de Gruyter Expositions in Mathematics 39

de Gruyter Expositions in Mathematics

Trigonometric Sums in Number Theory and Analysis

by

G. I. Arkhipov
V. N. Chubarikov
A. A. Karatsuba

Walter de Gruyter · Berlin · New York

Authors

Gennady I. Arkhipov
V. A. Steklov Mathematical Institute
Russian Academy of Sciences
8, Gubkina str.
119991, Moscow
Russia
e-mail: arkh@mi.ras.ru

Vladimir N. Chubarikov
Faculty of Mechanics and Mathematics
M. V. Lomonosov Moscow State University
Vorobjovy Gory
119899, Moscow
Russia
e-mail: chubarik@mech.math.msu.su

Anatoly A. Karatsuba
V. A. Steklov Mathematical Institute
Russian Academy of Sciences
8, Gubkina str.
119991, Moscow
Russia
e-mail: karacuba@mi.ras.ru

Title of the Russian original edition: Arkhipov, G. I.; Karatsuba, A. A.; Chubarikov, V. N.: Teoriya kratnykh trigonometricheskikh summ. *Publisher:* Nauka, Moscow 1987

Mathematics Subject Classification 2000: 11-02; 11D68, 11P05, 11P99

Key words: analytic number theory, multiple trigonometric sums, additive problems of number theory, Waring problem, Hilbert–Kamke problem, Artin problem, Hua problem, distribution of value of arithmetical functions

⊗ Printed on acid-free paper which falls within the guidelines of the ANSI to ensure permanence and durability.

Library of Congress Cataloging-in-Publication Data

Arkhipov, Gennadifi Ivanovich.
[Teorikila kratnykh trigonometricheskikh summ. English]
Trigonometric sums in number theory and analysis by / Gennady I. Arkhipov, Vladimir N. Chubarikov, Anatoly A. Karatsuba.
p. cm – (De Gruyter expositions in mathematics ; 39)
Includes bibliographical references and index.
ISBN 3-11-016266-0 (cloth : alk. paper)
1. Trigonometric sums. I. Chubarikov, Vladimir Nikolaevich. II. Karaktlsuba, Anatolifi Alekseevich. III. Title. IV. Series.
QA246.8.T75A75 2004
512.7–dc22 2004021280

ISBN 3-11-016266-0

Bibliographic information published by Die Deutsche Bibliothek

Die Deutsche Bibliothek lists this publication in the Deutsche Nationalbibliografie; detailed bibliographic data is available in the Internet at <http://dnb.ddb.de>.

Typesetting using the authors' TEX files: I. Zimmermann, Freiburg.
Printing and binding: Hubert & Co. GmbH & Co. KG, Göttingen.
Cover design: Thomas Bonnie, Hamburg.

Preface

The method of trigonometric sums was developed by I. M. Vinogradov in the first decades of the 20th century as a method for solving a wide range of problems in analytic number theory. The main problem in the study of trigonometric sums is to find an upper bound for the modulus of such sums. Presently, trigonometric sums with a single variable of summation have been studied rather completely, but many important problems still remain open, even in this area. In the theory of multiple trigonometric sums, to which the present monograph is primarily devoted, numerous new effects can be observed because there is a wide variety both of domains of the summation variables and of functions in the exponent.

In this monograph, the theory of multiple trigonometric sums is constructed systematically and several new applications of trigonometric sums and integrals in problems of number theory and analysis are described. At present, the theory of multiple trigonometric sums has reached the same degree of completion as the theory of one-dimensional trigonometric sums.

The first nine chapters of this translation are essentially identical with the Russian original of this book, which was published in 1987 by Nauka, Moscow. Chapters 10 to 12 are devoted to new results, and we hope that this English edition will be useful for a wide range of mathematicians. The reader can compare the original methods and the results that the authors obtained by these methods with the results presented in numerous papers after 1983. In particular, these new results concern estimates of trigonometric (oscillating) integrals and applications of the p-adic method in estimating trigonometric sums and in solving additive problems, including Waring's problem and Artin's conjecture on a local representation of zero by a form.

The authors wish to express their deep gratitude to the translator for very careful work with the manuscript of the Russian version of the book.

The authors

Basic Notation

We denote by $c, c_1, c_2, \ldots$ positive absolute constants which, in general, are different in different statements; $\varepsilon, \varepsilon_1, \varepsilon_2$ are positive arbitrarily small constants, and $\theta, \theta_1, \theta_2$ are complex-valued functions whose modulus does not exceed 1. For positive x, $\ln x = \log x$ is always the natural logarithm of the number x.

We shall use the standard notation of various mathematical symbols and number-theoretic functions without any special explanations.

For a real number α, the symbol $\|\alpha\|$ denotes the distance from α to the nearest integer number, i.e.,

$$\|\alpha\| = \min\big(\{\alpha\}, 1 - \{\alpha\}\big),$$

where $\{\alpha\}$ is the fractional part of α. The meaning of the symbol $\{\,\}$ should always be clear from the context (either the "fractional part" or the "braces").

For real numbers $\gamma_1, \ldots, \gamma_n, \delta_1, \ldots, \delta_n$, the relation

$$(\gamma_1, \ldots, \gamma_n) \equiv (\delta_1, \ldots, \delta_n) \pmod 1$$

means that all differences $\gamma_1 - \delta_1, \ldots, \gamma_n - \delta_n$ are integers.

For a positive A, the relation $B \ll A$ means that $|B| \le cA$; for positive A and B, the relation $A \asymp B$ means that $c_1 A \le B \le c_2 A$ as A becomes large.

If the limits of summation are not given under the summation sign, we shall assume that the summation is over all possible values of the variable of summation.

A system of Diophantine equations of the form

$$\sum_{j=1}^{2k} (-1)^j x_{1j}^{t_1} \ldots x_{rj}^{t_r} = 0, \quad 0 \le t_1 \le n_1, \ldots, 0 \le t_r \le n_r,$$

$$1 \le x_{1j} \le P_1, \ldots, 1 \le x_{rj} \le P_r, \quad j = 1, \ldots, 2k,$$

is said to be *complete*; if some of the equations are omitted in this system, the resulting system is said to be *incomplete*.

The range of values of the other parameters denoted by letters will be sufficiently clear from the text, and we sometimes do not make special mention of the range of values if it is clear from the context

New notation will be introduced in the course of the exposition; sometimes we shall recall notation that has already been used.

The statements and formulas are numbered separately in each chapter; auxiliary assertions are also numbered separately in each chapter. References to auxiliary assertions in the Appendix look as Lemma A.1, etc.

Contents

Introduction

In this monograph we give an exposition of the theory of trigonometric sums and its applications in number theory and analysis. This theory is based on the theory of multiple trigonometric sums developed by the authors in [2]–[12], [17]–[21], [23]–[34], [47]–[53]. By a *multiple trigonometric sum* we mean a sum of the form

$$S = \sum_{x_1=1}^{P_1} \sum_{x_r=1}^{P_r} \exp\{2\pi i F(x_1, \ldots, x_r)\}, \tag{1}$$

where $r \geq 1$, $P_1, \ldots, P_r$ are integers, and $F(x_1, \ldots, x_r)$ is a polynomial in r variables with real coefficients, i.e.,

$$F(x_1, \ldots, x_r) = \sum_{t_1=0}^{n_1} \cdots \sum_{t_r=0}^{n_r} \alpha(t_1, \ldots, t_r) x_1^{t_1} \ldots x_r^{t_r},$$

where $\alpha(t_1, \ldots, t_r)$ are real numbers.

When $r = 1$, such sums are called *Weyl sums*. The history of Weyl sums, their importance in mathematics, the methods available for studying them, and several results are presented in I. M. Vinogradov's monograph [165]. Hence here we shall not dwell in detail on the theory of these sums. We only give the general formulation, due to *I. M. Vinogradov*, of the problem of estimating Weyl sums. Such a sum has the form

$$S = \sum_{x=1}^{P} \exp\{2\pi i f(x)\},$$

where

$$f(x) = \alpha_1 x + \cdots + \alpha_n x^n.$$

It is easy to see that S is a periodic function with period 1 in each coefficient in $f(x)$. Hence it suffices to study S on the n-dimensional unit cube $0 \leq \alpha_1 < 1, \ldots, 0 \leq \alpha_n < 1$, which we denote by E. All points of E are divided into two sets E_1 and E_2 as follows:

$$E = E_1 \cup E_2.$$

At each point of E_1, I. M. Vinogradov obtains estimates for the sum S, and in many cases these estimates are best possible. The set E_1 itself consists of intervals and its measure is small:

$$\text{mes } E_1 = O(P^{-n(n+1)/2+1+2/n}).$$

At each point of E_2 there is an estimate for S which is uniform in appearance and has the form

$$S \ll P^{1-\rho}, \quad \rho = \rho(n) = \frac{c}{n^1 \ln n}; \tag{2}$$

the set E_2 has measure $1 - \text{mes } E_1$. Thus we know rather precise information about $|S|$ for any values of $\alpha_1, \ldots, \alpha_n$ in E. (For the exact statement of Vinogradov's theorem, see Section 3.1, Chapter 3, Theorem 3.2.)

By Ω we denote the m-dimensional unit cube in the m-dimensional Euclidean space, where $m = (n_1 + 1) \ldots (n_r + 1)$ and

$$0 \le \alpha(t_1, \ldots, t_r) < 1, \quad 0 \le t_1 \le n_1, \ldots, 0 \le t_r \le n_r.$$

Thus the coefficients in the polynomial $F(x_1, \ldots, x_r)$ of the multiple trigonometric sum (1) give a point in Ω. In our theory, we can also obtain estimates for $|S|$ on the entire Ω with an accuracy corresponding to the accuracy in the one-dimensional case. We divide the set Ω into two sets Ω_1 and Ω_2:

$$\Omega = \Omega_1 \cup \Omega_2.$$

On the point set Ω_1 whose measure is very small, we obtain an estimate for $|S|$, which in most cases is best possible. On the set Ω_2, we obtain a uniform estimate, which in the one-dimensional case corresponds to the estimate (2) (see Theorem 5.2 in Chapter 5 and Theorem 7.2 in Chapter 7).

Here it should be noted that in many applications it is necessary to have a uniform estimate for $|S|$ on the entire cube Ω_2. No estimate, no matter how sharp, on only a part of Ω_2 can enable one, for example, to obtain an asymptotic formula for the number of solutions of a complete system of equations (see Theorem 6.1 in Chapter 6).

The basis of our theory is the *mean value theorem*, i.e., the theorem that estimates the integral

$$J = J(\overline{P}; \overline{n}, k, r) = \int \cdots \int_{\Omega} \left| \sum_{x_1=1}^{P_1} \cdots \sum_{x_r=1}^{P_r} \exp\{2\pi i F(x_1, \ldots, x_r)\} \right|^{2k} dA$$

(see Theorem 4.2 in Chapter 4). The mean value theorem is proved by a p-adic method. It should be noted that the theory of multiple trigonometric sums is one of the motivations for the development of the p-adic method, but this motivation is the most important. For this reason, the book includes several problems of number theory which were solved by the p-adic method and for which this method was, in fact, created and developed, although these problems do not belong to the theory of multiple trigonometric sums.

By p-adic methods, we mean methods of analytic number theory which are based on the use of different properties of the system of residues of a power of a prime number p. The p-adic method used in this book was first invented in 1962–1966 (see [73], [77], [80], [81]). This technique has been further developed and improved (see [2]–[12], [17]–[21], [23]–[28], [30]–[34], [47]–[53], [86]) and, at present, includes several methods and considerations whose concepts are closely related to one another.

We list some of them that are most important:

1. the use of the circle method in the p-adic form,

2. the use of the p-adic analog of Vinogradov's u-numbers; the implementation of the Euler–Vinogradov "embedding principle" in the p-adic form for estimating the number of solutions of "Waring type" equations and congruences,

3. the lowering of the degree of a polynomial by shifting the argument (i.e., by dividing the values of the argument into progressions) by a number that is a multiple of some power of a prime number,

4. the recurrent reduction of additive problems for incomplete systems of residues modulo p^k to congruences for complete systems of residues and to problems of the same sort but with a fewer number of principal and nonprincipal parameters,

5. the use of regularity conditions for solutions to systems of equations and congruences in the p-adic form,

6. the use of variable parameters in the recurrence processes in items 2–4 and the optimization with respect to these parameters,

7. the passage from "jagged" systems to complete systems using a local p-adic variation in the unknowns,

8. the simultaneous use of several moduli of the form p^n that correspond to different prime numbers p,

9. the use of the idea of smoothing in the p-adic treatment,

10. the passages from polynomials to exponential functions and conversely in congruences, and

11. the p-adic and real methods for estimating the measure of the set of points at which the values of functions are small in terms of their parameters (coefficients, etc.) and for obtaining converse estimates of these parameters via the measure; the real interpretation of the methods and considerations given in items 2–4, 6, and 7.

Now let us discuss the contents of the monograph in more detail. Note that in each chapter, and sometimes in a section, we give necessary explanations of the results and of the methods used to obtain these results.

In Chapter 1, we study trigonometric integrals. We mainly find estimates from above for the moduli of such integrals. We note that integrals of this form are encountered not only in number theory, but also in mathematical analysis, mathematical statistics and probability theory, as well as in mathematical physics. As a consequence of these estimates, we obtain the complete solution of the Hua Loo-Keng problem stated in 1937 concerning the convergence exponent in the singular integral in Tarry's problem. We also give estimates from above for the convergence exponents in singular integrals in multidimensional analogs of Tarry's problem.

In Chapter 2, we study complete rational trigonometric sums. Based on estimates for such sums, we obtain upper bounds for the convergence exponents in the singular series in Tarry's problem and in multidimensional analogs of Tarry's problem.

In Chapter 3, we present two methods for estimating the mean values of Weyl sums: the "real" and "p-adic" methods. This chapter will help the reader to understand the main points in the theory of multiple trigonometric sums. In Section 3.1, we prove Vinogradov's mean value theorem and write Vinogradov's estimate for the Weyl sum. The mean value theorem is proved by using Vinogradov's original lemma on the "number of hits" and then estimating the number of solutions of the "one-sided" systems of equations. I. M. Vinogradov developed his method for estimating Weyl sums starting from the new method, which he constructed in 1934, for estimating the well-known Hardy–Littlewood function $G(n)$ in Waring's problem.

In Section 3.2, we consider one of the simplest versions of estimating $G(n)$ by Vinogradov's method. The "telescopic" construction of u-numbers in estimating $G(n)$ and the "telescopic" construction of "one-sided" systems of equations in the proof of the mean value theorem make the relation between the two Vinogradov's methods extremely clear.

In Section 3.3, we discuss an "analog of Waring's problem for congruences" and a p-adic method for solving this problem. In Section 3.4, we give a new p-adic proof of Vinogradov's mean value theorem. Moreover, the method studied in Section 3.4 corresponds to the method in Section 3.3 in the same way as the method considered in Section 3.1 depends on the method in Section 3.2. In Section 3.4, we also give estimates for trigonometric sums, which we shall use in the subsequent chapters. It should be noted that the theory of multiple trigonometric sums originates from this new p-adic method.

Finally, in Section 3.5 we present Yu. V. Linnik's method for proving Vinogradov's mean value theorem. We follow Linnik's paper written in 1943, which allows us to emphasize the common features of Linnik's method and the method given in Section 3.4, as well as distinctions between them.

In Chapter 4, we prove the main theorems in the theory of multiple trigonometric sums, namely, the mean value theorems. First, we prove the theorem for equivalent variables of summation (or the unknowns in the system of equations). This is the most important case in the theory and, at the same time, the simplest case. Then we prove the general mean value theorem.

In Chapter 5, we give estimates for multiple trigonometric sums. Here we prove lemmas on the multiplicity of intersection of regions of special form in multidimensional spaces. Based on these lemmas and the results obtained in Chapter 4, we obtain estimates for multiple sums.

In Chapter 6, some applications of the theory of multiple trigonometric sums are given. We consider problems of two types: asymptotic formulas for the number of solutions of systems of Diophantine equations and distributions of fractional parts of systems of polynomials in several variables.

There are many different problems in the multidimensional theory. Therefore, we restrict ourselves to the problems that are, in our opinion, most important and interesting.

In Chapter 7, we consider singular cases of the theory of multiple trigonometric sums. This singularity means that the difference between the limits of summation in a multiple trigonometric sum may be arbitrarily large.

In Chapter 8, we give a solution of the Hilbert–Kamke problem of representing natural numbers $N_1, \dots, N_n$ as sums of finitely many terms in natural numbers of the form $x, \dots, x^n$, respectively.

In Chapter 9, we use the p-adic method to solve two problems in number theory. In Section 9.1, we obtain a principally stronger result for Artin's problem of representing zero by values of a form in local fields. In Section 9.2, we give an estimate from above for $G(n)$ for large n; moreover, we give a simpler proof of the well-known Vinogradov's estimate and obtain a result that is even sharper.

In Chapter 10, we find some estimates for multiple trigonometric sums.

In Chapter 11, we consider an application of trigonometric sums in harmonic analysis.

In Chapter 12, we give some estimates for short Kloosterman sums.

Chapter 1

Trigonometric integrals

A *trigonometric integral* is defined to be an integral J of the form

$$J = \int_0^1 \cdots \int_0^1 \exp\{2\pi i F(x_1, \ldots, x_r)\} \, dx_1 \ldots dx_r,$$

where $F(x_1, \ldots, x_r)$ is a real function of r variables $x_1, \ldots, x_r$. Such integrals are used in analytic number theory, function theory, mathematical physics, probability theory, and mathematical statistics. One of the main problems concerning J is the problem of finding an upper bound for the modulus of J.

1.1 One-dimensional trigonometric integrals

In this section we prove exact estimates for integrals J in a certain class of functions $F(x)$. These auxiliary statements are also of independent interest; they and their analogs are used in different fields of mathematics.

Lemma 1.1. *Suppose that for* $0 < x < 1$ *a real function* $f(x)$ *has the nth-order derivative* $(n > 1)$ *and the inequality*

$$A \leq |f^{(n)}(x)|, \quad 0 < x < 1,$$

holds for some $A > 0$. *By E we denote the set of points in the interval* $0 < x < 1$ *such that*

$$|f'(x)| \leq B.$$

Then the measure $\mu = \mu(E)$ *of this set satisfies the estimate*

$$\mu = \mu(E) \leq (2n-2)(BA^{-1})^{1/(n-1)}.$$

Proof. The set E consists of intervals. We move the intervals of the set E together and thus form a single interval. Its length is μ. In this interval we choose n points such that the distance between them is equal to $\mu/(n-1)$. Then we move these integrals to their original places and thus obtain n points $x_1, x_2, \ldots, x_n$ in E such that

$$|x_k - x_j| \geq |k - j|\mu/(n-1).$$

Next, we consider the polynomial $g(x)$, which is the Lagrange interpolation polynomial corresponding to the function $f'(x)$ and the points of interpolation $x_1, x_2, \dots, x_n$,

$$g(x) = \sum_{\nu=1}^{n} f'(x_\nu) \frac{(x - x_1)\dots(x - x_{\nu-1})(x - x_{\nu+1})\dots(x - x_n)}{(x_\nu - x_1)\dots(x_\nu - x_{\nu-1})(x_\nu - x_{\nu+1})\dots(x_\nu - x_n)}.$$

The difference $F(x) = g(x) - f'(x)$ is $n - 1$ times differentiable and equal to zero at $x = x_1, x_2, \dots, x_n$. Hence, by Rolle's theorem, there exist points $\xi_1, \xi_2, \dots, \xi_{n-1}$,

$$x_1 < \xi_1 < x_2, \ x_2 < \xi_2 < x_3, \ \dots, \ x_{n-1} < \xi_{n-1} < x_n,$$

such that

$$F'(\xi_1) = \dots = F'(\xi_{n-1}) = 0.$$

Applying the same argument to the functions $F'(x)$, $F''(x)$, etc., in other words, applying Rolle's theorem to $F(x)$ subsequently $n - 1$ times, we find a point ξ with $0 < \xi < 1$ such that

$$F^{(n-1)}(\xi) = g^{(n-1)}(\xi) - f^{(n)}(\xi) = 0.$$

This relation implies

$$\frac{f^{(n)}(\xi)}{(n-1)!} = \sum_{\nu=1}^{n} \frac{f'(x_\nu)}{(x_\nu - x_1)\dots(x_\nu - x_{\nu-1})(x_\nu - x_{\nu+1})\dots(x_\nu - x_n)},$$

and the inequality

$$\frac{A}{(n-1)!} \le \frac{|f^{(n)}(\xi)|}{(n-1)!} \le B \sum_{\nu=1}^{n} \frac{1}{|(x_\nu - x_1)\dots(x_\nu - x_{\nu-1})(x_\nu - x_{\nu+1})\dots(x_\nu - x_n)|}$$

follows from the assumptions of the theorem and the properties of the points $x_1, \dots, x_n$.

Using the fact that

$$|x_k - x_j| \ge |k - j|\mu/(n-1),$$

we obtain

$$\frac{A}{(n-1)!} \le B \sum_{\nu=1}^{n} \frac{(n-1)^{n-1}}{\mu^{n-1}(\nu-1)!(n-\nu)!}$$

$$= \frac{B(n-1)^{n-1}}{(n-1)!\mu^{n-1}} \sum_{\nu=1}^{n} \frac{(n-1)!}{(\nu-1)!(n-\nu)!} = \frac{B(2n-2)^{n-1}}{(n-1)!\mu^{n-1}},$$

$$\mu^{n-1} \le (2n-2)^{n-1} B A^{-1}, \quad \mu \le (2n-1)(BA^{-1})^{1/(n-1)}.$$

The proof of the lemma is complete. □

We shall use Lemma 1.1 to estimate the trigonometric integral.

Lemma 1.2. *Suppose that for* $0 < x < 1$ *a real function* $f(x)$ *has the nth-order derivative* $(n > 1)$ *and the equality*

$$A \le |f^{(n)}(x)|, \quad 0 < x < 1,$$

holds for some $A > 0$. *Then the integral*

$$J = \int_0^1 \exp\{2\pi i f(x)\}\,dx$$

satisfies the estimate

$$|J| \le \min(1,\ 6nA^{-1/n}).$$

Proof. Representing J as the sum of integrals of the real and imaginary parts of the integrand function, we have

$$J = U + iV,$$

where

$$U = \int_0^1 \cos 2\pi f(x)\,dx, \quad V = \int_0^1 \sin 2\pi f(x)\,dx.$$

First, we consider the integral U. We divide the interval $0 < x < 1$ into two sets E_1 and E_2 as follows: the set E_1 consists of intervals such that the inequality

$$|f'(x)| \le B = 2^{-(n-1)/n} A^{1/n}$$

holds at each point of E_1, and the set E_2 consists of all other intervals. According to this partition, the integral U can be written as the sum of two terms:

$$U = U_1 + U_2,$$

where

$$U_1 = \int_{E_1} \cos 2\pi f(x)\,dx, \quad U_2 = \int_{E_2} \sin 2\pi f(x)\,dx.$$

Let μ be the sum of the lengths of the intervals that constitute E_1. Obviously, $|U_1| \le \mu$. Lemma 1.1 gives the following estimate for μ:

$$|U_1| \le \mu \le (2n-2)(BA^{-1})^{1/(n-1)}.$$

Let us find an upper bound for $|U_2|$. All the intervals in E_2 can be divided into at most $2n-2$ intervals in each of which the function $f'(x)$ is monotone and of constant sign. Indeed, the function $f''(x)$ can have at most $n-2$ zeros, since, otherwise, after Rolle's theorem is applied $n-2$ times to $f''(x)$, we would obtain a point ξ $(0 < \xi < 1)$ such that $f^{(n)}(\xi) = 0$, but this contradicts the assumption of the theorem that $|f^{(n)}(\xi)| \ge A > 0$. Hence $f'(x)$ has at most $n-1$ intervals on which it is monotone and has at most $2n-2$ intervals on which it is of constant sign. Suppose that $x_1 < x < x_2$ is one of such intervals and U_3 is the part of the integral U_2

corresponding to this interval. Without loss of generality, we consider only the case in which $f'(x)$ is an increasing function on this interval. By setting $f(x) = v$, $f(x_1) = v_1$, and $f(x_2) = v_2$, we readily obtain

$$U_3 = \int_{v_1}^{v_2} \cos 2\pi v \frac{dv}{f'(x)},$$

where $f'(x)$ is considered as a function of v. We use numbers of the form $0.5l + 0.25$, where l is integer, in the interior of the interval $v_1 \le v \le v_2$ to divide this interval into intervals whose lengths do not exceed 0.5. Then the integral U_3 can be represented as an alternating sum whose terms are monotonically decreasing in absolute value. Therefore, for some v_0 and σ such that $v_1 \le v_0 \le v_0 + \sigma \le v_2$ and $\sigma \le 0.5$, we have

$$|U_3| \le \int_{v_0}^{v_0+\sigma} \frac{dv}{f'(x)} = x'' - x', \quad v_0 = f'(x), \quad v_0 + \sigma = f(x'').$$

It follows from the Lagrange theorem on finite increments that

$$\sigma = f''(x) - f'(x) = f'(\xi)(x'' - x'), \quad x_1 \le x' \le \xi \le x'' \le x_2,$$

i.e.,

$$x'' - x' = \sigma\bigl(f'(\xi)\bigr)^{-1} \le (2B)^{-1}.$$

Hence,

$$\begin{aligned} |U_3| &\le (2B)^{-1}, \quad |U_2| \le (2n-2)(2B)^{-1} = (n-1)B^{-1}, \\ |U| \le |U_1| + |U_2| &\le (2n-2)(BA^{-1})^{1/(n-1)} + (n-1)B^{-1} \\ &\le 2(n-1)2^{(n-1)/n}A^{-1/n}. \end{aligned}$$

By a similar argument, we obtain the same upper bound for $|V|$. So we have

$$|J| \le 2\sqrt{2}\,2^{(n-1)/n}(n-1)A^{-1/n} < 6nA^{-1/n}.$$

The proof of the lemma is complete. □

From Lemma 1.2 we derive the following two consequences.

Corollary 1.1. *Suppose that a function $f(x)$ satisfies the assumptions of Lemma 1.2 on the interval $\alpha < x < \beta$. Then the following inequality holds:*

$$\left|\int_\alpha^\beta \exp\{2\pi i f(x)\}\,dx\right| \le \min(\beta - \alpha,\ 6nA^{-1/n}).$$

Proof. In this integral, we perform a change of the integration variable of the form $u = (x - \alpha)/(\beta - \alpha)$ and thus obtain

$$\int_\alpha^\beta \exp\{2\pi i f(x)\}\,dx = (\beta - \alpha)\int_0^1 \exp\{2\pi i g(u)\}\,du,$$

where $g(u) = f\big(u(\beta - \alpha) + \alpha\big)$. By assumption, we have

$$|g^{(n)}(u)| = (\beta - \alpha)^n \big|f^{(n)}\big(u(\beta - \alpha) + \alpha\big)\big| \geq (\beta - \alpha)^n A.$$

Therefore, applying Lemma 1.2, we arrive at the desired estimate

$$\left|\int_\alpha^\beta \exp\{2\pi i f(x)\}\, dx\right| = (\beta - \alpha)\left|\int_0^1 \exp\{2\pi i g(u)\}\, du\right|$$
$$\leq (\beta - \alpha) 6n \big((\beta - \alpha)^n A\big)^{-1/n} = 6nA^{-1/n}. \qquad \square$$

Corollary 1.2. *Suppose that $g(x)$ is a piecewise monotone continuous function, $\max_{0 \leq x \leq 1} |g(x)| = H$, the number of monotonicity intervals of the function $g(x)$ is equal to ρ, and $f(x)$ satisfies the conditions of Lemma* 1.2. *Then the integral*

$$I = \int_0^1 g(x) \exp\{2\pi i f(x)\}\, dx$$

satisfies the estimate

$$|I| \leq H \min(1,\ 24pnA^{-1/n}).$$

Proof. We divide the interval $0 < x < 1$ into intervals of monotonicity of the function $g(x)$. Let $x_1 \leq x \leq x_2$ be one of these intervals. Integrating by parts, we obtain

$$I_1 = \int_{x_1}^{x_2} g(x) \exp\{2\pi i f(x)\}\, dx = \int_{x_1}^{x_2} g(x)\, d\left(\int_0^x \exp\{2\pi i f(\xi)\}\, d\xi\right)$$
$$= g(x_2) \int_0^{x_2} \exp\{2\pi i f(\xi)\}\, d\xi - g(x_1) \int_0^{x_1} \exp\{2\pi i f(\xi)\}\, d\xi$$
$$- \int_{x_1}^{x_2} \left(\int_0^x \exp\{2\pi i f(\xi)\}\, d\xi\right) dg(x).$$

Passing to inequalities, we arrive at the estimate

$$|I_1| \leq 4H \max_{x_1 \leq y \leq x_2} \left|\int_0^y \exp\{2\pi i f(\xi)\}\, d\xi\right|.$$

Applying Corollary 1.1, we obtain

$$|I_1| \leq 24HnA^{-1/n}.$$

Thus we have

$$|I| \leq H \min(1,\ 24pnA^{-1/n}).$$

The proof of the corollary is complete. $\square$

Note that the estimate obtained in Lemma 1.2 is sharp in the parameters A and n. Indeed, let us choose $f(x) = \alpha x^n$ ($\alpha > 1$). Then $f^{(N)}(x) = n!\alpha$, and Lemma 1.2 implies the estimate

$$\left|\int_0^1 \exp\{2\pi i\alpha x^n\}\,dx\right| \le 6n(n!)^{-1/n}\alpha^{-1/n} \le 24\alpha^{-1/n}.$$

On the other hand, we have

$$\left|\int_0^1 \exp\{2\pi i\alpha x^n\}\,dx\right| \ge \left|\int_0^1 \cos 2\pi\alpha x^n\,dx\right| = U.$$

In the integral U, we perform a change of the integration variable of the form $y = \alpha x^n$ and thus obtain

$$|U| \ge \frac{1}{n}\alpha^{-1/n}\left|\int_0^{+\infty} \frac{\cos 2\pi y}{y^{1-1/n}}\,dy\right| - \frac{1}{n}\alpha^{-1/n}\left|\int_\alpha^{+\infty} \frac{\cos 2\pi y}{y^{1-1/n}}\,dy\right|,$$
$$\frac{1}{n}\alpha^{-1/n}\int_0^{+\infty} \frac{\cos 2\pi y}{y^{1-1/n}}\,dy = \alpha^{-1/n}\frac{\cos(\pi/2n)}{\sqrt[n]{2\pi}}\Gamma\left(1+\frac{1}{n}\right).$$

Now we estimate the remaining integral. Integrating one time by parts and passing to inequalities, we obtain

$$\int_\alpha^{+\infty} y^{-1+1/n}\cos 2\pi y\,dy = \frac{1}{2\pi}\alpha^{-1+1/n}\sin 2\pi\alpha + \frac{1}{2\pi}\left(1-\frac{1}{n}\right)\int_\alpha^{+\infty} y^{-2+1/n}\sin 2\pi y\,dy,$$
$$\left|\int_\alpha^{+\infty} y^{-1+1/n}\cos 2\pi y\,dy\right| \le \frac{1}{2\pi}\alpha^{-1+1/n} + \frac{1}{2\pi}\left(1-\frac{1}{n}\right)\int_\alpha^{+\infty} y^{-2+1/n}\,dy$$
$$= \frac{1}{2\pi}\alpha^{-1+1/n} + \frac{1}{2\pi}\alpha^{-1+1/n} = \frac{1}{\pi}\alpha^{-1+1/n}.$$

Hence we have

$$|U| \ge \alpha^{-1/n}\left\{\frac{\cos(\pi/2n)}{\sqrt[n]{2\pi}}\Gamma\left(1+\frac{1}{n}\right) - \frac{1}{\pi n}\alpha^{-1+1/n}\right\} \ge \frac{1}{8}\alpha^{-1/n}.$$

We have thus shown that the estimate obtained in Lemma 1.2 is sharp in the class of functions $f(x)$ under study. Nevertheless, this lemma does not completely solve the problem of estimating trigonometric integrals. Indeed, applying Lemma 1.2 to the integral J with the function

$$f(x) = \alpha x\left(x-\frac{1}{n}\right)\dots\left(x-\frac{n-1}{n}\right), \quad \alpha > 1,$$

we obtain

$$|J| \ll \alpha^{-1/n}$$

(the constant in $\ll$ depends on n). However, the integral J satisfies the exact estimate

$$|J| \ll \alpha^{-1/2}.$$

Let us prove this. Suppose that the polynomial $g(x)$ is given by the relation

$$g(x) = \alpha(x-\alpha_1)\dots(x-\alpha_n),$$

where $\alpha_1 < \alpha_2 < \dots < \alpha_n$ and $\delta = \min_{0<i<n}(\alpha_{j+1} - a_j)$. Then, for an arbitrary $g'(x)$, we have

$$g'(x) = n\alpha(x-\beta_1)\dots(x-\beta_{n-1}),$$

where

$$\alpha_1 < \beta_1 < \alpha_2 < \beta_2 < \dots < \alpha_{n-1} < \beta_{n-1} < \alpha_n, \tag{a}$$

$$|\alpha_j - \beta_k| > \delta/\ln(2e^2n - 3e^2). \tag{b}$$

Inequalities (a) follow from Rolle's theorem. Further, suppose, for instance, that $\min(\beta_k - \alpha_k, \alpha_{k+1} - \beta_k) = \beta_k - \alpha_k$ $(0 < k < n)$. Then $\alpha_{k+1} - \beta_k \geq \delta/2$,

$$0 = \frac{g'(\beta_k)}{g(\beta_k)} = \frac{1}{\beta_k - \alpha_1} + \dots + \frac{1}{\beta_k - \alpha_k} - \frac{1}{\alpha_{k+1} - \beta_k} - \dots - \frac{1}{\alpha_n - \beta_k},$$

$$\frac{1}{\beta_k - \alpha_k} = \frac{1}{\alpha_{k+1} - \beta_k} + \dots + \frac{1}{\alpha_n - \beta_k} - \frac{1}{\beta_k - \alpha_1} - \dots - \frac{1}{\beta_k - \alpha_{k-1}}$$

$$< \frac{1}{\delta/2} + \frac{1}{3\delta/2} + \dots + \frac{1}{(2k-3)\delta/2} < \frac{2}{\delta}\left(1 + \int_2^n \frac{du}{2u-3}\right)$$

$$- \delta^{-1}\ln(2e^2n \quad 3e^2),$$

which implies inequality (b).

Now we return to the polynomial

$$f(x) = \alpha x(x - 1/n)\dots\big(x - (n-1)/n\big), \quad \alpha > 1.$$

Twice applying (a) and (b), we obtain

$$f'(x) = n\alpha(x-\beta_1)\dots(x-\beta_{n-1}),$$

$$f''(x) = n(n-1)\alpha(x-\gamma_1)\dots(x-\gamma_{n-2}),$$

$$|\beta_j - \gamma_k| > \varkappa, \quad \varkappa = \big(n\ln^2(2e^2n - 3e^2)\big)^{-1}.$$

We assume that the set E_1 consists of points of the interval [0, 1] such that the distance between these points and the roots of the polynomial $f''(x)$ does not exceed $0.5\varkappa$. Then for $x \in E_1$, we have $|x - \gamma_k| > 0.5\varkappa$ $(k = 1, \dots, n-2)$ and

$$f''(x) = n(n-1)\alpha|x-\gamma_1|\dots|x-\gamma_{n-2}| > n(n-1)\alpha 2^{-n+2}\varkappa^{n-2}. \tag{c}$$

The other points $x \in [0, 1]$ form the set E_2. Then for $x \in E_2$ and some k $(1 \le k \le n-2)$, we have

$$|x - \gamma_k| \le 0.5\varkappa;$$

hence for any j $(j = 1, 2, \dots, n-1)$, we obtain

$$\begin{aligned} |x - \beta_j| &= |x - \gamma_k + \gamma_k - \beta_j| \ge |\beta_j - \gamma_k| - |x - \gamma_k| > 0.5\varkappa, \\ |f'(x)| &= n\alpha|x - \beta_1|\dots|x - \beta_{n-1}| > n\alpha 2^{-n+1}\varkappa^{n-1}. \end{aligned} \tag{d}$$

So, using inequalities (c) and (d) and Lemma 1.2 (obviously, Lemma 1.2 also holds for $n = 1$ if $f(x)$ is a polynomial), we obtain

$$\begin{aligned} J = \int_0^1 \exp\{2\pi i f(x)\}\,dx &= \int_{E_1} \exp\{2\pi i f(x)\}\,dx + \int_{E_2} \exp\{2\pi i f(x)\}\,dx \\ &\ll \alpha^{-1/2} + \alpha^{-1} \ll \alpha^{-1/2}. \end{aligned}$$

Theorem 1.1. *Suppose that $n \ge 1$, $\alpha_1, \dots, \alpha_n$ are real numbers, and*

$$f(x) = \alpha_n x^n + \dots + \alpha_1 x, \qquad \beta_r(x) = f^{(r)}(x)/r!, \quad r = 1, \dots, n,$$

$$H = H(\alpha_n, \dots, \alpha_1) = \min_{a \le x \le b} \sum_{r=1}^{n} |\beta_r(x)|^{1/r}.$$

Then the integral

$$J = \int_a^b \exp\{2\pi i f(x)\}\,dx$$

satisfies the estimate

$$|J| \le \min(b - a,\ 6en^3 H^{-1}).$$

Proof. First, we show that the interval $a < x < b$ can be covered by nonintersecting intervals $\Delta_1, \Delta_2, \dots, \Delta_m$, where $m \le 0.5(n^2 + n) - 1$, so that the inequality

$$|f^{(r)}(x)/r!| \ge (n^{-1}H)^r$$

holds on each Δ_j $(j = 1, 2, \dots, m)$ for some natural number r $(1 \le r \le n)$.

We realize this covering in n steps as follows (the last steps can be empty). For $k = 0, 1, \dots, n-1$, we consider the functions

$$\beta_{n-k}(x) = f^{(n-k)}(x)/(n-k)!$$

that are polynomials whose degree does not exceed k. First, we note that $\beta_{n-k}(x)$ has at most k intervals of monotonicity. Therefore, for any $D > 0$, the number of intervals such that $|\beta_{n-k}(x)| < D$ at each point of these intervals does not exceed k, while the number of intervals such that $|\beta_{n-k}(x)| \ge D$ at each point of these intervals does not exceed $k + 1$.

The first step: $k = 0$ and the function $\beta_n(x) = \alpha_n$. If

$$|\alpha_n| \geq (n^{-1}H)^n,$$

then we set $\Delta_1^{(1)} = (a, b]$, and thus complete the process of covering the interval (a, b). Assume the contrary, i.e., assume that

$$|\alpha_n| < (n^{-1}H)^n.$$

The second step: $k = 1$ and the function $\beta_{n-1}(x) = n\alpha_n x + \alpha_{n-1}$. If for any $x \in (a, b)$ we have the inequality

$$|\beta_{n-1}(x)| \geq (n^{-1}H)^{n-1}, \tag{1.1}$$

then we set $\Delta_1^{(2)} = (a, b]$ and thus complete the process of covering the interval (a, b). Assume the contrary, i.e., assume that there exist points $x \in (a, b)$ such that

$$|\beta_{n-1}(x)| < (n^{-1}H)^{n-1}.$$

The number of intervals at whose points the last inequality holds does not exceed 1. We denote these intervals by the symbol Δ'_2 and proceed to cover these intervals. The number of intervals at whose points inequality (1.1) holds does not exceed 2. We denote these intervals by the symbols $\Delta_1^{(2)}$ and $\Delta_2^{(2)}$ (they can also be empty).

Suppose that the kth step ($k < n$) is realized. Prior to making the $(k + 1)$st step, we have the point set Δ'_k consisting of $k - 1$ intervals such that the inequality

$$|\beta_{n-k+1}(x)| < (n^{-1}H)^{n-k+1}$$

holds at each point of these intervals. We proceed to cover the intervals that form the set Δ'_k. But the number of intervals such that the inequality

$$|\beta_{n-k+1}(x)| \geq (n^{-1}H)^{n-k+1}$$

holds at their points does not exceed k. We denote these intervals by the symbols $\Delta_1^{(k)}, \dots, \Delta_k^{(k)}$ (some of them can be empty).

The $k + 1$st step: the function $\beta_{n-k}(x)$ is a polynomial whose degree does not exceed k. If for any $x \in \Delta'_k$ we have the inequality

$$|\beta_{n-k}(x)| \geq (n^{-1}H)^{n-k}, \tag{1.2}$$

then we denote the intervals comprising Δ'_k by $\Delta_1^{(k+1)}, \dots, \Delta_k^{(k+1)}$ and thus complete the process of covering. Assume the contrary, i.e., assume that there exist $x \in \Delta'_k$ such that

$$|\beta_{n-k}(x)| < (n^{-1}H)^{n-k}.$$

We denote the set of points at which the last inequality holds by Δ'_{k+1}. The set Δ'_{k+1} consists of at most k intervals. We proceed to cover the intervals of the set Δ'_k. But the number of intervals at whose points inequality (2.2) holds does not exceed $k+1$. We denote these intervals by the symbols $\Delta_1^{k+1}, \ldots, \Delta_{k+1}^{k+1}$.

We show that the interval (a, b) is completely covered after the nth step (if it is covered earlier, we assume that the remaining steps are empty). Let $a < \xi < b$. Then, by the assumptions of the theorem, we have

$$H \leq \sum_{r=1}^{n} |\beta_r(\xi)|^{1/r},$$

i.e., for some r $(1 \leq r \leq n)$ we have the inequality

$$H \leq n|\beta_r(\xi)|^{1/r},$$

which can also be written as

$$|f^r(\xi)| \geq r!(n^{-1}H)^r.$$

Choosing the maximum value of such r and denoting it by the letter k, we see that ξ belongs to one of the intervals determined by the inequality

$$|\beta_k(x)| \geq (n^{-1}H)^k.$$

This proves that (a, b) is completely covered by the intervals $\Delta_j^{(k)}$, which we now denote by the symbols Δ_j $(j \leq m)$. The number m of the covering intervals does not exceed

$$2 + 3 + \cdots + n = 0.5(n^2 + n) - 1.$$

Now we can estimate J. We have the inequality

$$|J| \leq \sum_{j=1}^{m} \left| \int_{\Delta_j} \exp\{2\pi i f(x)\}\, dx \right|.$$

On each interval Δ_j the inequality

$$|f^{(r)}(x)| \geq r!(n^{-1}H)^r$$

holds for some r $(1 \leq r \leq n)$. Applying Corollary 1.1 of Lemma 1.2 (note that for $r = 1$, the lemma and its corollary hold in our case, because $f(x)$ is a polynomial and hence $f'(x)$ has at most $2n - 2$ intervals of monotonicity), we obtain the estimate

$$\left| \int_{\Delta_1} \exp\{2\pi i f(x)\}\, dx \right| \leq 6r(r!n^{-r}H^r)^{-1/r} \leq 6e(n^{-1}H)^{-1},$$

$$|J| \leq 6menH^{-1} \leq 6en^3H^{-1}.$$

The proof of the theorem is complete. □

The theorem proved above gives a correct (in the order of magnitude of H) estimate for the integral of a specific polynomial. More precisely, for any polynomial $f(x)$ on the integration interval $[a, b]$, it is possible to find a point c such that the trigonometric integral

$$J(c) = \int_a^c \exp\{2\pi i f(x)\}\, dx$$

has both upper and lower bounds of the order of T, where $T = \min(b-a,\ H^{-1})$ and H is the same as in Theorem 1.1. Let us prove this assertion.

The upper bound follows from Theorem 1.1, since for any point c, we have

$$\sum_{r=1}^{n} |\beta_r(x)|^{1/r} \geq H \quad \text{for} \quad \alpha \leq x \leq c \leq b.$$

Further, assume that the variable $\sum_{r=1}^{n} |\beta_r(x)|^{1/r}$ takes the value H at a point x_0 ($a \leq x_0 \leq b$). We represent the polynomial $f(x)$ in the form

$$f(x) = \sum_{r=0}^{n} \beta_r(x_0)(x - x_0)^r.$$

It follows from the relation

$$\sum_{j=1}^{n} |\beta_j(x_0)|^{1/j} = H$$

that

$$|\beta_r(x_0)|^{1/r} \leq H, \qquad r = 1, \ldots, n.$$

Therefore, if $|x - x_0| < \Delta = (33H)^{-1}$, then we have

$$|f(x) - f(x_0)| < \sum_{r=1}^{n} H^r \Delta^r = \sum_{r=1}^{n} 33^{-r} < 2^{-5},$$

and hence

$$\bigl|\exp\{2\pi i f(x)\} - \exp\{2\pi i f(x_0)\}\bigr| < 2\pi 2^{-5} < 4^{-1}.$$

Obviously, either the interval $[a, b]$ is contained in the interior of the interval $[x_0 - \Delta, x_0 + \Delta]$, or one of the intervals $[x_0 - \Delta, x_0]$ and $[x_0, x_0 + \Delta]$ is entirely contained in the interval $[a, b]$. In the first case we take the point b to be c. Then we have

$$\left|\int_a^b \exp\{2\pi i f(x)\}\, dx\right| \geq \left|\int_a^b \exp\{2\pi i f(x_0)\}\, dx\right| - \left|\int_a^b \bigl(\exp\{2\pi i f(x)\} - \exp\{2\pi i f(x_0)\}\bigr)\, dx\right|$$

$$\geq b-a-\frac{b-a}{4}=\frac{3}{4}(b-a),$$

i.e., $|J| \gg T$.

Let us consider the second case. Suppose that x_1 and x_2 are the left-hand and right-hand endpoints of one of the intervals $[x_0-\Delta, x_0]$ and $[x_0, x_0+\Delta]$ that belongs to $[a, b]$. Precisely as above, we obtain

$$\left|\int_{x_1}^{x_2} \exp\{2\pi i f(x)\}\,dx\right| > \frac{3}{4}\Delta.$$

If now we have

$$\left|\int_{a}^{x_1} \exp\{2\pi i f(x)\}\,dx\right| > \frac{1}{4}\Delta,$$

the we take the point x_1 to be c. Otherwise, we take x_2 to be c. Since

$$\left|\int_{a}^{x_2} \exp\{2\pi i f(x)\}\,dx\right| \geq \left|\int_{x_1}^{x_2} \exp\{2\pi i f(x)\}\,dx\right| - \left|\int_{a}^{x_1} \exp\{2\pi i f(x)\}\,dx\right| \geq \frac{\Delta}{2},$$

in both cases, we obtain

$$\left|\int_{a}^{c} \exp\{2\pi i f(x)\}\,dx\right| > \frac{1}{4}\Delta \gg T.$$

Thus we have proved the relations

$$T \ll |J(c)| \ll T,$$

as was stated above.

Theorem 1.1 can be generalized as follows.

Theorem 1.1′. *Suppose that a function $f(x)$ has the nth-order derivative on $[a, b]$ $(n > 1)$ and the number of intervals of monotonicity of its derivative does not exceed K. Then the following estimate holds*:

$$J \ll \min(b-a,\ H^{-1}),$$

where the constant in $\ll$ depends only on n and K.

Let us prove one more theorem concerning the upper bound for J that depends on the lower bound for the linear combination of the derivatives of the function $f(x)$.

Lemma 1.3. [1] *Let $0 < a < b$. If a real function $f(x)$ vanishes at $n+1$ points in the interval (a, b) and all zeros of the polynomial $a_0 + a_1x + \cdots + a_nx^n$ are real, then*

$$a_0 f(\xi) + a_1 f'(\xi) + a_2 f''(\xi) + \cdots + a_n f^{(n)}(\xi) = 0$$

at an interior point ξ of the interval (a, b).

[1] This lemma coincides with Problem 92 in Section 1, Chapter I, Part II, of the book [133]

Theorem 1.2. *Suppose that for* $0 < x < 1$, *a real function* $f(x)$ *has the nth-order* $(n > 1)$ *derivative, the number of intervals on which* $f'(x)$ *is monotone and of constant sign does not exceed* K, *real numbers* $a_1, a_2, \dots, a_n$ *satisfy the condition that all zeros of the polynomial* $a_1 + a_2 x + \dots + a_n x^{n-1}$ *are real, and* $a = \max(|a_1|, |a_2|, \dots, |a_n|)$. *Suppose also that the inequality*

$$\left|a_1 f'(x) + a_2 f''(x) + \dots + a_n f^{(n)}(x)\right| \geq A > 0$$

holds for all x from the interval $0 < x < 1$. *Then the following estimate holds*:

$$|J| = \left|\int_0^1 \exp\{2\pi i f(x)\}\, dx\right| \leq 24(Kan^{n-1})^{1/n} A^{-1/n}.$$

Proof. We have

$$J = U + iV,$$

where

$$U = \int_0^1 \cos 2\pi f(x)\, dx, \quad V = \int_0^1 \sin 2\pi f(x)\, dx.$$

First, we consider the integral U. We divide the interval $0 \leq x \leq 1$ into two sets E_1 and E_2. The set E_1 consists of intervals such that the inequality

$$|f'(x)| \leq B = \left(\frac{K}{8(n-1)}\right)^{(n-1)/n} a^{-1/n} A^{1/n}$$

holds at each point of these intervals. The set E_2 consists of all other points. Then we have

$$U = U_1 + U_2,$$

where

$$U_1 = \int_{E_1} \cos 2\pi f(x)\, dx, \quad U_2 = \int_{E_2} \cos 2\pi f(x)\, dx.$$

Let μ be the sum of the lengths of the intervals that constitute E_1. Obviously, $|U_1| \leq \mu$. Let us estimate μ from above. To this end, we choose n points $(0 \leq x_1 < \dots \leq x_n \leq 1)$ in E_1 so that

$$|x_k - x_j| \geq |k - j|\mu/(n-1).$$

Let $G(x)$ be a polynomial of the form (the interpolation Lagrange polynomial corresponding to $f'(x)$ with points of interpolation $x_1, \dots, x_n$):

$$g(x) = \sum_{\nu=1}^{n} f'(x_\nu) \frac{(x - x_1)\dots(x - x_{\nu-1})(x - x_{\nu+1})\dots(x - x_n)}{(x_\nu - x_1)\dots(x_\nu - x_{\nu-1})(x_\nu - x_{\nu+1})\dots(x_\nu - x_n)}.$$

Then the function $h(x) = f'(x) - g(x)$ is zero at n points and hence, by Lemma 1.3, there exists a number ξ $(0 < \xi < 1)$ such that

$$a_1 h(\xi) + \cdots + a_n h^{(n-1)}(\xi) = 0.$$

Hence we have

$$\left|a_1 f'(\xi) + \cdots + a_n f^{(n)}(\xi)\right| = \left|a_1 g(\xi) + \cdots + a_n g^{(n-1)}(\xi)\right| \geq A.$$

Let us estimate $|g^{(k)}(\xi)|$ for $k = 0, 1, \ldots, n-1$. We have

$$|g^{(k)}(\xi)| \leq \frac{(n-1)!}{(n-k-1)!} B \sum_{\nu=1}^{n} \frac{(n-1)^{n-1}}{(\nu-1)!(n-\nu)!\mu^{n-1}},$$

which implies

$$\begin{aligned} A &\leq \left|a_1 g(\xi) + \cdots + a_n g^{(n-1)}(\xi)\right| \leq a\big(|g(\xi)| + \cdots + |g^{(n-1)}(\xi)|\big) \\ &\leq aB\left(\frac{(n-1)!}{(n-1)!} + \cdots + \frac{(n-1)!}{0!}\right) \sum_{\nu=1}^{n} \frac{(n-1)^{n-1}}{(\nu-1)!(n-\nu)!\mu^{n-1}} \\ &\leq \mu^{1-n} e(n-1)^{n-1} 2^{n-1} aB, \quad \mu \leq 4(n-1)(aBA^{-1})^{1/(n-1)}. \end{aligned}$$

So, we have the following estimate for $|U_1|$:

$$|U_1| \leq \mu \leq 4(n-1)(aBA^{-1})^{1/(n-1)}.$$

Let us find an upper bound for $|U_2|$. The number of intervals on which the function $f'(x)$ is monotone and of constant sign does not exceed K. On each of these intervals, we consider the intervals from E_2. Let $x_1 \leq x \leq x_2$ be such an interval. In Lemma 1.2, we proved that

$$\left|\int_{x_1}^{x_2} \cos 2\pi f(x)\, dx\right| \leq (2B)^{-1}.$$

This implies the estimate $|U_1| \leq K(2B)^{-1}$, and thus we have

$$\begin{aligned} |U| \leq |U_1| + |U_2| &\leq 4(n-1)(aBA^{-1})^{1/(n-1)} + 0.5KB^{-1} \\ &\leq 2(8^{n-1}(n-1)^{n-1}Ka)^{1/n} A^{-1/n}. \end{aligned}$$

We shall obtain the same upper bound for $|V|$. Hence we have

$$|J| \leq 2\sqrt{2}\big(8^{n-1}(n-1)^{n-1}Ka\big)^{1/n} A^{-1/n} \leq 24(Kan^{n-1})^{1/n} A^{-1/n}.$$

The theorem is thereby proved. □

1.2 Singular integrals in Tarry's problem and related problems

We consider the following system of equations:

$$\begin{aligned} x_1 + \cdots + x_k &= y_1 + \cdots + y_k, \\ x_1^2 + \cdots + x_k^2 &= y_1^2 + \cdots + y_k^2, \\ &\vdots \\ x_1^n + \cdots + x_k^n &= y_1^n + \cdots + y_k^n, \end{aligned} \tag{1.3}$$

where the unknown variables $x_1, \dots, x_k, y_1, \dots, y_k$ are integers ranging from 1 to P ($P > 1$). We let $J_{k,n}(P)$ denote the number of solutions of this system of equation. The system of equations (1.3) is said to be *complete*. If some equations in system (1.3) are omitted, the system thus obtained is said to be *incomplete*. The problem of solving system (1.3) is called *Tarry's problem* (the history of this problem is discussed sufficiently complete in the review [70]). In 1938, using the powerful Vinogradov's method [165], [159], Hua Loo-Keng derived the following asymptotic formula for $J_{k,n}(P)$ as $P \to +\infty$:

$$J_{k,n}(P) = \sigma\theta_0 P^{2k-0.5(n^2+n)} + O(P^{2k-0.5(n^2+n)-\delta}), \tag{1.4}$$

where $\delta = \delta(n,k) > 0$, k is of order $n^2 \ln n$, σ is a *singular series*, and θ_0 is a *singular integral*.

We do not consider the singular series σ in detail, but only note that this series converges for some special relations between n and k and is a quantity that depends only on n and k. These and similar series will be studied in detail in Chapter 2.

The singular integral θ_0 has the form

$$\theta_0 = \int_{-\infty}^{+\infty} \cdots \int_{-\infty}^{+\infty} \left| \int_0^1 \exp\{2\pi i(\alpha_n x^n + \cdots + \alpha_1 x)\}\, dx \right|^{2k} d\alpha_n \dots d\alpha_1.$$

In [68] Hua Loo-Keng studied the conditions under which the series θ_0 converges. He proved that θ_0 converges for $2k > 0.5n^2 + n$; in the same paper it is also said that the problem of finding the exact value of the convergence exponent for the integral θ_0 remains open (see also [69]).

Definition 1.1. The *convergence exponent* of an improper integral

$$\theta' = \int_{-\infty}^{+\infty} \cdots \int_{-\infty}^{+\infty} |G(u_1, \dots, u_m)|^{2k}\, du_1 \dots du_m$$

is defined to be a number γ such that θ' converges for $2k > \gamma + \varepsilon$ and diverges for $2k < \gamma - \varepsilon$, where ε is an arbitrarily small number.

Here we prove a theorem stating that θ_0 converges for $2k > 0.5(n^2+n)+1$ and diverges for $2k \le 0.5(n^2+n)+1$, which completely solves the convergence exponent problem for the improper integral in Tarry's problem.

A formula similar to formula (1.4) is also valid for the number of solutions of an incomplete system of equations. The improper integral in this formula, which we denote by θ_0', differs from θ_0 only in that the corresponding monomials in the polynomial in the exponent in θ_0 are omitted. Here we also prove a theorem that completely explains for which values of k the integral θ_0' converges. Moreover, a significant difference between the convergence exponents of θ_0 and θ_0' is revealed.

Now we state and prove the following theorem about the convergence exponent of the integral θ_0.

Theorem 1.3. *The integral* $\theta_0 = \theta_0(k)$ *converges for* $2k > 0.5(n^2+n)+1$ *and diverges for* $2k \le 0.5(n^2+n)+1$.

Proof. **1.** We prove the first assertion of the theorem. First, we estimate the volume of the domain $\Omega = \Omega(\alpha_n, \dots, \alpha_1)$ of points $\alpha_n, \dots, \alpha_1$ at which the quantity H determined in Theorem 1.1 does not exceed P (P is a natural number). To this end, we set $u_r = r/P$ for $r = 1, 2, \dots, P$ and consider the domains $\Omega_r = \Omega_r(\alpha_n, \dots, \alpha_1)$ of points $(\alpha_n, \dots, \alpha_1)$ at which the inequality $|\beta_s(u_r)| \le 2^n P^s$, $s = 1, 2, \dots, n$, holds.

Let us find an upper bound for $\mu(\Omega_r)$, i.e., for the volume of Ω_r. We have

$$\mu(\Omega_r) = \int \cdots \int_{\Omega_r} d\alpha_n \dots d\alpha_1.$$

In this integral, we perform a change of the integration variables of the form

$$\alpha_s = \beta_s - \binom{s+1}{s} \beta_{s+1} u_\nu + \dots + (-1)^{n-s} \binom{n}{s} \beta_\nu^{n-s},$$

where β_s are new independent variables. The Jacobian of this transformation is equal to 1. Hence we have

$$\mu(\Omega_r) = \int_{|\beta_n| \le 2^n P^n} \cdots \int_{|\beta_1| \le 2^n P} d\beta_n \dots d\beta_1 = 2^{n^2+n} P^{0.5(n^2+n)}.$$

Now we show that if the inequality

$$H = H(\alpha_n, \dots, \alpha_1) \le P \tag{1.5}$$

holds at a point $(\alpha_n, \dots, \alpha_1)$, then $(\alpha_n, \dots, \alpha_1)$ belongs to the domain Ω for some r $(1 \le r \le P)$. Indeed, if inequality (10.5) holds, then for some ξ $(0 \le \xi \le 1)$ we have $|\beta_n(\xi)|^{1/s} \le P$ or $|\beta_n(\xi) \le P^s$ for each $s = 1, \dots, n$.

We take $r = [\xi P]$ and prove that $(\alpha_n, \dots, \alpha_1)$ belongs to Ω_r. Indeed, we set $y = u_r - \xi$, $u_r = r/P$, and obtain

$$|\beta_s(u_r)| = |\beta_s(\xi + y)| = \left|\beta_s(\xi) + \frac{1}{1!}\beta_s'(\xi)y + \dots + \frac{1}{(n-s)!}\beta_s^{(n-s)}(\xi)y^{n-s}\right|$$
$$\leq \sum_{k=0}^{n-s} \binom{s+k}{k} P^s < 2^n P^s.$$

So we have proved that each point $(\alpha_n, \dots, \alpha_1)$ that belongs to the domain Ω of points satisfying (10.5) belongs to the domain Ω_r for some r $(1 \leq r \leq P)$ and, moreover,

$$\mu(\Omega_r) = 2^{n^2+n} P^{0.5(n^2+n)}.$$

Hence

$$\mu(\Omega) \leq \sum_{r=1}^{P} \mu(\Omega_r) = 2^{n^2+n} P^{0.5(n^2+n)+1}.$$

We let $\pi(P)$ denote the set of points $(\alpha_n, \dots, \alpha_1)$ at which the inequality $P < H = H(\alpha_n, \dots, \alpha_1) \leq 2P$ holds. Then for the integral θ_0 we have the estimate

$$\theta_0 \leq \sum_{m=0}^{+\infty} \int\!\cdots\!\int_{\pi(2^m)} \left| \int_0^1 \exp\{2\pi i(\alpha_n x^n + \dots + \alpha_1 x)\}\, dx \right|^{2k} d\alpha_n \dots d\alpha_1$$
$$+ \int\!\cdots\!\int_{\Omega(1)} \left| \int_0^1 \exp\{2\pi i(\alpha_n x^n + \dots + \alpha_1 x)\}\, dx \right|^{2k} d\alpha_n \dots d\alpha_1.$$

We apply the estimate obtained in Theorem 1.1 to the integral

$$J = \int_0^1 \exp\{2\pi i(\alpha_n x^n + \dots + \alpha_1 x)\}\, dx,$$

where $(\alpha_n, \dots, \alpha_1)$ belongs to the domain $\pi(2^m)$ and trivially estimate the integral over the domain $\Omega(1)$. Moreover, estimating the volume of $\pi(P)$, we obtain

$$\theta_0 \leq (12en^3)^{2k} 2^{2n^2+4} \sum_{m=0}^{+\infty} 2^{m(0.5(n^2+n)+1-2k)} + 2^{n^2+n}.$$

The last series converges for $2k > 0.5(n^2+n)+1$, which proves the first assertion of the theorem.

2. For an integer $P \geq (600n)^{15n}$ and $r = 1, 2, \dots, P$, we consider the polynomials

$$f(x; r) = \alpha_n x^n + \alpha_{n-1} x^{n-1} + \dots + \alpha_1 x$$

with the highest-order coefficient α_n from the interval $P^n < \alpha_n \le (2P)^n$ and the coefficients $\alpha_{n-1}, \dots, \alpha_1$ determined by the relations

$$\begin{aligned}\alpha_n x^n &+ \alpha_{n-1}x^{n-1} + \dots + \alpha_1 x + \alpha_0 \\ &= \alpha_n(x-x_r)^n + \beta_{n-1}(x-x_r)^{n-1} + \dots + \beta_1(x-x_r),\end{aligned}$$

where $\beta_{n-1}, \dots, \beta_2, \beta_1$ are arbitrary numbers such that

$$|\beta_{n-1}| \le (c_1 P)^{n-1}, \dots, \ |\beta_2| \le (c_1 P)^2, \ |\beta_1| \le c_1 P;$$

here $c_1 = (600n)^{-9n}$ and, moreover, $x_r = 0.25 + r(2P)^{-1}$. We show that if $r_1 \ne r_2$, then $f(x; r_1) \ne f(x; r_2)$. Indeed, the coefficient α_{n-1} of the polynomial $f(x; r)$ is equal to

$$\alpha_{n-1} = -n\big(0.25 + r(2P)^{-1}\big)\alpha_n + \beta_{n-1}.$$

Therefore, if we let α'_{n-1} and α''_{n-1} respectively denote the coefficients α_{n-1} of the polynomials $f(x; r_1)$ and $f(x; r_2)$, then we have

$$|\alpha'_{n-1} - \alpha''_{n-1}| = \left|\frac{n}{2P}(r_2 - r_1)\alpha_n + \beta'_{n-1} - \beta''_{n-1}\right| \ge \frac{n}{2P}P^n - 2(c_1 P)^{n-1} > 0,$$

which means that $f(x; r_1) \ne f(x; r_2)$. By setting $P_m = (600n)^{15m}$, we obtain the following lower bound for θ_0:

$$\theta_0 > \sum_{m=n}^{+\infty}\sum_{r=1}^{P_m}\int_{P_m^n}^{(2P_m)^n}\int_{-(c_1P_m)^{n-1}}^{(c_1P_m)^{n-1}}\cdots\int_{-c_1P_m}^{c_1P_m}|J|^{2k}\,d\alpha_n d\beta_{n-1}\dots d\beta_1, \tag{1.6}$$

where

$$J = J(\alpha_n, \beta_{n-1}, \dots, \beta_1) = \int_0^1 \exp\{2\pi i f(x)\}\,dx,$$
$$f(x) = \alpha_n(x-x_r)^n + \beta_{n-1}(x-x_r)^{n-1} + \dots + \beta_1(x-x_r).$$

Let us find a lower bound for $|J|$. We set $\Delta = c/P$, $c = (600n)^{3n}$, and $P = P_m$. Then we have

$$J = \int_{-x_r}^{1-x_r} \exp\{2\pi i(\alpha_n x^n + \beta_{n-1}x^{n-1} + \dots + \beta_1 x)\}\,dx = J_1 + J_2 + J_3,$$

where

$$J_1 = \int_{-\Delta}^{\Delta} \exp\{2\pi i(\alpha_n x^n + \beta_{n-1}x^{n-1} + \dots + \beta_1 x)\}\,dx,$$
$$J_2 = \int_{-x_r}^{-\Delta} \exp\{2\pi i(\alpha_n x^n + \beta_{n-1}x^{n-1} + \dots + \beta_1 x)\}\,dx,$$

$$J_3 = \int_{\Delta}^{1-x_r} \exp\{2\pi i(\alpha_n x^n + \beta_{n-1}x^{n-1} + \cdots + \beta_1 x)\}\,dx.$$

First, we find upper bounds for $|J_2|$ and $|J_3|$. For any x from the intervals $-x_r \le x \le -\Delta$ and $\Delta \le x \le 1 - x_r$, we have the inequality

$$\left|\frac{d^{n-1}}{dx^{n-1}}(\alpha_n x^n + \beta_{n-1}x^{n-1} + \cdots + \beta_1 x)\right| = |n!\alpha_n x + (n-1)!\beta_{n-1}|$$
$$\ge n!\alpha_n|x| - (n-1)!|\beta_{n-1}| \ge n!P^n\Delta - (n-1)!(c_1 P)^{n-1} > 0.5cn!P^{n-1}.$$

Hence the quantity H determined in Theorem 1.1 does not exceed

$$(0.5cnP^{n-1})^{1/(n-1)} \ge c^{1/(n-1)}P.$$

Therefore, as a consequence of Theorem 1.1, we obtain the following estimates for $|J_2|$ and $|J_3|$:

$$|J_2| < 6en^3c^{-1/(n-1)}P^{-1}, \quad |J_3| < 6en^3c^{-1/(n-1)}P^{-1}.$$

Now let us calculate J_1. We have

$$J_1 = \int_{-\Delta}^{\Delta} \exp\{2\pi i\alpha_n x^n\}\,dx + \int_{-\Delta}^{\Delta} \Phi(x)\exp\{2\pi i\alpha_n x^n\}\,dx,$$

where

$$\Phi(x) = \exp\{2\pi i(\beta_{n-1}x^{n-1} + \cdots + \beta_1 x)\} - 1.$$

For $|x| \le \Delta$, we trivially obtain the following estimate for $|\Phi(x)|$:

$$\begin{aligned}
|\Phi(x)| &= 2\left|\sin\pi(\beta_{n-1}x^{n-1} + \cdots + \beta_1 x)\right| \\
&\le 2\pi\left(|\beta_{n-1}|\Delta^{n-1} + \cdots + |\beta_1|\Delta\right) \\
&\le 2\pi\left((c_1\Delta P)^{n-1} + \cdots + c_1\Delta P\right) \\
&< 2\pi\left(cc_1 + (cc_1)^2 + \cdots\right) \\
&= \frac{4\pi cc_1}{1 - cc_1}.
\end{aligned}$$

Hence we have

$$\left|\int_{-\Delta}^{\Delta} \Phi(x)\exp\{2\pi i\alpha_n x^n\}\,dx\right| < \frac{2\pi c^2 c_1}{1 - cc_1}P^{-1}.$$

Further, we note that

$$\int_{-\Delta}^{\Delta} \exp\{2\pi i\alpha_n x^n\}\,dx = \int_{-\infty}^{+\infty} \exp\{2\pi i\alpha_n x^n\}\,dx + R,$$

where

$$|R| \le 2\left|\int_{\Delta}^{+\infty} \exp\{2\pi i \alpha_n x^n\}\,dx\right|.$$

Performing a change of the integration variable of the form $u = \alpha_n x^n$ and following the usual reasoning (e.g., see the proof of Lemma 1.2), we obtain the following estimate for the last integral:

$$\frac{\sqrt{2}}{n}\alpha_n^{-1/n}(\alpha_n \Delta^n)^{-1+1/n} < \frac{\sqrt{2}}{n}\cdot\frac{1}{e^{n-1}}P^{-1},$$

i.e.,

$$|R| \le \frac{\sqrt{2}}{n}\cdot\frac{1}{e^{n-1}}P^{-1}.$$

Moreover, we have

$$\begin{aligned}\int_{-\infty}^{+\infty} \exp\{2\pi i \alpha_n x^n\}\,dx &= \alpha_n^{-1/n}\int_{-\infty}^{+\infty} \exp\{2\pi i u^n\}\,du\\ &= \alpha_n^{-1/n}\left(\int_{-\infty}^{+\infty} \cos 2\pi u^n\,du + i\int_{-\infty}^{+\infty} \sin 2\pi u^n\,du\right),\end{aligned}$$

$$\begin{aligned}\int_{-\infty}^{+\infty} \cos 2\pi u^n\,du &= \frac{2}{n\sqrt[n]{2\pi}}\int_0^{+\infty}\frac{\cos u}{u^{1-1/n}}\,du = \frac{\pi}{n\sqrt[n]{2\pi}\,\Gamma(1-1/n)\sin(\pi/2n)}\\ &= \frac{2\cos(\pi/2n)}{\sqrt[n]{2\pi}}\Gamma\left(1+\frac{1}{n}\right) = 2\varkappa(n),\end{aligned}$$

i.e.,

$$\left|\int_{-\infty}^{+\infty} \exp\{2\pi i \alpha_n x^n\}\,dx\right| \ge 2\varkappa(n)\alpha_n^{-1/n}.$$

Combining the above estimates, for $|J|$ we obtain the lower bound

$$\begin{aligned}|J| &\ge 2\varkappa(n)\alpha_n^{-1/n} - 12en^3c^{-1/(n-1)}P^{-1} - \frac{4\pi c^2 c_1}{1-cc_1}P^{-1} - \frac{2\sqrt{2}}{n}\cdot\frac{1}{e^{n-1}}P^{-1}\\ &\ge P^{-1}\left(\varkappa(n) - 12en^3c^{-1/(n-1)} - \frac{4\pi c^2 c_1}{1-cc_1} - \frac{2}{e^{n-1}}\right).\end{aligned}$$

Further, we have

$$\varkappa(n) = \frac{\cos(\pi/2n)}{\sqrt[n]{2\pi}}\Gamma\left(1+\frac{1}{n}\right) \ge \cos\frac{\pi}{4}\Gamma\left(2+\frac{1}{n}\right)\left(1+\frac{1}{n}\right) > \frac{\sqrt{2}}{2}$$

$$c = (600n)^{3n}, \quad c_1 = (600n)^{-9n},$$

$$\varkappa(n) - 12en^3c^{-1/(n-1)} - \frac{4\pi c^2 c_1}{1-cc_1} - \frac{2}{e^{n-1}} > \frac{1}{4}.$$

Thus we have proved that $|J| > (4P)^{-1}$. Substituting this estimate into formula (1.6), we obtain

$$\theta_0 > \sum_{m=n}^{+\infty} P_m P_m^n c_1^{1+2+\cdots+(n-1)} P_m^{1+2+\cdots+(n-1)} \left(\frac{1}{4P_m}\right)^{2k}$$
$$= c_1^{0.5(n^2-n)} 4^{-2k} \sum_{m=n}^{+\infty} (600n)^{15m(0.5(n^2+n)+1-2k)}.$$

It follows from this relation that the integral θ_0 diverges for $2k \le 0.5(n^2+n)+1$. The proof of the theorem is complete. □

Now we state and proof the following theorem about the singular integral θ_0' corresponding to the incomplete system of equations.

Theorem 1.4. *Suppose that natural numbers $r, \ldots, m, n$ satisfy the conditions $1 \le r < \cdots < m < n$ and $r + \cdots + m + n < 0.5(n^2+n)$,*

$$\theta_0' = \theta_0'(k) = \int_{-\infty}^{+\infty} \cdots \int_{-\infty}^{+\infty} \left| \int_0^1 \exp\{2\pi i(\alpha_n x^n + \alpha_m x^m + \cdots \right.$$
$$\left. \cdots + \alpha_r x^r)\} \, dx \right|^{2k} d\alpha_n d\alpha_m \ldots d\alpha_r.$$

Then the integral θ_0' converges for $2k > n + m + \cdots + r$ and diverges for $2k \le n + m + \cdots + r$.

Proof. Let P be a natural number, and let $u_\nu = \nu/P$ $(\nu = 1, \ldots, P)$. We define the domain $\Omega_\nu = \Omega_\nu(\alpha_n, \ldots, \alpha_r)$ as follows:

$$|\beta_s(u_\nu)| \le 2^n P^s, \quad s = 1, \ldots, n,$$

where $\beta_s = \beta_s(u_\nu)$ can be found from the relation

$$\alpha_n x^n + \alpha_m x^m + \cdots + \alpha_r x^r = \beta_n (x - u_\nu)^n + \beta_{n-1}(x - u_\nu)^{n-1} + \cdots + \beta_0;$$

in other words, we have

$$\alpha_n = \beta_n, \quad \alpha_{n-1} = \beta_{n-1} - n\beta_n u_\nu,$$
$$\vdots$$
$$\alpha_s = \beta_s - \binom{s+1}{s}\beta_{s+1}u_\nu + \cdots + (-1)^{n-s}\binom{n}{s}\beta_n u_\nu^{n-s},$$
$$\vdots$$
$$\alpha_1 = \beta_1 - \binom{2}{1}\beta_2 u_\nu + \cdots + (-1)^{n-1}\binom{n}{1}\beta_n u_\nu^{n-1},$$

and $\alpha_s = 0$ if $s \neq n, m, \dots, r$. Let s be the largest number for which $\alpha_1 = 0$. Then we express β_s in terms of $\beta_{s+1}, \dots, \beta_n$ and substitute this expression into the equation

$$\alpha_{s-1} = \beta_{s-1} - \binom{s}{s-1}\beta_s u_\nu + \cdots + (-1)^{n-s+1}\binom{n}{s-1}\beta_n u_\nu^{n-s+1}.$$

We perform this transformation for equations with $\alpha_s = 0$ $(s \neq n, m, \dots, r)$. Then we see that the previous system of equations for the unknown variables $\beta_n, \beta_{n-1}, \dots, \beta_1$ is equivalent to the following one:

$$\begin{aligned} \alpha_n = \beta_n, \quad \alpha_m &= \beta_m + a_{nm}\beta_n u_\nu^{n-m}, \\ &\vdots \\ \alpha_r &= \beta_r + \cdots + a_{mr}\beta_m u_\nu^{m-r} + a_{nr}\beta_n u_\nu^{n-r}, \end{aligned}$$

where $a_{nm}, \dots, a_{mr}, a_{nr}$ are some real constants. Since $n+m+\cdots+r < 0.5(n^2+n)$, we have $\alpha_s = 0$ for some s $(1 \le s \le n)$. Then the sth equation in the original system can be rewritten as

$$\begin{aligned} \beta_s - \binom{s+1}{s}\beta_{s+1}u_\nu + \cdots + (-1)^{n-1}\binom{n-1}{s}\beta_{n-1}u_\nu^{n-1-s} \\ = (-1)^{n-1}\binom{n}{s}\beta_n u_\nu^{n-s}. \end{aligned}$$

Hence, we have

$$\begin{aligned} |\beta_n| &\le \binom{n}{s}^{-1} u_\nu^{s-n} 2^n \left(P^s + \binom{s+1}{s} P^{s+1} u_\nu + \cdots + \binom{n-1}{s} P^{n-1} u_\nu^{n-1-s} \right) \\ &\le n 2^n P^{n-1} u_\nu^{-1} = n 2^n P^n \nu^{-1}. \end{aligned}$$

Let us find an upper bound for the volume of the domain Ω_ν. We have

$$\begin{aligned} \mu(\Omega_\nu) &= \int_{\Omega_\nu} \cdots \int d\alpha_n d\alpha_m \dots d\alpha_r \\ &= \int_{-n2^n P^n \nu^{-1}}^{n2^n P^n \nu^{-1}} \int_{-2^n P^m}^{2^n P^m} \cdots \int_{-2^n P^r}^{2^n P^r} d\beta_n d\beta_m \dots d\beta_r \\ &= n 2^{(n+1)l} \nu^{-1} P^{n+m+\cdots+r}, \end{aligned}$$

where l is the number of nonzero coefficients $\alpha_n, \alpha_m, \dots, \alpha_r$. Further, we show that if the inequality

$$H = H(\alpha_n, \alpha_m, \dots, \alpha_r) \le P$$

holds at a point $(\alpha_n, \alpha_m, \dots, \alpha_r)$, the point $(\alpha_n, \alpha_m, \dots, \alpha_r)$ belongs to Ω_ν for some ν $(1 \le \nu \le P)$. Indeed, from this inequality for some ξ $(1 \le \xi \le P)$ and each $s = 1, \dots, n$, we have

$$|\beta_s(\xi)|^{1/s} \le P, \quad \text{i.e.,} \quad |\beta_s(\xi)| \le P^s.$$

We set $\nu = [\xi P]$ and show that the point $(\alpha_n, \alpha_m, \dots, \alpha_r)$ belongs to Ω_ν. Let $y = u_\nu - \xi$. Then we have

$$|\beta_s(u_r)| = |\beta_s(\xi + y)| = \left|\beta_s(\xi) + \frac{1}{1!}\beta_s'(\xi)y + \dots + \frac{1}{(n-s)!}\beta_s^{(n-s)}(\xi)y^{n-s}\right|$$
$$\le \sum_{k=0}^{n-s} \binom{s+k}{k} P^s \le 2^n P^s, \quad s = 1, \dots, n.$$

So in the domain Ω of points at which the inequality $H \le P$ holds, each point $(\alpha_n, \alpha_m, \dots, \alpha_r)$ belongs to Ω for some ν $(1 \le \nu \le P)$. Hence, for the volume of the domain Ω, we obtain

$$\mu(\Omega) \le \sum_{\nu=1}^{P} \mu(\Omega_\nu) \le n2^{(n+1)l} P^{n+m+\dots+r}(\ln P + 1).$$

We let $\pi(P)$ denote the set of points $(\alpha_n, \alpha_m, \dots, \alpha_r)$ at which the inequality

$$P < H = H(\alpha_n, \alpha_m, \dots, \alpha_r) \le 2P$$

holds. Applying Theorem 1.1 that estimates trigonometric sums, we obtain

$$\theta_0' = \theta_0'(k) \le \sum_{m=0}^{+\infty} \int \dots \int_{\pi(2^m)} \left| \int_0^1 \exp\{2\pi i(\alpha_n x^n + \alpha_m x^m + \dots \right.$$
$$\left. \dots + \alpha_r x^r)\} \, dx \right|^{2k} d\alpha_n d\alpha_m \dots d\alpha_r + \mu(\Omega(1))$$

$$\le (6e \ln 2(n+1)^3)^{2k} n2^{(n+1)l} 2^{n+m+\dots+r} \sum_{s=0}^{+\infty} s2^{-s(2k-(n+m+\dots+r))} + n2^{(n+1)l}.$$

Hence the integral $\theta_0' = \theta_0'(k)$ converges for $2k > n + m + \dots + r$.

For integer $P \ge (600n)^{15n}$, we consider the domain of points $\alpha_n, \alpha_m, \dots, \alpha_r$ whose coordinates satisfy the inequalities

$$P^n < \alpha_n \le (2P)^n, \qquad |\alpha_m| \le (c_1 P)^m, \dots, |\alpha_r| \le (c_1 P)^r, \quad c_1 = (600n)^{-9n}.$$

Let us find a lower bound for the integral J for points in this domain:

$$J = \int_0^1 \exp\{2\pi i(\alpha_n x^n + \alpha_m x^m + \dots + \alpha_r x^r)\} \, dx = \int_0^\Delta + \int_\Delta^1,$$
$$\Delta = c/P, \quad c = (600n)^{3n}.$$

It follows from Theorem 1.1 that

$$\left| \int_\Delta^1 \exp\{2\pi f(x)\} \, dx \right| \le 6en^3 c^{-1/(n-1)} P^{-1},$$

$$f(x) = \alpha_n x^n + \alpha_m x^m + \cdots + \alpha_r x^r.$$

Indeed,

$$|\beta_{n-1}(x)| = \left|\frac{f^{(n-1)}(x)}{(n-1)!}\right|$$
$$= \left|n\alpha_n x + \frac{m(m-1)\dots(m-n+2)}{(n-1)!}\alpha_m x^{m-n+1}\right| \geq \frac{n}{2}\alpha_n \Delta,$$

$$|\beta_{n-1}(x)|^{-1/(n-1)} \leq \alpha_n^{-1/(n-1)}\Delta^{-1/(n-1)} \leq P^{-1}c^{-1/(n-1)}.$$

Next, we have

$$\int_0^{\Delta} \exp\{2\pi i(\alpha_n x^n + \alpha_m x^m + \cdots + \alpha_r x^r)\}\,dx$$
$$= \int_0^{\Delta} \exp\{2\pi i \alpha_n x^n\}\,dx + \int_0^{\Delta} \Phi(x)\exp\{2\pi i \alpha_n x^n\}\,dx,$$

where

$$|\Phi(x)| = 2|\sin\pi(\alpha_m x^m + \cdots + \alpha_r x^r)|$$
$$\leq 2\pi\left((c_1P)^m\left(\frac{c}{P}\right)^m + \cdots + (c_1P)^r\left(\frac{c}{P}\right)^r\right) \leq 2\pi\frac{(c_1c)^r}{1-c_1c} \leq \frac{2\pi c_1 c}{1-c_1c}.$$

Moreover, we have

$$\left|\int_0^{\Delta}\exp\{2\pi i\alpha_n x^n\}\,dx\right| \geq \left|\int_0^{+\infty}\exp\{2\pi i\alpha_n x^n\}\,dx\right| - \left|\int_{\Delta}^{+\infty}\exp\{2\pi i\alpha_n x^n\}\,dx\right|$$
$$\geq \frac{\cos(\pi/2n)}{\sqrt[n]{2\pi\alpha_n}}\Gamma\left(1+\frac{1}{n}\right) - \frac{\sqrt{2}}{n}c^{-1/(n-1)}P^{-1}.$$

Hence,

$$J > P^{-1}\left(\frac{\cos(\pi/2n)}{2\sqrt[n]{2\pi}}\Gamma\left(1+\frac{1}{n}\right) - \frac{\sqrt{2}}{n}\cdot\frac{1}{c^{n-1}} - 2\pi\frac{c_1c^2}{1-c_1c} - 6en^3c^{-1/(n-1)}\right)$$
$$\geq \frac{1}{10}P^{-1}.$$

We set $P_s = (600n)^{15s}$. Then

$$\theta_0' = \theta_0'(k)$$
$$> \sum_{s=n}^{+\infty}\int_{P_s^n}^{(2P_s)^n}\int_{-(c_1P_s)^m}^{(c_1P_s)^m}\cdots\int_{-(c_1P_s)^r}^{(c_1P_s)^r}\left|\int_0^1\exp\{2\pi i f(x)\}\,dx\right|^{2k} d\alpha_n d\alpha_m \dots d\alpha_r \geq$$

$$\geq \sum_{s=n}^{+\infty}(10P_s)^{-2k}P_s^n c_1^{m+\dots+r}P_s^{m+\dots+r} = 10^{-2k}c_1^{m+\dots+r}\sum_{s=n}^{+\infty}P_s^{-2k+n+m+\dots+r}$$

$$= 100^{-k}c_1^{m+\dots+r}\sum_{s=n}^{+\infty}(600n)^{15s(-2k+n+m+\dots+r)}.$$

It follows from this relation that $\theta_0'(k)$ diverges for $2k \leq n+m+\dots+r$. The proof of the theorem is complete. □

1.3 Multiple trigonometric integrals

In this section we derive several estimates for multiple trigonometric integrals. First, we obtain one more estimate for the one-dimensional trigonometric integral with a polynomial in the exponent.

Lemma 1.4. *Suppose that $f(x) = \alpha_n x^n + \dots + \alpha_1 x$, where $\alpha_n, \dots, \alpha_1$ are real numbers. Let the symbol α denote the maximum modulus of these numbers. Then the integral*

$$I = \int_0^1 \exp\{2\pi i f(x)\}\,dx$$

satisfies the estimate

$$|I| \leq \min(1,\ 32\alpha^{-1/n}).$$

Proof. We assume that $n > 1$ and $\alpha > (32)^n$, because, otherwise, the lemma is trivial. We have

$$I = U + iV,$$

where

$$U = \int_0^1 \cos 2\pi f(x)\,dx, \quad V = \int_0^1 \sin 2\pi f(x)\,dx.$$

Let us consider the integral U. We perform the following partition of the interval $0 \leq x \leq 1$ into two sets E_1 and E_2 each of which also consists of intervals: the set E_1 consists of the intervals such that the inequality

$$|f'(x)| \leq A = \left(\frac{n-1}{4e}\right)^{(n-1)/n}\alpha^{-1/n}$$

holds at each point of these intervals; the other intervals form the set E_2. Then we have

$$U = U_1 + U_2,$$

where

$$U_1 = \int_{E_1}\cos 2\pi f(x)\,dx, \quad U_2 = \int_{E_2}\cos 2\pi f(x)\,dx,$$

Let μ be the sum of lengths of the intervals comprising the set E_1. Obviously, we have $|U_1| \leq \mu$. Let us find an upper bound for μ. To this end, we move the intervals of the set E together and thus form a single interval (its length is μ). In this interval we choose n points such that the distance between them is equal to $\mu/(n-1)$ and then move these integrals to their original places. Thus we obtain n points $x_1, x_2, \ldots, x_n$ in E_1 such that

$$|x_k - x_j| \geq |k-j|\mu/(n-1).$$

For each $k = 1, \ldots, n$, we consider the relations

$$\alpha_1 + 2\alpha_2 x_k + \cdots + (r+1)\alpha_{r+1}x_k^r + \cdots + n\alpha_n x_k^{n-1} = f'(x_k)$$

as a linear system of equations for the unknowns $\alpha_1, 2\alpha_2, \ldots, n\alpha_n$. Let $\alpha = |\alpha_{r+1}|$, where $0 \leq r \leq n-1$. Then we have

$$(r+1)\alpha = |\Delta'/\Delta|,$$

where

$$\begin{vmatrix} 1 & x_1 & \ldots & x_1^{n-1} \\ 1 & x_2 & \ldots & x_2^{n-1} \\ \hdotsfor{4} \\ 1 & x_n & \ldots & x_n^{n-1} \end{vmatrix},$$

and the only difference between Δ' and Δ is that the $(r+1)$st column in Δ' is replaced by the column consisting of the right-hand sides $f'(x_1), \ldots, f'(x_n)$. Expanding Δ' with respect to the $(r+1)$st column, we obtain

$$|\Delta'| \leq \sum_{k=1}^{n} |f'(x_k)|\,|\Delta'_k|,$$

where Δ'_k is obtained from Δ by crossing out the $(r+1)$st column and the kth row. The quotient obtained by dividing Δ'_k by the $(n-1)$st-order Vandermonde determinant made up from the numbers $x_1, \ldots, x_{k-1}, x_{k+1}, \ldots, x_n$ is the $(n-1-r)$th elementary symmetric function of these numbers and does not exceed $\binom{n-1}{n-1-r} = \binom{n-1}{r}$. (Indeed, if we let s_m denote the mth elementary symmetric function of the numbers $z_1, z_2, \ldots, z_l$, then for each $k = 1, \ldots, l$ we have

$$0 = \prod_{\nu=1}^{l}(z_k - z_\nu) = z_k^l - s_1 z_k^{l-1} + s_2 z_k^{l-2} - \cdots + (-1)^l s_l,$$

i.e.,

$$(-1)^{l-1}s_l + z_k(-1)^{l-2}s_{l-1} + \cdots + z_k^{l-1} = z_k^l, \quad k = 1, \ldots, l,$$

and we find s_m from this system of equations.) Therefore,

$$(r+1)\alpha = \left|\frac{\Delta'}{\Delta}\right| \le \sum_{k=1}^{n} |f'(x_k)| \binom{n-1}{r} \Bigl(\prod_{\substack{j=1\\ j\ne k}}^{n} |x_k - x_j| \Bigr)^{-1}$$

$$\le A\binom{n-1}{r}(n-1)^{n-1}\mu^{-n+1}\sum_{k=1}^{n}\frac{1}{(k-1)!(n-k)!}$$

$$= A\left(\frac{2n-2}{\mu}\right)^{n-1}\frac{1}{r!(n-1-r)!},$$

$$\mu \le A^{-1/(n-1)} 4e\alpha^{-1/(n-1)} = 4e\left(\frac{n-1}{4e}\right)^{1/n}\alpha^{-1/n},$$

$$|U_1| \le 4e\left(\frac{n-1}{4e}\right)^{1/n}\alpha^{-1/n}.$$

Now we estimate $|U_2|$. All intervals comprising E_2 can be divided into at most $2n-2$ intervals on each of which the function $f'(x)$ is monotone and of constant sign. Let $x_1 \le x \le x_2$ be such an interval, and let I' be the corresponding part of the integral U_2. Without loss of generality, we assume that $f'(x)$ is an increasing function on this interval. Setting $f(x) = v$, $f(x_1) = v_1$, and $f(x_2) = v_2$, we readily obtain

$$I' = \int_{v_1}^{v_2} \cos 2\pi v \frac{dv}{f'(x)},$$

where $f'(x)$ is considered as a function of v.

In the interval $v_1 \le v \le v_2$ we take numbers of the form $0.5l + 0.25$ with integer l to divide this interval into subintervals whose lengths do not exceed 0.5. So the integral I' can be written as an alternating series. This implies that, for some v_0 and σ such that $v_1 \le v_0 \le v_0 + \sigma$, where $\sigma \le 0.5$, we have

$$|I'| \le \int_{v_0}^{v_0+\sigma} \frac{dv}{f'(x)} = x'' - x', \quad v_0 = f(x'), \quad v_0 + \sigma = f(x'').$$

By the Lagrange theorem on finite increments, we obtain

$$\sigma = f(x'') - f(x') = f'(\xi)(x'' - x'), \quad x_1 \le x' \le \xi \le x'' \le x_2,$$

i.e.,

$$x'' - x' = \frac{\sigma}{f'(\xi)} \le \frac{1}{2A}.$$

Hence we have

$$|U_2| \le (2n-2)\frac{1}{2A} = 4e\left(\frac{n-1}{4e}\right)^{1/n}\alpha^{-1/n}$$

and thus

$$|U| \le 8e\left(\frac{n-1}{4e}\right)^{1/n}\alpha^{-1/n}.$$

By a similar argument, we obtain the same upper bound for $|V|$. So we have

$$|I| \le 8e\sqrt{2}\left(\frac{n-1}{4e}\right)^{1/n}\alpha^{-1/n} < 32\alpha^{-1/n}.$$

The proof of the lemma is complete. □

Theorem 1.5. *Let*

$$\alpha = \max_{0\le t_1,\dots,t_r\le n} |\alpha(t_1,\dots,t_r)|, \quad \alpha(0,\dots,0) = 0,$$

$$I_r = \int_0^1 \cdots \int_0^1 \exp\{2\pi i F(x_1,\dots,x_r)\}\, dx_1 \dots dx_r,$$

where

$$F(x_1,\dots,x_r) = \sum_{t_1=0}^{n} \cdots \sum_{t_r=0}^{n} \alpha(t_1,\dots,t_r) x_1^{t_1} \dots x_r^{t_r}.$$

Then

$$|I_r| \le \min\left(1,\ 32^r \alpha^{-1/n} \ln^{r-1}(\alpha+2)\right).$$

Proof. The assertion of the theorem holds for $r = 1$ (see Lemma 1.4). We shall proceed by induction over the number of variables in the polynomial. We assume that the assertion of the theorem holds for $r-1$ variables and prove this assertion for r variables. Without loss of generality, we assume that the coefficient of the variable x_1 raised to a nonzero power has maximum modulus. Let $\alpha = |(\alpha(s_1,\dots,s_r))|$; then $s_1 = 0$. We set

$$F(x_1,\dots,x_r) = \sum_{t_1=0}^{n} \cdots \sum_{t_{r-1}=0}^{n} x_1^{t_1} \dots x_{r-1}^{t_{r-1}} \varphi_{t_1,\dots,t_{r-1}}(x_r),$$

$$t = [\ln(\alpha+1)] + 1, \qquad E_0 = \{x_r \,|\, |\varphi_{s_1,\dots,s_{r-1}}(x_r)| \le 1\},$$

$$E_k = \{x_r | \alpha^{(k-1)/t} < |\varphi_{s_1,\dots,s_{r-1}}(x_r)| < \alpha^{k/t}\}, \quad k = 1, 2, \dots, t-1,$$

$$E_t = \{x_r | \alpha^{(t-1)/t} < |\varphi_{s_1,\dots,s_{r-1}}(x_r)|\}.$$

By mes E_k we denote the sum of lengths of the intervals in E_k. The estimate

$$\text{mes}\{x \,|\, |f(x)| < A\} \le 4e(A\alpha^{-1})^{1/n}$$

was proved in Lemma 1.4. Hence we have

$$\text{mes}\, E_k \le 4e\alpha^{(-t+k)/(tn)}, \quad k = 0, 1, \dots, t.$$

The integral I_r satisfies the inequality

$$|I_r| \le \text{mes } E_0 + \sum_{k=1}^{t-1} \text{mes } E_k \max_{x_r \in E_k} \left| \int_0^1 \dots \int_0^1 \exp\{2\pi i F(x_1, \dots, x_r)\}\, dx_1 \dots dx_{r-1} \right|$$
$$+ \max_{x_r \in E_t} \left| \int_0^1 \cdots \int_0^1 \exp\{2\pi i F(x_1, \dots, x_r)\}\, dx_1 \dots dx_{r-1} \right|.$$

By the induction hypothesis, we have

$$\max_{x_r \in E_k} \left| \int_0^1 \cdots \int_0^1 \exp\{2\pi i F(x_1, \dots, x_r)\}\, dx_1 \dots dx_{r-1} \right|$$
$$\le \min\left(1,\ 32^{r-1} \alpha^{-(k-1)/(tn)} \ln^{r-2}(\alpha + 2)\right).$$

Hence

$$|I_r| \le 4e\alpha^{-1/n} + \sum_{k=1}^{t-1} 4e\alpha^{(-t+k)/(tn)} 32^{r-1} \alpha^{-(k-1)/(tn)} \ln^{r-2}(\alpha + 2)$$
$$+ 32^{r-1} \alpha^{-(t-1)/(tn)} \ln^{r-2}(\alpha + 2) \le 32^r \alpha^{-1/n} \ln^{r-1}(\alpha + 2).$$

Moreover, the trivial inequality $|I_r| \le 1$ is satisfied. Combining this estimate with the previous one, we arrive at the statement of the theorem. □

The following lemma shows the *accuracy* of the estimate obtained.

Lemma 1.5. *Let* $\alpha > 1$, *and let*

$$I_r(\alpha) = \int_0^1 \cdots \int_0^1 \exp\{2\pi i \alpha x_1^n \dots x_r^n\}\, dx_1 \dots dx_r.$$

Then we have the following upper bound:

$$|I_r(\alpha)| \ge \frac{1}{2\pi n^r (r-1)!} \alpha^{-1/n} (\ln \alpha)^{r-1}.$$

Proof. First, we note that

$$I_r(\alpha) = \frac{(-1)^{r-1}}{(r-1)!} \int_0^1 \exp\{2\pi i \alpha x^n\} (\ln x)^{r-1}\, dx.$$

We shall prove this formula by induction. This formula holds for $r = 1$. We assume that this formula also holds for $r - 1$ variables and prove it for r. By the induction hypothesis, we have

$$I_r(\alpha) = \int_0^1 I_{r-1}(\alpha x^n)\, dx = \int_0^1 \frac{(-1)^{r-2}}{(r-2)!} \left(\int_0^1 \exp\{2\pi i \alpha x^n y^n\} (\ln y)^{r-2}\, dy \right) dx.$$

After the change of variables $z = xy$, we obtain

$$I_r(\alpha) = \frac{(-1)^{r-2}}{(r-2)!}\int_0^1 \frac{dx}{x}\int_0^x \exp\{2\pi i\alpha z^n\}(\ln z - \ln x)^{r-2}\,dz.$$

We integrate the last integral by parts:

$$\begin{aligned}
I_r(\alpha) &= \frac{(-1)^{r-2}}{(r-2)!}\int_0^1 (d\ln x)\int_0^x \exp\{2\pi i\alpha z^n\} \\
&\qquad \times \left(\sum_{k=0}^{r-2}(-1)^k\binom{r-2}{k}(\ln x)^k(\ln z)^{r-2-k}\right)dz \\
&= \frac{(-1)^{r-2}}{(r-2)!}\sum_{k=0}^{r-2}\frac{(-1)^k}{k+1}\binom{r-2}{k}\int_0^1 d(\ln x)^{k+1}\int_0^x \exp\{2\pi i\alpha z^n\}(\ln z)^{r-2-k}\,dz \\
&= \frac{(-1)^{r-1}}{(r-2)!}\sum_{k=0}^{r-2}\frac{(-1)^k}{k+1}\binom{r-2}{k}\int_0^x \exp\{2\pi i\alpha z^n\}(\ln z)^{r-1}\,dz.
\end{aligned}$$

Since

$$\frac{1}{r-1}\sum_{k=0}^{r-2}(-1)^k\frac{r-1}{k+1}\binom{r-2}{k} = \frac{1}{r-1},$$

this implies the desired formula for r variables.

Let us find a lower bound for $|I_r(\alpha)|$ with $\alpha > 1$. Let $J = \operatorname{Im}(I_r(\alpha))$. Then we have

$$\begin{aligned}
J &= \frac{1}{(r-1)!}\int_0^1 \sin(2\pi\alpha y^n)\left(\ln\frac{1}{y}\right)^{r-1}dy \\
&= \frac{1}{n^r(r-1)!}\int_0^1 \sin(2\pi\alpha z)\left(\ln\frac{1}{z}\right)^{r-1}z^{-1+1/n}\,dz \\
&= \frac{1}{\pi\alpha n^r(r-1)!}\int_0^1 \left(\ln\frac{1}{z}\right)^{r-1}z^{-1+1/n}\,d(\sin^2(\pi\alpha z)) \\
&= -\frac{1}{\pi\alpha n^r(r-1)!}\int_0^1 \sin^2(\pi\alpha z)\,d\left(\left(\ln\frac{1}{z}\right)^{r-1}z^{-1+1/n}\right).
\end{aligned}$$

Moreover,

$$\begin{aligned}
J &= -\frac{1}{\pi\alpha n^r(r-1)!} \\
&\quad \times \int_{1/(2\alpha)}^{1+1/(2\alpha)} \cos^2(\pi\alpha z)\,d\left(\left(\ln\frac{1}{z-1/(2\alpha)}\right)^{r-1}\left(z-\frac{1}{2\alpha}\right)^{-1+1/n}\right).
\end{aligned}$$

Since $-\frac{d}{dz}\left(\left(\ln\frac{1}{z}\right)^{r-1}z^{-1+1/n}\right) \geq 0$ $(0 < z < 1)$ is a monotonically decreasing function, summing the expressions for J, we obtain

$$2J > -\frac{1}{\pi\alpha n^r(r-1)!}\int_{1/\alpha}^{1} d\left(\left(\ln\frac{1}{z}\right)^{r-1}z^{-1+1/n}\right) = \frac{1}{\pi n^r(r-1)!}\alpha^{-1/n}(\ln\alpha)^{r-1}.$$

Thus we have

$$|I_r(\alpha)| \geq J \geq \frac{1}{2\pi n^r(r-1)!}\alpha^{-1/n}(\ln\alpha)^{r-1}.$$

The proof of the lemma is complete. □

Theorem 1.6. *Let*

$$\alpha = \max_{t_1,\dots,t_r}|\alpha(t_1,\dots,t_r)|, \quad \alpha(0,\dots,0)=0,$$

$$I_r = \int_0^1\cdots\int_0^1 \exp\{2\pi i F(x_1,\dots,x_r)\}\,dx_1\dots dx_r,$$

where

$$F(x_1,\dots,x_r) = \sum_{t_1=0}^{n}\cdots\sum_{t_r=0}^{n}\alpha(t_1,\dots,t_r)x_1^{t_1}\dots x_r^{t_r}.$$

Then the integral I_r satisfies the estimate

$$|I_r| \leq \min\left(1,\ 32^r\alpha^{-1/n}\ln^{r-1}(\alpha+2)\right),$$

where $n = \max(n_1,\dots,n_r)$.

Proof. The polynomial $F(x_1,\dots,x_r)$ can be written as

$$F(x_1,\dots,x_r) = \sum_{t_1=0}^{n}\cdots\sum_{t_r=0}^{n}\beta(t_1,\dots,t_r)x_1^{t_1}\dots x_r^{t_r},$$

where the coefficients $\beta(t_1,\dots,t_r)$ are determined by the relations

$$\beta(t_1,\dots,t_r) = \begin{cases}\alpha(t_1,\dots,t_r) & \text{if } 0\leq t_1\leq n_1,\dots,0\leq t_r\leq n_r,\\ 0 & \text{otherwise.}\end{cases}$$

Suppose that $\beta = \max_{0\leq t_1,\dots,t_r\leq n}|\beta(t_1,\dots,t_r)|$. Then, obviously, we have $\beta=\alpha$. Now we use Theorem 1.5 to estimate I_r. We obtain

$$|I_r| \leq \min\left(1,\ 32^r\beta^{-1/n}\ln^{r-1}(\beta+2)\right) = \min\left(1,\ 32^r\alpha^{-1/n}\ln^{r-1}(\alpha+2)\right).$$

The theorem is thereby proved. □

Theorem 1.7. *Suppose that* $k = k_1 + \cdots + k_r$, $k_1, \ldots, k_r$ *are natural numbers,* $\nu = 1/k, \alpha_1, \ldots, \alpha_r$ *are real numbers, and the inequality*

$$\left| \frac{\partial^k f(x_1, \ldots, x_r)}{\partial x_1^{k_1} \ldots \partial x_r^{k_r}} \right| > H$$

holds for all points $(x_1, \ldots, x_r)$ $(0 \le x_1, \ldots, x_r \le 1)$.

Suppose also that there exist $v = \binom{k+r-1}{r-1}$ *directions* $(\alpha_1, \ldots, \alpha_r)$ *such that*

(a) *the number of monotonicity intervals of the derivative*

$$\left. \frac{\partial^k f(x_1 + \alpha_1 t, \ldots, x_r + \alpha_r t)}{\partial t^k} \right|_{t=0}$$

does not exceed m *on any of the intervals that contain any of the* v *directions* $(\alpha_1, \ldots, \alpha_r)$ *and lie in the cube* $0 \le x_1, \ldots, x_r \le 1$;

(b) *the modulus of the determinant of the matrix*

$$M = \left(\frac{k!}{s_1! \ldots s_r!} \alpha_1^{s_1} \ldots \alpha_r^{s_r} \right),$$

where $0 \le s_1, \ldots, s_r \le k$, $s_1 + \cdots + s_r = k$, *and the vectors* $(\alpha_1, \ldots, \alpha_r)$ *run over the* v *vectors mentioned above, is larger than* $R > 0$, *while the modulus of the algebraic complement of each element of the matrix* M *does not exceed* $T > 0$.

Then the integral

$$J = \int_0^1 \cdots \int_0^1 \exp\{2\pi i f(x_1, \ldots, x_r)\}\, dx_1 \ldots dx_r$$

satisfies the estimate

$$|J| \le 6kv^{2+\nu} m T^\nu R^{-\nu} H^{-\nu}.$$

Proof. We have the relations

$$\left. \frac{\partial^k f(x_1 + \alpha_1 t, \ldots, x_r + \alpha_r t)}{\partial t^k} \right|_{t=0} = \sum_{\substack{s_1=0 \\ s_1 + \cdots + s_r = k}}^{k} \cdots \sum_{s_r=0}^{k} \frac{k!}{s_1! \ldots s_r!} \alpha_1^{s_1} \ldots \alpha_r^{s_r} \frac{\partial^k f(x_1, \ldots, x_r)}{\partial x_1^{s_1} \ldots \partial x_r^{s_r}}$$

for each of the directions $(\alpha_1, \ldots, \alpha_n)$. Considering them as a system of linear equations for the unknowns

$$\frac{\partial^k f(x_1, \ldots, x_r)}{\partial x_1^{s_1} \ldots \partial x_r^{s_r}},$$

we find

$$\frac{\partial^k f(x_1, \ldots, x_r)}{\partial x_1^{s_1} \ldots \partial x_r^{s_r}} = \sum_{(\alpha_1, \ldots, \alpha_r)} \cdots \sum c(\alpha_1, \ldots, \alpha_r) \left. \frac{\partial^k f(x_1 + \alpha_1 t, \ldots, x_r + \alpha_r t)}{\partial t^k} \right|_{t=0},$$

where $\sum\limits_{(\alpha_1,\dots,\alpha_r)}\cdots\sum$ denotes the summation over all directions $(\alpha_1,\dots,\alpha_n)$ determined by the assumptions of the theorem; the coefficients $|c(\alpha_1,\dots,\alpha_r)|$ do not exceed TR^{-1}.

Since the inequality

$$\left|\frac{\partial^k f(x_1,\dots,x_r)}{\partial x_1^{k_1}\dots\partial x_r^{k_r}}\right| > H$$

holds for each point $(x_1,\dots,x_r)$ $(0 \le x_1,\dots,x_r \le 1)$, for any point $(x_1,\dots,x_r)$ there exists a direction $(\alpha_1,\dots,\alpha_r)$ such that

$$\left|\frac{\partial^k f(x_1+\alpha_1 t,\dots,x_r+\alpha_r t)}{\partial t^k}\bigg|_{t=0}\right| > v^{-1}T^{-1}RH.$$

Now we arrange the directions $(\alpha_1,\dots,\alpha_r)$ in some order and divide the cube $\Omega = \{(x_1,\dots,x_r) \mid 0 \le x_1,\dots,x_r \le 1\}$ into nonintersecting domains Ω_s $(s = 1,\dots,v)$.

The first domain Ω_1 consists of all points $(x_1,\dots,x_r)$ at which the modulus of the kth-order derivative in the first direction is larger than

$$v^{-1}T^{-1}RH;$$

the second domain Ω_2 consists of all points $(x_1,\dots,x_r)$ that do not belong to Ω_1 and at which the modulus of the kth-order derivative in the second direction is larger than the same value; the third domain Ω_3 consists of all points $(x_1,\dots,x_r)$ that do not belong to Ω_1 and Ω_2 and at which the modulus of the kth-order derivative in the third direction is larger than

$$v^{-1}T^{-1}RH,$$

etc. (some of the domains Ω_s can be empty).

Each interval parallel to the sth direction contains at most sm intervals from the set Ω_s. Indeed, each interval parallel to any of the v directions $(\alpha_1,\dots,\alpha_r)$ satisfying the assumptions of the theorem contains at most m intervals from Ω_1. By the construction of the set Ω_2, each interval (starting from the second) that is parallel to any of the v directions $(\alpha_1,\dots,\alpha_r)$ contains at most $2m$ intervals from Ω_2 (we throw away at most m intervals belonging to the set Ω_1 from at most m intervals lying in the corresponding monotonicity intervals of the kth-order derivative in some direction), etc. We write the integral J as

$$J = J_1 + \dots + J_v,$$

where

$$J_s = \int\limits_{\Omega_s}\cdots\int \exp\{2\pi i f(x_1,\dots,x_r)\}\,dx_1\dots dx_r$$

for $s = 1,\dots,v$.

Now let us estimate the integral J_s. We perform a linear orthogonal change of the integration variables so that the axis y_1 is parallel to the sth vector $(\alpha_1, \ldots, \alpha_r)$, while the other coordinate axes $y_2, \ldots, y_r$ are chosen so that the coordinate system $y_1, y_2, \ldots, y_r$ is orthogonal and oriented in the same way as the coordinate system $x_1, x_2, \ldots, x_r$. Under this change of variables, the domain Ω_s turns into the domain Ω'_s and $f(x_1, \ldots, x_r) = f_1(y_1, \ldots, y_r)$.

For each fixed point $(y_2, \ldots, y_r)$, we let $T(s; y_2, \ldots, y_r)$ denote the set of y_1 for which the point $(y_1, \ldots, y_r)$ belongs to Ω'_s. The set $T(s; y_2, \ldots, y_r)$ contains at most sm intervals. We let ω_s denote the range of the variables $y_2, \ldots, y_r$ corresponding to the points $(y_1, \ldots, y_r)$ belonging to the domain Ω'_s. Then we obtain

$$\begin{aligned}
|J_s| &= \left| \int \cdots \int_{\omega_s} \int_{T(s; y_2, \ldots, y_r)} \exp\{2\pi i f_1(y_1, \ldots, y_r)\}\, dy_1 \ldots dy_r \right| \\
&\le \int \cdots \int_{\omega_s} \left| \int_{T(s; y_2, \ldots, y_r)} \exp\{2\pi i f_1(y_1, \ldots, y_r)\}\, dy_1 \right| dy_2 \ldots dy_r \\
&\le \left| \int_{T(s; y_2^{(0)}, \ldots, y_r^{(0)})} \exp\{2\pi i f_1(y_1, y_2^{(0)}, \ldots, y_r^{(0)})\}\, dy_1 \right| \int \cdots \int_{\omega_s} dy_2 \ldots dy_r,
\end{aligned}$$

where $(y_2^{(0)}, \ldots, y_r^{(0)})$ is the point of the maximum modulus of the integral

$$\left| \int_{T(s; y_2, \ldots, y_r)} \exp\{2\pi i f_1(y_1, y_2, \ldots, y_r)\}\, dy_1 \right|.$$

Since the kth-order derivative of $f_1(y_1, \ldots, y_r)$ with respect to y_1 is larger than $v^{-1}T^{-1}RH$, it follows from the estimate for the integral of a single variable (see Lemma 1.2) that

$$|J_s| \le sm \cdot 6kv^{\nu}T^{\nu}R^{-\nu}H^{-\nu}.$$

Summing all estimates for J_s, we obtain

$$|J| \le |J_1| + \cdots + |J_v| \le \sum_{s=1}^{v} sm \cdot 6kv^{\nu}T^{\nu}R^{-\nu}H^{-\nu} \le 6kv^{2+\nu}mT^{\nu}R^{-\nu}H^{-\nu}.$$

The proof of the theorem is thus complete. □

Corollary 1.3. *Suppose that*

$$\left| \frac{\partial^k F(x_1, \ldots, x_r)}{\partial l^k} \right| \ge A > 0$$

in a direction l, $F(x_1, \ldots, x_r)$ satisfies the assumptions of Theorem 1.7, *and on any interval parallel to l and lying in the cube $0 \le x_1, \ldots, x_r \le 1$, the function $G(x_1, \ldots, x_r)$ is monotone and piecewise continuous and satisfies the condition*

$$|G(x_1, \ldots, x_r)| \le H.$$

Then the integral

$$J = \int_0^1 \cdots \int_0^1 G(x_1, \dots, x_r) \exp\{2\pi i F(x_1, \dots, x_r)\} \, dx_1 \dots dx_r,$$

satisfies the estimate

$$|J| \ll H A^{-1/k}.$$

Proof. We perform a linear orthogonal change of variables so that the axis y_1 is parallel to l, while the other coordinate axes are chosen so that the obtained and the original coordinate system are oriented in the same way. Under this change of variables, the unit cube $0 \le x_1, \dots, x_r \le 1$ turns into the domain Ω, $G(x_1, \dots, x_r) = G_1(y_1, \dots, y_r)$, and $F(x_1, \dots, x_r) = F_1(y_1, \dots, y_r)$. We obtain

$$J = \int \cdots \int_{\Omega} G_1(y_1, \dots, y_r) \exp\{2\pi i F_1(y_1, \dots, y_r)\} \, dy_1 \dots dy_r.$$

For fixed $y_2, \dots, y_r$, we let $T(y_2, \dots, y_r)$ denote the set of points $(y_1, \dots, y_r)$ from Ω. By ω we denote the range of the variables $y_2, \dots, y_r$. Integrating by parts, we obtain

$$\begin{aligned}
J &= \int \cdots \int_{\omega} \left(\int_{T(y_2,\dots,y_r)} G_1(y_1, \dots, y_r) \exp\{2\pi i F_1(y_1, \dots, y_r)\} \, dy_1 \right) dy_2 \dots dy_r \\
&= \int \cdots \int_{\omega} \left(\int_{T(y_2,\dots,y_r)} G_1(y_1, \dots, y_r) \right. \\
&\qquad \left. \times \frac{\partial}{\partial y_1} \left(\int_0^{y_1} \exp\{2\pi i F_1(\xi_1, y_2, \dots, y_r)\} \, d\xi_1 \right) dy_1 \right) dy_2 \dots dy_r \\
&= \int \cdots \int_{\omega} \left(\int_0^1 \exp\{2\pi i F_1(\xi_1, y_2, \dots, y_r)\} \, d\xi_1 \right) G_1(1, y_2, \dots, y_r) \, dy_2 \dots dy_r \\
&\quad - \int \cdots \int_{\omega} \left(\int_{T(y_2,\dots,y_r)} \left(\int_0^{y_1} \exp\{2\pi i F_1(\xi_1, y_2, \dots, y_r)\} \, d\xi_1 \right) \right. \\
&\qquad \left. \times G_1(y_1, \dots, y_r) \, dy_1 \right) dy_2 \dots dy_r.
\end{aligned}$$

Estimating the integral by Theorem 1.7 as

$$\left| \int_0^{y_1} \exp\{2\pi i F_1(\xi_1, y_2, \dots, y_r)\} \, d\xi_1 \right| \ll A^{-1/k},$$

we obtain the desired estimate. The proof of the corollary is complete. □

Theorem 1.8. *Suppose that* $n, k_1, \dots, k_r$ *are integers* ($n \geq 1$, $k_1, \dots, k_r \geq 0$, $k_1 + \cdots + k_r \leq n$), $\alpha(k_1, \dots, k_r)$ *are real numbers, and*

$$F(x_1, \dots, x_r) = \sum_{k=1}^{n} \sum_{\substack{k_1=0 \\ k_1+\cdots+k_r=k}}^{n} \cdots \sum_{k_r=0}^{n} \alpha(k_1, \dots, k_r) x_1^{k_1} \dots x_r^{k_r},$$

$$\beta(\overline{x}; \overline{k}) = \frac{1}{k_1! \dots k_r!} \cdot \frac{\partial^{k_1+\cdots+k_r} F(x_1, \dots, x_r)}{\partial x_1^{k_1} \dots \partial x_r^{k_r}},$$

$$H = \min_{0 \leq x_1, \dots, x_r \leq 1} \sum_{k=1}^{n} \sum_{\substack{k_1=0 \\ k_1+\cdots+k_r=k}}^{n} \cdots \sum_{k_r=0}^{n} |\beta(\overline{x}; \overline{k})|^{1/k}.$$

Then the integral

$$J = \int_0^1 \cdots \int_0^1 \exp\{2\pi i F(x_1, \dots, x_r)\} \, dx_1 \dots dx_r$$

satisfies the estimate

$$|J| \ll \min(1, H^{-1}),$$

where the constant in $\ll$ *depends only on* n *and* r.

Proof. We choose $v = \binom{n+r}{r} - 1$ directions $(\alpha_1, \dots, \alpha_r)$ so that the rank of each matrix

$$M_k = \left(\frac{(s_1 + \cdots + s_r)!}{s_1! \dots s_r!} \alpha_1^{s_1} \dots \alpha_r^{s_r} \right),$$

$$0 \leq s_1, \dots, s_r \leq n, \quad s_1 + \cdots + s_r = k, \quad 1 \leq k \leq n,$$

is maximal. In the matrix M_k we choose $v_k = \binom{k+r-1}{r-1}$ directions $(\alpha_1, \dots, \alpha_r)$ so that the determinant of the obtained submatrix is nonzero. In particular, we can take $(\alpha_1, \alpha_1^{n+1}, \dots, \alpha_1^{(n+1)^{r-1}})$, where α_1 runs through v distinct nonzero real numbers, to be the v directions with the required properties. Further, we have

$$\left. \frac{\partial^k F(x_1 + \alpha_1 t, \dots, x_r + \alpha_r t)}{\partial t^k} \right|_{t=0} = \sum_{\substack{k_1=0 \\ k_1+\cdots+k_r=k}}^{n} \cdots \sum_{k_r=0}^{n} \frac{k!}{k_1! \dots k_r!} \alpha_1^{k_1} \dots \alpha_r^{k_r} \frac{\partial^k F(x_1, \dots, x_r)}{\partial x_1^{k_1} \dots \partial x_r^{k_r}}.$$

We consider these v_k relations as a system of linear equations for the unknowns

$$\frac{1}{k_1! \dots k_r!} \cdot \frac{\partial^k F(x_1, \dots, x_r)}{\partial x_1^{k_1} \dots \partial x_r^{k_r}}.$$

Since the determinant of this system is not equal to zero, its modulus does not exceed some constant $c(n,r) > 0$. From the system of equations we find

$$\frac{1}{k_1!\dots k_r!}\cdot\frac{\partial^k F(x_1,\dots,x_r)}{\partial x_1^{k_1}\dots\partial x_r^{k_r}}$$
$$=\sum_{(\alpha_1,\dots,\alpha_r)}\cdots\sum c(\alpha_1,\dots,\alpha_r)\frac{\partial^k F(x_1+\alpha_1 t,\dots,x_r+\alpha_r t)}{\partial t^k}\bigg|_{t=0},$$

where $\sum_{(\alpha_1,\dots,\alpha_r)}\cdots\sum$ denotes the summation over the v_k directions mentioned above and the moduli of the coefficients $c(\alpha_1,\dots,\alpha_r)$ do not exceed some constant $c_1 = c_1(n,r) > 0$.

We divide the cube $0 \le x_1,\dots,x_r \le 1$ into nonintersecting domains $\omega_1,\dots,\omega_v$ so that at the points of ω_s some kth-order ($k < n+1$) partial derivative is larger than $(H/v)^k$ (some of the domains can be empty). To this end, we need to order the partial derivatives

$$\frac{\partial^k F(x_1,\dots,x_r)}{\partial x_1^{k_1}\dots\partial x_r^{k_r}},\qquad 0\le k_1,\dots,k_r\le n,\ k=k_1+\dots+k_r\le n.$$

First, we set $k=n$ and arrange the numbers in lexicographic order for $k_1+\dots+k_r = n$. Then we set $k = n-1$ and again arrange $(k_1,\dots,k_r)$ in lexicographic order, etc. The domain ω_1 consists of all points at which the lowest-order derivative is not less than $(H/v)^n$, the domain ω_2 consists of all points at which the next (in order) derivative is larger than $(H/v)^n$ and which do not belong to ω_1, etc.

We consider the domain ω_s for an arbitrary s. The corresponding partial derivative can be expressed in terms of directional derivatives. We divide the domain ω_s into nonintersecting domains $\omega_{s1},\dots,\omega_{sv}$ so that the domain ω_{s1} consists of all points at which the derivative along the first direction is not less than $(H/v)^k v^{-1}c_1^{-1}$, the domain ω_{s2} consists of all points at which the derivative along the second direction is not less than $(H/v)^k v^{-1}c_1^{-1}$ and which do not belong to ω_{s1}, etc. (some of the domains $\omega_{s\nu}$ can be empty). The intersection of the domain $\omega_{s\nu}$ with any straight line parallel to the νth direction contains at most n^{2v} intervals, since the number of solutions to the equations

$$\frac{\partial^k F(x_1+\alpha_1 t,\dots,x_r+\alpha_r t)}{\partial x_1^{k_1}\dots\partial x_r^{k_r}}=\left(\frac{H}{v}\right)^k,$$
$$\frac{\partial^k F(x_1+\alpha_1 t,\dots,x_r+\alpha_r t)}{\partial t^k}=\left(\frac{H}{v}\right)^k v^{-1}c_1^{-1}$$

for the unknown t does not exceed n, while the number of such equations is not less than $2v$. Let us estimate the integral

$$J'=\int\cdots\int_{\omega_{s\nu}}\exp\{2\pi i F(x_1,\dots,x_r)\}\,dx_1\dots dx_r.$$

We perform a linear orthogonal change of the integration variables so that the axis y_1 is parallel to the νth direction, while the other axes are directed so that the coordinate systems $x_1, \ldots, x_r$ and $y_1, \ldots, y_r$ are oriented in the same way. Under this change of variables, the domain $\omega_{s\nu}$ turns into the domain $\omega'_{s\nu}$ and $F(x_1, \ldots, x_r) = F'(y_1, \ldots, y_r)$. For each fixed point $(y_2, \ldots, y_r)$, we let $T(y_2, \ldots, y_r)$ denote the set of y_1 for which the point $y_1, y_2, \ldots, y_r$ belongs to $\omega'_{s\nu}$. The set $T(y_2, \ldots, y_r)$ consists of at most n^{2v} intervals. We let ω denote the range of variation of the variables $y_2, \ldots, y_r$. Then we obtain

$$\begin{aligned}|J'| &= \left|\int\cdots\int_{\omega'_{s\nu}} \exp\{2\pi i F_1(y_1, \ldots, y_r)\}\,dy_1 \ldots dy_r\right| \\ &\le \int\cdots\int_{\omega} \left|\int_{T(y_2,\ldots,y_r)} \exp\{2\pi i F_1(y_1, y_2, \ldots, y_r)\}\,dy_1\right| dy_2 \ldots dy_r \\ &\le \left|\int_{T(y_2^{(0)},\ldots,y_r^{(0)})} \exp\{2\pi i F_1(y_1, y_2^{(0)}, \ldots, y_r^{(0)})\}\,dy_1\right| \ll H^{-1}.\end{aligned}$$

Summing the obtained estimates over all domains $\omega_{s\nu}$ $(1 \le s, \nu \le v)$, we obtain $|J| \ll H^{-1}$. Since we always have $|J| \ll 1$, we arrive at the desired estimate. The proof of the theorem is complete. □

1.4 Singular integrals in multidimensional problems

The integrals studied in this section are closely related to the number of solutions of systems of Diophantine equations similar to system (1.3) in Section 1.2 but much more complicated (see Chapter 7). They have the form

$$\theta = \int_{-\infty}^{+\infty} \cdots \int_{-\infty}^{+\infty} \left|\int_0^1 \cdots \int_0^1 \exp\{2\pi i F(x_1, \ldots, x_r)\}\,dx_1 \ldots dx_r\right|^{2K} d\overline{\alpha}, \quad (1.7)$$

where

$$F(x_1, \ldots, x_r) = \sum_{t_1=0}^{n_1} \cdots \sum_{t_r=0}^{n_r} \alpha(t_1, \ldots, t_r) x_1^{t_1} \ldots x_r^{t_r}.$$

Precisely as in Section 1.2, some of the coefficients $\alpha(t_1, \ldots, t_r)$ can be identically zero. The problem of estimating the convergence exponent for integrals θ significantly depends on the polynomials $F(x_1, \ldots, x_r)$, i.e., on the systems of Diophantine equations to which θ correspond. Therefore, the methods for solving these problems are different.

Theorem 1.9. *Suppose that* $m = (n_1 + 1) \ldots (n_r + 1) - 1$, $\overline{\alpha}$ *is an* m*-dimensional vector whose coordinates are coefficients of the polynomial* $F(x_1, \ldots, x_r)$, *and* θ *is the singular integral* (1.7) *corresponding to* $F(x_1, \ldots, x_r)$. *Then* θ *converges for*

$$2K > nm, \quad n = \max(n_1, \ldots, n_r).$$

Proof. By Theorem 1.6, we have

$$|I_r| = \left| \int_0^1 \cdots \int_0^1 \exp\{2\pi i F(x_1, \ldots, x_r)\}\, dx_1 \ldots dx_r \right|$$
$$\leq \min\left(1,\ 32^r \alpha^{-1/n} \ln^{r-1}(\alpha+2)\right) \ll \min\left(1,\ \alpha^{\varepsilon - 1/n}\right),$$

where $\alpha = \max_{t_1,\ldots,t_r} |\alpha(t_1, \ldots, t_r)|$, $\alpha(0, \ldots, 0) = 0$, $\varepsilon > 0$ is an arbitrarily small fixed number, and thc constant in $\ll$ depends only on ε and r.

Further, we have

$$\min(1,\ \alpha^{\varepsilon-1/n})$$
$$= \min\left\{ \min\left(1, |\alpha(0, \ldots, 1)|^{\varepsilon-1/n}\right), \ldots, \min\left(1, |\alpha(n_1, \ldots, n_r)|^{\varepsilon-1/n}\right)\right\}$$
$$\leq \prod_{\substack{t_1=0 \\ t_1+\cdots+t_r\geq 1}}^{n} \cdots \prod_{t_r=0}^{n} \min\left(1,\ |\alpha(t_1, \ldots, t_r)|^{(-1+\varepsilon n)/(nm)}\right).$$

Hence

$$|I_r| \ll \prod_{\substack{t_1=0 \\ t_1+\cdots+t_r\geq 1}}^{n} \cdots \prod_{t_r=0}^{n} \min\left(1,\ |\alpha(t_1, \ldots, t_r)|^{(-1+\varepsilon n)/(nm)}\right),$$

$$\theta \ll \prod_{\substack{t_1=0 \\ t_1+\cdots+t_r\geq 1}}^{n} \cdots \prod_{t_r=0}^{n} \int_{-\infty}^{+\infty} \left(\min(1,\ |\alpha(t_1, \ldots, t_r)|^{(-1+\varepsilon n)/(nm)}\right)^{2K} d\alpha(t_1, \ldots, t_r).$$

Each of these single integrals has the form

$$\int_{-\infty}^{+\infty} \min\left(1,\ |\alpha|^{2K(-1+\varepsilon n)/(nm)}\right) d\alpha.$$

Since $\varepsilon > 0$ is an arbitrarily small fixed number, the last integral converges for

$$-2K/(nm) < -1, \quad 2K > nm.$$

This implies that θ converges for $2K > nm$. The proof of the theorem is complete. □

Theorem 1.10. *Suppose that the singular integral θ corresponds to a polynomial $F(x_1, \ldots, x_r)$ of the form*

$$F(x_1, \ldots, x_r) = \sum_{\substack{t_1=0 \\ t_1+\cdots+t_r\leq n}}^{n} \cdots \sum_{t_r=0}^{n} \alpha(t_1, \ldots, t_r) x_1^{t_1} \ldots x_r^{t_r}.$$

Then the integral θ converges for

$$2K > r\binom{n+r}{r+1} + r.$$

Proof. Let us estimate the volume of the domain $\Omega = \Omega(\overline{\alpha}; P)$ consisting of points $\alpha(k_1, \dots, k_r)$ $(0 \le k_1, \dots, k_r, k_1 + \dots + k_r \le n)$ at which the quantity H determined in Theorem 1.4 does not exceed P (P is a natural number). For $1 \le s_1, \dots, s_r \le P$ we set $\overline{u}(\overline{s}) = (s_1/P, \dots, s_r/P)$ and consider the domains $\Omega(\overline{s}) = \Omega(\overline{s}; \overline{\alpha})$ of points at which the inequalities

$$|\beta(\overline{k}; \overline{u}(\overline{s}))| \le (r+1)^n P^{k_1 + \dots + k_r}, \quad 0 \le k_1, \dots, k_r, \ k_1 + \dots + k_r \le n,$$

are satisfied.

Let us find an upper bound for the volume $\mu(\Omega(\overline{s}))$ of the domain $\Omega(\overline{s})$. We have

$$\mu(\Omega(\overline{s})) = \int \cdots \int_{\Omega(\overline{s})} d\overline{\alpha}.$$

To perform a change of variables in this integral, we find new variables from the relations

$$\begin{aligned} F(x_1 - u_1, \dots, x_r - u_r) &= \sum_{\substack{s_1=0 \\ 1 \le s_1 + \dots + s_r \le n}}^{n} \cdots \sum_{s_r=0}^{n} \beta(s_1, \dots, s_r)(x_1 - u_1)^{s_1} \dots (x_r - u_r)^{s_r} \\ &= \sum_{\substack{t_1=0 \\ t_1 + \dots + t_r \le n}}^{n} \cdots \sum_{t_r=0}^{n} \alpha(t_1, \dots, t_r) x_1^{t_1} \dots x_r^{t_r}, \end{aligned}$$

$$\begin{aligned} \alpha(t_1, \dots, t_r) = \sum_{s_1=t_1}^{n} \cdots \sum_{s_r=t_r}^{n} (-1)^{s_1 - t_1 + \dots + s_r - t_r} \binom{s_1}{t_1} \cdots \binom{s_r}{t_r} \beta(s_1, \dots, s_r) \\ \times u_1^{s_1 - t_1} \dots u_r^{s_r - t_r}, \quad 0 \le t_1, \dots, t_r, \ t_1 + \dots + t_r \le n. \end{aligned}$$

The Jacobian of this transformation is equal to 1. Hence we have

$$\mu(\Omega(\overline{s})) = \int \cdots \int_{\substack{|\beta(\overline{k})| < (r+1)^n P^{k_1 + \dots + k_r} \\ 0 \le k_1, \dots, k_r, \ k_1 + \dots + k_r \le n}} d\overline{\beta} = (2(r+1)^n)^v P^{\Delta}.$$

Here v is the number of coefficients $\beta(\overline{k})$ $(0 \le k_1, \dots, k_r, 1 \le k_1 + \dots + k_r \le n)$, $v = \binom{n+r}{r} - 1$, and the value of Δ is determined by the relation

$$\Delta = \sum_{\substack{t_1=0 \\ t_1 + \dots + t_r \le n}}^{n} \cdots \sum_{t_r=0}^{n} (t_1 + \dots + t_r) = \sum_{s=1}^{n} s \binom{s + r - 1}{s}.$$

It follows from the relations $\binom{s+r-1}{s} = \binom{s+r}{s} - \binom{s+r-1}{s-1}$ that

$$\Delta = \sum_{s=1}^{n} s\left(\binom{s+r}{s} - \binom{s+r-1}{s-1}\right) = n\binom{n+r}{n} - \binom{n+r}{n-1} = r\binom{n+r}{n-1}.$$

We show that if the inequality $H \leq P$ is satisfied at a point $\overline{\alpha}$, then this point belongs to $\Omega(\overline{s})$ for some $\overline{s} = (s_1, \ldots, s_r)$. It follows from this inequality that there exists a point $\overline{\xi} = (\xi_1, \ldots, \xi_r)$ $(0 \leq \xi_1, \ldots, \xi_r \leq 1)$ such that for each $k_1, \ldots, k_r \geq 0$, $1 \leq k_1 + \ldots k_r \leq n$, we have

$$|\beta(\overline{k}; \overline{\xi})|^{1/k} \leq P,$$

i.e.,

$$|\beta(\overline{k}; \overline{\xi})| \leq P^{k_1+\cdots+k_r}.$$

We choose $s_1 = [\xi_1 P], \ldots, s_r = [\xi_r P]$ and show that $\overline{\alpha}$ belongs to $\Omega(\overline{s})$. Let

$$\overline{y} = \overline{u}(\overline{s}) - \overline{\xi}, \quad \overline{u}(\overline{s}) = \left(\frac{s_1}{P}, \ldots, \frac{s_r}{P}\right).$$

Then, using the Taylor formula, we obtain

$$\begin{aligned} |\beta(\overline{k}; \overline{u}(\overline{s}))| &= |\beta(\overline{k}; \overline{\xi} + \overline{y})| \\ &= \Bigg|\beta(\overline{k}; \overline{\xi}) + \frac{1}{1!}\left(\beta'_{\xi_1}(\overline{k}; \overline{\xi})y_1 + \cdots + \beta'_{\xi_r}(\overline{k}; \overline{\xi})y_r\right) + \cdots \\ &\quad + \frac{1}{(n-s)!}\left(y_1\frac{\partial}{\partial \xi_1} + \cdots + y_r\frac{\partial}{\partial \xi_r}\right)^{n-s} \beta(\overline{k}; \overline{\xi})\Bigg| \leq (r+1)^n P^{k_1+\cdots+k_r}. \end{aligned}$$

Thus the point $\overline{\alpha}$ belongs to the domain Ω. Hence we have

$$\mu(\Omega) \leq \sum_{s_1=1}^{P} \cdots \sum_{s_r=1}^{P} \mu(\Omega(\overline{s})) = \left(2(r+1)^n\right)^v P^{\Delta+r}.$$

We let $\pi(P)$ denote the set of points $\overline{\alpha}$ at which $P < H \leq 2P$. Then for the integral θ we have

$$\begin{aligned} \theta = \sum_{m=0}^{+\infty} \int \cdots \int_{\pi(2^m)} \left| \int_0^1 \cdots \int_0^1 \exp\{2\pi i F(x_1, \ldots, x_r)\}\, dx_1 \ldots dx_r \right|^{2K} d\overline{\alpha} \\ + \int \cdots \int_{\Omega(\overline{\alpha}; 1)} \left| \int_0^1 \cdots \int_0^1 \exp\{2\pi i F(x_1, \ldots, x_r)\}\, dx_1 \ldots dx_r \right|^{2K} d\overline{\alpha}. \end{aligned}$$

Applying Theorem 1.8 to estimate the trigonometric integral, we obtain

$$\theta \ll \sum_{m=0}^{+\infty} (2(r+1)^n)^v 2^{\Delta+r} 2^{m(\Delta+r-2K)} + (2(r+1)^n)^v,$$

which implies that the integral θ converges for $2K > \Delta + r$. The proof of the theorem is complete. □

We note that the problem of the convergence exponent for singular integrals θ in multidimensional Tarry's problems remains open. Apparently, here each integral θ will have its own convergence exponent depending on the form of the corresponding polynomial $F(x_1, \ldots, x_r)$.

Concluding remarks on Chapter 1. 1. Lemma 1.1 is a generalization of Vinogradov's assertion (see Lemma 4 in Chapter II in [165]. In the special case $n = 1, 2$, Lemma 1.2 was proved in [145]. In the general case, this lemma was proved in [27], [28].

2. In Theorem 1.1, the modulus of the trigonometric integral of a polynomial is estimated via a positive function of the coefficients of this polynomial such that this function is invariant under shifts of the integration variable. This estimate is the best possible with respect to this function and the length of the integration interval.

3. Theorem 1.2 on estimating trigonometric integrals via the lower bound for a linear combination of the derivatives of the function contained in the exponent was generalized by I. A. Ikromov to the case of linear combinations with variable coefficients ([71]).

4. Theorem 1.3 ([27], [28]) gives a solution of the Hua Loo-Keng problem about the convergence exponent in the singular integral in Tarry's problem.

5. In Theorem 1.4, the convergence exponent in the singular integral is found for an incomplete systems of equations in Tarry's problem ([27], [28]).

6. Lemma 1.4 (in Section 1.3) was first proved by I. M. Vinogradov ([165]).

7. Theorems 1.5 and 1.9 were proved by V. N. Chubarikov ([47], [48]).

8. Theorems 1.7, 1.8, and 1.10 are contained in [28].

9. Estimates of trigonometric integrals are used in the mathematical theory of tomography, in harmonic analysis, etc. (see [135], [136], [137], [134]).

Chapter 2

Rational trigonometric sums

A *complete rational trigonometric sum modulo q* is defined to be a sum of the form

$$S = S(q, F(x_1, \dots, x_r)) = \sum_{x_1=1}^{q} \cdots \sum_{x_r=1}^{q} \exp\{2\pi i F(x_1, \dots, x_r)/q\}, \tag{2.1}$$

where q is a natural number and

$$F(x_1, \dots, x_r) = \sum_{t_1=0}^{n_1} \cdots \sum_{t_r=0}^{n_r} a(t_1, \dots, t_r) x_1^{t_1} \dots x_r^{t_r};$$

here $a(t_1, \dots, t_r)$ are integers.

In this chapter we shall study the following two problems:

(1) finding upper bounds for the moduli of such sums;

(2) finding the convergence exponents for singular series in Tarry's problem and its generalizations.

In most detail, we shall study the case of one-dimensional complete rational trigonometric sums modulo q, i.e., sums of the form

$$S(q, f(x)) = \sum_{x=1}^{q} \exp\{2\pi i f(x)/q\}, \tag{2.2}$$

where $f(x) = a_n x^n + \dots + a_1 x$ is a polynomial with integer coefficients. An upper bound sharp in order of increase of q was obtained for the absolute value of such a sum by Hua Loo-Keng in 1940 [70]. Here we present the derivation of this estimate given by Chen, Theorems 1 and 2 in [45], (see also [49]).

Next, we obtain exact values of the convergence exponents for singular series in Tarry's problem [68] and in its generalization to the case of "incomplete systems of equations." Finally, in the last sections, we obtain estimates for multiple complete trigonometric sums modulo q and present an upper bound for the convergence exponent of the singular series in the "multidimensional Tarry problem."

2.1 One-dimensional sums

In this section we estimate one-dimensional rational trigonometric sums modulo q. First, we prove some auxiliary lemmas.

Lemma 2.1. *Suppose that $f(x)$ is a polynomial with integer coefficients and $f(0) = 0$. Then the relation*

$$S\big(q_1, q_2, f(x)\big) = S\big(q_1, q_2^{-1} f(q_2 x)\big) S\big(q_2, q_1^{-1} f(q_1 x)\big)$$

holds for any coprime natural numbers q_1 and q_2.

Proof. Any residue x modulo $q_1 q_2$ can be uniquely represented as

$$x \equiv q_2 x_1 + q_1 x_2 \pmod{q_1 q_2},$$

where $1 \le x_1 \le q_1$ and $1 \le x_2 \le q_2$. This implies

$$f(x) \equiv f(q_2 x_1) + f(q_1 x_2) \pmod{q_1 q_2}.$$

Hence we have

$$\begin{aligned} S(q_1, q_2, f(x)) &= \sum_{x=1}^{q_1 q_2} \exp\{2\pi i f(x)/(q_1 q_2)\} \\ &= \sum_{x_1=1}^{q_1} \exp\{2\pi i q_2^{-1} f(q_2 x_1)/q_1\} \sum_{x_2=1}^{q_2} \exp\{2\pi i q_1^{-1} f(q_1 x_2)/q_2\} \\ &= S(q_1, q_2^{-1} f(q_2 x)) S(q_2, q_1^{-1} f(q_1 x)). \end{aligned}$$

The lemma is thereby proved. □

Lemma 2.2. *Suppose that $g(x)$ is a polynomial with integer coefficients and a is a root of $g(x)$ modulo p of multiplicity m. Suppose also that u is the largest power of p that divides all the coefficients of the polynomial*

$$h(x) = g(px + a).$$

Then the number of roots of the polynomial $p^{-u} h(x)$ modulo p, with their multiplicity taken into account, does not exceed m.

Proof. Since the residue a is a root of multiplicity m of the congruence $g(x) \equiv 0 \pmod{p}$, the polynomial $g(x)$ can be represented as

$$g(x) = (x - a)^m k(x) + p l(x),$$

where the degree of $l(x)$ is less than m and $(k(a), p) = 1$. This implies

$$p^{-u}h(x) = p^{-u}g(px + a) = p^{m-u}x^m k(px + a) + p^{1-u}l(px + a).$$

It follows from this relation that $m \geq n$. Hence the congruence $p^{-u}h(x) \equiv 0 \pmod{p}$ is equivalent to the congruence

$$p^{m-u}x^m k(a) + p^{1-u}l(px + a) \equiv 0 \pmod{p}.$$

Its degree does not exceed m, and hence the number of its solutions does not exceed m. The proof of the lemma is complete. □

Lemma 2.3. *Let $f(x) = a_n x^n + \cdots + a_1 x$, $p \nmid (a_n, \ldots, a_1)$, and let u be the highest power of p that is a divisor of all the coefficients of the polynomial*

$$g(x) = f(\lambda + px) - f(\lambda).$$

Then

$$1 \leq u \leq n.$$

Proof. Since the constant term in $g(x)$ is zero and all other its terms are divisible by p, we have $u \geq 1$.

Let τ be the largest number such that $(a_\tau, p) = 1$. Then the coefficients of x^τ in the polynomial $g(x)$ are divisible by p^τ but not divisible by $p^{\tau+1}$. Hence we have $u \leq \tau \leq n$. The lemma is thereby proved. □

Lemma 2.4. *Let $a > 1$, and let r, k be integers such that $1 \leq r \leq k$. We set*

$$M(r) = \max_{m_1+\cdots+m_r=k} \sum_{j=1}^{r} a^{m_j},$$

where $m_1, \ldots, m_r$ are positive integers. Then

$$M(r) \leq \max(ka, a^k).$$

Proof. Note that for $m \geq n \geq 1$ we have

$$a^m + a^n \leq a^{m+n-1} + a.$$

Indeed, for $a > 1$ we have $(a^m - a)(a^{n-1} - 1) > 0$ and hence

$$\sum_{j=1}^{r} a^{m_j} \leq a^{m_1+m_2-1} + \sum_{j=3}^{r} a^{m_j} + a \leq \cdots \leq a^{m_1+\cdots+m_r-r+1} + (r-1)a.$$

For $M(r)$ $(1 \leq r \leq k)$, this implies the upper bound

$$M(r) \leq (r-1)a + a^{k-r+1} = g(r).$$

Since $g''(r) = a^{k-r+1}(\log a)^2 > 0$, we obtain

$$M(r) \le \max\big(g(l),\ g(k)\big) = \max(ka,\ a^k).$$

The proof of the lemma is complete. □

Theorem 2.1. *Suppose that $n \ge 3$ is an integer, $f(x) = a_n x^n + \cdots + a_1 x + a_0$ is a polynomial with integer coefficients, $(a_n, \ldots, a_1, p) = 1$, p is a prime, and l is a natural number. Then*

$$|S(p^l, f(x))| \le c_1(n) p^{l(1-1/n)},$$

where

$$c_1(n) = \begin{cases} 1 & \text{if } p \ge (n-1)^{2n/(n-2)}, \\ n^{2/n} & \text{if } (n-1)^{2n/(n-2)} > p \ge (n-1)^{n/(n-2)}, \\ n^{3/n} & \text{if } (n-1)^{n/(n-2)} > p > n, \\ (n-1)n^{3/n} & \text{if } p \le n. \end{cases}$$

Proof. First we consider the case $p > n$. We write the estimate of the rational trigonometric sum with a prime denominator (see Lemma A.5, i.e., the Weil estimate) in the form

$$|S(p, f(x))| \le \min\big(p^{1/n},\ (n-1)p^{-0.5+1/n}\big)p^{1-1/n}.$$

This implies the estimate of the sum in Theorem 2.1 for $l = 1$ and $p > n$.

Now we suppose that $l \ge 2$. Let $\mu_1, \ldots, \mu_r$ be distinct roots of the congruence $f'(x) \equiv 0 \pmod p$, and let $m_1, \ldots, m_r$ be their multiplicities. We set $m_1 + \cdots + m_r = m$. Obviously, $0 \le m \le n-1$. Then we have

$$|S(p^l, f(x))| \le \sum_{v=1}^{p} |S_v|,$$

where

$$S_v = \sum_{\substack{0<x\le p^l \\ x\equiv v \pmod p}} \exp\{2\pi i f(x)/p^l\}.$$

We transform the sum S_v by using the substitution $x = y + p^{l-1}z$, where y and z take the values $y = 1, \ldots, p^{l-1}$ and $z = 0, \ldots, p-1$ independently. For $l \ge 2$ we obtain the relations

$$S_v = \sum_{\substack{0<y\le p^{l-1} \\ y\equiv v \pmod p}} \sum_{z=0}^{p-1} \exp\{2\pi i f(y + p^{l-1}z)/p^l\} = \tag{2.3}$$

$$= \sum_{\substack{0<y\le p^{l-1} \\ y\equiv v \pmod p}} \exp\{2\pi i f(y)/p^l\} \sum_{z=0}^{p-1} \exp\{2\pi i f'(y)z/p\}.$$

This yields that the relation $S_v = 0$ holds for $v \neq \mu_j \quad (j = 1, \ldots, r)$.

Let $\sigma_j \quad (j = 1, \ldots, r)$ be the largest power of p dividing all the coefficients of the polynomial

$$f(py + \mu_j) - f(\mu_j) = p^{\sigma_j} g_j(y).$$

Then it follows from the expansion

$$f(py+\mu_j) - f(\mu_j) = pyf'(\mu_j) + \frac{(py)^2 f''(\mu_j)}{2!} + \cdots + \frac{(py)^{m_j} f^{(m_j)}(\mu_j)}{m_j!} + \cdots$$

that $2 \le \sigma_j \le m_j + 1$.

Relation (2.3) with $1 \le j \le r$ and $l \ge 2$ implies

$$S_{\mu_j} = p \sum_{\substack{0<y\le p^{l-1} \\ y\equiv \mu_j \pmod p}} \exp\{2\pi i f(y)/p^l\} = p \sum_{y=0}^{p^{l-2}-1} \exp\{2\pi i f(\mu_j + py)/p^l\}. \quad (2.4)$$

We assume that $l > \sigma_j$. We estimate the sum S_{μ_j} by mathematical induction and thus obtain

$$\begin{aligned} |S_{\mu_j}| &= p\Big|\sum_{y=0}^{p^{l-2}-1} \exp\{2\pi i g_j(y)/p^{l-\sigma_j}\}\Big| = p^{\sigma_j - 1}\big|S(p^{l-\sigma_j}, g_j(y))\big| \quad (2.5)\\ &\le p^{\sigma_j-1+(l-\sigma_j)(1-1/n)} \max\big(1,\ \min(p^{1/n},\ (n-1)p^{-0.5+1/n})\big) \\ &= p^{l(1-1/n)-1+\sigma_j/n} \max\big(1,\ \min(p^{1/n},\ (n-1)p^{-0.5+1/n})\big). \end{aligned}$$

We assume that $l \le \sigma_j$. It follows from (2.4) that $|S_{\mu_j}| \le p^{l-1}$. Thus (2.5) also holds for $l \le \sigma_j$. This implies

$$|S(p^l, f(x))| \le \max\big(1,\ \min(p^{1/n},\ (n-1)p^{-0.5+1/n})\big) p^{l(1-1/n)} \sum_{j=1}^{r} p^{-1+\sigma_j/n}.$$

For $p \ge (n-1)^{n/(n-2)}$, by the inequalities $\sigma_j \le m_j + 1$ and $m_1 + \cdots + m_r = m \le n-1$ and Lemma 2.4, we obtain

$$\sum_{j=1}^{r} p^{-1+\sigma_j/n} \le p^{-1+1/n} \sum_{j=1}^{r} p^{m_j/n} \le p^{-1+1/n} \max\big((n-1)p^{1/n},\ p^{(n-1)/n}\big) \le 1.$$

Hence Theorem 2.1 is proved for $p \geq (n-1)^{n/(n-2)}$. For the case $n < p \leq (n-1)^{n/(n-2)}$ and $l \geq 2$, it suffices to obtain the estimate

$$|S(p^l, f(x))| \leq mp^{-1+3/n+l(1-1/n)}. \tag{2.6}$$

It follows from (2.4) that (2.6) holds for $l = 2$. Now we proceed by induction. We assume that (2.6) holds for $n < p \leq (n-1)^{n/(n-2)}$ and $2 \leq l \leq L$. We shall prove the estimate (2.6) for $n < p \leq (n-1)^{n/(n-2)}$ and $l = L+1$. Since we have $p > n$ and $2 \leq \sigma_j \leq m_j + 1$, we obtain the relation

$$g_j(y) \equiv p^{-\sigma_j}\Bigg(pf'(\mu_j) + \cdots + \frac{p^{m_j}y^{m_j-1}f^{(m_j)}(\mu_j)}{(m_j-1)!} + \frac{p^{m_j+1}y^{m_j}f^{(m_j+1)}(\mu_j)}{m_j!}\Bigg) \pmod p.$$

Hence the number of roots of the congruence $g_j'(y) \equiv 0 \pmod p$ does not exceed m_j.

If $\sigma_j < L$, then using (2.4), the inequality $\sigma_j \leq m_j + 1$, and the induction hypothesis (2.6) for $n < p \leq (n-1)^{n/(n-2)}$, we obtain

$$\begin{aligned}|S_{\mu_j}| &= p^{\sigma_j-1}\big|S(p^{L+1-\sigma_j}, g_j(y))\big| \leq m_j p^{\sigma_j-2+3/n+(L+1-\sigma_j)(1-1/n)} \\ &\leq m_j p^{-2+(m_j+1)/n+3/n} p^{(L+1)(1-1/n)}.\end{aligned} \tag{2.7}$$

If $\sigma_j \geq L$, then (2.4) implies

$$|S_{\mu_j}| \leq p^L. \tag{2.8}$$

We set $f_1(y) = yp^{1/n} - p^{y/n}$. Then we obtain

$$f_1'(y) = p^{1/n} - n^{-1}p^{y/n}\log p, \quad f_1''(y) \leq 0, \quad f_1(1) = 0.$$

It follows from the condition $n < p \leq (n-1)^{n/(n-2)}$ that $f_1'(1) \geq 0$ and $f_1(n-1) \geq 0$. Hence we have $f_1(y) \geq 0$ for $1 \leq y \leq n-1$. Since $1 \leq m_j \leq n-1$, we derive

$$m_j p^{1/n} \geq p^{m_j/n} \quad \text{for } n < p < (n-1)^{n/(n-2)}. \tag{2.9}$$

The following two cases are possible: (1) $L > \sigma_1 \geq \cdots \geq \sigma_r$; (2) $\sigma_1 \geq \cdots \geq \sigma_{r_1} \geq L > \sigma_{r_1+1} \geq \cdots \geq \sigma_r$. Suppose that conditions in case (1) are satisfied. Then from (2.7) $(1 \leq m_j \leq n-1)$ we obtain

$$\begin{aligned}\big|S(p^{L+1}, f(x))\big| &\leq p^{-2+3/n+(L+1)(1-1/n)}\sum_{j=1}^{r} m_j p^{(m_j+1)/n} \\ &\leq mp^{-1+3/n+(L+1)(1-1/n)}.\end{aligned}$$

Hence (2.6) holds in case (1). Now we assume that conditions in case (2) are satisfied. From (2.7) and (2.8) we obtain

$$|S(p^{L+1}, f(x))| \leq r_1 p^L + \sum_{j=r_1+1}^{r} m_j p^{(m_j+1)/n} p^{-2+3/n+(L+1)(1-1/n)}. \tag{2.10}$$

We set $m_1 + \cdots + m_{r_1} = M_1$ and $m_{r_1+1} + \cdots + m_r = M_2$. Then we have $M_1 + M_2 = m$. Since $m_j + 1 \ge \sigma_j \ge L$ for $j = 1, \ldots, r_1$, we have $M_1 + r_1 \ge r_1 L$, i.e., $L \le 1 + M_1/r_1$. Hence

$$L = (L+1)\Big(1-\frac{1}{n}\Big) - 1 + \frac{2}{n} + \frac{L-1}{n} \le (L+1)\Big(1-\frac{1}{n}\Big) - 1 + \frac{2}{n} + \frac{M_1}{r_1 n}. \quad (2.11)$$

We set $f_1(y) = yp^{M_1/(yn)}$ and calculate the derivatives

$$f_2'(y) = p^{M_r/(yn)}\Big(1 - \frac{M_r \log p}{yn}\Big), \quad f_2''(y) = p^{M_r/(yn)}\frac{M_1^2 \log^2 p}{y^3 n^2} \ge 0.$$

Hence for $1 \le r_1 \le M_1$ we obtain

$$r_1 p^{M_1/(r_1 n)} \le \max(p^{M_1/n},\ M_1 p^{1/n}) = A,$$

and (2.9) implies the inequality

$$A \le M_1 p^{1/n}. \quad (2.12)$$

It follows from (2.10) and (2.12) that

$$|S(p^{L+1}, f(x))| \le \Big(r_1 p^{M_1/(rn)} + \sum_{j=r_1+1}^{r} m_j p^{(m_j+1)/n-1+n}\Big) p^{(L+1)(1-1/n)-1+2/n}$$
$$\le (M_1 p^{1/n} + M_2 p^{1/n}) p^{(L+1)(1-1/n)-1+2/n} = mp^{(L+1)(1-1/n)-1+3/n}.$$

This implies that inequality (2.6) holds for $l = L+1$ and hence Theorem 2.1 holds for $n < p \le (n-1)^{n/(n-2)}$. The case $p > n$ has been studied completely.

Let $p \le n$. From the condition $p^t \,\|\, (na_n, \ldots, 2a_2, a_1)$, we find an integer t. Since $(a_n, \ldots, a_1, p) = 1$, we have $p^t \le n$. Suppose that $\mu_1, \ldots, \mu_r$ are distinct roots of the congruence $f'(x) \equiv 0 \pmod{p^{t+1}}$ $(0 \le x < p)$ and $m_1, \ldots, m_r$ are their multiplicities. Obviously, by setting $m_1 + \cdots + m_r = m$, we obtain $m \le n-1$. For $p \le n$ and $l \ge 1$, it suffices to prove that

$$|S(p^l, f(x))| \le n^{3/n} \max(1, m) p^{l(1-1/n)}. \quad (2.13)$$

Let $l < 2(t+1)$. Then it follows from the inequality $p^t \le n$ that

$$|S(p^l, f(x))| \le p^l = p^{l(1-1/n)} p^{l/n} \le p^{(2t+1)/n} p^{l(1-1/n)} \le n^{3/n} p^{l(1-1/n)}.$$

Now we assume that $l \ge 2(t+1)$ and transform the last sum by using the substitution $x = y + p^{l-t-1}z$ $(y = 1, \ldots, p^{l-t-1},\ z = 0, 1, \ldots, p^{t+1}-1)$. Then for $l \ge 2(t+1)$ we have

$$S_v = \sum_{\substack{0<y\le p^{l-t-1}\\ y\equiv v \pmod p}} \sum_{z=0}^{p^{t+1}-1} \exp\{2\pi i(f(y) + p^{l-t-1}zf'(y))/p^l\} = 0,$$

where $v \neq \mu_j \ (j = 1, \ldots, r)$.

Let $p^{\sigma_j} \parallel \big(f(py + \mu_j) - f(\mu_j)\big)$. Then we set

$$g_j(y) = p^{-\sigma_1}\big(f(py + \mu_j) - f(\mu_j)\big).$$

It follows from Lemma 2.3 that $1 \leq \sigma_j \leq n$. If $\sigma_j < l$, then we can apply the induction hypothesis (2.13) and Lemma 2.2. We obtain

$$\begin{aligned} |S_{\mu_j}| &= \Big| \sum_{\substack{x=1 \\ x \equiv m_j \pmod p}}^{p^l} \exp\{2\pi i f(x)/p^l\} \Big| = \Big| \sum_{y=0}^{p^{l-1}-1} \exp\{2\pi i g_j(y)/p^{l-\sigma_j}\} \Big| \\ &= p^{\sigma_j - 1} |S(p^{l-\sigma_j}, g_j(y))| \leq m_j n^{3/n} p^{\sigma_j - 1 + (l-\sigma_j)(1-1/n)} \qquad (2.14) \\ &= m_j n^{3/n} p^{l(1-1/n)+\sigma_j/n-1} \leq m_j n^{3/n} p^{l(1-1/n)}. \end{aligned}$$

We assume that $\sigma_j \geq l$. Then we have

$$|S_{\mu_j}| \leq p^{l-1} = p^{l(1-1/n)+l/n-1} \leq p^{l(1-1/n)+\sigma_j/n-1} \leq p^{l(1-1/n)}.$$

Thus we have proved (2.14) for $\sigma_j \geq l$. Further, we have the inequalities

$$|S(p^l, f(x))| \leq \sum_{j=1}^{r} m_j n^{3/n} p^{l(1-1/n)} = m n^{3/n} p^{l(1-1/n)} \leq (n-1) n^{3/n} p^{l(1-1/n)}.$$

The case $p \leq n$ is also studied. The proof of the theorem is complete. □

Theorem 2.2. *Suppose that $n \geq 3$ is an integer and $f(x) = a_n x^n + \cdots + a_1 x + a_0$ is a polynomial with integer coefficients, $(a_n, \ldots, a_1, q) = 1$, and q is a natural number. Then we have*

$$|S(q, f(x))| \leq c(n) q^{1-1/n},$$

where $c(n) = \exp\{4n\}$ for $n \geq 10$ and $c(n) = \exp\{nA(n)\}$ for $3 \leq n \leq 9$, $A(3) = 6.1$, $A(4) = 5.5$, $A(5) = 5$, $A(6) = 4.7$, $A(7) = 4.4$, $A(8) = 4.2$, and $A(9) = 4.05$.

Proof. Let $A = (n-1)^{2n/(n-2)}$, and let $B = (n-1)^{n/(n-2)}$. Then from Lemma 2.1 and Theorem 2.1 we obtain

$$\begin{aligned} |S(q, f(x))| &\leq ((n-1)n^{3/n})^{\pi(n)} (n^{3/n})^{\pi(B)-\pi(n)} (n^{2/n})^{\pi(A)-\pi(B)} q^{1-1/n} \\ &= (n-1)^{\pi(n)} n^{\pi(B)/n} n^{2\pi(A)/n} q^{1-1/n} = D q^{1-1/n}, \end{aligned}$$

where $\pi(x) = \sum_{p \leq x} 1$.

It is well known that the inequality

$$\pi(x) \leq 1.25x/\log x$$

holds for $x \geq 3$. It follows from this inequality that $D \leq \exp\{F(n)\}$, where

$$F(n) = 1.25n\left(\frac{\log(n-1)}{\log n} + \frac{(n-1)^{1+2/(n-2)}(n-2)\log n}{n^3\log(n-1)} + \frac{(n-1)^{2+4/(n-2)}(n-2)\log n}{n^3\log(n-1)}\right).$$

After simple calculations, we obtain the statement of the theorem. □

We will derive an estimate for the complete rational trigonometric sum modulo p^l, which depends on whether the coefficients of the polynomial in the exponent are divisible by powers of the prime p. We shall need this estimate to prove theorems on the convergence exponents of "singular series" in Tarry's problem and its generalizations.

Let $f(x) = a_1x + \cdots + a_nx^n$ be a polynomial with integer coefficients, $(a_n, \ldots, a_1, p) = 1$ $(n \geq 3)$, $w = [\log n/\log p]$, and $p^\tau \parallel (na_n, \ldots, 2a_2, a_1)$. Then $\tau \leq w$. We set

$$g(y) = f(y+\xi) = \sum_{s=0}^{n} b_s y^s,$$

where the coefficients of the polynomial $g(y)$ are given by the relations

$$b_n = a_n, \quad b_{n-1} = a_{n-1} + \binom{n}{1}a_n\xi,$$

$$\vdots$$

$$b_s = a_s + \binom{s+1}{s}a_{s+1}\xi + \cdots + \binom{n}{s}a_n\xi^{n-s},$$

$$\vdots$$

$$b_1 = a_1 + \binom{2}{1}a_2\xi + \cdots + \binom{n}{1}a_n\xi^{n-1}.$$

Note that $p^\tau \parallel (nb_n, \ldots, 2b_2, b_1)$. Now let $\xi = \xi_1, \ldots, \xi_m$ $(m \leq n)$ be roots of the congruence

$$p^{-\tau} f'(\xi) = p^{-\tau} b_1 \equiv 0 \pmod{p}. \tag{2.15}$$

For each root ξ of this congruence, we define the exponent $u_1 = u_1(\xi)$ as follows:

$$p^{u_1} \parallel (p^n b_n, \ldots, p^2 b_2, pb_1).$$

We choose a root of congruence (2.15) and then set

$$f(py+\xi) - f(\xi) = p^{u_1} f_1(y) = p^{u_1} \sum_{s=1}^{n} c_s y^s, \tag{2.16}$$

$$p^{u_1} c_s = p^s b_s, \quad s = 1, \ldots, n, \quad g_1(y) = f_1(y+\eta) = \sum_{s=0}^{n} d_s y^s.$$

Now we assume that $p^{\tau_1} \parallel (nd_n, \dots, 2d_2, d_1)$ and the numbers $\eta = \eta_1, \dots, \eta_r$ are roots of the congruence

$$p^{-\tau_1} f_1'(\eta) = p^{-\tau_1} d_1 \equiv 0 \pmod{p}. \tag{2.17}$$

As before, for each root η of congruence (2.17) we define the exponent $u_2 = u_2(\eta) = u_2(\xi, \eta)$ as follows:

$$p^{u_2} \parallel (p^n d_n, \dots, p^2 d_2, pd_1).$$

By $f_2(y)$ we denote a polynomial of the form

$$p^{u_2} f_2(y) = f_1(py + \eta) - f_1(\eta) = p^{u_2} \sum_{s=1}^{n} e_s y^s, \tag{2.18}$$

and by $g_2(y)$ we denote the polynomial $f_2(y+\xi)$. For the polynomial $g_2(y)$ we define the exponent u_3, etc. So we have a set of exponents $(u_1, u_2, \dots, u_t)$ corresponding to the set of roots $(\xi, \eta, \dots)$ of congruences (2.15) and (2.17). Moreover, from the inequalities $l - u_1 - \dots - u_{t-1} > 2w + 1$ and $l - u_1 - \dots - u_t \le 2w + 1$ we find the number $t = t(\xi, \eta, \dots)$. Then the following assertion readily follows from the definition and Lemma 2.2.

Lemma 2.5. *Suppose that $f(x)$ is a polynomial of degree n with integer coefficients that together with p are coprimes. Then the number of sets of exponents $(u_1, u_2, \dots)$ of the polynomial $f(x)$ does not exceed n.*

Lemma 2.6. *The following inequalities hold:*

$$n \ge u_1 \ge u_2 \ge \dots \ge u_t \ge 2.$$

Proof. Suppose that $s_1 = \max_{1 \le v \le n}\{v \mid (b_v, p) = 1\}$. Then, by the definition of c_{s_1}, we have $p^{s_1} b_{s_1} = p^{u_1} c_{s_1}$. Hence $u_1 \le s_1 \le n$. We assume that $s_2 = \max_{1 \le v \le n}\{v \mid (c_v, p) = 1\}$. Then $p^{s_2} b_{s_2} = p^{u_1} c_{s_2}$ $(s_1 \le u_1)$.

Further, since $b_1 \equiv 0 \pmod{p}$ and $pb_1 = p^{u_1} c_1$, we have $u_1 \ge 2$. It follows from the definition of d_v that $(d_{s_2}, p) = 1$ and $p^{u_2} \mid p^{s_2} d_{s_2}$ $(u_2 \le u_1)$. Since $d_1 \equiv 0 \pmod{p}$ and $p^{u_2} \parallel pd_1$, we have $u_2 \ge 2$. So

$$n \ge s_1 \ge u_1 \ge s_2 \ge u_2 \ge 2.$$

The lemma is thereby proved. □

Theorem 2.3. *Suppose that $n \ge 3$ is integer, $f(x) = a_n x^n + \dots + a_1 x$ is a polynomial with integer coefficients, $(a_n, \dots, a_1, p) = 1$, p is a prime, and l is a natural number. Let j be the least length of the set of exponents $(u_1, u_2, \dots)$ defined above. Then the following estimate holds:*

(a) $|S(p^l, f(x))| \le np^{l-j}$.

If, in addition, $s - 1 = u_1 + \cdots + u_j$, *then the following estimate holds*:

(b) $|S(p^l, f(x))| \le n^2 p^{l-j-1/2}$.

Proof. Let $p^\tau \parallel (na_n, \dots, 2a_2, a_1)$. Then we let $\xi^{(1)} = \xi_1^{(1)}, \dots, \xi_m^{(1)}$ denote distinct roots of the congruence

$$p^{-\tau} f'(\xi) \equiv 0 \pmod p, \quad 0 \le \xi < p.$$

Further, we assume that $l > 2w+1$, since, otherwise, $j = 0$ and the theorem obviously holds. We represent the sum $S(p', f(x))$ as

$$|S(p^l, f(x))| = \sum_{v=1}^{p} S_v,$$

where

$$S_v = \sum_{\substack{1 \le x \le p^l \\ x \equiv v \pmod p}} \exp\{2\pi i f(x)/p^l\}.$$

Substituting $x = y + p^{l-\tau+1} z$ $(y = 1, \dots, p^{l-\tau-1}, z = 0, 1, \dots, p^{\tau+1} - 1)$, we obtain

$$\begin{aligned} S_v &= \sum_{\substack{0<y\le p^{l-\tau-1} \\ y\equiv v \pmod p}} \sum_{z=0}^{p^{\tau+1}-1} \exp\left\{2\pi i \left(f(y) + p^{l-\iota-1} z f'(y)\right)/p^l\right\} \\ &= \sum_{\substack{0<y\le p^{l-\tau-1} \\ y\equiv v \pmod p}} \exp\{2\pi i f(y)/p^l\} \sum_{z=0}^{p^{\tau+1}\ 1} \exp\{2\pi i z f'(y)/p^{\tau+1}\}. \end{aligned}$$

Hence for $v \ne \xi_j^{(1)}$ $(j = 1, \dots, m)$, we have $S_v = 0$. In the case $v = \xi_j^{(1)}$ $(j = 1, \dots, m)$, taking into account the notation in (2.15)–(2.18), we obtain

$$\begin{aligned} S_v &= \sum_{y=1}^{p^{l-1}} \exp\left\{2\pi i \left(b_0(v) + b_1(v) p y + \cdots + b_n(v) p^n y^n\right)/p^l\right\} \\ &= \exp\{2\pi i b_0(v)/p^l\} p^{u_1-1} S(p^{l-u_1}, c_1 y + \cdots + c_n y^n). \end{aligned}$$

Next, following the preceding argument, we consider the sum

$$S(p^{l-u_1}, c_1 y + \cdots + c_n y^n).$$

Then, using the notation in (2.15)–(2.18), we obtain

$$S(p^l, f(x)) = \sum_{(\xi^{(1)},\xi^{(2)})} \exp\left\{2\pi i\left(\frac{b_0(\xi^{(1)})}{p^l} + \frac{d_0(\xi^{(2)})}{p^{l-u_1}}\right)\right\} p^{u_1+u_2-2} S(p^{l-u_1-u_2}, f_2(y)).$$

For each set of roots $(\xi^{(1)}, \xi^{(2)}, \dots)$ of congruences (2.15) and (2.17) and for the corresponding set of exponents, the number $t = t(\xi^{(1)}, \xi^{(2)}, \dots)$ is uniquely determined by the conditions

$$l - u_1 - \dots - u_{t-1} > 2w + 1, \quad l - u_1 - \dots - u_t \le 2w + 1.$$

Therefore, repeating the preceding argument appropriately many times, we obtain

$$S(p^l, f(x)) = \sum_{(\xi^{(1)},\dots,\xi^{(t)})} \exp\left\{2\pi i\left(\frac{b_0(\xi^{(1)})}{p^l} + \frac{d_0(\xi^{(2)})}{p^{l-u_1}} + \dots + \frac{g_0(\xi^{(t)})}{p^{l-u_1-\dots-u_{t-1}}}\right)\right\}$$
$$\times S(p^{l-u_1-\dots-u_t}, g_1 y + \dots + g_n y^n) p^{l-u_1-\dots-u_t-t}. \tag{2.19}$$

By Lemma 2.5, the number of sets $(\xi^{(1)}, \dots, \xi^{(t)})$ does not exceed n. Now we use the following trivial estimate of the sum:

$$|S(p^{l-u_1-\dots-u_t}, g_1 y + \dots + g_n y^n)| \le p^{l-u_1-\dots-u_t} \tag{2.20}$$

for $l - u_1 - \dots - u_t > 1$. If $l - u_1 - \dots - u_t = 1$, then we apply the Weil estimate (Lemma A.5):

$$|S(p, g_1 y + \dots + g_n y^n)| \le n\sqrt{p}. \tag{2.21}$$

After the substitution of inequalities (2.20) and (2.21) into (2.19), we obtain the desired assertions (a) and (b) of the theorem. The proof of the theorem is complete. □

2.2 Singular series in Tarry's problem and in its generalizations

Suppose that $n \ge 3$, $f(x) = (a_1/q_1)x + \dots + (a_n/q_n)x^n$, $(a_1, q_1) = \dots = (a_n, q_n) = 1$, and $q = q_1 \dots q_n$.

Definition 2.1. The *mean value σ of the complete rational trigonometric sum*

$$S(q, f(x)) = \sum_{x=1}^{q} \exp\{2\pi i f(x)\}$$

is defined by the expression

$$\sigma = \sum_{q_n=1}^{+\infty} \cdots \sum_{q_1=1}^{+\infty} \sum_{a_n=0}^{q_n-1}{}' \cdots \sum_{a_1=0}^{q_1-1}{}' |q^{-1} S(q, f(x))|^{2k},$$

where the prime on the summation sign means that a_s runs through the reduced system of residues modulo q_s $(s = n, \dots, 1)$. The series σ is also called the *singular series in Tarry's problem.*

We find the convergence exponent of the singular series σ. For this, we first perform several auxiliary transformations.

We write σ as

$$\sigma = \sum_{Q=1}^{+\infty} \underset{[q_n,\dots,q_1]=Q}{\sum_{q_n=1}^{+\infty} \cdots \sum_{q_1=1}^{+\infty}} \sum_{a_n=0}^{q_n-1}{}' \cdots \sum_{a_1=0}^{q_1-1}{}' \left|Q^{-1} S(Q, f(x))\right|^{2k}.$$

By Theorem 2.2, we have

$$\sigma \ll \sum_{Q=1}^{+\infty} Q^{n+\varepsilon-2k/n}.$$

Hence the series σ converges for $n - 2k/n < -1$, i.e., for $2k > n(n+1)$. Next, by Lemma 2.1, we rewrite the trigonometric sum $S(Q, f(x))$ with $Q = \prod_{p \nmid Q} p^{\alpha}$ and $Q_p = Q/p^{\alpha}$ in the form

$$S(Q, f(x)) = \prod_{p|Q} S(p^{\alpha}, Q_p^{-1} f(Q_p x)).$$

Since the series σ and σ_p converge absolutely, the latter relation implies $\sigma = \prod_p \sigma_p$, where σ_p is determined as

$$\sigma_p = 1 + \sum_{s=1}^{+\infty} A(p^s), \quad A(p^s) = \underset{p \nmid (a_n,\dots,a_1)}{\sum_{a_n=0}^{p^s-1} \cdots \sum_{a_1=0}^{p^s-1}} \left|p^{-s} S(p^s, a_n x^n + \cdots + a_1 x)\right|^{2k}.$$

We show that the infinite product $\prod_p \sigma_p$ converges for $2k > 0.5n(n+1)+2$. But first we prove a lemma concerning the arithmetic nature of the series σ_p.

Lemma 2.7. *The following relations hold*:

$$p^{-r(2k-n)} N(p^r) = \sum_{l=0}^{r} \underset{p \nmid (a_n,\dots,a_1)}{\sum_{a_n=0}^{p^l-1} \cdots \sum_{a_1=0}^{p^l-1}} \left|p^{-l} S(p^l, a_n x^n + \cdots + a_1 x)\right|^{2k}, \tag{2.22}$$

$$\sigma_p = \lim_{r\to\infty} p^{-r(2k-n)} N(p^r), \tag{2.23}$$

where the number $N(p^r)$ is the number of solutions of the congruences

$$x_1^h + \cdots + x_k^h \equiv y_1^h + \cdots + y_k^h \pmod{p^r}, \quad 1 \le h \le n,$$
$$1 \le x_1, \dots, x_k, y_1, \dots, y_k \le p^r.$$

Proof. Obviously, we have

$$N(p^r) = p^{-rn} \sum_{a_n=0}^{p^r-1} \cdots \sum_{a_1=0}^{p^r-1} |S(p^r, a_n x^n + \cdots + a_1 x)|^{2k}$$

$$= p^{-rn} \sum_{m=0}^{r} \sum_{\substack{a_n=0 \\ (a_n,\ldots,a_1,p^r)=p^m}}^{p^r-1} \cdots \sum_{a_1=0}^{p^r-1} |S(p^r, a_n x^n + \cdots + a_1 x)|^{2k}$$

$$= p^{-rn} \sum_{m=0}^{r} \sum_{\substack{b_n=0 \\ p\nmid(b_n,\ldots,b_1)}}^{p^{r-m}-1} \cdots \sum_{b_1=0}^{p^{r-m}-1} |p^m S(p^{r-m}, b_n x^n + \cdots + b_1 x)|^{2k}$$

$$= p^{2kr-rn} \sum_{l=0}^{r} \sum_{\substack{b_n=0 \\ p\nmid(b_n,\ldots,b_1)}}^{p^l-1} \cdots \sum_{b_1=0}^{p^l-1} |p^{-l} S(p^l, b_n x^n + \cdots + b_1 x)|^{2k}.$$

Relation (2.22) is proved. Passing to the limit as $r \to +\infty$ in this relation, we obtain (2.23). The proof of the lemma is complete. □

Theorem 2.4. *The singular series σ converges for $2k > 0.5n(n+1)+2$ and diverges for $2k \le 0.5n(n+1)+2$.*

Proof. First, let us find an upper bound for

$$A(p^s) = \sum_{\substack{a_n=0 \\ p\nmid(a_n,\ldots,a_1)}}^{p^s-1} \cdots \sum_{a_1=0}^{p^s-1} |p^{-s} S(p^s, a_n x^n + \cdots + a_1 x)|^{2k}.$$

We fix a solution $\xi = \xi^{(1)} + p\xi^{(2)} + \cdots + p^{j-1}\xi^{(j)}$ of the congruence $f'(x) \equiv 0 \pmod{p^l}$ defined before the statement of Lemma 2.5. A set of exponents $(u_1, \ldots, u_j)$ corresponds to this solution. We shall find an upper bound for the number of polynomials with this set of exponents. From the definition of the numbers $b_0, b_1, \ldots, b_n$, we have

$$f(x) = \sum_{s=1}^{n} a_s x^s = \sum_{s=0}^{n} b_s (x - \xi)^s,$$

where

$$
\begin{aligned}
a_n = b_n, \quad a_{n-1} &= b_{n-1} - \binom{n}{n-1} b_n \xi, \\
&\vdots \\
a_s &= b_s - \binom{s+1}{s} b_{s+1}\xi + \cdots + (-1)^{n-s}\binom{n}{s} b_n \xi^{n-s}, \qquad (2.24)\\
&\vdots \\
a_1 &= b_1 - \binom{2}{1} b_2 \xi + \cdots + (-1)^{n-1}\binom{n}{1} b_n \xi^{n-1}.
\end{aligned}
$$

Since $p^{u_1} \parallel (p^n b_b, \dots, pb_1)$, we have the relations

$$p^n b_n = p^{u_1} c_n, \dots, pb_1 = p^{u_1} c_1, \quad (c_n, \dots, c_1, p) = 1.$$

Hence we have $b_s = p^{u_1 - s} c_s$ for all $s \le u_1 - 1$. A system similar to (2.24) (see the definition before Lemma 2.5) uniquely determines the numbers $c_n, \dots, c_1$ in terms of $d_n, \dots, d_1$. From the definition of the exponent u_2 we have

$$p^n d_n = p^{u_2} e_n, \ \dots, \ pd_1 = p^{u_2} e_1, \quad (e_n, \dots, e_1, p) = 1.$$

Hence we obtain $d_s = p^{u_2 - s} e_s$ for $s \le u_2 - 1$. By writing similar relations for each exponent u_r $(r \le j)$ and fixing the coefficients with numbers $u_r, u_r + 1, \dots, u_{r-1} - 1$, for some constants $A_{u_1 - 1}, \dots, A_1$, we find

$$
\begin{aligned}
a_{u_1-1} &= p^{u_1-(u_1-1)} B_{u_1-1} + A_{u_1-1}, \\
&\vdots \\
a_{u_2-1} &= p^{u_1-(u_2-1)+u_2-(u_2-1)} B_{u_2-1} + A_{u_2-1}, \qquad (2.25)\\
&\vdots \\
a_1 &= p^{(u_1-1)+(u_2-1)+\cdots+(u_j-1)} B_1 + A_1.
\end{aligned}
$$

Since the coefficients $a_n, \dots, a_1$ of the polynomial $f(x)$ take values in the complete system of residues modulo p^s, it follows from relation (2.25) that the number of polynomials with the set of exponents $(u_1, \dots, u_j)$ does not exceed p^A,

$$A = ns - \frac{(u_1-1)u_1}{2} - \frac{(u_2-1)u_2}{2} - \cdots - \frac{(u_j-1)u_j}{2}.$$

Let $u_1 + \cdots + u_j = s_1$; then $s - 2w - 1 \le s_1 \le s$. We set $B = A - n(s - s_1)$ and show that the inequality $B \le jn(n+1)/2$ holds. Obviously, from (2.25) we have

$$
\begin{aligned}
B = s_1 n &- (u_1-1)u_1 - (u_2-1)u_2 - \cdots - (u_j-1)u_j \qquad (2.26)\\
&+ (u_1 - 1 + u_1 - 2 + \cdots + 1) + (u_2 - 1 + \cdots + 1) + \cdots \\
&+ (u_j - 1 + \cdots + 1)
\end{aligned}
$$

$$\begin{aligned}&= s_1(n-u_1+1)+(u_j-1)(s_1-u_1-u_2-\cdots-u_j)\\&\quad+(u_{j-1}-u_j)(s_1-u_1-\cdots-u_{j-1})+\cdots+(u_1-u_2)(s-u_1)\\&\quad+j(u_j-1+\cdots+1)+(j-1)(u_{j-1}-1+\cdots+u_j)+\cdots\\&\quad+(u_1-1+\cdots+u_2).\end{aligned}$$

Lemma 2.6 implies the inequalities $n \geq u_1 \geq u_2 \geq \cdots \geq u_j \geq 2$. Hence we have

$$\begin{aligned}&(u_{j-1}-u_j)(s_1-u_1-\cdots-u_{j-1})+(j-1)(u_{j-1}-1+\cdots+u_j)\\&\quad=(u_{j-1}-u_j)u_j+(j-1)(u_{j-1}-1+\cdots+u_j)\leq f(u_{j-1}-1+\cdots+u_j),\\&\quad\vdots\\&(u_1-u_2)(s_1-u_1)+(u_1-1+\cdots+u_2)\\&\quad=(u_1-u_2)(u_2+\cdots+u_j)+(u_1-1+\cdots+u_2)\\&\quad\leq(u_1-u_2)(j-1)u_2+(u_1-1+\cdots+u_2)\leq j(u_1-1+\cdots+u_2).\end{aligned}$$

Substituting these inequalities into (2.26), we obtain

$$B \leq s_1(n-u_1+1)+j\big((u_1-1)+(u_1-2)+\cdots+1\big).$$

Then we use the inequality $s_1 = u_1+\cdots+u_j \leq ju_1$ and obtain

$$\begin{aligned}B &\leq s_1(n-u_1+1)+j\big((u_1-1)+(u_1-2)+\cdots+1\big)\\&\leq j(n-u_1+1)u_1+j\big((u_1-1)+(u_1-2)+\cdots+1\big)\\&\leq j\big(n+(n-1)+\cdots+1\big)=jn(n+1)/2.\end{aligned}$$

Now we find an upper bound for the number of exponents $(u_1, u_2, \ldots, u_j)$ satisfying the conditions

$$n \geq u_1 \geq u_2 \geq \cdots \geq u_j \geq 2, \quad s \geq u_1+\cdots+u_j > s-2w-1.$$

Let e_n be the number of u_m equal to n; …; and let e_2 be the number of u_m equal to 2 $(1 \leq m \leq j)$. Then $ne_n+\cdots+2e_2 = s_1$ and the number of sets $(u_1, \ldots, u_j)$ coincides with the number of sets $(e_n, \ldots, e_2)$, since $n \geq u_1 \geq u_2 \geq \cdots \geq u_j \geq 2$. The first coordinate e_n can take at most $s/n+1$ values, …, the coordinate e_2 can take at most $0.5s+1$ values. Hence the number of sets $(e_n, \ldots, e_2)$ does not exceed s^n.

This means that the number of roots with j coordinates $(\xi^{(1)}, \ldots, \xi^{(j)})$ does not exceed p^j, the number of sets $(u_1, \ldots, u_j)$ does not exceed s^n, and the number of polynomials corresponding to the set of exponents $(u_1, \ldots, u_j)$ does not exceed

$$p^{0.5jn(n+1)+n(s-s_1)} \leq p^{0.5jn(n+1)+n(2w+1)}.$$

We divide all polynomials $f(x) = a_nx^n+\cdots+a_1x$ $(0 \leq a_n, \ldots, a_1 \leq p^s-1)$, $(a_n, \ldots, a_1, p) = 1$, into classes according to the length of the minimal set of exponents $(u_1, \ldots, u_j)$. The class A_j consists of all polynomials for which the minimal length of the set of exponents is equal to j $(j = 0, 1, \dots)$. By Theorem 2.3, for the polynomials contained in the class A_j, we have the estimate

$$|p^{-s} S(p^s, f(x))| \le np^{-j},$$

and the number of such polynomials does not exceed

$$s^n p^j p^{0.5jn(n+1)+n(2w+1)}.$$

Hence

$$A(p^s) \le \sum_{j_0 \le j} n^{2k} s^n p^{n(2w+1)} p^{(0.5n(n+1)+1-2k)j}, \tag{2.27}$$

where $j_0 = \max\big(1, (s - 2w - 1)/n\big)$.

Let us consider the case $p \le n$. If $s - 2w - 1 \ge n$, then (2.27) implies

$$A(p^s) \le n^{2k} s^n p^{n(2w+1)} p^{((s-2w-1)/n)(0.5n(n+1)+1-2k)}.$$

But if $s - 2w - 1 < n$, then (2.27) implies

$$A(p^s) \le n^{2k} s^n p^{n(2w+1)} (1 + p^{0.5n(n+1)+1-2k}).$$

Let $p > n$. Then $w = [\log n / \log p] = 0$ and s_1 is equal either to $s - 1$ or to s. If $s > n$, then formula (2.27) implies

$$\begin{aligned} A(p^s) &\le \sum_{j \ge (s-1)/n} n^{2k} s^n p^{n(2w+1)} p^{j(0.5n(n+1)+1-2k)} \\ &\le n^{2k} s^n p^{n(2w+1)} p^{((s-1)/n)(0.5n(n+1)+1-2k)}. \end{aligned}$$

Now if $p > n$, $2 \le s \le n$, then formula (2.27) implies the estimate

$$A(p^s) \le n^{2k} s^n p^{n(2w+1)} p^{0.5n(n+1)+1-2k}.$$

Finally, if $p > n$ and $s = 1$, then it follows from the Weil estimate (Lemma A.5) that (for $k > 0.25n(n+1) + 1 \ge n + 1$)

$$A(p) = \sum_{\substack{a_n=0 \\ p \nmid (a_n,\dots,a_1)}}^{p-1} \cdots \sum_{a_1=0}^{p-1} |p^{-1} S(p, a_n x^n + \cdots + a_1 x)|^{2k} \le p^n n^{2k} p^{-k} \le n^{2k} p^{-2}.$$

So for $P > n$ we have the estimate

$$\begin{aligned} \sigma_p - 1 = \sum_{s=1}^{+\infty} A(p^s) &\le n^{2k} p^{-2} + n^{2k+1+n} p^{n(2w+1)} p^{0.5n(n+1)+1-2k} \\ &+ \sum_{s>n} n^{2k} s^n p^{n(2w+1)} p^{((s-1)/n)(0.5n(n+1)+1-2k)} \end{aligned} \tag{2.28}$$

$$\ll p^{-2} + p^{0.5n(n+1)+1-2k+\varepsilon},$$

where $\varepsilon > 0$ is an arbitrarily small fixed number and $s^n \ll p^{\varepsilon(s-1)/n}$.

Let $p \le n$; then we have

$$\sigma_p = 1 + \sum_{s=1}^{+\infty} A(p^s) \le 1 + (n+2w)^{n+1} n^{2k} p^{n(2w+1)} (1 + p^{0.5n(n+1)+1-2k})$$
$$+ \sum_{s \ge n+2w+1} n^{2k} s^n p^{n(2w+1)} p^{((s-2w-1)/n)(0.5n(n+1)+1-2k)}. \tag{2.29}$$

For $2k > 0.5n(n+1)+1$, this formula implies the inequality $\sigma_p \ll 1$.

So, by the estimates (2.28) and (2.29), the series

$$\sigma = \prod_p \sigma_p = \prod_{p \le n} \sigma_p \prod_{p > n} \sigma_p$$

converges under the condition that $2k > 0.5n(n+1)+2$.

Let us prove that the series σ diverges for $2k \le 0.5n(n+1)+2$. Indeed, we have $\sigma > \sigma_1$, where

$$\sigma_1 = \sum_{p>n} \sum_{\substack{a_n=1 \\ (a_n,p)=1}}^{p^n} \cdots \sum_{\substack{(a_1,p)=1 \\ 1 \le a_1 \le p}} \sum_{c=0}^{p-1} \left| p^{-n} \sum_{x=1}^{p^n} \exp\left\{ 2\pi i \left(\frac{a_n}{p^n}(x+c)^n + \cdots + \frac{a_1}{p}(x+c) \right) \right\} \right|^{2k}.$$

For $p > n$, we have the equality

$$S_1 = \sum_{x=1}^{p^n} \exp\left\{ 2\pi i \left(\frac{a_n}{p^n} x^n + \cdots + \frac{a_1}{p} x \right) \right\} = p^{n-1}.$$

After the substitution $x = y + p^{n-1} z$ $(1 \le y \le p^{n-1}, 0 \le z \le p-1)$, we obtain

$$S_1 = \sum_{y=1}^{p^{n-1}} \exp\left\{ 2\pi i \left(\frac{a_n}{p^n} y^n + \cdots + \frac{a_1}{p} y \right) \right\} \sum_{z=0}^{p-1} \exp\left\{ 2\pi i \frac{n a_n z y^{n-1}}{p} \right\}$$
$$= p \cdot p^{n-2} = p^{n-1}.$$

Therefore, the series σ_1 satisfies the estimate

$$\sigma_1 = \sum_{p>n} \sum_{\substack{a_n=1 \\ (a_n,p)=1}}^{p^n} \cdots \sum_{\substack{a_1=1 \\ (a_1,p)=1}}^{p} \sum_{c=0}^{p-1} p^{-2k} > 2^{-n} \sum_{p>n} p^{0.5n(n+1)+1-2k}.$$

Since the series $\sum_{p>n} p^{-1}$ diverges, it follows from the last inequality that the series σ_1, as well as the series σ, diverges for $n(n+1)/2+1-2k \geq -1$, i.e., for $2k \leq n(n+1)/2+2$. The proof of the theorem is complete. □

Let $1 \leq m < r < \cdots < n$ be natural numbers, and let the number of numbers $m, r, \ldots, n$ be equal to l, $l \neq n$. Then the polynomial of degree n containing monomials of degrees $m, r, \ldots, n$ is said to be *jagged* (see [77]).

We consider the polynomial $f(x) = (a_m/q_m)x^m + \cdots + (a_n/q_n)x^n$ of degree $n \geq 3$, $(a_m, q_m) = \cdots = (a_n, q_n) = 1$, $q = q_m \ldots q_n$. We define the mean value of the complete rational trigonometric sum with jagged polynomial in the exponent as follows:

$$\sigma' = \sum_{q_n=1}^{+\infty} \cdots \sum_{q_m=1}^{+\infty} \sum_{\substack{a_n=0\\(a_n,q_n)=1}}^{q_n-1} \cdots \sum_{\substack{a_m=0\\(a_m,q_m)=1}}^{q_m-1} |q^{-1}S(q, qf(x))|^{2k}.$$

Similarly to the series σ, for $k > n(n+1)$, we represent the series σ' as an infinite product over all primes p from the series σ'_p, i.e., we represent it as $\sigma' = \prod_p \sigma'_p$, where

$$\sigma'_p = 1 + \sum_{s=1}^{+\infty} A_1(p^s),$$

$$A_1(p^s) = \sum_{\substack{a_n=0\\p\nmid(a_n,\ldots,a_m)}}^{p^s-1} \cdots \sum_{a_m=0}^{p^s-1} |p^{-s}S(p^s, a_n x^n + \cdots + a_m x^m)|^{2k}.$$

We show that the infinite product σ'_p converges for $2k > n + \cdots + r + m + 1$ and diverges for $2k \leq n + \cdots + r + m + 1$.

The statements and proofs of Lemmas 2.1 and 2.2 and of Theorems 2.1–2.3 are given in the form that is also suitable for jagged polynomials.

Theorem 2.5. *Suppose that* $1 \leq m < r < \cdots < n$ $(n \geq 4)$ *are natural numbers and the number of the numbers* $m, r, \ldots, n$ *is equal to* l, $l \neq n$. *Then the singular series* σ' *converges for* $2k > m + r + \cdots + n + 1$ *and diverges for* $2k \leq m + r + \cdots + n + 1$.

Proof. We fix a solution $\xi = \xi^{(1)} + p\xi^{(2)} + \cdots + p^{j-1}\xi^{(j)}$ of the system of congruences written before the statement of Lemma 2.5. To this solution there corresponds a set of exponents $u_1, \ldots, u_j$. Let us find an upper bound for the number of polynomials with this set of exponents $u_1, \ldots, u_j$.

As before, we assume that the numbers $b_0, b_1, \ldots, b_n$ are determined by the relation

$$f(x) = a_m x^m + \cdots + a_n x^n = b_0 + b_1(x-\xi) + \cdots + b_n(x-\xi)^n,$$

which can be written explicitly as

$$
\begin{aligned}
a_n = b_n, \quad a_{n-1} &= b_{n-1} - nb_n\xi, \\
&\vdots \\
a_s &= b_s - \binom{s+1}{s} b_{s+1}\xi + \cdots + (-1)^{n-s}\binom{n}{s} b_n \xi^{n-s}, \\
&\vdots \\
a_1 &= b_1 - \binom{2}{1} b_2\xi + \cdots + (-1)^{n-1}\binom{n}{1} b_n \xi^{n-1}.
\end{aligned}
\tag{2.30}
$$

Since $a_q = 0$ for $q \neq m, r, \dots, n$, the numbers b_q $(q \neq m, r, \dots, n)$ can be expressed in terms of $b_m, b_r, \dots, b_n$. We substitute the resulting expressions for b_q $(q \neq m, r, \dots, n)$ into the relations for $a_m, a_r, \dots, a_n$:

$$
\begin{aligned}
a_n &= b_n, \\
&\vdots \\
a_r &= b_r + \cdots + c_{2l} b_n \xi^{n-r}, \\
a_m &= b_m + c_{12} b_r \xi^{r-m} + \cdots + c_{1l} b_n \xi^{n-m},
\end{aligned}
\tag{2.31}
$$

where the coefficients $c_{12}, \dots, c_{1l}, \dots, c_{2l}$ are some integers. For convenience, from now on we denote the natural numbers $m, r, \dots, n$ as follows: $m = n_1, r = n_2, \dots, n = n_l$.

We will find necessary conditions on the coefficients of the polynomial $f(x) = a_m x^m + a_r x^r + \cdots + a_n x^n$ under which $f(x)$ has a given set of exponents $u_1, \dots, u_j$ and which allow us to estimate the number of polynomials with such exponents. By the definition of the integers $c_1, \dots, c_n$, we have

$$
p^n b_n = p^{u_1} c_n, \dots, p b_1 = p^{u_1} c_1, \quad (c_n, \dots, c_1, p) = 1. \tag{2.32}
$$

Lemma 2.2 implies the inequality $u_1 \leq n$. Let us consider the following two cases: (a) $u_1 < n$ and (b) $u_1 = n$. In case (a), we have $n_f < u_1 \leq n_{f+1}$ for some $f \leq l - 1$. We fix the values of $b_{n_l}, \dots, b_{n_{f+1}}$ and, instead of $b_{n_f}, \dots, b_{n_1}$, substitute their expressions in terms of $c_{n_f}, \dots, c_{n_1}$ into the system of equations (2.17): $b_{n_f} = c_{n_f} p^{u_1 - n_f}, \dots, b_{n_1} = c_{n_1} p^{u_1 - n_1}$.

The numbers $c_{n_1}, \dots, c_{n_f}$ can be uniquely expressed in terms of $d_{n_1}, \dots, d_{n_f}$ from a system similar to (2.17). Next, from Lemma 2.2 we have the inequality $u_2 \leq u_1$, and hence, for some $g \leq f$, we obtain $n_g < u_2 \leq n_{g+1}$. We fix the values of $d_{n_f}, \dots, d_{n_{g+1}}$. By definition, $d_1, \dots, d_n$ can be written as

$$
p^n d_n = p^{u_2} e_n, \ \dots, \ p d_1 = p^{u_2} e_1, \quad (e_n, \dots, e_1, p) = 1. \tag{2.33}
$$

Therefore, we have

$$
d_{n_g} = e_{n_g} p^{u_2 - n_g}, \ \dots, \ d_{n_1} = e_{n_1} p^{u_2 - n_1}.
$$

Now we define the variable t by the inequalities $u_t > n_1$ and $u_{t+1} \leq n_1$ and find the number h from the inequalities $n_h < u_t \leq u_{h+1}$. Hence for some $A_f, \dots, A_g, \dots, A_1$ we have

$$\begin{aligned} a_{n_f} &= p^{u_1 - n_f} B_f + A_f, \\ &\ \vdots \\ a_{n_g} &= p^{(u_1 - n_g) + (u_2 - n_g)} B_g + A_g, \\ &\ \vdots \\ a_{n_1} &= p^{(u_1 - n_1) + (u_2 - n_1) + \cdots + (u_t - n_1)} B_1 + A_1. \end{aligned}$$

Since the coefficients $a_{n_1}, \dots, a_{n_f}$ of the polynomial $f(x)$ run through the values of the complete system of residues modulo p^s, we see that the number of polynomials with exponents $u_1, \dots, u_j$ does not exceed p^A,

$$\begin{aligned} A = ls - fu_1 - gu_2 - \cdots - hu_t + (n_1 + \cdots + n_f) \\ + (n_1 + \cdots + n_g) + \cdots + (n_1 + \cdots + n_h) \end{aligned}$$

Let $u_1 + \cdots + u_j = s_1$; then we have $s - 2w - 1 \leq s_1 \leq s$. We show that $B = A - l(s - s_1)$ and $B \leq j(n_1 + \cdots + n_l - 1)$.

Obviously, we have the relations

$$\begin{aligned} B &= (l - f)s_1 + h(s_1 - u_1 - \cdots - u_t) + \cdots + (f - g)(s_1 - u_1) \\ &\quad + t(n_1 + \cdots + n_h) + \cdots + (n_{g+1} + \cdots + n_f) \\ &= (l - f)s_1 + h(u_{t+1} + \cdots + u_j) + \cdots + (f - g)(u_2 + \cdots + u_j) \\ &\quad + t(n_1 + \cdots + n_h) + \cdots + (n_{g+1} + \cdots + n_f). \end{aligned}$$

By the definition of $u_1, \dots, u_j$, we have

$$n_1 \geq u_{t+1} \geq \cdots \geq u_j, \quad n_{h+1} \geq u_t \geq \cdots \geq u_j, \quad n_{g+1} \geq u_2 \geq \cdots \geq u_j,$$

hence

$$\begin{aligned} B \leq (l - f)s_1 + h(j - t)n_1 + \cdots + (f - g)(j - 1)n_{g+1} \\ + t(n_1 + \cdots + n_h) + \cdots + (n_{g+1} + \cdots + n_f). \end{aligned}$$

Using the relations

$$\begin{aligned} h(j - t)n_1 + t(n_1 + \cdots + n_h) &\leq j(n_1 + \cdots + n_h), \\ &\ \vdots \\ (f - g)(j - 1)n_{g+1} + (n_{g+1} + \cdots + n_f) &\leq j(n_{g+1} + \cdots + n_f), \end{aligned} \tag{2.34}$$

we obtain $B \leq (l - f)s_1 + j(n_1 + \cdots + n_f)$.

Moreover, we have either the inequalities $s_1 = u_1 + \cdots + u_j \le jn_{f+1}$ or $s_1 \le j(n-1)$ if $f+1 = l$; $(l-f)s_1 \le (l-f)jn_{f+1} \le j(n_{f+1} + \cdots + n_l - 1)$.

We substitute the last inequality into (2.34) and thus obtain the estimate $B \le j(n_1 + \cdots + n_l - 1)$ stated above.

Further, we note that the number of the sets of exponents $(u_1, \ldots, u_j)$ satisfying the conditions

$$n \ge u_1 \ge \cdots \ge u_j \ge 2, \quad s \ge u_1 + \cdots + u_j \ge s - 2w - 1 \tag{2.35}$$

does not exceed s^n.

Starting from this fact and the estimate of B, in case (a), we see that the number of polynomials having j exponents $u_1, u_2, \ldots, u_j$ with conditions (2.35) does not exceed

$$s^n p^j p^{j(n_1+\cdots+n_l-1)+l(s-s_1)} = s^n p^{j(m+r+\cdots+n)+l(s-s_1)}.$$

Let us consider case (b), where $u_1 = n$. We first assume that $p > n$. Let $u_1 = \cdots = u_q = n$. We will show that $\xi^{(1)} = \xi^{(2)} = \cdots = \xi^{(q-1)} = 0$. From the definition of $b_n, b_{n-1}, \ldots, b_1$, we have

$$\begin{aligned} a_n &= b_n, \\ a_{n-1} &= b_{n-1} - nb_n\xi^{(1)}, \\ &\vdots \\ a_s &= b_s - \binom{s+1}{s} b_{s+1}\xi^{(1)} + \cdots + (-1)^{n-s}\binom{n}{s} b_n(\xi^{(1)})^{n-s}, \\ &\vdots \\ a_1 &= b_1 - \binom{2}{1} b_2\xi^{(1)} + \cdots + (-1)^{n-1}\binom{n}{1} b_n(\xi^{(1)})^{n-1}. \end{aligned} \tag{2.36}$$

Since $u_1 = n$, the values of $c_n, \ldots, c_1$ are determined as follows:

$$b_n = c_n, \quad b_{n-1} = pc_{n-1}, \quad \ldots \quad b_s = p^{n-s}c_s, \quad \ldots \quad b_1 = p^{n-1}c_1. \tag{2.37}$$

The values of $d_n, \ldots, d_1$ are determined similarly to the values of $b_n, \ldots, b_1$:

$$\begin{aligned} c_n &= d_n, \\ c_{n-1} &= d_{n-1} - nd_n\xi^{(2)}, \\ &\vdots \\ c_s &= d_s - \binom{s+1}{s} d_{s+1}\xi^{(2)} + \cdots + (-1)^{n-s}\binom{n}{s} d_n(\xi^{(2)})^{n-s}, \\ &\vdots \\ c_1 &= d_1 - \binom{2}{1} d_2\xi^{(2)} + \cdots + (-1)^{n-1}\binom{n}{1} d_n(\xi^{(2)})^{n-1}. \end{aligned} \tag{2.38}$$

Since the polynomial $f(x) = a_m x^m + a_r x^r + \cdots + a_n x^n$ is a jagged polynomial, there is an s such that $a_s = 0$; in other words, we have

$$(-1)^{n-s+1}\binom{n}{s} b_n (\xi^{(1)})^{n-s} = b_s - \binom{s+1}{s} b_{s+1}\xi^{(1)} + \cdots + (-1)^{n-s-1}\binom{n-1}{s} b_{n-1}(\xi^{(1)})^{n-s-1}. \tag{2.39}$$

It follows from relations (2.37) that $b_s, b_{s+1}, \ldots, b_{n-1}$ are divisible by p. Hence (2.39) implies that either b_n is divisible by p or $\xi^{(1)}$ is divisible by p. If we assume that $u_2 = n$, then $d_{n-1}, \ldots, d_1$ are divisible by p. In the case of $p \mid b_n$, because of the equalities $a_n = b_n = c_n = d_n$, we have $p \mid d_n$. This and (2.38) implies that $p \mid (c_n, c_{n-1}, \ldots, c_1)$, but $(c_n, \ldots, c_1, p) = 1$. Hence $\xi^{(1)}$ is divisible by p, but $0 \le \xi^{(i)} \le p-1$, and hence we have $\xi^{(1)} = 0$. In this case, relations (2.36) can be written as $a_n = b_n, \ldots, a_1 = b_1$, and hence we have $c_s = 0$. We can also treat $\xi^{(2)}, \ldots, \xi^{(q-1)}$ in a similar way. So we have proved that, in the case $u_1 = \cdots = u_q = n$, the variables $\xi^{(1)}, \ldots, \xi^{(q-1)}$ are zero.

As in case (a), we see that the number of polynomials having the solution $\xi = \xi^{(1)} + p\xi^{(2)} + \cdots + p^{j-1}\xi^{(j)}$ defined in Lemma 2.5 does not exceed p^A, where

$$\begin{aligned} A = {} & ls - (l-1)(u_1 + \cdots + u_q) - fu_{q+1} - \cdots - hu_t + \cdots \\ & + q(n_1 + \cdots + n_{l-1}) + (n_1 + \cdots + n_f) + \cdots + (n_1 + \cdots + n_h); \end{aligned}$$

As in casc (a), we let B denote the variable $A - l(s - s_1)$, where $s_1 = u_1 + \cdots + u_j$ and $s - 2w - 1 \le s_1 \le s$. We perform the transformations

$$\begin{aligned} B = {} & ls_1 - (l-1)(u_1 + \cdots + u_q) - fu_{q+1} - \cdots - hu_t \\ & + q(n_1 + \cdots + n_{l-1}) + (n_1 + \cdots + n_f) + \cdots + (n_1 + \cdots + n_h) \\ = {} & s_1 + h(s_1 - u_1 - \cdots - u_t) + \cdots + (l-1-f)(s_1 - u_1 - \cdots - u_q) \\ & + t(n_1 + \cdots + n_h) + \cdots + q(n_{f+1} + \cdots + n_{l-1}) \\ = {} & s_1 + h(u_{t+1} + \cdots + u_j) + \cdots + (l-1-f)(u_{q+1} + \cdots + u_j) \\ & + t(n_1 + \cdots + n_h) + \cdots + q(n_{f+1} + \cdots + n_{l-1}). \end{aligned}$$

Since $n_1 \ge u_{t+1} \ge \cdots \ge u_j, \ldots, n_{l-1} \ge u_{q+1} \ge \cdots \ge u_j$, we have

$$\begin{aligned} & h(u_{t+1} + \cdots + u_j) + t(n_1 + \cdots + n_h) \le h(j-t)n_1 + t(n_1 + \cdots + n_h) \\ & \quad \le j(n_1 + \cdots + n_h), \ldots, (l-1-f)(u_{q+1} + \cdots + u_j) + q(n_{f+1} + \cdots + n_{l-1}) \\ & \quad \le (l-1-f)(j-q-1)n_{f+1} + q(n_{f+1} + \cdots + n_{l-1}) \\ & \quad \le j(n_{f+1} + \cdots + n_{l-1}), \\ & s_1 = u_1 + \cdots + u_j = qn + u_{q+1} + \cdots + u_t + u_{t+1} + \cdots + u_j \le j(n-1) + q. \end{aligned}$$

Therefore, as in case (a), we see that the number of polynomials having all possible solutions $\xi^{(1)} + p\xi^{(2)} + \cdots + p^{j-1}\xi^{(j)}$ with the condition $u_1 = n$ does not exceed

$$s^{n-1}\sum_{q=1}^{j} p^{j-q}p^{A} = s^{n-1}\sum_{q=1}^{j} p^{j-q}p^{B+l(s-s_1)}$$

$$\leq s^{n-1}\sum_{q=1}^{j} p^{j-q}p^{j(m+r+\cdots+n)+q+l(s-s_1)}$$

$$\leq s^{n}p^{j(m+r+\cdots+n+1)+l(s-s_1)}.$$

In case (b), it remains to consider the case $p \leq n$. We estimate the number of polynomials corresponding to the solutions $\xi = \xi^{(1)} + p\xi^{(2)} + \cdots + p^{j-1}\xi^{(j)}$ similarly to case (a), but at the last step, estimating B, we use the inequality

$$(l-f)s_1 = (l-f)(u_1 + \cdots + u_j) \leq (l-f)jn_{f+1} \leq j(n_{f+1} + \cdots + n_l).$$

Then we can estimate the number K of polynomials as

$$K \leq s^{n}p^{j(m+r+\cdots+n+1)}.$$

Now let us estimate σ'_p, i.e., the p-adic density of the series σ',

$$\sigma'_p = 1 + \sum_{s=1}^{+\infty} A_1(p^s),$$

where

$$A_1(p^s) = \sum_{\substack{a_m=1 \\ p\nmid(a_m,\ldots,a_n)}}^{p^s} \cdots \sum_{a_n=1}^{p^s} \left| p^{-s}\sum_{x=1}^{p^s} \exp\{2\pi i(a_m x^m + \cdots + a_n x^n)/p^s\}\right|^{2k}.$$

First, we estimate $A_1(p^s)$. For polynomials $f(x) = a_m x^m + \cdots + a_n x^n$ with the set of exponents $(u_1, \ldots, u_j)$, Theorem 2.3 implies the estimate

$$\left| p^{-s}\sum_{x=1}^{p^s} \exp\{2\pi i(a_m x^m + \cdots + a_n x^n)/p^s\}\right| \leq np^{-j}.$$

Further, for $p > n$ the number K of such polynomials satisfies the inequality

$$K \leq s^{n}p^{j(m+r+\cdots+n)+l(s-s_1)},$$

and for $p \leq n$ this number satisfies the inequality

$$K \leq s^{n}p^{j(m+r+\cdots+n+1)+l(s-s_1)}.$$

It follows from the conditions $n \geq u_1 \geq \cdots \geq u_j \geq 2$ and $s \geq u_1 + \cdots + u_j \geq s - 2w - 1$ that j exceeds the largest of the two numbers $(s-2w-1)/n$ and 1. We estimate the variable $A_1(p^s)$ for $p > n$ and $p \leq n$ in different ways. First, we consider the case $p > n$. Then $w = [\log n / \log p] = 0$, and the variable s is equal either to $s-1$ or to s.

If, moreover, $s > n$, then Theorem 2.3 readily implies the inequality

$$A_1(p^s) \leq \sum_{j \geq s/n} s^n p^{j(m+r+\cdots+n)} n^{2k} p^{-2kj} + \sum_{j \geq (s-1)/n} s^n p^{j(m+r+\cdots+n)+l} n^{2k} p^{-2kj-k}$$
$$\leq 4n^{2k} s^n p^{(m+r+\cdots+n-2k)(s-1)/n}.$$

If $p > n$ and $2 \leq s \leq n$, then we have

$$A_1(p^s) \leq \sum_{j \geq 1} s^n p^{j(m+r+\cdots+n)} n^{2k} p^{-2kj} + \sum_{j \geq 1} s^n p^{j(m+r+\cdots+n)+l} n^{2k} p^{-2kj-k}$$
$$\leq 4n^{2k+n} p^{m+r+\cdots+n-2k}.$$

However, if $p > n$ and $s = 1$, then the Weil estimate (Lemma A.5) implies (for $2k > n + \cdots + r + m + 1 \geq 2l + 2$)

$$A_1(p) = \sum_{\substack{a_m=1 \\ p \nmid (a_m,\ldots,a_n)}}^{p} \cdots \sum_{a_n=1}^{p} \left| p^{-1} \sum_{x=1}^{p} \exp\{2\pi i (a_m x^m + \cdots + a_n x^n)/p\} \right|^{2k} \leq n^{2k} p^{l-k}.$$

Hence for $p > n$, we have the estimate

$$\sigma'_p - 1 = \sum_{s=1}^{+\infty} A_1(p^s) \leq n^{2k} p^{l-k} + 4n^{2k+n+1} p^{m+r+\cdots+n-2k}$$
$$+ 4n^{2k} \sum_{s>n} s^n p^{(m+r+\cdots+n-2k)(s-1)/n} \ll p^{l-k} + p^{m+r+\cdots+n-2k+\varepsilon},$$

where $\varepsilon > 0$ is an arbitrarily small number and $s^n \ll p^{\varepsilon(s-1)/n}$ as $s \to +\infty$.

Now we consider the case $p \leq n$. If $s - 2w - 1 \geq n$, then

$$A_1(p^s) \leq s^n p^{l(2w+1)} \sum_{j \geq (s-2w-1)/n} p^{j(m+r+\cdots+n+1-2k)} \leq 2ns^n e^{2nl} p^{m+r+\cdots+n+1-2k}.$$

For $p \leq n$ and $1 \leq s - 2w - 1$, we have

$$A_1(p^s) \leq s^n p^{l(2w+1)} \sum_{j \geq 1} p^{j(m+r+\cdots+n+1-2k)} \leq 2ns^n e^{2nl} p^{m+r+\cdots+n-2k}.$$

Therefore, for $p \leq n$ the series σ'_p converges for $m + r + \cdots + n + 1 - 2k < 0$, i.e., for $2k > m + r + \cdots + n + 1$. This implies that the series

$$\sigma' = \prod_p \sigma'_p = \prod_{p \leq n} \sigma'_p \prod_{p > n} \sigma'_p$$

converges for $2k > m + r + \cdots + n + 1$.

Let us prove that the series σ' diverges for $2k \le m + r + \cdots + n + 1$. Indeed, we have $\sigma' > \sigma$, where

$$\sigma_1 = \sum_{p>n} \sum_{\substack{a_m=1 \\ (a_m,p)=1}}^{p^m} \cdots \sum_{\substack{a_n=1 \\ (a_n,p)=1}}^{p^n} \left| p^{-n} \sum_{x=1}^{p^n} \exp\left\{ 2\pi i \left(\frac{a_n}{p^n} x^n + \cdots + \frac{a_m}{p^m} x^m \right) \right\} \right|^{2k}.$$

Further, for $p > n$ we have

$$S_1 = \sum_{x=1}^{p^n} \exp\left\{ 2\pi i \left(\frac{a_n}{p^n} x^n + \cdots + \frac{a_m}{p^m} x^m \right) \right\} = p^{n-1}.$$

Let us prove this relation. We can write each $1 \le x \le p^n$ as

$$x \equiv y + p^{n-1} z \pmod{p^n}, \quad 1 \le y \le p^{n-1},\ 1 \le z \le p.$$

Hence for $1 \le t \le n$, we have

$$x^t \equiv y^t + t p^{n-1} z y^{t-1} \pmod{p^n}.$$

Consequently,

$$S_1 = \sum_{y=1}^{p^{n-1}} \exp\left\{ 2\pi i \left(\frac{a_n}{p^n} y^n + \cdots + \frac{a_m}{p^m} y^m \right) \right\} \sum_{z=1}^{p} \exp\{2\pi i n a_n z y^{n-1}/p\} = p^{n-1}.$$

Therefore, we have the lower bound for the series σ_1:

$$\sigma_1 = \sum_{p>n} \sum_{\substack{a_m=1 \\ (a_m,p)=1}}^{p^m} \cdots \sum_{\substack{a_n=1 \\ (a_n,p)=1}}^{p^n} p^{-2k} \ge 2^{-l} \sum_{p>n} p^{m+r+\cdots+n-2k}.$$

But the series $\sum_{p>n} p^{-1}$ diverges and hence σ_1, as well as σ, diverges for $m + r + \cdots + n - 2k \ge -1$, i.e., for $2k \le m + r + \cdots + n + 1$. The proof of the theorem is complete. □

2.3 Multiple rational trigonometric sums

In this section, we obtain the upper bound for the modulus of the complete rational multiple trigonometric sum, i.e., for a sum of the form (2.1).

Lemma 2.8. *Let $F(x_1, \ldots, x_r)$ be a polynomial with integer coefficients, and let $F(0, \ldots, 0) = 0$. Then the following relation holds for any positive coprimes q_1 and q_2:*

$$S\big(q_1, q_2, F(x_1, \ldots, x_r)\big) = S\big(q_1, q_2^{-1} F(q_2 x_1, \ldots, q_2 x_r)\big) \times S\big(q_2, q_1^{-1} F(q_1 x_1, \ldots, q_1 x_r)\big).$$

Proof. If y_{ij} $(i = 1, \dots, r,\ j = 1, 2)$ runs through the complete system of residues modulo q_j, then $x_i = q_1 y_{i1} + q_2 y_{i2}$ $(i = 1, \dots, r)$ runs through the complete system of residues modulo q. Hence we obtain

$$F(x_1, \dots, x_r) = F(q_2 y_{11}, \dots, q_2 y_{r1}) + F(q_1 y_{12}, \dots, q_1 y_{r2}) \pmod{q}.$$

This congruence readily implies the statement of the lemma. The proof is complete. □

Lemma 2.9. *Suppose that $f(x) = a_0 + a_1 x + \dots + a_n x^n$ is a polynomial with integer coefficients, $(a_0, a_1, \dots, a_n, p) = 1$, and $N_p(\alpha, \beta)$ is the number of solutions of the congruence $f(x) \equiv 0 \pmod{p^\beta}$, $\alpha \geq \beta$, $1 \leq x \leq p^\alpha$. Then we have*

$$N_p(\alpha, \beta) \leq 3c_1(n) p^{\alpha - \beta/n},$$

where $c_1(n)$ is the constant in Theorem 2.1.

Proof. Without loss of generality, we assume that $(a_1, \dots, a_n, p) = 1$ (otherwise, the congruence $f(x) \equiv 0 \pmod{p^\beta}$ does not have solutions) and $n \geq 2$ (for $n = 1$ the estimate is trivial). Since the congruence $x \equiv x_1 \pmod{p^\beta}$ implies $f(x) \equiv f(x_1) \pmod{p^\beta}$, we have

$$N_p(\alpha, \beta) \leq p^{\alpha - 2\beta} \sum_{a=1}^{p^\beta} \sum_{x=1}^{p^\beta} \exp\{2\pi i a f(x)/p^\beta\}.$$

We divide the sum over a into $\beta + 1$ sums and collect together the sums over a for which $(a, p) = 1$ and $p \mid a$, but $p^2 \nmid a$, etc. Then we obtain

$$N_p(\alpha, \beta) \leq p^{\alpha - 2\beta} \sum_{k=0}^{\beta} S_k, \quad S_k = \sum_{\substack{a=1 \\ p^k \,\|\, a}}^{p^\beta} \sum_{x=1}^{p^\beta} \exp\{2\pi i a f(x)/p^\beta\}.$$

It follows from Theorem 2.1 ($a = a_1 p^k$ and $(a_1, p) = 1$) that

$$|S_k| \leq p^{\beta - k} \left| \sum_{x=1}^{p^\beta} \exp\{2\pi i a_1 f(x)/p^{\beta - k}\} \right| \leq c_1(n) p^{2\beta - k - (\beta - k)/n}.$$

Hence

$$\begin{aligned} N_p(\alpha, \beta) &\leq p^{\alpha - 2\beta} \sum_{k=0}^{\beta} |S_k| \leq c_1(n) p^{\alpha - \beta/n} \sum_{k=0}^{\beta} p^{-k + k/n} \\ &\leq c_1(n)(1 - p^{-1 + 1/n})^{-1} p^{\alpha - \beta/n}. \end{aligned}$$

The proof of the lemma is complete. □

Lemma 2.10. *Suppose that $n \geq 2$ is an integer and $F(x_1, \ldots, x_n)$ is a polynomial with integer coefficients*

$$F(x_1, \ldots, x_r) = \sum_{t_1=0}^{n} \cdots \sum_{t_r=0}^{n} a(t_1, \ldots, t_r) x_1^{t_1} \ldots x_r^{t_r},$$

where $(a(0, \ldots, 1), \ldots, a(n, \ldots, n), p) = 1$ and p is a prime. Then we have the estimate

$$|S(p^\alpha, F(x_1, \ldots, x_r))| \leq c_2(n) p^{r\alpha - \alpha/n},$$

where

$$c_2(n) = (3c_1(n))^r (\alpha + 1)^{r-1}.$$

Proof. We prove this lemma by induction on the number of variables in the polynomial $F(x_1, \ldots, x_r)$. For $r = 1$ the statement of the lemma holds (Theorem 2.1). We assume that the lemma holds for $r - 1$ variables and any α and prove it for r variables.

Let $(a(s_1, \ldots, s_r), p) = 1$. Without loss of generality, we can assume that $s_1 > 0$. We represent the polynomial $F(x_1, \ldots, x_r)$ as

$$F(x_1, \ldots, x_r) = \sum x_r = 1^{p^\alpha} = \sum_{t_1=0}^{n} \cdots \sum_{t_{r-1}=0}^{n} x_1^{t_1} \ldots x_{r-1}^{t_{r-1}} \varphi_{t_1, \ldots, t_{r-1}}(x_r).$$

Then

$$|S(p^\alpha, F(x_1, \ldots, x_r))| \leq \sum_{k=0}^{\alpha} \sum_{\substack{x_r=1 \\ p^k \parallel \varphi_{s_1, \ldots, s_r}(x_r)}}^{p^\alpha} \left| \sum_{x_1=1}^{p^\alpha} \cdots \sum_{x_{r-1}=1}^{p^\alpha} \exp\{2\pi i F(x_1, \ldots, x_r)/p^\alpha\} \right|.$$

By the induction hypothesis and by Lemma 2.9, we have

$$\begin{aligned} |S(p^\alpha, F(x_1, \ldots, x_r))| &\leq \sum_{k=0}^{\alpha} (3c_1(n))^{r-1} (\alpha+1)^{r-2} p^{(r-1)\alpha - (\alpha-k)/n} 3c_1(n) p^{\alpha - k/n} \\ &= (3c_1(n))^r (\alpha + 1)^{r-1} p^{r\alpha - \alpha/n}. \end{aligned}$$

The proof of the lemma is complete. □

Theorem 2.6. *Suppose that $n \geq 2$ is an integer, $n = \max(n_1, \ldots, n_r)$,*

$$F(x_1, \ldots, x_r) = \sum_{t_1=0}^{n_1} \cdots \sum_{t_r=0}^{n_r} a(t_1, \ldots, t_r) x_1^{t_1} \ldots x_r^{t_r}$$

is a polynomial with integer coefficients, and q is a natural number. Then

$$|S(q, F(x_1, \ldots, x_r))| \leq e^{7nr} 3^{r\nu(q)} (\tau(q))^{r-1} q^{r-1/n}, \tag{2.40}$$

where $\nu(q)$ is the number of distinct prime divisors of q and $\tau(q)$ is the number of divisors of q.

Proof. Let $q = p_1^{\alpha_1} \dots p_s^{\alpha_s}$ be the canonical decomposition of the number q. Then, since the sum $S(q, F(x_1, \dots, x_r))$ is multiplicative (Lemma 2.8), the estimate of $S(p^\alpha, F(x_1, \dots, x_r))$ in Lemma 2.10 implies that

$$\begin{aligned}|S(q, F(x_1, \dots, x_r))| &\le \prod_{p|q} (3c_1(n))^r (\tau(q))^{r-1} q^{r-1/n} \\ &\le \Big(\prod_{p|q} c_1(n)\Big)^r 3^{r\nu(q)} (\tau(q))^{r-1} q^{r-1/n} \le e^{7nr} 3^{r\nu(q)} (\tau(q))^{r-1} q^{r-1/n}.\end{aligned}$$

The proof of the theorem is complete. □

Lemma 2.11. *Suppose that $p \ge 3$ is a prime, m and n are natural numbers, $n > 1$, $(n, p) = 1$, $\alpha = mn$, $(a, p) = 1$, and*

$$S(p^\alpha, ax_1^n \dots x_r^n) = \sum_{x_1=1}^{p^\alpha} \dots \sum_{x_r=1}^{p^\alpha} \exp\{2\pi i a x_1^n \dots x_r^n / p^\alpha\}.$$

Then

$$S(p^\alpha, ax_1^n \dots x_r^n) \ge \frac{m^{r-1}}{(r-1)!} \left(1 - \frac{1}{p}\right)^{r-1} p^{r\alpha - m}.$$

Proof. We prove this lemma by induction. For $r = 1$ the statement holds (see [162], p. 270). We assume that it holds for $r - 1$ variables and prove it for r variables. We have

$$S(p^\alpha, ax_1^n \dots x_r^n) = \sum_{k=0}^{m-1} T_k + p^{r\alpha - m},$$

$$T_k = \sum_{x_1=1}^{p^\alpha} \dots \sum_{x_{r-1}=1}^{p^\alpha} \sum_{\substack{x_r=1 \\ p^k \,\|\, x_r}}^{p^\alpha} \exp\{2\pi i a x_1^n \dots x_r^n / p^\alpha\}.$$

By the induction hypothesis, we obtain

$$\begin{aligned} T_k &\ge \varphi(p^{\alpha-k}) p^{(r-1)kn} \frac{(m-k)^{r-2}}{(r-2)!} \left(1 - \frac{1}{p}\right)^{r-2} p^{(\alpha-kn)(r-1)-m+k} \\ &= \frac{(m-k)^{r-2}}{(r-2)!} \left(1 - \frac{1}{p}\right)^{r-1} p^{\alpha-k+(r-1)kn+(\alpha-kn)(r-1)-m+k} \\ &= \frac{(m-k)^{r-2}}{(r-2)!} \left(1 - \frac{1}{p}\right)^{r-1} p^{r\alpha-m}. \end{aligned}$$

Therefore,

$$S(p^\alpha, ax_1^n \dots x_r^n) \geq \left(1 - \frac{1}{p}\right)^{r-1} p^{r\alpha - m} \sum_{k=0}^{m-1} \frac{(m-k)^{r-2}}{(r-2)!}$$
$$\geq \frac{m^{r-1}}{(r-1)!}\left(1 - \frac{1}{p}\right)^{r-1} p^{r\alpha - m}.$$

The proof of the lemma is complete. □

2.4 Singular series in multidimensional problems

Suppose that $n \geq 2$, $n = \max(n_1, \dots, n_r)$, $F(x_1, \dots, x_r)$ is a polynomial with rational coefficients, and

$$F(x_1, \dots, x_r) = \sum_{t_1=0}^{n_1} \cdots \sum_{t_r=0}^{n_r} \frac{a(t_1, \dots, t_r)}{q(t_1, \dots, t_r)} x_1^{t_1} \dots x_r^{t_r},$$
$$\big(a(t_1, \dots, t_r), q(t_1, \dots, t_r)\big) = 1, \quad q(0, \dots, 0) = 1,$$
$$q = q(0, \dots, 1) \dots q(n_1, \dots, n_r), \quad m = (n_1 + 1) \dots (n_r + 1).$$

We consider a singular series σ of the form

$$\sigma = \sum_{q(n_1,\dots,n_r)=1}^{+\infty} \cdots \sum_{q(0,\dots,1)=1}^{+\infty} \sideset{}{'}\sum_{a(n_1,\dots,n_r)=0}^{q(n_1,\dots,n_r)-1} \cdots \sideset{}{'}\sum_{a(0,\dots,1)=0}^{q(0,\dots,1)-1} |q^{-1} S(q, qF(x_1, \dots, x_r))|^{2k},$$

where the prime on the summation signs means that

$$\big(a(n_1, \dots, n_r), q(n_1, \dots, n_r)\big) = 1, \ \dots, \ \big(a(0, \dots, 1), q(0, \dots, 1)\big) = 1.$$

The series σ is the mean value of complete multiple rational trigonometric sums.

Theorem 2.7. *The singular series σ converges for $2k > nm$.*

Proof. In the series σ, we collect all terms for which the numbers $q(t_1, \dots, t_r)$ $(0 \leq t_1 \leq n_1, \dots, 0 \leq t_r \leq n_r)$ have the same least common multiple equal to Q. Then Theorem 2.6 implies

$$\sigma \ll \sum_{Q=1}^{+\infty} \sigma(Q) Q^{(\varepsilon - 1/n)2k}, \tag{2.41}$$

where

$$\sigma(Q) = \sum_{\substack{q(n_1,\dots,n_r)=1 \\ [q(n_1,\dots,n_r),\dots,q(0,\dots,1)]=Q}}^{+\infty} \cdots \sum_{q(0,\dots,1)=1}^{+\infty} q(n_1, \dots, n_r), \dots, q(0, \dots, 1).$$

For $\sigma(Q)$, we have $\sigma(Q) \le Q^{m-1}(\tau(Q))^{m-1} \le c(\varepsilon_1)Q^{(1+\varepsilon_1)(m-1)}$, where $\varepsilon_1 > 0$ is an arbitrarily small fixed number. Substituting the estimate of $\sigma(Q)$ into (2.41), we see that the series σ converges for $2k > nm$. The theorem is thereby proved. □

Concluding remarks on Chapter 2. 1. Estimates for complete rational trigonometric sums with polynomials in the exponent were obtained by Hua Loo-Keng in 1940 [70]. Here Theorems 2.1 and 2.2 are proved following Chen's paper [45] (see also [144]).

2. We present the proofs of Theorems 2.3 and 2.4 closely following Hua Loo-Keng's paper [68].

3. Theorem 2.5 was proved by V. N. Chubarikov ([51]). Theorem 2.5 gives a solution to the problem of finding the convergence exponent of a singular series for the incomplete system of equations in Tarry's problem.

4. In connection with the problem of estimating complete trigonometric sums, A. A. Karatsuba ([83]) studied the accuracy as $p \to +\infty$, $n = n(p) \to +\infty$ of the Weil estimate for a complete trigonometric sum with an nth-degree polynomial in the exponent. The following theorem was proved:

For any ε, $0 < \varepsilon < 1/2$, there exists an infinite sequence of prime numbers p and a sequence of polynomials $f_n(x) = ax^n$, $(a, p) = 1$,

$$\frac{1}{2}\frac{p-1}{\log p}\log\frac{1}{\varepsilon} \le n \le \frac{p-1}{\log p}\log\frac{1}{\varepsilon},$$

such that

$$S(f_n) = \sum_{x=1}^{p} \exp\{2\pi i f_n(x)/p\} = \left(1 + \frac{2\pi\varepsilon}{1-\varepsilon}\theta\right)p, \quad \text{where } |\theta| \le 1.$$

The problem of finding similar estimates, if possible, for values of n less than $p/\log p$, say for n of order $\sqrt{p}$, was posed in [145]. V. M. Sidel'nikov [142] and V. I. Levenshtein [110] found that $S(f_n)$ is related to some problems in code theory. V. A. Zinov'ev and S. N. Litsyn [169] used the code-theoretical approach to solve the problem of estimating the accuracy of the Weil estimate. They proved that the Weil upper bound for complete trigonometric sums with a polynomial of degree n in the exponent is precise for n of order $\sqrt{Q}$ if these sums are considered in Galois fields F_Q, where $Q = p^m$, $m \ge 2$, and p is a fixed prime number.

L. A. Bassalygo, V. A. Zinov'ev, and S. N. Litsyn [38] found a relation between complete trigonometric sums with a polynomial in the exponent in Galois fields and the multiple trigonometric sums. They proved that the Weil estimate is exact already for $n \le \sqrt{Q}$.

Chapter 3

Weyl sums

In this chapter the letter S denotes a trigonometric sum of the form

$$S = S(\alpha_1, \dots, \alpha_n) = \sum_{x=1}^{P} \exp\{2\pi i f(x)\}, \tag{3.1}$$

where $f(x) = \alpha_1 x + \dots + \alpha_n x^n$ and $\alpha_1, \dots, \alpha_n$ are real numbers.

The sums S are called *Weyl sums*. This name was proposed by I. M. Vinogradov and became conventional.

3.1 Vinogradov's method for estimating Weyl sums

Vinogradov's method consists of the following two steps: first, one needs to reduce estimating an individual sum $S = S(\alpha_1, \dots, \alpha_n)$ to estimating the "mean" value of an even power of the modulus of S; second, one needs to estimate this "mean" value. The accuracy of averaging and estimating must satisfy some additional severe conditions.

First, we prove the *simplest properties of the sum S*.

Lemma 3.1. *The sum $S = S(\alpha_1, \dots, \alpha_n)$ treated as a function of the arguments $\alpha_1, \dots, \alpha_n$ is a periodic function in each of the arguments α_ν $(\nu = 1, \dots, n)$ with period equal to* 1.

Proof. We need to show that the congruence

$$(\alpha_1, \dots, \alpha_n) \equiv (\beta_1, \dots, \beta_n) \pmod 1 \tag{3.2}$$

implies the relation

$$S(\alpha_1, \dots, \alpha_n) = S(\beta_1, \dots, \beta_n). \tag{3.3}$$

For any integer x, (3.2) implies

$$\alpha_1 x + \dots + \alpha_n x^n \equiv \beta_1 x + \dots + \beta_n x^n \pmod 1,$$

and hence we have

$$\exp\{2\pi i(\alpha_1 x + \dots + \alpha_n x^n)\} = \exp\{2\pi i(\beta_1 x + \dots + \beta_n x^n)\},$$

which leads to (3.3). □

So, to know the behavior of all possible S, it suffices to know the behavior of S for which

$$(\alpha_1, \dots, \alpha_n) \in \Omega,$$

where Ω is the unit cube of the n-dimensional Euclidean space of the form $0 \le \alpha_1 < 1, \dots, 0 \le \alpha_n < 1$.

Definition 3.1. An integral J of the form

$$J = J(P; k, n) = \int_0^1 \cdots \int_0^1 |S(\alpha_1, \dots, \alpha_n)|^{2k} \, d\alpha_1 \dots d\alpha_n$$

is called the *mean value of the* $2k$*th power of the modulus of* S, or, briefly, the *mean value of* S.

The integral J is also called the *Vinogradov integral*. It is easy to see that J is equal to the number of solutions of the following system of equations in integers $x_1, \dots, x_{2k}$:

$$\begin{aligned} x_1 + \cdots + x_k &= x_{k+1} + \cdots + x_{2k}, \\ x_1^2 + \cdots + x_k^2 &= x_{k+1}^2 + \cdots + x_{2k}^2, \\ &\vdots \\ x_1^n + \cdots + x_k^n &= x_{k+1}^n + \cdots + x_{2k}^n, \\ 1 \le x_1, \dots&, x_{2k} \le P. \end{aligned} \tag{3.4}$$

The sum $S = S(\alpha_1, \dots, \alpha_n)$ is a continuous function of the arguments $\alpha_1, \dots, \alpha_n$, and therefore, any small variation in any of the arguments α_ν $(1 \le \nu \le n)$ results in small variations in S. More precisely, we state this property as Lemma 3.2.

Lemma 3.2. *Suppose that the inequalities*

$$|\alpha_1 - \beta_1| \le \Delta P^{-1}, \ \dots, \ |\alpha_n - \beta_n| \le \Delta P^{-n}$$

are satisfied for a given $\Delta > 0$. *Then*

$$S(\alpha_1, \dots, \alpha_n) = S(\beta_1, \dots, \beta_n) + 2\pi\theta n \Delta P, \quad |\theta| \le 1.$$

Proof. For any x $(1 \le x \le P)$, we have

$$|\beta_1 x + \cdots + \beta_n x^n - \alpha_1 x - \cdots - \alpha_n x^n| \le \Delta n$$

and, moreover, for a real φ, we have

$$|\exp\{2\pi i \varphi\} - 1| = 2|\sin \pi\varphi| \le 2\pi|\varphi|.$$

The lemma is thereby proved. □

Lemma 3.3. *The relation*

$$S = P_1^{-1} \sum_{y=1}^{P_1} \sum_{x=1}^{P} \exp\{2\pi i f(x+y)\} + 2\theta P_1, \quad |\theta| \leq 1,$$

holds for any integer P_1.

Proof. Let y be an integer $(1 \leq y \leq P_1)$. Then

$$S = \sum_{x=1}^{P} \exp\{2\pi i f(x)\} = \sum_{x=1}^{y} \exp\{2\pi i f(x)\} + \sum_{x=y+1}^{y+P} \exp\{2\pi i f(x)\} \tag{3.5}$$

$$- \sum_{x=1+P}^{y+P} \exp\{2\pi i f(x)\} = \sum_{x=1}^{P} \exp\{2\pi i f(x+y)\} + R,$$

where

$$R = R(y) = \sum_{x=1}^{y} \exp\{2\pi i f(x)\} - \sum_{x=1+P}^{y+P} \exp\{2\pi i f(x)\}.$$

The modulus of each term in these sums is equal to 1. Hence the modulus of R does not exceed $2y$. In other words, $R = 2\theta_1 y$, where $|\theta_1| \leq 1$. Summing both sides in (3.5) over y $(y = 1, \ldots, P_1)$, we prove the statement of the lemma. □

Definition 3.2. Suppose that $f_1(y), \ldots, f_{n-1}(y)$ are arbitrary real functions of an integer-valued argument y and $\Delta_1, \ldots, \Delta_{n-1}$ are arbitrary positive numbers that do not exceed 1. For each y $(y = 1, 2, \ldots, P_1)$, we consider domains $\Omega(y)$ of points in the $(n-1)$-dimensional Euclidean space of the form

$$(\gamma_1, \ldots, \gamma_{n-1}) \equiv \big(\{f_1(y)\} + \delta_1, \ldots, \{f_{n-1}(y)\} + \delta_{n-1}\big) \pmod 1,$$

where $|\delta_1| \leq \Delta_1, \ldots, |\delta_{n-1}| \leq \Delta_{n-1}$. For each point $(\alpha_1, \ldots, \alpha_{n-1})$ of the unit $(n-1)$-dimensional cube Ω, we let $g(\alpha_1, \ldots, \alpha_{n-1})$ denote the number of domains $\Omega(y)$ $(y = 1, \ldots, P_1)$ containing this point $(\alpha_1, \ldots, \alpha_{n-1})$. The number

$$G = \max_{(\alpha_1, \ldots, \alpha_{n-1}) \in \Omega} g(\alpha_1, \ldots, \alpha_{n-1})$$

is called the *multiplicity of intersection of the domains* $\Omega(y)$.

It follows from the definition of G that $1 \leq G \leq P_1$.

Using Lemmas 3.1–3.3, we reduce estimating the individual sum $S = S(\alpha_1, \ldots, \alpha_n)$ to estimating the mean value of the $2k$th power of the modulus of the trigonometric sum and then to estimating the multiplicity G of intersection of the domains $\Omega(y)$ defined by the polynomial $f(x) = \alpha_1 x + \cdots + \alpha_n x^n$.

Lemma 3.4. *Suppose that* $f(x) = \alpha_1 x + \cdots + \alpha_n x^n$, $0 < \Delta < 1$, $f_\nu(y) = (1/\nu!) f^{(\nu)}(y)$, $\Delta_\nu = \Delta P^{-\nu}$ $(\nu = 1, \ldots, n-1)$, $1 \le P_1 < P$, *and G is the multiplicity of intersection of the domains* $\Omega(y)$ *corresponding to given* $f_1(y), \ldots, f_{n-1}(y)$ *and* $\Delta_1, \ldots, \Delta_{n-1}$ $(y = 1, \ldots, P_1)$. *Then for any natural number k, the following inequality holds*:

$$|S| = |S(\alpha_1, \ldots, \alpha_n)| \le B + 2P_1 + 2\pi n \Delta P,$$

and moreover,

$$B^{2k} = (2\Delta)^{-n+1} P^{n(n-1)/2} P_1^{-1} G J(P; k, n-1).$$

Proof. By Lemma 3.3, we have

$$|S| \le W + 2P_1,$$

where

$$W = P_1^{-1} \sum_{y=1}^{P_1} \left| \sum_{x=1}^{P} \exp\{2\pi i f(x+y)\} \right|$$
$$= P_1^{-1} \sum_{y=1}^{P_1} \left| \sum_{x=1}^{P} \exp\{2\pi i (f_1(y)x + \cdots + f_{n-1}(y)x^{n-1} + \alpha_n x^n)\} \right|.$$

Suppose that $\delta_1, \ldots, \delta_{n-1}$ are arbitrary real numbers satisfying the conditions $|\delta_1| \le \Delta_1, \ldots, |\delta_{n-1}| \le \Delta_{n-1}$. Using Lemma 3.2, we find

$$W \le W_1 + 2\pi n \Delta P,$$

where

$$W_1 = P_1^{-1} \sum_{y=1}^{P_1} \left| \sum_{x=1}^{P} \exp\Big\{2\pi i \big((\{f_1(y)\} + \delta_1)x + \cdots \right.$$
$$\left. + (\{f_{n-1}(y)\} + \delta_{n-1})x^{n-1} + \alpha_n x^n\big)\Big\} \right|.$$

Raising W_1 to the power $2k$ and applying Hölder's inequality (Lemma A.1), we obtain

$$W_1^{2k} \le P_1^{-1} \sum_{y=1}^{P_1} \left| \sum_{x=1}^{P} \exp\Big\{2\pi i \big((\{f_1(y)\} + \delta_1)x + \cdots \right.$$
$$\left. + (\{f_{n-1}(y)\} + \delta_{n-1})x^{n-1} + \alpha_n x^n\big)\Big\} \right|^{2k}.$$

Then, integrating the last inequality over $-\Delta_\nu \le \delta_\nu \le \Delta_\nu$ with respect to δ_ν ($\nu = 1, \dots, n-1$) and recalling the definition of the multiplicity G of intersection of the domains $\Omega(y)$, we find

$$\begin{aligned} W^{2k} &\le 2^{-(n-1)}(\Delta_1 \dots \Delta_{n-1})^{-1} P_1^{-1} \\ &\quad \times \sum_{y=1}^{P_1} \int_{-\Delta_1}^{+\Delta_1} \cdots \int_{-\Delta_{n-1}}^{+\Delta_{n-1}} \Big| \sum_{x=1}^{P} \exp\big\{2\pi i \big((\{f_1(y)\} + \delta_1)x + \cdots \\ &\qquad + (\{f_{n-1}(y)\} + \delta_{n-1})x^{n-1} + \alpha_n x^n\big)\big\}\Big|^{2k} d\delta_1 \dots d\delta_{n-1} \\ &\le (2\Delta)^{-n+1} P^{n(n-1)/2} P_1^{-1} G \int_0^1 \cdots \int_0^1 \Big| \sum_{x=1}^{P} \exp\{2\pi i (\beta_1 x + \cdots \\ &\qquad + \beta_{n-1} x^{n-1} + \alpha_n x^n)\}\Big|^{2k} d\beta_1 \dots d\beta_{n-1} \\ &\le (2\Delta)^{-n+1} P^{n(n-1)/2} P_1^{-1} G J(P; k, n-1). \end{aligned}$$

The lemma is thereby proved. □

So we have reduced estimating $|S|$ to estimating the quantities G and J. For G, we have the trivial inequalities $1 \le G \le P_1$. Suppose that for a given polynomial $f(x) = \alpha_1 x + \cdots + \alpha_n x^n$ and the parameters given in Lemma 3.4, the value of G does not exceed $\Delta_0 P_1$, i.e.,

$$G \le \Delta_0 P_1, \quad \Delta_0 = P_1^{-c} < 1. \tag{3.6}$$

To what accuracy is it necessary to estimate J by using Lemma 3.4 in order to obtain a nontrivial estimate for $|S|$? There is another question. What is the best possible upper bound for $J = J(P; k, n)$? To answer these questions, we first consider the simplest properties of J and of some generalizations of J.

Lemma 3.5. *Suppose that $\lambda_1, \dots, \lambda_n$ are integers and $J_{kn}(\lambda_1, \dots, \lambda_n)$ is the number of solutions of the system of equations*

$$\begin{aligned} x_1 + \cdots - x_{2k} &= \lambda_1, \\ &\vdots \\ x_1^n + \cdots - x_{2k}^n &= \lambda_n, \\ 1 \le x_1, \dots, x_{2k} &\le P. \end{aligned} \tag{3.7}$$

Then the following relations hold:

$$\text{(a)} \quad J_{kn}(\lambda_1, \dots, \lambda_n) = \int_0^1 \cdots \int_0^1 \Big| \sum_{x \le P} \exp\{2\pi i (\alpha_1 x + \cdots + \alpha_n x^n)\}\Big|^{2k} \times \exp\{-2\pi i (\alpha_1 \lambda_1 + \cdots + \alpha_n \lambda_n)\}\, d\alpha_1 \dots d\alpha_n;$$

(b) $J_{kn}(\lambda_1, \dots, \lambda_n) \le J_{kn}(0, \dots, 0) = J(P; k, n) = J;$

(c) $\sum_{\lambda_1,\dots,\lambda_n} J_{kn}(\lambda_1, \dots, \lambda_n) = P^{2k};$

(d) $|\lambda_1| < kP, \ \dots, \ |\lambda_n| < kP^n;$

(e) $J = J(P; k, n) > (2k)^{-n} P^{2k-(n^2+n)/2};$

(f) *together with* $x_1, \dots, x_{2k}$, *the set of numbers* $x_1 + a, \dots, x_{2k} + a$ *is a solution of Eqs.* (3.4) *for any* a.

Proof. Assertion (a) becomes obvious if we raise the modulus of the integrand to the power $2k$ and integrate with respect to $\alpha_1, \dots, \alpha_n$; assertion (b) follows from the fact that the modulus of the integral does not exceed the modulus of the integrand; assertion (c) follows from the fact that the right-hand side of the relation is the number of all possible sets $x_1, \dots, x_{2k}$ of system (3.7), i.e., does not exceed P^{2k}; assertion (d) follows from the conditions on $x_1, \dots, x_{2k}$; assertion (e) follows from assertions (c), (b), and (d); assertion (f) can be proved by substituting the numbers $x_1+1, \dots, x_{2k}+a$ successively into the first, second, ..., last equations of system (3.4). The proof of the lemma is complete. □

It follows from assertion (e) in Lemma 3.5, i.e., from the estimate

$$J = J(P; k, n) > (2k)^{-n} P^{2k-(n^2+n)/2},$$

that the best possible estimate for J has the form

$$J = J(P; k, n) < c(n, k) P^{2k-(n^2+n)/2}, \tag{3.8}$$

where $c(n, k)$ is a positive constant depending only on n and k. The estimate (3.8) holds for k that are comparatively large as compared to n. Indeed, if (3.8) holds for $k \ge k_0 = k_0(n)$ and any $P \ge 1$, then the obvious inequality

$$J \ge P^{k_0}$$

implies

$$P^{k_0} < c(n, k_0) P^{2k_0-(n^2+n)/2},$$

i.e.,

$$1 < c(n, k_0) P^{k_0-(n^2+n)/2}.$$

Thus we have $k_0 \ge (n^2 + n)/2$, since for $k_0 < (n^2 + n)/2$ and $P \to +\infty$, the last inequality leads to a contradiction.

So we assume that (3.8) holds for $k \ge k_0$ and the estimate (3.6) holds for G. Then we obtain the following estimate for $|S|$ (applying Lemma 3.4, replacing n by $n - 1$ where it is necessary, and setting $P_1 = P^{1-c/(2k_0+n-1+c)}$ and $\Delta = P^{-c/(2k_0+n-1+c)}$):

$$|S| \le c_1(n, k_0) P^{1-c/(2k_0+n-1+c)}.$$

Obviously, this implies that, to obtain a more precise estimate of $|S|$, it is necessary to have (3.6) with $\Delta_0 = P_1^{-c}$ with $0 < c < 1$, where c is a constant, and to have (3.8) with the least possible value of $k_0 = k_0(n)$.

We also note that, instead of (3.8), it is possible to use a less precise inequality, namely, an inequality of the form

$$J < c(n,k)P^{2k-0.5(n^2+n)+\delta}, \tag{3.9}$$

where $\delta = \delta(n,k) > 0$ but satisfies the condition

$$\Delta_0 P^{\delta} < P^{-c_1}, \quad c_1 > 0.$$

Estimates of the form (3.8) and (3.9) are called *Vinogradov's mean value theorem.* They play a *fundamental role in Vinogradov's method for estimating Weyl sums.* Now we prove inequality (3.9). We shall follow [165].

First, we prove the original Vinogradov's lemma on the "number of hits," which sets the foundation of the mean value theorem.

Lemma 3.6. *Suppose that* $n > 2$, $P > (2n)^{4n}$, $H = (2n)^4$, *and* R *is the least number satisfying the condition* $HR \geq P$. *Finally, suppose that* $v_1, \dots, v_n$ *run through integers in the intervals*

$$X_1 < v_1 \leq Y_1, \ \dots, \ X_n < v_n \leq Y_n,$$

where, for some ω *such that* $0 \leq \omega < P$, *we have*

$$-\omega < X_1, \quad X_1 + R = Y_1, \quad Y_1 + R \leq X_2, \ \dots, \ X_n + R = Y_n, \quad Y_n \leq -\omega + P.$$

Then the number E_1 *of systems of values* $v_1, \dots, v_n$ *such that the sums* $V_1 = v_1 + \dots + v_n, \dots, V_n = v_1^n + \dots + v_n^n$ *lie respectively in some intervals of lengths*

$$1, \dots, P^{n-1} \tag{3.10}$$

satisfies the inequality

$$E_1 < \exp\{r(n) - 1\}H^{n(n-1)/2}, \quad r(n) = -\frac{n^2}{2}\ln n + \frac{3}{4}n^2 + \frac{3}{2}n.$$

Moreover, if $v_1', \dots, v_n'$ *run through the same values as* $v_1, \dots, v_n$ *(independently of the latter), then the number* E *of the cases where the differences* $V_1 - V_1', \dots, V_n - V_n'$ *lie respectively in some intervals of lengths*

$$P^{1-1/n}, \dots, P^{n(1-1/n)} \tag{3.11}$$

satisfies the inequality

$$E < 2\exp\{r(n)\}H^{n(n-3)/2}P^{(3n-1)/2}.$$

Proof. First, we estimate E_1. Let s be an integer such that $1 < s \le n$. If for given $v_{s+1}, \dots, v_n$ the sums $V_1, \dots, V_n$ lie respectively in intervals of lengths (3.10), then the sums $v_1 + \dots + v_s, \dots, v_1^s + \dots + v_s^s$ lie respectively in some intervals of lengths $1, \dots, P^{s-1}$.

Let $\eta_1, \dots, \eta_s$ and $\eta_1 + \xi_1, \dots, \eta_s + \xi_s$ be two sets of values of $v_1, \dots, v_s$ having this property and the least value η_s (hence $\xi_s > 0$). We obtain

$$\frac{(\eta_1 + \xi_1) - \eta_1}{\xi_1}\xi_1 + \dots + \frac{(\eta_s + \xi_s) - \eta_s}{\xi_s}\xi_s = \theta_0,$$
$$\vdots$$
$$\frac{(\eta_1 + \xi_1)^s - \eta_1^s}{s\xi_1}\xi_1 + \dots + \frac{(\eta_s + \xi_s)^s - \eta_s^s}{s\xi_s}\xi_s = \frac{\theta_{s-1}}{s}P^{s-1}$$

and thus derive

$$\Delta\xi_s - \Delta' = 0, \tag{3.12}$$

where

$$\Delta = \begin{vmatrix} \frac{(\eta_1+\xi_1)-\eta_1}{\xi_1} & \dots & \frac{(\eta_s+\xi_s)-\eta_s}{\xi_s} \\ \dots & \dots & \dots \\ \frac{(\eta_1+\xi_1)^s-\eta_1^s}{s\xi_1} & \dots & \frac{(\eta_s+\xi_s)^s-\eta_s^s}{s\xi_s} \end{vmatrix},$$

$$\Delta' = \begin{vmatrix} \frac{(\eta_1+\xi_1)-\eta_1}{\xi_1} & \dots & \frac{(\eta_{s-1}+\xi_{s-1})-\eta_{s-1}}{\xi_{s-1}} & \theta_0 \\ \dots & \dots & \dots & \dots \\ \frac{(\eta_1+\xi_1)^s-\eta_1^s}{s\xi_1} & \dots & \frac{(\eta_{s-1}+\xi_{s-1})^s-\eta_{s-1}^s}{s\xi_{s-1}} & \frac{\theta_{s-1}}{s}P^{s-1} \end{vmatrix}.$$

Next, we apply the following transformation to (3.12). We decompose both determinants in this relation with respect to the elements of the first column and, treating the result as the difference of values of some function of v_1 for $v_1 = \eta_1 + \xi_1$ and $v_1 = \eta_1$, apply the Lagrange formula. We obtain a new relation where the elements of the first column are replaced respectively by the numbers $1, \dots, x_1^{s-1}$ with some x_1 such that $X_1 < x_1 < Y_1$. Further, carrying our similar transformations for the second, third, ..., and penultimate columns and, finally, for the last column, but only in the first determinant, we obtain

$$\Delta_s\xi_s - \Delta_s' = 0,$$

$$\Delta_s = \begin{vmatrix} 1 & \dots & 1 \\ \dots & \dots & \dots \\ x_1^{s-1} & \dots & x_s^{s-1} \end{vmatrix}, \quad \Delta_s' = \begin{vmatrix} 1 & \dots & 1 & \theta_0 \\ \dots & \dots & \dots & \dots \\ x_1^{s-1} & \dots & x_{s-1}^{s-1} & \frac{\theta_{s-1}}{s}P^{s-1} \end{vmatrix},$$

$$X_1 < x_1 < Y_1, \ \dots, \ X_s < x_s < Y_s.$$

Hence we find

$$\Delta_s' = \sum_{r=0}^{s-1} \frac{\theta_r}{r+1} P^r U_r,$$

where U_r is the coefficient of x_s^r in the decomposition of

$$\Delta_s = (x_s - x_1)\dots(x_s - x_{s-1})\Delta_{s-1}$$

in powers of x_s, and hence it is equal to the product of Δ_{s-1} by the sum of products of the numbers $-x_1, \dots, -x_{s-1}$ taken till $s-1-r$. Therefore, we have

$$U_r \le \Delta_{s-1}\binom{s-1}{r}P^{s-1-r}, \quad \xi_s < \sum_{r=1}^{s-1} \frac{\binom{s-1}{r}P^{s-1}}{(r+1)(x_s - x_1)\dots(x_s - x_{s-1})}.$$

Because the inequality $x_{j+1} - x_j \ge (2t-1)R$ holds for $t \ge 1$, we have

$$\xi_s < \sum_{r=1}^{s} \frac{\binom{s}{r}H^{s-1}}{1\cdot 3\dots(2s-3)s} < \frac{(2^{s+1}-2)H^{s-1}}{3\dots(2s-1)} < L_s H^{s-1} - 1,$$

$$L_s = \frac{4}{(2-0.5)\dots(s-0.5)}.$$

Further,

$$\ln(2-0.5) + \dots + \ln(s-0.5) > \int_1^s \ln x\,dx = s\ln s - s + 1,$$

and therefore,

$$L_s < 4e^{s-1}s^{-s}.$$

So we have proved that, for $s > 1$ and given $v_{s+1}, \dots, v_n$, the number v_s can take only less than $4e^{s-1}s^{-s}H^{s-1}$ distinct values. Since, for given $v_2, \dots, v_n$, the number v_1 lies in an interval of length 1 and hence cannot take more than two distinct values, we have

$$E_1 < 2\prod_{s=2}^{n}(4e^{s-1}s^{-s}H^{s-1}) = 2\cdot 4^{n-1}(eH)^{n(n-1)/2}\prod_{s=2}^{n} s^{-s}.$$

Hence, because of the inequalities

$$\sum_{s=2}^{n} s\ln s > \int_1^n s\ln s\,ds > \frac{n^2}{2}\ln n - \frac{n^2}{4},$$

we obtain

$$E_1 < \exp\{r(n) - 1\}H^{n(n-1)/2}.$$

Further, since

$$\left(\frac{P^{1-1/n}}{1} + 1\right)\dots\left(\frac{P^{(n-1)(1-1/n)}}{P^{n-1}} + 1\right) < eP^{(n-1)/2},$$

the number E' of sets of values $v_1, \dots, v_n$ such that the sums $V_1, \dots, V_n$ lie respectively in some intervals of lengths (3.11) satisfies the inequality

$$E' < \exp\{r(n)\} H^{n(n-1)/2} P^{(n-1)/2}.$$

Finally, taking into account the fact that the number of all sets $v_1, \dots, v_n$ is less than $2P^n H^{-n}$, we obtain

$$E < 2\exp\{r(n)\} H^{n(n-3)/2} P^{(3n-1)/2}.$$

The proof of the lemma is complete. □

Theorem 3.1 (Vinogradov's mean value theorem). *Suppose that $\tau \geq 0$ is an integer, $k \geq n\tau$, and $P \geq 1$. Then*

$$J = J_k(P) = J_{kn}(P) \leq D_\tau P^{2k-\Delta(\tau)},$$

where

$$\Delta(\tau) = 0.5n(n+1)(1 - (1 - 1/n)^\tau), \quad D_\tau = (n\tau)^{6n\tau}(2n)^{4n(n+1)\tau}.$$

Proof. Obviously, it suffices to prove the theorem only for $k = n\tau$. For $\tau = 1$ and any P the theorem hold, since the integral $J_n(P)$ is equal to the number of solutions of the system of equations

$$\begin{gathered} x_1 + \cdots + x_n - x_{n+1} - \cdots - x_{2n} = 0, \\ \vdots \\ x_1^n + \cdots + x_n^n - x_{n+1}^n - \cdots - x_{2n}^n = 0, \\ 1 \leq x_i \leq P, \quad i = 1, \dots, 2n, \end{gathered}$$

which does not exceed

$$n!P^n \leq D_1 P^{2n-n}.$$

Moreover, for $\tau \geq 1$ and $P \leq D_\tau^{1/\Delta(\tau)}$, the theorem is trivial. Therefore, we shall consider only the case where $\tau \geq 1$ and $P > D_\tau^{1/\Delta(\tau)}$.

Let m and P_0 be natural numbers, and let the theorem be true for $\tau \leq m$ and $P \leq P_0$, as well as for $\tau \leq m+1$ and $P < P_0$. We shall prove that it is also true for $\tau \leq m+1$ and $P = P_0$. Thus, according to the principle of mathematical induction, it will be proved that the statement of the theorem is always true.

We set $k = n(m+1)$, $H = (2n)^4$, and $R = [PH^{-1} + 1]$. Then $P \leq RH$ and $J_k(P) \leq J_k(RH)$.

We transform the integrand in the integral $J_k(RH)$. First, we write

$$S = \sum_{x=1}^{RH} \exp\{2\pi i f(x)\} = \sum_{y=0}^{H-1} S(y),$$

where

$$S(y) = \sum_{z=1}^{R} \exp\{2\pi i f(z + Ry)\}, \quad f(x) = \alpha_1 x + \cdots + \alpha_n x^n.$$

Hence we have

$$S^k = \sum_{y_1=0}^{H-1} \cdots \sum_{y_k=0}^{H-1} S(y_1) \ldots S(y_k).$$

The set of numbers $y_1, \ldots, y_k$, as well as the product $S(y_1) \ldots S(y_k)$, is said to be *regular* if among the numbers $y_1, \ldots, y_k$ there are n numbers such that the modulus of the difference between any two of them does not exceed 1. The other sets and the corresponding products are said to be *irregular*. Now we set

$$S^k = W_1 + W_2,$$

where W_1 consists of regular products $S(y_1) \ldots S(y_k)$ and W_2 consists of irregular products. Then (see Lemma A.1) we have

$$J_k(RH) \le 2J_1 + 2J_2,$$

where

$$J_\mu = \int_0^1 \cdots \int_0^1 |W_\mu|^2 \, d\alpha_1 \ldots d\alpha_n, \quad \mu = 1, 2.$$

Let us estimate J_1. Applying Lemma A.1, we find

$$J_1 \le H^{2k} \max_{y_1, \ldots, y_k} \int_0^1 \cdots \int_0^1 |S(y_1) \ldots S(y_k)|^2 \, d\alpha_1 \ldots d\alpha_n.$$

We assume that the maximum is attained at the numbers $y_1, \ldots, y_n$ arranged so that $y_1 < \cdots < y_n$ and $y_{\nu+1} - y_\nu > 1$ $(\nu = 1, 2, \ldots, n-1)$. We divide the sum $S(y_\nu)$ $(\nu \ge n+1)$ into at most $t = [RP^{-1+1/n} + 1]$ small sums each of which has the summation interval of length $P^{1-1/n}$ or, perhaps, of length less than $P^{1-1/n}$ (the last sum). Then the product $S(y_{n+1}) \ldots S(y_k)$ can be represented as the sum of at most t^{k-n} terms of the form $S'(y_{n+1}) \ldots S'(y_k)$, where $S'(y_\nu)$ is one of the sums obtained by dividing the sum $S(y_\nu)$. Next, using the fact that the geometric mean of numbers does not exceed their arithmetic mean, we obtain

$$|S'(y_{n+1})|^2 \ldots |S'(y_k)|^2 \le \frac{|S'(y_{n+1})|^{2(k-n)} + \cdots + |S'(y_k)|^{2(k-n)}}{k-n}.$$

Hence

$$J_1 \le t^{2(k-n)} H^{2k} \int_0^1 \cdots \int_0^1 |S(y_1) \ldots S(y_n)|^2 |S'(y)|^{2(k-n)} \, d\alpha_1 \ldots d\alpha_n,$$

where y is one of the $y_{n+1}, \ldots, y_k$. But the last integral is equal to the number of solutions of the system of equations

$$(z_1 + Ry_1)^\nu + \cdots + (z_n + Ry_n)^\nu - (z_{n+1} + Ry_1)^\nu - \cdots - (z_{2n} + Ry_n)^\nu$$
$$= (z_{2n+1} + a)^\nu + \cdots - (z_{2k} + a)^\nu, \quad \nu = 1, 2, \ldots, n,$$

where $y_1, \ldots, y_n, a$ are fixed integers, $0 \le a = A + Ry < P$, $y_{\mu+1} - y_\mu > 1$ ($\mu = 1, 2, \ldots, n-1$), the unknowns $z_1, \ldots, z_{2n}$ vary from 1 to R, and the unknowns $z_{2n+1}, \ldots, z_{2k}$ vary from 1 to $P' \le P^{1-1/n}$. This system is equivalent to the following system (Lemma 3.5, (f)):

$$(z_1 + Ry_1 - a)^\nu + \cdots + (z_n + Ry_n - a)^\nu - (z_{n+1} + Ry_1 - a)^\nu - \cdots$$
$$- (z_{2n} + Ry_n - a)^\nu = z_{2n+1}^\nu + \cdots - z_{2k}^\nu, \quad \nu = 1, 2, \ldots, n.$$

Let J be the number of solutions of the last system of equations, and let $J'(\lambda_1, \ldots, \lambda_n)$ and $J''(\lambda_1, \ldots, \lambda_n)$ be the numbers of solutions of the systems

$$(z_1 + Ry_1 - a)^\nu + \cdots + (z_n + Ry_n - a)^\nu - (z_{n+1} + Ry_1 - a)^\nu - \cdots$$
$$- (z_{2n} + Ry_n - a)^\nu = \lambda_\nu, \quad \nu = 1, 2, \ldots, n,$$

and

$$z_{2n+1}^\nu + \cdots + z_{k+n}^\nu - z_{k+n+1}^\nu - \cdots - z_{2k}^\nu = \lambda_\nu, \quad \nu = 1, 2, \ldots, n.$$

Then we have

$$J = \sum_{\lambda_1, \ldots, \lambda_n} J'(\lambda_1, \ldots, \lambda_n) J''(\lambda_1, \ldots, \lambda_n).$$

Applying Lemma 3.5, (b), we obtain

$$J \le J''(0, \ldots, 0) \sum_{\lambda_1, \ldots, \lambda_n} J'(\lambda_1, \ldots, \lambda_n) \le J_{k-n}(P^{1-1/n}) \sum_{\lambda_1, \ldots, \lambda_n} J'(\lambda_1, \ldots, \lambda_n).$$

But the last sum is equal to the number of solutions of the system of inequalities

$$|(z_1 + Ry_1 - a)^\nu + \cdots + (z_n + Ry_n - a)^\nu - (z_{n+1} + Ry_1 - a)^\nu - \cdots$$
$$- (z_{2n} + Ry_n - a)^\nu| < (k-n) P^{\nu(1-1/n)}, \quad \nu = 1, 2, \ldots, n.$$

Applying the second assertion in Lemma 3.6, we obtain

$$\sum_{\lambda_1, \ldots, \lambda_n} J'(\lambda_1, \ldots, \lambda_n) < (2k)^n 2 \exp\{r(n)\} H^{n(n-3)/2} P^{(3n-1)/2}.$$

Combining these estimates, we arrive at the inequality

$$J_1 \le 2(2k)^n \exp\{r(n)\} (RP^{-1+1/n} + 1)^{2(k-n)}$$
$$\times H^{2k+n(n-3)/2} P^{(3n-1)/2} J_{k-n}(P^{1-1/n}).$$

By the induction hypothesis, we have

$$J_{k-n}(P^{1-1/n}) < D_m P^{(1-1/n)(2k-2n-\Delta(m))}.$$

Next, we find (using the fact that $P > D_{m+1}^{1/\Delta(m+1)}$)

$$\begin{aligned} k = n(m+1) > \Delta(m+1) &= 0.5n(n+1)(1-(1-1/n)^{m+1}) \\ &\leq 0.5(m+1)(n+1) \end{aligned}$$

and

$$P > (2n)^{8n} \quad \text{for } m \leq n; \quad P > (2n)^{8(m+1)} \quad \text{for } m > n;$$
$$\Delta(m+1) \leq 0.5n(n+1).$$

Therefore, we have

$$\begin{aligned} (RP^{-1+1/n}+1)^{2(k-n)} &\leq P^{2(k-n)/n} H^{-2(k-n)} (1+2P^{-1/n}H)^{2mn} \\ &\leq 2P^{2(k-n)/n} H^{-2(k-n)}, \end{aligned}$$

$$\begin{aligned} J_1 &\leq 2(2k)^n \exp\{r(n)\} 2P^{2(k-n)/n} H^{-2(k-n)} H^{2k+n(n-3)/2} P^{(3n-1)/2} \\ &\quad \times D_m P^{(1-1/n)(2k-2n-\Delta(m))} < 0.25 D_{m+1} P^{2k-\Delta(m+1)}. \end{aligned}$$

Now let us estimate J_2. Among the numbers $0, 1, \ldots, H-1$, an increasing series of $n-1$ numbers can be chosen in at most $H^{n-1}/(n-1)!$ ways. To each such series, there correspond $(2n-2)^k$ sets of $y_1, \ldots, y_k$. Hence the total number of irregular sets $y_1, \ldots, y_k$ does not exceed

$$\frac{H^{n-1}}{(n-1)!}(2n-2)^k = B.$$

Hence J_2 does not exceed

$$B^2 \int_0^1 \cdots \int_0^1 |S(y_1) \ldots S(y_k)|^2 \, d\alpha_1 \ldots d\alpha_n,$$

where $y_1, \ldots, y_k$ is the set for which the last integral takes its minimal value. Using again the inequality relating the arithmetic and geometric means, Lemma A.1, and the induction hypothesis, we obtain

$$\begin{aligned} J_2 &\leq B^2 \int_0^1 \cdots \int_0^1 |S(y)|^{2k} \, d\alpha_1 \ldots d\alpha_n \\ &= B^2 J_k(R) \leq B^2 D_{m+1} R^{2k-\Delta(m+1)} < 0.25 D_{m+1} P^{2k-\Delta(m+1)}. \end{aligned}$$

The statement of the theorem follows from the estimates for J_1 and J_2. The proof is complete. □

Choosing $l = l(n)$ in an appropriate way, one can obtain an arbitrarily small $\Delta(l)$. This, together with Vinogradov's lemma on estimating G, allows one to estimate the sum $S = S(\alpha_1, \ldots, \alpha_n)$ for all possible values of the coefficients $\alpha_1, \ldots, \alpha_n$.

The lemma on estimating G is stated as follows.

Lemma 3.7. *Suppose that $P \geq n^n$, $\nu = 1/n$, m is a positive integer that does not exceed P^{ν^2}, and to each integer y there corresponds its own point $(mY_{n-1}, \ldots, mY_1)$ determined by the expansion*

$$mf(x+y) - mf(y) = m\alpha_n x^n + mY_{n-1}x^{n-1} + \cdots + mY_1 x$$

of the polynomial $mf(x+y) - mf(y)$ in powers of x. To each $s = n, \ldots, 2$, we assign its own number $\tau_s = P^{s-0.5}$ and, moreover, represent a_s (this is always possible) in the form

$$\alpha_s = \frac{a_s}{q_s} + \frac{\theta_s}{q_s \tau_s}, \qquad (a_s, q_s) = 1,\ 0 < q_s \leq \tau_s.$$

We also let the symbol Q_0 denote the least common multiple of the numbers $q_n, \ldots, q_2$.

Let G be the number of points corresponding to the numbers y in the series $0, 1, \ldots, P-1$ which, by adding to their coordinates some numbers that numerically do not exceed

$$L_{n-1} = P^{-n+1}, \ \ldots, \ L_1 = P^{-1},$$

can be made congruent to the point corresponding to some definite number y_0 in the same series. Then for $Q \geq P^{0.5-0.4\nu}$ we have

$$G < mn^{2n-2} P^{0.5+0.4\nu}.$$

Proof. For the proof, see [165], p. 63. □

This lemma and Theorem 3.1 imply the general Vinogradov's estimate of the Weyl sum.

Theorem 3.2. *Let n be a constant number such that $n \geq 3$ and $\nu = n^{-1}$. We divide the points of the n-dimensional space into two classes: points of the first class and points of the second class. A point of the first class is defined to be a point*

$$\left(\frac{a_n}{q_n} + z_n, \ldots, \frac{a_1}{q_1} + z_1\right)$$

whose first summands are rational irreducible fractions with positive denominators whose least common multiple is a number $Q \leq P^{\nu}$ and whose second summands satisfy the condition

$$|z_s| \leq P^{-s+\nu}.$$

A point that is not a point of the first class is called a point of the second class. Then, by setting

$$\rho = \frac{1}{8n^2(\ln n + 1.5 \ln \ln n + 4.2)},$$

for $m \le P^{2\rho}$, *we have*

$$|T(m)| \le c(n) P^{1-\rho}, \quad c(n) = (2n)^{2n+2}\big(n(n+1)\ln \rho^{-1}\big)^{3/2}$$

for the points of the second class and, by setting

$$\delta_s = z_s P^s, \quad \delta_0 = \max\big(|\delta_n|, \dots, |\delta_1|\big),$$

for $m \le P^{4\nu^2}$, *we have*

$$|T(m)| \ll P(m, Q)^{\nu} Q^{-\nu+\epsilon_0}$$

for the points of the first class or, which is the same,

$$|T(m)| \ll P Q^{-\nu+\epsilon_0} \delta_0^{-\nu} \quad \text{if} \quad \delta_0 \ge 1,$$

where we introduced the notation

$$T(m) = \sum_{0<x\le P} \exp\{2\pi i m(\alpha_1 x + \cdots + \alpha_n x^n)\}.$$

Proof. For the proof, see [165], p. 66. □

Finally, from Theorem 3.1 and 3.2 we derive the "simplified upper bound" for J.

Theorem 3.3. *Suppose that* n *is a constant* ($n \ge 3$), k *is an integer, and*

$$k \ge [n^2(2\ln n + \ln \ln n + 4)].$$

Then the following estimate holds:

$$J = J_k(P) = J(P; k, n) \le c(n) P^{2k-0.5n(n+1)}.$$

Proof. For the proof of this Vinogradov's theorem, see [165], p. 70. □

In what follows (Theorem 3.9), we obtain a slightly more precise statement (where we estimate the value of $c(n)$).

3.2 An estimate of the function $G(n)$

The function $G(n)$ was introduced by Hardy and Littlewood while solving Waring's problem.

Definition 3.3. Let $n \geq 3$. Then $G(n)$ is equal to the least k for which the equation

$$x_1^n + \cdots + x_k^n = N$$

is solvable in natural numbers for any $N \geq N_0(n)$.

In 1919, Hardy and Littlewood found for $G(n)$ an upper bound of the form

$$G(n) \leq n2^{n-2}h, \qquad \lim_{n\to+\infty} h = 1,$$

increasing with n as a variable of order $n2^n$.

In 1934, I. M. Vinogradov developed a new method for estimating $G(n)$, which led him to the lemma on the "number of hits" and to a new method for estimating Weyl sums.

We present one of the simplest versions of estimating $G(n)$.

Theorem 3.4. *The function $G(n)$ satisfies the inequalities*

$$n < G(n) \leq 4n \ln n + 16n \ln \ln n + 8n.$$

Proof. We consider a sequence of numbers X of the form $X = P^n + P^{n-2}$, where $P \geq P_0(n)$ is a natural number. Since $[X^{1/n}] = P$, there is at most

$$P^k \leq P^n < X = P^n + P^{n-2}$$

natural numbers that do not exceed X and can be represented as the sum of $k \leq n$ natural terms of the form x^n. This implies the first statement of the theorem.

To prove the second statement, we consider the equation

$$x_1^n + \cdots + x_k^n + u_1^n + \cdots + u_m^n + u_{m+1}^n + \cdots + u_{2m}^n = N, \tag{3.13}$$

where $x_1, \ldots, x_k, u_1, \ldots, u_{2m}$ are natural numbers and

$$\begin{aligned}
P_1 &= 0.25N^{1/n} < u_1, & u_{m+1} &< 0.5N^{1/n} = 2P_1,\\
P_2 &= 0.5P_1^{1-1/n} < u_2, & u_{m+2} &< P_1^{1-1/n} = 2P_2,\\
&\vdots & &\vdots\\
P_m &= 0.5P_{m-1}^{1-1/n} < u_m, & u_{2m} &< P_{m-1}^{1-1/n} = 2P_m.
\end{aligned}$$

First, we have

$$4^{-n}N = P_1^n \le u_1^n + \dots + u_m^n + u_{m+1}^n + \dots + u_{2m}^n \le 4(2P_1)^n = 2^{-n+2}N.$$

Next, the equation

$$u_1^n + \dots + u_m^n = u_{m+1}^n + \dots + u_{2m}^n \tag{3.14}$$

has solutions only of the form

$$u_1 = u_{m+1},\ u_2 = u_{m+2},\ \dots,\ u_m = u_{2m}.$$

Indeed, if, for instance, $u_s \ne u_{m+s}$ $(s < m)$ and $u_1 = u_{m+1}, \dots, u_{s-1} = u_{m+s-1}$, then

$$|u_s^n - u_{m+s}^n| > nP_s^{n-1},$$
$$|u_{s+1}^n + \dots + u_m^n - u_{m+s+1} - \dots - u_{2m}^n| \le (2P_{s+1})^n = P_s^{n-1},$$

and relation (3.14) is impossible.

Let $I(N)$ be the number of solutions of Eq. (3.13). Then

$$I(N) = \int_0^1 S^k(\alpha)T_1^2(\alpha)\dots T_m^2(\alpha)\exp\{-2\pi i\alpha N\}\,d\alpha,$$

where

$$S(\alpha) = \sum_{0<x\le P} \exp\{2\pi i\alpha x^n\}, \quad T_j(\alpha) = \sum_{u_j} \exp\{-2\pi i\alpha u_j^n\},$$
$$P = N^{1/n}, \quad j = 1, \dots, m.$$

Following the partition procedure described in Theorem 3.2, we divide the points of the interval $[0, 1)$ into points of the first class E_1 and points of the second class E_2. The points of the first class are points of the form

$$\alpha = \frac{a}{q} + z, \quad (a, q) = 1, \quad 1 \le q \le P^{1/n}, \quad |z| \le P^{-n+1/n}.$$

All the other points of the interval $[0, 1)$ are points of the second class. We present the integral $I(N)$ as the sum of two integrals

$$I(N) = I_1(N) + I_2(N),$$
$$I_j(N) = \int_{E_j} S^k(\alpha)T_1^2(\alpha)\dots T_m^2(\alpha)\exp\{-2\pi i\alpha N\}\,d\alpha.$$

Let us find an upper bound for $|I_2(N)|$. By Theorem 3.2, for $\alpha \in E_2$, we have

$$|S(\alpha)| \le c(n)P^{1-\rho}, \quad \rho = \frac{1}{8n^2(\ln n + 1.5\ln\ln n + 4.2)},$$

and hence

$$|I_2(N)| \le c(n)P^{k(1-\rho)} \int_0^1 |T_1(\alpha)|^2 \dots |T_m(\alpha)|^2 \, d\alpha.$$

The last integral is equal to the number of solutions of Eq. (3.14), i.e., to the number of sets $u_1, \dots, u_m$, and does not exceed

$$P_1 P_2 \dots P_m \le c(n, m) N^{1-(1-\nu)^m}.$$

Hence

$$|I_2(N)| \le c(n, k) P_1 P_2 \dots P_m P^{k(1-\rho)}.$$

Let us find a lower bound for $I_1(N)$. By the definition of $I_1(N)$, we have

$$I_1(N) = \sum_{u_1} \cdots \sum_{u_{2m}} \int_{E_1} S^k(\alpha) \exp\{-2\pi i \alpha (N - u_1^n - \cdots - u_{2m}^n)\} \, d\alpha.$$

The number $N_1 = N - u_1^n - \cdots - u_{2m}^n$ lies between the bounds

$$(1 - 2^{-n+2})N \le N_1 \le (1 - 2^{-2n})N;$$

hence for the integral

$$\int_{E_1} S^k(\alpha) \exp\{-2\pi i \alpha N_1\} \, d\alpha$$

with $k \ge 4n$, we have the asymptotic formula

$$\int_{E_1} S^k(\alpha) \exp\{-2\pi i \alpha N_1\} \, d\alpha = \gamma \sigma(N_1) N_1^{k/n-1} + O(N_1^{k/n-1-1/(4n^3)}),$$

where

$$\gamma = \left(\Gamma\left(1 + \frac{1}{n}\right)\right)^k \left(\Gamma\left(\frac{k}{n}\right)\right)^{-1},$$

$$\sigma(N_1) = \sum_{q=1}^{+\infty} \sum_{\substack{0 \le a < q \\ (a,q)=1}} \left(q^{-1} \sum_{x=1}^{q} \exp\{2\pi i a x^n / q\}\right)^k \exp\{-2\pi i a N_1 / q\},$$

and moreover, $\sigma(N_1) \ge c(n, k) > 0$. This assertion is proved as in [10]. Hence for $N \ge N_0(n)$, we have the following estimate for $I_1(N)$:

$$I_1(N) \ge c(n, k) \sum_{u_1} \cdots \sum_{u_{2m}} N_1^{(k-n)/n} \ge c(n, k)(P_1 \dots P_m)^2 N^{(k-n)/n}.$$

For the inequality $I(N) = I_1(N) + I_2(N) > 0$ to hold, it suffices to have the inequality

$$(P_1 \dots P_m)^2 N^{(k-n)/n} \ge c(n, k) P_1 \dots P_m P^{k(1-\rho)}$$

or the inequality

$$N^{k/n-(1-\nu)^m} \geq c(n,k)N^{(k-k\rho)/n}.$$

The last inequality holds for $k = 8n$, $m \geq 2n \ln + 8n \ln \ln n$, and $N \geq N_0(n)$. So Eq. (3.13) is solvable, i.e., we have

$$G(n) \leq 4n \ln n + 16n \ln \ln n + 8n.$$

The proof of the theorem is complete. □

3.3 An analog of Waring's problem for congruences

In 1961 A. A. Karatsuba posed and solved a problem, which is called an *analog of Waring's problem for congruences*. This problem and the method used to solve it allowed one, first, to develop a new p-adic method for proving Vinogradov's mean value theorem and then to develop a general p-adic method that set the foundation of the theory of multiple trigonometric sums. In this section we consider this problem.

Let us consider the congruence

$$x_1^n + \cdots + x_t^n \equiv N \pmod{Q}, \quad 1 \leq x_1, \ldots, x_t \leq P. \tag{3.15}$$

We define a parameter r by the relation $r = \ln Q / \ln P$. We set $1 \leq r \leq n$. For $r > n$, $P \to +\infty$, $N \leq Q$, and fixed n and t, congruence (3.15) becomes an equation. Hence the larger r is, the "nearer" congruence (3.15) is to an equation. It follows from Section 3.2 that for $P \geq P_0(n)$ and $t \geq 4n \ln n + 16n \ln \ln n + 8n$, congruence (3.15) is solvable for any N. The problem of improving the lower bound for t arises. We give a sufficiently complete answer for a special form of moduli Q.

Theorem 3.5. *Suppose that m and r are natural numbers, $1 \leq r \leq \sqrt{n/3}$, p is a prime number, $p > n^6$, $Q = p^{2mnr}$, $P = p^{2mn}$, and $P \geq P_0(n)$. Then for $t \geq 4r + 4$, congruence* (3.15) *is solvable for any N. For $t < r$ there exist N such that congruence* (3.15) *does not have solutions.*

Proof. The second part of the theorem follows from the fact that the number of all possible values of the left-hand side in (3.15) does not exceed $P^t \leq P^{r-1} = p^{2mnr-2mn}$ and the number of all possible values of the right-hand side in (3.15) is exactly equal to $Q = p^{2mnr}$.

Let $W(N)$ be the number of solutions of (3.15). Then

$$W(N) = Q^{-1} \sum_{a=0}^{Q-1} S^t(a) \exp\{-2\pi i a N / Q\},$$

where

$$S(a) = \sum_{x=1}^{P} \exp\{2\pi i a x^n / Q\}.$$

Each positive a can be written as $a = bQp^{-\nu}$, $1 \le \nu \le 2mnr$, $0 < b < p^{\nu}$, and $(b, p) = 1$. Then

$$W(N) = P^t Q^{-1} + \sum_{\nu=1}^{2mnr} T(\nu),$$

where

$$T(\nu) = Q^{-1} \sum_{0<b<p^{\nu}}{}' \Big(\sum_{x=1}^{P} \exp\{2\pi i b x^n / p^{\nu}\} \Big)^t \exp\{-2\pi i b N / p^{\nu}\};$$

the prime on the sum over b means that $(b, p) = 1$.

By $W_0(N)$ we denote the sum

$$W_0(N) = P^t Q^{-1} + \sum_{\nu=1}^{2mn} T(\nu).$$

First, we prove that

$$W_0(N_1) > 0.5 P^{t_1} Q^{-1}$$

for any integer N_1 and $t = t_1 \ge 4$.

We note that the method used to prove this inequality can be treated as a discrete analog of the Hardy–Littlewood circle method in the form of the Vinogradov trigonometric sums (the partition of the sum over ν into two parts corresponds to the partition of the Waring's problem integration interval into the principal and additional intervals; small ν, i.e., $\nu \le 2mn$, correspond to principal intervals; so $W_0(N)$ gives the leading term of $W(N)$).

Since $P = p^{2mn}$ $(1 \le \nu \le 2mn)$, using the periodicity of the trigonometric sum in $T(\nu)$, we find

$$\sum_{x=1}^{P} \exp\{2\pi i b x^n / p^{\nu}\} = P p^{-\nu} \sum_{x=1}^{p^{\nu}} \exp\{2\pi i b x^n / p^{\nu}\}.$$

The well-known lemmas on complete trigonometric sums in Waring's problem (e.g., see [159], p. 270) easily imply the following formulas:

$$\sum_{x=1}^{p^{\nu}} \exp\{2\pi i b x^n / p^{\nu}\} = \begin{cases} p^{\nu - \nu/n} & \text{if } \nu \equiv 0 \pmod{n}, \\ p^{\nu-(\nu-1)/n-1} S(b, p) & \text{if } \nu \equiv 1 \pmod{n}, \\ p^{\nu - [\nu/n] - 1} & \text{if } \nu \not\equiv 0, 1 \pmod{n}; \end{cases}$$

$$S(b, p) = \sum_{x=1}^{p} \exp\{2\pi i b x^n / p\}, \quad |S(b, p)| < n\sqrt{p}. \tag{3.16}$$

We divide the sum over ν into the corresponding progressions:

$$\varkappa = \sum_{\nu=1}^{2mn} T(\nu) = \sum_{\nu_2=1}^{n} \sum_{\nu_1=0}^{2m-1} T(n\nu_1 + \nu_2) = B_1 + B_2 + B_3,$$

where

$$B_1 = \sum_{\nu_1=0}^{2m-1} T(n\nu_1 + 1), \quad B_2 = \sum_{\nu_2=2}^{n-1} \sum_{\nu_1=0}^{2m-1} T(n\nu_1 + \nu_2), \quad B_3 = \sum_{\nu_1=0}^{2m-1} T(n\nu_1 + n).$$

From (3.16) we find

$$T(n\nu_1 + 1) = P^{t_1} Q^{-1} p^{-(\nu_1+1)t_1} \sum_{0<b<p^\nu}{}' S^{t_1}(b, p) \exp\{-2\pi i b N_1/p^\nu\},$$

$$T(n\nu_1 + \nu_2) = P^{t_1} Q^{-1} p^{-(\nu_1+1)t_1} \sum_{0<b<p^\nu}{}' \exp\{-2\pi i b N_1/p^\nu\}, \quad 2 \le \nu_2 \le n.$$

Next, we have

$$\sum_{0<b<p^\nu}{}' \exp\{-2\pi i b N_1/p^\nu\} = \sum_{b=0}^{p^\nu-1} \exp\{-2\pi i b N_1/p^\nu\}$$
$$- \sum_{b=0}^{p^\nu-1} \exp\{-2\pi i b N_1/p^{\nu-1}\} = p^\nu \delta(N_1 p^{-\nu}) - p^{\nu-1} \delta(N_1 p^{-\nu+1}),$$

where $\delta(\xi) = 1$ if ξ is an integer and $\delta(\xi) = 0$ otherwise. This and the preceding formulas yield

$$B_2 + B_3 = \sum_{\nu_2=2}^{n} \sum_{\nu_1=0}^{2m-1} T(n\nu_1 + \nu_2)$$
$$= \sum_{\nu_1=0}^{2m-1} P^{t_1} Q^{-1} p^{-(\nu_1+1)t_1} \left(p^{n(\nu_1+1)} \delta(N_1 p^{-n(\nu_1+1)}) - p^{n\nu_1+1} \delta(N_1 p^{-n\nu_1-1})\right).$$

Let $N_1 = p^h N_2$ and $(N_2, p) = 1$. If $h \ge 2mn$, then $\delta(N_1 p^{-n(\nu_1+1)}) = \delta(N_1 p^{-n\nu-1}) = 1$ and

$$B_2 + B_3 = P^{t_1} Q^{-1} (p^n - p) p^{-t_1} \sum_{\nu_1=0}^{2m-1} p^{\nu_1(n-t_1)}$$
$$= P^{t_1} Q^{-1} (p^n - p) p^{-t_1} \frac{p^{2m(n-t_1)} - 1}{p^{n-t_1} - 1}.$$

Moreover, we always have

$$|T(n\nu_1+1)| < n^{t_1}P^{t_1}Q^{-1}p^{-(\nu_1+1)t_1+0.5t_1+n\nu_1+1},$$

$$|B_1| \le n^{t_1}P^{t_1}Q^{-1}p^{-0.5t_1+1}\sum_{\nu_1=0}^{2m-1} p^{(n-t_1)\nu_1} = n^{t_1}P^{t_1}Q^{-1}p^{-0.5t_1+1}\frac{p^{2m(n-t_1)}-1}{p^{n-t_1}-1}.$$

Hence

$$B_1+B_2+B_3 > P^{t_1}Q^{-1}\frac{p^{2m(n-t_1)}-1}{p^{n-t_1}-1}\left(p^{-t_1}(p^n-p)-n^{t_1}p^{-0.5t_1+1}\right).$$

It is easy to verify that, for $p > n^6$ $(n > 9)$ and any $t_1 \ge 4$,

$$B_1+B_2+B_3 > -0.5P^{t_1}Q^{-1}.$$

Let $h=0$. Then $B_2+B_3=0$ and $T(n\nu_1+1)=0$ for $\nu_1>0$, i.e.,

$$\begin{aligned} B_1+B_2+B_3 &= P^{t_1}Q^{-1}p^{-t_1}\sideset{}{'}\sum_{0<b<p} S^{t_1}(b,p)\exp\{-2\pi i bN_1/p\} \\ &> -n^{t_1}P^{t_1}Q^{-1}p^{-0.5t_1+1} > -0.5P^{t_1}Q^{-1}. \end{aligned}$$

Now let $1 \le h < 2mn$, $h = nh_1+h_2$, $0 < h_1 < 2m$, and $1 \le h_2 \le n$. Then

$$\varkappa - \sum_{\nu=1}^{h} T(\nu) + T(h+1) + \sum_{\nu=h+2}^{2mn} T(\nu).$$

Obviously, $T(\nu)=0$ for $\nu \ge h+2$. Next, for $\nu \le h$ we have

$$\begin{aligned} |T(n\nu_1+1)| &< P^{t_1}Q^{-1}p^{-(\nu_1+1)t_1+n\nu_1+1+t_1/2}, \\ T(n\nu_1+\nu_2) &= P^{t_1}Q^{-1}p^{-(\nu_1+1)t_1}(p^{\nu}-p^{\nu-1}), \quad \nu_2 \ne 1, \\ \sum_{\nu=1}^{h} T(\nu) &= \sum_{\nu_1=0}^{h_1-1}\sum_{\nu_2=1}^{n} T(n\nu_1+\nu_2) + \sum_{\nu_2=1}^{h_2} T(nh_1+\nu_2) \\ &> \sum_{\nu_1=0}^{h_1-1}\sum_{\nu_2=1}^{n} P^{t_1}Q^{-1}p^{-(\nu_1+1)t_1+n\nu_1+\nu_2}(1-p^{-1}) \\ &\quad - \sum_{\nu_1=0}^{h_1} P^{t_1}Q^{-1}p^{-(\nu_1+1)t_1+n\nu_1+1+t_1/2} \\ &\quad + \sum_{\nu_2=2}^{h_2} P^{t_1}Q^{-1}p^{-(h_1+1)t_1+nh_1+\nu_2}(1-p^{-1}). \end{aligned}$$

If $h_2 = n$, then

$$|T(h+1)| < n^{t_1} P^{t_1} Q^{-1} p^{-(h_1+1)t_1+nh_1+n+1-t_1/2};$$

if $h_2 < n$, then $T(h+1) = -P^{t_1} Q^{-1} p^{-(h_1+1)t_1+nh_1+h_2}$. So we always have

$$T(h+1) \geq -P^{t_1} Q^{-1} p^{-(h_1+1)t_1+nh_1+h_2}.$$

Hence

$$\varkappa > -P^{t_1} Q^{-1} p^{-(h_1+1)t_1+nh_1+h_2} + P^{t_1} Q^{-1} p^{-t_1} \sum_{\nu_1=0}^{h_1-1} p^{(n-t_1)\nu_1} \sum_{\nu_2=2}^{n} p^{\nu_2}(1-p^{-1})$$
$$- n^{t_1} P^{t_1} Q^{-1} p^{-0.5t_1+1} \sum_{\nu_1=0}^{h_1} p^{(n-t_1)\nu_1} + P^{t_1} Q^{-1}(1-p^{-1}) p^{-(h_1+1)t_1+nh_1} \sum_{\nu_2=2}^{h_2} p^{\nu_2}.$$

If $h_1 = 0$, then

$$\varkappa > P^{t_1} Q^{-1}(n^{t_1} p^{-0.5t_1+1} + p^{-t_1+1}) > -0.5 P^{t_1} Q^{-1};$$

if $h_1 \geq 1$, then

$$\varkappa > P^{t_1} Q^{-1} \left\{ \frac{(p^{(n-t_1)h_1} - 1)(p^{n-t_1} - p^{1-t_1})}{p^{n-t_1} - 1} \right.$$
$$\left. - n^{t_1} p^{-0.5t_1+1} \frac{p^{(n-t_1)(h_1+1)} - 1}{p^{n-t_1} - 1} - p^{(n-t_1)h_1-t_1+1} \right\}.$$

It is easy to verify that for all values of the parameters admissible by the assumptions of the lemma, the expression in braces does not exceed -0.5.

So $\varkappa > -0.5 P^{t_1} Q^{-1}$ and

$$W_0(N_1)) = P^{t_1} Q^{-1} + \varkappa > 0.5 P^{t_1} Q^{-1}.$$

We consider the following expression

$$W^*(N) = Q^{-1} \sum_{a=0}^{Q-1} \Big(\sum_{x=1}^{P} \exp\{2\pi i a x^n / Q\} \Big)^{t_1} \Big(\sum_{u} \exp\{2\pi i a u / Q\} \Big)^2$$
$$\times \Big(\sum_{u_0, v} \exp\{2\pi i a v^n u_0 / Q\} \Big) \exp\{-2\pi i a N / Q\},$$

where u, u_0, v run through natural numbers from the sets $\overline{U}$, $\overline{U}_0$, $\overline{V}$ each of which contains the sets U, U_0, V of distinct elements. As above, we represent the sum $W^*(N)$ as the sum of two terms:

$$W^*(N) = W_0^*(N) + W_1^*(N). \tag{3.17}$$

But $W_0^*(N)$ can also be written as

$$W_0^*(N) = \sum_{u,u_1}\sum_{u_0,v} Q^{-1}\sum_a \Big(\sum_{x=1}^{P}\exp\{2\pi iax^n/Q\}\Big)^{t_1}\exp\{-2\pi iaN_1/Q\},$$

where $N_1 = N - u - u_1 - v^n u_0$.

As already shown,

$$W_0(N_1) = Q^{-1}\sum_a \Big(\sum_{x=1}^{P}\exp\{2\pi iax^n/Q\}\Big)^{t_1}\exp\{-2\pi iaN_1/Q\} > 0.5P^{t_1}Q^{-1},$$

i.e.,

$$W_0^*(N) > 0.5U^2U_0VP^{t_1}Q^{-1}.$$

Let us construct the sets $\overline{U}$, $\overline{U}_0$, $\overline{V}$. We assume that $\xi_{\nu+1}$ run through the natural numbers that are not multiples of p and are not less than $p^{2mn-2m\nu}$ $(\nu = 0, 1, \dots, r-1)$. We let $A_{\nu+1}$ denote the set of $\xi_{\nu+1}$ for which $\xi_{\nu+1}^n$ are pairwise noncongruent modulo p^{2mn}. Since the congruence $\xi_{\nu+1}^n \equiv b \pmod{p^{2mn}}$ has at most n solutions for a fixed b coprime to p, the number of the numbers $\xi_{\nu+1}$ contained in $A_{\nu+1}$ is not less than

$$n^{-1}\varphi(p^{2mn-2m\nu}) \geq (2n)^{-1}p^{2mn-2m\nu}.$$

By $\overline{U}$ we denote the set of all numbers u of the form

$$u = \xi_1^n + (p^{2m}\xi_2)^n + \dots + (p^{2m(r-1)}\xi_r)^n, \quad \xi_\nu \in A_\nu;$$

the numbers u are pairwise noncongruent modulo $Q = p^{2mnr}$, and the number U of them is equal to

$$U = n^{-r}\prod_{\nu=0}^{r-1}\varphi(p^{2mn-2m\nu}) \geq (2n)^{-r}p^{2mnr-mr(r-1)} = (2n)^{-r}P^{r-r(r-1)/(2n)}.$$

Moreover, for $\nu = 0, 1, \dots, r-1$ we have

$$p^{2m\nu}\xi_{\nu+1} < p^{2mn} = P.$$

The set $\overline{U}_0$ is constructed similarly. We choose an integer $P_0 = p^{mn}$ and let $\zeta_{\nu+1}$ run through the values of natural numbers that are less than $p^{mn-m\nu}$ $(\nu = 0, 1, \dots, 2r-1)$ and not multiples of p. We let $B_{\nu+1}$ denote the set of $\zeta_{\nu+1}$ for which $\zeta_{\nu+1}^n$ are pairwise noncongruent modulo p^{mn}. By $\overline{U}_0$ we denote the set of all numbers u_0 of the form

$$u_0 = \zeta_1^n + (p^m\zeta_2)^n + \dots + (p^{m(2r-1)}\zeta_{2r})^n, \quad \zeta_{\nu+1} \in B_{\nu+1};$$

all u_0 are pairwise noncongruent modulo Q, we have the following lower bound for the number U_0 of them:

$$U_0 > (2n)^{-2r} p^{2mnr-mr(2r-1)} = (2n)^{-2r} P^{r-r(2r-1)/(2n)}.$$

We let the symbol $\overline{V}$ denote the set of all numbers from 1 to P_0 that are not multiples of p. For the number V of these numbers, we have the lower bound

$$V = P_0(1 - p^{-1}) > 0.5P_0.$$

We consider the product

$$v^n u_0 = (v\zeta_1)^n + (p^m v\zeta_2)^n + \cdots + (p^{m(2r-1)} v\zeta_{2r})^n.$$

Since $\zeta_{\nu+1} < p^{mn-m\nu}$, for the $(\nu + 1)$st bracket we have the upper bound

$$p^{m\nu} v\zeta_{\nu+1} < p^{mn} P_0 = P_0^2 = P.$$

After the sets $\overline{U}$, $\overline{U_0}$, $\overline{V}$ are constructed, it is easy to see that $W^*(N)$ in (3.17) is the number of representations of the number N modulo Q as the sum $t_1+2r+2r = t_1+4r$ of terms each of which is the nth power of a number less than P.

We estimate the sum

$$T(a) = \sum_{v,u_0} \exp\{2\pi i a v^n u_0/Q\}$$

under the condition that a belongs to the integration interval on which $W_1^*(N)$ is defined. It is easy to see that in this case

$$a/Q = b/p^\nu, \quad (b, p) = 1, \ \nu \geq 2mn + 1.$$

We define the integer s by the inequalities $smn < \nu \leq (s + 1)mn$. Then

$$u_0 \equiv u_0' = \zeta_1^n + (p^m\zeta_2)^n + \cdots + (p^{sm}\zeta_{s+1})^n \pmod{p^\nu}.$$

For a fixed u_0', the number u_0 congruent to u_0' is equal to

$$L \leq \prod_{\nu=s+1}^{2r-1} p^{mn-m\nu} = p^{mn(2r-s-1)-m(r(2r-1)-s(s+1)/2)}.$$

Hence, applying the Cauchy inequality, we find

$$|T(a)|^2 \leq U_0 L \sum_{u_0'<p^\nu} \Big|\sum_v \exp\{2\pi i b u_0' v^n/p^\nu\}\Big|^2 \leq \tag{3.18}$$

$$\leq U_0 L \sum_{a=1}^{p^\nu} \Big|\sum_v \exp\{2\pi i a v^n/p^\nu\}\Big|^2 \leq nU_0 L p^\nu V \leq$$

$$\leq nU_0Lp^{(s+1)mn}V \leq nU_0Vp^{2mnr-m(2r-1)}$$
$$= 2nU_0^2V^2P^{r(2r-1)/2n-(2r-1)/2n-1/2} < 2nU_0^2V^2P^{-1/3},$$
$$|T(a)| < \sqrt{2n}\,U_0VP^{-1/6}.$$

Finally, we estimate $W_1^*(N)$ using (3.18). We have

$$|W_1^*(N)| < \sqrt{2n}\,U_0VP^{-1/6}P^{t_1}Q^{-1}\sum_{a=0}^{Q-1}\Big|\sum_u \exp\{2\pi iau/Q\}\Big|^2$$
$$= \sqrt{2n}\,UU_0VP^{-1/6}P^{t_1}.$$

For the inequality $W^*(N) > 0$ to hold, it suffices to have $W_0^*(N) > W_1^*(N)$ or

$$0.5U^2U_0VP^{t_1}Q^{-1} > \sqrt{2n}\,UU_0VP^{t_1}P^{-1/6};$$

hence we obtain

$$U > 2\sqrt{2n}\,P^{r-1/6}. \tag{3.19}$$

But for U we had the estimate

$$U > (2n)^{-r}P^{r-r(r-1)/(2n)};$$

therefore, for $r \leq \sqrt{n/3}$ and $P \geq P_0(n)$ inequality (3.19) holds. The theorem is thereby proved. □

3.4 A new p-adic proof of Vinogradov's mean value theorem

I. M. Vinogradov, developing his method for estimating $G(n)$ and using the lemma on the "number of hits" to prove the mean value theorem, obtained a new method for estimating Weyl sums. Similarly, Karatsuba, starting from his own "analog of Waring's problem for congruences," arrived at a new p-adic method for proving Vinogradov's mean value theorem. As an analog the lemma on the "number of hits," he used Linnik's lemma on the number of solutions of a "complete system of congruences." A similar lemma was used earlier by Yu. V. Linnik in his p-adic proof of the mean value theorem. We shall discuss Linnik's method in detail in Section 3.5.

First, we prove the following three lemmas: Lemma 3.8 is an analog of the Bertrand postulate proved by P. L. Chebyshev; Lemma 3.9 gives the number of solutions of the complete system of congruences; and Lemma 3.10 gives the fundamental recursive inequality in the p-adic method.

Lemma 3.8. *For any natural n and $x \geq (2n)^2$, the interval $[x, 2x]$ contains at least n distinct prime numbers.*

Proof. For $x < 16$, the statement of the lemma is proved by a straightforward verification. Hence we assume that $x \geq 16$. First, we prove that

$$\psi(x) < x \ln 4. \tag{3.20}$$

For $x < 15$, this inequality is obvious. We assume that (3.20) holds for all y ($2 \leq y \leq x - 2$) and prove it for $y = x$. For any natural m we have

$$(4m)^{-1/2} < 2^{-2m}\binom{2m}{m} < (2m+1)^{-1/2}.$$

Since

$$\ln(k!) = \sum_{t \leq k} \ln t = \sum_{t \leq k}\sum_{d|t} \Lambda(d) = \sum_{d \leq k} \Lambda(d) \sum_{\substack{t \leq k \\ t=ud}} 1 = \sum_{d \leq k} \Lambda(d) \sum_{u \leq k/d} 1$$
$$= \sum_{u \leq k} \sum_{d \leq ku^{-1}} \Lambda(d) = \sum_{u \leq k} \psi(k/u), \quad \binom{2m}{m} = \frac{(2m)!}{(m!)^2},$$

we have

$$A = \ln\binom{2m}{m} = \psi(2m) - \psi(2m/2) + \psi(2m/3) - \cdots + \psi\big(2m/(2m-1)\big).$$

The function $\psi(x)$ does not decrease; hence we have

$$A \leq \psi(2m) - \psi(2m/2) + \psi(2m/3), \tag{3.21}$$
$$A \geq \psi(2m) - \psi(2m/2). \tag{3.22}$$

It follows from (3.22) that

$$\psi(2m) \leq A + \psi(m).$$

We assume that $2m$ is an even number that is the nearest to x. If $x = 2r + 1$, then $2m = 2r + 2$. By the induction hypothesis, we have (since $m \leq x - 2$)

$$\psi(2m) < 2m \ln 2 - \ln\sqrt{2m+1} + 2m \ln 2 \leq 2m \ln 4 - \ln\sqrt{2m+1}.$$

Next, if $2m < x$, then

$$\psi(x) = \psi(2m) < 2m \ln 4 < x \ln 4.$$

But if $2m \geq x$, then $2m \leq x + 1$; hence

$$\psi(x) \leq \psi(2m) \leq x \ln 4 + \ln 4 - \ln\sqrt{x+1}.$$

If $x \geq 15$, then $\ln 4 \leq \ln\sqrt{x+1}$. Hence $\psi(x) \leq x \ln 4$. So we have proved inequality (3.20). Next, from (3.21) we obtain

$$\psi(2m) - \psi(m) > A - \psi\left(\frac{2m}{3}\right) \geq m \ln 4 - \ln\sqrt{4m} - \frac{2m}{3}\ln 4 = \frac{m}{3}\ln 4 - \ln\sqrt{4m}.$$

We note that

$$\psi(2m) - \psi(m) = \sum_{m<n\le 2m} \Lambda(n) = \sum_{m<n\le 2m}{}_1 \Lambda(n) + \sum_{m<n\le 2m}{}_2 \Lambda(n),$$

where the symbol $\sum_1$ denotes the sum over prime numbers and the symbol $\sum_2$ denotes the sum over all other numbers.

In the second sum, to each number n for which $\Lambda(n) \ne 0$ we assign a prime p_n such that $p_n \,|\, n$. Obviously, $\Lambda(p_n) = \Lambda(n)$ and $p_n \le \sqrt{2m}$. This implies

$$\sum_{m<n\le 2m}{}_2 \Lambda(n) \le \psi(\sqrt{2m}) \le \sqrt{2m}\,\ln 4;$$

hence we have

$$S_1(m) = \sum_{m<n\le 2m}{}_1 \Lambda(n) = \sum_{m<p\le 2m} \ln p \ge \frac{m}{3}\ln 4 - \ln\sqrt{4m} - \sqrt{2m}\,\ln 4.$$

Next, since

$$\Lambda(n) \le 2m \quad \text{for } n \le 2m,$$

we have

$$\sum_{m<p\le 2m} 1 \ge \frac{m}{3}\cdot\frac{\ln 4}{\ln 2m} - 1 - \ln 4\frac{\sqrt{2m}}{\ln 2m} = S(m).$$

Now we show that $S(m) \ge \sqrt{m/2}$ for $m \ge 2^8$ or, which is the same,

$$f(m) = S(m) - \sqrt{m/2} \ge 0. \tag{3.23}$$

The derivative of the function $f(m)$ has the form

$$f'(m) = \frac{\ln 4}{3}\cdot\frac{1}{\ln 2m}\left(1 - \frac{1}{\ln 2m}\right) - \frac{\ln 4}{\sqrt{2m}\,\ln 2m}\left(1 - \frac{2}{\ln 2m}\right) - \frac{1}{2\sqrt{2m}}.$$

Hence

$$\begin{aligned} f'(m) &\ge \frac{1}{3\ln 2m} - \frac{1}{\sqrt{2m}\ln 2m} - \frac{1}{2\sqrt{2m}} \\ &= \frac{1}{\ln 2m}\left(\frac{1}{12} - \frac{1}{\sqrt{2m}}\right) + \left(\frac{1}{4\ln 2m} - \frac{1}{2\sqrt{2m}}\right). \end{aligned}$$

For $m \ge 2^8$ both terms in the last expression are positive (since $\ln 2m > 6$ and $2\ln 2m < \sqrt{2m}$ in this case). Therefore, for $m \ge 2^8$, the inequality $f'(m) > 0$ holds and hence the function $f(m)$ increases. So to prove inequality (3.23), it suffices to verify it for $m = 2^8$. But we have

$$f(2^8) = \frac{2^8}{3}\cdot\frac{2\ln 2}{9\ln 2} - 1 - 2^{4.5}\frac{2\ln 2}{9\ln 2} - 2^{3.5} > 0.$$

Inequality (3.23) is thereby proved.

Now we note that the interval $[x/2, x]$ contains all primes that are contained in the interval $(m, 2m]$. Moreover, if $x \geq (2n)^2$, then $2m > (2n)^2$.

Next, if $n \geq 12$, then $x \geq 4 \cdot 12^2$, $2m \geq 4 \cdot 12^2$, $m > 2 \cdot 12^2 > 2^8$, and hence the interval $[x/2, x]$ contains at least (see inequality (3.23)) $\sqrt{m/2} \geq n$ distinct prime numbers.

If $n \leq 11$, then for $x > 2 \cdot 2^8$ we have

$$\sqrt{m/2} > \sqrt{128} > 11,$$

and it follows from (3.23) that the interval $[x/2, x]$ contains at least n primes.

It remains to prove the lemma for $n \leq 11$ and $2^4 \leq x \leq 2^9$. Let k be a natural number ($6 \leq k \leq 50$). A straightforward verification using the table of primes (see [163], p. 166) shows that the interval $[0.25(k+1)^2,\ 0.5k^2]$ contains at least $(k+1)/4$ primes.

We let t denote the largest of the numbers k for which $k^2/2 \leq x$. Then

$$0.25(t+1)^2 \geq x/2, \quad 0.5t^2 \leq x.$$

Hence the interval $[0.25(t+1)^2, 0.5t^2]$ is completely contained in the interval $[x/2, x]$.

If $2^5 \leq x \leq 2^9$, then $2^4 \leq 0.25(t+1)^2 \leq 2^9$, $2^6 \leq (t+1)^2 \leq 2^{11}$, and $8 \leq t+1 < 50$. Hence the interval $[x/2, x]$ contains at least $0.25(t+1) > \sqrt{x/8}$ primes. This trivially implies the statement of the lemma. □

Lemma 3.9. *Suppose that* $1 \leq r \leq n$, p *is a prime* $(p > n)$, *and* $1 \leq P \leq p^r$. *Then the number* T *of solutions of the system of congruences*

$$\begin{aligned} x_1 + \cdots + x_n &\equiv y_1 + \cdots + y_n \pmod{p}, \\ x_1^2 + \cdots + x_n^2 &\equiv y_1^2 + \cdots + y_n^2 \pmod{p^2}, \\ &\vdots \\ x_1^n + \cdots + x_n^n &\equiv y_1^n + \cdots + y_n^n \pmod{p^n}, \end{aligned}$$
$$1 \leq x_1, \ldots, x_n, y_1, \ldots, y_n \leq P; \quad x_i \not\equiv x_j \pmod{p}, \quad i \neq j,$$

satisfies the estimate

$$T \leq n! p^{r(r-1)/2} P^n.$$

Proof. Obviously, we have $T \leq P^n T_1$, where T_1 is the number of solutions of such a system of congruences (for some fixed set of numbers $\lambda_1, \ldots, \lambda_n$):

$$\begin{aligned} x_1 + \cdots + x_n &\equiv \lambda_1 \pmod{p}, \\ x_1^2 + \cdots + x_n^2 &\equiv \lambda_2 \pmod{p^2}, \\ &\vdots \\ x_1^n + \cdots + x_n^n &\equiv \lambda_n \pmod{p^n}, \end{aligned} \tag{3.24}$$
$$1 \leq x_1, \ldots, x_n \leq p^r; \quad x_i \not\equiv x_j \pmod{p}, \quad i \neq j.$$

To estimate T_1, we represent x_t for each $t = 1, 2, \dots, n$ as

$$x_t = x_{1t} + px_{2t} + \cdots + p^{r-1}x_{rt}.$$

For a set $x_1, \dots, x_n$ to be a solution of system (3.24), it is necessary that the variables $x_{11}, \dots, x_{1n}$ satisfy the system of congruences

$$x_{11}^{\nu} + \cdots + x_{1n}^{\nu} \equiv \lambda_\nu \pmod p, \qquad \nu = 1, 2, \dots, n,$$

and the variables $x_{1s}, \dots, x_{ns}$ $(s = 2, \dots, r)$ satisfy their own system of linear congruences (for fixed $x_{11}, \dots, x_{1n}$)

$$x_{1s}(\nu x_{11}^{\nu-1}) + \cdots + x_{ns}(\nu x_{1s}^{\nu-1}) \equiv \lambda_{\nu s} \pmod p, \qquad \nu = s, \dots, r,$$

where $\lambda_{ss}, \dots, \lambda_{rs}$ are some integers.

The number of solutions of the first system does not exceed $n!$, since it follows from the elementary theory of symmetric functions that, for $p > n$ and fixed λ_ν, all solutions of this system are permutations of some unique solution. Next, because the variables x_t are pairwise noncongruent modulo p, the matrix of coefficients of each of the linear systems of congruences has the maximum rank. Hence the number of solutions of this system does not exceed p^s.

For T_1 and T, we obtain the estimates

$$T_1 \le n!p \cdot p^2 \dots p^{r-1} = n!p^{r(r-1)/2}, \quad T \le n!p^{r(r-1)/2}p^n.$$

The proof of the lemma is thus complete. □

Lemma 3.10. *Suppose that $k \ge n$, $1 \le r \le n$, and $P \ge 1$. Then in the interval $[P^{1/r}, 2P^{1/r}]$ there is a prime number p such that*

$$J = J(P; n, k) \le 4k^{2n} p^{2k-2n+r(r-1)/2} p^n J(P_1; n, k-n) + (2n)^{2kr} P^k, \qquad (3.25)$$

where $P_1 = Pp^{-1} + 1$.

Proof. If $P \le (4n^2)^r$, then, by setting $p = 2P^{1/r}$, we see that the second term in the inequality is less than P^{2k}, while the first term is always nonnegative, i.e., in this case inequality (3.25) becomes trivial. Therefore, we assume that $P \ge (4n^2)^r$. Then, by Lemma 3.8, on the interval $[P^{1/r}, 2P^{1/r}]$ there are at least r distinct primes. We choose some r primes on the interval $[P^{1/r}, 2P^{1/r}]$ and denote them by the letters $p_1, \dots, p_r$. Now, as above, we assume that

$$f(x) = \alpha_1 x + \cdots + \alpha_n x^n.$$

Then J can be represented as the multiple integral

$$(*) \quad J = J(P; n, k)$$
$$= \int_0^1 \cdots \int_0^1 \Big| \sum_{x_1 \le P} \cdots \sum_{x_k \le P} \exp\{2\pi i (f(x_1) + \cdots + f(x_k))\} \Big|^2 d\alpha_1 \dots d\alpha_n.$$

We divide all sets $\overline{x} = (x_1, \dots, x_k)$ into two classes A and B as follows: a set $\overline{x} = (x_1, \dots, x_k)$ belongs to the class A if among the numbers $p_1, \dots, p_r$ there is a number p^s such that among the numbers $x_1, \dots, x_k$ there are at least n numbers pairwise noncongruent modulo p; all other sets belong to the class B.

For brevity, we introduce a new notation (the explicit form of this notation is obvious) and then transform relation $(*)$ into the inequality

$$J = \int_\Omega \Big|\sum_{\overline{x}\in A} + \sum_{\overline{x}\in B}\Big|^2 d\Omega \le 2\int_\Omega \Big|\sum_{\overline{x}\in A}\Big|^2 d\Omega + 2\int_\Omega \Big|\sum_{\overline{x}\in B}\Big|^2 d\Omega = 2J_1 + 2J_2.$$

Let us estimate the integral J_1. The value of J_1 is the number of solutions of Eqs. (3.4) under the assumption that

$$\overline{x} = (x_1, \dots, x_k) \in A, \quad \overline{y} = (y_1, \dots, y_k) \in A.$$

We divide all sets $\overline{x} = (x_1, \dots, x_k) \in A$ into r sets $A_1, \dots, A_r$ as follows: the sets corresponding to their own number $p_s = p$ $(s = 1, \dots, r)$ belong to the same set; if a set corresponds to several p_s, then, for definiteness, we assume that it belongs to the set corresponding to the least p_s. Again we use the obvious notation and find

$$J_1 = \int_\Omega \Big|\sum_{\overline{x}\in A}\Big|^2 d\Omega = \int_\Omega \Big|\sum_{s=1}^{r}\sum_{\overline{x}\in A_s}\Big|^2 d\Omega \le r\sum_{s=1}^{r}\int_\Omega \Big|\sum_{\overline{x}\in A_s}\Big|^2 d\Omega = r\sum_{s=1}^{r} J_{1s}.$$

Let us estimate J_{1s}. The sum whose squared absolute value is the integrand in J_{1s} has the form

$$\sum_{\overline{x}\in A} = \sum_{x_1}\cdots\sum_{x_k},$$

and moreover, $\overline{x} = (x_1, \dots, x_k) \in A_s$, i.e., among the numbers $x_1, \dots, x_k$ there are n numbers pairwise noncongruent modulo p. We divide all sets in A_s into possibly intersecting classes as follows.

Let $t_1, \dots, t_n$ be natural numbers $(1 \le t_1 < t_2 < \cdots < t_n \le k)$. We let all sets $\overline{x} = (x_1, \dots, x_k)$ such that the numbers $x_1, \dots, x_k$ are pairwise noncongruent modulo p_s belong to the same class. Let R_1 and R_2 be two distinct classes from the set A_s. Renumbering the unknowns, we easily obtain

$$\int_\Omega \Big|\sum_{\overline{x}\in R_1}\Big|^2 d\Omega = \int_\Omega \Big|\sum_{\overline{x}\in R_2}\Big|^2 d\Omega.$$

Since, obviously, the total number of classes is $\binom{k}{n}$, denoting the set corresponding to $t_1 = 1,\ t_2 = 2, \dots, t_n = n$ by the symbol A_{1s}, we obtain

$$J_{1s} \le \binom{k}{n}^2 \int_\Omega \Big|\sum_{A_{1s}}\Big|^2 d\Omega \le \binom{k}{n}^2 \int_\Omega \Big|\sum_{x_1,\dots,x_n}{}'\Big|^2 \Big|\sum_{x\le P}\exp\{2\pi i f(x)\}\Big|^{2k-2n} d\Omega.$$

Here the prime on the first sum means that the sum is taken over the sets of numbers $x_1, \dots, x_k$ pairwise noncongruent modulo $p_s = p$. Dividing the second sum into p progressions with common difference p and applying Hölder's inequality, we obtain

$$J_{1s} \le \binom{k}{n}^2 \int_\Omega \Big| \sum_{x_1,\dots,x_n}{}' \Big|^2 \Big| \sum_{x \le P} \exp\{2\pi i f(x)\} \Big|^{2k-2n} d\Omega$$
$$\le \binom{k}{n}^2 p^{2k-2n-1} \sum_{y=1}^{p} \int_\Omega \Big| \sum_{x_1,\dots,x_n}{}' \Big|^2 \Big| \sum_{0 \le z \le Pp^{-1}} \exp\{2\pi i f(y+pz)\} \Big|^{2k-2n} d\Omega.$$

Let y_0 be the value of y for which the last integral takes its maximum. Using the fact that if $x_1^0, \dots, x_k^0, y_1^0, \dots, y_k^0$ is a solution of system (3.4), then $x_1^0 + a, \dots, x_k^0 + a, y_1^0 + a, \dots, y_k^0 + a$ is also a solution of system (3.4), we obtain the inequality

$$J_{1s} \le \binom{k}{n}^2 p^{2k-2n} \int_\Omega \Big| \sum_{x_1,\dots,x_n}{}'' \Big|^2 \Big| \sum_{0 \le z \le Pp^{-1}} \exp\{2\pi i f(pz)\} \Big|^{2k-2n} d\Omega,$$

where the symbol $\sum''$ means that the sum is taken over all sets of numbers $x_1, \dots, x_n$ that lie within the limits from $-y_0$ to $P - y_0$ and are pairwise noncongruent modulo p. The integral in the right-hand side is equal to the number of solutions of the following system of equations (we denote this number by J'):

$$\begin{aligned} x_1 + \dots + x_n - y_1 - \dots - y_n &= p(z_1 + \dots + z_{k-n} - v_1 - \dots - v_{k-n}), \\ x_1^2 + \dots + x_n^2 - y_1^2 - \dots - y_n^2 &= p^2(z_1^2 + \dots + z_{k-n}^2 - v_1^2 - \dots - v_{k-n}^2), \\ &\vdots \\ x_1^n + \dots + x_n^n - y_1^n - \dots - y_n^n &= p^n(z_1^n + \dots + z_{k-n}^n - v_1^n - \dots - v_{k-n}^n), \end{aligned}$$

where the unknowns $x_1, \dots, x_n$ and $y_1, \dots, y_n$ satisfy the condition stated above for the variables $x_1, \dots, x_n$, while the variables $z_1, \dots, z_{k-n}, v_1, \dots, v_{k-n}$ take all integer values from zero to Pp^{-1}.

We let the symbol $J'(Pp^{-1}; n, k-n; \Lambda)$, where $\Lambda = (\lambda_1, \dots, \lambda_n)$ is a set of integers, denote the number of solutions of the system of equations

$$z_1^\nu + \dots + z_{k-n}^\nu - v_1^\nu - \dots - v_{k-n}^\nu = \lambda_\nu, \quad \nu = 1, 2, \dots, n.$$

Obviously, we have the inequality

$$J'(Pp^{-1}; n, k-n; \Lambda) \le J(P_1; n, k-n)$$

(we use the fact that if a certain number is added to the solution of system (3.4), then a solution of system (3.4) is again obtained).

Let $D(\Lambda)$ be the number of solutions of the system of equations

$$x_1^\nu + \dots + x_n^\nu - y_1^\nu - \dots - y_n^\nu = p^\nu \lambda_\nu, \quad \nu = 1, 2, \dots, n.$$

Then, denoting the sum over all possible sets Λ by the symbol $\sum_{\Lambda}$, we obtain

$$J' = \sum_{\Lambda} D(\Lambda) J'(Pp^{-1}; n, k-n; \Lambda) \le J(P_1; n, k-n) \sum_{\Lambda} D(\Lambda) = J(P_1; n, k-n)T,$$

where T is the number of solutions of the system of congruences

$$x_1^{\nu} + \cdots + x_n^{\nu} - y_1^{\nu} - \cdots - y_n^{\nu} \equiv 0 \pmod{p^{\nu}}, \qquad \nu = 1, 2, \ldots, n,$$
$$1 \le x_1, \ldots, x_n, \; y_1, \ldots, y_n \le P.$$

The estimate in Lemma 3.9 can be applied to T:

$$T \le n! p^{0.5r(r-1)} P^n.$$

Hence, collecting the estimates obtained, we obtain the following inequality for J_1:

$$J_1 \le n! \binom{k}{n}^2 r^2 p^{2k-2n+0.5r(r-1)} P^n J(P_1; n, k-n),$$

where p denotes one of the numbers p_s for which the expression in the right-hand side takes its maximum.

Now we estimate J_2. The value of J_2 is the number of solutions of system of equations (3.4) under the assumption that $\overline{x} = (x_1, \ldots, x_n) \in B$ and $\overline{y} = (y_1, \ldots, y_n) \in B$. We find the upper bound for the number of elements $\overline{x} \in B$. Let p_s be one of the numbers $p_1, \ldots, p_r$. For each set $\overline{x} = (x_1, \ldots, x_n) \in B$, we consider the set $\overline{x}^{(s)}$ consisting of the remainders obtained by dividing the coordinates of $\overline{x}$ by p_s:

$$\overline{x}^{(s)} = (x_1^{(s)}, \ldots, x_k^{(s)}), \quad x_i \equiv x_i^{(s)} \pmod{p_s}, \; 0 \le x_i^{(s)} < p, \; i = 1, \ldots, k.$$

We let B_s denote the set of the sets $\overline{x}^{(s)}$ thus obtained. Let us estimate the number of elements of B_s. It does not exceed the number

$$\binom{p_s}{n-1}(n-1)^k.$$

Thus for each $\overline{x} - (x_1, \ldots, x_k) \in B$ we have obtained the system of congruences $\overline{x} \equiv x^{(s)} \pmod{p_s}$ $(s = 1, \ldots, r)$. It is possible to replace this system of congruences by a single congruence of the form $\overline{x} \equiv \overline{M} \pmod{p_1 \ldots p_r}$, where $\overline{M} = (M_1, \ldots, M_k)$ is a fixed set and $0 \le M_i < p_1 \ldots p_r$ $(i = 1, \ldots, k)$. Since each coordinate of $\overline{x}$ does not exceed $P < p_1 \ldots p_r$, the last congruence is equivalent to the equation $\overline{x} = \overline{M}$.

So the number of set B does not exceed the number of sets $\overline{M}$, i.e., the product

$$\binom{p_1}{n-1}(n-1)^k \ldots \binom{p_r}{n-1}(n-1)^k.$$

Now we estimate the number of $\overline{y} = (y_1, \dots, y_k)$ under the assumption that they satisfy system (3.4). If we fix $k - n$ of them, then the remaining numbers are uniquely determined with accuracy of the order of the terms, i.e., the number of all $y = (y_1, \dots, y_k)$ does not exceed $n!P^{k-n}$.

So we have

$$J_2 \le (n-1)^{kr} \binom{p_1}{n-1} \dots \binom{p_r}{n-1} n!P^{k-n} \le n^{2kr} P^{k-1}.$$

Combining the estimates for J_1 and J_2, we find the recursive formula

$$J \le 2J_1 + 2J_2 \le 2r^2 \binom{k}{n}^2 n!p^{2k-2n+0.5r(r-1)} P^n J(P_1; n, k-n) + (2n)^{2kr} P^{k-1}.$$

Making the right-hand side less sharp, we arrive at the statement of the lemma:

$$J \le 4k^{2n} p^{2k-2n+0.5r(r-1)} P^n J(P_1; n, k-n) + (2n)^{2kr} P^k.$$

□

Theorem 3.6 (The mean value theorem). *Let n, k, τ be natural numbers, and let*

$$\Delta(\tau) = 0.5(n^2 + n) - 0.5n^2(1 - 1/n)^\tau, \quad \varkappa(\tau) = 1.5(n+1)^2\tau.$$

Then the estimate

$$J = J(P; n, k) \le n^{2\Delta(\tau)n} 2^{\varkappa(\tau)} (8k)^{2n\tau} P^{2k-\Delta(\tau)}$$

holds for $k \ge n\tau$ and $P \ge 1$.

Proof. Without loss of generality, we can assume that $k = n\tau$ and $n \ge 2$. We proceed by induction on the parameter τ. For $\tau = 1$ the statement of the theorem holds, since in this case $k = n$, $\Delta(1) = n$, $\varkappa(1) = 1.5(n+1)^2$, and the estimate takes the form

$$J \le n^{2n^2} 2^{1.5(n+1)^2} (8n)^{2n} P^{2k-n},$$

which is somewhat less sharp than the estimate

$$J \le n!P^{2k-n},$$

which we can simply obtain.

Now we assume that the theorem holds for $\tau = m \ge 1$ and prove it for $\tau = m+1$. We estimate $J(P; n, n(m+1))$ by using the estimate in Lemma 3.10 with $r = n$. We obtain

$$J(P; n, n(m+1)) \le 4\big(n(m+1)\big)^{2n} p^{2n(m+1)+2n+0.5n(n-1)} \times P^n J(P_1; n, mn) + (2n)^{2n^2(m+1)} P^{n(m+1)}. \tag{3.26}$$

To estimate $J(P_1; n, nm)$, we use the estimate in the theorem with $\tau = m$:

$$J(P_1; n, nm) \le n^{2\Delta(m)n} 2^{\varkappa(m)} (8nm)^{2nm} P_1^{2nm-\Delta(m)}. \tag{3.27}$$

We substitute (3.27) into (3.26) and show that the estimate thus obtained is not less sharp than the estimate in the theorem with $\tau = m + 1$. We can assume that $P > (4k)^2$, since otherwise the estimate in the theorem is less sharp than the trivial estimate by P^{2k}. Indeed, we always have $\Delta(m + 1) \le n(m + 1)$, and hence for $P \le (4k)^2$, $k = n(m + 1)$, and $\tau = m + 1$, we have

$$P^{2k} \le (4k)^{2n(m+1)} \le (8k)^{2n(m+1)} n^{2n\Delta(m+1)} P^{2k-\Delta(m+1)}.$$

So let $P > (4k)^2$. Then we have

$$\begin{aligned} pP^{-1} &\le 2P^{-1+1/n} \le 2P^{-1/2} \le (2k)^{-1}, \\ P_1^{2k-2n-\Delta(m)} &= (Pp^{-1} + 1)^{2k-2n-\Delta(m)} \\ &\le P^{2k-2n-\Delta(m)} p^{-2k+2n+\Delta(m)} \left(1 + \frac{1}{2k}\right)^{2k} \\ &\le 3P^{2k-2n-\Delta(m)} p^{-2k+2n+\Delta(m)}. \end{aligned}$$

Hence the first term in the right-hand side of (3.26) does not exceed

$$\begin{aligned} &12k^{2n} n^{2\Delta(m)n} 2^{\varkappa(m)} (8nm)^{2nm} p^{\Delta(m)+0.5n(n-1)} P^{2k-n-\Delta(m)} \\ &\quad \le 12k^{2n} n^{2\Delta(m)n} 2^{\varkappa(m)+\Delta(m)+0.5n(n-1)} (8nm)^{2nm} \\ &\qquad \times P^{2k-n-\Delta(m)+\Delta(m)/n+0.5(n-1)} \\ &\quad \le 0.5n^{2\Delta(m+1)n} 2^{\varkappa(m+1)} (8k)^{2n(m+1)} P^{2k-\Delta(m+1)}, \end{aligned}$$

because it follows from the definition of $\Delta(\tau)$ and $\varkappa(\tau)$ that

$$\begin{aligned} \Delta(m + 1) &= \Delta(m) + n + 0.5 - (0.5n + \Delta(m)/n), \\ \varkappa(m + 1) &> \varkappa(m) + \Delta(m) + 0.5n(n - 1). \end{aligned}$$

Since we always have $\Delta(\tau) \le k$, we can assume that $P > (2n)^{2n}$. Otherwise, the product of the first two factors in the estimate in the theorem exceeds the contribution of lower P and the statement becomes trivial. So we have

$$\begin{aligned} P > (2n)^{2n}, \quad &\left((2n)^{-2n} P\right)^{k-\Delta(m+1)} \ge 1, \\ (2n)^{2kn} P^k ((2n)^{-2n} P)^{k-\Delta(m+1)} &\ge (2n)^{2kn} P^k, \end{aligned}$$

i.e.,

$$\begin{aligned} (2n)^{2kn} P^k &\le (2n)^{\Delta(m+1)n} P^{2k-\Delta(m+1)} \\ &\le 0.5n^{2\Delta(m+1)n} 2^{\varkappa(m+1)} (8k)^{2n(m+1)} P^{2k-\Delta(m+1)}. \end{aligned}$$

Thus we have obtained the desired estimate for $J(P; n, n\tau)$ with $\tau = m + 1$. The proof of the theorem is complete. □

From this theorem and estimates of trigonometric sums we obtain a "simplified upper bound" for J. Now let us study a slightly more general variable I. To this end, we consider the system of equations

$$\begin{aligned} x_1 + x_2 + \cdots + x_k &= N_1, \\ x_1^2 + x_2^2 + \cdots + x_k^2 &= N_2, \\ &\vdots \\ x_1^n + x_2^n + \cdots + x_k^n &= N_n, \end{aligned} \tag{3.28}$$

where $n \geq 3$, k, $N_1, \ldots, N_n$, P are natural numbers, and $x_1, x_2, \ldots, x_k$ are integer-valued unknowns such that $1 \leq x_1, \ldots, x_k \leq P$. We let I denote the number of solutions of this system.

Theorem 3.7. *For $k \geq n^2(4\ln n + 2\ln\ln n + 9)$ and $k \leq P^{0.1}$, the asymptotic formula holds:*

$$\begin{aligned} I &= I(P; n, k; N_1, \ldots, N_n) \\ &= \sigma\gamma P^{k-0.5n(n+1)} + \theta n^{30n^3} P^{k-0.5n(n+1)-(30(2+\ln n))^{-1}}, \end{aligned} \tag{3.29}$$

as well as the estimate $I \leq n^{30n^3} P^{k-0.5n(n+1)}$. Here θ ($|\theta| \leq 1$) is a real number and σ, γ are nonnegative numbers. The value of σ is equal to the sum of the "singular series in the Hilbert–Kamke problem," and the value of γ is equal to the value of the "singular integral in the Hilbert–Kamke problem."

Namely,

$$\sigma = \sum_{q_1=1}^{\infty} \cdots \sum_{q_n=1}^{\infty} \sum_{\substack{0 \leq a_1 < q_1 \\ (a_1, q_1)=1}} \cdots \sum_{\substack{0 \leq a_n < q_n \\ (a_n, q_n)=1}} q^{-k} V^k \exp\{2\pi i A\},$$

$$\gamma = \int_{-\infty}^{+\infty} \cdots \int_{-\infty}^{+\infty} W^k \exp\{-2\pi i B\}\, d\beta_1 \ldots d\beta_n,$$

where

$$q = q_1 \ldots q_n,$$

$$V = \sum_{x=1}^{q} \exp\left\{2\pi i\left(\frac{a_1 x}{q_1} + \cdots + \frac{a_n x^n}{q_n}\right)\right\}, \qquad A = \frac{a_1 N_1}{q_1} + \cdots + \frac{a_n N_n}{q_n},$$

$$W = \int_0^1 \exp\{2\pi i(\beta_1 x + \cdots + \beta_n x^n)\}\, dx, \qquad B = \frac{\beta_1 N_1}{P} + \cdots + \frac{\beta_n N_n}{P^n}.$$

The absolute convergence exponent of the singular series σ is $0.5n(n+1)+2$, while the absolute convergence exponent of the singular integral γ is equal to $0.5n(n+1)+1$.

First, we note that the absolute convergence exponents of the series σ and the integral γ were found in Chapters 1 and 2. Next, for I we have the trivial estimate $I \leq P^k$, because the number of all possible sets of integers $x_1, \ldots, x_k$ satisfying the inequalities $1 \leq x_1, \ldots, x_k \leq P$ is, obviously, equal to P^k. If the value of P in Theorem 3.7 is less than n^{40n}, then the estimate of the remainder term in the asymptotic formula in the theorem is less sharp than the above trivial estimate of I and this formula becomes meaningless. Therefore, in what follows, we assume that $P \geq n^{40n}$. We rewrite I as

$$I = \int_0^1 \cdots \int_0^1 S^k(A) \exp\{-2\pi i(\alpha_1 N_1 + \cdots + \alpha_n N_n)\}\, dA, \tag{3.30}$$

where A is a point of the n-dimensional space with coordinates $\alpha_1, \ldots, \alpha_n$ and $S(A)$ is the trigonometric sum,

$$S(A) = \sum_{x=1}^{P} \exp\{2\pi i f(x)\}, \quad f(x) = \alpha_1 x + \cdots + \alpha_n x^n.$$

Since the integrand function is periodic in $\alpha_1, \ldots, \alpha_n$ with period 1, the interval of integration with respect to the variables $\alpha_1, \ldots, \alpha_n$ in the multiple integral can be replaced by the domain Ω determined by the inequalities

$$-P^{0.3-s} < \alpha_s < 1 - P^{0.3-s}, \quad s = 1, 2, \ldots, n.$$

We let $\omega(a, q)$ denote the domain A of points satisfying the conditions

$$\alpha_s = \frac{a_s}{q_s} + z_s, \quad |z_s| < P^{0.3-s},$$
$$(a_s, q_s) = 1, \quad 0 \leq a_s < q_s, \quad s = 1, \ldots, n.$$

Here the numbers a_s and q_s are the sth coordinates of the sets of integers a and q.

Next, let Q be the least common multiple of the numbers $q_1, \ldots, q_n$. We divide all points of the domain Ω into two classes. All points of the domains $\omega(a, q)$ for which $Q \leq P^{0.3}$ belong to the first class Ω_1. All other points of the domain Ω belong to the second class Ω_2. We note that two distinct domains $\omega(a, q)$ and $\omega(a', q')$ whose points belong to the first class do not intersect. Indeed, we assume that such domains $\omega(a, q)$ and $\omega(a', q')$ have a common point A. Then for all s $(s = 1, \ldots, n)$ we have the relations

$$\alpha_s = \frac{a_s}{q_s} + z_s = \frac{a'_s}{q'_s} + z'_s;$$

in addition, $|z_s|, |z'_s| < P^{0.3-s}$ and for some s $(1 \leq s \leq n)$ we have $a_s/q_s \neq a'_s/q'_s$. Consequently,

$$\frac{a_s}{q_s} - \frac{a'_s}{q'_s} = z'_s - z_s.$$

Passing to inequalities, we obtain

$$P^{-0.6} \le \frac{1}{q_s q_s'} \le \left| \frac{a_s}{q_s} - \frac{a_s'}{q_s'} \right| = |z_s' - z_s| \le 2P^{0.3-s} \le 2P^{-0.7},$$

which is impossible, since $P^{0.1} > 2$ by the inequality $P \ge n^{40n}$.

Prior to proving the theorem, we prove two lemmas.

Lemma 3.11. *Suppose that a point A belongs to the domain Ω and its coordinates α_s $(s = 1, \dots, n)$ are represented in the form $\alpha_s = a_s/q_s + z_s$, where $(a_s, q_s) = 1$, $|z_s| \le \tau_s^{-1} = P^{0.5-s}$. Suppose that the least common multiple Q of the numbers $q_1, \dots, q_n$ satisfies the condition $Q \le P^{0.3}$. Then for the sum*

$$S(A) = \sum_{x=1}^{P} \exp\{2\pi i(\alpha_1 x + \cdots + \alpha_n x^n)\},$$

we have the asymptotic formula

$$S(A) = PQ^{-1}VW + R,$$

where

$$V = \sum_{x=1}^{Q} \exp\left\{2\pi i\left(\frac{a_1 x}{q_1} + \cdots + \frac{a_n x^n}{q_n}\right)\right\},$$

$$W = \int_0^1 \exp\{2\pi i(z_1 P x + \cdots + z_n P^n x^n)\}\,dx, \quad |R| \le 9nQ.$$

Proof. We represent the summation variable in the sum $S(A)$ as $x = Qy + t$, where t satisfies the condition $1 \le t \le Q$. Then $S(A) = S_1(A) + \theta Q$, where $|\theta| \le 1$,

$$S_1(A) = \sum_{t=1}^{Q} \sum_{y=0}^{P_1} \exp\{2\pi i f(Qy + t)\},$$

$$f(x) = \alpha_1 x + \cdots + \alpha_n x^n, \quad P_1 = [PQ^{-1}] + 1.$$

Since Q is a multiple of all numbers $q_1, \dots, q_n$, we have

$$\begin{aligned} f(Qy+t) &= \alpha_1(Qy+t) + \cdots + \alpha_n(Qy+t)^n \\ &= \frac{a_1}{q_1}(Qy+t) + \cdots + \frac{a_n}{q_n}(Qy+t)^n + z_1(Qy+t) + \cdots + z_n(Qy+t)^n \\ &= \frac{a_1}{q_1}t + \cdots + \frac{a_n}{q_n}t^n + z_1(Qy+t) + \cdots + z_n(Qy+t)^n + H, \end{aligned}$$

where H is an integer. Therefore,

$$S_1(A) = \sum_{t=1}^{Q} \exp\left\{2\pi i\left(\frac{a_1 t}{q_1} + \cdots + \frac{a_n t^n}{q_n}\right)\right\}$$
$$\times \sum_{y=0}^{P_1} \exp\left\{2\pi i\left(z_1(Qy+t) + \cdots + z_n(Qy+t)^n\right)\right\}.$$

Let $S(t)$ be the internal sum in the right-hand side of the last relation. Then

$$S_1(A) = \sum_{t=1}^{Q} e(t)S(t), \quad e(t) = \exp\left\{2\pi i\left(\frac{a_1 t}{q_1} + \cdots + \frac{a_n t^n}{q_n}\right)\right\}.$$

We set $\varphi(y) = z_1(Qy+t) + \cdots + z_n(Qy+t)^n$ and obtain

$$\left|\frac{d\varphi(y)}{dy}\right| = \left|\sum_{s=1}^{n} z_s(Qy+t)^{s-1} sQ\right| \le \sum_{s=1}^{n} \tau_s^{-1} P^{s-1} sQ$$
$$\le \sum_{s=1}^{n} sP^{0.5-s}P^{s-1}P^{0.3} \le 0.5n(n+1)P^{-0.2} \le 0.05,$$

because $|z_s| \le P^{0.5-s}$, $Q \le P^{0.3}$, and $Qy+t \le P$.

Next, since the derivative of the function $\varphi(y)$ is a polynomial of degree $n-1$, the interval $1 \le y \le P$ can be divided into at most $2n-2$ intervals on which this derivative is monotone and of constant sign. Hence the sum $S(t)$ can be divided into m sums ($m \le 2n-2$) so that each new sum satisfies the assumptions of Lemma A.2 with $\delta = 0.05$. Consequently, each such sum can be replaced by an integral so that the error does not exceed $3 + 2\delta/(1-\delta) < 4a$. Hence for some $|\theta_1| \le 1$ we obtain

$$S(t) = \int_0^{P_1} \exp\{2\pi i\varphi(y)\}\,dy + 8\theta_1(n-1).$$

We change the variable in the integral by setting $x = P^{-1}(Qy+t)$ and thus obtain

$$S(t) = PQ^{-1}\int_{tP^{-1}}^{P^{-1}(QP_1+t)} \exp\{2\pi i g(x)\}\,dx + 8\theta_1(n-1) = PQ^{-1}W + 8\theta_2 n,$$

where $g(x) = z_1Px + \cdots + z_nP^nx^n$ and $|\theta_2| \le 1$. Now, because $|V| \le Q$, we have

$$S_1(A) = \sum_{t=1}^{Q} e(t)S(t) = PQ^{-1}VW + 8\theta_2 nQ,$$

and hence

$$S(A) = S_1(A) + \theta Q = PQ^{-1}VW + R, \quad |R| \le 9nQ,$$

as required. The proof of the lemma is complete. □

Lemma 3.12. *Suppose that the point A belongs to the second class. Then the sum $S(A)$ satisfies the estimate*

$$|S(A)| \le (2n)^{2n+11} P^{1-\rho}, \quad \rho = \left(8n^2(\ln n + 1.5 \ln\ln n + 4.2)\right)^{-1}.$$

Proof. According to the well-known theorem in elementary number theory, each coordinate α_s of the point A can be represented as

$$\alpha_s = \frac{a_s}{q_s} + z_s, \quad (a_s, q_s) = 1, \quad 0 < q_s \le \tau_s = P^{s-0.5},$$
$$|z_s| \le (q_s \tau_s)^{-1}, \quad s = 1, \dots, n.$$

Let Q be the least common multiple of the numbers $q_1, \dots, q_n$, i.e., let $Q = [q_1, \dots, q_n]$. First, we assume that $Q \le P^{0.3}$. In this case, by Lemma 3.11, the sum $S(A)$ satisfies the asymptotic formula

$$S(A) = PQ^{-1}VW + R, \quad |R| \le 9nQ \le 9nP^{0.3}.$$

Since the point A belongs to the second class, we have $|z_s| \ge P^{0.3-s}$ for some s ($1 \le s \le n$). Therefore, using Lemma 1.4 (see Chapter 1) to estimate the integral

$$W = \int_0^1 \exp\{2\pi i g(x)\}\, dx, \quad g(x) = \sum_{s=1}^{n} z_s P^s x^s,$$

we obtain

$$|W| \le 2^5 P^{-0.3\nu}, \quad \nu = n^{-1},$$

because in this case we have

$$\max\left(|z_1|P, \dots, |z_n|P^n\right) \ge P^{0.3}.$$

Consequently,

$$|S(A)| \le PQ^{-1}|V|\,|W| + |R| \le 2^6 P^{1-0.3\nu} \le (2n)^{2n+11} P^{1-\rho},$$

i.e., the statement of the lemma is proved for $Q \le P^{0.3}$.

Now let $Q > P^{0.3}$, and let $Q_0 = [q_2, \dots, q_n]$. If it turns out that $Q_0 > P^{0.5-0.4\nu}$, then the desired result follows from Theorem 3.2, since $|S(A)|$ satisfies the estimate

$$|S(A)| \le cP^{1-\rho}, \quad \rho = \left(8n^2(\ln n + 1.5 \ln\ln n + 4.2)\right)^{-1},$$

where

$$c = c(n) = \left(n(n+1)\ln \rho^{-1}\right)^{3/2}(2n)^{2(n+1)} < (2n)^{2n+11};$$

hence $|S(A)| \le (2n)^{2n+11} P^{1-\rho}$, as stated in the lemma.

Now it remains to consider the case where $Q > P^{0.3}$, but $Q_0 < P^{0.5-0.4\nu}$. Here, as in the proof of Lemma 3.11, by setting

$$S(A) = \sum_{t=1}^{Q_0} \sum_{\substack{1\le x\le P \\ x\equiv t \pmod{Q_0}}} \exp\{2\pi i f(x)\},$$

we divide the interval of summation in the sum $S(A)$ into progressions. Then $S(A) = S_1(A) + R$, where

$$S_1(A) = \sum_{t=1}^{Q_0}\sum_{y=1}^{P_1} \exp\{2\pi i f(Q_0 y + t)\}, \quad P_1 = [PQ_0^{-1}] + 1, \ |R| \le Q_0.$$

The number Q_0 is a multiple of all $q_2, \dots, q_n$, but possibly not of q_1. Therefore,

$$\begin{aligned} f(Q_0 y + t) &= \sum_{s=1}^{n} \alpha_s (Q_0 y + t)^s = \sum_{s=1}^{n} \left(\frac{a_s}{q_s} + z_s\right)(Q_0 y + t)^s \\ &= \frac{by}{q_1} + \sum_{s=1}^{n} \frac{a_s t^s}{q_s} + \sum_{s=1}^{n} z_s (Q_0 y + t)^s + H, \end{aligned}$$

where b is the least in the absolute value residue of the number aQ_0 modulo q_1 and where H is an integer. Hence we have $|bq_1^{-1}| \le 0.5$ as well as

$$S_1(A) = \sum_{t=1}^{Q_0} \exp\left\{2\pi i \left(\frac{a_1 t}{q_1} + \dots + \frac{a_n t^n}{q_n}\right)\right\} S(t),$$

where

$$S(t) = \sum_{y=0}^{P_1} \exp\{2\pi i \varphi_t(y)\}, \quad \varphi_t(y) = \frac{by}{q_1} + \sum_{s=1}^{n} z_s (Q_0 y + t)^s.$$

We estimate the derivative of the function $\varphi_t(y)$ with respect to the variable y:

$$\begin{aligned} \left|\frac{d\varphi_t(y)}{dy}\right| &= \left|bq_1^{-1} + Q_0 \sum_{s=1}^{n} s z_s (Q_0 y + t)^{s-1}\right| \\ &\le 0.5 + P^{0.5-0.4\nu} \sum_{s=1}^{n} s P^{0.5-s} P^{s-1} \le 0.5 + n^2 P^{-0.4\nu} < 0.6, \end{aligned}$$

because $Q_0 < P^{0.5-0.4\nu}$, $|z_s| \le \tau_s^{-1} = P^{0.5-s}$, $P \ge n^{40n}$, which implies $n^2 P^{-0.4\nu} < n^{-8} < 0.1$. Dividing the range of the variable y into intervals on which the derivative of the function $\varphi_t(y)$ is monotone and of constant sign (the number of such intervals is at most $2n-2$) and applying Lemma A.2, we obtain

$$S(t) = \int_0^{P_1} \exp\{2\pi i \varphi_t(y)\}\,dy + R_1$$
$$= PQ_0^{-1} \int_0^1 \exp\{2\pi i g(x)\}\,dx + R_2 = PQ_0^{-1} W(t) + R_2,$$

where $|R_1| \le 12(n-1)$, $|R_2| \le 12n - 1$,

$$g(x) = -\frac{tb}{Q_0 q_1} + P\left(\frac{b}{Q_0 q_1} + z_1\right)x + \sum_{s=2}^{n} P^s z_s x^s.$$

Obviously, the variable $W(t)$ is determined by the last relation. The number b is integer and if $b \neq 0$, then the absolute value of the coefficient of the first power of the unknown in the polynomial $g(x)$ can be estimated as follows:

$$\left| P\left(\frac{b}{Q_0 q_1} + z_1\right)\right| \ge P\left|\frac{1}{Q_0 q_1} - \frac{1}{q_1 \tau_1}\right| \ge P q_1^{-1}(P^{0.4\nu - 0.5} - P^{-0.5})$$
$$\ge 0.5 P \tau_1^{-1} P^{0.4\nu - 0.5} = 0.5 P^{0.4\nu}, \qquad \nu = n^{-1}.$$

Therefore, applying Lemma 1.4 (see Chapter 1) to the integral $W(t)$, we obtain

$$|W(t)| \le 2^6 P^{-0.4\nu^2}.$$

This implies the estimate

$$|S(A)| \le |S_1(A)| + |R| \le \sum_{t=1}^{Q_0} |S(t)| + |R|$$
$$\le \sum_{t=1}^{Q_0} \left(PQ_0^{-1}|W(t)| + |R_1|\right) + |R| < 10^2 P^{1-0.4\nu^2},$$

and this estimate is much sharper than the desired estimate.

But if $b = 0$, then $Q_0 = Q$ and $Q_0 \ge P^{0.3}$. In this case the coefficients of the polynomial $g(x)$ are independent of t and we have the relations

$$W(t) = W, \quad S(A) = PQ^{-1} V W + R_3,$$
$$|R_3| \le |R| + \sum_{t=1}^{Q_0} |R_2| \le 12nQ \le 12n P^{0.3},$$

where V and W have the same values as in Lemma 3.11.

Now, estimating the sum V by Theorem 2.2 (see Chapter 2), we obtain the following estimate for $S(A)$:

$$|S(A)| \le 2e^{7n} P^{1-0.3\nu},$$

which is sharper than the desired estimate. So the proof of the lemma is complete. □

Proof of Theorem 3.7. According to the partition of the domain Ω into the classes Ω_1 and Ω_2, we can represent the integral I as the sum $I = I_1 + I_2$, where I_1 is the integral over the domain Ω_1 and I_2 is the integral over the domain Ω_2. So we have

$$I_1 = \int_{\Omega_1} S^k(A) \exp\{-2\pi i(\alpha_1 N_1 + \cdots + \alpha_n N_n)\}\, dA,$$
$$I_2 = \int_{\Omega_2} S^k(A) \exp\{-2\pi i(\alpha_1 N_1 + \cdots + \alpha_n N_n)\}\, dA.$$

In what follows, we derive an asymptotic formula for the integral I_1 and find an upper bound for the modulus of the integral I_2.

First, we consider the integral I_2. Let k_0, k_1, and k_2 be natural numbers such that $k_0 = k_1 + 2k_2 \le k$. Then we have

$$\begin{aligned}|I_2| &\le \left| \int_{\Omega_2} S^k(A) \exp\{-2\pi i(\alpha_1 N_1 + \cdots + \alpha_n N_n)\}\, dA \right. \\ &\le \int_{\Omega_2} |S^k(A)|\, dA \le P^{k-k_0} \max_{A\in\Omega_2} |S(A)|^{k_1} \int_{\Omega} |S(A)|^{2k_2}\, dA.\end{aligned}$$

Now we set $k_1 = n^2$ and $k_2 = n^2([\ln \rho^{-1}] + 1)$, where

$$\rho = \left(8n^2(\ln n + 1.5 \ln\ln n + 4.2)\right)^{-1}.$$

Applying Lemma 3.11 to estimate $\max_{\Omega_2} |S(A)|^{k_1}$ and estimating

$$\int_{\Omega} |S(A)|^{2k_2}\, dA$$

by Theorem 3.6, we obtain

$$\begin{aligned}I_2 &\le P^{k-k_0} \max_{A\in\Omega_2} |S(A)|^{n^2} \int_{\Omega} |S(A)|^{2k_2}\, dA \\ &\le P^{k-k_2} (2n)^{2n^3+11n^2} P^{n^2-n^2\rho}\, 2^{2n^2\tau}\, n^{2n\Delta} (8k_2)^{2n\tau} P^{2k_2-\Delta} = c_1 P^{c_2},\end{aligned}$$

where $\tau = k_2 n^{-1}$, $\Delta = 0.5n(n+1) - 0.5n^2(1-n^{-1})^{\tau}$, and c_1 and c_2 are obviously determined by the last relation.

Since $(1-n^{-1})^n < e^{-1}$, we have

$$(1-n^{-1})^{\tau} \le \rho, \quad \Delta \ge 0.5n(n+1) - 0.5n^2\rho.$$

In addition, since $n \ge 3$, we have

$$\begin{aligned}
&0.5n^2\rho > \big(30(2+\ln n)\big)^{-1},\\
&\tau = n[\ln \rho^{-1}] + n \le n\ln(8n^2(\ln n + 1.5\ln\ln n + 4.2)) \le n(2\ln n + \ln\ln n + 4),\\
&k_0 = k_1 + 2k_2 \le n^2(4\ln n + 2\ln\ln n + 9).
\end{aligned}$$

Hence

$$\begin{aligned}
c_1 &= (2n)^{2n^3+11n^2} 2^{2n^2\tau} n^{2n\Delta} (8k_2)^{2n\tau} < 0.5n^{30n^3},\\
c_2 &= k - k_0 + n^2 - n^2\rho + 2k_2 - 0.5n(n+1) + 0.5n^2\rho\\
&= k - 0.5n(n+1) - 0.5n^2\rho = k - 0.5n(n+1) - \big(30(2+\ln n)\big)^{-1}.
\end{aligned}$$

Thus we obtain the following final estimate for $|I_2|$:

$$|I_2| \le 0.5n^{30n^3} P^{k-0.5n(n+1)-(30(2+\ln n))^{-1}}. \tag{3.31}$$

Now we consider the integral I_1. Since the domains $\omega(a,q)$ do not intersect pairwise, the integral can be represented as the sum of integrals

$$I_1 = \sum_{a,q}^{P^{0.3}} I_{a,q},$$

where the sum $\sum_{a,q}^{P^{0.3}}$ is taken over all pairs of sets of integers (a,q) such that $(a_s, q_s) = 1$ $(0 \le a_s < q_s)$ for $s = 1, \dots, n$, $Q \le P^{0.3}$, and the integral $I_{a,q}$ is given by the relation

$$I_{a,q} = \int_{\omega(a,q)} S^k(A) \exp\{-2\pi i(\alpha_1 N_1 + \dots + \alpha_n N_n)\}\, dA.$$

Recall that Q is the least common multiple of the numbers $q_1, \dots, q_n$, i.e., $Q = [q_1, \dots, q_n]$. The trigonometric sum $S(A)$ in the last integral satisfies the assumptions of Lemma 3.11. Hence we have the following asymptotic formula for $S(A)$:

$$S(A) = PQ^{-1}VW + R,$$

where

$$V = \sum_{x=1}^{Q} \exp\left\{2\pi i\left(\frac{a_1 x}{q_1} + \dots + \frac{a_n x^n}{q_n}\right)\right\},$$

$$W = \int_0^1 \exp\{2\pi i(z_1 Px + \dots + z_n P^n x^n)\}\, dx, \quad |R| \le 9nQ \le 9nP^{0.3}.$$

Substituting this formula into the expression for $I_{a,q}$, we obtain

$$\begin{aligned}
I_{a,q} &= \int_{\omega(a,q)} S^k(A) \exp\{-2\pi i(\alpha_1 N_1 + \cdots + \alpha_n N_n)\}\, dA \\
&= \int_{\omega(a,q)} (PQ^{-1}VW + R)^k \exp\{-2\pi i(\alpha_1 N_1 + \cdots + \alpha_n N_n)\}\, dA \\
&= \int_{\omega(a,q)} (PQ^{-1}VW)^k \exp\{-2\pi i(\alpha_1 N_1 + \cdots + \alpha_n N_n)\}\, dA \\
&\quad + \sum_{s=0}^{k-1} \int_{\omega(a,q)} (PQ^{-1}VW)^s R^{k-s} \binom{k}{s} \exp\{-2\pi i(\alpha_1 N_1 + \cdots + \alpha_n N_n)\}\, dA \\
&= I_3 + \sum_{s=0}^{k-1} r_s
\end{aligned}$$

(the integrals I_3 and r_s $(s = 0, \ldots, k-1)$ are determined in an obvious way by the last relation).

Now we estimate each of the integrals r_s. For $s < 2n^2$, using the trivial estimates $P^{0.1} > 18n$, $|W| \le 1$, and $\binom{k}{s} < 2^k$, we obtain

$$\begin{aligned}
|r_s| &\le \int_{\omega(a,q)} \left| (PQ^{-1}VW)^s R^{k-s} \binom{k}{s} \right| dA \\
&\le P^s (9n)^{k-s} P^{0.3k-0.3s} \cdot 2^k \le (18n)^k P^{0.7s+0.3k} \le P^{0.7s+0.4k}.
\end{aligned}$$

For $s \ge 2n^2$, it follows from the estimates

$$\binom{k}{s} < k^{k-s}, \quad k \le P^{0.1}, \quad P^{0.1} > 9n,$$

that

$$\begin{aligned}
|r_s| &\le \int_{\omega(a,q)} \left| (PQ^{-1}VW)^s R^{k-s} \binom{k}{s} \right| dA \\
&\le |Q^{-1}V|^s P^s (9nP^{0.3})^{k-s} k^{k-s} \int_{\omega(a,q)} |W|^s\, dA \\
&\le |Q^{-1}V|^s P^{0.5(k+s)} \int_{\omega(a,q)} |W|^s\, dA.
\end{aligned}$$

Now we consider the integral

$$I_4 = \int_{\omega(a,q)} |W|^s\, dA.$$

We change the integration variables in this integral by setting

$$u_1 = z_1 P = \left(\alpha_1 - \frac{a_1}{q_1}\right) P, \ \ldots, \ u_n = z_n P^n = \left(\alpha_n - \frac{a_n}{q_n}\right) P^n.$$

We obtain

$$I_4 = P^{-0.5n(n+1)} \int_{-P^{0.3}}^{P^{0.3}} \cdots \int_{-P^{0.3}}^{P^{0.3}} \left| \int_0^1 \exp\{2\pi i(u_1 x + \cdots + u_n x^n)\}\, dx \right|^s du_1 \ldots du_n.$$

We divide the domain of integration over the variables $u_1, \ldots, u_n$ into the parts $\omega_0, \omega_1, \ldots, \omega_t, \ldots$ as follows. Suppose that $u_0 = \max(|u_1|, \ldots, |u_n|)$. For each value of the subscript t $(t = 1, \ldots, T)$, we define the domain ω_t by the condition

$$2^{t-1} < u_0 \leq 2^t.$$

The points $(u_1, \ldots, u_n)$ for which $u_0 \leq 1$ belong to the domain ω_0. We note that, starting from the number $T = [\log_2 P^{0.3}] + 1$, the domains ω_t are empty. Hence for the integral I_4 we have the estimate

$$I_4 \leq P^{-0.5n(n+1)} \sum_{t=0}^{T} \int_{\omega_1} \left| \int_0^1 \exp\{2\pi i(u_1 x + \cdots + u_n x^n)\} dx \right|^s du,$$

where $(u_1, \ldots, u_n)$ is a point of the domain ω_t. By Lemma 1.4 (see Chapter 1), for the integral

$$W_1 = \int_0^1 \exp\{2\pi i(u_1 x + \cdots + u_n x^n)\}\, dx,$$

we have the estimate

$$|W_1| < \min(1,\ 32 u_0^{-1/n}).$$

Since for $t = 1, \ldots, T$ the volume of the domain ω_t is, obviously, equal to $2^{(t+1)n} - 2^{tn}$, we use the above estimate for I_4 and, taking into account the inequality $s \geq 2n^2$, obtain

$$\begin{aligned} I_4 &\leq P^{-0.5n(n+1)} \Big(1 + \sum_{t=0}^{5n} (2^{(t+1)n} - 2^{tn}) + \sum_{t=5n+1}^{T} (2^{5-t/n})^s 2^{(t+1)n}\Big) \\ &\leq P^{-0.5n(n+1)} \Big(2^{(5n+1)n} + \sum_{t=5n+1}^{T} 2^{10n^2+n-tn}\Big) \leq P^{-0.5n(n+1)} \cdot 2^{5n^2+n+1} \\ &< 2^{6n^2} P^{-0.5n(n+1)}. \end{aligned}$$

Hence, for $s \geq 2n^2$, the integral r_s can be estimated as

$$|r_s| \leq 2^{6n^2} P^{0.5s+0.5k-0.5n(n+1)} |Q^{-1}V|^s.$$

So we have obtained the asymptotic formula

$$I_{a,q} = I_3 + R_1,$$

where

$$|R_1| \le 2n^2 P^{1.4n^2+0.4k} + 2^{6n^2} P^{0.5(k-n(n+1))} \sum_{s=2n^2}^{k-1} P^{0.5s} |Q^{-1}V|^s.$$

Hence for the integral I_1 we have the relation

$$I_1 = \sum_{a,q}^{P^{0.3}} I_{a,q} = \sum_{a,q}^{P^{0.3}} I_3 + \sum_{a,q}^{P^{0.3}} R_1.$$

We let the symbol $\sum_a'^{\,b}$ denote the summation over positive integers d that so not exceed b and run through the above system of residue modulo b. If $[q_1, \dots, q_n]$ denotes, as usual, the least common multiple of the numbers $q_1, \dots, q_n$, then the sum $\sum_{a,q}^{P^{0.3}}$ can be written as

$$\sum_{a,q}^{P^{0.3}} \cdots = \sum_{Q \le P^{0.3}} \sum_{\substack{q_1 \\ [q_1,\dots,q_n]=Q}} \cdots \sum_{q_n} \sum_{a_1}^{q_1}{}' \cdots \sum_{a_n}^{q_n}{}' \cdots .$$

Further, we set

$$I_5 = \sum_{a,q}^{P^{0.3}} I_3, \quad R_2 = \sum_{a,q}^{P^{0.3}}.$$

We estimate the variable R_2. For a fixed Q, we have

$$\sum_{\substack{q_1 \\ [q_1,\dots,q_n]=Q}} \cdots \sum_{q_n} \sum_{a_1}^{q_1}{}' \cdots \sum_{a_n}^{q_n}{}' 1 \le \sum_{q_1|Q} \varphi(q_1) \cdots \sum_{q_n|Q} \varphi(q_n) = Q^n. \tag{3.32}$$

Hence

$$\sum_{a,q}^{P^{0.3}} 1 \le \sum_{Q \le P^{0.3}} Q^n \le 1 + \int_1^{P^{0.3}} x^n\, dx \le P^{0.3(n+1)}.$$

Therefore, $|R_2| \le A + B \sum_1$, where

$$A = 2n^2 P^{1.4n^2+0.4k+0.3(n+1)}, \quad B = 2^{6n^2} P^{0.5(k-n(n+1))},$$

$$\sum\nolimits_1 = \sum_{Q \le P^{0.3}} \sum_{s=2n^2}^{k-1} \sum_{\substack{q_1 \\ [q_1,\dots,q_n]=Q}} \cdots \sum_{q_n} \sum_{a_1}^{q_1}{}' \cdots \sum_{a_n}^{q_n}{}' P^{0.5s} |Q^{-1}V|^s.$$

We divide the interval of summation over Q in the sum $\sum_1$ into two the parts: $Q \le \exp\{7n^2\}$ and $Q > \exp\{7n^2\}$. If it turns out that $P^{0.3} \le \exp\{7n^2\}$, then we assume that the second part is empty. We obtain $\sum_1 = \sum_2 + \sum_3$, where $\sum_2$ is the part of the sum $\sum_1$ corresponding to the first interval of summation and $\sum_3$ is the part of the sum $\sum_1$ corresponding to the second interval.

Using the trivial estimate $|Q^{-1}V| \leq 1$ and inequality (3.32), we estimate the sum

$$\sum\nolimits_2 \leq \sum_{s=2n^2}^{k-1} P^{0.5s} \sum_{Q \leq \exp\{7n^2\}} Q^n \leq \exp\{7n^2(n+1)\} P^{0.5k-0.5}.$$

In the sum $\sum_3$, we estimate $|Q^{-1}V|$ by using Theorem 2.2 (see Chapter 2). We obtain

$$\sum\nolimits_3 \leq \sum_{s=2n^2}^{k-1} P^{0.5s} \sum_{Q > \exp\{7n^2\}} Q^n (\exp\{7n\} Q^{-1/n})^s$$

$$\leq \sum_{s=2n^2}^{k-1} P^{0.5s} 0.5 \exp\{7n^2(n+1)\} \leq \exp\{7n^2(n+1)\} P^{0.5k-0.5}.$$

From the above estimates for the sums $\sum_2$ and $\sum_3$, we obtain the following estimate for the sum $\sum_1$:

$$\sum\nolimits_1 = \sum\nolimits_2 + \sum\nolimits_3 \leq 2\exp\{7n^2(n+1)\} P^{0.5k-0.5}.$$

Next, since $k \geq 9n^2$, we have

$$A = 2n^2 P^{1.4n^2+0.4k+0.3n+0.3} < P^{k-0.5n(n+1)-0.5},$$

$$|R_2| \leq A + B \sum\nolimits_1 \leq \exp\{8n^3\} P^{k-0.5n(n+1)-0.5}.$$

Now we consider the variable I_5. By definition, we have

$$I_5 = \sum_{a,q}^{P^{0.3}} I_3, \quad I_3 = \int_{\omega(a,q)} (PQ^{-1}VW)^k \exp\{-2\pi i(\alpha_1 N_1 + \cdots + \alpha_n N_n)\}\, dA.$$

We extend the integration over A in the integral I_3 to the entire space $\mathbb{R}^n$ by setting

$$I_6 = \int_{R^n} (PQ^{-1}VW)^k \exp\{-2\pi i(\alpha_1 N_1 + \cdots + \alpha_n N_n)\}\, dA.$$

Let $R_3 = I_5 - \sum_{a,q}^{P^{0.3}} I_6$. We estimate the variable R_3. Let I_7 be the difference between the integrals I_6 and I_3. Then

$$I_7 = I_6 - I_3 = \int_{\omega_1(a,q)} (PQ^{-1}VW)^k \exp\{-2\pi i(\alpha_1 N_1 + \cdots + \alpha_n N_n)\}\, dA,$$

where

$$\omega_1(a, q) = R^n \setminus \omega(a, q).$$

We estimate the integral I_7 as follows:

$$|I_7| \le \int_{\omega_1(a,q)} |PQ^{-1}VW|^k \, dA = |PQ^{-1}V|^k \int_{\omega_1(a,q)} |W|^k \, dA.$$

We shall estimate the integral $\int_{\omega_1(a,q)} |W|^k \, dA$ just in the same way as we estimated the integral I_4. First, we change the integration variables by setting

$$u_1 = z_1 P = \left(\alpha_1 - \frac{a_1}{q_1}\right) P, \ \ldots, \ u_n = z_n P^n = \left(\alpha_n - \frac{a_n}{q_n}\right) P^n.$$

We obtain

$$\int_{\omega_1(a,q)} |W|^k \, dA$$
$$= P^{-0.5n(n+1)} \underset{u_0 > P^{0.3}}{\int \cdots \int} \left| \int_0^1 \exp\{2\pi i (u_1 x + \cdots + u_n x^n)\} \, dx \right|^k du_1 \ldots du_n,$$

where $u_0 = \max(|u_1|, \ldots, |u_n|)$. We divide the domain of integration over the variables $u_1, \ldots, u_n$ into the parts ω_t $(t = T, T+1, \ldots)$ determined by the condition $2^{t-1} < u_0 \le 2^t$, where $T = [\log_2 P^{0.3}] + 1$. Using the estimate in Lemma 1.4 (see Chapter 1) for the integral

$$W_1 = \left| \int_0^1 \exp\{2\pi i (u_1 x + \cdots + u_n x^n)\} \, dx \right| \le 32 u_0^{-1/n},$$

we obtain (here $u = (u_1, \ldots, u_n)$)

$$\int_{u_0 > P^{0.3}} |W_1|^k \, du \le \sum_{t=T}^{+\infty} \int_{\omega_1} |W_1|^k \, du$$
$$\le \sum_{t=T-1}^{+\infty} (2^{5-t/n})^k 2^{(t+1)n} \le P^{-k/n} 2^{5k+k/n+1} < P^{-1}.$$

It follows from this estimate that

$$|R_3| = \left| I_5 - \sum_{a,q}^{P^{0.3}} I_6 \right| = \left| \sum_{a,q}^{P^{0.3}} I_7 \right| \le \sum_{a,q}^{P^{0.3}} |I_7|$$
$$\le \sum_{a,q}^{P^{0.3}} |PQ^{-1}V|^k \int_{\omega_1(a,q)} |W|^k \, dA \le P^{k-0.5n(n+1)-1} \sum_{a,q}^{P^{0.3}} |Q^{-1}V|^k.$$

Repeating the argument used to estimate the sum $\sum_{a,q}^{P^{0.3}} |Q^{-1}V|^s$ word for word, we obtain the following estimate for the sum $\sum_{a,q}^{P^{0.3}} |Q^{-1}V|^k$:

$$\sum_{a,q}^{P^{0.3}} |Q^{-1}V|^k \leq \exp\{7n^2(n+1)\}.$$

Hence

$$|R_3| = \left| I_5 - \sum_{a,q}^{P^{0.3}} I_6 \right| \leq \exp\{7n^2(n+1)\} P^{k-0.5n(n+1)-1}.$$

Now we extend the summation in the sum $\sum_{a,q}^{P^{0.3}} I_6$ to all natural numbers Q. We let the symbol $\sum_{a,q} I_6$ denote the series obtained after this change of the summation interval. We estimate the difference R_4 between these variables, i.e.,

$$R_4 = \sum_{a,q} I_6 - \sum_{a,q}^{P^{0.3}} I_6.$$

For this, we use the estimate in Lemma 1.4 (see Chapter 1) and relation (3.32) just as in estimating the variable R_2. We obtain

$$\begin{aligned}
|Q^{-1}V|^k &\leq |Q^{-1}V|^{2n^2} \leq \exp\{14n^3\} Q^{-2n}, \\
|R_4| &\leq \sum_{Q > P^{0.3}} \sum_{\substack{q_1 \\ [q_1,\ldots,q_n]=Q}} \cdots \sum_{q_n} \sum_{a_1}^{q_1}{}' \cdots \sum_{a_n}^{q_n}{}' |J_6| \\
&\leq P^k \int_{R^n} |W|^k \, dA \sum_{Q > P^{0.3}} \sum_{\substack{q_1 \\ [q_1,\ldots,q_n]=Q}} \cdots \sum_{q_n} \exp\{14n^3\} Q^{-2n} \sum_{a_1}^{q_1}{}' \cdots \sum_{a_n}^{q_n}{}' 1 \\
&\leq P^k \int_{R^n} |W|^k \, dA \cdot \exp\{14n^3\} \sum_{Q > P^{0.3}} Q^{-n} \\
&\leq (n-1) \exp\{14n^3\} P^{k-0.6} \int_{R^n} |W|^k \, dA.
\end{aligned}$$

The variable $\int_{R^n} |W|^k \, dA$ is estimated similarly to the integral I_4, only the summation over the parameter t is extended to the summation from zero to infinity. Therefore, we have

$$\int_{R^n} |W|^k \, dA \leq 2^{6n^2} P^{-0.5n(n+1)}. \tag{3.33}$$

We also note that the sum $\sum_{a,q} |Q^{-1}V|^k$ is estimated according to the same scheme as the sum $\sum_{a,q}^{P^{0.3}} |Q^{-1}V|^k$ and we have the same estimate for this sum:

$$\sum_{a,q} |Q^{-1}V|^k \le \exp\{7n^2(n+1)\}.$$

Substituting estimate (3.33) into the last inequality, we obtain

$$|R_4| \le \exp\{15n^3\} P^{k-0.5n(n+1)-0.6}.$$

At the same time, we have proved that the series $\sum_{a,q} I_6$ converges absolutely and, moreover,

$$\sum_{a,q} I_6 \le P^k \int_{R^n} |W|^k \, dA \sum_{a,q} |Q^{-1}V|^k \le \exp\{8n^3\} P^{k-0.5n(n+1)}. \qquad (3.34)$$

It follows from the above estimates for R_2, R_3, and R_4 that

$$\left| I_1 - \sum_{a,q} I_6 \right| = |R_2 + R_3 + R_4| \le \exp\{15n^3\} P^{k-0.5n(n+1)-0.5}.$$

Along with the above estimate for the integral I_2, this implies

$$\begin{aligned} R_5 &= \left| I - \sum_{a,q} I_6 \right| = \left| I_1 - \sum_{a,q} I_6 + I_2 \right| \le \left| I_1 - \sum_{a,q} I_6 \right| + |I_2| \\ &\le \exp\{15n^3\} P^{k-0.5n(n+1)-0.5} + 0.5n^{30n^3} P^{k-0.5n(n+1)-(30(2+\ln n))^{-1}} \\ &\le n^{30n^3} P^{k-0.5n(n+1)-(30(2+\ln n))^{-1}}. \end{aligned}$$

So for the variable I, we have obtained an asymptotic formula with the desired remainder term. Now, to complete the proof of Theorem 3.7, it suffices to show that its leading term can also be written in the form given in the statement of the theorem. But, first, we use this formula to estimate the integral I and thus to obtain the second assertion in the theorem.

From the estimate for the variable R_5 and inequality (3.34) we obtain

$$\begin{aligned} I \le R_5 + \left| \sum_{a,q} I_6 \right| &\le n^{30n^3} P^{k-0.5n(n+1)-(30(2+\ln n))^{-1}} \\ &\quad + \exp\{8n^3\} P^{k-0.5n(n+1)} \le n^{30n^3} P^{k-0.5n(n+1)}. \end{aligned}$$

Now we show that

$$\sum_{a,q} I_6 = \sigma \gamma P^{k-0.5n(n+1)}.$$

Indeed, let $q = q_1 \dots q_n$. Then we have

$$\sum_{Q=1}^{+\infty} \sum_{\substack{q_1 \quad\; q_n \\ [q_1,\dots,q_n]=Q}} \cdots \sum \cdots = \sum_{q_1=1}^{+\infty} \cdots \sum_{q_n=1}^{+\infty} \cdots .$$

Hence

$$\begin{aligned}\sum_{a,q} I_6 &= \sum_{Q=1}^{+\infty} \sum_{\substack{q_1 \quad\; q_n \\ [q_1,\dots,q_n]=Q}} \cdots \sum \sum_{a_1}^{q_1}{}' \cdots \sum_{a_n}^{q_n}{}' \int_{R^n} (PQ^{-1}VW)^k \\ &\quad \times \exp\{-2\pi i(\alpha_1 N_1 + \dots + \alpha_n N_n)\}\, dA \\ &= \sum_{q_1=1}^{+\infty} \cdots \sum_{q_n=1}^{+\infty} \sum_{a_1}^{q_1}{}' \cdots \sum_{a_n}^{q_n}{}' \int_{R^n} (PQ^{-1}VW)^k \\ &\quad \times \exp\{-2\pi i(\alpha_1 N_1 + \dots + \alpha_n N_n)\}\, dA.\end{aligned}$$

We consider the integral I_8 under the summation sign in the right-hand side of the last relation. We have

$$\begin{aligned}I_8 &= \int_{R^n} (PQ^{-1}VW)^k \exp\{-2\pi i(\alpha_1 N_1 + \dots + \alpha_n N_n)\}\, dA \\ &= (PQ^{-1}V)^k \exp\left\{ -2\pi i\left(\frac{a_1 N_1}{q_1} + \dots + \frac{a_n N_n}{q_n}\right)\right\} \\ &\quad \times \int_{-\infty}^{+\infty} \cdots \int_{-\infty}^{+\infty} W^k \exp\{-2\pi i(z_1 N_1 + \dots + z_n N_n)\}\, dz_1 \dots dz_n,\end{aligned}$$

because $\alpha_1 = a_1/q_1 + z_1, \dots, \alpha_n = a_n/q_n + z_n$. We change the integration variables by setting $\beta_1 = Pz_1, \dots, \beta_n = P^n z_n$ and obtain

$$\begin{aligned}&\int_{-\infty}^{+\infty} \cdots \int_{-\infty}^{+\infty} W^k \exp\{-2\pi i(z_1 N_1 + \dots + z_n N_n)\}\, dz_1 \dots dz_n \\ &\quad = \int_{-\infty}^{+\infty} \cdots \int_{-\infty}^{+\infty} \left(\int_0^1 \exp\{2\pi i(z_1 Px + \dots + z_n P^n x^n)\}\, dx \right)^k \\ &\quad \times \exp\{-2\pi i(z_1 N_1 + \dots + z_n N_n)\}\, dz_1 \dots dz_n \\ &\quad = P^{-0.5n(n+1)} \int_{-\infty}^{+\infty} \cdots \int_{-\infty}^{+\infty} \left(\int_0^1 \exp\{2\pi i(\beta_1 x + \dots + \beta_n x^n)\}\, dx \right)^k \\ &\quad \times \exp\left\{ -2\pi i\left(\frac{\beta_1 N_1}{P} + \dots + \frac{\beta_n N_n}{P^n}\right)\right\} d\beta_1 \dots d\beta_n = \gamma.\end{aligned}$$

Hence

$$I_8 = \gamma P^{k-0.5n(n+1)} (Q^{-1}V)^k \exp\left\{ -2\pi i\left(\frac{a_1 N_1}{q_1} + \dots + \frac{a_n N_n}{q_n}\right)\right\}.$$

Now we note that

$$Q^{-1}V = Q^{-1}\sum_{x=1}^{Q}\exp\left\{2\pi i\left(\frac{a_1x}{q_1}+\cdots+\frac{a_nx^n}{q_n}\right)\right\}$$
$$= q^{-1}\sum_{x=1}^{q}\exp\left\{2\pi i\left(\frac{a_1x}{q_1}+\cdots+\frac{a_nx^n}{q_n}\right)\right\}.$$

Therefore,

$$\sum_{a,q} I_6 = \sum_{q_1=1}^{+\infty}\cdots\sum_{q_n=1}^{+\infty}\sum_{a_1}^{q_1}{}'\cdots\sum_{a_n}^{q_n}{}'\gamma P^{k-0.5n(n+1)}$$
$$\times (Q^{-1}V)^k\exp\left\{2\pi i\left(\frac{a_1N_1}{q_1}+\cdots+\frac{a_nN_n}{q_n}\right)\right\}$$
$$= \gamma P^{k-0.5n(n+1)}\sum_{q_1=1}^{+\infty}\cdots\sum_{q_n=1}^{+\infty}\sum_{a_1}^{q_1}{}'\cdots\sum_{a_n}^{q_n}{}'$$
$$\times\left(q^{-1}\sum_{x=1}^{q}\exp\left\{-2\pi i\left(\frac{a_1x}{q_1}+\cdots+\frac{a_nx^n}{q_n}\right)\right\}\right)^k$$
$$\times\exp\left\{-2\pi i\left(\frac{a_1N_1}{q_1}+\cdots+\frac{a_nN_n}{q_n}\right)\right\} = \sigma\gamma P^{k-0.5n(n+1)},$$

as required. The proof of the theorem is complete. □

The argument used in the proof of Theorem 3.7 allows us to obtain a somewhat more general result. Moreover, the proof of this result differs from that of Theorem 3.7 only in the notation. In what follows, we state this result as Theorem 3.8 and assume that it has already been proved together with Theorem 3.7.

Theorem 3.8. *Let $k \geq n^2(4\ln n + 2\ln\ln n + 9)$, and let $k \leq P^{0.1}$. Then for the number I' of solutions of the Diophantine equations*

$$\sum_{m=1}^{l} x_m^s - \sum_{m=l+1}^{k} x_m^s = N_s, \quad s = 1,\ldots,n,$$

where $n \geq 3$, $l \leq k$, $1 \leq x_m \leq P$ $(m = 1,\ldots,k)$, the following asymptotic formula holds:

$$I' = \sigma'\gamma' P^{k-0.5n(n+1)} + \theta' n^{30n^3} P^{k-0.5n(n+1)-(30(2+\ln n))^{-1}},$$

as well as the estimate

$$I' \leq n^{30n^3} P^{k-0.5n(n+1)}.$$

Here $|\theta'| \le 1$ *and* σ' *and* γ' *are the singular series and the singular integral determined by the relations*

$$\sigma' = \sum_{q_1=1}^{+\infty} \cdots \sum_{q_n=1}^{+\infty} \sum_{a_1}^{q_1}{}' \cdots \sum_{a_n}^{q_n}{}' q^{-k} V^l \overline{V}^{k-l} \exp\{2\pi i A\},$$

$$\gamma' = \int_{-\infty}^{+\infty} \cdots \int_{-\infty}^{+\infty} W^l \overline{W}^{k-l} \exp\{2\pi i B\}\, d\beta_1 \dots d\beta_n.$$

The variables q, V, A, W *have the same meaning as in Theorem* 3.7.

As a corollary of Theorem 3.8, we obtain a *simplified estimate* in Vinogradov's mean value theorem for trigonometric sums.

Theorem 3.9. *Let the variable* J *be the mean value of the sum* $S(A)$, *namely, let*

$$J = J(P; k, n) = \int_0^1 \cdots \int_0^1 |S(A)|^{2k}\, dA,$$

where k is a natural number, $k \ge 0.5n^2(4\ln n + 2\ln\ln n + 9)$, *and the other notation has the same meaning as previously. The following estimate holds*:

$$J \le n^{30n^3} P^{2k-0.5n(n+1)}.$$

Proof. The statement of the theorem follows from the estimate of l' in Theorem 3.8. □

We point out the following fact. The problem of estimating J can be considered as the limit case of the problem of estimating the number K of solutions of a system of congruences of the form

$$\begin{gathered} x_1 + \cdots + x_k \equiv x_{k+1} + \cdots + x_{2k} \\ \vdots \\ x_1^n + \cdots + x_k^n \equiv x_{k+1}^n + \cdots + x_{2k}^n \\ 1 \le x_1, \dots, x_{2k} \le P. \end{gathered}$$

Indeed, if $Q > k(P^n - 1)$, then $K = J$. So the problem of estimating J is a problem in comparison theory for an incomplete ("short") system of residues. The method considered above reduced this problem to the problem of estimating T, i.e., to a problem in comparison theory for a complete system of residues. Hence, the use of the p-adic proof of the mean value theorem allows one to reduce estimating incomplete trigonometric sums and even the Weyl sums (that can be treated as incomplete trigonometric sums) to estimating complete trigonometric sums. This general consideration underlies the construction of the theory of multiple trigonometric sums.

3.5 Linnik's p-adic method for proving Vinogradov's mean value theorem

As already noted, the p-adic method for proving Vinogradov's mean value theorem (in a weaker form) was first proposed by Yu. V. Linnik. Here we state Linnik's theorem and outline the ideas underlying the proof.

Theorem 3.10. *Suppose that* $n \geq 3$, $\nu = 1/n$, $\sigma = 1 - \nu$, $t = [100n \ln n]$, *and* $Q_1 = P^{\nu}, Q_2 = P^{\nu\sigma}, \ldots, Q_j = P^{\nu\sigma^{j-1}}$, *where* $j = 1, 2, \ldots, t$. *Suppose that* $q_{j1}, q_{j2}, \ldots, q_{jQ'_j}$ *are all primes between* $0.5Q_j$ *and* Q_j *such that* $Q'_j > cQ_j/\ln Q_j$. *Suppose also that the variables* $x_1, \ldots, x_v$ *takes values of the form* $q_{1j_1}, q_{2j_2}, \ldots, q_{tj_t}$. *Then the number* V *of solutions of the system of equations*

$$x_1 + x_2 + \cdots + x_v = M_1,$$
$$\vdots$$
$$x_1^n + x_2^n + \cdots + x_v^n = M_n,$$

where $v = 32tn$, *satisfies the inequality*

$$V \ll P^{v - 0.5n(n+1) + n^{-50}}.$$

Outline of the proof (here we follow Linnik's paper [114]). **1.** We consider a 16-dimensional cube consisting of points of the form $(x_1, \ldots, x_{16n})$, where x_j are the variables used in the theorem. It is easy to see that $0 \leq x_j \leq P$. Let $q_{11}, \ldots, q_{1Q'_1}$ be the primes in the statement of the theorem. A point $M = (x_1, \ldots, x_{16n})$ is called a *singular point of the first order* if there exists precisely one number q_{1j} such that, among any $2n$ numbers $x_1, \ldots, x_{16n}$, there exist two numbers congruent to each other modulo q_{1j}. The number q_{1j} will be called the *modulus belonging to the point* M.

A point $M = (x_1, \ldots, x_{16n})$ is called a *singular point of the second order* if there exist precisely two moduli q_{1j} and q_{1k} $(j \neq k)$, and so on. All singular points whose order is larger than $m = [n/4]$ are said to be *essentially singular*, while the points of zero order are said to be *regular*. The set of all singular points of order j corresponding to given moduli $q_{11}, \ldots, q_{1j}$ will be denoted by the letter $G(q_{11}, \ldots, q_{1j})$.

2. The number V can be represented by the integral

$$V = \int_0^1 \cdots \int_0^1 \Big(\sum_{x_1}{}' \cdots \sum_{x_v}{}' \exp\big\{2\pi i(\alpha_1(x_1 + \cdots + x_v) + \cdots$$
$$+ \alpha_n(x_1^n + \cdots + x_v^n))\big\} \exp\big\{-2\pi i(\alpha_1 M_1 + \cdots + \alpha_n M_n)\big\}\Big)\, d\alpha_1 \ldots d\alpha_n$$
$$\leq \int_0^1 \cdots \int_0^1 \Big|\sum_x{}' \exp\{2\pi i f(x)\}\Big|^{32nt} d\alpha_1 \ldots d\alpha_n,$$

where the prime on the sum means that x takes values of the form

$$q_{1j_1}, \dots, q_{tj_t}, \quad f(x) = \alpha_1 x + \cdots + \alpha_n x^n.$$

By $S_{q_{1j}}$ we denote a sum of the form

$$\sum_x{}' \exp\{2\pi i f(x)\}$$

under the assumption that the first factor in the representation of x is equal to q_{1j}. Let $\sum'$ denote the summation over regular points. Then we have

$$\int_0^1 \cdots \int_0^1 \Big|\sum{}'\Big|^2 \Big|\sum_x{}' \exp\{2\pi i f(x)\}\Big|^{32n(t-1)} d\alpha_1 \dots d\alpha_n$$

$$\ll Q_1^{32n(t-1)-1} P^{32n} Q_1^{-0.5n(n+1)} \sum_{j=1}^{Q_1'} \int_0^1 \cdots \int_0^1 |S_{q_{1j}}|^{32n(t-1)} \, d\alpha_1 \dots d\alpha_n.$$

This estimate can be proved using Hölder's inequality and the following Lemma α.

Lemma α. *Suppose that q is a prime ($n! < q < P^\nu$), $0 < y_j \le P$ ($j = 1, \dots, n$), $y_i \not\equiv y_j \pmod q$ ($i \ne j$). Then the number W of solutions of the system of congruences*

$$\begin{aligned} y_1 + y_2 + \cdots + y_n &\equiv M_1 \pmod{q}, \\ y_1^2 + y_2^2 + \cdots + y_n^2 &\equiv M_2 \pmod{q^2}, \\ &\vdots \\ y_1^n + y_2^n + \cdots + y_n^n &\equiv M_n \pmod{q^n}, \end{aligned}$$

where $M_1, M_1, \dots, M_n$ are fixed numbers, satisfies the estimate

$$W \ll P^n q^{0.5n(n+1)}.$$

3. The number of points in the set $G(q_{11}, \dots, q_{1j})$ can be estimated using the following Lemma β.

Lemma β (V. A. Tartakovskii). *If $V(q_{11}, \dots, q_{1j})$ is the number of points in the set $G(q_{11}, \dots, q_{1j})$ ($j \le m$), then*

$$V(q_{11}, \dots, q_{1j}) \ll P^{16n} (q_{11} \dots q_{1j})^{-14n}.$$

4. By σ_j we denote a trigonometric sum of the form

$$\sum_{x_1}' \cdots \sum_{x_{16n}}' \exp\left\{2\pi i\big(f(x_1)+\cdots+f(x_{16n})\big)\right\},$$

where the sum is taken over all singular points of order j and $\sum(q_{11},\ldots,q_{1j})$ denotes a similar sum, but already over the set $G(q_{11},\ldots,q_{1j})$. Then we have

$$|\sigma_j|^2 \ll Q_1^j\Big(\Big|\sum(q_{11},\ldots,q_{1j})\Big|^2+\cdots+\Big|\sum(q_{1Q_1'-j+1},\ldots,q_{1Q_1'})\Big|^2\Big).$$

The Hölder inequality implies

$$\Big|\sum_x{}' \exp\{2\pi i f(x)\}\Big|^{32n(t-1)} \ll |S_{q_{11}}|^{32n(t-1)}+\cdots+|S_{q_{1j}}|^{32n(t-1)}$$
$$+\,Q_1^{32n(t-1)-1}\big(|S_{q_{1\,j+1}}|^{32n(t-1)}+\cdots+|S_{q_{1Q_1'}}|^{32n(t-1)}\big).$$

For $k>j$, we use Lemmas α and β to obtain

$$Q_1^j\int_0^1\cdots\int_0^1\Big|\sum(q_{11},\ldots,q_{1j})\Big|^2|S_{q_{1k}}|^{32n(t-1)}\,d\alpha_1\ldots d\alpha_n$$
$$\ll Q_1^{-j}P^{16n}Q_1^{-0.5n(n+1)}P^{16n}Q_1^{-14nj}\int_0^1\cdots\int_0^1|S_{q_{1k}}|^{32n(t-1)}\,d\alpha_1\ldots d\alpha_n$$
$$\ll \frac{1}{Q_1^j}P^{32n}Q_1^{-0.5n(n+1)}\int_0^1\cdots\int_0^1|S_{q_{1k}}|^{32n(t-1)}\,d\alpha_1\ldots d\alpha_n.$$

For $k\le j$, estimating the number of terms in $\sum(q_{11},\ldots,q_{1j})$ by Lemma β, we find

$$Q_1^j\int_0^1\cdots\int_0^1\Big|\sum(q_{11},\ldots,q_{1j})\Big|^2|S_{q_{1k}}|^{32n(t-1)}\,d\alpha_1\ldots d\alpha_n$$
$$\ll Q_1^{-j}Q_1^{32n(t-1)-1}P^{32n}Q_1^{-0.5n(n+1)}\int_0^1\cdots\int_0^1|S_{q_{1k}}|^{32n(t-1)}\,d\alpha_1\ldots d\alpha_n.$$

5. Let σ be a trigonometric sum,

$$\sigma=\sum_{x_1}''\cdots\sum_{x_{16n}}''\exp\{2\pi i(f(x_1)+\cdots+f(x_{16n}))\},$$

where the sum is taken over essentially singular points. The number of terms in this sum does not exceed

$$P^{16n}Q_1^{-14n[n/4]}Q^{[n/4]([n/4]-1)}.$$

Hence we have

$$\int_0^1 \cdots \int_0^1 |\sigma|^2 |S|^{32n(t-1)}\, d\alpha_1 \ldots d\alpha_n \ll Q_1^{32n(t-1)-1} P^{32n}$$

$$\times\, Q_1^{-28n[n/4]+2[n/4]([n/4]-1)} \sum_{j=1}^{Q_1'} \int_0^1 \cdots \int_0^1 |S_{q_{1j}}|^{32n(t-1)}\, d\alpha_1 \ldots d\alpha_n.$$

6. Since we have

$$|S|^{32n} \ll |\sigma_1|^2 + |\sigma_2|^2 + \cdots + |\sigma_m|^2 + |\sigma|^2 + \Big|\sum{}' \Big|^2,$$

collecting the above estimates, we find

$$V \ll Q_1^{32n(t-1)-1} P^{32n} Q_1^{-0.5n(n+1)} \sum_{j=1}^{Q_1'} \int_0^1 \cdots \int_0^1 |S_{q_{1j}}|^{32n(t-1)}\, d\alpha_1 \ldots d\alpha_n.$$

7. Applying the same argument to the sum S_{1j}, we pass to an inequality containing a power of $|S_{q_{1j},q_{2k}}|$ in the right-hand side and, continuing this procedure, arrive at the statement of the theorem.

A similar estimate can be obtained from this theorem for $J = J(P; n, k)$. To this end, in the trigonometric sum in the integrand of J, we must perform a shift of the summation variable of the form $x \to x + x'$, where x' takes values of the variables in the theorem. Applying Hölder's inequality, we can estimate a variable similar to V; then proceeding by iterations, we obtain the estimate

$$J = J(P; n, k) \ll P^{2k-0.5n(n+1)+n^{-50}}.$$

where $k \geq 16tn$ and $t = [100n \log n]$.

3.6 Estimate for Vinogradov's integral for k small relative to n^2

We now prove a generalization of Theorem 3.6 from which, as consequences, we obtain estimates of $J(P; n, k)$ for k small (large) relative to n^2.

Theorem 3.11. *Let $\tau, r_1, \ldots, r_\tau, m, k$ be natural numbers, where $\tau \geq 1$ and $1 = r_1 \leq r_2 \leq \cdots \leq r_\tau \leq n$. Further, set*

$$\Delta(\tau) = \Big(n - \frac{r_\tau - 1}{2}\Big) + \Big(1 - \frac{1}{r_\tau}\Big)\Big(n - \frac{r_{\tau-1} - 1}{2}\Big) + \cdots$$
$$+ \Big(1 - \frac{1}{r_\tau}\Big)\Big(1 - \frac{1}{r_{\tau-1}}\Big) \cdots \Big(1 - \frac{1}{r_2}\Big)\Big(n - \frac{r_1 - 1}{2}\Big),$$

$$\varkappa_\tau = \sum_{j=1}^{\tau} \left(r_j^2 + \Delta(j)\right).$$

Then the following estimate holds for $k \geq n\tau$ and $P \geq 1$:

$$J = J(P; n, k) \leq n^{2\Delta(\tau) r_\tau}\, 2^{\varkappa_\tau}\, (8k)^{2n\tau}\, P^{2k-\Delta(\tau)}.$$

Proof. Without loss of generality, we can assume that $k = n\tau$ and $n \geq 2$. We proceed by induction on the parameter τ. The assertion of the theorem holds for $\tau = 1$, since, in this case, we have $k = n$, $r_1 = 1$, $\varkappa_1 = n + 1$, and $\Delta(1) = n$, and the estimate has the form

$$J \ll 8^{n+1} n^{4n} P^{2k-n},$$

which is somewhat weaker than the following easily obtained estimate:

$$J \ll n!P^{2k-n}.$$

Now we assume that the theorem holds for $\tau = m \geq 1$ and prove that it holds for $\tau = m + 1$. We apply the estimate in Lemma 3.10 with $r = r_{m+1}$ to the number $J(P; n, n(m + 1))$. We assume that $r_{m+1} \geq 2$, since otherwise we have $r_1 = r_2 = \cdots = r_{m+1} \geq 1$ and $\Delta(m + 1) = n$, and our assertion becomes trivial. We obtain the inequality

$$\begin{aligned} J(P; n, n(m+1)) \leq 4k^{2n} R^{2k-2n+r_{m+1}(r_{m+1}-1)/2} P^n J(P_1; n, k-n) \qquad (3.35) \\ + (2n)^{2kr_{m+1}} P^k, \quad k = n(m+1). \end{aligned}$$

We apply the estimate in the theorem with $\tau = m$ to $J(P_1; n, k - n)$:

$$J(P_1; n, k-n) \leq n^{2\Delta(m) r_m} 2^{\varkappa_m} (8nm)^{2nm} P_1^{2k-2n-\Delta(m)}.$$

It remains to substitute this estimate into (3.35) and to show that the resulting estimate is no weaker than the estimate in the theorem for $\tau = m + 1$. We note that we can assume that $P > (4k)^2$, since otherwise the estimate in the theorem is weaker than the trivial estimate P^{2k}. In fact, we always have $\Delta(m + 1) \leq n(m + 1)$, and hence for $P \leq (4k)^2$ we have

$$P^{2k} \leq k^{2n(m+1)} n^{2r_{m+1}\Delta(m+1)} P^{2k-\Delta(m+1)}.$$

In this case

$$pP^{-1} \leq 2P^{-1+1/(r_{m+1})} \leq 2P^{-1/2} < (2k)^{-1},$$

and so

$$\begin{aligned} P_1^{2k-2n-\Delta(m)} &= (Pp^{-1} + 1)^{2k-2n-\Delta(m)} \\ &\leq P^{2k-2n-\Delta(m)} p^{-2k+2n+\Delta(m)} (1 + 1/(2k))^{2k} \\ &\leq 3P^{2k-2n-\Delta(m)} p^{-2k+2n+\Delta(m)}. \end{aligned}$$

Consequently, the first term on the right-hand side in (3.35) does not exceed

$$\begin{aligned}
&12k^{2n}n^{2\Delta(m)r_m}2^{\varkappa_m}(8nm)^{2nm}P^{\Delta(m)+r_{m+1}(r_{m+1}-1)/2}P^{2k-2n-\Delta(m)}\\
&\quad\le 12k^{2n}2^{\varkappa_m+\Delta(m)+r_{m+1}(r_{m+1}-1)/2}n^{2\Delta(m)r_m}(8nm)^{2nm}\\
&\qquad\times P^{2k-\left(\Delta(m)+n+1/2-(r_{m+1}/2+\Delta(m)/r_{m+1})\right)}\\
&\quad\le \frac{1}{2}n^{2\Delta(m+1)r_{m+1}}2^{\varkappa_{m+1}}(8k)^{2n(m+1)}P^{2k-\Delta(m+1)},
\end{aligned}$$

since it follows from the definition of $\Delta(\tau)$ and $\varkappa_\tau$ that

$$\Delta(m+1)=\Delta(m)+n+\frac{1}{2}-\left(\frac{r_{m+1}}{2}+\frac{\Delta(m)}{r_{m+1}}\right),$$

$$\varkappa_{m+1}>\varkappa_m+\Delta(m)+\frac{r_{m+1}(r_{m+1}-1)}{2}.$$

Now we show that the second term in (3.35) does not also exceed

$$\frac{1}{2}n^{2\Delta(m+1)r_{m+1}}2^{\varkappa_{m+1}}(8k)^{2n(m+1)}P^{2k-\Delta(m+1)}.$$

Since we always have $\Delta(m+1)\le k$, we can assume that $P>(2n)^{2r_{m+1}}$, because otherwise the first factor in the estimate in the theorem exceeds the lower bound in P, and the assertion becomes trivial. Thus we have

$$P>(2n)^{2r_{m+1}},\quad \left((2n)^{-2r_{m+1}}P\right)^{k-\Delta(m+1)}\ge 1,$$

$$(2n)^{2kr_{m+1}}P^k\left((2n)^{-2r_{m+1}}P\right)^{k-\Delta(m+1)}\ge (2n)^{2kr_{m+1}}P^k,$$

i.e.

$$(2n)^{2kr_{m+1}}P^k\le n^{2\Delta(m+1)r_{m+1}}P^{2k-\Delta(m+1)},$$

$$\frac{1}{2}(2n)^{2\Delta(m+1)r_{m+1}}P^k\le\frac{1}{2}n^{2\Delta(m+1)r_{m+1}}2^{\varkappa_{m+1}}(8k)^{2n(m+1)}P^{2k-\Delta(m+1)}.$$

We have thereby obtained the desired estimate for $J(P;n,n\tau)$ with $\tau=m+1$. The proof of the theorem is complete. □

Now let us estimate $J(P;n,k)$ for k relatively small with respect to n^2.

Theorem 3.12. *Let $r_1=1$, and let $r_{m+1}=[\sqrt{2mn}]$ for all m in the interval $1\le m\le n/2$. Then*

$$\Delta(m)>mn(1-\sqrt{8m/(9n)}).$$

Proof. We note that

$$\Delta(1)=n,\quad \Delta(m)\le mn=\frac{1}{2}(r_{m+1}+\theta)^2.$$

Since

$$\Delta(m+1) = \Delta(m) + n + \frac{1}{2} - \left(\frac{r_{m+1}}{2} + \frac{\Delta(m)}{r_{m+1}}\right),$$

it follows that

$$\Delta(m+1) - \Delta(m) > n - \sqrt{2mn},$$

because

$$\frac{r_{m+1}}{2} + \frac{\Delta(m)}{r_{m+1}} \leq \frac{r_{m+1}}{2} + \frac{(r_{m+1}+\theta)^2}{2r_{m+1}} = \frac{r_{m+1}}{2} + \frac{r_{m+1}}{2} + \theta + \frac{\theta^2}{2r_{m+1}} < \sqrt{2mn} + \frac{1}{2}.$$

Therefore,

$$\sum_{s=1}^{m-1} \big(\Delta(s+1) - \Delta(s)\big) = \Delta(m) - \Delta(1) = \Delta(m) - n > n(m-1) - \sum_{s=1}^{m-1} \sqrt{2sn}$$

$$> n(m-1) - \int_0^m \sqrt{2sn}\, ds = n(m-1) - \sqrt{2n}\,\frac{2}{3} m\sqrt{m}.$$

We hence finally obtain

$$\Delta(m) > mn(1 - \sqrt{8m/(9n)}\,),$$

as required. □

Corollary 3.1. *For every* ε, $0 < \varepsilon < 1/2$, *and* $k = mn$, $k \leq \varepsilon^2 n^2$, *the following estimate holds*:

$$J \ll P^{k(1+\varepsilon)}.$$

Proof. Obviously, it suffices to show that $\Delta(m)$ in the relation $J \ll P^{2k-\Delta(m)}$ satisfies the inequality

$$\Delta(m) \geq k(1-\varepsilon) = mn(1-\varepsilon).$$

It follows from Theorem 3.12 that

$$\Delta(m) \geq mn(1 - \sqrt{8m/(9n)}\,).$$

But since $k = mn \leq \varepsilon^2 n^2$, we have

$$m/n \leq \varepsilon^2, \quad \sqrt{8m/(9n)} \leq \varepsilon\sqrt{8/9} < \varepsilon.$$

Hence

$$\Delta(m) \geq mn(1 - \sqrt{8m/(9n)}\,) > mn(1-\varepsilon) = k(1-\varepsilon),$$

as required. □

Before estimating $J(P; n, k)$ for large values of k, we prove an auxiliary assertion.

Lemma 3.13. *The quantities r_m, $m = 1, \ldots, \tau$, in Theorem* 3.11 *can be chosen so that*

$$\Delta(m+1) - \Delta(m) > n - \sqrt{2\Delta(m)}.$$

Proof. We set $r_1 = 1$ and $r_{m+1} = \sqrt{2\Delta(m)} - \theta, 0 \le \theta < 1$. Then

$$\begin{aligned} \Delta(m) &= \frac{1}{2}(r_{m+1} + \theta)^2, \\ \Delta(m+1) - \Delta(m) &= n + \frac{1}{2} - \left(\frac{r_{m+1}}{2} + \frac{(r_{m+1} + \theta)^2}{2r_{m+1}}\right) \\ &= n + \frac{1}{2} - \sqrt{2\Delta(m)} - \frac{\theta^2}{2r_{m+1}} > n - \sqrt{2\Delta(m)}. \end{aligned}$$

The proof of the lemma is complete. □

We consider the function $\varphi(y)$, $0 \le y < 1$, defined as

$$\varphi(y) = \int_0^y \frac{t\,dt}{1-t} = -y + \ln\frac{1}{1-y}.$$

This function increases from zero to infinity with the argument increasing from zero to one. It is monotone along with its derivatives. We define a function $z(x)$ by the equation $n\varphi(\sqrt{2z}/n^2) = x$.

Theorem 3.13. *If the quantities r_m are chosen as in Lemma* 3.13*, then*

$$\Delta(m) \ge z(m).$$

Proof. We consider the function $m(\Delta)$ that is defined on the range of values of $\Delta(m)$ and is inverse to the latter function, i.e., $m(\Lambda(m)) - m$.

To prove the theorem, it suffices to show that $x(\Delta) \ge m(\Delta)$, where $x(z)$ is the inverse of $z(x)$. The function $m(\Delta(m))$ increases by 1 with m increasing by 1. Since the theorem follows immediately from the definition for $m = 0$, to prove the theorem completely, we need to show that

$$R = x(\Delta(m+1)) - x(\Delta(m)) \ge 1.$$

Applying Lagrange's theorem on finite increments, we obtain

$$R = \big(\Delta(m+1) - (\Delta(m)\big)x'\big(\alpha\Delta(m+1) + (1-\alpha)\Delta(m)\big), \quad 0 \le \alpha \le 1.$$

Since $x'(z)$ is monotonically increasing, it follows from Lemma 3.13 that

$$\begin{aligned} R &> \big(\Delta(m+1) - \Delta(m)\big)x'(\Delta(m)) \\ &> n\big(n - \sqrt{2\Delta(m)}\big)\frac{\sqrt{2\Delta(m)/n^2}}{1 - \sqrt{2\Delta(m)/n^2}}\sqrt{\frac{2}{n^2}}\frac{1}{2\sqrt{\Delta(m)}} = 1, \end{aligned}$$

as required. The proof of the theorem is complete. □

Theorem 3.13 enables us immediately to choose τ for the required lower bound in Vinogradov's theorem, and hence in many cases this theorem is convenient for applications.

Suppose that we would like to obtain the lower bound for $\Delta(\tau) \geq \alpha n^2$. Theorem 3.13 shows that to do this it suffices to take τ to be the least integer such that $\tau \geq n\varphi(\sqrt{\alpha})$. For example, if $\alpha = 1/4$, then $\tau = [n(\ln 2 - 1/2)] + 1$, and if $\alpha = \varepsilon^2$, then $\tau = [n(-\ln(1-\varepsilon) - \varepsilon)] + 1$.

In conclusion, we note that, in general, the fundamental theorem can be used to obtain results which are somewhat sharper than Theorems 3.12 and 3.13. However, in that case both the statements and the computations involved in the proofs become more complicated. For small values of n, one can successively choose r_m in the optimal way.

In particular, in principle, this allows one to estimate trigonometric sums by Vinogradov's method more precisely than by Weyl's method as soon as $n \geq 11$.

Corollary 3.2. *If in Theorem* 3.11 *we set* $r_2 = r_3 = \cdots = r_\tau = n$, *then we obtain the estimate*

$$J(P; n, k) \leq n^{2\Delta(\tau)n}\, 2^{\varkappa}\, (8k)^{2n\tau}\, P^{2k-\Delta},$$

where

$$\Delta = \frac{n(n+1)}{2} - \frac{n^2}{2}\left(1 - \frac{1}{n}\right)^{\tau},$$

$$\varkappa = n^2\tau + \frac{n(n+1)}{2}\tau - \frac{n^2(n-1)}{2}\left(1 - \left(1 - \frac{1}{n}\right)^{\tau}\right) < \frac{3(n+1)^2\tau}{2},$$

i.e., we obtain the statement of Theorem 3.6.

Concluding remarks on Chapter 3. 1. The contents of this chapter was discussed in detail in the Introduction. We only note that the new p-adic method appeared after A. A. Karatsuba (see [72], [74], [75], [73], [76]) studied rational trigonometric sums with denominator equal to the power of a prime. The sums considered by A. G. Postnikov [138], where he considered the boundary of zeros of the Dirichlet L-functions with character whose modulus is equal to the power of a prime, turned out to be especially interesting. The class of such sums and of their generalizations was studied in [76], where they were called the *L-sums*.

2. In number theory, problems modulo the power of a fixed prime were studied by S. M. Rozin [140], Yu. V. Kashirskii [105], M. B. Barban, Yu. V. Linnik, and N. G. Chudakov [37], V. N. Chubarikov [46], and M. M. Petechuk [132].

3. An analog of Waring's problem for congruences described in Section 3.3 led to a local analog of the Hardy–Littlewood hypothesis stating that $G(n) = O(n)$ (see [99], [101], [102]).

Suppose that m is a natural number, $m \geq m_1 > 0$, E_m is a complete system of residues modulo m, $A \subseteq E_m$, $\|A\|$ is the number of elements in A, and $\|A\| \geq 2$. A set A is called a *basis* of E_m of order $k = k(A)$ if each $\ell \in E_m$ can be represented as

$$x_1 + \cdots + x_k \equiv \ell \pmod{m}, \qquad x_1, \ldots, x_k \in A,$$

and there exists an $\ell_1 \in E_m$ such that

$$x_1 + \cdots + x_k \not\equiv \ell_1 \pmod{m}$$

for any $x_1, \ldots, x_{k-1} \in A$. A set A is said to be *regular* (*c-regular*) modulo m if it is a basis of E_m of order $k = k(A)$ and there exists an absolute constant such that

$$k \leq c \log m / \log \|A\|.$$

The function $k = k(A)$ is precisely a local analog of the Hardy–Littlewood function $G(n)$. If A is a regular set, then this analog has a right upper bound (the function $G(n)$ itself does not have such an upper bound), since the inequality

$$k = k(A) > \log m / \log \|A\|$$

holds trivially.

The following assertion can be regarded as a local analog of Waring's problem and the Hardy–Littlewood hypothesis on $G(n)$: for any $\varepsilon > 0$ and any natural number n, there exists an $m_1 = m_1(\varepsilon; n) > 0$ such that the number set $A = \{x^n, 1 \leq x \leq m^\varepsilon\}$ is regular modulo m for any $m \geq m_1$.

The hypothetical estimate $G(n) = O(n)$ readily implies a local analog of Waring's problem for any modulus m. In [99] a local analog of Waring's problem was proved for moduli equal to the powers of fixed prime numbers. In [101] regular sets with special moduli related to $\operatorname{ind} x$ were considered.

4. A problem similar to the problem of the existence of regular sets, but in a more general form, is known in the literature as Rohrbach's problem for finite groups (see M. B. Natanson's paper [125]).

5. If an arbitrary set A is considered, then (as Yu. Belyi noted in his letter to A. A. Karatsuba) for any $\varepsilon > 0$ and $\delta > 0$, for each $m \geq m_1 = m_1(\varepsilon, \delta) = (2\varepsilon^{-1} + 2)^{(1+\delta)/\delta}(\varepsilon^{-1} + 1)$, there exists a set $A \subseteq E_m$, $\|A\| \leq m^\varepsilon$, such that

$$k = k(A) \leq (1 + \delta)\frac{\log m}{\log \|A\|}.$$

The set mentioned by Yu. Belyi consists of the so-called k-regular numbers x, $x = 1, 2, \ldots, m$, $k = [\varepsilon^{-1}]$ (a number x is said to be *k-regular* if the congruences $e_1 \equiv \cdots \equiv e_t \pmod{k}$ hold for this number written in the binary number system, $x = 2^{e_1} + \cdots + 2^{e_t}$).

6. At the end of Section 3.4, we noted that the p-adic proof of the mean value theorem and its further use in estimating the Weyl sums, in fact, reduces estimating

the incomplete (short) trigonometric sums to estimating complete trigonometric sums. This was realized by A. A. Karatsuba in [77], [78], [79], [80], [82], where he estimated the number of solutions to systems of congruences and to incomplete systems of equations.

7. The results obtained by the p-adic method, in particular, the asymptotic formulas for the number of solutions to Diophantine equations of Waring type in numbers having small prime divisors, were delivered by A. A. Karatsuba at the International Congress of Mathematicians in Vancouver [87] (see also [88]).

8. Vinogradov's mean value theorem found numerous applications in various problems in analytic number theory. We mention only the monographs by K. Chandrasekharan [44], by S. M. Voronin and A. A. Karatsuba [166], and by A. A. Karatsuba [98], where some applications of this theorem in the theory of the Riemann zeta function are given, as well as some generalizations of this theorem to algebraic number fields obtained by Y. Eda [61] and I. M. Kozlov [109].

9. Theorems 3.7 and 3.8 were proved by G. I. Arkhipov in [8].

10. Theorem 3.9 was proved by G. I. Arkhipov in [6]; this is a refined version of the theorem proved by A. A. Karatsuba in [86].

11. The statements presented in Section 3.6 were proved by G. I. Arkhipov and A. A. Karatsuba in [17], [18].

12. Some estimates obtained by O. V. Tyrina in [146] make the statements of theorems in Section 3.6 more precise.

Chapter 4

Mean value theorems for multiple trigonometric sums

In this chapter, we use the p-adic method to prove two fundamental theorems in the theory of multiple trigonometric sums. In Theorem 4.1, the variables of summation are equivalent. This restriction simplifies both the statement of the theorem and its proof and allows us to concentrate our attention on the key points of the method.

4.1 The mean value theorem for the multiple trigonometric sum with equivalent variables of summation

Precisely as in the particular (one-dimensional) case $r = 1$, the mean value of the $2k$th power of the modulus of an r-multiple ($r \geq 1$) trigonometric sum gives the number of solutions of a system of Diophantine equations. The system is written as follows: it is symmetric, i.e., its left- and right-hand sides differ only in the names of the variables; the system consists of $m = (n_1 + 1) \dots (n_r + 1)$ equations; the left-hand side of each equation contains k terms of the form $x^u y^v \dots z^w$, where $u, v, \dots, w$ are nonnegative integers, and the number of them does not exceed n (so the number of all possible sets $u, v, \dots, w$ is equal to $m = (n_1 + 1) \dots (n_r + 1)$). All terms in one equation have a fixed set of exponents $u, v, \dots, w$. To avoid introducing extra letters, we reindex the variables as follows: we shall write x_1 instead of x, x_2 instead of $y, \dots,$ and x_r instead of z. To distinguish the terms in equations, we introduce the second subscript on the variables. This subscript is just the number of the term in the equation. Moreover, we number the terms in the left-hand side by even second subscripts and the terms in the right-hand side by odd subscripts. We also denote the exponents $u, v, \dots, w$ by the letters $t_1, t_2, \dots, t_r$. So our system of equations can be written as

$$\sum_{j=1}^{2k} (-1)^j x_{1j}^{t_1} \dots x_{rj}^{t_r} = 0, \quad 0 \leq t_1 \leq n_1, \ \dots, \ 0 \leq t_r \leq n_r. \tag{4.1}$$

Now we assume that each unknown x_{ij} takes all integer values from 1 to P. If we denote the number of solutions of this system of equations by J, then we see that J depends on $P, n_1, \dots, n_r, k, r$, where P is the main parameter. In all our estimates, we assume that $P \to +\infty$, while $n_1, \ldots, n_r, k, r$ are constant. However, our goal is to obtain estimates that are also uniform in $n_1, \ldots, n_r, k, r$. This means that these parameters can increase together with P, but, as we shall see later, not too fast.

Precisely as in the one-dimensional case, it is easy to see that

$$J = J(P; \overline{n}, k, r) = \int_{\Omega} \cdots \int \left| \sum_{x_r=1}^{P} \cdots \sum_{x_r=1}^{P} \exp\{2\pi i F(x_1, \ldots, x_r)\} \right|^{2k} d\Omega,$$

where $\overline{n} = (n_1, \ldots, n_r)$,

$$F(x_1, \ldots, x_r) = \sum_{t_1=0}^{n_1} \cdots \sum_{t_r=0}^{n_r} \alpha(t_1, \ldots, t_r) x_1^{t_1} \ldots x_r^{t_r}.$$

Here Ω is an m-dimensional ($m = (n_1 + 1) \ldots (n_r + 1)$) cube of the form

$$0 \le \alpha(t_1, \ldots, t_r) < 1, \quad 0 \le t_1 \le n_1, \ \ldots, \ 0 \le t_r \le n_r.$$

Our goal is to obtain an estimate of the integral $J(P; \overline{n}, k, r)$ so that this estimate be sharp in the main parameter P for the correct order of the parameter k, (about the exact estimate, see below).

To realize this goal, we use the same p-adic method as in Section 3.4, Chapter 3. We note that at present we cannot obtain the desired result for the integral $J(P; \overline{n}, k, r)$ in any other way.

First, we outline the scheme for proving the mean value theorem for multiple sums. A preliminary analysis shows that, using the p-adic method, we can reduce estimating the variable J to estimating the number of solutions of some systems of congruences and to estimating J_1, where J_1 is a variables of the same nature as J, but with a fewer number of parameters. In this case, we can vary the parameters of the system of congruences.

However, this is not sufficient for obtaining the desired estimate of J, because the number of solutions of the system of congruences will be too large. However, it should be noted that if not all the values of the set of variables are admitted in the integral J, then we obtain several conditions on the unknown variables in the system of congruences, and we hope that these conditions decrease the number of its solutions till some admissible value. Moreover, it is also necessary to estimate the integral J over the remaining set of values of the sets of variables. In the one-dimensional case, the required condition can be imposed rather simply (see the partition of solutions of the system of equations into sets of the first and second kind in Section 3.4).

The main difficulty consists precisely in finding such a condition in the multidimensional case. Here the role of this condition is played by the condition that the set of variables in the integral J is regular, which we introduce later.

The further argument significantly repeats the proof of the one-dimensional theorem, although, of course, the corresponding calculations are more cumbersome and sometimes require other technical solutions.

4.1.1 Definitions

For convenience, we introduce several new abbreviations. We arrange the terms determining $F(x_1, \dots, x_r)$, i.e., the monomials

$$\alpha(t_1, \dots, t_r) x_1^{t_1} \dots x_r^{t_r},$$

in ascending order of the numbers $t_1+(n_1+1)t_2+\dots+(n_1+1)\dots(n_{r-1}+1)t_r$. By A we denote the vector whose coordinates are $\alpha(t_1, \dots, t_r)$ in the same order as they enter $F(x_1, \dots, x_r)$. By $S(A)$ we denote the multiple trigonometric sum in the integrand of $J(P; \overline{n}, k, r)$. We arrange the integers $\lambda(t_1, \dots, t_r)$ for $0 \le t_1 \le n_1, \dots, 0 \le t_r \le n_r$ in the same order as $\alpha(t_1, \dots, t_r)$. By Λ we denote the vector composed of $\lambda(t_1, \dots, t_r)$ arranged in this order.

We consider the system of equations similar to Eqs. (4.1), but with arbitrary not necessarily zero right-hand sides:

$$\sum_{j=1}^{2k} (-1)^j x_{1j}^{t_1} \dots x_{rj}^{t_r} = \lambda(t_1, \dots t_r), \quad 0 \le t_1 \le n_1, \dots, 0 \le t_r \le n_r, \tag{4.2}$$

$$1 \le x_{1j}, \dots, x_{rj} \le P, \quad j = 1, 2, \dots, 2k.$$

We denote the number of solutions of this system by $J(P; \overline{n}, k, r; \Lambda)$. As pointed out above, system (4.1) is said to be *complete*, while a system of equations similar to (4.1), but without several equations, is said to be *incomplete*.

Definition 4.1. If $\overline{x} = (a_1, \dots, a_s)$ and $\overline{y} = (b_1, \dots, b_s)$ are two vectors with integer coordinates, then the congruence $\overline{x} \equiv \overline{y} \pmod q$ means that $a_i \equiv b_i \pmod q$ $(i = 1, \dots, s)$.

Definition 4.2. We consider the matrix

$$M = (x_{1j}^{t_1} \dots x_{rj}^{t_r}), \quad 0 \le t_1 \le n_1, \dots, 0 \le t_r \le n_r, \ j = 1, 2, \dots, k,$$

(so the matrix M has $m = (n_1 + 1) \dots (n_r + 1)$ columns and k rows). We shall say that the matrix M *corresponds to the vectors* $\overline{x}_1 = (x_{11}, \dots, x_{r1}), \dots, \overline{x}_k = (x_{1k}, \dots, x_{rk})$ and, conversely, the vectors $\overline{x}_1, \dots, \overline{x}_k$ are said to be *corresponding to the matrix* M.

Definition 4.3. Let k be a natural number. A set of vectors $\overline{x}_1, \dots, \overline{x}_k$ is said to be *regular modulo* q if the rank (modulo q) of the matrix M corresponding to these vectors is maximal. Otherwise, the above set is said to be *singular*.

We note that if $k \ge m$ and the vectors $\overline{x}_1, \dots, \overline{x}_k$ are regular modulo q, then the rank modulo q of the matrix M corresponding to these vectors is equal to m.

For brevity, the solutions of system (4.1) that are regular (singular) sets are also said to be *regular* (*singular*).

4.1.2 Simple lemmas

Here we state and prove two simple lemmas.

Lemma 4.1. *The following relations hold*:

(a) $J = J(P; \overline{n}, k, r; \Lambda) = \int \dots \int_{\Omega} |S(A)|^{2k} \exp\{-2\pi i A \times \Lambda\} \, dA;$

(b) $J = J(P; \overline{n}, k, r; \Lambda) \le J(P; \overline{n}, k, r) \le P^{2k_1 r} J(P; \overline{n}, k - k_1, r);$

(c) $\sum_{\Lambda} J(P; \overline{n}, k, r; \Lambda) = P^{2kr};$

(d) $|S(A)|^{2k} = \sum_{\Lambda} J(P; \overline{n}, k, r; \Lambda) \exp\{2\pi i A \times \Lambda\};$

(e) $J(P; \overline{n}, k, r) \ge (2k)^{-m} P^{2kr - 0.5m(n_1 + \dots + n_r)}.$

Proof. For integer λ, we have

$$\int_0^1 \exp\{2\pi i \alpha \lambda\} d\alpha = \begin{cases} 1 & \text{if } \lambda = 0, \\ 0 & \text{if } \lambda \ne 0. \end{cases}$$

This relation implies assertion (a) if we raise the absolute value of the integrand to the power $2k$ and integrate over Ω; assertion (b) follows from the fact that the absolute value of the integral does not exceed the value of the integral of the absolute value of the integrand; assertion (c) follows from the fact that the left-hand side of the relation is the number of all possible sets $\overline{x}_1, \dots, \overline{x}_{2k}$ of system (4.1), i.e., it is equal to P^{2kr}; to prove assertion (d) we first raise the sum $S(A)$ to the power $2k$ and then collect similar terms with $\exp\{2\pi i A \times A\}$; assertion (e) follows from assertions (b) and (c).□

Lemma 4.2. (a) *If the vectors* $\overline{x}_1, \dots, \overline{x}_{2k}$ *form a solution of system* (4.1), *then for any vector* $\overline{a} = (a_1, \dots, a_r)$, *the vectors* $\overline{x}_1 + \overline{a}, \dots, \overline{x}_{2k} + \overline{a}$ *also form a solution of system* (4.1).

(b) *If the vectors* $\overline{x}_1, \dots, \overline{x}_k$ *form a regular* (*singular*) *set modulo* q, *then for any vector* $\overline{a} = (a_1, \dots, a_r)$, *the vectors* $\overline{x}_1 + \overline{a}, \dots, \overline{x}_{2k} + \overline{a}$ *also form regular* (*singular*) *set modulo* q.

Proof. (a) Let $\overline{x}_j = (x_{1j}, \dots, x_{rj}$ be a solution of Eqs. (4.1). Removing the parentheses, we find

$$\sum_{j=1}^{2k} (-1)^j (x_{1j} + a_1)^{t_1} \dots (x_{rj} + a_r)^{t_r}$$

$$= \sum_{j=1}^{2k} (-1)^j \sum_{v_1=0}^{t_1} \binom{t_1}{v_1} a_1^{v_1} x_{1j}^{t_1 - v_1} \cdots \sum_{v_r=0}^{t_r} \binom{t_r}{v_r} a_r^{v_r} x_{rj}^{t_r - v_r} =$$

$$= \sum_{v_1=0}^{t_1} \cdots \sum_{v_r=0}^{t_r} \binom{t_1}{v_1} a_1^{v_1} \cdots \binom{t_r}{v_r} a_r^{v_r} \sum_{j=1}^{2k} (-1)^j x_{1j}^{t_1-v_1} \cdots x_{rj}^{t_r-v_r} = 0.$$

The proof of assertion (a) is complete.

Remark 4.1. Assertion (a) remains valid if congruences modulo any arbitrary value of q are considered instead of (4.1).

(b) If the vectors $\overline{x}_1, \ldots, \overline{x}_k$ form a singular set, then the rows of the matrix M corresponding to this set are linearly dependent modulo q. Since the matrix M has a special form, this statement is equivalent to the existence of a polynomial in r variables, $F(\overline{y}) = F(y_1, \ldots, y_r)$, such that the coefficient of the highest-order (in lexicographic order) term in this polynomial is equal to 1 and the congruence

$$F(\overline{x}_s) \equiv 0 \pmod{q} \tag{4.3}$$

holds for any $\overline{x}_s$ $(s = 1, \ldots, k)$ from the set mentioned above.

In this case, the degree of the polynomial does not exceed n_i in each variable. Obviously, the polynomial $G(\overline{y}) = F(\overline{y} - \overline{a})$ has the same coefficient of the highest-order (in lexicographic order) term as the polynomial $F(\overline{y})$, but relation (4.3) implies

$$G(\overline{x}_s + \overline{a}) \equiv 0 \pmod{q},$$

i.e., the vectors $\overline{x}_s + \overline{a}, \ldots, \overline{x}_k + \overline{a}$ also form a singular set.

If the original set is regular, then it remains to be regular under the shift by $\overline{a}$, and performing the shift by $-\overline{a}$, we return to the original set. Assertion (b) is proved. □

Lemma 4.3. *Let q be a prime number, and let T_0 be the number of solutions of the system of congruences*

$$\sum_{j=1}^{2m} (-1)^j y_{1j}^{t_1} \cdots y_{rj}^{t_r} \equiv 0 \pmod{q},$$

$$0 \le t_1 \le n_1, \ldots, 0 \le t_r \le n_r, \quad m = (n_1 + 1) \ldots (n_r + 1),$$

where each unknown variable y_{ij} runs through the values of the complete system of residues modulo q. Then T_0 satisfies the estimate

$$T_0 \le (m-1)! q^{2mr-m+1}.$$

Proof. We write the variable T_0 as

$$T_0 = q^{-m} \sum_A \Big| \sum_{y_1=1}^{q} \cdots \sum_{y_r=1}^{q} \exp\{2\pi i F_A(y_1, \ldots, y_r)/q\} \Big|^{2m},$$

where

$$F_A(y_1, \dots, y_r) = \sum_{t_1=0}^{n_1} \cdots \sum_{t_r=0}^{n_r} a(t_1, \dots, t_r) x_1^{t_1} \dots x_r^{t_r};$$

here A is an integer-valued set consisting of numbers $a(t_1, \dots, t_r)$; $\sum_A$ denotes summation over all sets A that are different modulo q. We make the following change of summation variables:

$$\begin{aligned} y_1 = z_1, \quad y_2 &= z_2 + z_1^{n_1+1}, \\ y_3 &= z_3 + z_1^{(n_1+1)(n_2+1)}, \\ &\vdots \\ y_r &= z_r + z_1^{(n_1+1)(n_2+1)\dots(n_{r-1}+1)}. \end{aligned}$$

If the coordinates y_i of the vector $(y_1, \dots, y_r)$ run through complete systems of residues modulo q, then the coordinates z_j of the vector $(z_1, \dots, z_r)$ also run through complete systems of residues modulo q, and conversely. Therefore, we have

$$T_0 = q^{-m} \sum_A \Bigg| \sum_{z_1=1}^{q} \cdots \sum_{z_r=1}^{q} \exp\{2\pi i F_A(z_1, z_2 + z_1^{n_1+1}, \dots, \\ z_r + z_1^{(n_1+1)\dots(n_{r-1}+1)})/q)\} \Bigg|^{2m}.$$

Applying Hölder's inequality (Lemma A.1), we obtain

$$T_0 \le q^{(r-1)(2m-1)} \sum_{z_2=1}^{q} \cdots \sum_{z_r=1}^{q} V, \tag{4.4}$$

where

$$V = q^{-m} \sum_A \Bigg| \sum_{z=1}^{q} \exp\{2\pi i F_A(z, z_2 + z^{n_1+1}, \dots, z_r + z^{(n_1+1)\dots(n_{r-1}+1)})/q\} \Bigg|^{2m}.$$

The variable V is equal to the number of solutions of the following system of congruences (for fixed $z_2, \dots, z_r$):

$$\sum_{j=1}^{2m} (-1)^j z_{1j}^{t_1} (z_2 + z_{1j}^{n_1+1})^{t_2} \dots (z_r + z_{1j}^{(n_1+1)\dots(n_{r-1}+1)})^{t_r} \equiv 0 \pmod{q}, \tag{4.5}$$

$$0 \le t_1 \le n_1, \ 0 \le t_2 \le n_2, \ \dots, \ 0 \le t_r \le n_r.$$

By Lemma 4.2 (a), together with the solution $\overline{x}_j = (x_{1j}, \dots, x_{rj})$ $(j = 1, \dots, 2m)$ of system (4.5), where $x_{1j} = z_{1j}$, $x_{2j} = z_{1j}^{n_1+1}, \dots, x_{rj} = z_{1j}^{(n_1+1)\dots(n_{r-1}+1)}$, the set of vectors $\overline{x}_1 + \overline{a}, \dots, \overline{x}_{2m} + \overline{a}$, where $\overline{a} = (0, -z_2, \dots, -z_r)$, is also a solution of system (4.5), and conversely, i.e., system (4.5) is equivalent to the system

$$\sum_{j=1}^{2m} (-1)^j z_{1j}^{t_1+t_2(n_1+1)+\cdots+t_r(n_1+1)\dots(n_{r-1}+1)} \equiv 0 \pmod q.$$

However, since for $0 \le t_1 \le n_1, \dots, 0 \le t_r \le n_r$, the sum

$$t_1 + t_2(n_1+1) + \cdots + t_r(n_1+1)\dots(n_{r-1}+1)$$

turns without repetitions through all integer values in the interval $0 \le t \le m-1$, we can rewrite the last sum of congruences as

$$\sum_{j=1}^{2m} (-1)^j y_j^t \equiv 0 \pmod q, \quad 0 \le t \le m-1. \tag{4.6}$$

We prove that the number of solutions of system (4.6) does not exceed $(m-1)!q^{m+1}$. We arbitrarily fix variables with odd numbers and a variable with the number $2m$. Then, for some $\lambda_1, \dots, \lambda_{m-1}$, the remaining $m-1$ variables $y_2, y_4, \dots, y_{2m-2}$ satisfy the system

$$\sum_{j=1}^{m-1} y_{2j}^t \equiv \lambda_t \pmod q, \quad 1 \le t \le m-1. \tag{4.7}$$

If $q \ge m$, then system (4.7) has at most $(m-1)!$ solutions (see the proof of Lemma 3.9 in Section 3.4, Chapter 4). But if $q < m$, then the number of solutions of system (4.7) can be estimated as follows: we omit all the congruences for which $t \ge q$. Obviously, the number of solutions can only increase after this. To estimate the number of solutions of this system, we use the last estimate of the variable V (where the values of the parameters are changed appropriately). We find

$$V \le (q-1)!q^{2m-q+1} \le (m-1)!q^{m+1}.$$

Substituting this estimate into (4.4), we obtain the desired estimate,

$$T_0 \le (m-1)!q^{2mr-m+1}.$$

The proof of the lemma is complete. □

4.1.3 Lemma on the number of solutions of a complete system of congruences

In this section we prove a fundamentally important lemma on complete systems of congruences. For simplicity, we assume that $n_1 = \cdots = n_r = n$. The case of arbitrary $n_1, \dots, n_r$ will be studied in Section 4.3 in general situation.

Lemma 4.4. *Let p be a prime, and let T be the number of solutions of the system of congruences*

$$\sum_{j=1}^{2m}(-1)^j x_{1j}^{t_1}\dots x_{rj}^{t_r} \equiv 0 \pmod{p^{t_1+\dots+t_r}}, \quad 0 \le t_1,\dots,t_r \le n, \tag{4.8}$$

where $B \le x_{sj} < B + p^{rn}$ $(s = 1,\dots,r;\ j = 1,\dots,2m)$ and the vectors $\overline{x}_j = (x_{1j},\dots,x_{rj})$ $(j = 2,4,\dots,2m)$ satisfy the regularity condition modulo p. Then

$$T \le m!\,p^{2mr^2n-0.5rnm}.$$

Remark 4.2. In general, the proof of the lemma on complete systems is as follows. First, each vector is represented p-adically in the form

$$\overline{x}_j = \overline{x}_{j0} + p\overline{x}_{j1} + \dots + p^{rn-1}\overline{x}_{j\,rn-1}.$$

Next, the fact that $\overline{x}_1,\dots,\overline{x}_{2m}$ satisfy (4.8) is used to derive necessary conditions on the coordinates of the vector $\overline{x}_j$ written p-adically, i.e., conditions on the vectors $\overline{x}_{j\nu}$ $(\nu = 0,1,\dots,rn-1)$. For each fixed $\nu \ge 1$, these conditions say that the vectors $\overline{x}_{j\nu}$ satisfy a certain system of linear congruences, where the rank of the matrix of coefficients of the system is maximal, since the set $\overline{x}_2,\dots,\overline{x}_{2m}$ is regular modulo p.

The fact that the rank of the matrix of coefficients is maximal allows us to estimate the number T_ν, i.e., the number of admissible sets $\overline{x}_{1\nu},\dots,\overline{x}_{2m\nu}$ for $\nu \ge 1$. But if $\nu = 0$, then T_ν can be estimated by using Lemma 4.3. Since we have the inequality

$$T \le T_0T_1\dots T_{r\,n-1},$$

we obtain the final result multiplying together the estimates for T_ν for all ν.

Proof. The unknowns in the system of congruences (4.8) run through a complete system of residues modulo p^{nr}, the regularity condition modulo p is independent of which representatives of the residue classes modulo p are taken, and the congruences in (4.8) are taken modulo p^s $(s \le nr)$. Hence the number of solutions of the complete system of congruences (4.8) is independent of precisely what the integers run through a complete system of residues modulo p^{nr}. So we can set $B = 0$. We shall further assume that $p > n$. In the case $p \le n$, there are no solutions to (4.8) that satisfy the regularity condition modulo p.

In fact, if $n \ge p$, then the matrix M has a row $(x_{12}, x_{14},\dots,x_{12m})$ and a row $(x_{12}^p, x_{14}^p,\dots,x_{12m}^p)$; obviously, these rows are linearly dependent modulo p, so that M has less than maximal rank modulo p.

We write each unknown in the form $x_{sj} = x_{sj0} + px_{sj1} + \dots + p^{rn-1}x_{sjr\,n-1}$, $0 \le x_{sj0},\dots,x_{sjr\,n-1} \le p-1$ $(s = 1,\dots,r,\ j = 1,\dots,2m)$ and find necessary conditions that are satisfied by the variables $x_{sj\nu}$. The congruences in (4.8) for which

$t_1 + \cdots + t_r \geq 1$ are satisfied modulo p. Since $x_{sj} \equiv x_{sj0} \pmod{p}$, it follows that the unknowns x_{sj0} satisfy the system of congruences

$$\sum_{j-1}^{2m} (-1)^j x_{1j0}^{t_1} \cdots x_{rj0}^{t_r} \equiv 0 \pmod{p}, \tag{4.9}$$

where $0 \leq t_1, \ldots, t_r \leq n$, $t_1 + \cdots + t_r \geq 1$, and the unknowns $\overline{x}_{20}, \ldots, \overline{x}_{2m0}$ form a regular set. We let T_0 denote the number of solutions of this system.

Let us estimate T_0. We omit the regularity conditions on the unknowns in (4.9). Obviously, $T_0 \leq T'$, where, by Lemma 4.3, the value of T' does not exceed $(m - 1)!p^{2mr-m+1}$, i.e.,

$$T_0 \leq (m-1)!p^{2mr-m+1}.$$

Let now $\nu \geq 1$. We set

$$u_{sj\nu} = \sum_{\mu=0}^{\nu} p^{\mu} x_{sj\mu}.$$

For a fixed ν ($1 \leq \nu \leq m-1$), we consider the system of congruences (we denote this system by the symbol W_ν)

$$\sum_{j=1}^{2m} (-1)^j u_{1j\nu}^{t_1} \cdots u_{rj\nu}^{t_r} \equiv \lambda(t_1, \ldots, t_r) \pmod{p^{\nu-1}},$$

where $t_1 + \cdots + t_r \geq \nu + 1$, $0 \leq t_1, \ldots, t_r \leq n$, and moreover, the unknowns $\overline{u}_{j\nu}$ satisfy the regularity condition modulo p. We let $T(W_\nu)$ denote the number of its solutions.

It is obvious that if the unknowns $\overline{x}_j$ satisfy system (4.8), then the unknowns $\overline{u}_{j\nu}$ satisfy the system W_ν. Next, we fix an arbitrary solution of the system $W_{\nu-1}$ and find conditions that must be satisfied by the unknowns $\overline{u}_{j\nu}$ in this case. We have

$$u_{sj\nu} = u_{sj\,\nu-1} + p^{\nu} x_{sj\nu}.$$

Hence

$$u_{1j\nu}^{t_1} \cdots u_{rj\nu}^{t_r} \equiv u_{1j\,\nu-1}^{t_1} \cdots u_{rj\,\nu-1}^{t_r}$$
$$+ p^{\nu} \sum_{s=1}^{r} t_s u_{1j\,\nu-1}^{t_1} \cdots u_{s-1\,j\,\nu-1}^{t_{s-1}} u_{sj\,\nu-1}^{t_s-1} u_{s+1\,j\,\nu-1}^{t_{s+1}} \cdots u_{rj\,\nu-1}^{t_r} x_{sj\nu} \pmod{p^{\nu+1}},$$

where we have the corresponding term in the last sum to be zero for $t_s = 0$. Consequently, modulo p^{ν}, we have the system of congruences

$$\sum_{j=1}^{2m} (-1)^j u_{1j\nu}^{t_1} \cdots u_{rj\nu}^{t_r} \equiv \lambda(t_1, \ldots, t_r) \pmod{p^{\nu}},$$

where $t_1 + \cdots + t_r \geq \nu + 1$ and $0 \leq t_1, \ldots, t_r \leq n$. Therefore, for some fixed $\lambda'(t_1, \ldots, t_r)$, the system W_ν is equivalent to the system of linear congruences

$$\sum_{j=1}^{2m}(-1)^j \sum_{s=1}^{r} t_s \Big(\prod_{q=1}^{r} u_{qj\,\nu-1}^{t_q}\Big) u_{sj\nu-1}^{-1} x_{sj\nu} \equiv \lambda'(t_1, \ldots, t_r) \pmod p,$$

where $t_1 + \cdots + t_r \geq \nu + 1$ and $0 \leq t_1, \ldots, t_r \leq n$. We let T_ν denote the number of solutions to this system. Then

$$T(W_\nu) \leq T_\nu T(W_{\nu-1}).$$

Thus if we are given an estimate for $T(W_{\nu-1})$, then to estimate $T(W_s)$, it suffices to estimate T_ν. To do this, we construct r subsystems of congruences from the system of congruences W_ν. The first subsystem includes those congruences for which $t_1 \geq \nu+1$ and $t_2 = \cdots = t_r = 0$. The second subsystem includes those congruences for which $t_1+t_2 \geq \nu+1$, $t_2 \neq 0$, and $t_3 = \cdots = t_r = 0$. The $(r-1)$st subsystem includes those for which $t_1 + t_2 + \cdots + t_{r-1} \geq \nu + 1$, $t_{r-1} \neq 0$, and $t_r = 0$, and the rth subsystem includes those for which $t_1 + \cdots + t_r \geq \nu + 1$ and $t_r \neq 0$.

We let $R_r(\nu)$ denote the number of solutions in integers $t_1, \ldots, t_r$ of the inequalities $t_1 + \cdots + t_r \geq \nu$ and $0 \leq t_1, \ldots, t_r \leq n$.

We note that the first subsystem consists of $R_1(\nu + 1)$ congruences, the second consists of $R_2(\nu + 1) - R_1(\nu + 1)$ congruences, the $(r - 1)$st subsystem consists of $R_{r-1}(\nu + 1) - R_{r-2}(\nu + 1)$ congruences, and the rth subsystem consists of $R_r(\nu + 1) - R_{r-1}(\nu + 1)$ congruences.

We shall estimate T_ν as follows. For the first subsystem of congruences, we estimate the number of its solutions $x_{1j\nu}$ $(j = 1, \ldots, 2m)$. Next, we fix the $x_{1j\nu}$ and for the second subsystem, we find an estimate for the number of its solutions $x_{2j\nu}$ $(j = 1, \ldots, 2m)$. We next fix $x_{1j\nu}, x_{2j\nu}, \ldots, x_{s-1\,j\nu}$ $(j = 1, \ldots, 2m)$ and find the number of solutions of the sth subsystem. Let us consider the first system of congruences

$$\sum_{j=1}^{2m}(-1)^j x_{1j0}^{t-1} x_{1j\nu} \equiv \lambda'(t, 0, \ldots, 0) \pmod p, \qquad n \geq t \geq \nu + 1.$$

Because of the regularity condition modulo p, the congruences of this subsystem form a system of linearly independent congruences modulo p, i.e., the matrix of its coefficients has maximal rank modulo p. Hence, we can find $u = R_1(\nu + 1)$ indices $1 \leq j_1 < j_2 < \cdots < j_u \leq 2m$ such that the determinant of the matrix

$$\begin{vmatrix} x_{1j_10}^{\nu+1} & x_{1j_20}^{\nu+1} & \cdots & x_{1j_u0}^{\nu+1} \\ \cdots & \cdots & \cdots & \cdots \\ x_{1j_10}^{n} & x_{1j_20}^{n} & \cdots & x_{1j_u0}^{n} \end{vmatrix}$$

is not congruent to zero modulo p. Thus, by adding certain values from a complete system of residues modulo p to the unknowns $x_{1j\nu}$ $(j \neq j_s, s = 1, \ldots, u)$ in the first subsystem, we uniquely determine $x_{1j_1\nu}, \ldots, x_{1j_u\nu}$ $(j = 1, \ldots, 2m)$. This implies that the number of solutions of the first subsystem does not exceed $p^{2m-R_1(\nu+1)}$.

Suppose that we have found $x_{1j\nu}, \ldots, x_{s-1\,j\nu}$ $(j = 1, \ldots, 2m)$. We estimate the number of solutions $x_{sj\nu}$ of the sth subsystem. For some $\lambda''(t_1, \ldots, t_s, 0, \ldots, 0)$, this subsystem is equivalent to the system of congruences

$$\sum_{j=1}^{2m}(-1)^j x_{1j0}^{t_1} \ldots x_{s-1j0}^{t_{s-1}} x_{sj0}^{t_s-1} x_{sj\nu} \equiv \lambda''(t_1, \ldots, t_s, 0, \ldots, 0) \pmod p,$$

$$0 \le t_1, \ldots, t_s \le n, \quad t_1 + \cdots + t_s \ge \nu + 1, \quad t_s \ne 0.$$

Because of the regularity condition modulo p, we find that the congruences in this system form a set of linearly independent congruences modulo p. Since the number of congruences in this system is equal to $R_s(\nu+1) - R_{s-1}(\nu+1)$, it follows that the number of its solutions does not exceed $p^{2m-R_s(\nu+1)-R_{s-1}(\nu+1)}$. Consequently, T_ν, which is the number of solutions of the νth system of congruences, does not exceed

$$\begin{aligned} T_\nu &\le p^{2m-R_1(\nu+1)} p^{2m-R_2(\nu+1)+R_1(\nu+1)} \ldots p^{2m-R_r(\nu+1)+R_{r-1}(\nu+1)} \\ &\le p^{2mr-R_r(\nu+1)}. \end{aligned}$$

Earlier, it was shown that $T \le T_0 T_1 \ldots T_{r\,n-1}$, which implies

$$T \le m! p^{2mr-m+1} \prod_{\nu=1}^{r\,n-1} p^{2mr-R_r(\nu+1)}.$$

We note that $R_r(1) = m - 1$. Hence

$$T \le m! p^{2mr^2n-\Delta}, \quad \Delta = \sum_{\nu=1}^{rn} R_r(\nu).$$

We let $R_r^*(\nu)$ denote the number of solutions of the equation $t_1 + \cdots + t_r = \nu$ in integers $0 \le t_1, \ldots, t_r \le n$. Then

$$\Delta = \sum_{\nu=1}^{rn} R_r(\nu) = \sum_{\nu=1}^{rn}\sum_{k=\nu}^{rn} R_r^*(k).$$

Changing the order of summation, we obtain

$$\Delta = \sum_{k=1}^{rn} k R_r^*(k) \sum_{\nu=1}^{k} 1 = \sum_{k=1}^{rn} k R_r^*(k).$$

From the definition of $R_r^*(k)$ we have

$$\Delta = \sum_{k=1}^{rn} k R_r^*(k) = \sum_{t_1=0}^{n} \cdots \sum_{t_r=0}^{n} (t_1 + \cdots + t_r) = 0.5rnm.$$

Thus the proof of Lemma 4.3 is complete. □

4.1.4 The fundamental lemma

In this section we prove the fundamental lemma, which then readily implies the mean value theorem. In the lemma, we obtain a recurrence inequality which is the basis of the p-adic method in the class of problems under study.

Fundamental lemma. *Suppose that $n \geq 2$, $r \geq 1$, $k \geq 2m$, and $P \geq 1$. Then there exists a number p in the interval $[P^{1/(nr)}, 2P^{1/(nr)}]$ such that*

$$J(P; n, k, r) \leq 2k^{2m} p^{2mr^2n+2rk-0.5rnm-2rm} J(P_1; n, k-m, r) + (2^r rn)^{2rnk} P^{2rk-k}/8,$$

where $P_1 = Pp^{-1} + 1$.

Before giving a formal proof of this lemma, we outline it.

1. We divide the sets of vector solutions $\overline{x}_1, \dots, \overline{x}_{2k}$ of system (4.1) into two classes. The first class includes the solutions for which the sets $\overline{x}_1, \overline{x}_3, \dots, \overline{x}_{2k-1}$ and the sets $\overline{x}_2, \overline{x}_4, \dots, \overline{x}_{2k}$ satisfy the regularity condition modulo p for at least one $p = p_s$, where p_s $(s = 1, \dots, rn)$ are pairwise distinct prime numbers that are larger than $P^{1/(rn)}$ and do not exceed $2P^{1/(rn)}$ (these prime numbers exist if $P > (2nr)^{2nr}$; but if P does not exceed $(2nr)^{2nr}$, then the inequality in the lemma becomes trivial because of the second term). The second class includes all of the other solutions of (4.1). It is convenient to carry out the partition of the solutions into two classes using the representation of J as the square of the modulus of the kth power of a multiple trigonometric sum (see formula (4.10) below).

2. We estimate the number of solutions to (4.1) belonging to the first class. To do this, we use successive transformations to reduce everything to estimating the number of solutions of (4.1), but with fewer values of the parameters P and k.

The first step in the transformation consists in reducing everything to estimating the number of solutions of (4.1) satisfying the condition that the first m vectors among the $\overline{x}_1, \overline{x}_3, \dots, \overline{x}_{2k-1}$ and the first m vectors among the $\overline{x}_2, \overline{x}_4, \dots, \overline{x}_{2k}$ satisfy the regularity condition modulo p. All of the other solutions in the first class are obtained from these by permuting these m vectors among the k places possible for them. By the same token, it suffices to estimate the number of solutions of (4.1) with the above condition and then to multiply the resulting estimate by $\binom{k}{m}^2$ in order to obtain an upper bound for the number of solutions of (4.1) in the first class.

The second step in the transformations is the following. If all of the remaining $k - m$ vectors in the left- and in the right-hand side of (4.1) have coordinates that are multiples of p, then it is clear from (4.1) that the first m vectors $\overline{x}_1, \overline{x}_3, \dots, \overline{x}_{2m-1}$ and $\overline{x}_2, \overline{x}_4, \dots, \overline{x}_{2m}$ must satisfy the system of congruences in Lemma 4.3, which is precisely our goal. In order to obtain what we need, we partition into arithmetic progressions with difference p the integers in the interval of variation of the coordinates of the last $k - m$ vectors corresponding to the left- and right-hand sides of (4.1), i.e., we represent the vectors $\overline{x}_j$ $(j = 2m + 1, \dots, 2k - 1, 2k)$ in the form $\overline{x}_j =$

$\overline{y}_j + p\overline{z}_j$, where the coordinates of the vector $\overline{y}_j$ vary from 1 to p, and the coordinates of the vector $\overline{z}_j$ vary from 0 to Pp^{-1}. Next, if we bring the summation over $\overline{y}_j$ $(j = 2m+1, \ldots, 2k-1, 2k)$ outside the absolute value sign and apply Hölders's inequality, we find that all of the $\overline{x}_j$ $(j \geq 2m+1)$ will have the form $\overline{x}_j = \overline{a} + p\overline{z}_j$, where $\overline{a}$ is some fixed vector. We can now apply Lemma 4.2 (a) concerning shifts. In this case the regularity condition modulo p for the first m vectors in the left- and right-hand sides of (4.1) will not be disturbed, while the last $k-m$ vectors will become multiples of p.

The third step is rather obvious. If we move the terms in (4.1) corresponding to $\overline{x}_j - \overline{a}$ $(j = 1, \ldots, 2m)$ to the left and the terms corresponding to $p\overline{z}_j$ $(j = 2m+1, \ldots, 2k-1, 2k)$ to the right, then in the right-hand side, we obtain a product of powers of p by terms in parentheses. Each of the terms in parentheses has the same form as the left-hand side of (4.1) except that, instead of $2kr$ variables, there are $2(k-m)r$ variables, and they run through the nonnegative integers up to the number Pp^{-1}. If each of the terms in parentheses takes all possible integer values, then the resulting system is a system of congruences corresponding to the first $2m$ vectors, i.e., the complete system of congruences given in Lemma 4.3. But for fixed values of terms in parentheses, the maximal number of unknowns is obtained if all of the terms in parentheses vanish (see Lemma 4.1 (b)). If we take this maximal values outside the summation, which is over all values of the terms in parentheses, we arrive at the product of two factors: the first factor is the number of solutions of (4.1) with $2(k-m)$ vectors of unknowns, where the coordinates of these vectors vary from 1 to $P_1 = Pp^{-1} + 1$; the second factor is the number of solutions to the complete system of congruences given in Lemma 4.3.

So we can estimate the number of solutions of (4.1) which belong to the first class.

3. We estimate the number of solutions of system (4.1) which belong to the second class. We immediately note that they will be few in number (the second term in the fundamental lemma). By definition, the second class includes the solutions of (4.1) for which the vectors $\overline{x}_1, \overline{x}_3, \ldots, \overline{x}_{2k-1}$ (or $\overline{x}_2, \overline{x}_4, \ldots, \overline{x}_{2k}$) do not satisfy the regularity condition modulo $p = p_s$ for a single $s = 1, \ldots, rn$. This means that for any $p = p_s$ $(1 \leq s \leq rn)$ the matrix M corresponding to the vectors $\overline{y}_1 = \overline{x}_1$, $\overline{y}_2 = \overline{x}_3, \ldots, \overline{y}_k = \overline{x}_{2k-1}$, has rank modulo p lower than m. This fact, in turn, means that the rows of this matrix are linearly dependent modulo $p = p_s$, i.e., there exist numbers $c_1, \ldots, c_m$ not all congruent to zero modulo $p = p_s$ such that the linear combinations of the rows of M with coefficients $c_1, \ldots, c_m$ are congruent to zero modulo $p = p_s$. The numbers $c_1, \ldots, c_m$ themselves depend on $p = p_s$ and on the set of vectors $\overline{y}_1, \overline{y}_2, \ldots, \overline{y}_k$.

We find necessary conditions satisfied by the vectors $\overline{y}_1(p), \overline{y}_2(p), \ldots, \overline{y}_k(p)$ that are congruent to the vectors $\overline{y}_1, \overline{y}_2, \ldots, \overline{y}_k$ modulo $p = p_s$ $(s = 1, \ldots, rn)$. We partition all of these vectors into sets corresponding to a fixed choice of the numbers $c_1, \ldots, c_m$ (this last set depends only on $p = p_s$ $(1 \leq s \leq rn)$). We put the vectors $\overline{y}_1(p), \overline{y}_2(p), \ldots, \overline{y}_k(p)$ in the set $B(c_1, \ldots, c_m)$ if the linear combinations of the rows of the matrix M corresponding to $\overline{y}_1(p), \overline{y}_2(p), \ldots, \overline{y}_k(p)$ with coefficients

$c_1, \ldots, c_m$ are congruent to zero modulo p. Clearly, the sets $B(c_1, \ldots, c_m)$ can be empty, and they can overlap. We find an estimate from above for the number of elements in $B(c_1, \ldots, c_m)$. Multiplying our estimate by $2p^{m-1}$, which is not less than the number of all sets $c_1, \ldots, c_m$ (it should be noted that we can assume that $c_1 = 0$ or 1 without loss of generality), we obtain an upper bound for the number of sets of vectors $\overline{y}_1(p), \overline{y}_2(p), \ldots, \overline{y}_k(p)$.

The number of elements in $B(c_1, \ldots, c_m)$ does not exceed the number of solutions of the following system of congruences: the linear combination of the columns of the matrix M corresponding to $\overline{y}_1(p), \overline{y}_2(p), \ldots, \overline{y}_k(p)$ with coefficients $c_1, \ldots, c_m$ is congruent to zero modulo p, where the variables in the congruence, i.e., the coordinates $\overline{y}_i(p)$, take values from a complete system of residues modulo p. Each of the congruences in the resulting system is independent of the others, since it includes unknowns in the columns. By the same token, the number of solutions of the system does not exceed the product of the numbers of solutions to all separate congruences. An individual congruence has the following form: a polynomial in r variables of degree that does not exceed rn, which is not identically zero modulo $p = p_s$ (the numbers $c_1, \ldots, c_m$ are not all zero modulo $p = p_s$), is congruent to zero modulo $p = p_s$.

The number of solutions of such a congruence does not exceed nrp_s^{r-1}, i.e., the number of elements in $B(c_1, \ldots, c_m)$ does not exceed $(nrp_s^{r-1})^k$, and the number of vectors $\overline{y}_1(p), \overline{y}_2(p), \ldots, \overline{y}_k(p)$ does not exceed $2p_s^{m-1}(nrp_s^{r-1})^k$ (recall that we take $c_1 = 0$ or 1, while $c_1, \ldots, c_m$ are arbitrary).

Thus we have proved that the vectors $\overline{y}_1, \ldots, \overline{y}_k$ are congruent to the vectors $\overline{y}_1(p), \ldots, \overline{y}_k(p)$ modulo $p = p_s$ $(s = 1, \ldots, rn)$ and the number of all possible vectors $\overline{y}_1(p), \ldots, \overline{y}_k(p)$ does not exceed $2p_s^{m-1}(nrp_s^{r-1})^k$ for each $p = p_s$. All of these rn systems of linear vector congruences can be replaced by a single one modulo $p_1, \ldots, p_{rn}$, where the number of right-hand sides in this single system cannot be larger than $\prod_{s=1}^{rn}(2p_s^{m-1}(nrp_s^{r-1})^k)$.

The coordinates of the left-hand sides of the resulting system of congruences do not exceed $P < p_1, \ldots, p_{rn}$, the coordinates of the right-hand sides do not exceed $p_1, \ldots, p_{rn}$, and the congruences themselves are considered modulo $p_1, \ldots, p_{rn}$ and have the form $\overline{y}_i \equiv \overline{a} \pmod{p_1, \ldots, p_{rn}}$ $i = 1, \ldots, k$, i.e., this system of congruences is equivalent to a system of linear vector equations. By the same token, we find that the number of vectors $\overline{y}_1 = \overline{x}_1, \overline{y}_2 = \overline{x}_3, \ldots, \overline{y}_k = \overline{x}_{2k-1}$ which satisfy (4.1) and belong to the second class does not exceed $\prod_{s=1}^{rn}(2p_s^{m-1}(nrp_s^{r-1})^k)$.

Proof of the Fundamental Lemma. The proof will be given in subsections corresponding to the steps in the outline.

We first exclude the trivial cases. If $P \leq (2nr)^{2nr}$, then we obtain $p = 2P^{1/(nr)}$. In that case, the second term in the inequality in the lemma exceeds P^{2rk}, while the first term is always nonnegative. Hence the lemma becomes trivial.

Thus we assume that $P > (2nr)^{2nr}$.

1. If $P > (2nr)^{2nr}$, the interval $[P^{1/(nr)}, 2P^{1/(nr)}]$ contains at least rn distinct prime numbers (Lemma 3.8, Chapter 3). We take rn of such prime numbers, and denote them by the letters $p_1, \dots, p_{rn}$. Recall that, for convenience, we have introduced the following abbreviated notation: we let $\overline{x}_j$ denote the vector $\overline{x}_j = (x_{1j}, \dots, x_{rj})$ $(j = 1, \dots, 2k)$. If $F(x, y, \dots, z)$ is a polynomial in r variables $x, y, \dots, z$, then $F(\overline{x}_j) = F(x_{1j}, \dots, x_{rj})$. We can write the number $J(P; n, k, r)$ as the integral

$$J = J(P; n, k, r) = \int_{\Omega}\cdots\int \Big| \sum_{\overline{x}_1} \sum_{\overline{x}_3} \cdots \sum_{\overline{x}_{2k-1}} \exp\big\{2\pi i\big(F(\overline{x}_1) + F(\overline{x}_3) + \cdots + F(\overline{x}_{2k-1})\big)\big\}\Big|^2 d\Omega, \tag{4.10}$$

where the summation is over the vectors $\overline{x}_1, \overline{x}_3, \dots, \overline{x}_{2k-1}$ whose coordinates take integer values and vary from 1 to P.

We divide all of the vectors $\overline{x}_1, \overline{x}_3, \dots, \overline{x}_{2k-1}$ into two classes A and B: the class A includes those satisfying the regularity condition modulo $p = p_s$ for at least one value of s $(1 \le s \le rn)$, while the class B includes the remaining vectors. Recall that a set of vectors $(\overline{x}_1, \overline{x}_3, \dots, \overline{x}_{2k-1})$ satisfies the regularity condition modulo p if the corresponding matrix M $(0 \le t_1, \dots, t_r \le n)$

$$M = \big(x_{11}^{t_1} \dots x_{r1}^{t_r},\ x_{13}^{t_1} \dots x_{r3}^{t_r},\ \dots,\ x_{1\,2k-1}^{t_1} \dots x_{r\,2k-1}^{t_r}\big),$$

which consists of k columns and $m = (n+1)^r$ rows, has maximal rank, which in our case is $m \le k$, modulo p. In accordance with the partition into classes, we rewrite (4.10) as follows, where we use the obvious abbreviated notation:

$$J = \int_{\Omega}\cdots\int \Big|\sum_A + \sum_B\Big|^2 d\Omega.$$

From this we arrive at the inequality (we apply the Cauchy inequality)

$$J \le 2J_1 + 2J_2,$$

where

$$J_1 = \int_{\Omega}\cdots\int \Big|\sum_A\Big|^2 d\Omega, \quad J_2 = \int_{\Omega}\cdots\int \Big|\sum_B\Big|^2 d\Omega.$$

2. Let us estimate J_1. We partition the class A into rn disjoint classes $A_1, \dots, A_{rn}$ as follows: the vectors $\overline{x}_1, \overline{x}_3, \dots, \overline{x}_{2k-1}$ belong to the class A_s if they satisfy the regularity condition modulo p_s and do not belong to the classes $A_1, \dots, A_{s-1}$. We transform J_1 as follows (applying Hölder's inequality):

$$J_1 = \int_{\Omega}\cdots\int \Big|\sum_{s=1}^{rn}\sum_{A_s}\Big|^2 d\Omega \le rn \sum_{s=1}^{rn} \int_{\Omega}\cdots\int \Big|\sum_{A_s}\Big|^2 d\Omega \le (rn)^2 J_0,$$

where J_0 denotes the largest value attained by integrals of the form

$$\int_{\Omega}\cdots\int \Big|\sum_{A_s}\Big|^2 d\Omega, \quad 1 \le s \le rn.$$

Thus, the summation in the last sum is over those vectors $\overline{x}_1, \overline{x}_3, \ldots, \overline{x}_{2k-1}$, which satisfy the regularity condition for some modulus $p = p_s$ and which do not belong to the classes $A_1, \ldots, A_{s-1}$. The number J_0 can only increase if we remove the assumption that $\overline{x}_1, \overline{x}_3, \ldots, \overline{x}_{2k-1}$ do not belong to $A_1, \ldots, A_{s-1}$; in addition, everywhere below (when estimating J_0) we omit the index of p_s and only use the fact that $p = p_s$ lies on the interval $[P^{1/(rm)}, 2P^{1/(rn)}]$. Thus we must estimate J_0:

$$J_0 = \int\limits_{\Omega}\cdots\int \Big|\sum_A\Big|^2 d\Omega,$$

where A now denotes the class of sets of vectors $(\overline{x}_1, \overline{x}_3, \ldots, \overline{x}_{2k-1})$ satisfying the regularity condition modulo p.

First step. Because the vectors $\overline{x}_1, \overline{x}_3, \ldots, \overline{x}_{2k-1}$ satisfy the regularity condition modulo p (in other words, because the matrix M corresponding to $\overline{y}_1 = \overline{x}_1, \overline{y}_2 = \overline{x}_3, \ldots, \overline{y}_k = \overline{x}_{2k-1}$ has rank m), there exist m columns in the matrix M that are linearly independent modulo p. To these m columns, there correspond m vectors $\overline{x}_{j_1}, \ldots, \overline{x}_{j_m}$ that, by definition, satisfy the regularity condition modulo p (we shall say that to the vectors $\overline{x}_1, \overline{x}_3, \ldots, \overline{x}_{2k-1}$ in A there correspond vectors $\overline{x}_{j_1}, \ldots, \overline{x}_{j_m}$ if $\overline{x}_{j_1}, \ldots, \overline{x}_{j_m}$ satisfy the regularity condition modulo p). We consider the elements of A to which the first m vectors of $\overline{x}_1, \overline{x}_3, \ldots, \overline{x}_{2m-1}$ correspond. It is obvious that the other elements of the class A differ from these only by the indices of the vectors to which they correspond. Since m elements can be put in k places in $\binom{k}{m}$ ways, we have

$$J_0 \le \binom{k}{m}^2 \int\limits_{\Omega}\cdots\int \Big|\sum_A{}'\Big|^2 d\Omega, \tag{4.11}$$

where the prime on the sum in the integrand denotes summation over the elements of A which correspond to the vectors $\overline{x}_1, \overline{x}_3, \ldots, \overline{x}_{2m-1}$. The integral in the right-hand side can only increase if we sum over those $\overline{x}_1, \overline{x}_3, \ldots, \overline{x}_{2k-1}$ for which the first m vectors satisfy the regularity condition modulo p and the other vary arbitrarily. In the rest of the argument, we shall make this assumption.

Second step. The sum in the integral in (4.11) has the following form:

$$\Big|\sum_A{}'\Big|^2 = \Big|\sum_{\overline{x}_1}\cdots\sum_{\overline{x}_{2m-1}}\sum_{\overline{x}_{2m+1}}\cdots\sum_{\overline{x}_{2k-1}}\Big|^2, \tag{4.12}$$

where no restrictions are placed on summation over $\overline{x}_{2m+1}, \ldots, \overline{x}_{2k-1}$. We partition into arithmetic progressions with difference p the intervals over which the variables vary; these variables are the coordinates of the vectors $\overline{x}_j$ $(j = 2m+1, \ldots, 2k-1)$. In other words, we represent $\overline{x}_j$ $(j = 2m+1, \ldots, 2k-1)$ in the form $\overline{x}_j = \overline{y}_j + p\overline{z}_j$, where the coordinates of the vectors $\overline{y}_j$ vary from 1 to p, and those of the vectors $\overline{z}_j$ vary from 0 to Pp^{-1} (below we shall also write $\overline{x} = \overline{y} + p\overline{z}$).

If we assume that the coordinates of $\overline{z}_j$ take on all integer values from 0 to Pp^{-1}, then the integral in the right-hand side in (4.11) can only increase; we make this assumption. If we take the summation over $\overline{y}_j$ outside the absolute value sign in (4.12) and apply Hölder's inequality, we obtain the following inequality (with the obvious abbreviated notation):

$$\Big|\sum_A{}'\Big|^2 \le \Big|\sum_{\overline{x}_1}\cdots\sum_{\overline{x}_{2m-1}}\Big|^2\Big|\sum_{\overline{y}}\sum_{\overline{z}}\Big|^{2(k-m)}$$
$$\le p^{2r(k-m)-r}\sum_{\overline{y}}\Big|\sum_{\overline{x}_1}\cdots\sum_{\overline{x}_{2m-1}}\Big|^2\Big|\sum_{\overline{z}}\Big|^{2(k-m)}.$$

We obtain the following estimate for J_0:

$$J_0 \le \binom{k}{m}^2 p^{2r(k-m)-r}\sum_{\overline{y}}\int\cdots\int_{\Omega}\Big|\sum_{\overline{x}_1}\cdots\sum_{\overline{x}_{2m-1}}\Big|^2\Big|\sum_{\overline{z}}\Big|^{2(k-m)}d\Omega$$
$$\le \binom{k}{m}^2 p^{2r(k-m)}\max_{\overline{y}}\int\cdots\int_{\Omega}\Big|\sum_{\overline{x}_1}\cdots\sum_{\overline{x}_{2m-1}}\Big|^2\Big|\sum_{\overline{z}}\Big|^{2(k-m)}d\Omega.$$

Suppose that this last maximum is attained for $\overline{y} = \overline{y}^{(0)} = (y_1^{(0)}, \dots, y_r^{(0)})$, so that the above integral is equal to the number of solutions of the following system of equations:

$$\sum_{j=1}^{2m}(-1)^j x_{1j}^{t_1}\dots x_{rj}^{t_r} = \sum_{j=2m+1}^{2k}(-1)^j(y_1^{(0)}+pz_{1j})^{t_1}\dots(y_r^{(0)}+pz_{rj})^{t_r},$$
$$0 \le t_1, \dots, t_r \le n,$$

where the vectors $\overline{x}_1, \overline{x}_3, \dots, \overline{x}_{2m-1}$ corresponding to the left-hand side of this system of equations satisfy the regularity condition modulo p, the unknowns z_{ij} $(i = 1, \dots, r;$ $j = 2m+1, \dots, 2k)$ take arbitrary integer values from 0 to Pp^{-1}, and $y_1^{(0)}, \dots, y_r^{(0)}$ are fixed integers. By Lemma 4.2 (a), we can perform a shift of the unknowns in this system by the numbers $y_1^{(0)}, \dots, y_r^{(0)}$. We rewrite the system as follows:

$$\sum_{j=1}^{2m}(-1)^j(x_{1j}-y_1^{(0)})^{t_1}\dots(x_{rj}-y_r^{(0)})^{t_r} \tag{4.13}$$
$$= p^{t_1+\dots+t_r}\sum_{j=2m+1}^{2k}(-1)^j z_{1j}^{t_1}\dots z_{rj}^{t_r}, \quad 0 \le t_1, \dots, t_r \le n.$$

Third step. We let $J'(P_1; n, k-m, r; \Lambda)$ denote the number of solutions of the system of equations

$$\sum_{j=2m+1}^{2k}(-1)^j z_{1j}^{t_1}\dots z_{rj}^{t_r} = \lambda(t_1, \dots, t_r), \quad 0 \le t_1, \dots, t_r \le n,$$

where $\lambda(t_1, \dots, t_r)$ are certain fixed integers, Λ is the set of $\lambda(t_1, \dots, t_r)$, and the unknowns z_{ij} vary in the range indicated above. By Lemma 4.1 (a), we have the inequality

$$J'(P_1; n, k-m, r; \Lambda) \leq J'(P_1; n, k-m, r; \overline{0}),$$

where $\overline{0}$ denotes the set Λ in which all the $\lambda(t_1, \dots, t_r)$ are zero. Using Lemma 4.2 (a), we shift the interval of variation of the variables z_{ij} by 1 to the right; then we find that

$$J'(P_1; n, k-m, r; \overline{0}) = J(P_1; n, k-m, r).$$

Next, we let $J'(\Lambda)$ denote the number of solutions of the system

$$\sum_{j=1}^{2m} (-1)^j (x_{1j} - y_1^{(0)})^{t_1} \dots (x_{rj} - y_r^{(0)})^{t_r} = p^{t_1+\dots+t_r} \lambda(t_1, \dots, t_r),$$

$$0 \leq t_1, \dots, t_r \leq n,$$

and we let T denote the number of solutions of the system

$$\sum_{j=1}^{2m} (-1)^j (x_{1j} - y_1^{(0)})^{t_1} \dots (x_{rj} - y_r^{(0)})^{t_r} \equiv 0 \pmod{p^{t_1+\dots+t_r}}, \qquad (4.14)$$

$$0 \leq t_1, \dots, t_r \leq n,$$

where all terms have been defined earlier. From the definition of the congruences it follows that $\sum_\Lambda J'(\Lambda) = T$.

Furthermore, the number of solutions of (4.13) is equal to

$$\sum_\Lambda J(P_1; n, k-m, r; \Lambda)\, J'(\Lambda),$$

and this, in turn, does not exceed $J(P_1; n, k-m, r)T$.

By Lemma 4.2 (b), the sets of unknowns in the system of congruences (4.14) satisfy the regularity condition modulo p, i.e., Lemma 4.4 on complete systems of congruences can be used to estimate T. Consequently, we have

$$T \leq m! p^{2mr^2n - 0.5rnm},$$

$$J_0 \leq \binom{k}{m}^2 p^{2r(k-m)} T J(P-1; n, k-m, r),$$

$$J_1 \leq (rn)^2 J_0 \leq \frac{(rn)^2}{m!} k^{2m} p^{2mr^2n - 0.5rnm + 2kr - 2rm} J(P_1; n, k-m, r)$$

$$\leq k^{2m} p^{2mr^2n + 2rk - 0.5rnm - 2rm} J(P_1; n, k-m, r).$$

3. Let us estimate J_2. To do this, we estimate the number of elements in the class B, i.e., the number of terms in the trigonometric sum the square of whose modulus is contained in the integral J_2. The vectors $\overline{x}_1, \overline{x}_3, \dots, \overline{x}_{2k-1}$ belong to the class B; this means that the matrix M corresponding to these vectors has rank modulo $p = p_s$ less than m for any $s = 1, 2, \dots, rn$, i.e., the rows of M are linearly dependent modulo $p = p_s$ $(1 \le s \le rn)$. In other words, for any set $\overline{x}_1, \overline{x}_3, \dots, \overline{x}_{2k-1}$ in B and for any $p = p_s$ $(1 \le s \le rn)$, there exist integers $c_1, \dots, c_m$, not all congruent to zero modulo p, such that the linear combinations of the rows of M with these numbers $c_1, \dots, c_m$ as coefficients are congruent to zero modulo p.

Note that:

(1) the numbers $c_1, \dots, c_m$ can take any values in a complete set of residues modulo p;

(2) the numbers $c_1, \dots, c_m$ depend on $\overline{x}_1, \overline{x}_3, \dots, \overline{x}_{2k-1}$ and on $p = p_s$;

(3) we can assume that the number c_1 takes one of two values, namely 0 or 1, since the relations in which $c_1, \dots, c_m$ appear are homogeneous in $c_1, \dots, c_m$;

(4) if we let $\overline{x}(p)$ denote the vector obtained by taking the least nonnegative residue modulo p of the coordinates of the vector $\overline{x}$, then regularity (or singularity) of the set $\overline{x}_1, \overline{x}_3, \dots, \overline{x}_{2k-1}$ modulo p implies regularity (singularity) of the set $\overline{x}_1(p), \overline{x}_3(p), \dots, \overline{x}_{2k-1}(p)$ modulo p.

To estimate the number of elements in B, we proceed as follows.

First step. For each s $(1 \le s \le rn)$, we estimate the number of sets $\overline{x}_1(p_s), \overline{x}_3(p_s), \dots, \overline{x}_{2k-1}(p_s)$. For each set of integers $c_1, \dots, c_m$, where $c_1 = 0$ or 1 and not all of these numbers are congruent to 0 modulo p_s, we let $B(c_1, \dots, c_m)$ denote the sets $\overline{x}_1(p_s), \overline{x}_3(p_s), \dots, \overline{x}_{2k-1}(p_s)$ for which the linear combinations of the rows of the corresponding matrix M having coefficients $c_1, \dots, c_m$ are congruent to zero modulo p_s. Let us estimate the number of elements in $B(c_1, \dots, c_m)$. If the sets $\overline{x}_1(p_s), \overline{x}_3(p_s), \dots, \overline{x}_{2k-1}(p_s)$ belong to $B(c_1, \dots, c_m)$, then their coordinates

$$x_{ij}(p_s), \quad i = 1, \dots, r, \quad j = 1, 3, \dots, 2k-1,$$

satisfy the congruences

$$\sum_{t_1=0}^{n} \cdots \sum_{t_r=0}^{n} c(t_1, \dots, t_r) x_{1j}^{t_1}(p_s) \dots x_{rj}^{t_r}(p_s) \equiv 0 \pmod{p_s},$$
$$j = 1, 3, \dots, 2k-1,$$

where $c(t_1, \dots, t_r)$ are the same as $c_1, \dots, c_m$ except with a different indexing, which is more convenient in our argument.

Each of these k congruences is independent of the others, i.e., the unknowns in the congruences do not overlap. The left-hand side of one of the congruences is a polynomial in r variables whose coefficients are not all congruent to zero modulo p_s and whose degree does not exceed rn. Consequently, the single congruence has no more than rnp_s^{r-1} solutions. Hence we see that the number of elements in $B(c_1, \dots, c_m)$ does not exceed $(rnp_s^{r-1})^k$ and the number of sets $\overline{x}_1(p_s), \overline{x}_3(p_s), \dots, \overline{x}_{2k-1}(p_s)$ does not exceed $2p_s^{m-1}(rnp_s^{r-1})^k$.

Second step. By definition, if the sets $\overline{x}_1, \overline{x}_3, \ldots, \overline{x}_{2k-1}$ belong to B, then for each $s = 1, 2, \ldots, rn$, the vectors $\overline{x}_j(p_s)$ satisfy the congruences

$$\overline{x}_j \equiv \overline{x}_j(p_s) \pmod{p_s}, \quad j = 1, 3, \ldots, 2k-1.$$

If the right-hand sides of the congruences are fixed, we can replace the congruences modulo $p_1, p_2, \ldots, p_{rn}$ by a single congruence modulo $p_1, \ldots, p_{rn}$ having the form

$$\overline{x}_j \equiv \overline{a}_j(p_s) \pmod{p_1 \ldots p_{rn}}, \quad j = 1, 3, \ldots, 2k-1, \tag{4.15}$$

where the right-hand side $\overline{a}_j$ of the resulting congruence is uniquely determined by the right-hand sides $\overline{x}_j(p_s)$ $(s = 1, \ldots, rn)$ and the coordinates of $\overline{a}_j$ are residues modulo $p_1, \ldots, p_{rn}$, nonnegative and less than $p_1, \ldots, p_{rn}$. The number of possible right-hand sides $\overline{a}_j$ is no larger than

$$U = \prod_{s=1}^{rn} \left(2p_s^{m-1}(rnp_s^{r-1})^k\right).$$

Since $p_s \le 2P^{1/(rn)}$ $(s = 1, \ldots, rn)$, it follows that

$$U \le 2^{rn(m+k(r-1))}(rn)^{rnk} P^{kr-k+m-1}.$$

Each coordinate of the vector $\overline{x}_j$ does not exceed $P < p_1 \ldots p_{rn}$. Hence, congruences (4.15) are equivalent to the relations

$$\overline{x}_j = \overline{a}_j, \quad j = 1, 3, \ldots, 2k-1,$$

i.e., there are no more than U elements in B.

Consequently (recall that $k \ge 2m$), we obtain the following estimate for J_2:

$$J_2 \le U^2 \le 2^{-2mrn}(2^r rn)^{2rnk} P^{2kr-2k+2m-2} < \frac{1}{16}(2^r rn)^{2rnk} P^{2kr-k}.$$

The above estimates for J_1 and J_2 give us the statement of the fundamental lemma:

$$J \le 2k^{2m} p^{2mr^2n+2rk-0.5rnm-2rm} J(P_1; , n, k-m, r) + \frac{1}{8}(2^r rn)^{2rnk} P^{2rk-k}. \quad \square$$

4.1.5 The mean value theorem

In this subsection we prove Theorem 4.1 on the mean value of the $2k$th power of the modulus of an r-fold trigonometric sum.

Theorem 4.1. *Suppose that $\tau \ge 0$ is an integer, $k \ge m\tau$, and $P \ge 1$. Then the following estimate holds for $J = J(P; n, k, r)$:*

$$J \le k^{2m\tau} 4^{mr^2n\tau}(nr)^{2nr\Delta(\tau)} P^{2rk-0.5rnm+\delta(\tau)},$$

where

$$\delta(\tau) = 0.5rnm(1 - 1/(rn))^\tau, \quad \Delta(\tau) = 0.5rnm - \delta(\tau).$$

Proof. We show that the theorem holds for $\tau = 0$ and $\tau = 1$. For $\tau = 0$ and $k \geq 0$, we have

$$\delta(\tau) = 0.5rnm, \quad \Delta(\tau) = 0,$$

and the estimate in the theorem takes the form $J \leq P^{2rk}$, which is always the case. For $\tau = 1$ and $k \geq m$, we have

$$\delta(\tau) = 0.5rnm - 0.5m, \quad \Delta(\tau) = 0.5m,$$

and the estimate in the theorem takes the form

$$J \leq k^{2m} 4^{mr^2 n} (nr)^{mrn)} P^{2kr-0.5m}.$$

We show that in this case J satisfies an even sharper estimate. From Lemma 4.1 (b) $(k \geq m)$ we have

$$J(P; n, k, r) \leq P^{2r(k-m)} J(P; , n, m, r).$$

We take a prime q in the interval $[P, 2P]$ and note that $J(P; n, m, r)$ does not exceed the number of solutions of a system of congruences modulo q having the same form as (4.1) with $k = m$ and in which the unknowns take values in a complete set of residues modulo q. The number of solutions of such a system of congruences does not exceed the number T_0 in Lemma 4.3, i.e.,

$$J(P; n, m, r) \leq m! q^{2mr-m+1} \leq m! 2^{2mr-m+1} P^{2mr-m+1},$$

$$J(P; n, k, r) \leq m! 2^{2mr-m+1} P^{2kr-m+1} \leq k^{2m} 4^{mr^2 n} (nr)^{mnr} P^{2kr-0.5m}.$$

Thus, it remains to prove the theorem for $\tau \geq 2$ and $k \geq m\tau$.

Again from Lemma 4.1 (b) $(k \geq m\tau)$, we have

$$J(P; n, k, r) \leq P^{2r(k-m\tau)} J(P; n, m\tau, r).$$

If we prove the theorem for $J(P; n, m\tau, r)$, then it also follows for $J(P; n, k, r)$.

We assume that the statement of the theorem holds for $J(P; n, m\tau, r)$ $(\tau \geq 1)$, and we prove it for $J(P; n, m(\tau+1), r)$. We apply the fundamental lemma to $J(P; n, m(\tau+1), r)$:

$$J(P; n, m(\tau+1), r) \leq 2(m(\tau+1))^{2m} p^{2mr^2 n + 2rm(\tau+1) - 0.5rnm - 2rm}$$
$$\times J(P_1; n, m\tau, r) + \frac{1}{8}(2^r rn)^{2rnm(\tau+1)} P^{2rm(\tau+1) - m(\tau+1)},$$

where $P_1 = Pp^{-1} + 1$. We apply the induction assumption to $J(P_1; n, m\tau, r)$:

$$J(P_1; n, m\tau, r) \leq (m\tau)^{2m\tau} 4^{mr^2 n\tau} (nr)^{2nr\Delta(\tau)} P_1^{2rm\tau - 0.5rnm + \delta(\tau)}.$$

Substituting this estimate into the previous one, we obtain

$$J(P; n, m(\tau+1), r) \leq W_1 + W_2,$$

where

$$W_1 = 2(m(\tau+1))^{2m}(m\tau)^{2m\tau}4^{mr^2n\tau}(nr)^{2nr\Delta(\tau)} \times p^{2mr^2n+2mr\tau-0.5rnm}P_1^{2rm\tau-0.5rnm+\delta(\tau)},$$

$$W_2 = \frac{1}{8}(2^r rn)^{2rmn(\tau+1)}P^{2rm(\tau+1)-m(\tau+1)}.$$

It remains to show that $W_1 \le 0.5W_0$ and $W_2 \le 0.5W_0$, where

$$W_0 = (m(\tau+1))^{2m(\tau+1)}4^{mr^2n(\tau+1)}(nr)^{2nr\Delta(\tau+1)}p^{2rm(\tau+1)-\Delta(\tau+1)}.$$

We first show that $W_1 \le 0.5W_0$. First, we can assume that $P \ge (2rm\tau)^2$, since otherwise the theorem is trivial. Furthermore,

$$P_1^{\varkappa} < 3P^{\varkappa}p^{-\varkappa},$$

where $\varkappa = 2rm\tau - 0.5rnnm + \delta(\tau)$, since

$$(P_1pP^{-1})^{\varkappa} = (1+pP^{-1})^{\varkappa} < e < 3,$$

which holds because $p < P^{1/2}$ and $\varkappa < 2rm\tau \le P^{1/2}$.

Consequently, setting $\varkappa_1 = 2mr^2n + 2rm\tau - 0.5rnm$, we have

$$p^{\varkappa_1}P^{\varkappa} \le 3p^{\varkappa_1-\varkappa}P^{\varkappa} \le 3\cdot 2^{\varkappa_1-\varkappa}P^{\varkappa+(\varkappa_1-\varkappa)/(rn)}.$$

But $\varkappa + (\varkappa_1-\varkappa)/(rn) = 2rm(\tau+1) - \Delta(\tau+1)$ and $2^{\varkappa_1-\varkappa} = 4^{mr^2n}$, and hence

$$p^{\varkappa_1}P^{\varkappa} \le 3\cdot 4^{mr^2n}P^{2rm(\tau+1)-\Delta(\tau+1)}.$$

From this we have

$$\begin{aligned}
W_1 &= 2(m(\tau+1))^{2m}(m\tau)^{2m\tau}4^{mr^2n\tau}(nr)^{2nr\Delta(\tau)}p^{2mr^2n+2rm\tau-0.5rnm} \\
&\quad\times P_1^{2rm\tau-0.5rnm+\delta(\tau)} \\
&= 2(m(\tau+1))^{2m}(m\tau)^{2m\tau}4^{mr^2n\tau}(nr)^{2nr\Delta(\tau)}p^{\varkappa_1}P_1^{\varkappa} \\
&\le 0.5(m(\tau+1))^{2m(\tau+1)}4^{mr^2n(\tau+1)}(nr)^{2nr\Delta(\tau+1)}P^{2rm(\tau+1)-\Delta(\tau+1)} \\
&\quad\times 12\left(\frac{\tau}{\tau+1}\right)^{2m\tau} \\
&= 0.5W_0\cdot 12\left(\frac{\tau}{\tau+1}\right)^{2m\tau} \le 0.5W_0\cdot 0.1875 < 0.5W_0.
\end{aligned}$$

Thus the inequality $W_1 \le 0.5W_0$ has been established.

We now show that we also have $W_2 \le 0.5W_0$. Indeed, we can assume that $P > (nr)^{2nr}$, since otherwise the theorem holds trivially. Furthermore, $m(\tau+1) - \Delta(\tau+1) \ge 0$ and hence

$$\begin{aligned}(nr)^{2nr\Delta(\tau+1)}&P^{2rm(\tau+1)-\Delta(\tau+1)}\\&= (nr)^{2nrm(\tau+1)}P^{2rm(\tau+1)-m(\tau+1)}\left((nr)^{2nr}P^{-1}\right)^{\Delta(\tau+1)-m(\tau+1)}\\&> (nr)^{2rm(\tau+1)-m(\tau+1)}P^{2rm(\tau+1)-m(\tau+1)} = 8\cdot 4^{-r^2nm(\tau+1)}W_2.\end{aligned}$$

From this we have

$$\begin{aligned}W_0 &= (m(\tau+1))^{2m(\tau+1)}4^{mr^2n(\tau+1)}(nr)^{2nr\Delta(\tau+1)}P^{2rm(\tau+1)-\Delta(\tau+1)}\\&> 8^{-1}(m(\tau+1))^{2m(\tau+1)}W_2 > 2W_2.\end{aligned}$$

The last inequality now trivially implies that $W_2 < 0.5W_0$. Hence the proof of the theorem is complete. □

Remark 4.3. It is obvious that Theorem 4.1 remains true if the unknowns in (4.1) are subjected to any additional restrictions.

As a supplement to the theorem, we now show to what extent the size of the parameter k in this theorem is the correct one. As an example, we examine how small k can be in order for the lower bound in the estimate of the integral to be equal to $\Delta(rn)$. Recall that, to obtain such a bound, we must take $k \ge m\tau = mrn$ from our theorem.

Thus let k be such that $J(P;n,k,r) \ll P^{2kr-\Delta(rn)}$ for all $P > 1$. Among all the solutions of (4.1), the number of which is expressed by the integral $J(P;n,k,r)$, we consider those for which $x_{sr} = 0$ for all $s = 1,\dots,2k$. There will obviously be $J(P;n,k,r-1)$ such solutions and

$$J(P;n,k,r-1) \le J(P;n,k,r) \tag{4.16}$$

for all admissible values of the parameters.

In Lemma 4.1 it was proved that

$$J(P;n,k,r-1) \gg P^{2k(r-1)-0.5(r-1)n(n+1)^{r-1}}.$$

Hence in order that (4.16) hold for all P, it is necessary that

$$2kr - \Delta(rn) \ge 2k(r-1) - 0.5(r-1)n(n+1)^{r-1}.$$

We hence conclude that we must have

$$\begin{aligned}2k &\ge \Delta(rn) - \frac{(r-1)n(n+1)^{r-1}}{2} = \frac{rnm}{2}\left(1-\left(1-\frac{1}{rn}\right)^{rn}\right) - \frac{(r-1)nm}{2(n+1)}\\&\ge \frac{rnm}{2}\left(1-\frac{1}{e}-\frac{1}{n+1}\right) \ge \frac{rnm}{6}.\end{aligned}$$

This implies that, in order to obtain the required lower bound, the parameter k must in any case be chosen no less than $rnm/12$, as compared to rnm in our theorem. This gives us grounds for saying that the order of k in the theorem estimating the integral $J(P; n, k, r)$ is correct with respect to all parameters.

4.2 The mean value theorem for multiple trigonometric sums of general form

Here we find an upper bound for the mean value of the $2k$th power of the modulus of a multiple trigonometric sum with the summation variables $x_1, \dots, x_r$ varying in the range $1 \le x_1 \le P_1, \dots, 1 \le x_r \le P_r$. Since the unknowns in the complete system of equations (for the notation, see below), as well as the summation variables in a multiple trigonometric sum, are not equivalent, we shall estimate the contribution of each unknown taking this fact into account. In the mean value theorem, using the main principle of the p-adic method, we reduce estimating the number of solutions of a complete system to estimating the number of solutions of the same system in which the unknowns vary in a smaller range as before, but the original "degree" of nonequivalence is preserved. To realize this consideration, we must prove several statements about the complete system of congruences and the recurrence inequality that are more general than those in Section 4.1. The parts of the proofs that coincide with those given in Section 4.1 we perform without detailed explanations.

The complete system of equations has the form

$$\sum_{j=1}^{2k}(-1)^j x_{1j}^{t_1}\dots x_{rj}^{t_r} = 0, \quad 0 \le t_1 \le n_1, \dots, 0 \le t_r \le n_r, \tag{4.17}$$

where the unknowns vary within the limits

$$1 \le x_{1j} \le P_1,\ 1 \le x_{2j} \le P_2,\ \dots,\ 1 \le x_{rj} \le P_r.$$

In what follows, without loss of generality, we can assume that $1 < P_1 = \min(P_1, P_2, \dots, P_r)$.

We let $J = J(P; n, k, r)$ denote the number of solutions of system (4.17) (for brevity, we sometimes write $\overline{P}$, which means that we have the vector $\overline{P} = (P_1, P_2, \dots, P_r)$).

If in system (4.17) some equations are omitted, such a system is said to be *incomplete*.

We let $F(x_1, \dots, x_r)$ denote a polynomial of the form

$$F(x_1, \dots, x_r) = F_A(x_1, \dots, x_r) = \sum_{t_1=0}^{n_1}\cdots\sum_{t_r=0}^{n_r}\alpha(t_1, \dots, t_r)x_1^{t_1}\dots x_r^{t_r},$$

where $\alpha(t_1, \dots, t_r)$ are real numbers, the monomials $\alpha(t_1, \dots, t_r)x_1^{t_1}\dots x_r^{t_r}$ are arranged in ascending order of numbers $t_1 + (n_1+1)t_2 + (n_1+1)(n_2+1)t_3 + \cdots +$

$(n_1+1)\dots(n_{r-1}+1)t_r$, A is the vector whose coordinates are the coefficients of the polynomial $F(x_1,\dots,x_r)$ in the same order as they enter $F(x_1,\dots,x_r)$.

Let

$$S_t(A) = \sum_{x_1 \le P_1} \cdots \sum_{x_r \le P_r} \exp\{2\pi i t F(x_1,\dots,x_r)\}, \quad S(A) = S_1(A).$$

In the m-dimensional Euclidean space, by Ω we denote the unit m-dimensional cube ($m = (n_1+1)(n_2+1)\dots(n_r+1)$) of the form

$$0 \le \alpha(t_1,\dots,t_r) < 1, \quad 0 \le t_1 \le n_1, \dots, 0 \le t_r \le n_r.$$

4.2.1 Lemma on the complete system of congruences

The following lemma is one of the two fundamental lemmas in the p-adic proof of the mean value theorem for multiple trigonometric sums.

Lemma 4.5. *Suppose that $\mu_1,\dots,\mu_r$ are arbitrary natural numbers, $0 \le t_1 \le n_1,\dots,0 \le t_r \le n_r$, $m = (n_1+1)\dots(n_r+1)$, $n = n_1+\dots+n_r$, $\varkappa = \mu_1 n_1 + \dots + \mu_r n_r$, p is a prime number, and T is the number of solutions of the congruences*

$$\sum_{j=1}^{2m} (-1)^j x_{1j}^{t_1} \dots x_{rj}^{t_r} \equiv 0 \pmod{p^{\mu_1 t_1+\dots+\mu_r t_r}},$$

$$0 \le t_1 \le n_1,\dots,0 \le t_r \le n_r,$$

where each of the unknowns $x_{1j},\dots,x_{rj}$ of the system takes $p^\varkappa$ successive values, and $B_s \le x_{sj} < B_s + p^\varkappa$ $(s = 1,\dots,r,\ j = 1,\dots,2m)$ Suppose also that the vectors $\overline{x}_j = (x_{1j},\dots,x_{rj})$, where $j = 2,4,\dots,2m$, satisfy the regularity condition modulo p. Then T satisfies the estimate

$$T \le m! p^{\varkappa(2mr-0.5m)}.$$

Proof. Using the fact that $\varkappa \ge \mu_1 t_1 + \dots + \mu_r t_r$, i.e., each unknown runs through the complete system of residues, we can set $B_s = 0$ $(s = 1,\dots,r)$. Moreover, we can assume that $p > \max(n_1,\dots,n_r)$, since in the case $p \le \max(n_1,\dots,n_r)$, there are no solutions satisfying the regularity condition modulo p. We represent each unknown x_{sj} as

$$x_{sj} = x_{sj0} + p x_{sj1} + \dots + p^{\varkappa-1} x_{sj\,\varkappa-1},$$

where

$$0 \le x_{sj0}, x_{sj1},\dots,x_{sj\,\varkappa-1} \le p-1, \quad s = 1,\dots,r;\ j = 1,\dots,2m,$$

and find necessary conditions which are satisfied by the variables $x_{sj\nu}$. Considering all the congruences in the system modulo p and using the fact that $x_{sj} \equiv x_{sj0} \pmod p$, we find the conditions on x_{sj0}:

$$\sum_{j=1}^{2m}(-1)^j x_{1j0}^{t_1}\dots x_{rj0}^{t_r} \equiv 0 \pmod p,$$
$$0 \le t_1 \le n_1,\ \dots,\ 0 \le t_r \le n_r, \quad \mu_1 t_1 + \dots + \mu_r t_r \ge 1.$$

Let T_0 be the number of solutions of this system. Then, by Lemma 4.3 with q replaced by p, we have the estimate

$$T_0 \le (m-1)!p^{2mr-m+1}.$$

The further argument completely coincides with the corresponding argument in Section 4.1.3, but, instead of the conditions $t_1+\dots+t_r \ge \nu$ and $0 \le t_1, \dots, t_r \le n$, one must consider the conditions $\mu_1 t_1+\dots+\mu_r t_r \ge \nu$ and $0 \le t_1 \le n_1, \dots, 0 \le t_r \le n_r$.

We let $R_r(\nu)$ denote the number of solutions in integers $t_1, \dots, t_r$ of the inequalities $\mu_1 t_1+\dots+\mu_r t_r \ge \nu, 0 \le t_1 \le n_1, \dots, 0 \le t_r \le n_r$. Then for T we have the estimate

$$T \le m!p^{2mr-m+1}\prod_{\nu=1}^{\varkappa-1} p^{2mr-R_r(\nu+1)}.$$

Since $R_r(1) = m-1$, we also have

$$T \le m!p^{2mr\varkappa-R}, \quad R = \sum_{\nu=1}^{\varkappa} R_r(\nu).$$

If $R_r'(\nu)$ is the number of solutions of the equation $\mu_1 t_1 + \dots + \mu_r t_r = \nu$ in integers $0 \le t_1 \le n_1, \dots, 0 \le t_r \le n_r$, then we have the relations

$$\begin{aligned} R &= \sum_{\nu=1}^{\varkappa} R_r(\nu) = \sum_{\nu=1}^{\varkappa}\sum_{k=\nu}^{\varkappa} R_r'(k) = \sum_{k=\nu}^{\varkappa} R_r'(k)\sum_{\nu=1}^{k} 1 \\ &= \sum_{k=1}^{\varkappa} kR_r'(k) = \sum_{t_1=0}^{n_1}\dots\sum_{t_r=0}^{n_r}(\mu_1 t_1 + \dots + \mu_r t_r) = 0.5m\varkappa, \end{aligned}$$

which implies the statement of the lemma. □

4.2.2 Recurrence inequality

Lemma 4.6. *We consider the sets consisting of $k \ge 2n$ vectors $(\overline{x}_1, \overline{x}_2, \dots, \overline{x}_k)$ such that each of the vectors $\overline{x}$ in this set has r coordinates, i.e.,*

$$\overline{x} = (y_1, \dots, y_r),$$

where $y_1, \dots, y_r$ *are natural numbers,* $1 \le y_1 \le P_1, \dots, 1 \le y_r \le P_r$ *(hence if* $P_1, \dots, P_r$ *are integers, then the total number of sets is* $P_1^k \dots P_r^k$*), and* $P_1 = \min(P_1, \dots, P_r)$.

Suppose also that $\varkappa \ge 1$ *is a natural number,* $\varkappa\gamma = 1$*, and* $p_1, \dots, p_\varkappa$ *are distinct primes such that*

$$0.5P_1^\gamma \le p_j \le P_1^\gamma, \quad j = 1, \dots, \varkappa.$$

We divide all these sets into two classes A *and* B*. Class* A *consists of the sets satisfying the regularity condition modulo* $p = p_j$ *at least for a single value of* j $(1 \le j \le \varkappa)$*; all the other sets belong to class* B*. Then the number of sets in class* B *does not exceed*

$$D = n^{k\varkappa} 2^{kr+\varkappa} (P_1 \dots P_r)^k (p_1 \dots p_\varkappa)^{-k+m-1}.$$

Proof. We reduce estimating D to estimating (already carried out in Section 4.1) the number of vector sets in the second class, which was defined in the same section. As shown above, we have

$$\overline{x} \equiv \overline{y} \pmod q$$

(i.e., the vector $\overline{x}$ is congruent to the vector $\overline{y}$ modulo q) if their corresponding coordinates are congruent modulo q. It is easy to see that if there are two vector sets $(\overline{x}_1, \overline{x}_2, \dots, \overline{x}_k)$ and $(\overline{y}_1, \overline{y}_2, \dots, \overline{y}_k)$ and the vectors from the second set are congruent to the corresponding vectors from the first set modulo $p_1, \dots, p_k$, then these sets belong to the same class (either A or B). Thus

$$D \le \left(P_1(p_1, \dots, p_\varkappa)^{-1} + 1\right)^k \dots \left(P_r(p_1, \dots, p_\varkappa)^{-1} + 1\right)^k V,$$

where V is the number of sets $(\overline{y}_1, \overline{y}_2, \dots, \overline{y}_k)$ consisting of k vectors $\overline{y}$ such that each of their coordinates is an integer strictly less than $p_1 \dots p_\varkappa$. But we have already proved (see Section 4.1) that V satisfies the inequality

$$V \le \prod_{j=1}^{\varkappa} (2p_j^{m-1}(np_j^{r-1})^k),$$

where $n = n_1 + \dots + n_r$. This implies

$$D \le n^{k\varkappa} 2^{kr+\varkappa} (P_1 \dots P_r)(p_1 \dots p_\varkappa)^{-k+m-1},$$

as required. □

Lemma 4.7. *Let* $P_1 = \min(P_1, P_2, \dots, P_r)$*. For* $P_1 > 1$ *and each* $s = 1, \dots, r$*, we determine natural numbers* μ_1 *and* ν_s *from the relations*

$$-\frac{1}{2} \le \frac{\ln P_s}{\ln P_1} - \mu_s < \frac{1}{2}, \quad -1 < \frac{\ln P_s}{\ln P_1} - \nu_s \le 0.$$

We also set

$$\varkappa = \mu_1 n_1 + \cdots + \mu_r n_r, \quad \varkappa\gamma = 1,$$

where $n_1, \ldots, n_r$ are natural numbers. Let p be a real number in the interval $0.5P_1^\gamma \le p \le P_1^\gamma$. We determine numbers Q_s $(s = 1, \ldots, r)$ by the relations

$$Q_s = P_s p^{-\mu_s} + 1.$$

Then for all $s = 1, \ldots, r$ the following relations hold:

$$1. \quad Q_1 \le Q_s; \qquad 2. \quad \ln Q_s \le \nu_s \ln Q_1.$$

Proof. **1.** If $\mu_s = 1$ for some s $(1 \le s \le r)$, then we obtain

$$Q_1 = P_1 p^{-1} + 1 \le P_s p^{-1} + 1 = Q_s.$$

Let now $\mu_s > 1$. Then it follows from the definition of μ_s that

$$\ln P_s / \ln P_1 \ge \mu_s - 1/2, \quad P_s \ge P_1^{\mu_s - 1/2},$$

and hence

$$P_s p^{-\mu_s} \ge P_1^{\mu_s - 1/2} p^{-\mu_s} = (P_1 p^{-1})^{\mu_s} P_1^{-1/2} \ge P_1^{3/2} p^{-2}.$$

Since $\varkappa = \mu_1 n_1 + \cdots + \mu_r n_r \ge 2$, $\gamma \le 1/2$, we have

$$p \le P_1^\gamma \le P_1^{1/2}, \quad P_1^{3/2} p^{-2} \ge P_1 p^{-1}, \quad P_s p^{-\mu_s} \ge P_1 p^{-1},$$
$$Q_s = P_s p^{-\mu_s} + 1 \ge P_1 p^{-1} + 1 = Q_1.$$

The first assertion in the lemma is proved.

2. We first note that for $x \ge y > 1$, we have the inequality

$$\ln(x+1)/\ln(y+1) \le \ln x / \ln y. \tag{4.18}$$

Indeed, in this case we have the relation $x = y^{1+\alpha}$ for some $\alpha \ge 0$. Therefore, inequality (4.18) is equivalent to the inequality

$$\ln(1 + y^{1+\alpha})/\ln(1+y) \le 1 + \alpha,$$

which, in turn, follows from the inequality

$$(1+y)^{1+\alpha} \ge 1 + y^{1+\alpha}.$$

Using (4.18), we successively obtain

$$\frac{\ln Q_s}{\ln Q_1} = \frac{\ln(P_s p^{-\mu_s} + 1)}{\ln(P_1 p^{-1} + 1)} \le \frac{\ln P_s p^{-\mu_s}}{\ln P_1 p^{-1}} = \frac{\ln P_s - \mu_s \ln p}{\ln P_1 - \ln p}.$$

Let $\ln P_s/\ln P_1 = \mu_s + \alpha_s$, where $-0.5 \le \alpha_s < 0.5$. Then

$$\ln P_s = \mu_s \ln P_1 + \alpha_s \ln P_1,$$
$$\frac{\ln Q_s}{\ln Q_1} \le \frac{\mu_s(\ln P_1 - \ln p) + \alpha_s \ln P_1}{\ln P_1 - \ln p} = \mu_s + \frac{\alpha_s \ln P_1}{\ln P_1 p^{-1}}.$$

If $\alpha_s \le 0$, then $\mu_s - 1 < \mu_s + \alpha_s \le \mu_s$, i.e., by definition, we have $\nu_s = \mu_s$. Hence

$$\frac{\ln Q_s}{\ln Q_1} \le \mu_s + \frac{\alpha_s \ln P_1}{\ln P_1 p^{-1}} \le \nu_s.$$

If $0 < \alpha_s < 0.5$, then $\mu_s < \ln P_s/\ln P_1 = \mu_s + \alpha_s < \mu_s + 1 = \nu_s$. Further,

$$\frac{\ln P_1}{\ln P_1 p^{-1}} = \frac{\ln P_1 p^{-1} + \ln p}{\ln P_1 p^{-1}} = 1 + \frac{\ln p}{\ln P_1 p^{-1}} \le 2.$$

Hence we obtain

$$\frac{\ln Q_s}{\ln Q_1} \le \mu_s + \frac{\alpha_s \ln P_1}{\ln P_1 p^{-1}} \le \mu_s + 2\alpha_s < \mu_s + 1 = \nu_s.$$

The second assertion is the lemma is also proved. □

The following lemma is the second fundamental lemma in the p-adic proof of the mean value theorem and is called the *lemma on the recurrence inequality*.

Lemma 4.8. *Suppose that $r > 1$, $n_1, \dots, n_r$ are natural numbers, $k \ge 2m$, and $\min(P_1, \dots, P_r) = P_1 > 1$. For each $s = 1, \dots, r$, we determine natural numbers μ_s and ν_s by the relations*

$$-0.5 \le \frac{\ln P_s}{\ln P_1} - \mu_s < 0.5, \quad -1 < \frac{\ln P_s}{\ln P_1} - \nu_s \le 0.$$

We also set

$$\varkappa = \mu_1 n_1 + \dots + \mu_r n_r, \quad \gamma\varkappa = 1, \quad \mu = \mu_1 + \dots + \mu_r.$$

Then there exists a number p in the interval $[0.5P_1^\gamma, P_1^\gamma]$ such that the following inequality holds:

$$J(\overline{P}; \overline{n}, k) \le k^{2m}\varkappa^2 2^{2mr}(P_1 \dots P_r)^{2m} p^{2\mu(k-m)-m\varkappa/2} J(\overline{Q}; \overline{n}, k-m)$$
$$+ 2^{-\varkappa}(4\varkappa)^{2k\varkappa}(P_1 \dots P_r)^{2k} P_1^{-k},$$

where $\overline{P} = (P_1, \dots, P_r)$, $\overline{Q} = (Q_1, \dots, Q_r)$, and the variables $Q_1, \dots, Q_r$ are determined by the relations

$$Q_s = P_s p^{-\mu_s} + 1, \quad s = 1, 2, \dots, r,$$

and also satisfy the conditions

$$Q_1 = \min(Q_1, Q_2, \dots, Q_r), \quad \ln Q_s \le \nu_s \ln Q_1, \quad s = 1, \dots, r.$$

Proof. We assume that $P_1 > (4\varkappa)^{2\varkappa}$ because, otherwise, by setting $p = P_1^{\gamma}$, we see that the second term in the right-hand side in the assertion of the lemma is larger than $(P_1 \dots P_r)^{2k}$. By Lemma 3.8 (see Chapter 3), the interval $[0.5P_1^{\gamma}, P_1^{\gamma}]$ contains at least $\varkappa$ prime numbers, which we denote by $p_1, \dots, p_\varkappa$. The further argument repeats the argument in Section 4.1, and we successively obtain the estimates

$$J \le 2J_1 + 2J_2, \quad J_1 \le \varkappa^2 J_3, \quad J_3 \le \binom{k}{m}^2 J_4,$$
$$J_4 = \int_{\Omega}\cdots\int \Big|\sum_{\overline{x}_1}{}' \cdots \sum_{\overline{x}_{2m-1}}{}'\Big|^2 \Big|\sum_{\overline{x}}\Big|^{2(k-m)} d\Omega,$$

where the prime on the distinguished repeated sum denotes summation over the sets $\overline{x}_1, \overline{x}_3, \dots, \overline{x}_{2m-1}$ satisfying the regularity condition modulo p. Now the summation over $\overline{x} = (x_1, \dots, x_r)$ in the last sum can be represented as the summation over arithmetic progressions whose form depends on the number of the coordinate of the vector $\overline{x}$ as follows: if x_s is the sth coordinate of $\overline{x}$, then

$$x_s = y_s + p^{\mu_s} z_s, \quad 0 \le y_s < p^{\mu_s}, \quad 0 \le z_s < P_s p^{-\mu_s}, \quad s = 1, \dots, r.$$

The rest of the argument in Section 4.1 concerning the estimate of J_4 is preserved. Applying Lemma 4.5 when necessary, we obtain the final inequality

$$J_4 \le p^{2(\mu_1+\cdots+\mu_r)(k-m)} m!(P_1 p^{-\varkappa} + 1)^{2m}$$
$$\dots (P_r p^{-\varkappa} + 1)^{2m} p^{2mr\varkappa - 0.5m\varkappa} J(\overline{Q}; \overline{n}, k-m),$$
$$J_1 \le \varkappa^2 \frac{k^{2m}}{m!} 2^{2mr} p^{2(\mu_1+\cdots+\mu_r)(k-m)-0.5m\varkappa} (P_1 \dots P_r)^{2m} J(\overline{Q}; \overline{n}, k-m).$$

We see that the estimate obtained for J_1 corresponds to the first term in the statement of the lemma.

Let us estimate J_2. The value of J_2 does not exceed the squared number U of terms from class B. But class B contains sets of vectors $\overline{x}_1, \overline{x}_3, \dots, \overline{x}_{2k-1}$ that are singular modulo p for $p = p_j$ $(1 \le j \le \varkappa)$. By Lemma 4.6, the number U does not exceed

$$n^{k\varkappa} 2^{kr+\varkappa} (P_1 \dots P_r)^k (p_1 \dots p_\varkappa)^{m-k-1}.$$

Hence J_2 satisfies the estimate

$$J_2 \le U^2 \le n^{2k\varkappa} 2^{2kr+2\varkappa} (P_1 \dots P_r)^{2k} (p_1 \dots p_\varkappa)^{-2k+2m-2}$$
$$\le n^{2k\varkappa} 2^{2kr+4\varkappa+2k\varkappa-2m\varkappa} (P_1 \dots P_r)^{2k} P_1^{-2k+2m-2}.$$

Hence, for $k \ge 2m$, we obtain the estimate

$$J_2 \le 2^{-2\varkappa} (4\varkappa)^{2k\varkappa} (P_1 \dots P_r)^{2k} P_1^{-k},$$

which corresponds to the second term in the statement of the lemma. The estimates obtained imply the desired estimate in the lemma.

The last part of the statement of the lemma, namely, the relations

$$Q_1 = \min(Q_1, \dots, Q_r), \quad \ln Q_s \le \nu_s \ln Q_1, \quad s = 1, \dots, r,$$

follow from the definition of Q_s and Lemma 4.7. □

4.2.3 The mean value theorem

Theorem 4.2. *Suppose that $\tau \ge 0$ is an integer and $n_1, \dots, n_r$ are natural numbers. Then for $k_1 \ge k = m\tau$, the variable $J = J(\overline{P}; \overline{n}, k_1)$ satisfies the estimate*

$$J \le k^{2m\tau} \varkappa^{4\varkappa^2 \Delta(\tau)} 2^{8m\varkappa\tau} (P_1 \dots P_r)^{2k_1} P^{-\varkappa\Delta(\tau)},$$

where $\varkappa = n_1\nu_1 + \dots + n_r\nu_r$, $\gamma\varkappa = 1$, $m = (n_1+1)\dots(n_r+1)$, and

$$\Delta(\tau) = 0.5m(1-(1-\gamma)^\tau), \quad P = (P_1^{n_1} \dots P_r^{n_r})^\gamma.$$

Here $\nu_1, \dots, \nu_r$ are natural numbers such that

$$-1 < \frac{\ln P_s}{\ln P_1} - \nu_s \le 0, \quad s = 1, \dots, r.$$

Proof. It sufficcs to prove the theorem for $k_1 = k = m\tau$. If $\tau = 0$, then the statement of the theorem becomes trivial. Suppose that $\tau = 1$ and $k = m$. We choose a prime number q in the interval $P_1 \le q \le 2P_1$. Let T be the number of solutions of the system of congruences modulo q corresponding to the system of equations (4.17) (instead of Eqs. (4.17), we consider the system of congruences modulo q with the same conditions on the unknowns). Obviously, we have

$$J \le T \le (P_1 q^{-1} + 1)^{2m} \dots (P_r q^{-1} + 1)^{2m} T_0,$$

where T_0 is the variable estimated in Lemma 4.3. We obtain

$$\begin{aligned} &T_0 \le m! q^{2mr-m+1}, \\ &J \le m! 2^{4mr} (P_1 \dots P_r)^{2m} P_1^{-m+1} \le m^{2m} 2^{8m\varkappa} (P_1 \dots P_r)^{2m} P^{-0.5m}. \end{aligned}$$

The last inequality becomes obvious if we recall that

$$n_1 + n_2 \frac{\ln P_2}{\ln P_1} + \dots + n_r \frac{\ln P_r}{\ln P_1} \le n_1\nu_1 + \dots + n_r\nu_r = \varkappa, \quad \varkappa\gamma = 1, \quad \varkappa \ge r.$$

Thus we see that the statement of the theorem holds for $\tau = 0$ and $\tau = 1$.

We assume that the statement of the theorem holds for $\tau = s$ and prove it for $\tau = s+1$. Since $s \geq 1$, $s+1 \geq 2$, and $k = m(s+1) \geq 2m$, we can use Lemma 4.8 to estimate $J = J(\overline{P}; \overline{n}, k)$:

$$J \leq k^{2m} \varkappa^2 2^{2mr} (P_1 \dots P_r)^{2m} p^{2\mu(k-m)-0.5m\varkappa} J(\overline{Q}; \overline{n}, k-m) \qquad (4.19)$$
$$+ 2^{-\varkappa} (4\varkappa)^{2k\varkappa} (P_1 \dots P_r)^{2k} P_1^{-k},$$

where $\mu = \mu_1 + \dots + \mu_r$ and $\mu_1, \dots, \mu_r$ are natural numbers determined by the conditions

$$-0.5 \leq \frac{\ln P_j}{\ln P_1} - \mu_j < 0.5, \quad j = 1, \dots, r,$$
$$0.5 P_1^{\gamma} \leq p \leq P_1^{\gamma}, \quad \overline{Q} = (Q1, \dots, Q_r), \quad Q_s = P_s p^{-\mu_s} + 1, \quad s = 1, \dots, r.$$

We apply the induction assumption to estimate $J(\overline{Q}; \overline{n}, k-m)$. For this, we note that $Q_1 = \min(Q_1, \dots, Q_r) > 1$. Next, we determine natural numbers $\nu_1', \dots, \nu_r'$ from the relations

$$-1 < \ln Q_j / \ln Q_1 - \nu_j' \leq 0, \quad j = 1, \dots, r,$$

and set $\varkappa_1 = n_1 \nu_1' + \dots + n_r \nu_r'$ and $\varkappa_1 \gamma_1 = 1$. We note that Lemma 4.7 readily implies the estimates

$$\ln Q_j / \ln Q_1 \leq \nu_j, \quad \text{i.e.} \quad \nu_j' \leq \nu_j, \quad j = 1, \dots, r,$$

Hence $\varkappa_1 \leq \varkappa$ and $\gamma \leq \gamma_1$.

So it follows from the induction assumption that ($k - m = ms$, $\tau = s$)

$$J(\overline{Q}; \overline{n}, k-m) \leq (ms)^{2ms} \varkappa_1^{4\varkappa_1^2 \Delta_1(s)} 2^{8ms\varkappa_1} (Q_1 \dots Q_r)^{2ms} Q^{-\varkappa_1 \Delta_1(s)},$$

where $\Delta_1(s) = 0.5m(1 - (1-\gamma_1)^s)$ and $Q = (Q_1^{n_1} \dots Q_r^{n_r})^{\gamma_1}$.

From (4.19) and the last inequality, we find

$$J \leq W_1 + W_2,$$

where

$$W_1 = (m(s+1))^{2m} \varkappa^2 2^{2mr} (ms)^{2ms} \varkappa_1^{4\varkappa_1^2 \Delta_1(s)} 2^{8ms\varkappa_1}$$
$$\times (P_1 \dots P_r)^{2m} p^{2\mu ms - 0.5m\varkappa} (Q_1 \dots Q_r)^{2ms} Q^{-\varkappa_1 \Delta_1(s+1)},$$
$$W_2 = 2^{-\varkappa} (4\varkappa)^{2m(s+1)\varkappa} (P_1 \dots P_r)^{2m(s+1)} P_1^{-k}.$$

It remains to prove that

$$W_1 \leq 0.5W, \quad W_2 \leq 0.5W,$$

where

$$W = (m(s+1))^{2m(s+1)} \varkappa^{4\varkappa^2 \Delta(s+1)} 2^{8m(s+1)\varkappa} (P_1 \dots P_r)^{2m(s+1)} P^{-\varkappa \Delta(s+1)}$$

(W is the right-hand side in the inequality in the statement of the theorem). As already noted, we have

$$P_1^{n_1} \dots P_r^{n_r} \le P_1^{\varkappa}. \tag{4.20}$$

Moreover, $\Delta(s+1) = 0.5m(1-(1-\gamma)^{s+1}) \le 0.5m(s+1)\gamma$. Hence

$$(P_1^{n_1} \dots P_r^{n_r})^{-\Delta(s+1)} \ge P^{-0.5m(s+1)},$$

i.e., the lowering obtained in the theorem is not better than $P_1^{-0.5m(s+1)}$, and hence if $P_1 \le (m(s+1))^4$, then, taking the first factor in W into account, we obtain $W > (P_1 \dots P_r)^{2m(s+1)}$. The statement of the theorem thus becomes trivial. Hence we assume

$$P_1 > (m(s+1))^4 \ge (2ms)^2.$$

Moreover, we can assume that $P_1^{n_1} \dots P_r^{n_r} > \varkappa^{4\varkappa^2}$. Otherwise, the estimate of J becomes trivial because of the second factor in the statement of the theorem. Thus we can assume that $P_1 \ge \varkappa^{4\varkappa}$.

Now we estimate W_1. By the definition of the variables $Q_1, \dots, Q_r$, we obtain

$$(Q_1 \dots Q_r)^{2ms} \le (P_1 \dots P_r)^{2ms}(1+pP_1^{-1})^{2ms} \dots (1+pP_r^{-1})^{2ms} p^{-2\mu ms}.$$

Next, $\gamma \le 0.5$, $p \le \sqrt{P}$, and $P_j \ge P_1$ $(j = 1, \dots, r)$. For $\mu_j = 1$, we have the estimates

$$p^{\mu_j} P_j^{-1} = pP_j^{-1} \le P_1^{-0.5};$$

for $\mu_j \ge 2$, the inequality $P_j \ge P_1^{\mu_j - 0.5}$ implies

$$p^{\mu_j} P_j^{-1} \le (pP_1^{-1})^{\mu_j} P_1^{0.5} \le P_1^{-0.5\mu_j+0.5} \le P_1^{-0.5}.$$

So we always have (because $P_1 \ge (2ms)^2$) $p^{\mu_j} P_j^{-1} \le P_1^{-0.5} \le 1/(2ms)$. Therefore, we obtain

$$(1+p^{\mu_j} P_j^{-1})^{2ms} \le (1+1/(2ms))^{2ms} \le 3,$$
$$(Q_1 \dots Q_r)^{2ms} \le 3^r (P_1 \dots P_r)^{2ms} p^{-2\mu ms}.$$

We again use the definition of Q_j to obtain

$$P_j p^{-\mu_j} < Q_j,$$
$$(Q_1^{n_1} \dots Q_r^{n_r})^{-\Delta_1(s)} \le (P_1^{n_1} \dots P_r^{n_r})^{-\Delta_1(s)} p^{(n_1\mu_1+\dots+n_r\mu_r)\Delta_1(s)},$$
$$(P_1 \dots P_r)^{2m} p^{2\mu ms - 0.5m\varkappa} (Q_1 \dots Q_r)^{2ms} (Q_1^{n_1} \dots Q_r^{n_r})^{-\Delta_1(s)}$$
$$\le 3^r (P_1 \dots P_r)^{2m(s+1)} (P_1^{n_1} \dots P_r^{n_r})^{-\Delta_1(s)} p^{-0.5m\varkappa+(n_1\mu_1+\dots+n_r\mu_r)\Delta_1(s)}.$$

Now we show that

$$\begin{aligned}
&3^r(m(s+1))^{2m}\varkappa^2 2^{2mr}(ms)^{2ms}\varkappa_1^{4\varkappa_1^2\Delta_1(s)}2^{8ms\varkappa_1}\\
&\quad\times(P_1\dots P_r)^{2m(s+1)}(P_1^{n_1}\dots P_r^{n_r})^{-\Delta_1(s)}p^{-0.5m\varkappa+(n_1\mu_1+\dots+n_r\mu_r)\Delta_1(s)}\\
&\quad\le 0.5(m(s+1))^{2m(s+1)}\varkappa^{4\varkappa^2\Delta(s+1)}2^{8m(s+1)\varkappa}(P_1\dots P_r)^{2m(s+1)}p^{-\varkappa\Delta(s+1)}.
\end{aligned}$$

We increase the left-hand side of the last inequality replacing $\varkappa_1$ by $\varkappa$. It is easy to see that

$$3^r(m(s+1))^{2m}\varkappa^2 2^{2mr}(ms)^{2ms}2^{8ms\varkappa}\le 0.5(m(s+1))^{2m(s+1)}2^{8m(s+1)\varkappa}.$$

Indeed, after cancellation (with some roughening of the left-hand side), we obtain the relation

$$3^r\varkappa^2 2^{2mr}\le 2^{8m\varkappa-1},$$

which is always trivial (one must only have in mind that $\varkappa\ge n_1+\dots+n_r\ge r$). Now it remains to prove the inequality

$$\begin{aligned}
&\varkappa^{4\varkappa^2\Delta_1(s)}(P_1^{n_1}\dots P_r^{n_r})^{-\Delta_1(s)}p^{-0.5m\varkappa+(n_1\mu_1+\dots+n_r\mu_r)\Delta_1(s)}\\
&\quad\le\varkappa^{4\varkappa^2\Delta(s+1)}(P_1^{n_1}\dots P_r^{n_r})^{-\Delta(s+1)},
\end{aligned}$$

or the equivalent inequality

$$(P_1^{n_1}\dots P_r^{n_r}\varkappa^{-4\varkappa^2})^{\Delta(s+1)-\Delta_1(s)}\le p^{0.5m\varkappa-(n_1\mu_1+\dots+n_r\mu_r)\Delta_1(s)}. \tag{4.21}$$

It follows from the definition of μ_j that $\mu_j\le\nu_j$ $(j=1,\dots,r)$. Hence we have $\varkappa=n_1\nu_1+\dots+n_r\nu_r\ge n_1\mu_1+\dots+n_r\mu_r$. Moreover,

$$\Delta_1(s)=0.5m\big(1-(1-\gamma_1)^s\big)\le 0.5m.$$

Hence

$$0.5m\varkappa-(n_1\mu_1+\dots+n_r\mu_r)\Delta_1(s)>0.$$

Next, as noted above, we have

$$P_1^{n_1}\dots P_r^{n_r}\varkappa^{-4\varkappa^2}>1.$$

Therefore, if $\Delta(s+1)-\Delta_1(s)\le 0$, then inequality (4.21) holds trivially. Let $\Delta(s+1)-\Delta_1(s)>0$. We have the inequalities

$$P_1^{n_1}\dots P_r^{n_r}\le P_1^{\varkappa},\quad P_1\le 2^{\varkappa}p^{\varkappa},\quad P_1^{n_1}\dots P_r^{n_r}\varkappa^{-4\varkappa^2}\le 2^{\varkappa^2}\varkappa^{-4\varkappa^2}p^{\varkappa^2}<p^{\varkappa^2},$$
$$p^{0.5m\varkappa-(n_1\mu_1+\dots+n_r\mu_r)\Delta_1(s)}\ge p^{0.5m\varkappa-\varkappa\Delta_1(s)}.$$

Now we prove the inequality

$$p^{\varkappa^2(\Delta(s+1)-\Delta_1(s))}\le p^{0.5m\varkappa-\varkappa\Delta_1(s)}$$

or the equivalent inequality $\varkappa\big(\Delta(s+1) - \Delta_1(s)\big) \le 0.5m - \Delta_1(s)$. We successively obtain ($\varkappa\gamma = 1$)

$$0.5m\varkappa\big((1-\gamma_1)^s - (1-\gamma)^{s+1}\big) \le 0.5m(1-\gamma_1)^s,$$
$$(1-\gamma_1)^s - (1-\gamma)^{s+1} \le \gamma(1-\gamma_1)^s, \quad (1-\gamma_1)^s(1-\gamma) \le (1-\gamma)^{s+1},$$
$$\gamma \le \gamma_1.$$

Thus we have proved the inequality $W_1 \le 0.5W$.

The inequality $W_2 \le 0.5W$ can be proved much simpler. If

$$(4\varkappa)^{2m(s+1)\varkappa} \le 2^{8m(s+1)\varkappa} P_1^{m(s+1)} (P_1^{n_1} \dots P_r^{n_r})^{-\Delta(s+1)},$$

then the desired relation holds trivially. But, as already noted, we have

$$(P_1^{n_1} \dots P_r^{n_r})^{\Delta(s+1)} \le P_1^{0.5m(s+1)}, \quad P_1 > \varkappa^{4\varkappa}.$$

Hence we obtain the obvious inequality

$$2^{4m(s+1)\varkappa}\varkappa^{2m(s+1)\varkappa} \le 2^{8m(s+1)\varkappa}\varkappa^{2m(s+1)\varkappa}.$$

The proof of the theorem is complete. □

Remark 4.4. The statement of the theorem remains valid if the unknowns in system (4.17) are subjected to any additional restrictions.

4.2.4 On the accuracy of the estimate in the mean value theorem

We note that the estimate in Theorem 4.2 is correct in the order of magnitude of increasing variables $P_1, \dots, P_r$. Indeed, following the argument similar to the argument in the proof of Lemma 4.1 (e), it is possible to prove the inequality

$$J(\overline{P}; \overline{n}, k) \ge (2k)^{-m}(P_1 \dots P_r)^{2k}(P_1^{n_1} \dots P_r^{n_r})^{-0.5m}.$$

On the other hand, for $k = cm\varkappa \log m\varkappa$, where $c > 1$ is a constant, i.e., for the k for which the upper bound for $J(\overline{P}; \overline{n}, k)$ is usually used in applications, Theorem 4.2 implies the inequality

$$J(\overline{P}; \overline{n}, k) \le \exp\{c_1 m\varkappa^2 \log m\varkappa\}(P_1 \dots P_r)^{2k}(P_1^{n_1} \dots P_r^{n_r})^{-0.5m+\delta},$$

where $c_1 > 0$ is a constant, δ does not exceed $\gamma/2$, and $\delta \to 0$ with increasing c.

Moreover, essentially using Theorem 4.2 for the above-mentioned values of k, we obtain an asymptotic formula for $J(\overline{P}; \overline{n}, k)$ (see Theorems 6.1 and 6.2 in Chapter 6).

We also show that the variable $k = k(\tau) = m\tau$ in the mean value theorem has a correct order of increase in the variables $n_1, \dots, n_r$. More precisely, we show that it is impossible to set $k(\tau) = 0.1m\tau$ instead of $k(\tau) = m\tau$ in the statement of

Theorem 4.2. The outline of the proof coincides, in general, with the outline of the corresponding argument in Section 4.5.1. We assume that, for all $s = 1, 2, \ldots, r$, $\ln P_s / \ln P_1 \leq \alpha(\overline{n}, r)$, where $\alpha(\overline{n}, r) \geq 1$ is a constant depending only on $\overline{n}$ and r.

Let $k_2(\tau)$ be the least positive integer such that the estimate

$$J(\overline{P}; \overline{n}, k_1) \ll (P_1 \ldots P_r)^{2k_1} (P_1^{n_1} \ldots P_r^{n_r})^{-\Delta(\tau)} \tag{4.22}$$

holds for all $P_1, \ldots, P_r$ and all $k_1 \geq k_2(\tau)$. Hereafter, we assume that the constant in Vinogradov's sign depends only on $\overline{n}$ and r. By the condition $\ln P_s / \ln P_1 \leq \alpha(\overline{n}, r)$, the variable $\varkappa$ in the statement of Theorem 4.1 satisfies the inequality $\varkappa \ll 1$. Hence Theorem 4.1 implies that the estimate (4.22) holds for $k_1 \geq k(\tau)$. Hence we have $k_2 = k_2(\tau) \leq k(\tau)$.

Since the theorem is proved for $k(\tau) = m\tau$, we have $k_2 \leq m\tau$. Our goal is to prove the inequality

$$k_2 = k_2(\tau) > 0.1k(\tau) = 0.1m\tau \quad \text{for any } \tau \ll 1.$$

By $m = m_r$ we denote the value of $(n_1 + 1)(n_2 + 1) \ldots (n_r + 1)$ and by m_{r-1} we denote the value of $(n_2 + 1) \ldots (n_r + 1)$. We consider the solutions of the system

$$\sum_{j=1}^{2k_2} (-1)^j x_{1j}^{t_1} \ldots x_{rj}^{t_r} = 0, \quad 0 \leq t_1 \leq n_1, \ \ldots, \ 0 \leq t_r \leq n_r,$$

that contain the unknowns $x_{1j} = 1$, $j = 1, \ldots, 2k_2$. We denote the number of such solutions by J_{r-1}. Obviously, it coincides with the number of solutions of the system

$$\sum_{j=1}^{2k_2} (-1)^j x_{2j}^{t_2} \ldots x_{rj}^{t_r} = 0, \quad 0 \leq t_2 \leq n_2, \ \ldots, \ 0 \leq t_r \leq n_r,$$

where the unknowns vary in the limits $1 \leq x_{2j} \leq P_2, \ \ldots, \ 1 \leq x_{rj} \leq P_r$, $j = 1, \ldots, 2k_2$.

As already noted, since $k_2 \ll 1$, the variable J_{r-1} has the lower bound

$$J_{r-1} \gg (P_2 \ldots P_r)^{2k_2} (P_2^{n_2} \ldots P_r^{n_r})^{-0.5m_{r-1}}. \tag{4.23}$$

Moreover, we have the trivial relation $J_r = J(\overline{P}; \overline{n}, k_2) \geq J_{r-1}$. This and inequalities (4.22) and (4.23) imply

$$(P_1 \ldots P_r)^{2k_2} (P_1^{n_1} \ldots P_r^{n_r})^{-\Delta(\tau)} \gg J \gg (P_2 \ldots P_r)^{2k_2} (P_2^{n_2} \ldots P_r^{n_r})^{-0.5m_{r-1}}.$$

It follows from the last inequality that

$$2k_2 \geq \Delta(\tau)\left(n_1 + \frac{\ln P_2}{\ln P_1} n_2 + \cdots + \frac{\ln P_r}{\ln P_1} n_r\right)$$

$$- 0.5m_{r-1}\left(n_2 \frac{\ln P_2}{\ln P_1} + \cdots + n_r \frac{\ln P_r}{\ln P_1}\right).$$

We transform the right-hand side and thus obtain

$$2k_2 \geq \left(n_1 + \frac{\ln P_2}{\ln P_1} n_2 + \cdots + \frac{\ln P_r}{\ln P_1} n_r\right)$$
$$\times (\Delta(\varkappa) - 0.5m_{r-1}) + 0.5(m_r - m_{r-1}) > 0.2m_r\varkappa = 0.2m\varkappa$$

for $n_1 \geq 2$ and $\tau = \varkappa$. Hence $k_2 = k_2(\varkappa) > 0.1m\varkappa = 0.1k(\varkappa)$, as required.

Concluding remark on Chapter 4. The mean value theorem for trigonometric sums of arbitrary multiplicity was proved by G. I. Arkhipov and V. N. Chubarikov [11], [10]. The statement of this theorem contained a new result even for the case of one-dimensional sums, i.e., for the case of the classical Vinogradov's mean value theorem. The improved results in these papers were obtained by induction on the set of two parameters one of which it the length of the summation interval and the other is the "accuracy" of averaging [11]. Simultaneously, a less precise result was obtained by S. B. Stechkin by the method of successive iterations [143].

Chapter 5

Estimates for multiple trigonometric sums

In this chapter we obtain a general estimate for the trigonometric sum $S(A)$ introduced in Section 4.2. We divide all points $\alpha(t_1, \dots, t_r)$ with the condition $0 \le \alpha(t_1, \dots, t_r) < 1$, where $t_1 + \dots + t_r \ge 1$, into two classes depending on their approximation by fractions. We obtain a uniform, rather sharp estimate of $|S(A)|$ for points of the second class, which comprise the overwhelming majority of points. We obtain an estimate which in many cases is best possible for points of the first class. In deriving the estimate of $|S(A)|$ for points of the first class, we shall use the estimates obtained in Chapters 1 and 2 for multiple trigonometric integrals and complete multiple trigonometric sums. In addition, we shall also need a generalization of van der Korput's lemma to the multidimensional case.

In deriving the estimate of $|S(A)|$ for points of the second class, we shall need the theorems on the multiplicity of the intersection of regions that we prove in Section 5.1.

5.1 Theorems on the multiplicity of intersection of multidimensional regions

The theorems in this section give an upper bound for the multiplicity of regions. We shall use this bound to estimate the trigonometric sums of general type.

We now explain the logical interrelation of the results in this section.

In Chapter 3 we describe in detail the general scheme for reducing an estimate of an individual trigonometric sum to an estimate of its mean value. In particular, in this reduction a point of the m-dimensional space whose coordinates are the coefficients of the polynomial $F(\overline{x})$ in the exponent of the multiple trigonometric sum is enclosed in a rectangular region ω in such a way that the absolute value of the trigonometric sum is almost the same throughout the region. If the interval of summation is shifted by a vector $\overline{y} = (y_1, \dots, y_r)$, then the coefficients of the polynomial $F(\overline{x})$ in the exponent of the trigonometric sum change and become themselves some polynomials in $\overline{y}$. In the argument below, we shall denote these polynomials by $B(\overline{t}) = B(t_1, \dots, t_r) = B(\overline{t}; \overline{y})$. The condition that the region $\omega = \Omega(\overline{y}_1)$ intersects with a fixed region $\Omega(y_0)$ essentially means that the respective coefficients of $B(\overline{t}; \overline{y}_1)$ and $B(\overline{t}; \overline{y}_0)$ are close to one another modulo 1 for all $\overline{t} = (t_1, \dots, t_r)$. In other words, if these regions intersect, then some congruences nonlinear in $\overline{y}$ hold modulo 1.

In order to use these congruences to obtain an upper bound for the number G (the number of distinct vectors $\overline{y}_1$ satisfying these congruences, i.e., the number of regions that intersect the fixed region), we use a technique of I. M. Vinogradov, which enables us, from the above system of nonlinear congruences, to derive a system of congruences that are linear in $\overline{y}_1$.

Unlike the one-dimensional case, this system contains several unknowns. Hence, to realize Vinogradov's technique for obtaining an estimate of the correct order for G in dependence on the value of the least common multiple of the denominators in rational approximations of the coefficients of $F(\overline{x})$, we must use some additional considerations.

The method by means of which a nonlinear system can be reduced to a linear system splits into two parts. The first step consists in finding for the $B(\overline{t}; \overline{y}_1)$ a particular representation in terms of the linear forms contained in the other coefficients, i.e., in the other polynomials $B(\overline{t}; \overline{y}_1)$. We think that every polynomial is the sum of a constant term, a linear form, a quadratic form, a cubic form, and so on. At the second step, we use this representation to derive congruences in linear forms from the system of nonlinear congruences for the coefficients of $B(\overline{t}; \overline{y})$.

In the one-dimensional case, the first step is trivial, and the main difficulty is with the second step. In the multidimensional case which we shall study here, both steps are of approximately equal difficulty. The first of these steps is studied in Lemma 5.1 and the second in Lemma 5.2.

In Lemma 5.1, to obtain the required representation of the coefficients in terms of linear forms, we construct a system of polynomials with nonnegative integer coefficients by means of generating functions; then the required representation is demonstrated by direct transformations.

The proof of Lemma 5.2 is to a large extent similar to that in the one-dimensional case. The multidimensionality yields new parameters; hence, one must use the multidimensional induction and, in order to preserve the accuracy, once again select numerical factors.

To estimate G, we prove the following three assertions. First, we recall some old notions and introduce new ones.

Let $F(x_1, \dots, x_r)$ be a polynomial introduced in Section 4.2, i.e.,

$$F(x_1, \dots, x_r) = \sum_{t_1=0}^{n_1} \cdots \sum_{t_r=0}^{n_r} \alpha(t_1, \dots, t_r) x_1^{t_1} \dots x_r^{t_r}.$$

We define a function $B = B(u_1, \dots, u_r) = B(\overline{u})$ by the relation

$$F(x_1 + y_1, \dots, x_r + y_r) - F(x_1 + z_1, \dots, x_r + z_r)$$

$$= \sum_{t_1=0}^{n_1} \cdots \sum_{t_r=0}^{n_r} \alpha(t_1, \dots, t_r)\big((x_1 + y_1)^{t_1} \dots (x_r + y_r)^{t_r} - (x_1 + z_1)^{t_1} \dots (x_r + z_r)^{t_r}\big)$$

$$= \sum_{t_1=0}^{n_1} \cdots \sum_{t_r=0}^{n_r} B(t_1, \ldots, t_r) x_1^{t_1} \ldots x_r^{t_r}.$$

Hence the function $B(t_1, \ldots, t_r)$ depends not only on $t_1, \ldots, t_r$ but also on $y_1, \ldots, y_r, z_1, \ldots, z_r$ and (by definition) can be written as

$$B(u_1, \ldots, u_r) = \sum_{t_1=u_1}^{n_1} \sum_{t_r=u_r}^{n_r} \alpha(t_1, \ldots, t_r) \binom{t_1}{u_1} \cdots \binom{t_r}{u_r} \times (y_1^{t_1-u_1} \ldots y_r^{t_r-u_r} - z_1^{t_1-u_1} \ldots z_r^{t_r-u_r}).$$

Next, we set $u = u_1 + \cdots + u_r$, $v = v_1 + \cdots + v_r$, $n = n_1 + \cdots + n_r$, and $t = t_1 + \cdots + t_r$. We define a function $A = A(u_1, \ldots, u_r; s) = A(\overline{u}; s)$ as the sth degree form in the polynomial $B(t_1, \ldots, t_r)$; in other words,

$$A(u_1, \ldots, u_r; s) = A(\overline{u}; s) = \sum_{\substack{v_1=u_1 \\ v=s+u}}^{n_1} \cdots \sum_{v_r=u_r}^{n_r} \alpha(v_1, \ldots, v_r) \binom{v_1}{u_1} \cdots \binom{v_r}{u_r} \times (y_1^{v_1-u_1} \ldots y_r^{v_r-u_r} - z_1^{v_1-u_1} \ldots z_r^{v_r-u_r}).$$

It follows from the definitions of B and A that

$$B(\overline{u}) = \sum_{s=0}^{n-u} A(\overline{u}; s).$$

Moreover, we note that

$$A(\overline{u}; 1) = \sum_{j=1}^{r} (u_j + 1) \alpha(u_1, \ldots, u_j + 1, \ldots, u_r)(y_j - z_j).$$

Lemma 5.1. *There exist polynomials $H(\overline{u}; \overline{v}; s)$ in the unknowns $y_1, \ldots, y_r$, $z_1, \ldots, z_r$ such that*

$$A(\overline{u}; s) \frac{1}{u_1! \ldots u_r! s!} \sum_{\substack{v_1=u_1 \\ v=s-1+u}}^{n_1} \cdots \sum_{v_r=u_r}^{n_r} v_1! \ldots v_r! H(\overline{u}; \overline{v}; s) A(\overline{v}; 1);$$

the sum of the coefficients of each of the polynomials $H(\overline{u}; \overline{v}; s)$ does not exceed sr^{s-1}, and the sum of powers of the variables y_j, z_j $(j = 1, \ldots, r)$ contained in any monomial in the polynomial H does not exceed $v_j - u_j$ $(j = 1, \ldots, r)$.

Proof. We define a function $g(w_1, \dots, w_r; s)$ which will be the generating function of the polynomials $H(\overline{u}; \overline{v}; s)$ by the relation

$$g(w_1, \dots, w_r; s) = w_1^{u_1} \dots w_r^{u_r} \frac{(y_1 w_1 + \dots + y_r w_r)^s - (z_1 w_1 + \dots + z_r w_r)^s}{(y_1 w_1 + \dots + y_r w_r) - (z_1 w_1 + \dots + z_r w_r)}$$
$$= \sum_{\substack{v_1 = u_1 \\ v = s-1+u}}^{+\infty} \dots \sum_{v_r = u_r}^{+\infty} H(\overline{u}; \overline{v}; s) w_1^{v_1} \dots w_r^{v^r}. \tag{5.1}$$

The definition of $g(w_1, \dots, w_r; s)$ implies

$$g(w_1, \dots, w_r; s)$$
$$= w_1^{u_1} \dots w_r^{u_r} \sum_{t=0}^{s-1} (y_1 w_1 + \dots + y_r w_r)^t (z_1 w_1 + \dots + z_r w_r)^{s-t-1}. \tag{5.2}$$

The coefficient of the monomial $w_1^{v_1} \dots w_r^{v_r}$ in the right-hand side of this relation is the polynomial $H(\overline{u}; \overline{v}; s)$ in the variables $y_1, \dots, y_r, z_1, \dots, z_r$. Hence it is easy to see that the coefficients of the polynomial $H(\overline{u}; \overline{v}; s)$ are integers. If we now set $w_1 = \dots = w_r = 1$, $y_1 = \dots = y_r = 1$, and $z_1 = \dots = z_r = 1$ in (5.2), then the right-hand side of this relation is, first, equal to the sum of all coefficients of all monomials $H(\overline{u}; \overline{v}; s)$, where $v_1 \geq u_1, \dots, v_r \geq u_r$, and, second, is equal to the number

$$\sum_{t=0}^{s-1} r^t r^{s-t-1} = s r^{s-1}.$$

Therefore, the sum of the coefficients of each monomial $H(\overline{u}; \overline{v}; s)$, $v_1 \geq u_1$, $\dots, v_r \geq u_r$, does not exceed $s r^{s-1}$. Moreover, the sum of powers of the variables y_j, z_j contained in each monomial in the polynomial $H(\overline{u}; \overline{v}; s)$ is equal to $v_j - u_j$ $(j = 1, \dots, r)$. Indeed, to this end, we set $z_j = y_j$ in the right-hand side of (5.2). Then the power of y_j before $w_j^{v_j}$ will be the required sum. But it is easy to see that in this case the power of y_j before $w_j^{v_j}$ is equal to $v_j - u_j$.

We transform the product

$$g(w_1, \dots, w_r; s)\big((y_1 w_1 + \dots + y_r w_r) - (z_1 w_1 + \dots + z_r w_r)\big), \tag{5.3}$$

first using the second part of relation (5.1) and then the first part. Equating the coefficients of $w_1^{v_1} \dots w_r^{v_r}$ to one another, we obtain the desired statement of the lemma.

So, from relation (5.1), we have obtained the following system of relations (here we change the order of summation and change the summation variable):

$$
\begin{aligned}
&g(w_1,\dots,w_r;s)((y_1-z_1)w_1+\dots+(y_r-z_r)w_r)\\
&=\sum_{\substack{v_1\ge u_1\\ v=s-1+u}}\cdots\sum_{v_r\ge u_r} H(\overline{v};\overline{u};s)w_1^{v_1}\dots w_r^{v_r}\sum_{j=1}^{r}(y_j-z_j)w_j\\
&=\sum_{\substack{v_1\ge u_1\\ v=s-1+u}}\cdots\sum_{v_r\ge u_r} H(v_1,\dots,v_r;\overline{u};s)\\
&\quad\times\sum_{j=1}^{r}(y_j-z_j)w_1^{v_1}\dots w_{j-1}^{v_{j-1}}w_j^{v_j+1}w_{j+1}^{v_{j+1}}\dots w_r^{v_r}\\
&=\sum_{j=1}^{r}\sum_{v_1\ge u_1}\cdots\sum_{v_{j-1}\ge u_{j-1}}\ \sum_{v_j+1\ge u_j+1}\ \sum_{v_{j+1}\ge u_{j+1}}\cdots\sum_{v_r\ge u_r}\\
&\qquad\quad {\scriptstyle v_1+\dots+v_{j-1}+(v_j+1)+v_{j+1}+\dots+v_r=s+u_1+\dots+u_r}\\
&\quad\times H\big(v_1,\dots,v_{j-1},v_j+1-1,v_{j+1},\dots,v_r;\overline{u};s\big)\\
&\quad\times(y_j-z_j)w_1^{v_1}\dots w_{j-1}^{v_{j-1}}w_j^{v_j+1}w_{j+1}^{v_{j+1}}\dots w_r^{v_r}\\
&=\sum_{j=1}^{r}\sum_{v_1\ge u_1}\cdots\sum_{v_{j-1}\ge u_{j-1}}\ \sum_{\substack{v_j\ge u_j+1\\ v=s+u}}\ \sum_{v_{j+1}\ge u_{j+1}}\cdots\sum_{v_r\ge u_r}\\
&\quad\times H\big(v_1,\dots,v_{j-1},v_j-1,v_{j+1},\dots,v_r;\overline{u};s\big)\\
&\quad\times(y_j-z_j)w_1^{v_1}\dots w_{j-1}^{v_{j-1}}w_j^{v_j}w_{j+1}^{v_{j+1}}\dots w_r^{v_r}.
\end{aligned}
$$

We show that the summation over v_j in the last sum can start from u_j. A new term equal to $(y_j-z_j)w_1^{v_1}\dots w_r^{v_r}$ multiplied by $H(v_1,\dots,v_{j-1},u_j-1,v_{j+1},\dots,v_r;\overline{u};s)$ appears in this sum for $v_j=u_j$. But, according to definition (5.1), the last factor is the coefficient of the monomial $w_1^{v_1}\dots w_j^{u_j-1}\dots w_r^{v_r}$ in the decomposition of $g=g(w_1,\dots,w_r;s)$ in powers of $w_1^{\varkappa_1}\dots w_r^{\varkappa_r}$. All the monomials contained in g have the degree in the variable w_j no less than u_j, because g is the product of $w_1^{u_1}\dots w_r^{u_r}$ by some polynomial in the same variables, namely, by

$$\frac{(y_1w_1+\dots+y_rw_r)^s-(z_1w_j+\dots+z_rw_r)^s}{(y_1w_1+\dots+y_rw_r)-(z_1w_1+\dots+z_rw_r)}.$$

Therefore, the factor $H(v_1,\dots,v_{j-1},u_j-1,v_{j+1},\dots,v_r;\overline{u};s)$ is equal to zero. Hence we can start the summation in the last sum from $v_j=u_j$ and rewrite this sum as

$$\sum_{j=1}^{r}\sum_{v_1\ge u_1}\cdots\sum_{\substack{v_j\ge u_j\\ v=s+u}}\cdots\sum_{v_r\ge u_r} H\big(v_1,\dots,v_j-1,\dots,v_r;\overline{u};s\big)\qquad(5.4)$$
$$\times(y_j-z_j)w_1^{v_1}\dots w_r^{v_r}$$

$$= \sum_{\substack{v_1 \geq u_1 \\ v=s+u}} \cdots \sum_{v_r \geq u_r} w_1^{v_1} \dots w_r^{v_r}$$
$$\times \sum_{j=1}^{r} (y_j - z_j) H(v_1, \dots, v_j - 1, \dots, v_r; \overline{u}; s).$$

Now we use the right-hand side of (5.1) to transform product (5.2). First, by the Newton binomial formula, we have

$$(y_1 w_1 + \dots + y_r w_r)^s = \sum_{k_1+\dots+k_r=s} \frac{s!}{k_1! \dots k_r!} y_1^{k_1} \dots y_r^{k_r} w_1^{k_1} \dots w_r^{k_r}.$$

Therefore, after obvious transformations, we obtain

$$\begin{aligned}
&g(w_1, \dots, w_r; s)((y_1 - z_1)w_1 + \dots + (y_r - z_r)w_r) \\
&\quad = w_1^{u_1} \dots w_r^{u_r} \left((y_1 w_1 + \dots + y_r w_r)^s - (z_1 w_1 + \dots + z_r w_r)^s\right) \\
&\quad = w_1^{u_1} \dots w_r^{u_r} \sum_{k_1+\dots+k_r=s} \frac{s!}{k_1! \dots k_r!} \\
&\qquad \times w_1^{k_1} \dots w_r^{k_r} (y_1^{k_1} \dots y_r^{k_r} - z_1^{k_1} \dots z_r^{k_r}) \\
&\quad = \sum_{k_1+\dots+k_r=s} \frac{s!}{k_1! \dots k_r!} w_1^{k_1+u_1} \dots w_r^{k_r+u_r} (y_1^{k_1} \dots y_r^{k_r} - z_1^{k_1} \dots z_r^{k_r}) \\
&\quad = \sum_{\substack{v_1 \geq u_1 \\ v=s+u}} \cdots \sum_{v_r \geq u_r} \frac{s!}{(v_1 - u_1)! \dots (v_r - u_r)!} \\
&\qquad \times w_1^{v_1} \dots w_r^{v_r} (y_1^{v_1-u_1} \dots y_r^{v_r-u_r} - z_1^{v_1-u_1} \dots z_r^{v_r-u_r})
\end{aligned} \tag{5.5}$$

(in the penultimate sum, we made a change of the summation variables of the form $k_1 + u_1 = v_1, \dots, k_r + u_r = v_r$).

Comparing the coefficients of $w_1^{v_1} \dots w_r^{v_r}$ in the last sums in (5.4) and (5.5), we obtain

$$\frac{s!}{(v_1 - u_1)! \dots (v_r - u_r)!} (y_1^{v_1-u_1} \dots y_r^{v_r-u_r} - z_1^{v_1-u_1} \dots z_r^{v_r-u_r})$$
$$= \sum_{j=1}^{r} (y_j - z_j) H(v_1, \dots, v_j - 1, \dots, v_r; \overline{u}; s),$$

or

$$y_1^{v_1-u_1} \dots y_r^{v_r-u_r} - z_1^{v_1-u_1} \dots z_r^{v_r-u_r}$$
$$= \frac{(v_1 - u_1)! \dots (v_r - u_r)!}{s!} \sum_{j=1}^{r} (y_j - z_j) H(v_1, \dots, v_j - 1, \dots, v_r; \overline{u}; s).$$

We first substitute this identity into the expression for $A(\overline{u}; s)$ and use the relation

$$\binom{v}{u} = \frac{v!}{u!(v-u)!}.$$

Then, again using the relation $H(v_1, \dots, u_j - 1, \dots, v_r; \overline{u}; s) = 0$, we change the order of summation and apply the formula for $A(\overline{u}; 1)$. We obtain

$$A(\overline{u}; s) = \sum_{\substack{v_1=u_1 \\ v=s+u}}^{n_1} \cdots \sum_{v_r=u_r}^{n_r} \alpha(v_1, \dots, v_r) \frac{v_1!}{u_1!(v_1-u_1)!} \cdots \frac{v_r!}{u_r!(v_r-u_r)!}$$

$$\times \frac{(v_1-u_1)! \dots (v_r-u_r)!}{s!} \sum_{j=1}^{r} (y_j - z_j) H(v_1, \dots, v_j - 1, \dots, v_r; \overline{u}; s)$$

$$= \frac{1}{u_1! \dots u_r! s!} \sum_{\substack{v_1=u_1 \\ v=s+u}}^{n_1} \cdots \sum_{v_r=u_r}^{n_r} v_1! \dots v_r! \alpha(v_1, \dots, v_r)$$

$$\times \sum_{j=1}^{r} (y_j - z_j) H(v_1, \dots, v_j - 1, \dots, v_r; \overline{u}; s)$$

$$= \frac{1}{u_1! \dots u_r! s!} \sum_{\substack{v_1=u_1 \\ v_1+\dots+(v_j-1)+\dots+v_r=s-1+u_1+\dots+u_r}}^{n_1} \cdots \sum_{v_j-1=u_j-1}^{n_j} \cdots \sum_{v_r=u_r}^{n_r}$$

$$\times v_1! \dots (v_j - 1 + 1)! \dots v_r! \alpha(v_1, \dots, v_j - 1 + 1, \dots, v_r)$$

$$\times \sum_{j=1}^{r} (y_j - z_j) H(v_1, \dots, v_j - 1, \dots, v_r; \overline{u}; s)$$

$$= \frac{1}{u_1! \dots u_r! s!} \sum_{v_1=u_1}^{n_1} \cdots \sum_{\substack{v_j=u_j-1 \\ v=s-1+u}}^{n_j} \cdots \sum_{v_r=u_r}^{n_r} v_1! \dots (v_j+1)! \dots v_r!$$

$$\times \alpha(v_1, \dots, v_j + 1, \dots, v_r) \sum_{j=1}^{r} (y_j - z_j) H(v_1, \dots, v_j, \dots, v_r; \overline{u}; s) =$$

$$= \frac{1}{u_1! \dots u_r! s!} \sum_{\substack{v_1=u_1 \\ v=s-1+u}}^{n_1} \cdots \sum_{v_r=u_r}^{n_r} v_1! \dots v_r! (v_j + 1)$$

$$\times \alpha(v_1, \dots, v_j + 1, \dots, v_r) \sum_{j=1}^{r} (y_j - z_j) H(v_1, \dots, v_r; \overline{u}; s) =$$

$$= \sum_{\substack{v_1=u_1 \\ v=s-1+u}}^{n_1} \cdots \sum_{v_r=u_r}^{n_r} H(\overline{v};\overline{u};s) \sum_{j=1}^{r} \frac{v_1!\dots v_r!}{u_1!\dots u_r!s!}(v_j+1)(y_j-z_j)$$

$$\times \alpha(v_1,\dots,v_j+1,\dots,v_r)$$

$$= \frac{1}{u_1!\dots u_r!s!} \sum_{\substack{v_1=u_1 \\ v=s-1+u}}^{n_1} \cdots \sum_{v_r=u_r}^{n_r} v_1!\dots v_r! H(\overline{v};\overline{u};s) A(\overline{v};1).$$

The proof of the lemma is complete. □

Lemma 5.2. *We let L_1 denote the number of solutions of the system of inequalities*

$$\|B(u_1,\dots,u_r)\| \le P_1^{-u_1}\dots P_r^{-u_r}, \tag{5.6}$$
$$u_1 = 0,1,\dots,n_1;\ \dots;\ u_r = 0,1,\dots,n_r,$$
$$n = n_1+\dots+n_r, \quad u = u_1+\dots+u_r, \quad 1 \le u \le n-1,$$

under the assumption that the unknowns $y_1,\dots,y_r$ run respectively through the integers in the intervals $[-Y_1,Y_1],\dots,[-Y_r,Y-r]$, $z_1,\dots,z_r$ are fixed integers from the same intervals, $Y_1 \le P_1,\dots,Y_r \le P_r$, and L_2 denotes the number of solutions under the same conditions of the linear system

$$\left\| \frac{n!}{(u+1)!}\cdot\frac{(n+1)!}{(u+2)!}A(u_1,\dots,u_r;1) \right\| \tag{5.7}$$
$$\le \frac{n!}{(u+1)!}\cdot\frac{(n+1)!}{(u+2)!}(4rn^2)^{n-u-1}P_1^{-u_1}\dots P_r^{-u_r},$$
$$u_1 = 0,1,\dots,n_1;\dots;u_r = 0,1,\dots,n_r, \quad 1 \le u \le n-1.$$

Then the following inequality holds:

$$L_1 \le L_2.$$

Proof. We divide all inequalities in (5.7) into groups of inequalities E_μ ($\mu = 0,1,\dots,n-1$). Each group E_μ contains the inequalities for which the sum $u = u_1+\dots+u_r$ has the same value equal to μ. We show that each solution of system (5.6) satisfies all inequalities in E_μ for any μ, which readily implies the statement of the lemma.

We prove the last assertion as follows: at the first step, we prove that each solution of system (5.6) is also a solution of the system of inequalities consisting of the inequalities in (5.6) and of the inequalities from the group E_μ with the maximal value $\mu = u_0 = n-1$. At the second step, we prove that each solution of system (5.6) is also a solution of the system of inequalities consisting of the inequalities in (5.7), the inequalities from the group E_{u_0}, and the inequalities from the group E_{u_0-1}, etc., till E_1. In other words, we proceed by induction on the parameter μ.

Let $\mu = u_0 = n - 1$. For this value of the parameter μ, the group E_μ consists of r linear inequalities. Indeed, in this case the equation

$$u = u_1 + \cdots + u_r = n - 1 = n_1 + \cdots + n_r - 1,$$
$$0 \le u_1 \le n_1, \ldots, 0 \le u_r \le n_r,$$

has solutions $u_1 = n_1 - 1, u_2 = n_2, \ldots, u_r = n_r; \ldots; u_1 = n_1, u_2 = n_2, \ldots, u_r = n_r - 1$; moreover,

$$\begin{aligned} &B(n_1, \ldots, n_{j-1}, n_j - 1, n_{j+1}, \ldots, n_r) \\ &\quad = A(n_1, \ldots, n_{j-1}, n_j - 1, n_{j+1}, \ldots, n_r; 1) = n_j \alpha(n_1, \ldots, n_r)(y_j - z_j) \end{aligned}$$

and the coefficients of $A(n_1, \ldots, n_{j-1}, n_j - 1, n_{j+1}, \ldots, n_r; 1)$ in the inequalities in E_{u_0} are equal to 1. Besides, the right-hand sides of the inequalities in E_{u_0} exceed in magnitude the right-hand sides of the inequalities in system (5.6) with the corresponding indices. Thus the induction assumption holds for $\mu = u_0$. Now we assume that the desired assertion is proved for $\mu = k + 1$ and prove it for $\mu = k$. Let

$$u = u_1 + \cdots + u_r = \mu = k.$$

By the definition of $B(u_1, \ldots, u_r)$, we have

$$B(u_1, \ldots, u_r) = \sum_{s=1}^{n-u} A(\overline{u}; s).$$

Hence

$$A(\overline{u}; 1) = B(\overline{u}) - \sum_{s=1}^{n-u} A(\overline{u}; s).$$

Multiplying both sides of this relation by $\frac{n!}{(u+1)!} \cdot \frac{(n+1)!}{(u+2)!}$ and using the expression for $A(\overline{u}; s)$ obtained in Lemma 5.1, we arrive at

$$\begin{aligned} \frac{n!}{(u+1)!} \cdot \frac{(n+1)!}{(u+2)!} A(\overline{u}; 1) &= \frac{n!}{(u+1)!} \cdot \frac{(n+1)!}{(u+2)!} B(\overline{u}) \\ &- \sum_{s=2}^{n-u} \sum_{\substack{v_1=u_1 \\ v=s-1+u}}^{n_1} \cdots \sum_{v_r=u_r}^{n_r} \frac{v_1! \ldots v_r!}{u_1! \ldots u_r!} \cdot \frac{(v+2)!}{s!(u+1)!} \cdot \frac{(v+1)!}{(u+2)!} \\ &\times H(\overline{v}; \overline{u}; s) \frac{n!}{(v+1)!} \cdot \frac{(n+1)!}{(v+2)!} A(\overline{v}; 1). \end{aligned} \tag{5.8}$$

Next, for any integer $y_1, \ldots, y_r, z_1, \ldots, z_r$, the variable

$$\frac{v_1! \ldots v_r!}{u_1! \ldots u_r!} \cdot \frac{(v+2)!}{s!(u+1)!} \cdot \frac{(v+1)!}{(u+2)!} H(\overline{v}; \overline{u}; s)$$

is integer because the numbers

$$\frac{v_1!\dots v_r!}{u_1!\dots u_r!},\quad \frac{(v+2)!}{s!(u+1)!},\quad \frac{(v+1)!}{(u+2)!}$$

are integer for $s > 1$, $n_1 \geq v_1 \geq u_1, \dots, n_r \geq v_r \geq u_r$, and $v = v_1 + \dots + v_r = s - 1 + u_1 + \dots + u_r = s - 1 + u$ and, by Lemma 5.1, $H(\overline{v}; \overline{u}; s)$ is a polynomial in $y_1, \dots, y_r, z_1, \dots, z_r$ with integer coefficients.

Now we note that if $\alpha = \beta \pmod 1$, then $\|\alpha\| = \|\beta\|$ and for any integer d, the inequality $\|d\alpha\| \leq d\|\alpha\|$ holds.

Passing from (5.8) to a congruence modulo 1 and using this remark, we obtain the inequality

$$\begin{aligned}\left\|\frac{n!}{(u+1)!}\cdot\frac{(n+1)!}{(u+2)!}A(\overline{u};1)\right\| &\leq \frac{n!}{(u+1)!}\cdot\frac{(n+1)!}{(u+2)!}\|B(\overline{u})\| \\ &+\sum_{s=2}^{n-u}\sum_{\substack{v_1=u_1\\ v=s-1+u}}^{n_1}\cdots\sum_{v_r=u_r}^{n_r}\frac{v_1!\dots v_r!}{u_1!\dots u_r!}\cdot\frac{(v+2)!}{s!(u+1)!}\cdot\frac{(v+1)!}{(u+2)!} \\ &\times H(\overline{v};\overline{u};s)\left\|\frac{n!}{(v+1)!}\cdot\frac{(n+1)!}{(v+2)!}A(\overline{v};1)\right\|.\end{aligned}\tag{5.9}$$

In the last sum, we have $v = v_1 + \dots + v_r = s - 1 + u = s - 1 + u_1 + \dots + u_r \geq 1 + u_1 + \dots + u_r = 1 + u > k$. Therefore, applying the induction assumption to the variables

$$\left\|\frac{n!}{(v+1)!}\cdot\frac{(n+1)!}{(v+2)!}A(\overline{v};1)\right\|,$$

we see that they do not exceed

$$\frac{n!}{(v+1)!}\cdot\frac{(n+1)!}{(v+2)!}(4rn^2)^{n-v-1}P_1^{-v_1}\dots P_r^{-v_r}.$$

It follows from the assumptions of the lemma that the value of $\|B(\overline{u})\|$ does not exceed $P_1^{-v_1}\dots P_r^{-v_r}$. Moreover, by Lemma 5.1, the sum of the coefficients of the polynomial $H(\overline{v}; \overline{u}; s)$ does not exceed sr^{s-1}, and the sum of the powers of the variables y_j, z_j in each monomial is equal to $v_j - u_j$. Hence

$$|H(\overline{v};\overline{u};s)| \leq sr^{s-1}P_1^{v_1-u_1}\dots P_r^{v_r-u_r}.$$

Substituting the above estimates into inequality (5.9), we obtain

$$\begin{aligned}\left\|\frac{n!}{(u+1)!}\cdot\frac{(n+1)!}{(u+2)!}A(\overline{u};1)\right\| &\leq \frac{n!}{(u+1)!}\cdot\frac{(n+1)!}{(u+2)!}P_1^{-u_1}\dots P_r^{-u_r} \\ &+\sum_{s=2}^{n-u}\sum_{\substack{v_1=u_1\\ v=s-1+u}}^{n_1}\cdots\sum_{v_r=u_r}^{n_r}\frac{v_1!\dots v_r!}{u_1!\dots u_r!}\cdot\frac{(v+2)!}{s!(u+1)!}\cdot\frac{(v+1)!}{(u+2)!}\end{aligned}$$

$$\times\, sr^{s-1} P_1^{v_1-u_1} \dots P_r^{v_r-u_r} \frac{n!}{(v+1)!} \cdot \frac{(n+1)!}{(v+2)!} (4rn^2)^{n-v-1} P_1^{-v_1} \dots P_r^{-v_r}$$

$$= \frac{n!}{(u+1)!} \cdot \frac{(n+1)!}{(u+2)!} P_1^{-u_1} \dots P_r^{-u_r} (1+\varkappa),$$

where

$$\varkappa = \sum_{s=2}^{n-u} \frac{r^{s-1}}{(s-1)!} \sum_{\substack{v_1=u_1 \\ v=s-1+u}}^{n_1} \cdots \sum_{v_r=u_r}^{n_r} \frac{v_1! \dots v_r!}{u_1! \dots u_r!} (4rn^2)^{n-v-1}$$

$$= \sum_{s=2}^{n-u} \frac{r^{s-1}}{(s-1)!} (4rn^2)^{n-v-s} \sum_{\substack{v_1=u_1 \\ v=s-1+u}}^{n_1} \cdots \sum_{v_r=u_r}^{n_r} \frac{v_1! \dots v_r!}{u_1! \dots u_r!}.$$

Let us find an upper bound for $\varkappa$. By the Newton binomial formula, we have

$$(y_1 + \dots + y_r)^{s-1} = \sum_{k_1=0}^{s-1} \cdots \sum_{k_r=0}^{s-1} \frac{(s-1)!}{k_1! \dots k_r!} y_1^{k_1} \dots y_r^{k_r},$$

and hence

$$\sum_{\substack{v_1=u_1 \\ v=s-1+u}}^{n_1} \cdots \sum_{v_r=u_r}^{n_r} \frac{v_1! \dots v_r!}{u_1! \dots u_r!} = \sum_{\substack{k_1=0 \\ k_1+\dots+k_r=s-1}}^{n_1-u_1} \cdots \sum_{k_r=0}^{n_r-u_r} \frac{(k_1+u_1)! \dots (k_r+u_r)!}{u_1! \dots u_r!}$$

$$\le \sum_{\substack{k_1=0 \\ k_1+\dots+k_r=s-1}}^{s-1} \cdots \sum_{k_r=0}^{s-1} n_1^{k_1} \dots n_r^{k_r} \le (n_1 + \dots + n_r)^{s-1} = n^{s-1}.$$

Thus for $\varkappa$ we obtain

$$\varkappa \le \sum_{s=2}^{n-u} \frac{r^{s-1}}{(s-1)!} (4rn^2)^{n-u-s} n^{s-1} \le (4rn^2)^{n-u-1} \sum_{s=2}^{n-u} \frac{r^{s-1}}{(s-1)!} (4rn^2)^{-s+1} n^{s-1}$$

$$< (4rn^2)^{n-u-1} \sum_{s=2}^{+\infty} \frac{1}{(s-1)! 4^{s-1}} = (4rn^2)^{n-u-1} (\sqrt[4]{e} - 1).$$

Hence, recalling the formulas $u = u_1 + \dots + u_r < u_0 = n_1 + \dots + n_r - 1 = n - 1$, $r > 1$, $u = u_1 + \dots + u_r \ge 1$, we obtain the estimate

$$1 + \varkappa < 1 + (4rn^2)^{n-u-1} (\sqrt[4]{e} - 1) < (4rn^2)^{n-u-1},$$

as required. Thus we have proved the induction assumption for $\mu = k$, which implies that this assumption holds for all μ. The proof of the lemma is complete. □

Now we state and prove the following theorem on the upper bound for the multiplicity of intersection of multidimensional regions.

Theorem 5.1. *For all $t_1, \dots, t_r$ such that $0 \le t_1 \le n_1, \dots, 0 \le t_r \le n_r$, the numbers $\tau(t_1, \dots, t_r)$ are determined by the relations*

$$\tau(t_1, \dots, t_r) = P_1^{t_1} \dots P_r^{t_r} \Delta, \quad \Delta = (P_1^{n_1} \dots P_r^{n_r})^{-(3\varkappa)^{-1}}.$$

Let the coefficients $\alpha(t_1, \dots, t_r)$ of the polynomial $F(x_1, \dots, x_r)$ have the form

$$\alpha(t_1, \dots, t_r) = \frac{a(t_1, \dots, t_r)}{q(t_1, \dots, t_r)} + \frac{\theta(t_1, \dots, t_r)}{q(t_1, \dots, t_r)\tau(t_1, \dots, t_r)},$$

where the integers $\alpha(t_1, \dots, t_r)$ and $q(t_1, \dots, t_r)$ satisfy the conditions

$$\big(a(t_1, \dots, t_r), q(t_1, \dots, t_r)\big) = 1, \quad 0 < q(t_1, \dots, t_r) \le \tau(t_1, \dots, t_r),$$

and the absolute values of the real numbers $\theta(t_1, \dots, t_r)$ do not exceed 1. We let Q_0 denote the least common multiple of the numbers $q(t_1, \dots, t_r)$ such that $t = t_1 + \dots + t_r \ge 2$. We also determine the variables

$$c(t_1, \dots, t_r) = c(t_1, \dots, t_r; \overline{y})$$

by the relations

$$F(x_1 + y_1, \dots, x_r + y_r) - F(x_1, \dots, x_r) = \sum_{t_1=0}^{n_1} \dots \sum_{t_r=0}^{n_r} c(t_1, \dots, t_r) x_1^{t_1} \dots x_r^{t_r}.$$

We determine the region $\Omega = \Omega(y_1, \dots, y_r)$ of points $\gamma(t_1, \dots, t_r)$ by the conditions

$$\begin{gathered} \|\gamma(t_1, \dots, t_r) - c(t_1, \dots, t_r)\| < 0.5 P_1^{-t_1} \dots P_r^{-t_r}, \\ 0 \le t_1 \le n_1, \dots, 0 \le t_r \le n_r, \\ 1 \le t = t_1 + \dots + t_r \le n - 1 = n_1 + \dots + n_r - 1, \end{gathered}$$

so that the integers $y_1, \dots, y_r$ can take any values in the intervals $-Y_1 \le y_1 \le Y_1 \le P_1, \dots, -Y_r \le y_r \le Y_r \le P_r$. We choose one of the regions $\Omega = \Omega(y_1, \dots, y_r)$, in other words, we consider the region $\Omega_0 = \Omega(z_1, \dots, z_r)$ corresponding to some of the numbers $z_1, \dots, z_r$ from the above intervals. Let G be the number of regions intersecting with Ω_0. Then G satisfies the estimate

$$G \le (rn)^{4n} P_1 \dots P_r(\Delta + Q_0^{-1}).$$

Proof. First, we exclude the trivial cases of the theorem. A trivial estimate of G is the number of all possible values of $y_1, \dots, y_r$, i.e.,

$$(2Y_1 + 1) \dots (2Y_r + 1) \le (2P_1 + 1) \dots (2P_r + 1);$$

the right-hand side is larger than $P_1, \ldots, P_r$. Therefore, in order the two estimates in the theorem be nontrivial, it is necessary to satisfy each of the following inequalities:

$$P_1 \ldots P_r > (rn)^{4n} P_1 \ldots P_r \Delta, \quad P_1 \ldots P_r > (rn)^{4n} P_1 \ldots P_r Q_0^{-1}.$$

From the first inequality, we obtain the condition on P_1:

$$\Delta^{-1} > (rn)^{4n}, \quad P_1 > (rn)^{12n};$$

from the second inequality, we obtain the condition on Q_0:

$$Q_0 > (rn)^{4n}.$$

Thus, in proving the theorem, we assume that

$$P_1 > (rn)^{12n}, \quad Q_0 > (rn)^{4n}.$$

We use Lemma 5.2 to reduce estimating G to estimating the number of solutions of a system of linear inequalities, which, in turn, will be estimated using Lemma A.4.

If the regions $\Omega = \Omega(y_1, \ldots, y_r)$ and $\Omega_0 = \Omega(z_1, \ldots, z_r)$ intersect, then they have at least one common point γ with the coordinates $\gamma(t_1, \ldots, t_r)$ $(0 \le t_1 \le n_1, \ldots, 0 \le t_r \le n_r, 1 \le t = t_1 + \cdots + t_r \le n - 1 = n_1 + \cdots + n_r - 1)$. This implies that the following inequalities hold simultaneously for each of the numbers $\gamma(t_1, \ldots, t_r)$:

$$\|\gamma(t_1, \ldots, t_r) - c(t_1, \ldots, t_r; \overline{y})\| < 0.5 P_1^{-t_1} \ldots P_r^{-t_r},$$
$$\|\gamma(t_1, \ldots, t_r) - c(t_1, \ldots, t_r; \overline{z})\| < 0.5 P_1^{-t_1} \ldots P_r^{-t_r}.$$

Therefore, for $c(t_1, \ldots, t_r; \overline{y})$ and $c(t_1, \ldots, t_r; \overline{z})$, we have the inequalities

$$\|c(t_1, \ldots, t_r; \overline{y}) - c(t_1, \ldots, t_r; \overline{z})\| < P_1^{-t_1} \ldots P_r^{-t_r}.$$

Using the notation of Lemma 5.2, we rewrite the last relations as

$$\|B(t_1, \ldots, t_r)\| < P_1^{-t_1} \ldots P_r^{-t_r}. \tag{5.10}$$

Thus we are under the assumptions of Lemma 5.2, and we must estimate the number L_1 of solutions of the system of inequalities (5.10) under the condition that the unknowns $y_1, \ldots, _r$ take values of the integers from the corresponding intervals $-Y_1 \le y_1 \le Y_1, \ldots, -Y_r \le y_r \le Y_r$ and $z_1, \ldots, z_r$ are fixed numbers in the same intervals. Applying Lemma 5.2, we see that $L = G$ does not exceed L_2, where L_2 is the number of solutions of the following linear system of inequalities:

$$\begin{aligned} &\left\| \frac{n!}{(t+1)!} \cdot \frac{(n+1)!}{(t+2)!} A(t_1, \ldots, t_r; 1) \right\| \\ &\quad \le \frac{n!}{(t+1)!} \cdot \frac{(n+1)!}{(t+2)!} (4rn^2)^{n-t-1} P_1^{-t_1} \ldots P_r^{-t_r}, \end{aligned} \tag{5.11}$$

$$t_1 = 0, 1, \dots, n_1; \ \dots; \ t_r = 0, 1, \dots, n_r,$$
$$n = n_1 + \dots + n_r, \quad t = t_1 + \dots + t_r, \quad 1 \le t \le n - 1.$$

Next, we estimate the number L_2 in different ways depending on the values of $q(t_1, \dots, t_r)$ (which are the denominators of the rational approximations to $\alpha(t_1, \dots, t_r)$). First, we estimate L_2 under the condition that there exists a $q(t_1, \dots, t_r)$ $(t_1 = n_1, \dots, t_r = n_r, u_1 + \dots + u_r = u \ge 2)$ such that

$$q(u_1, \dots, u_r) \ge \Delta^{-1}.$$

Since $u_1 + \dots + u_r \ge 2$, there exists a $u_k \ge 1$. We consider the inequality in system (5.11) that corresponds to the variables $t_1, \dots, t_r$ satisfying the conditions $t_1 = u_1, \dots, t_{k-1} = u_{k-1}, t_k = u_k - 1, t_{k+1} = u_{k+1}, \dots, t_r = u_r$. This inequality has the form

$$\left\| \frac{n!}{u!} \cdot \frac{(n+1)!}{(u+1)!} \left(\sum_{\substack{j=1 \\ j \ne k}}^{r} (u_j + 1)\alpha(u_1, \dots, u_j + 1, \dots, u_r)(y_j - z_j) \right.\right.$$
$$\left.\left. + u_k \alpha(u_1, \dots, u_k, \dots, u_r)(y_k - z_k) \right) \right\|$$
$$\le \frac{n!}{u!} \cdot \frac{(n+1)!}{(u+1)!} (4rn^2)^{n-u} P_1^{-u_1} \dots P_r^{-u_r} P_k.$$

For fixed $y_1, \dots, y_{k-1}, y_{k+1}, \dots, y_r$, to estimate the number L_3 of the numbers y_k satisfying this inequality, we use Lemma A.4 (first, we write the chosen α as $\alpha = a/q + \theta/(q\tau)$):

$$L_3 \le (\lambda Y m + m + 2V)(2Yq^{-1} + 1),$$

where

$$m = \frac{n!}{u!} \cdot \frac{(n+1)!}{(u+1)!} u_k, \quad \lambda = \tau^{-1}(u_1, \dots, u_r), \quad Y = 2Y_k + 1,$$
$$V = \frac{n!}{u!} \cdot \frac{(n+1)!}{(u+1)!} (4rn^2)^{n-u} P_1^{-u_1} \dots P_r^{-u_r} P_k q, \quad q = q(u_1, \dots, u_r).$$

After simple calculations, from this inequality we obtain the estimate $L_2 \le L_3(2Y_1 + 1) \dots (2Y_r + 1)(2Y_k + 1)^{-1} \le (rn)^{4n} P_1 \dots P_r \Delta$, which corresponds to this case of the statement of the lemma.

Now we estimate L_2 under the assumption that each $q(t_1, \dots, t_r)$ does not exceed Δ^{-1}. By $Q(t_1, \dots, t_r)$ we denote the least common multiple of the numbers $q(t_1 + 1, t_2, \dots, t_r), q(t_1, t_2 + 1, \dots, t_r), \dots, q(t_1, t_2, \dots, t_r + 1)$. The following two cases are possible:

(1) there exists a $Q(t_1, \dots, t_r)$ $(t_1 = u_1, \dots, t_r = u_r)$ such that

$$Q(u_1, \dots, u_r) \ge \Delta^{-1};$$

(2) for all $t_1, \ldots, t_r$ the following inequality holds:

$$Q(t_1, \ldots, t_r) \geq \Delta^{-1}.$$

(1). In this case we consider the inequality in system (5.11) with the indices $t_1 = u_1, \ldots, t_r = u_r$. It has the form

$$\left\| \frac{n!}{(u+1)!} \cdot \frac{(n+1)!}{(u+2)!} \sum_{j=1}^{r} (u_j+1)\alpha(u_1, \ldots, u_j+1, \ldots, u_r)(u_j - z_j) \right\| \tag{5.12}$$
$$\leq \frac{n!}{(u+1)!} \cdot \frac{(n+1)!}{(u+2)!} (4rn^2)^{n-u-1} P_1^{-u_1} \ldots P_r^{-u_r}.$$

In this relation, we pass from the numbers $\alpha(u_1, \ldots, u_j+1, \ldots, u_r)$ to their rational approximations. By the condition of the lemma, we have

$$\alpha(u_1, \ldots, u_j+1, \ldots, u_r) = \frac{a(u_1, \ldots, u_j+1, \ldots, u_r)}{q(u_1, \ldots, u_j+1, \ldots, u_r)} + \frac{\theta(u_1, \ldots, u_j+1, \ldots, u_r)}{q(u_1, \ldots, u_j+1, \ldots, u_r)\tau(u_1, \ldots, u_j+1, \ldots, u_r)}.$$

We replace each α in the left-hand side of (5.12) by the rational fraction a/q. The absolute value of the remainder thus obtained does not exceed the number

$$\frac{n!}{(u+1)!} \cdot \frac{(n+1)!}{(u+2)!} \sum_{j=1}^{r} (u_j+1) \frac{4Y_j}{q(u_1, \ldots, u_j+1, \ldots, u_r)\tau(u_1, \ldots, u_j+1, \ldots, u_r)}$$
$$\leq \frac{4n!}{(u+1)!} \cdot \frac{(n+1)!}{(u+2)!} \sum_{j=1}^{r} (u_j+1) P_j P_1^{-u_1} \ldots P_r^{-u_r} \Delta^{-1}$$
$$< \frac{8((n+1)!)^2}{(u+1)!(u+2)!} P_1^{-u_1} \ldots P_r^{-u_r} \Delta^{-1}.$$

Therefore, the number of solutions of inequality (5.12) does not exceed the number of solutions of the inequality

$$\left\| \frac{n!}{(u+1)!} \cdot \frac{(n+1)!}{(u+2)!} \sum_{j=1}^{r} (u_j+1) \frac{a(u_1, \ldots, u_j+1, \ldots, u_r)}{q(u_1, \ldots, u_j+1, \ldots, u_r)} (y_j - z_j) \right\| \tag{5.13}$$
$$\leq \frac{8((n+1)!)^2}{(u+1)!(u+2)!} P_1^{-u_1} \ldots P_r^{-u_r} \Delta^{-1}$$
$$+ \frac{n!}{(u+1)!} \cdot \frac{(n+1)!}{(u+2)!} (4rn^2)^{n-u-1} P_1^{-u_1} \ldots P_r^{-u_r}$$
$$< \frac{((n+1)!)^2}{(u+1)!(u+2)!} (4rn^2)^{n-u} P_1^{-u_1} \ldots P_r^{-u_r} \Delta^{-1} = A^{-1}.$$

We transform the last inequality as follows: we divide the common factors out of the numerator and the denominator of the fractions before $y_j - z_j$ and note that only the terms

$$\frac{n!}{(u+1)!}\cdot\frac{(n+1)!}{(u+2)!}(u_j+1) \quad \text{and} \quad q(u_1,\dots,u_j+1,\dots,u-r)$$

can have common factors.

The obtained denominators q_j of the irreducible fractions satisfy the inequalities

$$q_j \le q(u_1,\dots,u_j+1,\dots,u_r) \le \frac{n!}{(u+1)!}\cdot\frac{(n+1)!}{(u+2)!}(u_j+1)q_j.$$

Moreover, the common least multiple of the numbers $q_1,\dots,q_r$, which we denote by Q_1, is no less than

$$(u_1+1)^{-1}\dots(u_r+1)^{-1}\frac{(u+1)!}{n!}\cdot\frac{(u+2)!}{(n+1)!}Q.$$

We represent each y_j from the interval $-Y_j \le y_j \le Y_j$ in the form $y_j = q_j y_j' + x_j$, where $0 \le x_j < q_j$. Then inequality (5.13) can be written as

$$\left\|\sum_{j=1}^{r}\frac{b_j x_j}{q_j} + \alpha\right\| \le A^{-1}, \tag{5.14}$$

where α is a real number and $(b_j, q_j) = 1$ $(j = 1,\dots,r)$. Therefore, if L_4 is the number of solutions of the last inequality for the unknowns $x_1,\dots,x_r$, then the number of solutions of inequality (5.13) and hence G do not exceed

$$L_4(2Y_1q_1^{-1}+1)\dots(2Y_rq_r^{-1}+1).$$

Let us estimate L_4. Each unknown x_j takes a value $0, 1, \dots, q_j - 1$, the numbers b_j and q_j are coprimes, and the function $\|x\|$ is periodic with period 1, hence we can write x_j instead of $b_j x_j$ in (5.14), i.e., (5.14) can be written as

$$\left\|\frac{x_1}{q_1}+\dots+\frac{x_r}{q_r}+\alpha\right\| \le A^{-1}. \tag{5.15}$$

We consider one of the fractions x_j/q_j, which we denote by x/q. We assume that the canonical decomposition of the number q into prime divisors has the form $q = p_1^{\alpha_1}\dots p_s^{\alpha_s}$. Then the numbers x/q and $z_1/p_1^{\alpha_1}+\dots+z_s/p_s^{\alpha_s}$ take the same values modulo 1 when running through the complete system of residues modulo q, and the numbers $z_1,\dots,z_s$ run independently through the complete systems of residues modulo $p_1^{\alpha_1},\dots,p_s^{\alpha_s}$, respectively. Let p^α be one of the factors in the canonical decomposition of the number Q_1. Since Q_1 is the least common multiple of the numbers $q_1,\dots,q_r$, then p^α enters the canonical decomposition of one of the q_j. Representing

x_j/q_j as the sum of the terms $z_1/p_1^{\alpha_1} + \cdots + z_s/p_s^{\alpha_s}$, we reduce inequality (5.15) to the inequality

$$\left\| \frac{z_1}{p_1^{\alpha_1}} + \cdots + \frac{z_s}{p_s^{\alpha_s}} + \beta \right\| \le A^{-1}.$$

In this inequality, $p_1^{\alpha_1} \ldots p_s^{\alpha_s} = Q_1$ is the canonical decomposition of Q_1 into prime divisors and β is the sum of the remaining terms and the number α. The sum β takes exactly $q_1 \ldots q_r Q_1^{-1}$ values. The last inequality can be transformed once again to the form

$$\|z Q_1^{-1} + \beta\| \le A^{-1},$$

where z runs through the complete system of residues modulo Q_1. The number of solutions of this inequality does not exceed $q_1 \ldots q_r Q_1^{-1}(1 + 2Q_1 A^{-1})$. Hence, recalling the lower bounds for Q_1 and Q and the explicit expression of A and performing simple calculations for G, we obtain the inequality

$$G \le q_1 \ldots q_r Q_1^{-1}(1+2Q_1 A^{-1})(2Y_1 q_1^{-1}+1) \ldots (2Y_r q_r^{-1}+1) \le (rn)^{4n} P_1 \ldots P_r \Delta,$$

which proves the statement of the lemma in case (1).

(2) In this case, the proof is rather similar to that in case (1), except for several details. In all inequalities in system (5.11) we pass from the numbers $\alpha(t_1, \ldots, t_j + 1, \ldots, t_r)$ to their rational approximations in the same way as in case (1). Repeating word for word the argument till formula (5.13) for each inequality in the system, we see that G does not exceed the number L_5 of solutions to the system of inequalities

$$\begin{aligned}
&\left\| \sum_{j=1}^{r} \frac{b(t_1, \ldots, t_j + 1, \ldots, t_r)}{q_1(t_1, \ldots, t_j + 1, \ldots, t_r)} (y_j - z_j) \right\| \\
&\quad \le \frac{4n!}{(t+1)!} \cdot \frac{(n+1)!}{(t+2)!} P_1^{-t_1} \ldots P_r^{-t_r} \Delta^{-1} \sum_{j=1}^{r} \frac{t_j + 1}{q(t_1, \ldots, t_j + 1, \ldots, t_r)} \\
&\quad + \frac{n!}{(t+1)!} \cdot \frac{(n+1)!}{(t+2)!} (4rn^2)^{n-t-1} P_1^{-t_1} \ldots P_r^{-t_r}, \\
&\quad t_1 = 0, 1, \ldots, n_1, \;\; \ldots, \;\; t_r = 0, 1, \ldots, n_r, \\
&n = n_1 + \cdots + n_r, \quad t = t_1 + \cdots + t_r, \quad 1 \le t \le n - 1.
\end{aligned} \tag{5.16}$$

The value of the right-hand side of each of these inequalities does not exceed $0.5Q^{-1}(t_1, \ldots, t_r)$, hence the system of inequalities (5.16) is equivalent to the system of congruences

$$\sum_{j=1}^{r} \frac{b(t_1, \ldots, t_j + 1, \ldots, t_r)}{q_1(t_1, \ldots, t_j + 1, \ldots, t_r)} (y_j - z_j) \equiv 0 \pmod{1}. \tag{5.17}$$

Further, let $Q_1(t_1, \ldots, t_r)$ be the least common multiple of the numbers

$$q_1(t_1 + 1, t_2, \ldots, t_r), q_1(t_1, t_2 + 1, \ldots, t_r), \ldots, q_1(t_1, t_2, \ldots, t_r + 1),$$

and let $Q_{1j}(t_1, \ldots, t_r)$ be determined by the relations

$$Q_1(t_1, \ldots, t_r) = q_1(t_1, \ldots, t_j + 1, \ldots, t_r) Q_{1j}(t_1, \ldots, t_r).$$

Then system (5.17) is equivalent to the system of congruences

$$\sum_{j=1}^{r} Q_{1j}(t_1, \ldots, t_r) b(t_1, \ldots, t_j + 1, \ldots, t_r)(y_j - z_j) \equiv 0 \qquad (5.18)$$
$$(\text{mod } Q_1(t_1, \ldots, t_r)).$$

We note that the total set of numbers

$$Q_{11}(t_1, \ldots, t_r) b(t_1 + 1, \ldots, t_r), \ldots, Q_{1r}(t_1, \ldots, t_r) b(t_1, \ldots, t_r + 1), Q_1(t_1, \ldots, t_r)$$

is a set of coprimes.

Let Q be the least common multiple of the numbers $Q_1(t_1, \ldots, t_r)$, $t_1 = 0, 1, \ldots, n_1, \ldots, t_r = 0, 1, \ldots, n_r, t_1 + \cdots + t_r \geq 2$, and let $Q = p_1^{\alpha_1} \ldots p_s^{\alpha_s}$ be the canonical decomposition of Q into prime divisors. Then for each $p_k^{\alpha_k}$ $(k = 1, \ldots, s)$, there exists a $Q_1(t_1^{(k)}, \ldots, t_r^{(k)})$ $(t_1^{(k)} + \cdots + t_r^{(k)} \geq 2)$ multiple of it and hence the following congruences are satisfied:

$$\sum_{j=1}^{r} Q_{1j}(t_1^{(k)}, \ldots, t_r^{(k)}) b(t_1^{(k)}, \ldots, t_j^{(k)} + 1, \ldots, t_r^{(k)})(y_j - z_j) \equiv 0 \qquad (5.19)$$
$$(\text{mod } p_k^{\alpha_k}), \quad k = 1, \ldots, s,$$

and at least for one of j $(1 \leq j \leq r)$ the number

$$Q_{1j}(t_1^{(k)}, \ldots, t_r^{(k)}) b(t_1^{(k)}, \ldots, t_j^{(k)} + 1, \ldots, t_r^{(k)}) = Q_{1jk} b_{jk}$$

is not multiple of p_k (this follows from the above remark that r of such products are coprimes modulo $Q_1(t_1, \ldots, t_r)$).

We consider the largest natural number μ such that

$$p_1^{\alpha_1} \ldots p_\mu^{\alpha_\mu} \leq Y = Y_1$$

and consider congruences (5.19) for $k = 1, \ldots, \mu$. We find the numbers R_k form the relations

$$p_k^{\alpha_k} R_k = p_1^{\alpha_1} \ldots p_\mu^{\alpha_\mu}.$$

For each r_j $(0 \leq r_j < p_1^{\alpha_1} \ldots p_\mu^{\alpha_\mu})$ the number of solutions of the congruences

$$y_j - z_j \equiv r_j \pmod{p_1^{\alpha_1} \ldots p_\mu^{\alpha_\mu}}$$

does not exceed the number $2Y_j(p_1^{\alpha_1} \ldots p_\mu^{\alpha_\mu})^{-1} + 1$. Representing each r_j as

$$r_j \equiv R_1 r_{1j} + \cdots + R_\mu r_{\mu j} \pmod{p_1^{\alpha_1} \ldots p_\mu^{\alpha_\mu}},$$

where $0 \le r_{kj} < p_k^{\alpha_k}$, we pass from system (5.19) to the system

$$\sum_{j=1}^{r} Q_{1jk} b_{jk} (R_1 r_{1j} + \cdots + R_\mu r_{\mu j}) \equiv 0 \pmod{p_k^{\alpha_k}}, \quad k = 1, \ldots, \mu. \qquad (5.20)$$

If L_6 is the number of solutions of the last system of congruences for the unknowns $r_{1j}, \ldots, r_{\mu j}$, then

$$L_5 \le L_6 \prod_{j=1}^{r} \left(2Y_j (p_1^{\alpha_1} \ldots p_\mu^{\alpha_\mu})^{-1} + 1\right).$$

In turn, system (5.20) is equivalent to the system

$$\sum_{j=1}^{r} Q_{1jk} b_{jk} R_k r_{kj} \equiv 0 \pmod{p_k^{\alpha_k}}, \quad k = 1, \ldots, \mu,$$

because each $R_1, \ldots, R_\mu$ except R_k is a multiple of $p_k^{\alpha_k}$. The number of solutions of each of the congruences in this system does not exceed $p_k^{\alpha_k(r-1)}$ (this is a linear congruence with r unknowns running through the complete systems of residues modulo $p_k^{\alpha_k}$, and the coefficient of at least one of the unknowns and its absolute value are coprimes). Hence

$$L_6 \le \prod_{k=1}^{\mu} p_k^{\alpha_k(r-1)} = (p_1^{\alpha_1} \ldots p_\mu^{\alpha_\mu})^{r-1}.$$

Combining the estimates L_6 and L_5 for G, we obtain the inequality

$$\begin{aligned} G \le L_5 &\le (p_1^{\alpha_1} \ldots p_\mu^{\alpha_\mu})^{r-1} \prod_{j=1}^{r} \left(2Y_j (p_1^{\alpha_1} \ldots p_\mu^{\alpha_\mu})^{-1} + 1\right) \\ &\le 3^r Y_1 \ldots Y_r (p_1^{\alpha_1} \ldots p_\mu^{\alpha_\mu})^{-1}. \end{aligned}$$

If $\mu = s$, then

$$p_1^{\alpha_1} \ldots p_\mu^{\alpha_\mu} = Q, \quad Q \ge \frac{Q_0}{n!(n+1)!}$$

and

$$G \le 3^r Y_1 \ldots Y_r n!(n+1)! Q_0^{-1} \le 3^r n!(n+1)! P_1 \ldots P_r Q_0^{-1}.$$

If $\mu < s$, then, by the definition of μ, we have

$$Y < p_1^{\alpha_1} \ldots p_\mu^{\alpha_\mu} p_{\mu+1}^{\alpha_{\mu+1}}.$$

Since each $p_j^{\alpha_j}$ divides one of the numbers $q(t_1, \ldots, q_r)$ that does not exceed Δ^{-1} in the situation under study, we have

$$p_{\mu+1}^{\alpha_{\mu+1}} \le \Delta^{-1}, \quad p_1^{\alpha_1} \ldots p_\mu^{\alpha_\mu} \ge Y\Delta$$

and hence

$$G \leq 3^r Y_2 \dots Y_r \Delta < 3^r P_1 \dots P_r \Delta.$$

Combining the two estimates for G, we also obtain the statement of the theorem in case (2). The proof of the theorem is now complete. □

5.2 Estimates for multiple trigonometric sums

As already noted in the Introduction, multiple trigonometric sums have several distinctive features that form a significant distinction between multiple trigonometric sums and one-dimensional sums. One of such distinctions is the variety of regions in which both the principal and nonprincipal parameters can vary. Now we consider trigonometric sums whose summation variables belong to an r-dimensional parallelepiped of the form $1 \leq x_1 \leq P_1, \dots, 1 \leq x_r \leq P_r$.

We divide the points of the cube Ω (for the definition of Ω and the sum $S(A)$, see Section 4.2) into two classes Ω_1 and Ω_2. To this end, we first determine the region $\Omega(a, q)$ in the following way: the region $\Omega(a, q)$ contains a point A with coordinates $\alpha(t_1, \dots, t_r)$ if

$$\alpha(t_1, \dots, t_r) = \frac{a(t_1, \dots, t_r)}{q(t_1, \dots, t_r)} + \beta(t_1, \dots, t_r),$$

where

$$0 \leq a(t_1, \dots, t_r) < q(t_1, \dots, t_r), \quad \big(a(t_1, \dots, t_r), q(t_1, \dots, t_r)\big) = 1,$$

and

$$|\beta(t_1, \dots, t_r)| \leq P_1^{-t_1} \dots P_r^{-t_r} P^{0.1}, \quad 0 \leq t_1 \leq n_1, \dots, 0 \leq t_r \leq n_r.$$

By Q we denote the least common multiple of the numbers $q(t_1, \dots, t_r)$. Thus to each region $\Omega(a, q)$, there corresponds its own Q. The first class Ω_1 contains the regions $\Omega(a, q)$ for which $Q < P^{0.1}$. The second class Ω_2 contains all other points of the cube Ω. We estimate the sum $S(A)$ depending on what class the point A belongs.

Lemma 5.3. *Let*

$$\tau(t_1, \dots, t_r) = P_1^{t_1} \dots P_r^{t_r} P^{-1/3},$$

and let the coordinates $\alpha(t_1, \dots, t_r)$ of a point $A \in \Omega$ be written as

$$\alpha(t_1, \dots, t_r) = \frac{a(t_1, \dots, t_r)}{q(t_1, \dots, t_r)} + \frac{\theta(t_1, \dots, t_r)}{q(t_1, \dots, t_r)\, \tau(t_1, \dots, t_r)},$$
$$\big(a(t_1, \dots, t_r), q(t_1, \dots, t_r)\big) = 1 \quad 1 \leq q(t_1, \dots, t_r) \leq \tau(t_1, \dots, t_r),$$
$$|\theta(t_1, \dots, t_r)| \leq 1, \quad 0 \leq t_1 \leq n_1, \dots, 0 \leq t_r \leq n_r.$$

By Q_0 we denote the least common multiple of the numbers $q(t_1, \dots, t_r)$ such that $t_1 + \cdots + t_r \geq 2$. Then the following estimate holds for $Q_0 \geq P^{1/6}$:

$$|S(A)| \leq 2^{32\varkappa} P_1 \dots P_r P^{-\rho},$$

where $\rho = (32m\varkappa \log 8m\varkappa)^{-1}$.

Proof. We take the numbers $Y_1 = [P_1 P^{-\rho}], \dots, Y_r = [P_r P^{-\rho}]$; in the sum $S(A)$ we shift the summation variables $x_1, \dots, x_r$ by $y_1 \leq Y_1, \dots, y_r \leq Y_r$ and sum over all natural numbers $y_1, \dots, y_r$ within these limits. We obtain the inequality

$$|S(A)| \leq W + r2^r P_1 \dots P_r P^{-\rho},$$

where

$$W = (Y_1 \dots Y_r)^{-1} \sum_{y_1=1}^{Y_1} \cdots \sum_{y_r=1}^{Y_r} \left| \sum_{x_1=1}^{P_1} \cdots \sum_{x_r=1}^{P_r} \exp\{2\pi i F_A(x_1 + y_1, \dots, x_r + y_r)\} \right|.$$

We also take $\tau = [\varkappa \log 6m\varkappa]$, $k = m\tau$, raise W to the power $2k$, and then apply Hölder's inequality. We obtain

$$W^{2k} \leq (Y_1 \dots Y_r)^{-1} \sum_{y_1=1}^{Y_1} \cdots \sum_{y_r=1}^{Y_r} |S|^{2k}, \tag{5.21}$$

where

$$S = \sum_{x_1=1}^{P_1} \cdots \sum_{x_r=1}^{P_r} \exp\{2\pi i F_A(x_1 + y_1, \dots, x_r + y_r)\}.$$

We write the polynomial $F_A(x_1 + y_1, \dots, x_r + y_r)$ in powers of the unknowns $x_1, \dots, x_r$. Then, applying the above notation, we obtain

$$F_A(x_1 + y_1, \dots, x_r + y_r) = F_B(x_1, \dots, x_r),$$

where the set of coefficients B depends on $y_1, \dots, y_r$ (these coefficients themselves are polynomials in $y_1, \dots, y_r$). As previously, we consider the set of coefficients B as a point with coordinates $B(t_1, \dots, t_r)$ in the m-dimensional space. By $\omega = \omega(y_1, \dots, y_r)$ we denote the region of points β with coordinates $\beta(t_1, \dots, t_r)$ in the m-dimensional space satisfying the conditions

$$\|\beta(t_1, \dots, t_r) - B(t_1, \dots, t_r)\| < 0.5 P_1^{-t_1} \dots P_r^{-t_r} P^{-\rho},$$
$$0 \leq t_1 \leq n_1, \ \dots, \ 0 \leq t_r \leq n_r.$$

For any point $\beta \in \omega$, we have the chain of relations

$$S = \sum_{x_1=1}^{P_1} \cdots \sum_{x_r=1}^{P_r} \exp\{2\pi i F_B(x_1, \dots, x_r)\} =$$

$$= \sum_{x_1=1}^{P_1} \cdots \sum_{x_r=1}^{P_r} \exp\{2\pi i F_\beta(x_1, \dots, x_r)\} + R,$$

$$|R| \le \sum_{x_1=1}^{P_1} \cdots \sum_{x_r=1}^{P_r} \big| \exp\{2\pi i F_B(x_1, \dots, x_r)\} - \exp\{2\pi i F_\beta(x_1, \dots, x_r)\} \big|$$
$$\le 2\pi m P_1 \dots P_r P^{-\rho},$$
$$|S|^{2k} \le 2^{2k} \big(|S_\beta|^{2k} + (2\pi m P_1 \dots P_r P^{-\rho})^{2k} \big).$$

Substituting the last estimate into (5.21) and integrating both sides of the obtained inequality over the region ω (note that only the first term in the right-hand side depends on β over which we integrate), we obtain

$$W^{2k} \le 2^{2k} (Y_1, \dots, Y_r)^{-1} P^{0.5m\varkappa + m\rho} I + (4\pi m P_1 \dots P_r P^{-\rho})^{2k},$$

where

$$I = \sum_{y_1=1}^{Y_1} \cdots \sum_{y_r=1}^{Y_r} \int \cdots \int_{\omega} |S_\beta|^{2k} \, d\overline{\beta}.$$

Let G be the maximal multiplicity of intersection of the regions $\omega = \omega(y_1, \dots, y_r)$ with a fixed region $\omega(z_1, \dots, z_r)$. Then

$$I \le GJ, \quad J = \int \cdots \int_{\Omega} |S(A)|^{2k} \, dA.$$

By Theorem 4.2 in Chapter 4, we have the estimate

$$J \le k^{2m\tau} \varkappa^{4\varkappa^2 \Delta(\tau)} 2^{8m\varkappa\tau} (P_1 \dots P_r)^{2k} (P_1^{n_1} \dots P_r^{n_r})^{\Delta(\tau)}.$$

Next, by Theorem 5.1, G satisfies the estimate

$$G \le (rn)^{4n} P_1 \dots P_r (P^{-1/3} + Q_0^{-1}).$$

From the above estimates, we obtain

$$W^{2k} \le 2^{2k} (Y_1, \dots, Y_r)^{-1} P^{0.5m\varkappa + m\rho} GJ + (4\pi m P_1 \dots P_r P^{-\rho})^{2k}$$
$$\le 2 \cdot 2^{2k} (Y_1, \dots, Y_r)^{-1} (rn)^{4n} k^{2m\tau} \varkappa^{4\varkappa^2 \Delta(\tau)} 2^{8m\varkappa\tau} (P_1 \dots P_r)^{2k+1}$$
$$\times P^{-1/6 + 0.5m\varkappa + m\rho - \varkappa\Delta(\tau)} + (4\pi m P_1 \dots P_r P^{-\rho})^{2k}.$$

Since $\tau = [\varkappa \log 6m\varkappa]$ and $k = m\tau$, we have

$$(Y_1, \dots, Y_r)^{-1} P_1 \dots P_r P^{-1/6 + 0.5m\varkappa + m\rho - \varkappa\Delta(\tau)}$$
$$\le 2^r P^{-1/6 + (m+r)\rho + 0.5m\varkappa - \varkappa\Delta(\tau)} \le 2^r P^{-1/16},$$

because

$$
\begin{aligned}
Y_1 &\geq 0.5P_1P^{-\rho}, \ \ldots, \ Y_r \geq 0.5P_rP^{-\rho}, \\
0.5m\varkappa - \varkappa\Delta(\tau) &= 0.5m\varkappa(1-\gamma)^{\tau} \leq 0.5m\varkappa(1-\gamma)^{\varkappa \log 6m\varkappa - 1} \\
&\leq 0.5m\varkappa \exp\left\{ -\left(\frac{1}{\varkappa} + \frac{1}{2\varkappa^2}\right)(\varkappa \log 6m\varkappa - 1)\right\} \\
&= \frac{1}{12}\exp\left\{ -\frac{\log 6m\varkappa}{2\varkappa} + \frac{1}{\varkappa} + \frac{1}{2\varkappa^2}\right\} < \frac{1}{12}, \\
&-\frac{\log 6m\varkappa}{2\varkappa} + \frac{1}{\varkappa} + \frac{1}{2\varkappa^2} < 0, \\
(m+r)\rho &= \frac{m+r}{32m\varkappa \log 8m\varkappa} \leq \frac{1}{16\varkappa \log 8m\varkappa} < \frac{1}{96}.
\end{aligned}
$$

Extracting the $2k$th root, we obtain

$$
\begin{aligned}
|W| &\leq 2^{3+0.5(r+1)k^{-1}+2nk^{-1}\log_2(rn)+2\log_2 k + 2\varkappa^2 k^{-1}\Delta(\tau)\log_2\varkappa + 4\varkappa} \\
&\quad \times P_1 \ldots P_r P^{-2(32m\tau)^{-1}} \leq 2^{19\varkappa} P_1 \ldots P_r P^{-\rho}, \\
\rho &= (32m\varkappa \log 8m\varkappa)^{-1},
\end{aligned}
$$

because

$$
\begin{gathered}
\Delta(\tau) = 0.5m(1-(1-\gamma)^{\tau}) \leq 0.5m, \quad \varkappa \leq 2^{\varkappa}, \\
m = (n_1+1)\ldots(n_r+1) \leq 2^n \leq 2^{\varkappa}, \quad m + \varkappa + 1 \leq 2^{2\varkappa}, \\
2\varkappa^2 k^{-1}\Delta(\tau)\log_2\varkappa \leq 2\varkappa \frac{3\log_2\varkappa}{\log 8m\varkappa} \leq 4\varkappa, \\
2\log_2 k \leq 2\log_2(m\varkappa \log 8m\varkappa) \leq 2\log_2 m + 2\log_2\varkappa + 2\log_2\log 8m\varkappa \leq 8\varkappa, \\
\log_2(\log 8 + \log m + \log\varkappa) \leq \log_2(m+\varkappa+1) \leq 2\varkappa, \\
3 + \frac{r+1+4n\log_2(rn)}{2k} \leq 3 + \frac{6n\log_2(rn)}{m\varkappa \log 8m\varkappa} \leq \frac{3\log_2(rn)}{4\log 8m\varkappa} + 3 \leq 5.
\end{gathered}
$$

The proof of the lemma is complete. □

Lemma 5.4. *Suppose that $F(x_1, \ldots, x_r)$ is a real differentiable function for $0 \leq x_j \leq P_j$, $P_j \leq P$ $(j = 1, \ldots, r)$, inside the interval of variation of the variables, the function $\partial F(x_1, \ldots, x_r)/\partial x_j$ is piecewise monotone and of constant sign in each of the variables x_j $(j = 1, \ldots, r)$ for any fixed values of the other variables, and the number of intervals of monotonicity and constant sign does not exceed s. Next, let the inequalities*

$$
\left|\frac{\partial F(x_1, \ldots, x_r)}{\partial x_j}\right| \leq \delta, \quad j = 1, \ldots, r,
$$

hold for $0 < \delta < 1$. *Then*

$$\sum_{x_1=0}^{P_1} \cdots \sum_{x_r=0}^{P_r} \exp\{2\pi i F(x_1, \dots, x_r)\}$$
$$= \int_0^{P_1} \cdots \int_0^{P_r} \exp\{2\pi i F(x_1, \dots, x_r)\}\, dx_1 \dots dx_r$$
$$+ \theta_1 r s P^{r-1}\big(3 + 2\delta/(1-\delta)\big), \quad |\theta_1| \le 1.$$

Proof. To prove this lemma, we successively apply Lemma A.2. For fixed values of the variables $x_1, \dots, x_r$, by Lemma A.2, we have

$$\sum_{x_1=0}^{P_1} \exp\{2\pi i F(x_1, \dots, x_r)\} = \int_0^{P_1} \exp\{2\pi i F(x_1, \dots, x_r)\}\, dx_1$$
$$+ \theta s\big(3 + 2\delta/(1-\delta)\big).$$

We sum over the other variables both sides of the relation

$$\sum_{x_1=0}^{P_1} \cdots \sum_{x_r=0}^{P_r} \exp\{2\pi i F(x_1, \dots, x_r)\}$$
$$= \sum_{x_2=0}^{P_2} \cdots \sum_{x_r=0}^{P_r} \int_0^{P_1} \exp\{2\pi i F(x_1, \dots, x_r)\}\, dx_1 + \theta P^{r-1} s\big(3 + 2\delta/(1-\delta)\big).$$

We again apply Lemma A.2 to

$$\sum_{x_3=0}^{P_3} \cdots \sum_{x_r=0}^{P_r} \int_0^{P_1} \left(\sum_{x_2=0}^{P_2} \exp\{2\pi i F(x_1, \dots, x_r)\} \right) dx_1$$

over the variable x_2 for fixed other variables $x_1, x_3, \dots, x_r$.

Precisely in the same way, we deal with the remaining variables $x_3, \dots, x_r$. Collecting together all relations obtained, we arrive at the statement of the lemma. The proof of Lemma 5.4 is complete. □

Now we can prove the lemma on the estimates of the trigonometric sums $S(A)$ for A belonging to the class Ω_1.

Lemma 5.5. *The following estimate holds for the points A of the first class Ω_1:*

$$|S(A)| \le 2(5n^{2n})^{r\nu(Q)} (\tau(Q))^{r-1} P_1 \dots P_r Q^{-\nu}.$$

Moreover, if we set

$$\delta(t_1, \dots, t_r) = P_1^{t_1} \dots P_r^{t_r} \beta(t_1, \dots, t_r), \quad \delta = \max_{t_1, \dots, t_r} |\delta(t_1, \dots, t_r)|,$$

then for $\delta > 1$, the following estimate holds:

$$|S(A)| \le 2^{6r}(5n^{2n})^{r\nu(Q)}(\tau(Q))^{r-1}P_1 \dots P_r(\delta Q)^{-\nu}\big(\ln(\delta+2)\big)^{r-1}.$$

Proof. In the sum $S(A)$, we make a change of summation variables of the form

$$x_j = Q\xi_j + \eta_j, \quad 1 \le \eta_j \le Q,$$
$$-\eta_j Q^{-1} < \xi_j \le (P_j - \eta_j)Q^{-1}, \quad j = 1, \dots, r.$$

Then the sum $S(A)$ will have the form

$$S(A) = \sum_{\eta_1=1}^{Q} \cdots \sum_{\eta_r=1}^{Q} \exp\{2\pi i F_a(\eta_1, \dots, \eta_r)\} W(\eta_1, \dots, \eta_r),$$

where

$$F_a(\eta_1, \dots, \eta_r) = \sum_{t_1=0}^{n_1} \cdots \sum_{t_r=0}^{n_r} \frac{a(t_1, \dots, t_r)}{q(t_1, \dots, t_r)} \eta_1^{t_1} \dots \eta_r^{t_r},$$

$$W(\eta_1, \dots, \eta_r) = \sum_{\xi_1} \cdots \sum_{\xi_r} \exp\{2\pi i F_\beta(Q\xi_1 + \eta_1, \dots, Q\xi_r + \eta_r)\},$$

$$F_\beta(Q\xi_1 + \eta_1, \dots, Q\xi_r + \eta_r) = \sum_{t_1=0}^{n_1} \cdots \sum_{t_r=0}^{n_r} \beta(t_1, \dots, t_r)(Q\xi_1 + \eta_1)^{t_1} \dots (Q\xi_r + \eta_r)^{t_r}.$$

For any j $(1 \le j \le r)$, we have

$$\left|\frac{\partial F_\beta}{\partial \xi_j}\right| = \Bigg|\sum_{t_1=0}^{n_1} \cdots \sum_{t_r=0}^{n_r} \beta(t_1, \dots, t_r) t_j Q (Q\xi_1 + \eta_1)^{t_1} \dots \dots (Q\xi_j + \eta_j)^{t_j - 1} \dots (Q\xi_r + \eta_r)^{t_r}\Bigg|$$
$$\le \sum_{t_1=0}^{n_1} \cdots \sum_{t_r=0}^{n_r} P_1^{-t_1} \dots P_r^{-t_r} P^{0.1} t_j Q P_1^{t_1} \dots P_j^{t_j-1} \dots P_r^{t_r}$$
$$= 0.5 n_j m P_j^{-1} P^{0.1} Q \le 0.5.$$

Therefore, we can apply Lemma 5.4 to the sum $W(\eta_1, \dots, \eta_r)$. Hence

$$W(\eta_1, \dots, \eta_r) = \int_{-\eta_1 Q^{-1}}^{(P_1-\eta_1)Q^{-1}} \cdots \int_{-\eta_r Q^{-1}}^{(P_r-\eta_r)Q^{-1}} \exp\{2\pi i F_\beta(Q\xi_1 + \eta_1, \dots, Q\xi_r + \eta_r)\}\, d\xi_1 \dots d\xi_r +$$

$$
\begin{aligned}
&+ 2\theta_2 rn P_2 \dots P_r Q^{-r+1}\\
&= Q^{-r} \int_0^{P_1} \cdots \int_0^{P_r} \exp\{2\pi i F_\beta(x_1, \dots, x_r)\}\, dx_1 \dots dx_r\\
&+ 2\theta_2 rn P_2 \dots P_r Q^{-r+1}, \quad |\theta_2| \le 1.
\end{aligned}
$$

In the last integral we make a change of integration variables of the form $x_j \to P_j x_j$. Recalling the definition of the variables $\delta(t_1, \dots, t_r)$, we obtain the relation

$$
\begin{aligned}
W(\eta_1, \dots, \eta_r) = P_1 \dots P_r Q^{-r} \int_0^1 \cdots \int_0^1 \exp\{2\pi i F_\delta(x_1, \dots, x_r)\}\, dx_1 \dots dx_r\\
+ 2\theta_2 rn P_2 \dots P_r Q^{-r+1},
\end{aligned}
$$

where

$$
F_\delta(x_1, \dots, x_r) = \sum_{t_1=0}^{n_1} \cdots \sum_{t_r=0}^{n_r} \delta(t_1, \dots, t_r) x_1^{t_1} \dots x_r^{t_r}.
$$

For the sum $S(A)$, we find

$$
S(A) = P_1 \dots P_r UV + 2\theta_3 rn P_2 \dots P_r Q,
$$

where

$$
U = Q^{-r} \sum_{\eta_1=1}^{Q} \cdots \sum_{\eta_r=1}^{Q} \exp\{2\pi i F_a(\eta_1, \dots, \eta_r)\},
$$

$$
V = \int_0^1 \cdots \int_0^1 \exp\{2\pi i F_\delta(x_1, \dots, x_r)\}\, dx_1 \dots dx_r.
$$

To estimate U, we apply Theorem 2.2 in Chapter 2 and, to estimate V, we apply Theorem 1.6 in Chapter 1:

$$
|U| \le (5n^{2n})^{r\nu(Q)} (\tau(Q))^{r-1} Q^{-\nu}, \quad |V| \le \min\left(1,\ 32^r \delta^{-\nu} \ln^{r-1}(\delta + 2)\right).
$$

Substituting these estimates into the formula for $S(A)$, we obtain the statement of the lemma. □

Lemma 5.6. *Let a point A belong to the second class Ω_2, and let $Q_0 < P^{1/6}$. Then the following estimate holds for $S(A)$:*

$$
\begin{aligned}
|S(A)| \le (5n^{2n})^{r\nu(Q_0)} (\tau(Q_0))^{r-1} P_1 \dots P_r P^{-\nu/10}\\
+ 2^{8r} (r\nu^{-1})^{r-1} P_1 \dots P_r P^{-\nu/16},
\end{aligned}
$$

where $\nu(Q)$ is the number of distinct prime divisors of Q, $\tau(Q)$ is the number of divisors of Q, and $\nu \max(n_1, \dots, n_r) = 1$.

Proof. We divide the intervals of summation over $x_1, \dots, x_r$ into arithmetic progressions with difference Q_0 and transform the sum $S(A)$ as in Lemma 5.5. We obtain the relation

$$S(A) = Q_0^{-r} P_1 \dots P_r W + 16\theta_4 nr P_2 \dots P_r Q_0,$$

where

$$W = \sum_{\eta_1=1}^{Q_0} \cdots \sum_{\eta_r=1}^{Q_0} \exp\left\{2\pi i \left(\sum_{\substack{t_1=0\\ t_1+\dots+t_r\geq 2}}^{n_1} \cdots \sum_{t_r=0}^{n_r} \frac{a(t_1, \dots, t_r)}{q(t_1, \dots, t_r)} \eta_1^{t_1} \dots \eta_r^{t_r} \right)\right\}$$

$$\times \int_0^1 \cdots \int_0^1 \exp\{2\pi i \Phi(x_1, \dots, x_r)\}\, dx_1 \dots dx_r,$$

$$\Phi(x_1, \dots, x_r) = \sum_{\substack{t_1=0\\ t_1+\dots+t_r\geq 2}}^{n_1} \cdots \sum_{t_r=0}^{n_r} \delta(t_1, \dots, t_r) x^{t_1} \dots x^{t_r}$$

$$+ \sum_{j=1}^{r} \left(\left(\frac{a'(0, \dots, 1, \dots, 0)}{q(0, \dots, 1, \dots, 0)} + Q_0 \beta(0, \dots, 1, \dots, 0) \right) \frac{P_j x_j - \eta_j}{Q_0} \right.$$

$$\left. + \eta_j \beta(0, \dots, 1, \dots, 0) \right), \qquad \delta(t_1, \dots, t_r) = P_1^{t_1} \dots P_r^{t_r} \beta(t_1, \dots, t_r);$$

the symbol $(0, \dots, 1, \dots, 0)$ means that 1 stands in the jth place and 0 stand in all other places. The variable $a'(0, \dots, 1, \dots, 0)$ is determined by the congruence

$$a(0, \dots, 1, \dots, 0) Q_0 \equiv a'(0, \dots, 1, \dots, 0) \pmod{q(0, \dots, 1, \dots, 0)}$$

with the condition that $|a'(0, \dots, 1, \dots, 0)| \leq 0.5 q(0, \dots, 1, \dots, 0)$.

Let now $Q_0 \neq Q$, and hence let $Q_0 < Q$. Then there exists a j $(1 \leq j \leq r)$ such that the relation

$$a'(0, \dots, 1, \dots, 0) \not\equiv 0 \pmod{q(0, \dots, 1, \dots, 0)}$$

holds, i.e., $|a'(0, \dots, 1, \dots, 0)| \geq 1$. Hence the absolute value of the coefficient of x_j in the polynomial $\Phi(x_1, \dots, x_r)$ is no less than

$$\frac{P_j}{Q_0} \left(\frac{1}{q(0, \dots, 1, \dots, 0)} - \frac{Q_0}{q(0, \dots, 1, \dots, 0)\tau(0, \dots, 1, \dots, 0)} \right)$$

$$\geq \frac{P_j}{2Q_0 q(0, \dots, 1, \dots, 0)} \geq 0.5 P^{1/6}.$$

Applying Theorem 1.6 in Chapter 1 to the integral in the sum W, we obtain the estimate

$$|W| \leq Q_0^r 2^{5r+1} (0.5 P^{1/6})^{-\nu} (\ln P)^{r-1}.$$

Since $\ln P \le 12\nu^{-1}(r-1)P^{\nu/(12(r-1))}$ for any $P \ge 1$, we have

$$|W| \le 2^{5r+2}(\nu^{-1}r)^{r-1}Q_0^r P^{-\nu/12}.$$

So in the case under study, we obtain the estimate

$$\begin{aligned}|S(A)| &\le 2^{5r+2}(\nu^{-1}r)^{r-1}P_1\dots P_r P^{-\nu/12} + 16nr P_2\dots P_r Q_0\\ &< 2^{8r}(r\nu^{-1})^{r-1}P_1\dots P_r P^{-\nu/12}.\end{aligned}$$

Let $Q_0 = Q$, then $S(A)$ can be written as

$$S(A) = P_1\dots P_r UV + 16\theta_4 nr P_2\dots P_r Q_0,$$

where, as in Lemma 5.5,

$$U = Q^{-r}\sum_{x_1=1}^{Q}\cdots\sum_{x_r=1}^{Q}\exp\{2\pi i F_a(x_1,\dots,x_r)\},$$

$$V = \int_0^1\cdots\int_0^1 \exp\{2\pi i F_\delta(x_1,\dots,x_r)\}\,dx_1\dots dx_r.$$

Since the point A belongs to the second class Ω_2, we have either $Q \ge P^{0.1}$ or $\delta \ge P^{0.1}$. If $Q \ge P^{0.1}$, then by Theorem 2.6 in Chapter 2, U satisfies the estimate

$$|U| \le (5n^{2n})^{r\nu(Q_0)}(\tau(Q_0))^{r-1}Q_0^{-\nu} \le (5n^{2n})^{r\nu(Q_0)}(\tau(Q_0))^{r-1}P^{-\nu/10}.$$

If $\delta \ge P^{0.1}$, then by Theorem 1.6, for V we have the estimate

$$|V| \le 2^{5r}P^{0.1\nu}\left(\ln(P^{0.1}+2)\right)^{r-1} \le 2^{8r-3}(r\nu^{-1})P^{-\nu/16}.$$

These two estimates imply the statement of the lemma. The proof of the lemma is complete. □

In the following theorem we give an estimate for the trigonometric sum $S(A)$ on the entire unit cube Ω.

Theorem 5.2. *Suppose that A is a point of the first class Ω_1. Then the following estimate holds*:

$$|S(A)| \le 2(5n^{2n})^{r\nu(Q)}(\tau(Q))^{r-1}P_1\dots P_r Q^{-\nu}, \quad \nu\max(n_1,\dots,n_r) = 1.$$

Moreover, if

$$\delta(t_1,\dots,t_r) = P_1^{t_1}\dots P_r^{t_r}\beta(t_1,\dots,t_r), \quad \delta = \max_{t_1,\dots,t_r}|\delta(t_1,\dots,t_r)|,$$

then the following estimate holds for $\delta > 1$:

$$|S(A)| \leq 2^{6r}(5n^{2n})^{r\nu(Q)}(\tau(Q))^{r-1}P_1 \dots P_r(\delta Q)^{-\nu}\big(\ln(\delta+2)\big)^{r-1}.$$

Suppose that A is a point of the second class Ω_2. *We set*

$$\tau(t_1, \dots, t_r) = P_1^{t_1} \dots P_r^{t_r} P^{-1/3}$$

and write the coordinates $\alpha(t_1, \dots, t_r)$ *of the point A as*

$$\alpha(t_1, \dots, t_r) = \frac{a(t_1, \dots, t_r)}{q(t_1, \dots, t_r)} + \frac{\theta(t_1, \dots, t_r)}{q(t_1, \dots, t_r)\tau(t_1, \dots, t_r)},$$
$$(a(t_1, \dots, t_r), q(t_1, \dots, t_r)) = 1, \quad 1 \leq q(t_1, \dots, t_r) \leq \tau(t_1, \dots, t_r),$$
$$|\theta(t_1, \dots, t_r)| \leq 1, \quad 0 \leq t_1 \leq n_1, \dots, 0 \leq t_r \leq n_r.$$

Let Q_0 *denote the least common multiple of the numbers* $q(t_1, \dots, t_r)$ *under the condition that* $t_1 + \dots + t_r \geq 2$. *Then the following estimate holds for* $Q_0 \geq P^{1/6}$:

$$|S(A)| \leq 2^{32\varkappa} P_1 \dots P_r P^{-\rho},$$

where $\rho = (32m\varkappa \log 8m\varkappa)^{-1}$.

But if $Q_0 \leq P^{1/6}$, *then* $S(A)$ *satisfies the estimate*

$$|S(A)| \leq (5n^{2n})^{r\nu(Q_0)}(\tau(Q_0))^{r-1}P_1 \dots P_r P^{-0.1\nu} + 2^{8r}(r\nu^{-1})^{r-1}P_1 \dots P_r P^{-\nu/16}.$$

Proof. This assertion follows from Lemmas 5.3, 5.5, and 5.6. □

Concluding remark on Chapter 5. The results considered in Section 5.1 were obtained by G. I. Arkhipov in [2], [3], [4] for polynomials in two variables. In the general case, this result was obtained by G. I. Arkhipov and V. N. Chubarikov [11], [10].

Chapter 6

Several applications

In this chapter we apply the estimates obtained for multiple trigonometric sums to several problems in number theory. These problems can be divided into two groups: the first group deals with asymptotic formulas for the number of solutions of complicated systems of Diophantine equations; the second group deals with distributions of fractional parts of polynomials or sets of polynomials (joint distributions). Multidimensional problems, in contrast to one-dimensional problems, have many specific characteristics.

One of such specific properties is the existence of a variety of domains in which the variable parameters can vary. We mean both the principal parameters such as $x_1, \ldots, x_r$ (for example, $x_1, \ldots, x_r$ belong to a parallelepiped, ellipsoid, generalized ball, etc.) and the nonprincipal parameters such as $t_1, \ldots, t_r$. For example, it is possible to consider $F(x_1, \ldots, x_r)$ of the form

$$F(x_1, \ldots, x_r) = \sum_{t_1+\cdots+t_r \leq n} \cdots \sum \alpha(t_1, \ldots, t_r)\, x_1^{t_1} \ldots x_r^{t_r}.$$

6.1 Systems of Diophantine equations

In this section we obtain asymptotic formulas for the mean value of a power of the modulus of a multiple trigonometric sum, which, as mentioned above, give the number of solutions of a complicated system of Diophantine equations. The main problem here is to derive such formulas for the least possible power to which we raise the moduli of the sums. The solution of this problem is the main result of this section.

In the case of one-dimensional trigonometric sums, this problem is called *Tarry's problem.*

6.1.1 An asymptotic formula for the mean value of a multiple trigonometric sum

In the problem studied here, we derive an asymptotic formula using the mean value theorem for multiple trigonometric sums (Section 4.2, Chapter 4) and the estimates from above for the modulus of multiple trigonometric sums (Section 5.2, Chapter 5), for the multiple trigonometric integral, and for multiple complete rational trigonometric sums (Chapters 1 and 2).

We shall use the notation introduced in Section 5.2, Chapter 5.

Lemma 6.1. *Suppose that A is a point of the first class Ω_1. Then the sum $S(A)$ satisfies the relation*

$$S(A) = P_1 \dots P_r UV + O(P_2 \dots P_r Q),$$

where

$$U = Q^{-r} \sum_{x_1=1}^{Q} \cdots \sum_{x_r=1}^{Q} \exp\{2\pi i F_a(x_1, \dots, x_r)\},$$

$$F_a(x_1, \dots, x_r) = \sum_{t_1=0}^{n_1} \cdots \sum_{t_r=0}^{n_r} \frac{a(t_1, \dots, t_r)}{q(t_1, \dots, t_r)} x_1^{t_1} \dots x_r^{t_r},$$

$$V = \int_0^1 \cdots \int_0^1 \exp\{2\pi i F_\beta(x_1, \dots, x_r)\}\, dx_1 \dots dx_r,$$

$$F_\beta(x_1, \dots, x_r) = \sum_{t_1=0}^{n_1} \cdots \sum_{t_r=0}^{n_r} \beta(t_1, \dots, t_r) P_1^{t_1} \dots P_r^{t_r} x_1^{t_1} \dots x_r^{t_r}.$$

Proof. This formula was obtained in Chapter 5 in the proof of Lemma 5.5 in estimating the trigonometric sum $S(A)$ at points of the first class Ω_1. □

Lemma 6.2. *Suppose that $k > 2\nu^{-1}m$ and*

$$J_1 = \int \cdots \int_{\Omega_1} |S(A)|^{2k}\, dA.$$

Then the variable J_1 satisfies the asymptotic formula

$$J_1 = \sigma\theta(P_1 \dots P_r)^{2k} P^{-0.5m\varkappa} + O\big((P_1 \dots P_r)^{2k} P^{-0.5m\varkappa - 0.1}\big),$$

where

$$\theta = \int_{-\infty}^{+\infty} \cdots \int_{-\infty}^{+\infty} \left| \int_0^1 \cdots \int_0^1 \exp\{2\pi i F_A(x_1, \dots, x_r)\}\, dx_1 \dots dx_r \right|^{2k} dA,$$

$$\sigma = \sum_{q(0,\dots,1)=1}^{+\infty} \cdots \sum_{q(n_1,\dots,n_r)=1}^{+\infty} \sum_{\substack{a(0,\dots,1)=1 \\ (a(0,\dots,1),q(0,\dots,1))=1}}^{q(0,\dots,1)} \cdots \sum_{\substack{a(n_1,\dots,n_r)=1 \\ (a(n_1,\dots,n_r),q(n_1,\dots,n_r))=1}}^{q(n_1,\dots,n_r)} |U(a,q)|^{2k},$$

$$U(a,q) = q^{-r} \sum_{x_1=1}^{q} \cdots \sum_{x_r=1}^{q} \exp\{2\pi i F_a(x_1, \dots, x_r)\},$$

$$q = q(0, \dots, 1) \dots q(n_1, \dots, n_r).$$

Proof. The domain Ω_1 consists of nonintersecting domains $\Omega(a, q)$ for which the least common multiple of the numbers $q(0, \dots, 1), \dots, q(n_1, \dots, n_r)$, equal to Q, does not exceed $P^{0.1}$. Therefore, the variable J_1 can be written as

$$J_1 = \sum_{Q \le P^{0.1}} \sum_{\substack{q(0,\dots,1)\ge 1 \\ [q(0,\dots,1),\dots,q(n_1,\dots,n_r)]=Q}} \cdots \sum_{q(n_1,\dots,n_r)\ge 1} \sum_{\substack{a(0,\dots,1)=1 \\ (a(0,\dots,1),q(0,\dots,1))=1}}^{q(0,\dots,1)} \cdots \sum_{\substack{a(n_1,\dots,n_r)=1 \\ (a(n_1,\dots,n_r),q(n_1,\dots,n_r))=1}}^{q(n_1,\dots,n_r)} J_3,$$

where

$$J_3 = \int \cdots \int_{\Omega(a,q)} |S(A)|^{2k} \, dA.$$

By ω we denote the domain of sets β whose coordinates $\beta(t_1, \dots, t_r)$ satisfy the inequalities

$$|\beta(t_1, \dots, t_r)| \le P_1^{-t_1} \dots P_r^{-t_r} P^{0.1}, \quad 0 \le t_1 \le n_1, \dots, 0 \le t_r \le n_r.$$

By Lemma 6.1, we can write the integral J_3 as

$$\begin{aligned} J_3 = {} & (P_1 \dots P_r)^{2k} |U(a, q)|^{2k} \int \cdots \int_{\omega} |V|^{2k} \, d\beta \\ & + O\big((P_2 \dots P_r)^{2k} Q^{2k} P^{-0.5m\varkappa + 0.1m}\big) \\ & + O\Big(P_1^{2k-1} (P_2 \dots P_r)^{2k} Q |U(a, q)|^{2k-1} \int \cdots \int_{\omega} |V|^{2k-1} \, d\beta\Big). \end{aligned}$$

We perform the change of integration variables

$$\gamma(t_1, \dots, t_r) = P_1^{t_1} \dots P_r^{t_r} \beta(t_1, \dots, t_r), \quad 0 \le t_1 \le n_1, \dots, 0 \le t_r \le n_r.$$

After this change, the domain ω becomes the domain ω_1 determined by the inequalities

$$|\gamma(t_1, \dots, t_r)| \le P^{0.1}, \quad 0 \le t_1 \le n_1, \dots, 0 \le t_r \le n_r,$$

and the integral J_3 takes the form

$$\begin{aligned} J_3 = {} & (P_1 \dots P_r)^{2k} P^{-0.5m\varkappa} |U(a, q)|^{2k} \int \cdots \int_{\omega_1} |V_1(\gamma)|^{2k} \, d\gamma \\ & + O\big((P_2 \dots P_r)^{2k} Q^{2k} P^{-0.5m\varkappa + 0.1m}\big) \\ & + O\Big(P_1^{2k-1} (P_2 \dots P_r)^{2k} Q P^{-0.5m\varkappa} |U(a, q)|^{2k-1} \int \cdots \int_{\omega_1} |V_1(\gamma)|^{2k-1} \, d\gamma\Big), \end{aligned}$$

where $V_1(\gamma) = V$.

Next, by Theorem 1.6 (Chapter 1), the integral

$$\int_{-\infty}^{+\infty} \cdots \int_{-\infty}^{+\infty} |V_1(\gamma)|^{2k} \, d\gamma$$

converges for $k > 0.5\nu^{-1}m$. Therefore, if we trivially bound $|V_1(\gamma)|$ above by unity, then we find the following expression for J_3:

$$\begin{aligned} J_3 = {} & (P_1 \dots P_r)^{2k} P^{-0.5m\varkappa} |U(a,q)|^{2k} \int_{-\infty}^{+\infty} \cdots \int_{-\infty}^{+\infty} |V_1(\gamma)|^{2k} \, d\gamma \\ & - (P_1 \dots P_r)^{2k} P^{-0.5m\varkappa} |U(a,q)|^{2k} \underset{R^m \setminus \omega_1}{\int \cdots \int} |V_1(\gamma)|^{2k} \, d\gamma \\ & + O\big((P_2 \dots P_r)^{2k} Q^{2k} P^{-0.5m\varkappa + 0.1m}\big) \\ & + O\Big((P_1 \dots P_r)^{2k} P_1^{-1} P^{-0.5m\varkappa} Q |U(a,q)|^{2k-1}\Big). \end{aligned}$$

To estimate the integral of $|V_1(\gamma)|^{2k}$ over the points that do not belong to ω_1, we rewrite the estimate of the trigonometric integral $V_1(\gamma)$ obtained in Theorem 1.6 in the more convenient form

$$\begin{aligned} |V_1(\gamma)| & \ll \min\big(1, \, |\gamma(0,\dots,1)|^{-\nu+\varepsilon}, \dots, |\gamma(n_1,\dots,n_r)|^{-\nu+\varepsilon}\big) \\ & \ll \min\big(1, \, |\gamma(0,\dots,1)|^{(-\nu+\varepsilon)/m}\big) \dots \min\big(1, \, |\gamma(n_1,\dots,n_r)|^{(-\nu+\varepsilon)/m}\big), \end{aligned}$$

where $\varepsilon > 0$ is an arbitrary small constant, the constant in $\ll$ depends only on $\overline{n}$ and r, and $\nu \max(n_1,\dots,n_r) = 1$. Since for the points γ that do not belong to ω_1, the absolute value of at least one of the coordinates $\gamma(t_1,\dots,t_r)$ is larger than $P^{0.1}$, we have

$$\underset{R^m \setminus \omega_1}{\int \cdots \int} |V_1(\gamma)|^{2k} \, d\gamma \le \sum_{t_1=0}^{n_1} \cdots \sum_{\substack{t_r=0 \\ t_1+\cdots+t_r \ge 1}}^{n_r} R(t_1,\dots,t_r),$$

where

$$\begin{aligned} R(t_1,\dots,t_r) = {} & \int_{-\infty}^{+\infty} d\gamma(0,\dots,1) \dots \int_{|\gamma(t_1,\dots,t_r)| > P^{0.1}} d\gamma(t_1,\dots,t_r) \dots \\ & \cdots \int_{-\infty}^{+\infty} |V_1(\gamma)|^{2k} \, d\gamma(n_1,\dots,n_r). \end{aligned}$$

Applying the above estimate of $V_1(\gamma)$, we obtain

$$R(t_1,\dots,t_r) \ll \int_{|\gamma| > P^{0.1}} \min(1, \, |\gamma|^{2k(-\nu+\varepsilon)/m}) \, d\gamma \ll P^{0.1(1+2k(-\nu+\varepsilon)/m)},$$

where, as before, the constant in $\ll$ depends only on $n_1, \dots, n_r$ and r. Hence,

$$\int \cdots \int_{R^m \setminus \omega_1} |V_1(\gamma)|^{2k} \, d\gamma \ll P^{0.1(1+2k(-\nu+\varepsilon)/m)}.$$

This and the estimate for multiple rational trigonometric sums (Theorem 2.6) imply the following formula for J_3:

$$\begin{aligned} J_3 &= \theta (P_1 \dots P_r)^{2k} P^{-0.5m\varkappa} |U(a, q)|^{2k} \\ &\quad + O\big((P_1 \dots P_r)^{2k} P^{-0.5m\varkappa} (P_1^{-2k} P_1^{0.1m} Q^{2k} \\ &\quad + P_1^{-1} Q^{1-(2k-1)\nu+(2k-1)\varepsilon} + Q^{2k(-\nu+\varepsilon)} P^{0.1(1+2k(-\nu+\varepsilon)/m)})\big). \end{aligned}$$

Substituting this formula for J_3 into the expression for J_1, we obtain

$$\begin{aligned} J_1 &= \theta (P_1 \dots P_r)^{2k} P^{-0.5m\varkappa} \sum_{Q \le P^{0.1}} \sum_{q(0,\dots,1) \ge 1} \cdots \sum_{\substack{q(n_1,\dots,n_r) \ge 1 \\ [q(0,\dots,1),\dots,q(n_1,\dots,n_r)] = Q}} \\ &\quad \times \sum_{\substack{a(0,\dots,1)=1 \\ (a(0,\dots,1),q(0,\dots,1))=1}}^{q(0,\dots,1)} \cdots \sum_{\substack{a(n_1,\dots,n_r)=1 \\ (a(n_1,\dots,n_r),q(n_1,\dots,n_r))=1}}^{q(n_1,\dots,n_r)} |U(a, q)|^{2k} \\ &\quad + O\Big((P_1 \dots P_r)^{2k P^{-0.5m\varkappa}} \sum_{Q \le P^{0.1}} (P_1^{-2k} P_1^{0.1m} Q^{2k+m+m\varepsilon} \\ &\quad + P_1^{-1} Q^{m+1-\nu(2k-1)+m\varepsilon+2k\varepsilon} + Q^{m+2k(-\nu+\varepsilon)+m\varepsilon} P^{0.1(1+2k(-\nu+\varepsilon)/m)})\Big). \end{aligned}$$

It follows from the estimate for the trigonometric sum $U(a, q)$ (see Theorem 2.6) that the singular series σ converges for $k > 0.5m\nu^{-1}$. Hence for the integral J_1, we have

$$\begin{aligned} J_1 &= \sigma\theta (P_1 \dots P_r)^{2k} P^{-0.5m\varkappa} - \theta (P_1 \dots P_r)^{2k} P^{-0.5m\varkappa} \\ &\quad \times \sum_{Q > P^{0.1}} \sum_{q(0,\dots,1) \ge 1} \cdots \sum_{\substack{q(n_1,\dots,n_r) \ge 1 \\ [q(0,\dots,1),\dots,q(n_1,\dots,n_r)] = Q}} \\ &\quad \times \sum_{\substack{a(0,\dots,1)=1 \\ (a(0,\dots,1),q(0,\dots,1))=1}}^{q(0,\dots,1)} \cdots \sum_{\substack{a(n_1,\dots,n_r)=1 \\ (a(n_1,\dots,n_r),q(n_1,\dots,n_r))=1}}^{q(n_1,\dots,n_r)} |U(a, q)|^{2k} + R, \end{aligned}$$

where

$$\begin{aligned} |R| &\ll (P_1 \dots P_r)^{2k} P^{-0.5m\varkappa} \big(P_1^{-2k} P^{0.2m+0.2k+0.1m\varepsilon+0.1} \\ &\quad + P_1^{-1} + P^{0.1(2+m-2k\nu-2k\nu m^{-1}+\varepsilon(m+2k+2km^{-1}))}\big). \end{aligned}$$

Hence, for $k \geq 2m\nu^{-1}$, the value of $|R|$ does not exceed

$$\ll (P_1 \dots P_r)^{2k} P^{-0.5m\varkappa - 0.1}.$$

Next, we use the fact that the numbers $q(t_1, \dots, t_r)$ are divisors of Q. Hence the number of sets $(q(0, \dots, 1), \dots, q(n_1, \dots, n_r))$ for which their least common multiple is equal to Q does not exceed $(\tau(Q))^m \ll Q^{m\varepsilon}$. By Theorem 2.6, this fact and the estimate of $|U(a, q)|$ imply (for $k > 2m\nu^{-1}$)

$$\begin{aligned} J_1 &= \sigma\theta(P_1 \dots P_r)^{2k} P^{-0.5m\varkappa} \\ &+ O\Big((P_1 \dots P_r)^{2k} P^{-0.5m\varkappa} \sum_{Q > P^{0.1}} Q^{m+m\varepsilon-2k\nu+2k\varepsilon}\Big) \\ &+ O\big((P_1 \dots P_r)^{2k} P^{-0.5m\varkappa-0.1}\big). \end{aligned}$$

Since

$$\sum_{Q > P^{0.1}} Q^{m-2k\nu+\varepsilon(m+2k)}) \ll P^{0.1(1+m-2k\nu+\varepsilon(m+2k))} \ll P^{-0.1},$$

the preceding relation yields the desired asymptotic formula for J_1. The proof of the lemma is complete. □

Lemma 6.3. *Let*

$$J_2 = \int \cdots \int_{\Omega_2} |S(A)|^{2k} \, dA.$$

Then the following estimate holds for $k \geq 4m\varkappa \log 16m\varkappa$:

$$J_2 \ll e^a (P_1 \dots P_r)^{2k} P^{-0.5m\varkappa - \rho_1},$$

where

$$\rho_1 = (32\varkappa \log 8m\varkappa)^{-1}, \quad a = 64m\varkappa^2 \log 16m\varkappa + 32m\varkappa \log^2 16m\varkappa.$$

Proof. We set $k_1 = m$, $k_2 = m\tau$, and $\tau = [\varkappa \log 16m\varkappa^2 + \varkappa \log\log 8m\varkappa] + 1$. Obviously, it suffices to prove the statement of the lemma for $k = k_1 + k_2$. We have the inequality

$$J_2 \leq DJ, \tag{6.1}$$

where

$$D = \max_{A \in \Omega_2} |S(A)|^{2k_1}, \quad J = J(\overline{P}; \overline{n}, k_2) = \int \cdots \int_{\Omega} |S(A)|^{2k} \, dA.$$

Let us estimate J from above. Theorem 4.2 (Chapter 4) implies the inequality

$$J \leq k_2^{2m\tau} \varkappa^{4\varkappa^2 \Delta(\tau)} 2^{8m\varkappa\tau} (P_1 \dots P_r)^{2k_2} P^{-0.5m\varkappa+\delta},$$

where

$$\delta = 0.5m\varkappa(1-\gamma)^\tau \le 0.5m\varkappa \exp\{-\log 16m\varkappa^2 - \log\log 8m\varkappa\}$$
$$= (32\varkappa \log 8m\varkappa)^{-1} = \rho_1.$$

Further, we have

$$J = \exp\{16m\varkappa \log^2 16m\varkappa + 36m\varkappa^2 \log 16m\varkappa\}(P_1 \dots P_r)^{2k_2} P^{-0.5m\varkappa+\rho_1},$$

because

$$k_2^{2m\tau}\varkappa^{4\varkappa^2\Delta(\tau)}2^{9m\varkappa\tau} \le \exp\{4m\varkappa \log(16m\varkappa^2)\log(2m\varkappa \log 16m\varkappa^2)\}$$
$$\times \exp\{2\varkappa^2 m \log\varkappa\}\exp\{16m\varkappa^2 \log 16m\varkappa^2\}$$
$$\le \exp\{16m\varkappa \log^2 16m\varkappa + 36m\varkappa^2 \log 16m\varkappa\}.$$

Theorem 5.2 estimating the trigonometric sum $S(A)$ for $Q \ge P^{1/6}$ implies the estimate

$$|S(A)|^{2k_1} \le 2^{64m\varkappa}(P_1 \dots P_r)^{2m} P^{-2\rho_1}$$

and, for $Q < P^{1/6}$, the estimate

$$|S(A)|^{2k_1} \ll (P_1 \dots P_r)^{2m} P^{-0.1\nu m},$$

where the constant in $\ll$ depends only on $\overline{n}$ and r. Hence for D we have the estimate

$$D \ll 2^{64m\varkappa}(P_1 \dots P_r)^{2m} P^{-2\rho_1}.$$

Substituting the estimates of J and D into inequality (6.1), we obtain the estimate for J_2 stated in the lemma. The proof of the lemma is complete. □

Theorem 6.1. *Let* $k \ge 4m\varkappa \log 16m\varkappa$, *and let*

$$J = \int \cdots \int_{\Omega} |S(A)|^{2k}\, dA.$$

Then the following asymptotic formula holds:

$$J = \sigma\theta(P_1 \dots P_r)^{2k} P^{-0.5m\varkappa} + O\big(e^a (P_1 \dots P_r)^{2k} P^{-0.5m\varkappa-\rho_1}\big),$$

where

$$\rho_1 = (32\varkappa \log m\varkappa)^{-1}, \quad a = 64m\varkappa^2 \log 16m\varkappa + 32m\varkappa \log^2 16m\varkappa,$$

and the constant in the sign O depends only on $\overline{n}$ and r.

Proof. This assertion follows from Lemmas 6.2 and 6.3. □

In Chapter 8 we shall need Theorem 6.2, which is close in content to Theorem 6.1. Here we only state this theorem because their proofs coincide word for word.

Theorem 6.2. *Let* $k \geq 8m\varkappa \log 16m\varkappa$, *and let*

$$J = \int \cdots \int_{\Omega} S^k(A) \exp\{-2\pi i(A \times N)\} \, dA,$$

where

$$(A \times N) = \sum_{t_1=0}^{n_1} \cdots \sum_{t_r=0}^{n_r} \alpha(t_1, \ldots, t_r) \, N(t_1, \ldots, t_r);$$

here $N(t_1, \ldots, t_r)$ *are arbitrary natural numbers. Then the following asymptotic formula holds*:

$$J = \sigma\theta(P_1 \ldots P_r)^k P^{-0.5m\varkappa} + O\big(e^a (P_1 \ldots P_r)^k P^{-0.5m\varkappa - \rho_1}\big),$$

where

$$\sigma = \sum_{q(0,\ldots,1)=1}^{+\infty} \cdots \sum_{q(n_1,\ldots,n_r)=1}^{+\infty}$$

$$\times \sum_{\substack{a(0,\ldots,1)=1 \\ (a(0,\ldots,1),q(0,\ldots,1))=1}}^{q(0,\ldots,1)} \cdots \sum_{\substack{a(n_1,\ldots,n_r)=1 \\ (a(n_1,\ldots,n_r),q(n_1,\ldots,n_r))=1}}^{q(n_1,\ldots,n_r)} (U(a,q))^k \exp\left\{-2\pi i\left(\frac{a}{q} \times N\right)\right\},$$

$$\left(\frac{a}{q} \times N\right) = \sum_{t_1=0}^{n_1} \cdots \sum_{t_r=0}^{n_r} \frac{a(t_1, \ldots, t_r)}{q(t_1, \ldots, t_r)} N(t_1, \ldots, t_r),$$

$$\theta = \int_{-\infty}^{+\infty} \cdots \int_{-\infty}^{+\infty} \left(\int_0^1 \cdots \int_0^1 \exp\{2\pi i F_A(x_1, \ldots, x_r)\} \, dx_1 \ldots dx_r\right)^k$$
$$\times \exp\{-2\pi i(A \times M)\} \, dA,$$

$$(A \times M) = \sum_{t_1=0}^{n_1} \cdots \sum_{t_r=0}^{n_r} \alpha(t_1, \ldots, t_r) \, M(t_1, \ldots, t_r),$$

$$M(t_1, \ldots, t_r) = N(t_1, \ldots, t_r) P_1^{-t_1} \ldots P_r^{-t_r},$$

$$\rho_1 = (32\varkappa \log 8m\varkappa)^{-1}, \quad a = 64m\varkappa^2 \log 16m\varkappa + 32m\varkappa \log^2 16m\varkappa,$$

and the constant in the sign O *depends only on* $\overline{n}$ *and* r.

6.1.2 Multiple trigonometric sums with summation domains of special form

Now we consider a summation domain somewhat more complicated than the parallelepiped and show how multiple sums can be estimated in this case, what asymptotic formulas for the number of solutions of the complete system of equations can be obtained, etc. Since the proofs of the theorems mainly coincide with those performed above, we shall concentrate only on the parts of argument that characterize the case under study. We shall deal with the domain E_r that has the form

$$e_1x_1^{s_1} + \dots + e_rx_r^{s_r} \le P_0, \quad 1 \le x_1 \le P_1, \dots, 1 \le x_r \le P_r,$$

where $e_1, \dots, e_r$ are equal to either 0 or 1, $s_1, \dots, s_r$ are natural numbers such that $s_1 \le n_1, \dots, s_r \le n_r$, and the numbers $P_1, \dots, P_r$ satisfy the condition: if $e_j = 1$, then $P_j^{s_j} = P_0$.

Let $\chi(x_1, \dots, x_r)$ be the characteristic function of the domain E_r, i.e.,

$$\chi(x_1, \dots, x_r) = \begin{cases} 1 & \text{if } (x_1, \dots, x_r) \in E_r, \\ 0 & \text{if } (x_1, \dots, x_r) \notin E_r. \end{cases}$$

We shall prove two auxiliary lemmas.

Lemma 6.4. *Let* $1 \le y_1 \le P_1, \dots, 1 \le y_r \le P_r$. *Then the sum*

$$R = \sum_{x_1=1}^{P_1+y_1} \cdots \sum_{x_r=1}^{P_r+y_r} |\chi(x_1, \dots, x_r) - \chi(x_1 - y_1, \dots, x_r - y_r)|$$

satisfies the estimate

$$R \le 2^r(y_1P_2 \dots P_r + \dots + P_1 \dots P_{r-1}y_r).$$

Proof. Since we have the inequality

$$\begin{aligned} R \le \sum_{x_1=1}^{P_1+y_1} \cdots \sum_{x_r=1}^{P_r+y_r} \big(&|\chi(x_1, \dots, x_r) - \chi(x_1 - y_1, x_2, \dots, x_r)| \\ &+ |\chi(x_1 - y_1, x_2, x_3, \dots, x_r) - \chi(x_1 - y_1, x_2 - y_2, x_3, \dots, x_r)| + \cdots \\ &+ |\chi(x_1 - y_1, \dots, x_{r-1} - y_{r-1}, x_r) - \chi(x_1 - y_1, \dots, x_{r-1} - y_{r-1}, x_r - y_r)|\big), \end{aligned}$$

it suffices, for fixed $z_1, \dots, z_{q-1}, z_{q+1}, \dots, z_r$, to estimate from above the number R_q of z_q such that $\chi_q = |\chi(z_1, \dots, z_{q-1}, z_q, z_{q+1}, \dots, z_r) - \chi(z_1, \dots, z_{q-1}, z_q - y_q, z_{q+1}, \dots, z_r)| = 1$. Obviously, $\chi_q = 1$ for $1 \le z_q \le y_q$ and for z_q such that $\chi(z_1, \dots, z_{q-1}, z_q, z_{q+1}, \dots, z_r) = 0$ and $\chi(z_1, \dots, z_{q-1}, z_q - y_q, z_{q+1}, \dots, z_r) = 1$.

In the case $e_q = 0$, the last conditions holds for $P_q + 1 \le z_q \le P_q + y_q$. But if $e_q = 1$, then z_q is determined by the inequalities

$$e_1 z_1^{s_1} + \dots + e_q z_q^{s_q} + \dots + e_r z_r^{s_r} > P_0,$$
$$e_1 z_1^{s_1} + \dots + e_q (z_q - y_q)^{s_q} + \dots + e_r z_r^{s_r} \le P_0.$$

Hence we have $A < z_q \le A + y_q$, where

$$A = (P_0 - e_1 z_1^{s_1} - \dots - e_{q-1} z_{q-1}^{s_{q-1}} - e_{q+1} z_{q+1}^{s_{q+1}} - \dots - e_r z_r^{s_r})^{1/s_q}.$$

Hence $R_q \le 2y_q$. This implies the inequality

$$\begin{aligned} R &\le 2y_1(P_2 + y_2)\dots(P_r + y_r) + \dots + (P_1 + y_1)\dots(P_{r-1} + y_{r-1})2y_r \\ &\le 2^r(y_1 P_2 \dots P_r + \dots + P_1 \dots P_{r-1}\, y_r). \end{aligned}$$

The proof of the lemma is complete. □

Lemma 6.5. *Suppose that the function $\chi_1(x_1, \dots, x_r)$ is determined by the relations*

$$\chi_1(x_1, \dots, x_r) = \begin{cases} 1 & \text{if } e_1 x_1^{s_1} + \dots + e_r x_r^{s_r} \le 1,\ x_1 > 0, \dots, x_r > 0, \\ 0 & \text{otherwise}, \end{cases}$$

where $e_1, \dots, e_r$ are equal either to 0 or to 1, $s_1, \dots, s_r$ are natural numbers satisfying the conditions that $s_1 \le n_1, \dots, s_r \le n_r$, and ν^{-1} is equal to the largest of the numbers $n_1, \dots, n_r$. Then the integral

$$V = \int_0^1 \cdots \int_0^1 \chi_1(x_1, \dots, x_r) \exp\{2\pi i F(x_1, \dots, x_r)\}\, dx_1 \dots dx_r,$$

where, as above,

$$F(x_1, \dots, x_r) = \sum_{t_1=0}^{n_1} \cdots \sum_{t_r=0}^{n_r} \alpha(t_1, \dots, t_r)\, x_1^{t_1} \dots x_r^{t_r},$$
$$\alpha = \max_{t_1, \dots, t_r} |\alpha(t_1, \dots, t_r)|, \quad \alpha(0, \dots, 0) = 0,$$

satisfies the estimate

$$|V| \le \min(1,\ 2^{12r} \alpha^{-\nu} (\ln(\alpha + 2))^r).$$

Proof. If $e_1 = \dots = e_r = 0$, then we obtain the desired statement by setting $n = \nu^{-1}$ in Theorem 1.6 (Chapter 1). Therefore, we assume that $e_j = 1$ for some j $(1 \le j \le r)$. If the value of α is small, $\alpha \le 2^{12r\nu^{-1}}$, then the estimate in the lemma becomes trivial. We assume that $\alpha > 2^{12r\nu^{-1}}$. Now we set $r = 1$ and $\Delta = \alpha^{-1}$

in Lemma A.3 and take the function $\psi(x)$ from this lemma. Then, according to the properties of the function $\psi(x)$, we obtain the relations

$$\psi(x) = 1 \quad \text{if} \quad \Delta < x \le 1 - \Delta \pmod 1,$$
$$0 \le \psi(x) \le 1 \quad \text{if} \quad 0 \le x \le \Delta \pmod 1 \quad \text{or} \quad 1 - \Delta \le x \le 1 \pmod 1,$$
$$\psi(x) = 1 - \Delta + \sum_{m=1}^{+\infty} g_m \exp\{2\pi i m x\} + h_m \exp\{-2\pi i m x\},$$

and moreover,

$$\max(|g_m|, |h_m|) < (\pi m)^{-1} \quad \text{if} \quad 1 \le m \le \Delta^{-1},$$
$$\max(|g_m|, |h_m|) < (\pi^2 \Delta m^2)^{-1} \quad \text{if} \quad m > \Delta^{-1}.$$

We use the properties of the function $\psi(x)$ to write the integral V as

$$V = V_0 + \theta_1(V_1 + V_2),$$

where

$$V_0 = \int_0^1 \cdots \int_0^1 \psi(e_1 x_1^{s_1} + \cdots + e_r x_r^{s_r}) \exp\{2\pi i F(x_1, \ldots, x_r)\}\, dx_1 \ldots dx_r,$$
$$V_1 = \underset{e_1 x_1^{s_1} + \cdots + e_r x_r^{s_r} \le \Delta}{\int_0^1 \cdots \int_0^1} dx_1 \ldots dx_r, \quad V_2 = \underset{1-\Delta \le e_1 x_1^{s_1} + \cdots + e_r x_r^{s_r} \le 1}{\int_0^1 \cdots \int_0^1} dx_1 \ldots dx_r,$$
$$|\theta_1| \le 1.$$

Now we estimate V_1 and V_2. For this, we consider the integral $V_3 = V_3(\Lambda)$,

$$V_3 = \underset{x^a + \cdots + y^b + \cdots + z^d \le \Lambda}{\int_0^1 \cdots \int_0^1} dx \ldots dy \ldots dz,$$

where $a, \ldots, b, \ldots, d$ are natural numbers and $0 < \Lambda \le 1$. A change of integration variables implies

$$\begin{aligned} V_3 &= \frac{1}{a \ldots b \ldots d} \Lambda^{a^{-1} + \cdots + b^{-1} + \cdots + d^{-1}} \\ &\quad \times \underset{u + \cdots + v + \cdots + w \le 1}{\int_0^{\Lambda^{-1}} \cdots \int_0^{\Lambda^{-1}}} u^{-1+a^{-1}} \ldots v^{-1+b^{-1}} \ldots w^{-1+d^{-1}}\, du \ldots dv \ldots dw \\ &= (a \ldots b \ldots d)^{-1} \Lambda^{a^{-1} + \cdots + b^{-1} + \cdots + d^{-1}} \\ &\quad \times \underset{u + \cdots + v + \cdots + w \le 1}{\int_0^1 \cdots \int_0^1} u^{-1+a^{-1}} \ldots v^{-1+b^{-1}} \ldots w^{-1+d^{-1}}\, du \ldots dv \ldots dw. \end{aligned}$$

The last integral is the well-known Dirichlet integral (e.g., see [90], p. 58) and is equal to

$$\frac{\Gamma(a^{-1})\dots\Gamma(b^{-1})\dots\Gamma(d^{-1})}{\Gamma(a^{-1}+\dots+b^{-1}+\dots+d^{-1})}\cdot\frac{1}{a^{-1}+\dots+b^{-1}+\dots+d^{-1}}.$$

Thus we have

$$V_3 = c\,\Lambda^{a^{-1}+\dots+b^{-1}+\dots+d^{-1}},$$

where

$$c=\frac{1}{a\dots b\dots d}\cdot\frac{1}{a^{-1}+\dots+b^{-1}+\dots+d^{-1}}\cdot\frac{\Gamma(a^{-1})\dots\Gamma(b^{-1})\dots\Gamma(d^{-1})}{\Gamma(a^{-1}+\dots+b^{-1}+\dots+d^{-1})}\le 2.$$

Now we let the letters $a,\dots,b,\dots,d$ denote the s_j for which $e_j=1$. Then we obtain the relations

$$\begin{aligned}V_1=V_3(\Delta),\quad V_2&=V_3(1)-V_3(1-\Delta)=c\big(1-(1-\Delta)^{a^{-1}+\dots+b^{-1}+\dots+d^{-1}}\big)\\&=c\Delta(a^{-1}+\dots+b^{-1}+\dots+d^{-1})\,\xi^{a^{-1}+\dots+b^{-1}+\dots+d^{-1}-1},\end{aligned}$$

where $1-\Delta\le\xi\le 1$. Thus (recall that $\Delta=\alpha^{-1}$ and $\alpha>2^{12r}$) we have

$$V_1\le c\Delta^{\nu},\quad V_2\le 2cr\Delta.$$

Now we shall estimate V_0. For this, we expand the function $\psi(x)$ in the Fourier series and pass to the inequalities

$$|V_0|\le(1-\Delta)|I_r|+\sum_{m=1}^{+\infty}\big(|g_m|\,|\Phi_1(m)|+|h_m|\,|\Phi_2(m)|\big),$$

where

$$\begin{aligned}I_r&=\int_0^1\cdots\int_0^1\exp\{2\pi i F(x_1,\dots,x_r)\}\,dx_1\dots dx_r,\\ \Phi_1(m)&=\int_0^1\cdots\int_0^1\exp\{2\pi i(F(x_1,\dots,x_r)\\&\qquad+m(e_1x_1^{s_1}+\dots+e_rx_r^{s_r}))\}\,dx_1\dots dx_r,\\ \Phi_2(m)&=\int_0^1\cdots\int_0^1\exp\{2\pi i(F(x_1,\dots,x_r)\\&\qquad-m(e_1x_1^{s_1}+\dots+e_rx_r^{s_r}))\}\,dx_1\dots dx_r.\end{aligned}$$

By Theorem 1.6 in Chapter 1, we have

$$|I_r|\le 32^r\alpha^{-\nu}\ln^{r-1}(\alpha+2).$$

Further, if $m \geq 2\Delta^{-1} = 2\alpha$, then the coefficient largest in absolute value in the polynomials contained in the exponential in the integrals $\Phi_1(m)$ and $\Phi_2(m)$ is larger than or equal to $m/2$, because the coefficient largest in absolute value of $F(x_1, \ldots, x_r)$ does not exceed $\alpha \leq m/2$ and $e_j = 1$.

Now we apply the estimate in Theorem 1.6. For $m \geq 2\Delta^{-1}$, we obtain

$$|\Phi_1(m)| \leq 32^r (m/2)^{-\nu} \ln^{r-1}(m+2), \quad i = 1, 2;$$

$$|g_m|\,|\Phi_1(m)| + |h_m|\,|\Phi_2(m)| \leq \frac{2 \cdot 32^r}{\pi^2 \Delta m^2} \left(\frac{m}{2}\right)^{-\nu} \ln^{r-1}(m+2).$$

Now let $1 \leq m \leq 2\Delta^{-1}$. In this case the largest coefficient of the polynomials under study is no less than $|m - \Delta^{-1}| = |m - \alpha|$ and does not exceed $m + \alpha$. We again apply Theorem 1.6 for $m \leq \Delta^{-1} - 1$ and $m \geq \Delta^{-1} + 2$ and find

$$|\Phi_i(m)| \leq 32^r |m - \Delta^{-1}|^{-\nu} \big(\log(m + \Delta^{-1} + 2)\big)^{r-1}, \quad i = 1, 2;$$

$$|g_m|\,|\Phi_1(m)| + |h_m|\,|\Phi_2(m)| \leq \frac{2 \cdot 32^r}{\pi m} \log^{r-1}(2\alpha + 2)\, |m - \Delta^{-1}|^{-\nu}.$$

Substituting the obtained estimates into the inequality for V_0 and, in the case $\Delta^{-1} - 1 < m < \Delta^{-1} + 2$, estimating $\Phi_i(m)$ trivially, we obtain the estimate

$$\begin{aligned}
|V_0| &\leq 32^r \alpha^{-\nu} \ln^{r-1}(\alpha + 2) \\
&\quad + \sum_{1 \leq m \leq \Delta^{-1} - 1} \frac{2 \cdot 32^r}{\pi m} |m - \Delta^{-1}|^{-\nu} \log^{r-1}(2\alpha + 2) \\
&\quad + \sum_{\Delta^{-1} - 1 < m < \Delta^{-1} + 2} \frac{2}{\pi m} + \sum_{m \geq \Delta^{-1} + 2} \frac{2 \cdot 32^r}{\pi \Delta m^2} \left(\frac{m}{2}\right)^{-\nu} \log^{r-1}(m+2) \\
&\leq 2^{10r} \alpha^{-\nu} \log^r(\alpha + 2).
\end{aligned}$$

This and the estimates for the integrals V_1 and V_2 imply the statement of the lemma. The proof is complete. □

Now we let $T(A)$ denote the trigonometric sum where the summation variables belong to the domain E_r. If, as before, $\chi(x_1, \ldots, x_r)$ denotes the characteristic function of the domain E_r, then the sum $T(A)$ can be written as

$$T(A) = \sum_{x_1=1}^{P_1} \cdots \sum_{x_r=1}^{P_r} \chi(x_1, \ldots, x_r) \exp\{2\pi i F_A(x_1, \ldots, x_r)\}.$$

Recall that the domain E_r is given by the inequalities

$$e_1 x_1^{s_1} + \cdots + e_r x_r^{s_r} \leq P_0, \quad 1 \leq x_1 \leq P_1, \ldots, 1 \leq x_r \leq P_r,$$

where $e_1, \dots, e_r$ are equal either to 0 or 1, $s_1, \dots, s_r$ are natural numbers ($s_1 \le n_1, \dots, s_r \le n_r$), and the numbers $P_1, \dots, P_r$ satisfy the condition that if $e_j = 1$, then $P_j^{s_j} = P_0$.

We shall prove the following theorem on the estimate of the sum $T(A)$ for points A from the cube Ω.

Theorem 6.3. *Suppose that a point A belongs to the class Ω_1. Then the estimate $|T(A)| \le 2(5n^{2n})^{r\nu(Q)}\,(\tau(Q))^{r-1}\,P_1 \dots P_r Q^{-\nu}$ holds. Moreover, if we set*

$$\delta(t_1, \dots, t_r) = P_1^{t_1} \dots P_r^{t_r} \beta(t_1, \dots, t_r), \quad \delta = \max_{t_1, \dots, t_r} |\delta(t_1, \dots, t_r)|,$$

then for $\delta > 1$ the following estimate holds:

$$|T(A)| \le 2^{16r}(5n^{2n})^{r\nu(Q)}(\tau(Q))^{r-1} P_1 \dots P_r (\delta Q)^{-\nu}(\ln(\delta + 2))^r.$$

Suppose that a point A belongs to the second class Ω_2. As above, we set

$$\tau(t_1, \dots, t_r) = P_1^{t_1} \dots P_r^{t_r} P^{-1/3}$$

and represent the coordinates $\alpha(t_1, \dots, t_r)$ of the point A as

$$\alpha(t_1, \dots, t_r) = \frac{a(t_1, \dots, t_r)}{q(t_1, \dots, t_r)} + \frac{\theta(t_1, \dots, t_r)}{q(t_1, \dots, t_r)\,\tau(t_1, \dots, t_r)},$$
$$(a(t_1, \dots, t_r), q(t_1, \dots, t_r)) = 1, \quad 1 \le q(t_1, \dots, t_r) \le \tau(t_1, \dots, t_r),$$
$$|\theta(t_1, \dots, t_r)| \le 1, \quad 0 \le t_1 \le n_1, \dots, 0 \le t_r \le n_r.$$

By Q_0 we denote the least common multiple of the numbers $q(t_1, \dots, t_r)$ satisfying the condition $t_1 + \dots + t_r \ge 2$. Then for $Q_0 \ge P^{1/6}$ the following estimate holds:

$$|T(A)| \le 2^{32\varkappa} P_1 \dots P_r P^{-\rho}, \quad \rho = (32m\varkappa \log(8m\varkappa))^{-1}.$$

But if $Q_0 < P^{1/6}$, then for $T(A)$ the following estimate holds:

$$|T(A)| \le (5n^{2n})^{r\nu(Q_0)}(\tau(Q_0))^{r-1} P_1 \dots P_r P^{-\nu/10} + 2^{16r}(r\nu^{-1})^{r-1} P_1 \dots P_r P^{-\nu/16}.$$

Proof. Suppose that a point A belongs to Ω_1. Then, repeating the argument of Lemma 5.5 in Chapter 5, we obtain

$$T(A) = \sum_{\eta_1=1}^{Q} \dots \sum_{\eta_r=1}^{Q} \exp\{2\pi i F_a(\eta_1, \dots, \eta_r)\}\, W_1(\eta_1, \dots, \eta_r),$$

where

$$F_a(\eta_1, \dots, \eta_r) = \sum_{t_1=0}^{n_1} \dots \sum_{t_r=0}^{n_r} \frac{a(t_1, \dots, t_r)}{q(t_1, \dots, t_r)} \eta_1^{t_1} \dots \eta_r^{t_r},$$

$$W_1 = W_1(\eta_1, \dots, \eta_r) = \sum_{\xi_1} \cdots \sum_{\xi_r} \chi(Q\xi_1 + \eta_1, \dots, Q\xi_r + \eta_r)$$
$$\times \exp\{2\pi i F_\beta(Q\xi_1 + \eta_1, \dots, Q\xi_r + \eta_r)\},$$
$$F_\beta(Q\xi_1 + \eta_1, \dots, Q\xi_r + \eta_r)$$
$$= \sum_{t_1=0}^{n_1} \cdots \sum_{t_r=0}^{n_r} \beta(t_1, \dots, t_r)(Q\xi_1 + \eta_1)^{t_1} \dots (Q\xi_r + \eta_r)^{t_r}.$$

Moreover, the summation is performed over $\xi_1, \dots, \xi_r$ such that the point $(Q\xi_1 + \eta_1, \dots, Q\xi_r + \eta_r)$ belongs to E_r.

Now we show that, with high accuracy, it is possible to replace the sum $W_1(\eta_1, \dots, \eta_r)$ by an integral. For this, we first consider the sum

$$G = \sum_{\xi_r} \chi(Q\xi_1 + \eta_1, \dots, Q\xi_r + \eta_r) \exp\{2\pi i F_\beta(Q\xi_1 + \eta_1, \dots, Q\xi_r + \eta_r)\},$$

where the summation variable ξ_r varies in the interval $[A_r, B_r]$. In the case $e_r = 0$, the variables A_r and B_r are determined by the relations $A_r = -\eta_r Q^{-1}$ and $B_r = (P_r - \eta_r)Q^{-1}$. But if $e_r = 1$, then A_r is determined as the least number and B_r is determined as the largest number that satisfy the inequalities

$$e_1(Q\xi_1 + \eta_1)^{s_1} + \cdots + e_r(QA_r + \eta_r)^{s_r} \geq 1, \qquad QA_r + \eta_r > 0,$$
$$e_1(Q\xi_1 + \eta_1)^{s_1} + \cdots + e_r(QB_r + \eta_r)^{s_r} \leq P_0, \quad QB_r + \eta_r \leq P_r.$$

We note that $A_r \geq -\eta_r Q^{-1}$ and $B_r \leq (P_r - \eta_r)Q^{-1}$ both in the first and in the second case. Further, as already shown,

$$\left| \frac{\partial F_\beta(Q\xi_1 + \eta_1, \dots, Q\xi_r + \eta_r)}{\partial \xi_r} \right| \leq 0.5,$$

and the number of intervals on which the function

$$F_1(\xi_r) = \frac{\partial F_\beta(Q\xi_1 + \eta_1, \dots, Q\xi_r + \eta_r)}{\partial \xi_r}$$

is monotone and of constant sign does not exceed $2n_r$. Hence to the sum G_1, it is possible to apply Lemma 5.4 (Chapter 5) with $r = 1$.

We obtain the relation

$$G_1 = \sum_{A_r \leq \xi_r \leq B_r} \exp\{2\pi i F_\beta(Q\xi_1 + \eta_1, \dots, Q\xi_r + \eta_r)\}$$
$$= \int_{A_r}^{B_r} \exp\{2\pi i F_\beta(Q\xi_1 + \eta_1, \dots, Q\xi_r + \eta_r)\}\, d\xi_r + 16\theta_1 n_r.$$

By the definition of $\chi(x_1, \dots, x_r)$ and by the inequalities $-\eta_r Q^{-1} \le A_r \le \xi_r \le B_r \le (P_r - \eta_r)Q^{-1}$, we can rewrite the last integral as

$$G_1 = \int_{-\eta_r Q^{-1}}^{(P_r-\eta_r)Q^{-1}} \chi(Q\xi_1 + \eta_1, \dots, Q\xi_r + \eta_r) \\ \times \exp\{2\pi i F_\beta(Q\xi_1 + \eta_1, \dots, Q\xi_r + \eta_r)\}\, d\xi_r + 16\theta_1 n_r.$$

Substituting G_1 into the sum W_1 and changing the order of integration and summation, we obtain

$$W_1 = \sum_{\xi_1} \cdots \sum_{\xi_{r-2}} \int_{-\eta_r Q^{-1}}^{(P_r-\eta_r)Q^{-1}} G_2\, d\xi_r + 16\theta_2 n_r P_1 \dots P_{r-1} Q^{-r+1},$$

where

$$G_2 = \sum_{\xi_{r-1}} \chi(Q\xi_1 + \eta_1, \dots, Q\xi_r + \eta_r) \exp\{2\pi i F_\beta(Q\xi_1 + \eta_1, \dots, Q\xi_r + \eta_r)\}$$

and the summation is taken over integer ξ_{r-1} from the interval $[A_{r-1}, B_{r-1}]$. Here A_{r-1} is the least number and B_{r-1} is the largest number that satisfy the inequalities

$$e_1(Q\xi_1 + \eta_1)^{s_1} + \dots + e_{r-1}(QA_{r-1} + \eta_{r-1})^{s_{r-1}} + e_r(Q\xi_r + \eta_r)^{s_r} \ge 1, \\ QA_{r-1} + \eta_{r-1} > 0, \\ e_1(Q\xi_1 + \eta_1)^{s_1} + \dots + e_{r-1}(QB_{r-1} + \eta_{r-1})^{s_{r-1}} + e_r(Q\xi_r + \eta_r)^{s_r} \le P_0, \\ QB_{r-1} + \eta_{r-1} \le P_{r-1}.$$

Performing the argument similarly to that for the sum G_1, we can replace the sum G_2 by an integral so that the error does not exceed $16n_{r-1}$. Therefore, for W_1 we have the relation

$$W_1 = \sum_{\xi_1} \cdots \sum_{\xi_{r-2}} \int_{-\eta_{r-1} Q^{-1}}^{(P_{r-1}-\eta_{r-1})Q^{-1}} d\xi_{r-1} \int_{-\eta_r Q^{-1}}^{(P_r-\eta_r)Q^{-1}} \chi(Q\xi_1 + \eta_1, \dots, Q\xi_r + \eta_r) \\ \times \exp\{2\pi i F_\beta(Q\xi_1 + \eta_1, \dots, Q\xi_r + \eta_r)\}\, d\xi_r \\ + 16\theta_3(n_{r-1} Q^{-r+1} P_1 \dots P_{r-2} P_r + n_r Q^{-r+1} P_1 \dots P_{r-1}).$$

Continuing the argument similarly to that for the variables ξ_r and ξ_{r-1}, we obtain the expression

$$W_1(\eta_1, \dots, \eta_r) = \int_{-\eta_1 Q^{-1}}^{(P_1-\eta_1)Q^{-1}} \cdots \int_{-\eta_r Q^{-1}}^{(P_r-\eta_r)Q^{-1}} \chi(Q\xi_1 + \eta_1, \dots, Q\xi_r + \eta_r) \\ \times \exp\{2\pi i F_\beta(Q\xi_1 + \eta_1, \dots, Q\xi_r + \eta_r)\}\, d\xi_1 \dots d\xi_r \\ + 16\theta_4 n P_2 \dots P_r Q^{-r+1}.$$

We set $\chi_1(x_1,\ldots,x_r) = \chi(P_1x_1,\ldots,P_rx_r)$. Then we have

$$\chi_1(x_1,\ldots,x_r) = \begin{cases} 1 & \text{if } e_1x_1^{s_1} + \cdots + e_rx_r^{s_r} \le 1, \\ & 0 \le x_1 \le 1,\ldots,0 \le x_r \le 1, \\ 0 & \text{otherwise,} \end{cases}$$

In the integral for $W_1(\eta_1,\ldots,\eta_r)$, we perform the change of variables

$$P_1x_1 = Q\xi_1 + \eta_1, \;\ldots,\; P_rx_r = Q\xi_r + \eta_r,$$

and thus obtain

$$W_1(\eta_1,\ldots,\eta_r) = P_1\ldots P_rQ^{-r}V + 16\theta_4 nP_2\ldots P_rQ^{-r+1},$$

where

$$V = \int_0^1\cdots\int_0^1 \chi_1(x_1,\ldots,x_r)\exp\{2\pi i F_\delta(x_1,\ldots,x_r)\}\,dx_1\ldots dx_r.$$

After these transformations, the sum $T(A)$ takes the form

$$T(A) = P_1\ldots P_rUV + 16\theta_4 nP_2\ldots P_rQ,$$

where

$$U = Q^{-r}\sum_{\eta_1=1}^{Q}\cdots\sum_{\eta_r=1}^{Q}\exp\{2\pi i F_a(\eta_1,\ldots,\eta_r)\}.$$

To estimate the sum U, as above, we apply Theorem 2.6 (Chapter 2). To estimate the integral V, we use Lemma 6.5. This readily implies the statement of the theorem for points A of the first class Ω_1.

Suppose that A belongs to the second class Ω_2. We first consider the case $Q_0 \ge P^{1/6}$. We shift the domain of summation over $x_1,\ldots,x_r$ in the sum $T(A)$ by integer numbers $y_1,\ldots,y_r$ such that $1 \le y_1 \le P_1,\ldots,1 \le y_r \le P_r$. We obtain

$$\begin{aligned} T(A) = &\sum_{x_1=y_1+1}^{P_1+y_1}\cdots\sum_{x_r=y_r+1}^{P_r+y_r}\chi(x_1-y_1,\ldots,x_r-y_r)\exp\{2\pi i F_A(x_1,\ldots,x_r)\} \\ &+\sum_{x_1=1}^{P_1+y_1}\cdots\sum_{x_r=1}^{P_r+y_r}\big(\chi(x_1,\ldots,x_r)-\chi(x_1-y_1,\ldots,x_r-y_r)\big) \\ &\times\exp\{2\pi i F_A(x_1,\ldots,x_r)\}. \end{aligned}$$

Using Lemma 6.4 and summing over $y_1 \le Y_1,\ldots,y_r \le Y_r$, where $Y_1 = [P_1P^{-\rho}],\ldots,Y_r = [P_rP^{-\rho}]$, we obtain

$$|T(A)| \le W + r2^rP_1\ldots P_rP^{-\rho},$$

$$W \le (Y_1 \dots Y_r)^{-1} \sum_{y_1=1}^{Y_1} \cdots \sum_{y_r=1}^{Y_r}$$

$$\times \left| \sum_{x_1=1}^{P_1} \cdots \sum_{x_r=1}^{P_r} \chi(x_1, \dots, x_r) \exp\{2\pi i F_A(x_1 + y_1, \dots, x_r + y_r)\} \right|.$$

Next, repeating the argument of Lemma 5.3 in Chapter 5 word for word, we obtain the inequality

$$W^{2k} \le 2^{2k}(Y_1 \dots Y_r)^{-1} P^{0.5m\varkappa + m\rho} I_0 + (4\pi m P_1 \dots P_r P^{-\rho})^{2k},$$

where

$$I_0 = \sum_{y_1=1}^{Y_1} \cdots \sum_{y_r=1}^{Y_r} \int \cdots \int_{\omega} |T_\beta|^{2k} d\beta,$$

$$T_\beta = \sum_{x_1=1}^{P_1} \cdots \sum_{x_r=1}^{P_r} \chi(x_1, \dots, x_r) \exp\{2\pi i F_\beta(x_1, \dots, x_r)\}.$$

If, as before, G is the minimal multiplicity of the intersection of the domains ω with a chosen domain $\omega(z_1, \dots, z_r)$, then

$$I_0 \le GI, \quad I = \int \cdots \int_{\Omega} |T(A)|^{2k} \, dA.$$

The inequality $I \le J$ follows from the remark to Theorem 4.2 (Chapter 4). Hence we have $I_0 \le GJ$. Therefore, in the case under study, performing the same argument as in Lemma 5.3 (Chapter 5) for the sum $S(A)$, for $T(A)$ we obtain the same estimate as for $S(A)$.

In the remaining case $Q_0 \le P^{1/6}$, the sum $T(A)$ can be estimated similarly to the sum $S(A)$ in Lemma 5.6 (Chapter 5) by replacing, if necessary, the estimates for the integral

$$\int_0^1 \cdots \int_0^1 \exp\{2\pi i F_\delta(x_1, \dots, x_r)\} \, dx_1 \dots dx_r$$

by the estimates for the integral in Lemma 6.5. The proof of the theorem is complete. □

We use Theorem 4.2 (Chapter 4) and the estimates in Theorem 6.3 to obtain an asymptotic formula for the mean value of a power of the modulus of the multiple trigonometric sum $T(A)$.

Theorem 6.4. *Let* $k \ge 4m\varkappa \log 16m\varkappa$, *and let*

$$I_0 = \int \cdots \int_{\Omega} |T(A)|^{2k} \, dA.$$

Then the following asymptotic formula holds:

$$J_0 = \sigma\theta_0(P_1 \dots P_r)^{2k} P^{-0.5m\varkappa} + O\big(e^a (P_1 \dots P_r)^{2k} P^{-0.5m\varkappa-\rho_1}\big),$$

where

$$\rho_1 = (32\varkappa \log 8m\varkappa)^{-1}, \quad a = 64m\varkappa^2 \log 16m\varkappa + 32m\varkappa \log^2 16m\varkappa,$$

$$\theta_0 = \int_{-\infty}^{+\infty} \cdots \int_{-\infty}^{+\infty} \bigg| \int_0^1 \cdots \int_0^1 \chi_1(x_1, \dots, x_r) \times \exp\{2\pi i F_A(x_1, \dots, x_r)\}\, dx_1 \dots dx_r \bigg|^{2k} dA$$

and the singular series σ *is the same as in Theorem* 6.2.

Proof. The proof is similar to that of Theorem 6.2. We divide the points of the cube Ω into two classes: Ω_1 and Ω_2. For the sum $T(A)$ in the case in which A belongs to Ω_1, a formula similar to that obtained in Lemma 6.1 was proved in Theorem 6.3. Next, we can derive an asymptotic formula for the integral over the points of the first class repeating the argument in Lemma 6.2 word for word, but replacing the estimates for the trigonometric integral in Theorem 1.6 by the estimates for the integral in Lemma 6.5. The integral over the points of the second class can be obtained similarly to the proof of Lemma 6.3, but in this case the estimates in Theorem 5.2 (Chapter 5) are replaced by the estimates in Theorem 6.3. □

6.2 Fractional parts of polynomials

Theorems on the uniform distribution of the fractional parts of polynomials in several variables also give an application of estimates for multiple trigonometric sums. We note that a necessary condition for the joint distribution of the fractional parts of several polynomials is the condition that they be linearly independent over a set of integers, provided that these numbers vary in intervals of a sufficiently small length as compared with the intervals in which the variables of the polynomial vary. This fact is taken into account in the statements of theorems (about the partition of sets of polynomials into classes see below).

6.2.1 Joint distributions

We introduce the following notation. Let $f_j(x_1, \dots, x_r)$ be polynomials in r variables with real coefficients,

$$f_j(x_1, \dots, x_r) = \sum_{t_1=0}^{n_1} \cdots \sum_{t_r=0}^{n_r} \alpha_j(t_1, \dots, t_r)\, x_1^{t_1} \dots x_r^{t_r}.$$

Further, we assume that $m = (n_1 + 1) \dots (n_r + 1)$, $P_1, \dots, P_r$ are positive numbers, $1 < P_1 = \min(P_1, \dots, P_r) = P$, $\Delta = P^{-2\rho}$, $\rho = (32m\varkappa \log 8m\varkappa)^{-1}$, $\varkappa$ is introduced in Theorem 5.2 in Chapter 5, $d_1, \dots, d_s$ are integers, and $|d_j| \le \Delta^{-1}$ $(j = 1, \dots, s)$.

We define real numbers B by the relations

$$B = B(t_1, \dots, t_r; d_1, \dots, d_s) = d_1\alpha_1(t_1, \dots, t_r) + \dots + d_s\alpha_s(t_1, \dots, t_r).$$

Let a and q be integers, and let

$$B = \frac{a}{q} + z, \quad q \ge 1, \quad (a, q) = 1, \quad |z| \le (q\tau)^{-1},$$
$$\tau = \tau(t_1, \dots, t_r) = P_1^{t_1} \dots P_r^{t_r} P^{-1/3}.$$

We assume that, for fixed $d_1, \dots, d_s$, the number $Q = Q(d_1, \dots, d_s)$ is the least common multiple of the numbers $q = q(t_1, \dots, t_r)$, $t_1 + \dots + t_r \ge 1$, $0 \le t_1 \le n_1, \dots, 0 \le t_r \le n_r$. We set

$$Q_0 = \min_{d_1,\dots,d_s} Q(d_1, \dots, d_s),$$
$$\delta = \delta(d_1, \dots, d_s) = \max_{t_1,\dots,t_r} P_1^{t_1} \dots P_r^{t_r} |z(t_1, \dots, t_r)|,$$

where $t_1 + \dots + t_r \ge 1, 0 \le t_1 \le n_1, \dots, 0 \le t_r \le n_r$, and

$$\delta_0 = \min_{d_1,\dots,d_s} \delta(d_1, \dots, d_s).$$

We divide the sets of polynomials $(f_1, \dots, d_s)$ with coefficients

$$0 \le \alpha_j(t_1, \dots, t_r) < 1, \quad j = 1, \dots, s, \quad 0 \le t_1 \le n_1, \dots, 0 \le t_r \le n_r,$$

into two classes E_1 and E_2. The first class E_1 contains sets of polynomials $(f_1, \dots, f_s)$ for which $Q_0 \le P^{0.5}$. The second class E_2 contains all other sets of polynomials $(f_1, \dots, f_s)$.

Theorem 6.5. *Suppose that $D(\sigma_1, \dots, \sigma_s)$ is the number of integer-valued sets $x_1, \dots, x_r$ satisfying the conditions*

$$\{f_1(x_1, \dots, x_r)\} < \sigma_1, \quad \dots, \quad \{f_s(x_1, \dots, x_r)\} < \sigma_s,$$
$$1 \le x_1 \le P_1, \dots, 1 \le x_r \le P_r.$$

We represent $D(\sigma_1, \dots, \sigma_s)$ as

$$D(\sigma_1, \dots, \sigma_s) = P_1 \dots P_r\sigma_1 \dots \sigma_s + \lambda(\sigma_1, \dots, \sigma_s).$$

Then

$$\lambda(\sigma_1, \dots, \sigma_s) \ll P_1 \dots P_r\Delta_1,$$

where Δ_1 *is determined as*

$$\Delta_1 = \exp\{32\varkappa\}P^{-\rho_1}, \quad \rho_1 = (32m\varkappa \log 8m\varkappa)^{-1}$$

for a set of polynomials of the second class and as $\Delta_1 = (Q_0\delta_0)^{-\nu+\varepsilon}$ *for a set of polynomials* $(f_1, \ldots, f_s)$ *of the first class.*

Proof. Let $S(d_1, \ldots, d_s)$ be a multiple trigonometric sum, and let

$$S(d_1, \ldots, d_s) = \sum_{x_1=1}^{P_1} \cdots \sum_{x_r=1}^{P_r} \exp\left\{2\pi i \sum_{j=1}^{s} d_j f_j(x_1, \ldots, x_r)\right\}.$$

In view of the partition of sets of polynomials into two classes E_1 and E_2, by Theorem 5.2 (Chapter 5), we have

$$|S(d_1, \ldots, d_s)| \ll P_1 \ldots P_r \Delta_0,$$

where $|d_1|, \ldots, |d_s| < \Delta^{-1}$, and the quantity Δ_0 for the sets of the second class is determined as

$$\Delta_0 = \exp\{32\varkappa\}P^{-\rho_0}, \quad \rho_0 = (32m\varkappa \log 8m\varkappa)^{-1},$$

and for the sets of the first class as

$$\Delta_0 = (Q_0\delta_0)^{-1/n+\varepsilon_3}.$$

Without loss of generality, we assume that $\Delta_0 \le 0.1$. We consider the periodic function $\psi_j(x)$ with period 1 introduced in Lemma A.3. Let α_j and β_j be arbitrary real numbers such that $\Delta_0 \le \beta_j - \alpha_j \le 1 - \Delta_0$ $(1 \le j \le s)$. In this lemma, we set $r = 1$. Then we have

$$\begin{aligned}
&\psi_j(f_j(x_1, \ldots, x_r)) = 1\\
&\quad \text{if} \quad \alpha_j + 0.5\Delta_0 \le f_j(x_1, \ldots, x_r) < \beta_j - 0.5\Delta_0 \pmod 1,\\
&0 \le \psi_j(f_j(x_1, \ldots, x_r)) \le 1\\
&\quad \text{if} \quad \alpha_j - 0.5\Delta_0 < f_j(x_1, \ldots, x_r) < \alpha_j + 0.5\Delta_0 \pmod 1\\
&\quad \text{or} \quad \beta_j - 0.5\Delta_0 < f_j(x_1, \ldots, x_r) < \beta_j + 0.5\Delta_0 \pmod 1,\\
&\psi_j(f_j(x_1, \ldots, x_r)) = 0\\
&\quad \text{if} \quad \beta_j + 0.5\Delta_0 < f_j(x_1, \ldots, x_r) < 1 + \alpha_j - 0.5\Delta_0 \pmod 1,
\end{aligned}$$

$$\begin{aligned}
&\psi_j(f_j(x_1, \ldots, x_r)) = \beta_j - \alpha_j + \sum_{d=1}^{+\infty} \big(g_d \exp\{2\pi i d f_j(x_1, \ldots, x_r)\}\\
&\qquad\qquad + h_d \exp\{-2\pi i d f_j(x_1, \ldots, x_r)\}\big),\\
&\max(|g_d|, |h_d|) \le (\pi d)^{-1} \qquad \text{if} \quad 1 \le d \le \Delta_0^{-1},
\end{aligned}$$

$$\max(|g_d|, |h_d|) \le (\pi^2 \Delta_0 d^2)^{-1} \quad \text{if} \quad d > \Delta_0^{-1}.$$

We set

$$U = U(\alpha_1, \beta_1, \dots, \alpha_s, \beta_s) = \sum_{x_1=1}^{P_1} \dots \sum_{x_r=1}^{P_r} \psi_1(f_1(x_1, \dots, x_r)) \dots \psi_s(f_s(x_1, \dots, x_r)).$$

Then we have the relation

$$U = P_1 \dots P_r(\beta_1 - \alpha_1) \dots (\beta_s - \alpha_s) + H,$$

where

$$H \ll \sum_{0 \le |d_1| \le \Delta_0^{-1}} \cdots \sum_{0 \le |d_s| \le \Delta_0^{-1}}{}' \frac{|S(d_1, \dots, d_s)|}{\overline{d}_1 \dots \overline{d}_s}$$
$$+ \sum_{k=1}^{s} \sum_{0 \le |d_1| \le \Delta_0^{-1}} \cdots \sum_{\Delta_0^{-1} < |d_k| \le \Delta^{-1}} \cdots \sum_{0 \le |d_s| \le \Delta_0^{-1}} \frac{|S(d_1, \dots, d_s)|}{\Delta_0^2 d_k^2 \overline{d}_1 \dots \overline{d}_{k-1} \overline{d}_{k+1} \dots \overline{d}_s}$$
$$+ \sum_{k=1}^{s} \sum_{0 \le |d_1| < \Delta^{-1}} \cdots \sum_{|d_k| > \Delta^{-1}} \cdots \sum_{0 \le |d_s| \le \Delta^{-1}} \frac{P_1 \dots P_r}{\Delta_0^2 d_k^2 \overline{d}_1 \dots \overline{d}_{k-1} \overline{d}_{k+1} \dots \overline{d}_s}$$

and $\overline{d} = \max(1, |d|)$; here the prime on the summation sign means that the summation is taken over all $d_a, \dots, d_s$ mentioned above except for $d_1 = \dots = d_s = 0$.

Since, for $|d_1|, \dots, |d_s| < \Delta^{-1}$, the sum $S(d_1, \dots, d_s)$ satisfies the estimate

$$|S(d_1, \dots, d_s)| \ll P_1 \dots P_r \Delta_0,$$

we have

$$H \ll \sum_{0 \le |d_1| \le \Delta_0^{-1}} \cdots \sum_{0 \le |d_s| \le \Delta_0^{-1}}{}' \frac{P_1 \dots P_r \Delta_0}{\overline{d}_1 \dots \overline{d}_s}$$
$$+ \sum_{k=1}^{s} \sum_{0 \le |d_1| \le \Delta_0^{-1}} \cdots \sum_{\Delta_0^{-1} \le |d_k| \le \Delta^{-1}} \cdots \sum_{0 \le |d_s| \le \Delta_0^{-1}} \frac{P_1 \dots P_r \Delta_0}{\Delta_0^2 d_k^2 \overline{d}_1 \dots \overline{d}_k}$$
$$+ \sum_{k=1}^{s} \sum_{|d_1| \le \Delta^{-1}} \cdots \sum_{|d_k| > \Delta^{-1}} \cdots \sum_{|d_s| \le \Delta^{-1}} \frac{P_1 \dots P_r}{\Delta_0 d_k^2 \overline{d}_1 \dots \overline{d}_{k-1} \overline{d}_{k+1} \dots \overline{d}_s},$$

which implies $H \ll P_1 \dots P_r \Delta_1$. Hence $D(\sigma_1, \dots, \sigma_s)$ satisfies the above formula, and $\lambda(\sigma_1, \dots, \sigma_s)$ satisfies the estimate

$$|\lambda(\sigma_1, \dots, \sigma_s)| \ll P_1 \dots P_r \Delta_1.$$

The proof of the theorem is complete. □

We consider an example illustrating Theorem 6.5.

Example 6.1. Let $F_1(x, y)$ and $F_2(x, y)$ be two polynomials with real coefficients in which the powers of each of the variables do not exceed n $(n \geq 4)$. Suppose that $0 \leq s, t \leq n$, $s + t \geq 1$, $P \gg 1$, and the coefficient of the expression $x^s y^t$ in the first polynomial is equal to $\sqrt{2}$, while the corresponding coefficient in the second polynomial is equal to $\sqrt{3}$. Further, let $D(\sigma_1, \sigma_2)$ be the number of sets of integers x and y such that the inequalities

$$\{F_1(x, y)\} < \sigma_1, \quad \{F_2(x, y)\} < \sigma_2, \quad 1 \leq x, y \leq P,$$

are satisfied simultaneously. Then we have the asymptotic formula

$$D(\sigma_1, \sigma_2) = P^2\sigma_1\sigma_2 + O(P^2\Delta),$$

where $\Delta = P^{-\rho}$, $\rho = \gamma(n^3 \ln n)^{-1}$, and $\gamma > 0$ is an absolute constant.

Indeed, it follows from the partition of points of the unit $(n + 1)^2$-dimensional cube into points of the first and second classes (see the definition of the partition of points into classes at the beginning of this section (Theorem 6.5)) that any point corresponding to the polynomials $F_1(x, y)$ and $F_2(x, y)$ belongs to the second class. Let us prove this assertion.

To this end, we consider the number $\alpha = m_1\sqrt{2} + m_2\sqrt{3} \neq 0$. We set $\tau = P^{t+s-1/4}$. It is known that there exist integer coprimes a and q such that

$$\left|\alpha - \frac{a}{q}\right| \leq \frac{1}{q^\tau}, \quad q \leq \tau. \tag{6.2}$$

Since α is an irrational algebraic number of degree no larger than 4, it follows from the Liouville theorem that for all a and q, $(a, q) = 1$, we have

$$\left|\alpha - \frac{a}{q}\right| > \frac{1}{c(\alpha)q^4},$$

where $c(\alpha) = \max_{|\xi-\alpha|\leq 1} |f'(\xi)|$ and $f(\xi)$ is a polynomial that is irreducible over the field of rational numbers and has a root $\alpha = m_1\sqrt{2}+m_2\sqrt{3}$. (If $m_1 \neq 0$ and $m_2 \neq 0$, then $f(\xi) = (\xi - m_1\sqrt{2} - m_2\sqrt{3})(\xi - m_1\sqrt{2} + m_2\sqrt{3})(\xi + m_1\sqrt{2} - m_2\sqrt{3})(\xi + m_1\sqrt{2} + m_2\sqrt{3})$.)

For integers m_1 and m_2 whose absolute values do not exceed $P^{2/(n+1)^2}$, the constant $c(\alpha)$ can be bounded above as follows: $c(\alpha) \leq 1000P^{6/(n+1)^2}$. Then the number q in (6.2) is larger than

$$0.1\,P^{(t+s)/3-1/12-2/(n+1)^{-2}} > P^{2/(n+1)^{-2}}.$$

Hence the point corresponding to $F_1(x, y)$ and $F_2(x, y)$ belongs to the second class, and the above asymptotic formula holds for $D(\sigma_1, \sigma_2)$.

Theorem 6.6. *Let $D_1(\sigma_1, \dots, \sigma_s)$ be the number of sets of integers $x_1, \dots, x_r$ satisfying the conditions*

$$\{f_1(x_1, \dots, x_r)\} < \sigma_1, \quad \dots, \quad \{f_s(x_1, \dots, x_r)\} < \sigma_s, \quad (x_1, \dots, x_r) \in E_r$$

(E_r is determined as in Theorem 6.3). Represent $D_1(\sigma_1, \dots, \sigma_s)$ in the form

$$D_1(\sigma_1, \dots, \sigma_s) = V(P_0)\sigma_1 \dots \sigma_s + \lambda(\sigma_1, \dots, \sigma_s),$$

where $V(P_0)$ is the volume of the domain E_r. Then

$$|\lambda(\sigma_1, \dots, \sigma_s)| \ll P_1 \dots P_r \Delta_1.$$

where Δ_1 is the same as in Theorem 6.5.

Proof. The proof is the same as that of Theorem 6.5. We only use the estimates from Theorem 6.3 where it is necessary. □

6.2.2 Distribution of the fractional parts of polynomials

We only formulate three theorems on the fractional parts of polynomials in several variables. Their proofs are based on Theorem 5.2 in Chapter 5, Theorem 6.2, and Lemma A.3 and do not differ from the proof of Theorem 6.5.

Theorem 6.7. *Let $D(\sigma)$ be the number of sets of integers $x_1, \dots, x_r$ satisfying the conditions*

$$\{F(x_1, \dots, x_r)\} < \sigma, \quad 1 \le x_1 \le P_1, \dots, 1 \le x_r \le P_r.$$

Represent $D(\sigma)$ in the form

$$D(\sigma) = P_1 \dots P_r \sigma + \lambda(\sigma).$$

Then

$$\lambda(\sigma) \ll P_1 \dots P_r \Delta_1,$$

where Δ_1 for the polynomial of the second class Ω_2 is determined as

$$\Delta_1 = \exp\{32\varkappa\} P^{-\rho_1}, \quad \rho_1 = (32 m\varkappa \log 8m\varkappa)^{-1},$$

and for the polynomial of the first class as

$$\Delta_1 = (Q\delta)^{-\nu+\varepsilon}.$$

Theorem 6.8. *Let $D_1(\sigma)$ be the number of sets of integers $x_1, \dots, x_r$ satisfying the conditions*

$$\{F(x_1, \dots, x_r)\} < \sigma, \quad (x_1, \dots, x_r) \in E_r$$

(E_r is determined as in Theorem 6.3). *Represent $D_1(\sigma)$ in the form*

$$D_1(\sigma) = V(P_0)\sigma + \lambda(\sigma),$$

where $V(P_0)$ is the volume of the domain E_r. Then

$$\lambda(\sigma) \ll P_1 \dots P_r \Delta_1,$$

where Δ_1 is the same as in Theorem 6.5.

Theorem 6.9. *The following relation holds*:

$$\sum_{x_1=1}^{P_1} \cdots \sum_{x_r=1}^{P_r} \{F(x_1, \dots, x_r)\} = 0.5 P_1 \dots P_r + O(P_1 \dots P_r \Delta_1),$$

where Δ_1 is the same as in Theorem 6.7 *and the constant in the sign O depends only on $n_1, \dots, n_r$ and r.*

Concluding remarks on Chapter 6. The results considered in this chapter were obtained by the authors and published in [25], [26], [30], [31] and [32].

Chapter 7

Special cases of the theory of multiple trigonometric sums

In this chapter we obtain estimates for multiple trigonometric sums for which the ranges of variation of the principal parameters are essentially different. In fact, such sums are sums of lesser multiplicity. In deriving the main results, we use the estimates for trigonometric sums obtained in Chapter 5 and the estimates for multiple trigonometric integrals obtained in Chapter 1. Finally, special cases of the theory are based on the mean value theorem for multiple trigonometric sums (see Chapter 4). This once again confirms the conjecture that the main case of the theory of multiple trigonometric sums is the case of sums with equivalent variables of summation.

In Section 7.1, we prove a theorem on the estimate for a double trigonometric sum, and in Section 7.2, we prove a theorem on the estimate for an r-fold trigonometric sum ($r \geq 2$). We prove these theorem by induction on the number of variables, and the results of Section 7.1 form the base of induction.

In Section 7.3, we derive an asymptotic formula for the number of solutions of the complete system of Diophantine equations with unknowns for which the ranges of variation are essentially different.

7.1 Double trigonometric sums

7.1.1 Main notions and auxiliary lemmas

We introduce the main notions and state auxiliary lemmas necessary for further studies.

Let n and P be natural numbers, and let $\tau(t) = P^{t-1/6}$ $(t = 1, \ldots, n)$. By Ω we denote the unit cube in the n-dimensional Euclidean space with coordinates $\alpha(t)$ determined by the conditions

$$-\tau^{-1}(t) \leq \alpha(t) < 1 - \tau^{-1}(t), \quad t = 1, \ldots, n.$$

Let $f(t)$ be a polynomial with real coefficients $\alpha(t)$. By S we denote the trigonometric sum

$$S = S(A) = \sum_{x=1}^{P} \exp\{2\pi i f(x)\},$$

where A is a point in the cube Ω whose coordinates are the numbers $\alpha(t)$, i.e., the coefficients of the polynomial $f(x)$.

We divide all the points of the cube Ω into two classes. A point A with coordinates $\alpha(t)$ $(1 \le t \le n)$ belongs to the first class if $\alpha(t)$ can be represented as

$$\alpha(t) = \frac{a(t)}{q(t)} + \beta(t), \quad \big(a(t), q(t)\big) = 1, \quad 0 \le a(t) < q(t),$$
$$|\beta(t)| \le P^{-t+0.1}, \quad t = q, \dots, n,$$

and, moreover, the least common multiple of the numbers $q(1), \dots, q(n)$, i.e., $Q = [q(1), \dots, q(n)]$, does not exceed $P^{0.1}$. The other points of the cube Ω belong to the second class Ω_2.

In what follows, we shall often apply the well-known Dirichlet theorem on the approximation of a real number by a rational fraction in the following statement.

Theorem 7.1 (Dirichlet). *For any real number $\tau \ge 1$ and any real number α, there exists a natural number q $(1 \le q \le \tau)$ and a natural number α coprime to q such that*

$$\alpha = \frac{a}{q} + \beta, \quad \textit{where} \quad |\beta| \le (q\tau)^{-1}. \tag{7.1}$$

A representation of the number α in the form (7.1) is called the *D-approximation* of α corresponding to τ.

Lemma 7.1. (a) *Let a point A belong to the first class Ω_1. Then the following estimate holds*:

$$|S(A)| \ll PQ^{-1/n+\varepsilon}; \tag{7.2}$$

moreover, if we set

$$\delta(t) = P^t\beta(t), \quad \delta = \max_{1 \le t \le n} |\delta(t)|,$$

then for $\delta > 1$, the following estimate holds:

$$|S(A)| \ll PQ^{-1/n+\varepsilon}; \tag{7.3}$$

the constant in $\ll$ depends only on n and ε.

Suppose that a point A belongs to the second class Ω_2. Then the following inequality holds:

$$|S(A)| \ll P^{1-\rho_1}, \tag{7.4}$$

where $\rho_1 = c_1(n^2 \ln n)^{-1}$ and the constant in $\ll$ depends only on n.

(b) *Suppose that $n \ge 2$, $P \ge 1$, $f(x)$ is a polynomial with real coefficients $\alpha(t)$, and*

$$f(x) = \alpha(1)x + \cdots + \alpha(n)x^n.$$

Consider the D-approximation of the number $\alpha(t)$ $(1 \le t \le n)$ corresponding to $\tau(t) = P^{t-1/6}$, i.e., consider the relations

$$\alpha(t) = \frac{a(t)}{q(t)} + \beta(t), \quad l \le q(t) \le \tau(t),$$
$$\big(a(t), q(t)\big) = 1, \quad |\beta(t)| \le \big(q(t), \tau(t)\big)^{-1}.$$

Let Q be the least common multiple of the numbers $q(1), \dots, q(n)$. Then for $Q > P^{0.1}$ the following estimate holds:

$$|S(A)| = \left|\sum_{x=1}^{P} \exp\{2\pi i f(x)\}\right| \ll P^{1-\rho_1}, \quad \rho_1 = c_1(n^2 \ln n)^{-1}.$$

where the constant in $\ll$ depends only on n.

Proof. Assertion (a) is a consequence of Theorem 5.2 in Chapter 5 for $r = 1$.

Let us prove assertion (b). If $A = (\alpha(1), \dots, \alpha(n))$ is a point of the second class in Lemma 7.1 (a), then the above estimate holds for $|S(A)|$. Let A be a point of first class in Lemma 7.1 (a). Then, by the definition of points of the first class, we can represent each $\alpha(t)$ as

$$\alpha(t) = \frac{a_0(t)}{q_0(t)} + \beta_0(t), \quad \big(a_0(t), q_0(t)\big) = 1, \quad |\beta_0(t)| \le P^{-t+0.1},$$
$$Q_0 = [q_0(1), \dots, q_0(n)] \le P^{0.1}.$$

There exists a t such that the denominator $q(t)$ in the D-approximation of $\alpha(t)$ is not equal to $q_0(t)$ (otherwise, the least common multiples of the numbers $q(t)$ and $q_0(t)$ would coincide, which is not the case). We show that $q_0(t)$ and hence the least common multiple Q_0 is larger than $0.5P^{1/15}$. Indeed, we have the inequalities

$$\frac{1}{q(t)q_0(t)} \le \left|\frac{a(t)}{q(t)} - \frac{a_0(t)}{q_0(t)}\right| \le |\beta(t)| + |\beta_0(t)| \le P^{-t+0.1} + P^{-1+1/6}q^{-1}(t),$$
$$\frac{1}{q_0(t)} \le q(t)P^{-t+0.1} + P^{-t+1/6} \le P^{-1/15} + P^{-5/6} \le 2P^{-1/15},$$
$$q_0(t) \ge 0.5P^{1/15}.$$

Now, to estimate $|S(A)|$, we apply Lemma 7.1 (a) (note that A belongs to the class Ω_1 and $Q_0 \ge 0.5P^{1/15}$) and obtain

$$|S(A)| \ll P^{1-1/(30n)} \ll P^{1-\rho_1},$$

as required. □

Lemma 7.2. *Suppose that* σ *is an arbitrary number such that* $0 < \sigma \leq 1$ *and the number* $D(\sigma)$ *of values of the variable* x *contained in the interval* $1 \leq x \leq P$ *under the condition that* $\{f(x)\} < \sigma$ *can be represented as*

$$D(\sigma) = P\sigma + \lambda(\sigma).$$

Then

(1) *if the point* A *belongs to the second class* Ω_2 *in Lemma* 7.1 (a), *then*

$$\lambda(\sigma) \ll P^{1-\rho_1};$$

(2) *if the point* A *belongs to the first class* Ω_1 *in Lemma* 7.1 (a), *then*

$$\lambda(\sigma) \ll PQ^{-1/n+\varepsilon},$$

and for $\delta > 1$

$$\lambda(\sigma) \ll P(Q\delta)^{-1/n+\varepsilon};$$

(3) *if the point* A *satisfies the conditions of Lemma* 7.1 (b), *then*

$$\lambda(\sigma) \ll P^{1-\rho_1}.$$

The constants in $\ll$ *depend only on* n *and* ε.

Proof. This lemma is a simple consequence of Theorem 6.5 (Chapter 6) for $s = r = 1$ and of Lemma 7.1 (a). □

In the preceding lemmas on the estimates of one-dimensional trigonometric sums, we introduced notions such as the cube Ω, the domain Ω_1 of points of the first class, and domain Ω_2 of points of the second class. To study multiple trigonometric sums, we must introduce similar notions. For them, we shall use the same notation, and it will be clear from the context in what sense these notions are used.

Now we introduce the notation. Let n_1, n_2, P_1, and P_2 be natural numbers,

$$P_1 \leq P_2; \quad m = (n_1+1)(n_2+1); \quad n = \max(n_1, n_2), \quad n \geq 2;$$
$$\tau(t_1, t_2) = P_1^{t_1-1/6} P_2^{t_2},$$

where $0 \leq t_1 \leq n_1, 0 \leq t_2 \leq n_2, t_1 + t_2 \geq 1$. By Ω we denote the unit cube in the m-dimensional Euclidean space with coordinates $\alpha(t_1, t_2)$ that is determined by the conditions

$$-\tau^{-1}(t_1, t_2) \leq \alpha(t_1, t_2) < 1 - \tau^{-1}(t_1, t_2), \quad \alpha(0,0) = 0,$$
$$0 \leq t_1 \leq n_1, \quad 0 \leq t_2 \leq n_2, \quad t_1 + t_2 \geq 1.$$

A *multiple* (more precisely, a *double*) trigonometric sum is defined to be the sum

$$S = S(A) = \sum_{x_1=1}^{P_1} \sum_{x_2=1}^{P_2} \exp\{2\pi i F(x_1, x_2)\},$$

where $F(x_1, x_2)$ is a polynomial with real coefficients $\alpha(t_1, t_2)$,

$$F(x_1, x_2) = \sum_{t_1=0}^{n_1} \sum_{t_2=0}^{n_2} \alpha(t_1, t_2) x_1^{t_1} x_2^{t_2}.$$

In this case we assume that a point A from the m-dimensional space belongs to the cube Ω and its coordinates are the coefficients $\alpha(t_1, t_2)$ of the polynomial $F(t_1, t_2)$.

We divide the cub Ω into classes Ω_1 and Ω_2. The first class contains points A with coordinates $\alpha(t_1, t_2)$ that satisfy the relations

$$\alpha(t_1, t_2) = \frac{a(t_1, t_2)}{q(t_1, t_2)} + \beta(t_1, t_2), \quad \big(a(t_1, t_2), q(t_1, t_2)\big) = 1,$$
$$0 \le a(t_1, t_2) < q(t_1, t_2), \quad |\beta(t_1, t_2)| \le P_1^{-t_1+0.1} P_2^{-t_2},$$

and for which the least common multiple Q of all the numbers $q(t_1, t_2)$ does not exceed $P_1^{0.1}$. All other points of the cube Ω belong to the second class Ω_2.

Lemma 7.3. *The points of the first class satisfy the estimate*

$$|S(A)| \ll P_1 P_2 Q^{-1/n+\varepsilon};$$

if, moreover, we set

$$\delta(t_1, t_2) = P_1^{t_1} P_2^{t_1}(t_1, t_2), \quad \delta = \max_{t_1, t_2} |\delta(t_1, t_2)|,$$

then for $\delta > 1$ the estimate

$$|S(A)| \ll P_1 P_2 Q^{-1/n+\varepsilon}$$

also holds; the constants in $\ll$ depend only on n and ε.

Proof. The statement of the lemma readily follows from Lemma 5.5 in Chapter 5. □

Lemma 7.4. *Suppose that a point A belongs to the second class Ω_2 and ν is a natural number from the interval*

$$-1 < \frac{\ln P_2}{\ln P_1} - \nu \le 0.$$

Further, we set $\varkappa = n_1 + \nu n_2$. Then we have

$$|S(A)| \ll \exp\{32\varkappa\} P_1 P_2 P_1^{-\rho_2}, \quad \rho_2 = (32 m\varkappa \ln 8m\varkappa)^{-1},$$

where the constant in $\ll$ depends only on n.

Proof. This assertion follows from Theorem 5.2 in Chapter 5. □

7.1.2 Lemmas

In proving the first main lemma, we shall use the auxiliary Lemmas 7.5–7.7 from this section. Prior to stating these lemmas, we introduce several necessary notions.

Suppose that $g(x)$ is an nth degree polynomial with real coefficients $\alpha_n, \alpha_{n-1}, \dots, \alpha_1, \alpha_0$,

$$g(x) = \alpha_n x^n + \alpha_{n-1}x^{n-1} + \cdots + \alpha_1 x + \alpha_0.$$

For each ν $(1 \le \nu \le n)$, setting $\tau_\nu = P_2^{\nu-1/6}$, we consider the D-approximations of the coefficients α_ν corresponding to τ_ν:

$$\alpha_\nu = \frac{a_\nu}{q_\nu} + \beta_\nu,$$

where $(a_\nu, q_\nu) = 1$, $1 \le q_\nu \le \tau_\nu$, $|\beta_\nu| \le (q_\nu \tau_\nu)^{-1}$. These approximations are called the *first D-approximation* of the numbers α_ν $(\nu = 1, \dots, n)$.

Let Q be the least common multiple of the denominators q_ν of the rational approximations of α_ν, i.e., let $Q = [q_n, q_{n-1}, \dots, q_1]$.

Further, we consider some other approximations of the same numbers α_ν $(1 \le \nu \le n)$. For this, we set

$$\tau_\nu^* = P_1^{s-1/6} P_2^\nu,$$

where s is a fixed natural number that does not exceed n. Then, by the Dirichlet theorem, we have

$$\alpha_\nu = \frac{a_\nu^*}{q_\nu^*} + \beta_\nu^*,$$

where $(a_\nu^*, q_\nu^*) = 1$, $1 \le q_\nu^* \le \tau_\nu^*$, $|\beta_\nu^*| \le (q_\nu^* \tau_\nu^*)^{-1}$, and $1 \le \nu \le n$. These approximations are said to be the *second D-approximation* of the numbers α_ν $(\nu = n, \dots, 1)$.

Now, using the introduced notation, we state the following lemma.

Lemma 7.5. *Suppose that the value of Q does not exceed $H_2 = P_1^{10n^2\rho}$ and, moreover, there exists a natural number t that does not exceed n and satisfies the condition*

$$\tau_t < q_t^* \le \tau_t^*.$$

Setting $\Delta = P_1^{-s+2\rho n}$, we consider all intervals of the form

$$\left[\frac{A}{B} - \Delta, \frac{A}{B} + \Delta\right], \tag{7.5}$$

where A and B are natural numbers, $1 \le B \le P_1^{2\rho n} = H_1$, and $(A, B) = 1$. Denote by Y the number of y $(1 \le y \le P_2)$ for which the fractional part of the polynomial $g(y)$ lies at least in one of these intervals. Then under the conditions

$$1 \le c_1(n) \le P_1, \quad \ln P_2 \ge 1.2(n+1)\ln P, \quad \rho \le 0.02n^{-2},$$

the variable Y satisfies the inequality

$$Y \le c_1(n) P_2 P_1^{-\rho}.$$

Proof. Without loss of generality, we assume that H_1 is an integer. Then the number of intervals (7.5) is equal to

$$\Phi(H_1) = \sum_{r=1}^{H_1} \varphi(r) \le H_1^2.$$

We define a periodic function $\chi(x)$ with period 1 by the relations

$$\chi(x) = \begin{cases} 1 & \text{if } |x| \le \Delta, \\ (2\Delta - |x|)\Delta^{-1} & \text{if } \Delta < |x| \le 2\Delta, \\ 0 & \text{if } 2\Delta < |x| \le 0.5. \end{cases}$$

The Fourier series of the function $\chi(x)$ has the form

$$\chi(x) = \Delta + \sum_{m=-\infty}^{+\infty}{}' c(m) \exp\{2\pi i m x\}, \tag{7.6}$$

where the prime on the summation sign means that $c(0) = 0$; moreover,

$$|c(m)| \le \min\left(\Delta, (\Delta m^2)^{-1}\right), \quad m = \pm 1, \pm 2, \dots .$$

Suppose that s_μ $(\mu = 1, \dots, \Phi(H_1))$ are the centers of the intervals (7.5), i.e., let $s_\mu = AB^{-1}$. We define new functions $\chi_\mu(x)$ by the relations

$$\chi_\mu(x) = \chi(x + s_\mu) = \chi(x + AB^{-1}).$$

Then for Y we find the estimate

$$Y \le Z = \sum_{\mu \le \Phi(H_1)} \sum_{y \le P_2} \chi_\mu(g(y)).$$

We use expansion (7.6) to obtain

$$Z = P_2 \Phi(H_1)\Delta + Z_1,$$

where

$$Z_1 = \sum_{\mu \le \Phi(H_1)} \sum_{m=-\infty}^{+\infty}{}' c(m) \sum_{y \le P_2} \exp\{2\pi i (g(y) + s_\mu)\}.$$

We let M and M_1 denote

$$M = \Delta^{-1}, \quad M_1 = \Delta^{-1} H_1^2 P_1^{\rho}.$$

Then, applying the corresponding estimate of $|c(m)|$, we obtain the inequality

$$|Z| \leq \Phi(H_1)\Delta \sum_{1\leq m<M} \left|\sum_{y\leq P_2} \exp\{2\pi img(y)\}\right| \tag{7.7}$$
$$+ \Phi(H_1)\Delta^{-1} \sum_{M\leq m<M_1} m^{-2}\left|\sum_{y\leq P_2} \exp\{2\pi img(y)\}\right| + P_2P_1^{-\rho}.$$

To prove the lemma, it remains to estimate the modulus of the trigonometric sum

$$T(m) = \sum_{y\leq P_2} \exp\{2\pi img(y)\}.$$

Recall that by Q we denoted the least common multiple of the denominators of the rational fractions approximating the numbers α_ν $(\nu = n, \dots, 1)$ in the first of the D-approximations and, by the assumptions of the lemma, Q does not exceed $P_1^{10n^2\rho}$. We represent each natural number $y \leq P_2$ in the form

$$y = Qu + v, \quad 1 \leq v \leq Q, \quad (1-v)Q^{-1} \leq u \leq (P_2 - v)Q^{-1},$$

and note that the polynomial $g(y)$ satisfies the relation

$$g(Qu + v) \equiv F(v) + G(Qu + v) + \alpha_0 \pmod 1,$$

where

$$F(v) = \sum_{\nu=1}^{n} \frac{a_\nu}{q_\nu} v^\nu, \quad G(y) = \sum_{\nu=1}^{n} \beta_\nu y^\nu.$$

Hence for $|T(m)|$ we obtain the inequality

$$|T(m)| \leq Q|T_1(m)|, \tag{7.8}$$

where $T_1(m)$ denotes a trigonometric sum over u of the form

$$T_1(m) = \sum_{1\leq Qu+v\leq P_2} \exp\{2\pi imG(Qu + v)\};$$

here v is a fixed natural number $(1 \leq v \leq Q)$.

We show that for any m from the interval $1 \leq m \leq M$, it is possible to replace the trigonometric sum $T_1(m)$ by an integral with an appropriate accuracy. For this, we give an estimate from above for the absolute value of the derivative with respect to u of the polynomial $mG(Qu + v)$. We use the fact that P_2 is much larger than P_1 (more precisely, $\ln P_2 \geq 1.2(n + 1) \ln P_1$) to obtain

$$\left|m\frac{d}{du}G(Qu + v)\right| = \left|m\sum_{\nu=1}^{n} \nu\beta_\nu(Qu + v)^{\nu-1}Q\right|$$

$$\le M_1 \sum_{\nu=1}^{n} \nu \tau_\nu^{-1} P_1^{\nu-1} Q \le n^2 M_1 P_1^{-5/6} \le 0.5.$$

This implies that Lemma A.2 can be applied to $T_1(m)$, i.e.,

$$T_1(m) = \int_{-vQ^{-1}}^{(P_2-v)Q^{-1}} \exp\{2\pi i m G(Qu+v)\}\,du + O(1)$$
$$= Q^{-1}\int_0^{P_2} \exp\{2\pi i m G(u)\}\,du + O(1).$$

Substituting this formula into (7.8), we obtain

$$|T(m)| \le \left| \int_0^{P_2} \exp\{2\pi i m G(u)\}\,du \right| + c_1 Q. \tag{7.9}$$

We denote the last integral by J, perform a change of the variable of integration of the form $u = P_2 z$, and find

$$J = P_2 \int_0^1 \exp\{2\pi i m h(z)\}\,dz,$$

where

$$h(z) = G(P_2 z) = \sum_{\nu=1}^{n} \beta_\nu P_2^\nu z^\nu = \sum_{\nu=1}^{n} \delta_\nu z^\nu.$$

Let us bound $\delta = \max_{1\le\nu\le n} |\delta_\nu|$ from below. For this, we use the second condition of the lemma stating that there exists a natural number t that does not exceed n and satisfies the condition $\tau_t < q_t^* \le \tau_t^*$. We consider the first and the second D-approximation of the number α_t:

$$\alpha_t = \frac{a_t}{q_t} + \beta_t, \quad \alpha_t = \frac{a_t^*}{q_t^*} + \beta_t^*.$$

Then we have $q_t^* > \tau_t \ge q_t$; hence $q_y \ne q_t^*$ and

$$|\beta_t| = \left| \frac{a_t}{q_t} - \frac{a_t^*}{q_t^*} - \beta_t^* \right| \ge \frac{1}{q_t q_t^*} - |\beta_t^*| \ge (Q\tau_t^*)^{-1} - (q_t^*\tau_t^*)^{-1} \ge 0.5(Q\tau_t^*)^{-1}.$$

Therefore,

$$\delta \ge 0.5(Q\tau_t^*)^{-1} P_2^t \ge 0.5 H_2^{-1} P_1^{-s+1/6}.$$

Applying Theorem 1.6 (Chapter 1) to J, we obtain

$$|J| \le c_2 P_2 m^{-1/n} H_2^{1/n} P_1^{s/n-1/(6n)}.$$

It follows from this estimate and relations (7.7) and (7.9) that

$$|Z_1| \le c_3\Big\{H_1^2\Delta \sum_{1\le m<M} (P_2 m^{-1/n} H_2^{1/n} P_1^{s/n-1/(6n)} + H_2)$$
$$+ H_1^2\Delta^{-1} \sum_{M\le m<M_1} (P_2 m^{-2-1/n} H_2^{1/n} P_1^{s/n-1/(6n)} + H_2 m^{-2}) + P_2 P_1^{-\rho}\Big\}$$
$$\le c_4 P_2 P_1^{-\rho}.$$

This and the formula for Z imply the statement of the lemma. □

Now we consider n polynomials $g_\nu(y)$ of the form

$$g_\nu(y) = \frac{a_\nu}{q_\nu} + \sum_{s=0}^{n} \beta_\nu(s) y^s, \qquad \nu = 1, 2, \ldots, n,$$

where

$$(a_\nu, q_\nu) = 1, \quad 1 \le q_\nu \le \tau_\nu, \quad \tau_\nu = P_2^\nu P_1^{-1/6},$$
$$|\beta_\nu(s)| \le \tau_\nu^{-1}(s), \quad \tau_\nu(s) = P_1^{s-1/6} P_2^\nu, \quad 0 \le s \le n.$$

Let Q_1 be the least common multiple of the numbers $q_1, \ldots, q_n$, i.e., let

$$Q_1 = [q_1, \ldots, q_n]. \tag{7.10}$$

For each natural number y that does not exceed P_1, we consider the D-approximations of real numbers $g_\nu(y)$ corresponding to $\tau_\nu = P_2^{\nu=1/6}$, in other words, we consider the relations

$$g_\nu(y) = \frac{a_\nu(y)}{q_\nu(y)} + \beta_\nu(y),$$

where $\big(a_\nu(y), q_\nu(y)\big) = 1$, $1 \le q_\nu(y) \le \tau_\nu$, and $|\beta_\nu(y)| \le \big(q_\nu(y)\tau_\nu\big)^{-1}$.

By $Q_1(y)$ we denote the least common multiple of the numbers $q_\nu(y)$ $(\nu = 1, \ldots, n)$.

Lemma 7.6. *Suppose that the number Q_1 is larger than $P_1^{0.05-5n^3\rho}$. Let Y be the number of natural numbers y $(1 \le y \le P_1)$ for which the relations*

$$|\beta_\nu(y)| \le \Delta_\nu = P_2^{-\nu} P_1^{2\rho n}, \quad Q_1(y) \le H = P_1^{20\rho n^4}, \quad \nu = 1, \ldots, n,$$

are satisfied. Then the variable Y satisfies the estimate

$$Y \ll P_1^{1-\rho}, \quad \rho = c(n^4 \ln n)^{-1},$$

where the constant in $\ll$ depends only on n.

Proof. In the n-dimensional space, we consider the set Ω_0 of points $g(y) = (g_1(y), \dots, g_n(y))$ $(y = 1, 2, \dots, P_1)$ whose coordinates $g_\nu(y)$ satisfy the assumptions of the lemma. We show that, first, the set Ω_0 can intersect only the domain Ω_1 of the first class. Indeed, by the definition of the domain Ω_1 of the first class, each of its points $\alpha = (\alpha_1, \dots, \alpha_n)$ has the form

$$\alpha_\nu = \frac{b_\nu}{h_\nu} + z_\nu, \quad 1 \le h_\nu \le \tau_\nu, \quad (b_\nu, h_\nu) = 1,$$
$$|z_\nu| \le (h_\nu \tau_\nu)^{-1}, \quad \nu = 1, \dots, n, \quad [h_1, \dots, h_n] \le H.$$

Therefore, the distance between the corresponding coordinates of the centers of two domains Ω_1 and Ω_2 of the first class is no less than H^{-2} and hence the distance between the corresponding coordinates of the domains is no less than

$$H^{-2} - 2\tau_\nu^{-1} \ge 0.5H^{-2}, \quad \nu = 1, \dots, n.$$

The difference between any coordinate of a point in the set Ω_0, say, the νth coordinate, and the νth coordinate of the point $(a_1/q_1, \dots, a_\nu/q_\nu, \dots, a_n/q_n)$ is no less than

$$\zeta = \sum_{s=0}^{n} |\beta_\nu(s)| P_1^s \le (n+1)\tau_\nu^{-1}(0).$$

This implies that if Ω_n intersects any domain of the first class, then this can be only a single domain. Therefore, if $Y \ne 0$, then

$$\frac{a_\nu(y)}{q_\nu(y)} = \frac{b_\nu}{h_\nu}, \quad \left| g_\nu(y) - \frac{b_\nu}{h_\nu} \right| \le \Delta_\nu, \quad \nu = 1, \dots, n,$$

for all y satisfying the assumptions of the lemma.

We let $B_\nu(y)$ denote the polynomial

$$B_\nu(y) = \sum_{s=0}^{n} \beta_\nu(s) y^s.$$

Then we can rewrite the last relations as

$$\left| B_\nu(y) - \frac{b_\nu}{h_\nu} + \frac{a_\nu}{q_\nu} \right| \le \Delta_\nu, \quad \nu = 1, \dots, n.$$

Since $Q \ge P_1^{0.05 - 5n^3\rho}$, there exists a q_μ such that $q_\mu \ge P_1^{0.02n^{-1}} > H$; therefore, $a_\mu/q_\mu \ne b_\mu/h_\mu$.

We give an estimate from above for the number Y_1 of y satisfying the inequality

$$\left| B_\nu(y) - \frac{b_\nu}{h_\nu} + \frac{a_\nu}{q_\nu} \right| \le \Delta_\mu. \tag{7.11}$$

We proceed as in Lemma 7.5 and introduce the function

$$\psi(x) = \chi\left(x + \frac{a_\nu}{q_\nu} - \frac{b_\nu}{h_\nu}\right),$$

where $\chi(x)$ is the function defined in Lemma 7.5. Then

$$Y_1 \le \sum_{1\le y\le P_2} \psi\big(B_\mu(y)\big) = Y_2.$$

Expanding the function $\psi(x)$ into the Fourier series and passing to inequalities, we obtain

$$Y_2 \le c_2\left(P_1\Delta + \sum_{1\le m<M} \Delta|T(m)| + \sum_{M\le m<M_1} \Delta^{-1}m^{-2}|T(m)| + P_1^{1-\rho}\right), \qquad (7.12)$$

where

$$T(m) = \sum_{1\le y\le P_2} \exp\{2\pi i m B_\mu(y)\}, \quad \Delta = \Delta_\mu, \quad M = \Delta^{-1}, \quad M_1 = MP_1^\rho.$$

The further argument is similar to that in Lemma 7.5. From the sum $T(m)$ we pass to an integral and use Theorem 1.6 in Chapter 1 to estimate this integral (first, we bound δ from below).

First, we estimate the modulus of the derivative of the polynomial $mB_\mu(y)$. We have

$$\left|m\frac{d}{dy}B_\mu(y)\right| \le |m|\sum_{s=1}^{n} s\tau_\mu^{-1}(s)P_1^{s-1} < n^2P_1^{9\rho n^3-5/6} \le 0.5.$$

Applying Lemma A.2, we find

$$T(m) = \int_0^{P_1} \exp\{2\pi i m B_\mu(y)\}\,dy + O(1).$$

In the last integral we perform a change of the variable of integration of the form $P_1x = y$ and pass to estimates. We obtain

$$|T(m)| \le P_1\left|\int_0^1 \exp\{2\pi i m A_\mu(x)\}\,dx\right| + c_3,$$

where $A_\mu(x) = \sum_{s=1}^n \delta_s x^s$ and $\delta_s = \beta_\mu(s)P_1^s$.

Now we bound $\delta = \max_{1\le\nu\le n}|\delta_\nu|$ from below. First, we successively obtain the inequalities

$$\left|\frac{a_\mu}{q_\mu} - \frac{b_\mu}{h_\mu} + \sum_{s=0}^{n}\beta_\mu(s)\right| \ge \left|\frac{a_\mu}{q_\mu} - \frac{b_\mu}{h_\mu}\right| - \sum_{s=0}^{n}|\beta_\mu(s)|$$

$$\geq (q_\mu h_\mu)^{-1} - (q_\mu \tau_\mu(0))^{-1} - \sum_{s=1}^{n} \tau^{-1}(s) \geq (4H\tau_\mu(0))^{-1}.$$

For some y, this and (7.11) imply

$$\left| \sum_{s=0}^{n} \beta_\mu(s)(y^s - 1) \right| \geq \left| \frac{a_\mu}{q_\mu} - \frac{b_\mu}{h_\mu} + \sum_{s=0}^{n} \beta_\mu(s) \right| - \left| \frac{a_\mu}{q_\mu} - \frac{b_\mu}{h_\mu} + \sum_{s=0}^{n} \beta_\mu(s) y^s \right|$$
$$\geq (4H\tau_\mu(0))^{-1} - \Delta_\mu \geq (8H\tau_\mu(0))^{-1}.$$

Therefore, we have

$$(8H\tau_\mu(0))^{-1} \leq \left| \sum_{s=0}^{n} \beta_\mu(s)(y^s - 1) \right| \leq \sum_{s=1}^{n} |\delta_s| \frac{y^s - 1}{P_1^s} \leq n\delta,$$
$$\delta \geq (8nH\tau_\mu(0))^{-1}.$$

We apply Theorem 1.6 (Chapter 1) to the trigonometric integral and find the following estimate for the sum $T(m)$:

$$|T(m)| \leq c_4 P_1 \min\left(1, |m|^{-1/n} H^{1/n} \tau_\mu^{1/n}(0)\right) + c_4. \tag{7.13}$$

It follows from (7.12) and (7.13) that

$$Y_2 \leq c_5 \Bigg(P_1 \Delta + \Delta \sum_{1 \leq m < M} m^{-1/n} H^{1/n} \tau_\mu^{1/n}(0)$$
$$+ \Delta^{-1} \sum_{M \leq m < M_1} m^{-2-1/n} H^{1/n} \tau_\mu^{1/n}(0) + P_1^{1-\rho} \Bigg) \leq c_6 P_1^{1-\rho},$$
$$Y \leq Y_1 \leq Y_2 \leq c_6 P_1^{1-\rho},$$

as required. □

To prove the first main lemma, we need one more lemma, namely, Lemma 7.7, whose statement, in fact, differs from that of Lemma 7.6 only in the values of several parameters. But nevertheless, to prove this lemma, we must introduce several new definitions. Since the proof of this lemma does not differ from that of Lemma 7.6 except for the notation, we omit it here.

Let n polynomials $g_\nu(y)$ be given,

$$g_\nu(y) = \frac{a_\nu}{q_\nu} + \sum_{s=0}^{n} \beta_\nu(s) y^s, \quad \nu = 1, 2, \ldots, n,$$

where $(a_\nu, q_\nu) = 1$, $1 \leq q_\nu \leq \tau_\nu = P_1^{\nu - 1/6}$, $|\beta_\nu(s)| \leq \tau_\nu^{-1}(s)$, and $\tau_\nu(s) = P_1^{\nu-1/6} P_2^s$.

We denote the least common multiple of the numbers $q_1, \dots, q_n$ by Q_2. For each natural number $y \le P_2$, we consider the D-approximations of the numbers $g_\nu(y)$ corresponding to $\tau_\nu = P_1^{\nu-1/6}$, i.e., we consider the relations

$$g_\nu(y) = \frac{a_\nu(y)}{q_\nu(y)} + \beta_\nu(y),$$

where $\big(a_\nu(y), q_\nu(y)\big) = 1$, $1 \le q_\nu(y) \le \tau_\nu$, and $|\beta_\nu(y)| \le (q_\nu(y)\tau_\nu)^{-1}$.
We set $Q_2(y) = [q_1(y), q_2(y), \dots, q_n(y)]$.

Lemma 7.7. *Suppose that Q_2 is larger than $P^{0.05-5\rho n^3}$ and Y is the number of natural numbers y from the interval $1 \le y \le P_2$ for which the following conditions hold:*

$$|\beta_\nu(y)| \le \Delta_\nu = P_1^{-\nu+2\rho n}, \quad Q_2(y) \le H = P_1^{20\rho n^4}, \quad \nu = 1, \dots, n.$$

Then Y satisfies the estimate

$$Y \ll P_2 P_1^{-\rho}, \quad \rho = c(n^4 \ln n)^{-1},$$

where the constant in $\ll$ depends only on n.

In Lemma 7.9, from an estimate for the trigonometric sum $S(A)$, we derive an estimate for the sum $S_q(A)$ in which the summation variable x takes values from an arithmetic progression modulo $q \le P^{0.05/n}$, where P is the length of the summation interval in $S(A)$. To this end, we need the following lemma.

Lemma 7.8. *Suppose that the coordinates $\alpha(t)$ of a point A can be represented as*

$$\alpha(t) = \frac{a(t)}{q(t)} + \beta(t), \quad \delta(t) = P^t \beta(t), \quad \delta = \max_{1\le t\le n} |\delta(t)|, \quad 0 \le t \le n,$$

where $a(t)/q(t)$ is an irreducible fraction and the polynomial $f(x)$ has the form

$$f(x) = \alpha(n)x^n + \cdots + \alpha(1)x + \alpha(0).$$

Consider the polynomial $g(x)$ defined by the relation

$$g(x) = f(x+y) = \sum_{t=0}^{n} \alpha_0(t)x^t,$$

where y is an integer such that $|y| \ll P$; we set $Q = [q(1), \dots, q(n)]$ and $Q_0 = [q_0(1), \dots, q_0(n)]$. Then for $0 \le t \le n$, the coefficients $\alpha_0(t)$ of the polynomial $g(x)$ can be represented as

$$\alpha_0(t) = \frac{a_0(t)}{q_0(t)} + \beta_0(t)$$

and, moreover, $Q = Q_0$ and $\delta \ll \delta_0 \ll \delta$, where δ_0 is determined similarly to δ, but the numbers $\beta_0(t)$ are taken instead of the numbers $\beta(t)$, and the constants in $\ll$ depend only on n.

Proof. We represent the polynomial $f(x)$ as the sum

$$f(x) = f_1(x) + f_2(x),$$

where $f_1(x)$ has coefficients $a(t)/q(t)$ and $f_2(x)$ has numbers $\beta(t)$ as the coefficients. By setting

$$g_1(x) = f_1(x + y), \quad g_2(x) = f_2(x + y),$$

we obtain

$$g(x) = g_1(x) + g_2(x).$$

For $a_0(t)/q_0(t)$, we take the coefficients of the polynomial $g_1(x)$. For the numbers $\beta_0(t)$, we take the coefficients of $g_2(x)$. In the polynomials $f_1(x)$ and $g_1(x)$, the coefficients of all powers except the constant term will have the common denominator. Then we obtain

$$f_1(x) = Q^{-1} f_3(x) + a(0)q^{-1}(0), \quad g_1(x) = Q_0^{-1} g_3(x) + a_0(0)q_0^{-1}(0),$$

where $f_3(x)$ and $g_3(x)$ are polynomials with integer coefficients without constant terms. Let

$$g_4(x) = g_1(x) - a_0(0)g_0^{-1}(0) = Q^{-1} f_3(x + y) + R, \tag{7.14}$$

where R is a rational number. The polynomial $g_4(x)$ has rational coefficients, and the least common multiple of their denominators is equal to Q_0, while the constant term in $g_4(x)$ is zero. On the other hand, since y is an integer, the coefficients of the polynomial $g_5(x)$, where

$$g_5(x) = f_3(x + y),$$

are integers. This implies that the least common multiple of all coefficients in the polynomial $Q^{-1}g_5(x)$ divides the number Q_0.

If we set $x = 0$ in (7.14), then we see that the constant term in the polynomial $Q^{-1}g_5(x)$ is equal to R, and the polynomial $g_4(x)$ can be obtained from the polynomial $Q^{-1}g_5(x)$ by omitting the constant term. In this case, the least common multiple of the denominators of all coefficients in the polynomial $Q^{-1}g_5(x)$ can only be decreased only by an integer factor. Hence the number Q_0 is a divisor of the number Q. If in this argument we replace the polynomial $f_3(x)$ by $g_3(x)$ and the polynomial $g_1(x)$ by $f_1(x)$, then we prove that Q also divides Q_0. Hence $Q = Q_0$ and the first statement of the lemma is proved.

Now let us prove the second statement of the lemma. We set

$$f_6(x/P) = f_2(x), \quad g_6(x/P) = g_2(x).$$

Obviously, the coefficients of the polynomials $f_6(x)$ and $g_6(x)$ are respectively the numbers $\delta(t)$ and $\delta_0(t)$ $(t = 0, 1, \dots, n)$. Moreover, the relation $g_2(x) = f_2(x + y)$ implies $g_6(z) = f_6(z + y/P)$. Hence we have

$$g_6(z) = \sum_{t=0}^{n} \delta_0(t)z^t = \sum_{t=0}^{n} \delta(t)(z + yP^{-1})^t.$$

Opening the brackets in the right-hand side of the last relation and taking the inequality $|yP^{-1}| \ll 1$ into account, we obtain $\delta_0 \ll \delta$. Interchanging the polynomials $g_2(x)$ and $f_2(x)$ in this argument, we see that $\delta \ll \delta_0$ and hence $\delta \ll \delta_0 \ll \delta$. The proof of Lemma 7.8 is thereby complete. □

By $S_q(A)$ we denote a trigonometric sum of the form

$$S_q(A) = \sum_{x=1}^{P}{}' \exp\{2\pi i f(x)\},$$

where, as before, $f(x)$ is a polynomial with coefficients $\alpha(t)$ that simultaneously are the coordinates of a point $A \in \Omega$ and the prime on the summation sign means that the summation is taken not over the entire interval, but over a progression of the form $x = zq - y$, where q is a natural number and y satisfies the inequality $0 \leq y < q$.

Lemma 7.9. *Suppose that q satisfies the inequality*

$$q^n m \leq P^{0.05}.$$

Then in the notation of Lemma 7.1 (a) *we have the following estimates*:

(1) *If A is a point of the second class, then*

$$|S_q(A)| \ll P^{1-0.5\rho_1} q^{-1}.$$

(2) *If A is a point of the first class, $\delta \leq P^{0.04}$, and $Q \leq P^{0.07}$, then*

(a) $|S_q(A)| \ll Pq^{-1}Q_1^{-1/n+\varepsilon}$,

and for $\delta \geq 1$,

(b) $|S_q(A)| \ll Pq^{-1}(Q_1\delta)_1^{-1/n+\varepsilon}$,

where $Q_1 = Q/(Q, q^n)$.

The constants in $\ll$ depend only on n and ε.

(3) *For the remaining points A of the first class, the estimate in item* (1) *holds.*

Proof. Without loss of generality, we can assume that $P \equiv 0 \pmod q$.

1. First, we consider the case $y = 0$. Then, obviously, for all $t = 1, \dots, n$ we have

$$q^t \alpha(t) = \alpha_0(t), \tag{7.15}$$

where $\alpha_0(t)$ are coefficients of the polynomial $f_0(x) = f(qx)$.

Now we note that the partition of the points of the cube Ω into classes Ω_1 and Ω_2 depends on the value of the parameter P.

We consider the point A with coordinates $(\alpha_0(1), \dots, \alpha_0(n))$ and the corresponding trigonometric sum

$$S_0(A_0) = \sum_{x=1}^{P_0} \exp\{2\pi i f_0(x)\},$$

where $P_0 = Pq^{-1}$. In the above notation, we have

$$S_q(A) = S_0(A_0).$$

We shall estimate the sum $S_0(A_0)$ depending on the class to which the point A_0 belongs with respect to the parameter P_0 and the point A belongs with respect to the parameter P. If the point A belongs to the second class, then the point A_0 can belong both to the second and to the first class. In the first case, which is the most simple, the desired estimate for $S_0(A_0)$ readily follows from Lemma 7.1 (a) because $P_0 > P^{0.5}$. Now we assume that the point A belongs to the second class and the point A_0 belongs to the first class. In this case, the coordinates of the point A_0 can be represented as

$$\alpha_0(t) = \frac{a_0(t)}{q_0(t)} + \beta_0(t),$$

where $(a_0(t), q_0(t)) = 1$, $0 \le a_0(t) < q_0(t)$, $|\beta_0(t)| \le P^{-t+0.1}$, and $Q_0 = [q_0(1), \dots, q_0(n)]$ does not exceed $P_0^{0.1}$. If this representation contained $Q_0 \le P^{0.05}$, then, by (7.15) and the inequalities

$$q^t \le q^n \le P^{0.05}, \quad |\beta_0(t)|q^{-1} \le P_0^{-t+0.1}q^{-1} \le P^{-t+0.1},$$

the point A would belong to the first class, which is impossible. Hence we have $Q_0 > P^{0.05}$ and Lemma 7.1 (a) implies the desired estimate for the sum $S_0(A_0)$. Thus, if the point A belongs to the second class, then the statement of the lemma is proved for $y = 0$.

Now let the point A belong to the first class. If in this case we have $Q \le P^{0.07}$ and $\delta \le P^{0.05}$, then again using relation (7.15) to estimate Q_0 and δ_0 in terms of Q and $\beta(t)$, we prove that the point A_0 also belongs to the first class. To estimate the sum $S_0(A_0)$, we use Lemma 7.1 (a) and obtain the inequalities in item (2) of the statement of the lemma. Moreover, in the case $Q \le P^{0.07}$ and $P^{0.04} \le \delta \le P^{0.05}$, we obtain a stronger estimate. But if either $Q > P^{0.07}$ or $\delta > P^{0.05}$, then again the point A_0 can belong both to the first and to the second class. In the first of these cases, we estimate $S_0(A_0)$ by Lemma 7.1 (a) using (7.2) for $Q > P^{0.07}$ and (7.3) for $\delta > P^{0.05}$. The second case is studied similarly to the case considered at the beginning of the proof where A and A_0 belong to the second class. Thus the statement of the lemma is proved for $Y = 0$.

2. We use Lemma 7.8 to reduce the case of progressions with constant term $y \neq 0$ to the case $y = 0$. We change the variable of summation in the sum $S_q(A)$ by setting $x = z - y$ and define the polynomial $f_1(z)$ by the relation $f_1(z) = f(z - y)$. Then the sum $S_q(y)$ satisfies the relation

$$|S_q(A)| = |S_q(A_1)|,$$

where

$$S_q(A_1) = \sum_{z=1}^{P}{}' \exp\{2\pi i(f_1(z) - f_1(0))\};$$

the prime on the summation sign means that the summation is taken over integer z that are multiples of q, while the coordinates of the point A are the coefficients of the polynomial $f_1(z) - f_1(0)$.

The sum $S_q(A_1)$ is a sum of the same form as $S_q(A)$, but it is generated by a different point of the n-dimensional space. The progression over which the summations is performed in $S_q(A_1)$ satisfies the assumption of item (1) in this lemma. Hence all the results obtained above can be applied to $S_q(A_1)$. It is only necessary to study all possibilities for the points A and A_1 to belong to the first and second classes. We shall consider these cases.

(a) The points A and A_1 belong to the second class. In this case, the desired estimate follows from the results in item (1).

(b) The points A and A_1 belong to different classes. In this case, it follows from Lemma 7.8 that

$$\delta \gg P^{0.1}$$

for a point of the first class, and the process of estimating the sum $S_q(A)$ is reduced to item (1).

(c) Both points A and A_1 belong to the first class. Then it follows from Lemma 7.8 that the value of the parameter Q is the same for both points and the ratio of values of the parameter δ is bounded above and below by some constants. If we have

$$Q < P^{0.07}, \quad \delta \le P^{0.04}$$

for the point A, then the point A_1 satisfies the inequalities

$$Q < P^{0.07}, \quad \delta \ll P^{0.04} < P^{0.05},$$

and the desired estimate can be obtained from item (1). But if we have either $Q > P^{0.07}$ or $\delta > P^{0.04}$ for the point A, then the point A_1 satisfies either

$$Q > P^{0.07} \quad \text{or} \quad \delta \gg P^{0.04};$$

this last case can again be reduced to item (1). The proof of Lemma 7.9 is complete. □

7.1.3 The first main lemma

In this section we prove the first main lemma on estimating the double trigonometric sum on points of the second class.

In fact, this lemma contains all characteristic features of the theory in question.

The first main lemma. *Let $F(x_1, x_2)$ be a polynomial with real coefficients $\alpha(t_1, t_2)$ of the form*

$$F(x_1, x_2) = \sum_{t_1=0}^{n_1} \sum_{t_2=0}^{n_2} \alpha(t_1, t_2) x_1^{t_1} x_2^{t_2}, \quad \alpha(0, 0) = 0,$$

let P_1 and P_2 be natural numbers, $P_1 \le P_2$, and let $P_1 \to +\infty$. Consider the D-approximations of the numbers $\alpha(t_1, t_2)$ corresponding to $P_1^{t_1-1/6} P_2^{t_2}$:

$$\alpha(t_1, t_2) = \frac{a(t_1, t_2)}{q(t_1, t_2)} + \beta(t_1, t_2), \quad \big(a(t_1, t_2), q(t_1, t_2)\big) = 1,$$
$$1 \le q(t_1, t_2) \le \tau(t_1, t_2), \quad |\beta(t_1, t_2)| \le \big(q(t_1, t_2)\tau(t_1, t_2)\big)^{-1},$$
$$0 \le t_1 \le n_1, \quad 0 \le t_2 \le n_2.$$

Denote by Q the least common multiple of the numbers $q(t_1, t_2)$. Then for $Q > P_1^{0.1}$ the trigonometric sum

$$S = S(A) = \sum_{x_1=1}^{P_1} \sum_{x_2=1}^{P_2} \exp\{2\pi i F(x_1, x_2)\}$$

satisfies the estimate

$$|S| \le c P_1 P_2 P_1^{-\rho},$$

where $c = c(n_1, n_2) > 0$, $\rho = \gamma(n^4 \ln n)^{-1}$, and $\gamma > 0$ is an absolute constant.

Outline of the proof of the first main lemma. We can rewrite our sum S as

$$S = S(A) = \sum_{x_2=1}^{P_2} \sum_{x_1=1}^{P_1} \exp\{2\pi i (A_0 + A_1 x_1 + \cdots + A_{n_1} x_1^{n_1})\},$$

where

$$A_s = f_s(x_2) = \sum_{t=0}^{n_2} \alpha(s, t) x_2^t.$$

If the point $(A_1, \dots, A_{n_1})$ belongs to the second class Ω_2 in Lemma 7.1 (a) with respect to the parameter P_1, then this lemma can be applies to the sum over x_1. Suppose that the point $(A_1, \dots, A_n)$ belongs to the first class Ω_1. If the least common multiple Q or the value of δ in Lemma 7.1 (a) is "large," then we again can apply this lemma. However, if both Q and δ are "small," but the fractional parts of $f_s(x_2)$ are uniformly distributed at least for a single s $(1 \le s \le n_2)$, then the number of points x_2 possessing this property is "small" and the corresponding part of the sum S can be estimated trivially by the number of terms. But if the fractional parts of $f_s(x_2)$ are not uniformly distributed, then here (using the fact that P_1 and P_2 are essentially distinct) we again can show that the number of such x_2 is "not large." If P_1 and P_2 do not differ significantly, then the desired estimate for $|S|$ follows from Theorem 5.2 in Chapter 5.

The main points in the proof are the following.

1. First, we note that the desired statement contains two essentially different cases. The first of these cases is conditionally said to be "*two-dimensional*," and the second is said to be "*one-dimensional*."

On the plane, we consider points with integer coordinates (t_1, t_2) which are the indices of the coefficients $\alpha(t_1, t_2)$ of the polynomial $F(x_1, x_2)$ contained in the exponential. These points lie in a rectangular of the form $0 \le t_1 \le n_1, 0 \le t_2 \le n_2$. We divide the points of this rectangular into three classes E_0, E_1, and E_2. The class E_1 contains all the points (t_1, t_2) lying on the ordinate axis (the $0t_2$-axis); the class E_2 contains all the points (t_1, t_2) lying on the abscissa axis (the $0t_1$-axis); and the class E_0 contains all other points of the rectangular. By Q_j $(j = 0, 1, 2)$ we denote the least common multiple of the denominators of the rational fractions in the D-approximations of the coefficients $\alpha(t_1, t_2)$ corresponding to $\tau(t_1, t_2)$ provided that $(t_1, t_2) \in E_j$, i.e., $Q_j = \text{l.c.m.}(q(t_1, t_2))$, where $(t_1, t_2) \in E_j$ $(j = 0, 1, 2)$ and

$$\alpha(t_1, t_2) = \frac{a(t_1, t_2)}{q(t_1, t_2)} + \beta(t_1, t_2), \quad \big(a(t_1, t_2), q(t_1, t_2)\big) = 1,$$

$$q(t_1, t_2) \le \tau(t_1, t_2) = P_1^{t_1 - 1/6} P_2^{t_2}, \quad |\beta(t_1, t_2)| \le \big(q(t_1, t_2)\tau(t_1, t_2)\big)^{-1}.$$

Clearly, we have $Q = [Q_0, Q_1, Q_2]$.

2. Let $Q_0 > P_1^{10\rho n^3}$. We shall conditionally say that this case is "two-dimensional." Then we can estimate the sum as follows. We write the polynomial $F(x_1, x_2)$ as

$$F(x_1, x_2) = \sum_{s=0}^{n_1} f_s(x_2) x_1^s = \sum_{s=0}^{n_1} A(s) x_1^s.$$

Next, there exists a polynomial $f_s(x_2)$ for which the value $Q(s)$ of the least common multiple of the denominators of the rational fractions in the D-approximations of the coefficients $\alpha(s, t)$, corresponding to $\tau(s, t) = P_1^{s-1/6} P_2^t$ $(1 \le t \le n_2)$, is larger than $P_1^{10\rho n}$. The fractional parts of the polynomial $f_s(x_2)$ $(1 \le x_2 \le P_2)$ can be either uniformly distributed or not uniformly distributed. Let us consider the possible cases.

3. If all denominators of the fractions in the D-approximations of the coefficients $\alpha(s, t)$ $(1 \le t \le n_2)$, corresponding to $\tau(s, t)$, do not exceed $\tau_2(s) = P_2^{s-1/6}$ $(s = 1, \dots, n_2)$, then the fractional parts of the polynomial $f_s(x_2)$ are uniformly distributed.

For each x_2 $(1 \le x_2 \le P_2)$, we consider the points $(A(1), \dots, A(n_1))$ and the D-approximations of the coordinates $A(\nu)$ $(1 \le \nu \le n_1)$ corresponding to $\tau_1(s) = P_1^{s-1/6}$. By $Q(x_2)$ we denote the least common multiple of the denominators of the fractions in this D-approximation. If $Q(x_2)$ is sufficiently large, then the part of the double sum corresponding to this x_2 can be estimated rather well as a one-dimensional sum with respect to x_1. But if $Q(x_2)$ is small, then the estimate for the one-dimensional sum can be rather bad. But the number Y of values of the variable x_2 for which $Q(x_2)$ is small is also small, since the fractional parts of the polynomial $f_s(x_2)$ are uniformly distributed. Hence the part of the two-dimensional sum corresponding to all such values of x_2 can be estimated rather well. This already implies the desired estimate for S.

4. Now we assume that in some coefficients $\alpha(s,t)$ $(1 \le t \le n_2)$ in the polynomial $f_s(x_2)$, the denominators of the fractions in their D-approximations corresponding to $\tau(s,t)$ are larger than $\tau_2(t)$. Then we take new D-approximations of these coefficients which already correspond to $\tau_2(t)$. If the least common multiple of the rational fractions in the new D-approximations of the coefficients $\alpha(s,t)$ $(1 \le t \le n_2)$ is larger than some value (we choose it in the course of the proof), then the fractional parts of $f_s(x_2)$ $(1 \le x_2 \le P_2)$ are uniformly distributed and to estimate the sum S in this case, we must repeat the argument of item 3.

5. We assume that the least common multiple of the fractions in the new D-approximations of the coefficients of the nonzero powers of the variable x_2 in the polynomial $f_s(x_2)$ does not exceed H. In this case, we consider the point $M(x_2) = (\{f_1(x_2)\}, \dots, \{f_{n_1}(x_2)\})$ and, for each fixed x_2, we take the D-approximations of its coordinates $\{f_\nu(x_2)\}$ $(\nu = 1, \dots, n_2)$ corresponding to the numbers $\tau_1(\nu) = P_1^{\nu-1/6}$. If the least common multiple of the denominators of the fractions in these last approximations is sufficiently large, then the inner sum over x_1 in the double sum S can be estimated rather well. Otherwise, we can obtain both good and bad estimates for the sum over x_1. The values of the variable x_2 for which the sum over x_1 is estimated badly are conditionally said to be "*bad*." The number of such variables will be denoted by Y. It turns out that a good estimate can be obtained for Y (in fact, obtaining this estimate is the contents of Lemma 7.5). Hence the desired estimate for the sum S can be obtained. Thus we have completely discussed item 5 and, together with it, the entire "two-dimensional" case.

It should be noted that the main difficulty in the "two-dimensional" case is contained in estimating the variable Y. This estimate is significantly based on the use of the inequality

$$\ln P_2 \ge 1.2(n+1)\ln P_1$$

(see Lemma 7.5).

6. Now let $Q_0 < P_1^{10\rho n^3}$. Then either Q_1 or Q_2 is large. This case is conditionally said to be "one-dimensional." Now we consider only the case of large Q_1, because if Q_2 is large, then the outline of the argument coincides with that in the case of large Q_1, only instead of Lemma 7.6, we use Lemma 7.7 in the appropriate places. We partition the summation over x_2 into arithmetic progressions with difference Q_0. After simple transformations, we obtain the estimate

$$|S| \le \sum_{x_1=1}^{P_1} \sum_{z_2=1}^{Q_0} \left| \sum_{y_2} \exp\{2\pi i \Psi(x_1, Q_0 y_2 + z_2)\} \right| = T_1,$$

where the inner sum is taken over the variable y_2 satisfying the conditions $1 \le Q_0 y_2 + z_2 \le P_2$, the polynomial $\Psi(x_1, x_2)$ has the form

$$\Psi(x_1, x_2) = \sum_{\nu=1}^{n_2} g_\nu(x_1) x_2^\nu, \quad g_\nu(x_1) = \frac{a(0,\nu)}{q(0,\nu)} + \sum_{t=0}^{n_1} \beta(t,\nu) x_1^t,$$

and the variables $a(0,\nu)$, $q(0,\nu)$, and $\beta(t,\nu)$ $(1 \le \nu \le n_2)$ are defined in item 1.

We take the D-approximations of the coordinates $g_\nu(x_1)$ of the point

$$(\{g_1(x_1)\}, \ldots, \{g_{n_2}(x_1)\})$$

corresponding to $\tau_2(\nu) = P_2^{\nu-1/6}$. If the least common multiple of the denominators of the fractions in these D-approximations is sufficiently large, then we can apply Lemma 7.9 to the inner sum over y_2. Otherwise, the sum S is estimated according to the scheme given in item 5, only we use Lemma 7.6 instead of Lemma 7.5. Thus the "one-dimensional" case has been discussed completely.

Proof. We shall follows the above plan.

1. Suppose that Q_0 is the least common multiple of the numbers $q(t_1,t_2)$ satisfying the conditions $t_1 \ge 1$ and $t_2 \ge 1$, Q_1 is the least common multiple of the numbers $q(0,t_2)$ $(t_2 \ge 1)$, and Q_2 is the least common multiple of the numbers $q(t_1,0)$ $(t_1 \ge 1)$. By the assumptions of the lemma, we have $Q = [Q_0, Q_1, Q_2] \ge P_1^{0.1}$ We shall separately consider the two cases: the case of large Q_0 and the case of small Q_0.

2. Let $Q_0 \ge P_1^{10\rho n^3}$. For each s $(1 \le s \le n_1)$, by $Q(s)$ we denote the least common multiple of the numbers $q(s,t)$ $(1 \le t \le n_2)$. By the definitions of $Q(s)$ and Q_0, we have

$$Q_0 = [Q(1), \ldots, Q(n_1)],$$

and hence there exists an s $(1 \le s \le n_1)$ for which

$$Q(s) \ge Q_0^{1/n} \ge P_1^{10\rho n^2}.$$

Depending on the value of $Q(s)$, we consider the following three subcases of item 2.

(a) $P_1^{10\rho n^2} \le Q(s) < P_2^{0.1}$;

(b) $P_2^{0.1} \le Q(s)$ and the inequality

$$q(s,t) \le P_2^{t-1/6}$$

holds for $1 \le t \le n_2$;

(c) $P_2^{0.1} \le Q(s)$ and there exists a t $(1 \le r \le n_2)$ such that

$$P_2^{t-1/6} < q(s,t).$$

Before studying these subcases, we write the polynomial $F(x_1,x_2)$ as

$$F(x_1,x_2) = \sum_{t_1=0}^{n_1}\sum_{t_2=0}^{n_2} \alpha(t_1,t_2)x_1^{t_1}x_2^{t_2} = \sum_{s=0}^{n_2} f_s(x_2)x_1^s,$$

where

$$f_s(x_2) = \sum_{t=0}^{n_2} \alpha(s,t)x_2^t.$$

For a given s, we take the number

$$\tau_1(s) = P^{s-1/6}$$

and consider the D-approximations of the fractional parts of the polynomial $f_s(x_2)$ $(1 \le x_2 \le P_2)$ corresponding to $\tau_1(s)$, i.e., we consider relations of the form

$$\{f_s(x_2)\} = \frac{b}{r} + \frac{\theta}{r\tau_1(s)}, \tag{7.16}$$

where $(b, r) = 1$, $1 \le r \le \tau_1(s)$, and $|\theta| \le 1$.

3. (a) We represent the sum $S = S(A)$ as three terms:

$$S = S_1 + S_2 + S_3,$$

where

$$S_j = \sum_{x_2 \le P_2}^{(j)} \sum_{x_1=1}^{P_1} \exp\{2\pi i F(x_1, x_2)\}, \quad j = 1, 2, 3,$$

and the domain of summation over the variable x_2 in each of the sums S_j is its own and is determined as follows. We consider the inner sum over x_1:

$$\begin{aligned} S(x_2) &= \sum_{x_1=1}^{P_1} \exp\{2\pi i F(x_1, x_2)\} \qquad (7.17)\\ &= \sum_{x_1=1}^{P_2} \exp\{2\pi i (A_0 + A_1 x_1 + \cdots + A_{n_1} x^{n_1})\}, \end{aligned}$$

where the numbers $a_0, A_1, \ldots, A_{n_1}$ depend on x_2 and $A_s = \{f_s(x_2)\}$.

If the point $(A_1, \ldots, A_{n_1}) = (\{f_1(x_2)\}, \ldots, \{f_{n_1}(x_2)\})$ of the n_1-dimensional space is a point of the second class with respect to the parameter P_1 defined in Lemma 7.1 (a), then the corresponding x_2 belongs to the sum S_1. But if this point is a point of the first class and its sth coordinate $A_s = \{f_s(x_2)\}$ satisfies relation (7.16) with $r > H = P_1^{2\rho n}$, then the corresponding x_2 belongs to the sum S_2. All other x_2 belong to the sum S_3.

For the values of x_2 contained in the sum S_1, the point $(A_1, \ldots, A_{n_1})$ is a point of the second class, and hence Lemma 7.1 (a) implies the estimate

$$|S(x_2)| \ll P_1^{1-\rho_1} \ll P_1^{1-\rho}, \quad \rho_1 = c(n^2 \ln n)^{-1}.$$

But for the values of x_2 contained in the sum S_2, the least common multiple of the denominators of the rational fractions in representation (7.16) is larger than H, and hence, by Lemma 7.1 (a), we have

$$|S(x_2)| \ll P_1 H^{-1/n+\varepsilon} \ll P_1^{1-\rho}, \quad \rho_1 = c(n^4 \ln n)^{-1}.$$

Now we give an estimate from above for the number Y of values of x_2 contained in the sum S_3. By Lemma 7.2 on the distribution of the fractional parts of the polynomial $f_s(x_2)$, the number Γ of values of x_2 for which $\{f_s(x_2)\}$ belongs to the interval $[b/r - 1/(r\tau_1(s)), b/r + 1/(r\tau_1(s))]$ can be estimated as

$$\Gamma \ll P_2\big((r\tau_1(s))^{-1} + P_1^{-5\rho n}\big).$$

Indeed, the polynomial $f_s(x_2)$ has the form

$$f_s(x_2) = \sum_{t=0}^{n_2} \alpha(s,t)x_2^t,$$

where, by the assumptions of the lemma, the coefficients $\alpha(s,t)$ can be written as

$$\alpha(s,t) = \frac{a(s,t)}{q(s,t)} + \beta(s,t),$$
$$|\beta(s,t)| \le \big(q(s,t)\tau(s,t)\big)^{-1} \le P_1^{-s+1/6}P_2^{-t} \le P_2^{-t},$$

and the least common multiple $Q(s)$ of the denominators $q(s,t)$ does not exceed $P_2^{0.1}$, i.e., the point $(\alpha(s,1),\dots,\alpha(s,n_2))$ belongs, with respect to the parameter P_2, to the first class defined in Lemma 7.1 (a). Hence to estimate the number of the fractional parts of the polynomial $f_s(x_2)$ contained in this interval, we can use Lemma 7.2, item (2), with $Q = Q(s) > P_1^{10\rho n^2}$. The number of such intervals with $r \le H$ does not exceed $\sum_{r\le H}\varphi(r) \le H^2$. Hence,

$$Y \le P_2H^2\big((\tau_1(s))^{-1} + P_1^{-5\rho n}\big) \ll P_2P_1^{-\rho n} \ll P_2P_1^{-\rho}.$$

We substitute the estimates obtained for $|S(x_2)|$ into the sums S_1 and S_2 and trivially estimate the sum S_3 by the number of terms. We obtain

$$|S| \le |S_1| + |S_2| + |S_3| \ll P_1P_2P_1^{-\rho}.$$

(b) We shall proceed as in case (a). We represent the sum S as

$$S = S_4 + S_5 + S_6,$$

where

$$S_j = \sum_{x_2\le P_2}^{(j)} \sum_{x_1=1}^{P_1} \exp\{2\pi i F(x_1,x_2)\}, \quad j = 4,5,6,$$

and the domain of summation over the variable x_2 in each of the sums is its own and is determined as follows. We consider representation (7.17) of the inner sum over x_1. If the point $(A_1,\dots,A_{n_1})$ is a point of the second class with respect to the parameter P_1 defined in Lemma 7.1 (a), then the corresponding x_2 belongs to the sum S_4. If this point is a point of the first class, but its sth coordinate $A_s = \{f_s(x_2)\}$ satisfies relation (7.16) with $r \geq H = P_2^{0.25\rho_1}$, then the corresponding x_2 belongs to the sum S_5. All other x_2 belong to the sum S_6.

If the value of x_2 belongs to the sum S_4, then, by Lemma 7.1 (a), for the sum $S(x_2)$ we have

$$|S(x_2)| \ll P_1^{1-\rho_1} \ll P_1^{1-\rho},$$

where $\rho_1 = c_1(n^2 \ln n)^{-1}$ and $\rho = c(n^4 \ln n)^{-1}$.

Now we estimate $|S(x_2)|$ for the values of x_2 contained in S_5. Relation (7.16) implies

$$A_s = \frac{b}{r} + \frac{\theta}{r\tau_1(s)}.$$

If $r \geq P_1^{0.1}$, then we can use Lemma 7.1 (b) to estimate $|S(x_2)|$ and thus obtain

$$|S(x_2)| \ll P_1^{1-\rho_1}.$$

Now let $r < P_1^{0.1}$ (we also assume that $H_1 = P_2^{0.25\rho_1} < P_1^{0.1}$, since, otherwise, $r \geq H_1 \geq P_1^{0.1}$, and this case we have just studied). Since the point $(A_1, \ldots, A_{n_1})$ belongs to the first class, we have

$$A_\nu = \frac{a_\nu}{q_\nu} + \beta_\nu, \quad |\beta_\nu| \leq P_1^{-\nu+0.1}, \quad \nu = 1, \ldots, n_1,$$

and $q = [q_1, \ldots, q_{n_1}] \leq P_1^{0.1}$. Let us prove that $a_s/q_s = b/r$.

Indeed, otherwise, we would have the inequalities

$$\frac{1}{rq_s} \leq \left|\frac{b}{r} - \frac{a_s}{q_s}\right| \leq |\beta_s| + r^{-1}\tau_1^{-1}(s) \leq 2P_1^{-5/6}, \quad r < P_1^{0.1}, \quad q_s \leq q \leq P_1^{0.1},$$

which contradict each other. Hence we have $q_s = r$ and $q \geq q_s = r > H_1$.

We use Lemma 7.1 (a) to estimate $|S(x_2)|$ and thus find

$$|S(x_2)| \ll P_1 H_1^{-1/n+\varepsilon} \ll P_1^{1-\rho}.$$

Next, as in case (a), by Lemma 7.2, the number of numbers x_2 contained in the sum S_6 does not exceed

$$\Gamma \ll P_2 H_1^2 \left(\tau_1^{-1}(s) + P_2^{-\rho_1}\right), \quad \rho_1 = c_1(n^2 \ln n)^{-1}.$$

Indeed, the polynomial $f_s(x_2)$ has the form

$$f_s(x_2) = \sum_{t=0}^{n_2} \alpha(s,t) x_2^t,$$

where the coefficients $\alpha(s,t)$ can be represented as

$$\alpha(s,t) = \frac{a(a,t)}{q(s,t)} + \beta(s,t),$$

$$|\beta(s,t)| \le \big(q(s,t)\tau(s,t)\big)^{-1} \le \big(q(s,t)\tau_2(t)\big)^{-1}, \quad q(s,t) \le \tau_2(t) = P_2^{t-1/6};$$

and moreover, $Q(s) = [q(s,1), \dots, q(s,n_2)] \ge P_2^{0.1}$. Therefore, by Lemma 7.2, item (3), the number of the fractional parts of $f_s(x_2)$ contained in the interval

$$\left[\frac{b}{r} - \frac{1}{r\tau_1(s)}, \frac{b}{r} + \frac{1}{r\tau_1(s)}\right],$$

does not exceed $\Gamma_1 \ll P_2\big((r\tau_1(s))^{-1} + P_2^{-\rho_1}\big)$, and the number of such intervals does not exceed H_1^2.

Thus we have

$$|S| \le |S_4| + |S_5| + |S_6| \ll P_1 P_2 P_1^{-\rho}.$$

4. Now we consider case (c). In this case we have $Q(s) \ge P_2^{0.1}$, and there exists a t_0 $(1 \le t_0 \le n_2)$ such that the denominator $q(s,t_0)$ of the fraction in the D-approximation of the number $\alpha(s,t_0)$, corresponding to $\tau(s,t_0)$, satisfies the condition

$$\tau_2(t_0) = P_2^{t_0-1/6} < q(s,t_0) \le \tau(s,t_0).$$

Now let us consider new D-approximations of the numbers $\alpha(s,t)$ for all t $(1 \le t \le n_2)$ corresponding to $\tau_2(t) = P_2^{t-1/6}$, i.e., the relations

$$\alpha(s,t) = \frac{d(t)}{h(t)} + \frac{\theta}{h(t)\tau_2(t)}, \tag{7.18}$$
$$\big(d(t), h(t)\big) = 1, \quad 1 \le h(t) \le \tau_2(t), \quad |\theta| \le 1.$$

If in this approximation of the numbers $\alpha(s,t)$, it turns out that the least common multiple of the numbers $h(t)$ is larger than $P_2^{0.1}$, then the sum S is, in fact, estimated in the same way as in case (b). We represent the sum S as

$$S = S_7 + S_8 + S_9,$$

where

$$S_j = \sum_{x_2 \le P_2}^{(j)} \sum_{x_1=1}^{P_1} \exp\{2\pi i F(x_1,x_2)\}, \quad j = 7, 8, 9,$$

and the domain of summation over the variable x_2 in each of the sums is its own and is determined as follows. We consider representation (7.17) of the inner sum over x_1. If the point $(A_1, \dots, A_{n_1})$ is a point of the second class with respect to the parameter P_1 defined in Lemma 7.1 (a), then the corresponding x_2 belongs to the sum S_7. If this point is a point of the first class, but its sth coordinate $A_s = \{f_s(x_2)\}$ satisfies relation (7.16) with $r \ge H_1 = P_2^{0.25\rho_1}$, then the corresponding x_2 belongs to the sum S_8. All other x_2 belong to the sum S_9. The sums S_7 and S_8 are estimated precisely in the same way as the sums S_4 and S_5, respectively.

We give an estimate from above for the number Y of values of x_2 contained in the sum S_9. For this, we consider the D-approximations of the coefficients $\alpha(s,t)$ $(1 \le t \le n_2)$ of the polynomial $f_s(x_2)$ corresponding to $\tau_2(t) = P_2^{t-1/6}$, i.e., relations (7.18). Since the least common multiple of the numbers $h(t)$ is larger than $P_2^{0.1}$, by Lemma 7.2, item (3), the number of the fractional parts of $f_s(x_2)$ in the interval

$$\left[\frac{b}{r} - \frac{1}{r\tau_1(s)}, \frac{b}{r} + \frac{1}{r\tau_1(s)}\right]$$

does not exceed $\Gamma_1 \ll P_2\big((r\tau_1(s))^{-1} + P_2^{-\rho_1}\big)$, while the number of such intervals does not exceed H_1^2. Hence for the variable Y, we have the estimate

$$Y \ll P_2 H_1^2 \big(\tau_1^{-1}(s) + P_2^{-\rho_1}\big) \ll P_2 P_1^{-\rho}.$$

Thus

$$|S| \ll P_1 P_2 P_1^{-\rho}.$$

Now we assume that in the new D-approximation the least common multiple of the numbers $h(t)$ does not exceed $P_2^{0.1}$. We denote this least common multiple by $Q_1(s)$. Here the following two cases are possible: $P_2^{0.1} \ge Q_1(s) > H_2 = P_1^{10n^2\rho}$ and $Q_1(s) \le H_2$. First, we consider the case $Q_1(s) > H_2$. We represent the fractional parts of the polynomial $f_s(x_2)$ as

$$\{f_s(x_2)\} = \frac{b}{r} + \frac{\theta}{r\tau_1(s)}, \tag{7.19}$$
$$(b,r) = 1, \quad 1 \le r \le \tau_1(s) = P_1^{s-1/6}, \quad |\theta| \le 1.$$

In other words, we consider the D-approximations of the numbers $\{f_s(x_2)\}$ corresponding to $\tau_1(s) = P_1^{s-1/6}$ for each x_2 $(1 \le x_2 \le P_2)$. We divide the sum S into three sums:

$$S = S_{10} + S_{11} + S_{12},$$

where

$$S_j = \sum_{x_2 \le P_2}^{(j)} \sum_{x_1=1}^{P_1} \exp\{2\pi i F(x_1,x_2)\}, \quad j = 10, 11, 12,$$

and the domain of summation over the variable x_2 in each of the sums is its own and is determined as follows. We consider representation (7.17) of the inner sum over x_1. If the point $(A_1,\ldots,A_{n_1}) = (\{f_1(x_2)\},\ldots,\{f_{n_1}(x_2)\})$ is not a point of the second class with respect to the parameter P_1 defined in Lemma 7.1 (a), then the corresponding x_2 belongs to the sum S_{10}. If this point is a point of the first class, defined in Lemma 7.1 (a), and moreover, $r > K = P_1^{2\rho n}$ in representation (7.19), then the corresponding x_2 belongs to the sum S_{11}. All other x_2 belong to the sum S_{12}. In other words, the sum S_{12} contains all x_2 for which the denominators r in representation (7.19) do not exceed K.

We use Lemma 7.1 (a) and representation (7.17) to estimate the sum S_{10} as

$$|S(x_2)| \ll P_1^{1-\rho_1} \ll P_1^{1-\rho}, \quad |S_{10}| \ll P_1 P_2 P_1^{-\rho}.$$

For S_{11}, we estimate the inner sum as follows. If $r \geq P_1^{0.1}$, then, by Lemma 7.4, we have

$$|S(x_2)| \ll P_1^{1-\rho_1}.$$

But if $r < P_1^{0.1}$, then $|S(x_2)|$ can be estimated in the same way as in the sum S_5 for $r < P_1^{0.1}$, i.e., for the coordinates of the point $(A_1, \ldots, A_{n_1})$, we consider the relations

$$A_\nu = \frac{a_\nu}{q_\nu} + \beta_\nu, \quad (a_\nu, q_\nu) = 1, \quad |\beta_\nu| \ll P_1^{-\nu+0.1}, \quad \nu = 1, \ldots, n_1,$$

$$q = [q_1, \ldots, q_{n_1}] < P_1^{0.1},$$

which hold because this point belongs to the first class with respect to the parameter P_1. We show that $a_s/q_s = b/r$. This implies that $q \geq q_s = r > K$. Next, we apply Lemma 7.1 (a) to the sum $S(x_2)$ and find

$$|S(x_2)| \ll P_1 K^{-1/n+\varepsilon} \ll P_1^{1-\rho}.$$

Thus we have

$$|S_{11}| \ll P_1 P_2 P_1^{-\rho}.$$

Now we consider the sum S_{12}. Estimating this sum by the number of terms, we obtain

$$|S_{12}| \leq Y P_1,$$

where Y is the number of values of x_2 for which the fractional part of the polynomial $f_s(x_2)$ belongs at least to one of the intervals of the form

$$\left[\frac{b}{r} - \frac{1}{r\tau_1(s)}, \frac{b}{r} + \frac{1}{r\tau_1(s)}\right], \quad (b, r) = 1, \quad r \leq K. \tag{7.20}$$

A point $(\alpha(s, 1), \ldots, \alpha(s, n_2))$ from the n_2-dimensional cube Ω is generated by the coefficients $\alpha(s, t)$ of the polynomial $f_s(x_2)$. We first assume that this point belongs to the second class with respect to the parameter P_2. Then, by Lemma 7.2, item (1), the number Γ of the fractional parts of the polynomial $f_s(x_2)$ in the interval (7.20) does not exceed

$$\Gamma_1 \ll P_2\big((r\tau_1(s))^{-1} + P_2^{-\rho_1}\big), \quad \rho_1 = c_1(n^2 \ln n)^{-1}.$$

Now we assume that this point belongs to the first class. This means that for the numbers $\alpha(s, t)$ $(1 \leq t \leq n_2)$, we have the representations

$$\alpha(s, t) = \frac{a_t}{q_t} + \beta_t, \quad |\beta_t| \ll P_1^{-t+0.1}, \quad (a_t, q_t) = 1,$$

$$q = [q_1, \ldots, q_{n_2}] < P_2^{0.1}.$$

At the same time, for the numbers $\alpha(s,t)$ $(1 \le t \le n_2)$, we have relations (7.18) for $H_2 < Q_1(s) < P_2^{0.1}$, i.e.,

$$\alpha(s,t) = \frac{d(t)}{h(t)} + \frac{\theta}{h(t)\tau_2(t)}, \qquad \tau_2(t) = P_2^{t-1/6},$$
$$Q_1(s) = [h(1), \ldots, h(n_2)], \qquad H_2 < Q_1(s) \le P_2^{0.1}.$$

This implies that the relations $a_t/q_t = d(t)/h(t)$ hold for all t $(1 \le t \le n_2)$.

Indeed, let the relation

$$a_t/q_t \ne d(t)/h(t)$$

hold for some t $(1 \le t \le n_2)$. Then, on the one hand, we have

$$\left| \frac{a_t}{q_t} - \frac{d(t)}{h(t)} \right| \ge \frac{1}{q_t h(t)} \ge P_2^{-0.2},$$

since q_t and $h(t)$ satisfy the inequalities $q_t \le q \le P_2^{0.1}$ and $h(t) < Q_1(s) \le P_2^{0.1}$. On the other hand, we have

$$\left| \frac{a_t}{q_t} - \frac{d(t)}{h(t)} \right| \le |\beta_t| + \frac{1}{h(t)\tau_2(t)} \le P_2^{-t+0.1} + P_2^{-t+1/6} \le 2P_2^{-t+1/6},$$

since $|\beta_t| \le P_2^{-t+0.1}$ and $1 \le h(t) \le \tau_2(t) = P_2^{t-1/6}$.

The estimates obtained for $|a_t/q_t - d(t)/h(t)|$ contradict each other. Hence we have $a_t/q_t = d(t)/h(t)$ for all t $(1 \le t \le n_2)$. This implies $q = Q_1(s)$ and hence $q < H_2$. Now we estimate Γ by using Lemma 7.2, item (2), as

$$\Gamma_1 \ll P_2\big((r\tau_1(s))^{-1} + H_2^{-1/n+\varepsilon}\big).$$

So, for Γ in both cases, we have the estimate

$$\Gamma_1 \ll P_2\big((r\tau_1(s))^{-1} + P_2^{-\rho_1} + H_2^{-1/n+r}\big) \ll P_2 H_2^{-1/(2n)}.$$

Since the number of intervals (7.20) does not exceed K^2, we have

$$Y \ll K^2\Gamma \ll P_2 K^2 H_2^{-1/(2n)} \ll P_2 P_1^{-\rho}.$$

Hence, for $Q_1(s) > H_2$, we have

$$|S| \le |S_{10}| + |S_{11}| + |S_{12}| \ll P_1 P_2 P_1^{-\rho}.$$

5. Now we consider the case $Q_1(s) \le H_2$. First, we estimate the sum $S = S(A)$ under the assumption that $\ln P_2 \le 1.2(n+1)\ln P_1$. If the point A with coordinates

$\alpha(t_1, t_2)$ $(0 \le t_1 \le n_1, 0 \le t_2 \le n_2, t_1 + t_2 \ge 1)$ that are the coefficients of the polynomial

$$F(x_1, x_2) = \sum_{t_1=0}^{n_1} \sum_{t_2=0}^{n_2} \alpha(t_1, t_2) x_1^{t_1} x_2^{t_2}$$

belongs to the second class Ω_2, then, by Lemma 7.4, since $\varkappa < c_2 n^2$, the sum $S = S(A)$,

$$S = \sum_{x_1=1}^{P_1} \sum_{x_2=1}^{P_2} \exp\{2\pi i F(x_1, x_2)\},$$

can be estimated as follows: $|S| \ll P_1 P_2 P_1^{-\rho}$, where $\rho = c(n^4 \ln n)^{-1}$.

But if the point A belongs to the first class, then the definition of points of the first class implies the relations

$$\alpha(t_1, t_2) = \frac{a_0(t_1, t_2)}{q_0(t_1, t_2)} + \beta_0(t_1, t_2),$$
$$\big(a_0(t_1, t_2), q_0(t_1, t_2)\big) = 1, \quad |\beta_0(t_1, t_2)| \le P_1^{-t_1+0.1} P_2^{-t_2},$$

and the number q_0, which is the least common multiple of $q_0(t_1, t_2)$ $(0 \le t_1 \le n_1, 0 \le t_2 \le n_2, t_1 + t_2 \ge 1)$, is less than $P_1^{0.1}$. Recall that Q is the least common multiple of the denominators $q(t_1, t_2)$ of the fractions in the D-approximations of $\alpha(t_1, t_2)$ $(0 \le t_1 \le n_1, 0 \le t_2 \le n_2, t_1 + t_2 \ge 1)$ corresponding to $\tau(t_1, t_2) = P_1^{t_1 - 1/6} P_2^{t_2}$. By the assumptions of the lemma, Q is larger than $P_1^{0.1}$. Hence there exists a set (t_1, t_2) such that $q(t_1, t_2) \ne q_0(t_1, t_2)$.

We show that $q_0 \ge 0.5 P_1^{1/15}$. Indeed, we have the system of inequalities

$$\frac{1}{q(t_1, t_2) q_0(t_1, t_2)} \le \left| \frac{a(t_1, t_2)}{q(t_1, t_2)} - \frac{a_0(t_1, t_2)}{q_0(t_1, t_2)} \right| \le |\beta(t_1, t_2)| + |\beta_0(t_1, t_2)|$$
$$\le P_1^{-t_1+0.1} P_2^{-t_2} + P_1^{-t_1+1/6} P_2^{-t_2} q^{-1}(t_1, t_2),$$
$$q_0^{-1}(t_1, t_2) \le q(t_1, t_2) P_1^{-t_1+0.1} P_2^{-t_2} + P_1^{-t_1+1/6} P_2^{-t_2} \le 2 P_1^{-1/(15)},$$
$$q_0 \ge q_0(t_1, t_2) \ge 0.5 P_1^{1/(15)}.$$

Applying Lemma 7.3, we find

$$|S(A)| \ll P_1 P_2 P_1^{-1/(30n)} \ll P_1 P_2 P_1^{-\rho}.$$

We have estimated the sum S in the case $\ln P_2 \le 1.2(n+1) \ln P_1$. Therefore, we assume that $\ln P_2 > 1.2(n+1) \ln P_1$.

We introduce the notation

$$\{f_s(x_2)\} = \frac{b}{r} + \beta, \quad (b, r) = 1, \quad 1 \le r \le \tau_1(s), \tag{7.21}$$

$$\tau_1(s) = P_1^{s-1/6}, \quad |\beta| \le \left(r\tau_1(s)\right)^{-1}, \quad \delta = P_1^s|\beta|.$$

We divide the sum S into three sums:

$$S = S_{13} + S_{14} + S_{15},$$

where

$$S_j = \sum_{x_2 \le P_2}^{(j)} \sum_{x_1 \le P_1} \exp\{2\pi i F(x_1, x_2)\}, \quad j = 13, 14, 15,$$

and the domain of summation over the variable x_2 in each of the sums is its own and is determined as follows. We consider representation (7.17) of the inner sum over x_1. If the point $(A_1, \dots, A_{n_1}) = (\{f_1(x_2)\}, \dots, \{f_{n_1}(x_2)\})$ is a point of the second class with respect to the parameter P_1 defined in Lemma 7.1 (a), then the corresponding x_2 belongs to the sum S_{13}. If this point is a point of the first class, and moreover, $r > K = P_1^{2n\rho}$ or $\delta > K$ in representation (7.21), then the corresponding x_2 belongs to the sum S_{14}. Finally, all the other x_2 belong to the sum S_{15}. In other words, the sum S_{15} contains all x_2 for which the denominator r and the value of δ in representation (7.21) do not exceed K.

For the sums $S(x_2)$ contained in S_{13}, by Lemma 7.1 (a), we have the estimate

$$|S(x_2)| \ll P_1^{1-\rho_1}, \quad \rho_1 = c_1(n^2 \ln n)^{-1};$$

this implies

$$|S_{13}| \ll P_1 P_2 P_1^{-\rho}.$$

Now we consider the sum S_{14}. If r is larger than $P_1^{0.1}$, then the least common multiple of the denominators of the fractions in the D-approximations of the coordinates A_ν of the point $(A_1, \dots, A_{n_1})$ corresponding to $\tau_1(\nu)$ is larger than $P_1^{0.1}$ and, for the sums $S(x_2)$ contained in S_{14}, by Lemma 7.1 (b), we have the estimate

$$|S(x_2)| \ll P_1^{1-\rho_1}.$$

Now we assume that r does not exceed $P_1^{0.1}$. For the sums $S(x_2)$ contained in S_{14}, the point $(A_1, \dots, A_{n_1}) = (\{f_1(x_2)\}, \dots, \{f_{n_1}(x_2)\})$ belongs to the first class. This means that

$$A_\nu = \frac{a_\nu}{q_\nu} + \beta_\nu, \quad |\beta_\nu| \le P_1^{-\nu+0.1}, \quad (a_\nu, q_\nu) = 1,$$
$$1 \le \nu \le n_1, \quad q = [q_1, \dots, q_{n_1}] < P_1^{0.1}.$$

Thus, as in the case of the sum S_5, we obtain $a_s/q_s = b/r$ and hence $q \ge q_s = r$ and $\beta_s = \beta$. Therefore, by the definition of the sum S_{14}, we have $q > K$ or $|\beta_s|P_1^s = \delta > K$.

To estimate the sum $S(x_2)$ contained in S_{14}, we use Lemma 7.1 (a) and obtain

$$|S(x_2)| \ll P_1 K^{-1/n+\varepsilon} \ll P_1^{1-\rho}.$$

Thus for all sums $S(x_2)$ in S_{14}, we have the estimate $|S(x_2)| \ll P_1^{1-\rho}$ and hence $|S_{14}| \ll P_1 P_2 P_1^{-\rho}$.

Now we consider the sum S_{15}. We trivially estimate each of the sums $S(x_2)$ contained in S_{15} and obtain $|S_{15}| \le Y P_1$, where Y is the number of values of x_2 $(1 \le x_2 \le P_2)$ for which the fractional part of the polynomial $f_s(x_2)$ belongs to at least one of the intervals of the form $[b/r - P_1^{-s+2\rho n}, b/r + P_1^{-s+2\rho n}]$, where $(b,r) = 1$ and $r \le K = P_1^{2\rho n}$.

Since $\ln P_2 \ge 1.2(n+1)\ln P_1$ and $\rho \le 0.02 n^{-2}$, by Lemma 7.5, we have $Y \ll P_2 P_1^{-\rho}$. Hence $|S_{15}| \ll P_1 P_2 P_1^{-\rho}$.

Hence for $Q_1(s) < H_2$, we obtain

$$|S| \le |S_{13}| + |S_{14}| + |S_{15}| \ll P_1 P_2 P_1^{-\rho};$$

thus the statement of the lemma is proved for $Q_0 > P_1^{10n^3\rho}$.

6. Now we consider the case $Q_0 \le P_1^{10n^3\rho}$. Since $Q = [Q_0, Q_1, Q_2] > P_1^{0.1}$, we have either $Q_1 > P_1^{0.05-5n^3\rho}$ or $Q_2 > P_1^{0.05-5n^3\rho}$.

First, we assume that $Q_1 > P_1^{0.05-5n^3\rho}$. Then we represent x_1 and x_2 as

$$x_1 = Q_0 y_1 + z_1, \quad 0 < z_1 \le Q_0, \quad (1 - z_1)Q_0^{-1} < y_1 \le (P_1 - z_1)Q_0^{-1},$$
$$x_2 = Q_0 y_2 + z_2, \quad 0 < z_2 \le Q_0, \quad (1 - z_2)Q_0^{-1} < y_2 \le (P_2 - z_2)Q_0^{-1}.$$

We write the D-approximations of the coefficients $\alpha(t_1,t_2)$ of the polynomial $F(x_1,x_2)$ corresponding to $\tau(t_1,t_2) = P_1^{t_1-1/6}P_2^{t_2}$:

$$\alpha(t_1,t_2) = \frac{a(t_1,t_2)}{q(t_1,t_2)} + \beta(t_1,t_2), \quad \big(a(t_1,t_2), q(t_1,t_2)\big) = 1,$$
$$1 \le q(t_1,t_2) \le \tau(t_1,t_2), \quad |\beta(t_1,t_2)| \le \big(q(t_1,t_2)\tau(t_1,t_2)\big)^{-1}.$$

Hence the polynomial $F(x_1,x_2)$ can be represented as

$$F(Q_0y_1 + z_1, Q_0y_2 + z_2) = \sum_{t_1=0}^{n_1}\sum_{t_2=0}^{n_2} \alpha(t_1,t_2)(Q_0y_1+z_1)^{t_1}(Q_0y_2+z_2)^{t_2}$$
$$\equiv \Phi(z_1,z_2) + \Psi_1(Q_0y_1+z_1) + \Psi(Q_0y_1+z_1, Q_0y_2+z_2) \pmod 1,$$

where

$$\Phi(z_1,z_2) = \sum_{t_1=0}^{n_1}\sum_{t_2=0}^{n_2} \frac{a(t_1,t_2)}{q(t_1,t_2)} z_1^{t_1} z_2^{t_2}, \quad \Psi_1(x) = \sum_{t_1=0}^{n_1} \alpha(t_1,0)x^{t_1},$$
$$\Psi(x_1,x_2) = \sum_{t_2=0}^{n_2} x_2^{t_2}\left(\frac{a(0,t_2)}{q(0,t_2)} + \sum_{t_1=0}^{n_1} \beta(t_1,t_2)x_1^{t_1}\right).$$

We transform the trigonometric sum S as follows:

$$\begin{aligned} S &= \sum_{x_1=1}^{P_1}\sum_{x_2=1}^{P_2} \exp\{2\pi i F(x_1, x_2)\} \\ &= \sum_{z_1=1}^{Q_0}\sum_{y_1}\sum_{z_2=1}^{Q_0} \exp\{2\pi i(\Phi(z_1, z_2) + \Psi_1(z_1, z_2))\} \\ &\quad \times \sum_{y_2} \exp\{2\pi i \Psi(Q_0 y_1 + z_1, Q_0 y_2 + z_2)\}. \end{aligned}$$

The summation over y_1 is performed in the limits $-z_1 Q_0^{-1} < y_1 \le (P_1 - z_1)Q_0^{-1}$ and over y_2 in the limits $-z_2 Q_0^{-1} < y_2 \le (P_2 - z_2)Q_0^{-1}$. It follows from the last relation that

$$\begin{aligned} |S| &\le \sum_{z_1=1}^{Q_0}\sum_{y_1}\sum_{z_2=1}^{Q_0} \left| \sum_{y_2} \exp\{2\pi i \Psi(Q_0 y_1 + z_1, Q_0 y_2 + z_2)\} \right| \\ &= \sum_{x_1=1}^{P_1}\sum_{z_2=1}^{Q_0} \left| \sum_{y_2} \exp\{2\pi i \Psi(x_1, Q_0 y_2 + z_2)\} \right| = T_1. \end{aligned}$$

Next, we shall estimate the sum T_1. We write the polynomial $\Psi(x_1, x_2)$ as

$$\Psi(x_1, x_2) = \sum_{\nu=1}^{n_2} g_\nu(x_1) x_2^\nu = \sum_{\nu=1}^{n_2} B_\nu(x_1) x_2^\nu,$$

$$g_\nu(x_1) = \frac{a(0, \nu)}{q(0, \nu)} + \sum_{t_1=0}^{n_1} \beta(t_1, \nu) x_1^{t_1} = B_\nu.$$

We consider the D-approximations of the fractional parts of the polynomials $g_\nu(x_1)$ ($\nu = 1, \dots, n_2$) corresponding to $\tau_2(\nu) = P_2^{\nu - 1/6}$:

$$\{g_\nu(x_1)\} = \frac{a_\nu(x_1)}{q_\nu(x_1)} + \beta_\nu(x_1), \quad \big(a_\nu(x_1), q_\nu(x_1)\big) = 1,$$
$$1 \le q_\nu(x_1) \le \tau_2(\nu), \quad |\beta_\nu(x_1)| \le \big(q_\nu(x_1)\tau_2(\nu)\big)^{-1}.$$

By $Q_1(x_1)$ we denote the least common multiple of $q_1(x_1), \dots, q_{n_2}(x_1)$ and by δ the largest of $|\beta_\nu(x_1)| P_2^\nu$ $(1 \le \nu \le n_2)$. We divide the sum T_1 into three sums:

$$T_1 = S_{16} + S_{17} + S_{18},$$

where

$$S_j = {\sum_{x_1 \le P_1}}' \sum_{z_2=1}^{Q_0} |S(x_1, z_2)|, \quad j = 16, 17, 18,$$

$$S(x_1, z_2) = \sum_{y_2} \exp\{2\pi i \Psi(x_1, Q_0 y_2 + z_2)\},$$

the summation over the variable y_2 is performed within the limits $-z_2 Q_0^{-1} < y_2 \le (P_2 - z_2) Q_0^{-1}$, and the domain of summation over the variable x_1 in each of the sums S_{16}, S_{17}, S_{18} is its own and is determined as follows. If the point $(B_1, \dots, B_{n_2})$ is a point of the second class with respect to the parameter P_2, then the corresponding x_1 belongs to the sum S_{16}. If this point is a point of the first class, and $Q_1(x_1) \ge H = P_1^{20n^4\rho}$ or $\delta > P_1^{2n\rho}$, then the corresponding x_1 belongs to the sum S_{14}. Finally, all the other x_1 belong to the sum S_{18}. In other words, the sum S_{18} contains all x_1 for which $Q_1(x_1) < H = P_1^{20n^4\rho}$ and the inequality $|\beta_\nu(x_1)| P_2^\nu \le P_1^{2n\rho}$ holds for all ν $(1 \le \nu \le n_2)$.

For the sums $S(x_1, z_2)$ contained in S_{16}, by Lemma 7.9, item (1), we have the estimate

$$|S(x_1, z_2)| \ll P_2^{1-0.5\rho_1} Q_0^{-1} \ll P_2 Q_0^{-1} P_1^{-\rho}.$$

Hence

$$|S_{16}| \ll P_1 P_2 P_1^{-\rho}.$$

Now we pass to estimating the sum S_{17}. Since the point $(B_1, \dots, B_{n_2})$ belongs to the first class, we can represent its coordinates B_ν as

$$B_\nu = \frac{b_\nu}{r_\nu} + \beta_\nu, \quad (b_\nu, r_\nu) = 1, \quad |\beta_\nu| \le P_2^{-\nu+0.1}, \quad \nu = 1, \dots, n_2, \tag{7.22}$$

$$r = \lfloor r_1, \dots, r_{n_2} \rfloor \le P_2^{0.1}.$$

We first consider the case $Q_1(x_1) \le P_2^{0.1}$. Here we have $q_\nu(x_1) \ne r_\nu$ for some ν $(1 \le \nu \le n_2)$. This implies the inequalities

$$\frac{1}{q_\nu(x_1) r_\nu} \le \left| \frac{a_\nu(x_1)}{q_\nu(x_1)} - \frac{b_\nu}{r_\nu} \right| \le |\beta_\nu| + |\beta_\nu(x_1)| \le P_2^{-\nu+0.1} + q_\nu^{-1}(x_1) P_2^{-\nu+1/6},$$

$$r_\nu^{-1} \le q_\nu(x_1) P_2^{-\nu+0.1} + P_2^{-\nu+1/6} \le 2P_2^{-1/15}.$$

Hence we have $r \ge r_\nu \ge 0.5 P_2^{1/15}$. But if the inequalities $0.5 P_2^{1/15} \le r \le P_2^{0.07}$ hold for r, then, by Lemma 7.9, item (2), (a), we obtain

$$|S(x_1, z_2)| \ll P_2 Q_0^{-1} R^{-1/n+\varepsilon}, \qquad R = r/(r, Q_0^n).$$

Since

$$(r, Q_0^n) \le P_1^{10n^4\rho} \le P_1^{0.01}, \quad r \ge P_2^{0.05},$$

we can estimate our sum as

$$|S(x_1, z_2)| \ll P_2 Q_0^{-1} P_1^{-\rho}.$$

But if the inequality $r > P_2^{0.07}$ holds for r, then, by Lemma 7.9, item (3), we obtain the following estimate for the same sum:

$$S(x_1, z_2)| \ll P_2^{1-0.5\rho_1} Q_0^{-1} \ll P_2 Q_0^{-1} P_1^{-\rho}.$$

Now we consider the case $Q_1(x_1) < P_2^{0.1}$. Since the point $(B_1, \dots, B_n)$ belongs to the first class, relations (7.22) holds for this point. Then, as in the case of the sum S_{12}, the relations $a_\nu(x_1)/q_\nu(x_1) = b_\nu/r_\nu$ hold for all ν $(1 \le \nu \le n_2)$. Hence we have

$$Q_1(x_1) = r, \quad \beta_\nu(x_1) = \beta_\nu, \quad \delta_\nu = \beta_\nu P_2^\nu.$$

If $P_2^{0.07} \ge r = Q_1(x_1) \ge H = P_1^{20n^4\rho}$, then, by Lemma 7.9, item (2), (a), the sum $S(x_1, z_2)$ satisfies the estimate

$$|S(x_1, z_2)| \ll P_2 Q_0^{-1} R^{-1/n+\varepsilon}, \quad R = r/(r, Q_0^n).$$

Since

$$(r, Q_0^n) \le Q_0^n \le P_1^{10n^4\rho}, \quad r \ge P_1^{20n^4\rho},$$

then $R > P_2^{10n^4\rho}$ and hence

$$|S(x_1, z_2)| \ll P_2 Q_0^{-1} P_1^{-\rho}.$$

If $r > P_2^{0.07}$ or $\delta > P_2^{0.04}$, then, by Lemma 7.9, item (3), we have

$$|S(x_1, z_2)| \ll P_2^{1-0.5\rho_1} Q_0^{-1} \ll P_2 Q_0^{-1} P_1^{-\rho}.$$

If $P_1^{2n\rho} \le \delta \le P_2^{0.04}$, then, by Lemma 7.9, item (2), (b), we have the estimate

$$|S(x_1, z_2)| \ll P_2 Q_0^{-1} \delta^{-1/n+\varepsilon} \ll P_2 Q_0^{-1} P_1^{-\rho}.$$

Thus if $Q_1(x_1) \ge H$ or the value $|\beta_s(x_2)|$ is larger than $P_1^{-s} P_1^{2n\rho}$ for some s $(1 \le s \le n_2)$, then we have the estimate

$$|S(x_1, z_2)| \ll P_2 Q_0^{-1} P_1^{-\rho},$$

which implies

$$|S_{17}| \ll P_1 P_2 P_1^{-\rho}.$$

Now we estimate the sum S_{18}. We trivially estimate each of the sums $S(x_1, z_2)$ contained in S_{18} and obtain $|S_{18}| \le Y P_2$, where Y is the number of values of x_1 $(1 \le x_1 \le P_1)$ for which

$$|\beta_\nu(x_1)| \le \Delta_\nu = P_1^{-\nu} P_1^{2n\rho}, \quad 1 \le \nu \le n_2, \quad Q_1(x_1) \le H = P_1^{20n^4\rho},$$

and the point $(B_1, \dots, B_{n_2})$ belongs to the first class. By Lemma 7.6, we have $Y \ll P_1^{1-\rho}$. Hence,

$$|S_{18}| \ll P_1 P_2 P_1^{-\rho}, \quad |S| \le |S_{16}| + |S_{17}| + |S_{18}| \ll P_1 P_2 P_1^{-\rho}.$$

Thus the statement of the lemma is proved in the case $Q_1 > P_1^{0.05-5n^3\rho}$.

Now let $Q_2 > P_1^{0.05-5n^3\rho}$. As in the preceding case, we partition the variables of summation into arithmetic progressions with difference Q_0 and obtain

$$|S| \le \sum_{x_2=1}^{P_2} \sum_{z_1=1}^{Q_0} \left| \sum_{y_1} \exp\{2\pi i \Psi_1(Q_0 y_1 + z_1, x_2)\} \right| = T_2,$$

where

$$|\Psi_1(x_1, x_2) = \sum_{t_1=1}^{n_1} x_1^{t_1} \left(\frac{a(t_1, 0)}{q(t_1, 0)} + \sum_{t_2=0}^{n_2} \beta(t_1, t_2) x_2^{t_2} \right).$$

We write the polynomial $\Psi_1(x_1, x_2)$ in the form

$$\Psi_1(x_1, x_2) = \sum_{\nu=1}^{n_1} g_\nu(x_2) x_1^\nu = \sum_{\nu=1}^{n_1} B_\nu x_1^\nu,$$

$$B_\nu = g_\nu(x_2) = \frac{a(\nu, 0)}{q(\nu, 0)} + \sum_{t_2=0}^{n_2} \beta(\nu, t_2) x_2^{t_2},$$

and take the D-approximations of the fractional parts of $g_\nu(x_2)$ $(\nu = 1, \dots, n)$ corresponding to $\tau_1(\nu) = P_1^{\nu-1/6}$:

$$\{g_\nu(x_2)\} = \frac{a_\nu(x_2)}{q_\nu(x_2)} + \beta_\nu(x_2), \quad \big(a_\nu(x_2), q_\nu(x_2)\big) = 1,$$
$$1 \le q_\nu(x_2) \le \tau_1(\nu), \quad |\beta_\nu(x_2)| \le \big(q_\nu(x_2)\tau_1(\nu)\big) = 1.$$

By $Q_2(x_2)$ we denote the least common multiple of the numbers $q_1(x_2), \dots, q_{n_1}(x_2)$ and by δ the largest value of

$$\delta_\nu = |\beta_\nu(x_2)| P_1^\nu, \quad \nu = 1, \dots, n_1.$$

We divide the sum T_2 into three sums:

$$T_2 = S_{19} + S_{20} + S_{21}, \tag{7.23}$$

where

$$S_j = \sum_{x_2}^{(j)} \sum_{z_1=1}^{Q_0} |S(z_1, x_2)|, \quad j = 19, 20, 21,$$

$$S(z_1, x_2) = \sum_{y_1} \exp\{2\pi i \Psi(Q_0 y_1 + z_1, x_2)\},$$

the summation over the variable y_1 is performed within the limits $(1 - z_1)Q_0^{-1} < y_1 \le (P_1 - z_1)Q_0^{-1}$, and the domain of summation over the variable x_2 in each of the sums S_{19}, S_{20}, S_{21} is its own and is determined as follows. If the point $(B_1, \dots, B_{n_1})$ is a point of the second class with respect to the parameter P_1, then the corresponding x_2 belongs to the sum S_{19}. If this point is a point of the first class, and $Q_2(x_2) \ge H = P_1^{20n^4\rho}$ or $\delta > P_1^{2n\rho}$, then the corresponding x_2 belongs to the sum S_{20}. Finally, all the other x_2 belong to the sum S_{21}. In other words, the sum S_{21} contains all x_2 for which $Q_2(x_2) < H = P_1^{20n^4\rho}$ and the inequality

$$|\beta_s(x_2)|P_1^s < P_1^{2n\rho}$$

holds for all s $(1 \le s \le n_1)$.

Each of the sums in expression (7.23) can be estimated in the same way as the corresponding sum in the preceding case, only in estimating the sum S_{21} we apply Lemma 7.7 instead of Lemma 7.6 used to estimate S_{17}. The proof of the main lemma is complete. □

7.1.4 Estimate for the double trigonometric sum

In this section we estimate the double trigonometric sum $S(A)$ for all points A of the unit m-dimensional cube Ω.

Theorem 7.2. *Suppose that a point A belongs to the first class Ω_1. Then the following estimate holds*:

$$|S(A)| \ll P_1 P_2 Q^{-1/n+\varepsilon}.$$

Moreover, if we set

$$\delta(t_1, t_2) = P_1^{t_1} P_2^{t_2} \beta(t_1, t_2), \quad \delta = \max_{t_1,t_2} |\delta(t_1, t_2)|,$$

then the estimate

$$|S(A)| \ll P_1 P_2 (Q\delta)^{-1/n+\varepsilon}$$

holds for $\delta > 1$.

Suppose that a point A belongs to the second class Ω_2. Then the following estimate holds:

$$|S(A)| \ll P_1 P_2 P_1^{-\rho}, \quad \rho = c(n^4 \ln n)^{-1}.$$

The constants in $\ll$ depend only on n and ε.

Proof. For points A of the first class Ω_1, the statement of the theorem follows from Lemma 7.3. We prove this theorem for points of the second class Ω_2. We consider the D-approximations of coefficients $\alpha(t_1, t_2)$ of the polynomial $F(x_1, x_2)$ corresponding to $\tau(t_1, t_2) = P_1^{t_1-1/6} P_2^{t_2}$, i.e., we consider the relations

$$\alpha(t_1, t_2) = \frac{a_0(t_1, t_2)}{q_0(t_1, t_2)} + \beta(t_1, t_2), \quad \big(a_0(t_1, t_2), q_0(t_1, t_2)\big) = 1,$$
$$1 \le q_0(t_1, t_2) \le \tau(t_1, t_2), \quad |\beta_0(t_1, t_2)| \le \big(q_0(t_1, t_2)\tau(t_1, t_2)\big)^{-1},$$
$$0 \le t_1 \le n_1, \quad 0 \le t_2 \le n_2, \quad t_1 + t_2 \ge 1.$$

By q_0 we denote the least common multiple of the numbers $q_0(t_1, t_2)$ $(0 \le t_1 \le n_1, 0 \le t_2 \le n_2, t_1 + t_2 \ge 1)$ and by δ_0 the value

$$\delta_0 = \max_{t_1, t_2} |\beta_0(t_1, t_2)| P_1^{t_1} P_2^{t_2}.$$

By condition, A is a point of the second class. This means that either q_0 or δ_0 are large; more precisely, the following two cases are possible:

(a) $q_0 \ge P_1^{0.1}$;

(b) $q_0 < P_1^{0.1}$ and $\delta_0 \ge P_1^{0.1}$.

First, we consider case (a). Since $q_0 \ge P_1^{0.1}$, the first main lemma implies the following estimate for the sums $S = S(A)$:

$$|S| \ll P_1 P_2 P_1^{-\rho}, \quad \rho = c(n^4 \ln n)^{-1}.$$

Now we consider case (b). In this case, the sum $S(A)$ can be estimated similarly to the sum $S(A)$ for points A of the first class (see the proof of Lemma 5.5 in Chapter 5).

To this end, we partition the summation over the variables x_1 and x_2 in the trigonometric sum $S(A)$ into arithmetic progressions with difference q_0.

It turns out that, with a good accuracy, the parts of the sum $S(A)$ corresponding to the same progressions can be replaced by trigonometric integrals. Next, after simpler transformations, we see that the sum $S(A)$ can be approximated well by the product of the trigonometric series by a double complete rational trigonometric sum with denominator q_0. Now to obtain the desired estimate of the sum $S(A)$, it suffices to estimate the trigonometric integral. Let us follow the above reasoning.

We partition the full summation over x_1 and x_2 into progressions with difference q_0, i.e., we perform a change of the summation variables of the form

$$x_j = q_0 y_j + z_j, \quad 0 < z_h j \le q_0,$$
$$-z_j q_0^{-1} < y_j \le (P_j - z_j) q_0^{-1}, \quad j = 1, 2.$$

Then we can write the sum $S = S(A)$ as

$$S(A) = \sum_{z_1=1}^{q_0} \sum_{z_2=1}^{q_0} \exp\{2\pi i \Phi(z_1, z_2)\} W(z_1, z_2),$$

where

$$\Phi(z_1, z_2) = \sum_{t_1=0}^{n_1} \sum_{t_2=0}^{n_2} \frac{a_0(t_1, t_2)}{q_0(t_1, t_2)} z_1^{t_1} z_2^{t_2},$$

$$W(z_1, z_2) = \sum_{y_1} \sum_{y_2} \exp\{2\pi i F_\beta(q_0 y_1 + z_1, q_0 y_2 + z_2)\},$$

$$F_\beta(x_1, x_2) = \sum_{t_1=0}^{n_1} \sum_{t_2=0}^{n_2} \beta_0(t_1, t_2) x_1^{t_1} x_2^{t_2}.$$

We replace the sum $W(z_1, z_2)$ by the trigonometric integral. For this, we first estimate from above the partial derivatives of the polynomial $F_\beta(q_0y_1+z_1, q_0y_2+z_2)$ with respect to the variables y_1 and y_2. We obtain

$$\left|\frac{\partial F_\beta(q_0 y_1 + z_1, q_0 y_2 + z_2)}{\partial y_1}\right|$$
$$= \left|\sum_{t_1=1}^{n_1} \sum_{t_2=0}^{n_2} \beta_0(t_1, t_2) t_1 q_0 (q_0 y_1 + z_1)^{t_1-1} (q_0 y_2 + z_2)^{t_2}\right|$$
$$\le \sum_{t_1=1}^{n_1} \sum_{t_2=0}^{n_2} q_0^{-1}(t_1, t_2) P_1^{-t_1+1/6} P_2^{-t_2} t_1 q_0 P_1^{t_1-1} P_2^{t_2}$$
$$\le (n+1)^3 q_0 P_1^{-5/6} \le 0.5, \qquad \left|\frac{\partial F_\beta}{\partial y_2}\right| \le 0.5.$$

Next, applying Lemma 5.4 (Chapter 5) to the sum $W(z_1, z_2)$, we obtain

$$W(z_1, z_2) = \int_{-z_1q_0^{-1}}^{(P_1-z_1)q_0^{-1}} \int_{-z_2q_0^{-1}}^{(P_2-z_2)q_0^{-1}} \exp\{2\pi i F_\beta(q_0 y_1 + z_1, q_0 y_2 + z_2)\}\, dy_1 dy_2$$
$$+O(P_2 q_0^{-1}) = q_0^{-2} \int_0^{P_1} \int_0^{P_2} \exp\{2\pi i F_\beta(x_1, x_2)\}\, dx_1 dx_2 + O(P_2 q_0^{-1})$$
$$= P_1 P_2 q_0^{-2} \int_0^1 \int_0^1 \exp\{2\pi i F_\delta(x_1, x_2)\}\, dx_1 dx_2 + O(P_2 q_0^{-1}),$$

$$F_\beta(x_1, x_2) = \sum_{t_1=0}^{n_1} \sum_{t_2=0}^{n_2} \delta_0(t_1, t_2) x_1^{t_1} x_2^{t_2}, \qquad \delta(t_1, t_2) = P_1^{t_1} P_2^{t_2} \beta_0(t_1, t_2).$$

Now, estimating the trigonometric integral

$$I = \int_0^1 \int_0^1 \exp\{2\pi i F_\delta(x_1, x_2)\}\, dx_1 dx_2$$

by Theorem 1.6 (Chapter 1), we obtain $|I| \le 32^r \delta_0^{-1/n} \ln^{r-1}(\delta_0 + 2) \ll P_1^{-1/(20n)}$ because we have $\delta_0 \ge P_1^{0.1}$ in case (b). Hence

$$|W(z_1, z_2)| \ll P_1 P_2 q_0^{-2} P_1^{-1/(20n)}.$$

This implies the estimate

$$|S| \leq \sum_{z_1=1}^{q_0} \sum_{z_2=1}^{q_0} |W(z_1, z_2)| \ll P_1 P_2 P_1^{-\rho}$$

for the sum S. The proof of the theorem is complete. □

7.2 r-fold trigonometric sums

In the preceding section, we estimated the double trigonometric sum for all points of the unit cube Ω. Our further goal is to derive a similar estimate for the r-fold sum for any natural number r. For this, we need several auxiliary lemmas generalizing the statements of Sections 7.1.1 and 7.1.2 to the case of sums of arbitrary multiplicity. It should be noted that many points in the proofs of the lemmas coincide for multiple and double sums. To avoid repetitions, we here, if possible, will refer to the corresponding argument in the proofs of similar statements in Sections 7.1.1 and 7.1.2.

7.2.1 Auxiliary lemmas

We introduce new notions and notation. Let $n_1, \dots, n_r, P_1, \dots, P_r$ be natural numbers: $P_1 \leq P_2 \leq \dots \leq P_r$, $m = (n_1+1)\dots(n_r+1)$, $n = \max(n_1, \dots, n_r)$, $n \geq 2$; let

$$\tau(t_1, \dots, t_t) = P_1^{t_1 - 1/6} P_2^{t_2} \dots P_r^{t_r},$$

where $0 \leq t_1 \leq n_1, \dots, 0 \leq t_r \leq n_r$, $t_1 + \dots + t_r \geq 1$.

Now by $\Omega = \Omega(r)$ we denote the unit cube in the m-dimensional Euclidean space, $m = m_r$. Suppose that the coordinates $\alpha(t_1, \dots, t_r)$ of points A of this cube are determined by the conditions

$$0 \leq \alpha(0, \dots, 0) < 1, \quad -\tau^{-1}(t_1, \dots, t_r) \leq \alpha(t_1, \dots, t_r) < 1 - \tau^{-1}(t_1, \dots, t_r),$$
$$0 \leq t_1 \leq n_1, \dots, 0 \leq t_r \leq n_r, \quad t_1 + \dots + t_r \geq 1.$$

A *multiple* or *r-fold trigonometric sum* is defined to be the sum

$$S = S(A) = \sum_{x_1=1}^{P_1} \dots \sum_{x_r=1}^{P_r} \exp\{2\pi i F(x_1, \dots, x_r)\},$$

where

$$F(x_1, \dots, x_r) = \sum_{t_1=0}^{n_1} \dots \sum_{t_r=0}^{n_r} \alpha(t_1, \dots, t_r) x_1^{t_1} \dots x_r^{t_r};$$

the coordinates of A are the coefficients of the polynomial $F(x_1, \dots, x_r)$. Since the modulus of the sum $S = S(A)$ is independent of the value of the constant term $\alpha(0, \dots, 0)$ in the polynomial $F(x_1, \dots, x_r)$, we set it to be zero in what follows.

We divide the points of the cube Ω into two classes Ω_1 and Ω_2. The first class Ω_1 contains points A whose coordinates satisfy the conditions:

(a) $\quad \alpha(t_1,\dots,t_r) = \dfrac{a(t_1,\dots,t_r)}{q(t_1,\dots,t_r)} + \beta(t_1,\dots,t_r),$

$\big(\alpha(t_1,\dots,t_r), q(t_1,\dots,t_r)\big) = 1, \quad 0 \le \alpha(t_1,\dots,t_r) < q(t_1,\dots,t_r),$

$\quad |\beta(t_1,\dots,t_r)| \le P_1^{-t_1+0.1} P_2^{-t_2} \dots P_r^{-t_r};$

(b) the least common multiple Q of all $q(t_1,\dots,t_2)$ does not exceed $P_1^{0.1}$.

The second class Ω_2 contains the other points of the cube Ω. In what follows, we obtain a uniform estimate for the trigonometric sum $S(A)$ on points of the second class. We note that the derivation of this estimate splits into two significantly different cases depending on the D-approximations of the coordinates of the point A. According to this, the class Ω_2 splits into two domains ω_1 and ω_2. Let us determine them.

We consider the D-approximations of the coordinates $\alpha(t_1,\dots,t_r)$ of a point $A \in \Omega_2$ corresponding to $\tau(t_1,\dots,t_r)$, i.e.,

$$\alpha(t_1,\dots,t_r) = \frac{a(t_1,\dots,t_r)}{q(t_1,\dots,t_r)} + \beta(t_1,\dots,t_r)$$
$$\big(\alpha(t_1,\dots,t_r), q(t_1,\dots,t_r)\big) = 1,$$
$$0 \le \alpha(t_1,\dots,t_r) < q(t_1,\dots,t_r) \le \tau(t_1,\dots,t_r),$$
$$|\beta(t_1,\dots,t_r)| \le \big(q(t_1,\dots,t_r), \tau(t_1,\dots,t_r)\big)^{-1},$$
$$0 \le t_1 \le n_1, \dots, 0 \le t_r \le n_r, \quad t_1 + \dots + t_r \ge 1.$$

Let Q be the least common multiple of all the numbers $q(t_1,\dots,t_r)$ and let

$$\delta(t_1,\dots,t_r) = \beta(t_1,\dots,t_r) P_1^{t_1} \dots P_r^{t_r}.$$

We set $\delta = \max_{t_1,\dots,t_r} |\delta(t_1,\dots,t_r)|$. A point A belongs to the domain ω_2 if $Q \ge P_1^{0.1}$ and to the domain ω_1 if $Q < P_1^{0.1}$, but $\delta \ge P_1^{0.1}$. We note that the other points of the cube Ω form exactly the first class Ω_1.

Lemma 7.10. *The points A of the first class satisfy the estimates*:

(a) $|S(A)| \ll P_1 \dots P_r Q^{-1/n+\varepsilon}$;

(b) $|S(A)| \ll P_1 \dots P_r (Q\delta)^{-1/n+\varepsilon}$ *for* $\delta \ge 1$;

the constants in $\ll$ *depend only on* n, r, *and* ε.

Proof. This lemma in a somewhat more precise statement was proved in Chapter 5 (Lemma 5.5). □

Lemma 7.11. *The points A from the domain* ω_1 *satisfy the estimate*

$$|S(A)| \ll P_1 \dots P_r \delta^{-1/n+\varepsilon} \ll P_1 \dots P_r P_1^{-0.05n^{-1}};$$

the constant in $\ll$ *depends only on* n, r, *and* ε.

Proof. This lemma is, in fact, similar to that in item (b) of the preceding Lemma 7.10. Its proof is a word for word repetition of the argument in Lemma 5.5 (Chapter 5). □

Lemma 7.12. *Suppose that a point A belongs to the domain ω_2. For $s = 2, \dots, r$, by ν_s we denote the natural number from the interval*

$$-1 < \frac{\ln P_s}{\ln P_1} - \nu_s \le 0.$$

Let $\varkappa = n_1 + \nu_2 n_2 + \cdots + \nu_r n_r$. Then the sum $S(A)$ satisfies the estimate

$$|S(A)| \ll \exp\{32\varkappa\} P_1 \dots P_r P_1^{-\rho}, \quad \rho = c(m\varkappa \log m\nu k)^{-1};$$

the constant in $\ll$ depends only on n and r.

Proof. This statement follows from Theorem 5.2 in Chapter 5. □

Lemma 7.13. *Let $D(\sigma)$ be the number of integer-valued sets $(x_1, \dots, x_r)$ satisfying the conditions*

$$\{F(x_1, \dots, x_r)\} < \sigma, \quad 1 \le x_1 \le P_1, \ \dots, \ 1 \le x_r \le P_r.$$

We represent $D(\sigma)$ in the form

$$D(\sigma) = \sigma P_1 \dots P_r + \lambda(\sigma).$$

Then in the notation of Lemmas 7.10 and 7.12, the following estimates hold:

(1) *If the point A belongs to the first class Ω_1, then*

(a) $|\lambda(\sigma)| \ll P_1 \dots P_r Q^{-1/n+\varepsilon}$;

(b) $|S(A)| \ll P_1 \dots P_r (Q\delta)^{-1/n+\varepsilon}$ *for* $\delta > 1$.

(2) *If the point A belongs to the domain ω_1 from Ω_1, then*

$$|\lambda(\sigma)| \ll P_1 \dots P_r Q^{-1/n+\varepsilon} \ll P_1 \dots P_r P_1^{-0.05n^{-1}}.$$

(3) *If the point A belongs to the domain ω_2 from Ω_2, then*

$$|\lambda(\sigma)| \ll P_1 \dots P_r P_1^{-\rho} \exp\{32\varkappa\}.$$

The constants in $\ll$ depend only on n, r, and ε.

Proof. The statement of the lemma follows from Theorem 6.5 in Chapter 6 for $s = 1$. □

Lemmas 7.14–7.18 given below generalize the lemmas given in Section 7.1.2 to the case of polynomials in arbitrarily many variables.

We introduce the notation. Let s be a natural number, $s < r$. We set

$$F(x_1, \dots, x_r) = \sum_{t_1=0}^{n_1} \cdots \sum_{t_s=0}^{n_s} g_{t_1,\dots,t_s}(x_{s+1}, \dots, x_r) x_1^{t_1} \dots x_s^{t_s}.$$

Hence, for $0 \le t_1 \le n_1, \dots, 0 \le t_s \le n_s$, we have

$$g_{t_1,\dots,t_s}(x_{s+1}, \dots, x_r) = \sum_{t_{s+1}=0}^{n_{s+1}} \cdots \sum_{t_r=0}^{n_r} \alpha(t_1, \dots, t_r) x_{s+1}^{t_{s+1}} \dots x_r^{t_r}.$$

Suppose that the number s is determined by the relations

$$\begin{aligned} P_r^{5/6} &\le P_1^{n_1+1} \dots P_{r-1}^{n_{r-1}}, \\ &\vdots \\ P_{s+2}^{5/6} &\le P_1^{n_1+1} \dots P_{s+1}^{n_{s+1}}, \\ P_{s+1}^{5/6} &\le P_1^{n_1+1} \dots P_s^{n_s}. \end{aligned} \tag{7.24}$$

We have already found the numbers

$$\tau = \tau(t_1, \dots, t_r) = P_1^{t_1-1/6} P_2^{t_1} \dots P_r^{t_r}.$$

Now we determine the numbers

$$\eta = \eta(t_{s+1}, \dots, t_r) = P_{s+1}^{t_{s+1}-1/6} P_{s+2}^{t_{s+2}} \dots P_r^{t_r}.$$

and consider two D-approximations for each of the numbers $\alpha(t_1, \dots, t_r)$ respectively corresponding to $\tau(t_1, \dots, t_r)$ and $\eta(t_{s+1}, \dots, t_r)$:

$$\alpha(t_1, \dots, t_r) = \frac{a(t_1, \dots, t_r)}{q(t_1, \dots, t_r)} + \beta(t_1, \dots, t_r) = \frac{a_0(t_1, \dots, t_r)}{q_0(t_1, \dots, t_r)} + \beta_0(t_1, \dots, t_r)$$

where

$$\begin{aligned} |\beta(t_1, \dots, t_r)| &\le \big(q(t_1, \dots, t_r)\tau(t_1, \dots, t_r)\big)^{-1} = (q\tau)^{-1}, \\ |\beta_0(t_1, \dots, t_r)| &\le \big(q_0(t_1, \dots, t_r)\eta(t_{s+1}, \dots, t_r)\big)^{-1} = (q_0\eta)^{-1}. \end{aligned}$$

Let $Q_0 = Q_0(t_1, \dots, t_r)$ be the least common multiple of the numbers $q_0 = q_0(t_1, \dots, t_r)$ under the conditions $0 \le t_{s+1} \le n_{s+1}, \dots, 0 \le t_r \le n_r$, $t_{s+1} + \dots + t_r \ge 1$.

Lemma 7.14. *Suppose that, for some set $(t_1, \dots, t_s)$, the value of Q_0 does not exceed $H_0 = P_1^{10n^2\rho}$ and, for the same set, there is a set $(t_1, \dots, t_s, t_{s+1}, \dots, t_r)$ such that $\eta < q \le \tau$.*

We set $\Delta = P_1^{-t_1} \dots P_s^{-t_s} P_1^{2n\rho}$ *and consider all intervals of the form*

$$\left[\frac{A}{B} - \Delta, \frac{A}{B} + \Delta\right], \tag{7.25}$$

where A and B are integers,

$$0 \le A, \quad 1 \le B \le H_1 = P_1^{2n\rho}, \quad (A, B) = 1.$$

Let Y denote the number of sets $(x_{s+1}, \dots, x_r)$ *for which the fractional parts of the polynomial g belong to one of the intervals* (7.25) *for* $1 \le x_{s+1} \le P_{s+1}, \dots, 1 \le x_r \le P_r$. *Then for* $\rho \le 0.02n^{-2}$ *we have*

$$Y \ll P_{s+1} \dots P_r P_1^{-\rho}.$$

Proof. Repeating the beginning of the proof of Lemma 7.5 word for word, we obtain the inequality

$$Y \le Z = \sum_{\mu \le \Phi(H_1)} \sum_{x_{s+1} \le P_{s+1}} \cdots \sum_{x_r \le P_r} \chi_\mu\big(g(x_{s+1}, \dots, x_r)\big).$$

From expansion (7.6) of the function $\chi(x)$ into the Fourier series, we have

$$Z = P_{s+1} \dots P_r \Phi(H_1)\Delta + Z_1,$$

where

$$Z_1 = \sum_{\mu \le \Phi(H_1)} \sum_{m=-\infty}^{+\infty}{}' c(m) \times \sum_{x_{s+1} \le P_{s+1}} \cdots \sum_{x_r \le P_r} \exp\{2\pi i m s_\mu\} \exp\{2\pi i m g(x_{s+1}, \dots, x_r)\}.$$

Suppose that, as in Lemma 7.5, $M = \Delta^{-1}$, $M_1 = \Delta^{-1} H_1^2 P_1^\rho$, and $T(m)$ is a multiple trigonometric sum,

$$T(m) = \sum_{x_{s+1} \le P_{s+1}} \cdots \sum_{x_r \le P_r} \exp\{2\pi i m g(x_{s+1}, \dots, x_r)\}.$$

Then, using the estimates for the Fourier coefficients $c(m)$, we obtain

$$|Z_1| \le \Phi(H_1)\Delta \sum_{1 \le m < M} |T(m)| + \Phi(H_1)\Delta^{-1} \sum_{M \le m < M_1} m^{-2}|T(m)| + P_{s+1} \dots P_r P_1^{-\rho}. \tag{7.26}$$

Now we estimate the sum $T(m)$. By the assumptions of the lemma, the value of Q_0 does not exceed H_0. We divide the summation interval for each variable $x_{s+1}, \dots, x_r$ into arithmetic progressions with difference Q_0. We obtain

$$\begin{gathered} x_\nu = Q_0 u_\nu + v_\nu, \quad 1 \le v_\nu \le Q_0, \\ (1 - v_\nu) Q_0^{-1} \le u_\nu \le (P_\nu - v_\nu) Q_0^{-1}, \quad s < \nu \le r. \end{gathered} \tag{7.27}$$

The polynomial $g(x_{s+1}, \dots, x_r) = g_{t_1,\dots,t_s}(x_{s+1}, \dots, x_r)$ satisfies the relation

$$\begin{aligned} & g(Q_0 u_{s+1} + v_{s+1}, \dots, Q_0 u_r + v_r) \equiv F(v_{s+1}, \dots, v_r) \\ & \quad + G(Q_0 u_{s+1} + v_{s+1}, \dots, Q_0 u_r + v_r) + \alpha(t_1, \dots, t_s, 0, \dots, 0) \pmod 1, \\ & F(v_{s+1}, \dots, v_r) = \sum_{\substack{t_{s+1}=0 \\ t_{s+1}+\dots+t_r \ge 1}}^{n_{s+1}} \cdots \sum_{t_r=0}^{n_r} \frac{a_0(t_{s+1}, \dots, t_r)}{q_0(t_{s+1}, \dots, t_r)} v_{s+1}^{t_{s+1}} \dots v_r^{t_r}, \\ & G(y_{s+1}, \dots, y_r) = \sum_{\substack{t_{s+1}=0 \\ t_{s+1}+\dots+t_r \ge 1}}^{n_{s+1}} \cdots \sum_{t_r=0}^{n_r} \beta_0(t_{s+1}, \dots, t_r) y_{s+1}^{t_{s+1}} \dots y_r^{t_r}. \end{aligned}$$

Hence we have the estimate $|T(m)| \le Q^{r-s} |T_1(m)|$, where

$$T_1(m) = \sum_{u_{s+1}} \cdots \sum_{u_r} \exp\{2\pi i m G(Q_0 u_{s+1} + v_{s+1}, \dots, Q_0 u_r + v_r)\};$$

here the summation over $u_{s+1}, \dots, u_r$ is taken within the limits given in (7.27).

We estimate from above the absolute value of the partial derivatives of the polynomial $mG(Q_0 u_{s+1} + v_{s+1}, \dots, Q_0 u_r + v_r)$ with respect to u_ν $(s < \nu \le r)$ for $m \le M$. Using inequalities (7.24), we obtain

$$\begin{aligned} & \left| m \frac{\partial}{\partial u_\nu} G(Q_0 u_{s+1} + v_{s+1}, \dots, Q_0 u_r + v_r) \right| \\ & \quad = \left| m \sum_{t_{s+1}=0}^{n_{s+1}} \cdots \sum_{t_\nu=1}^{n_\nu} \cdots \sum_{t_r=0}^{n_r} t_\nu \beta_0(t_{s+1}, \dots, t_r) \right. \\ & \qquad \left. \times (Q_0 u_{s+1} + v_{s+1})^{t_{s+1}} \dots (Q_0 u_\nu + v_\nu)^{t_\nu - 1} \dots (Q_0 u_r + v_r)^{t_r} Q_0 \right| \\ & \quad \le M_1 \sum_{t_{s+1}=0}^{n_{s+1}} \cdots \sum_{t_\nu=1}^{n_\nu} \cdots \sum_{t_r=0}^{n_r} t_\nu \eta^{-1}(t_{s+1}, \dots, t_r) P_{s+1}^{t_{s+1}} \dots P_\nu^{t_\nu - 1} Q_0 \\ & \quad \le \frac{m_r n_\nu}{2 m_s} M_1 P_{s+1}^{1/6} P_\nu^{-1} Q_0 \le 0.5. \end{aligned}$$

Hence, following Lemma 5.4 (Chapter 5), we can replace the sum $T_1(m)$ by the integral as follows:

$$T_1(m) = \int_{-v_{s+1}Q_0^{-1}}^{(P_{s+1}-v_{s+1})Q_0^{-1}} \dots \int_{-v_r Q_0^{-1}}^{(P_r-v_r)Q_0^{-1}} \exp\{2\pi i m G(Q_0 u_{s+1} + v_{s+1}, \dots,$$

$$\dots, Q_0 u_r + v_r)\}\, du_{s+1} \dots du_r + O(P_{s+2}Q_0^{-1} \dots P_r Q_0^{-1})$$

$$= Q_0^{-r+s} \int_0^{P_{s+1}} \dots \int_0^{P_r} \exp\{2\pi i m G(u_{s+1}, \dots, u_r)\}\, du_{s+1} \dots du_r$$

$$+ O(P_{s+2} \dots P_r Q_0^{-r+s+1}).$$

Thus we obtain the estimate

$$|T(m)| \le \left| \int_0^{P_{s+1}} \dots \int_0^{P_r} \exp\{2\pi i m G(u_{s+1}, \dots, u_r)\}\, du_{s+1} \dots du_r \right|$$

$$+ c_1 P_{s+1} \dots P_r Q_0.$$

We perform a change of variables. By setting $u_{s+1} = P_{s+1}z_{s+1}, \dots, u_r = P_r z_r$, we obtain

$$J = P_{s+1} \dots P_r \int_0^1 \dots \int_0^1 \exp\{2\pi i m H(z_{s+1}, \dots, z_r)\}\, dz_{s+1} \dots dz_r,$$

where

$$H(z_{s+1}, \dots, z_r) = G(P_{s+1}z_{s+1}, \dots, P_r z_r)$$

$$= \sum_{\substack{t_{s+1}=0 \\ t_{s+1}+\dots+t_r \ge 1}}^{n_{s+1}} \dots \sum_{t_r=0}^{n_r} \beta_0(t_{s+1}, \dots, t_r) P_{s+1}^{t_{s+1}} \dots P_r^{t_r} z_{s+1}^{t_{s+1}} \dots z_r^{t_r}$$

$$= \sum_{\substack{t_{s+1}=0 \\ t_{s+1}+\dots+t_r \ge 1}}^{n_{s+1}} \dots \sum_{t_r=0}^{n_r} \delta_0(t_{s+1}, \dots, t_r) z_{s+1}^{t_{s+1}} \dots z_r^{t_r}.$$

To estimate the integral J, it is necessary to give an estimate from below for

$$\delta_0 = \max_{t_{s+1}, \dots, t_r} |\delta_0(t_{s+1}, \dots, t_r)|.$$

By the assumptions of the lemma, there exists a set $(t_1, \dots, t_r)$ for which $\eta < q \le \tau$. For the variable $\alpha(t_1, \dots, t_r)$ corresponding to this set, we consider the D-approximations corresponding to the parameters τ and η. We obtain

$$\alpha(t_1, \dots, t_r) = \frac{a}{q} + \beta = \frac{a_0}{q_0} + \beta_0, \quad q_0 \le \eta < q \le \tau,$$

and hence $q \neq q_0$.

Now we give an estimate from below for β_0:

$$|\beta_0| = \left|\frac{a_0}{q_0} - \frac{a}{q} - \beta\right| \geq \frac{1}{qq_0} - |\beta| \geq \frac{1}{Q_0\tau} - \frac{1}{q\tau} \geq 0.5(Q_0\tau)^{-1}.$$

The last inequality implies the following estimate for δ_0:

$$\begin{aligned}\delta_0 &> |\beta_0| P_{s+1}^{t_{s+1}} \dots P_r^{t_r} \geq 0.5(Q_0\tau)^{-1} P_{s+1}^{t_{s+1}} \dots P_r^{t_r} \\ &\geq 0.5 H_0^{-1} P_1^{-t_1+1/6} P_2^{-t_2} \dots P_s^{-t_s}.\end{aligned}$$

Now, to estimate J for $1 \leq m \leq M$, we apply Theorem 1.6 (Chapter 1):

$$|J| \ll P_{s+1} \dots P_r m^{-1/(2n)} H_0^{1/(2n)} P_1^{t_1/(2n)-1/(12n)} P_2^{t_2/(2n)} \dots P_s^{t_s/(2n)}.$$

Successively substituting this estimate first into the sum $T(m)$ and then into formula (7.26), we obtain

$$\begin{aligned}|Z_1| &\ll \Phi(H_1)\Delta \sum_{1\leq m\leq M} \big(P_{s+1} \dots P_r (m^{-1} H_0 P^{t_1-1/6} P_2^{t_2} \dots P_s^{t_s})^{1/(2n)} \\ &\qquad + P_{s+2} \dots P_r H_0\big) \\ &\quad + \Phi(H_1)^{-1}\Delta \sum_{1\leq m\leq M} \big(P_{s+1} \dots P_r m^{-2} (m^{-1} H_0 P^{t_1-1/6} P_2^{t_2} \dots P_s^{t_s})^{1/(2n)} \\ &\qquad + P_{s+2} \dots P_r H_0 m^{-2}\big) + P_{s+1} \dots P_r P_1^{-\rho} \\ &\ll P_{s+1} \dots P_r P_1^{-\rho}.\end{aligned}$$

This implies the estimate for Y given in the lemma. The proof of the lemma is complete. □

Let $1 \leq s < r$. We consider the polynomial

$$\Psi(x_1, \dots, x_r) = \sum_{\substack{t_{s+1}=0 \\ t_{s+1}+\dots+t_r \geq 1}}^{n_{s+1}} \cdots \sum_{t_r=0}^{n_r} G(t_{s+1}, \dots, t_r) x_{s+1}^{t_{s+1}} \dots x_r^{t_r}.$$

In this formula the variables $G = G(t_{s+1}, \dots, t_r)$ are polynomials in the variables $x_1, \dots, x_r$ of the form

$$\begin{aligned}G &= g_{t_{s+1},\dots,t_r}(x_1, \dots, x_s) \\ &= \frac{a(0, \dots, 0, t_{s+1}, \dots, t_r)}{q(0, \dots, 0, t_{s+1}, \dots, t_r)} + \sum_{t_1=0}^{n_1} \cdots \sum_{t_s=0}^{n_s} \beta(t_1, \dots, t_r) x_1^{t_1} \dots x_s^{t_s},\end{aligned}$$

and moreover,

$$\begin{gathered}\big(a(0,\dots,0,t_{s+1},\dots,t_r),q(0,\dots,0,t_{s+1},\dots,t_r)\big)=1,\\ 1\le q(0,\dots,0,t_{s+1},\dots,t_r)\le\tau(0,\dots,0,t_{s+1},\dots,t_r),\\ |\beta(0,\dots,0,t_{s+1},\dots,t_r)|\le\big(q(0,\dots,0,t_{s+1},\dots,t_r)\tau(0,\dots,0,t_{s+1},\dots,t_r)\big)^{-1},\\ |\beta(t_1,\dots,t_r)|\le\tau^{-1}(t_1,\dots,t_r),\quad \tau(t_1,\dots,t_r)=P_1^{t_1-1/6}P_2^{t_2}\dots P_r^{t_r},\\ 0\le t_1\le n_1,\dots,0\le t_r\le n_r,\quad t_{s+1}+\dots+t_r\ge 1.\end{gathered}$$

Let Q_1 be the least common multiple of the numbers $q(0,\dots,0,t_{s+1},\dots,t_r)$ for $0\le t_{s+1}\le n_{s+1},\dots,0\le t_r\le n_r$, $t_{s+1}+\dots+t_r\ge 1$. For each set $(x_1,\dots,x_s)$ such that $1\le x_1\le P_1,\dots,1\le x_s\le P_s$, the D-approximations of the fractional parts of the polynomial $g=g_{t_{s+1},\dots,t_r}(x_1,\dots,x_s)$ corresponding to $\eta(t_{s+1},\dots,t_r)=P_{s+1}^{t_{s+1}-1/6}P_{s+2}^{t_{s+2}}\dots P_r^{t_r}$ are considered. In other words, we consider the relations

$$\{g\}=\frac{a_1(t_{s+1},\dots,t_r)}{q_1(t_{s+1},\dots,t_r)}+\beta_1(t_{s+1},\dots,t_r),$$

where

$$\begin{gathered}\big(a_1(t_{s+1},\dots,t_r),q_1(t_{s+1},\dots,t_r)\big)=1,\quad 1\le q_1(t_{s+1},\dots,t_r)\le\eta(t_{s+1},\dots,t_r),\\ |\beta(t_{s+1},\dots,t_r)|\le\big(q_1(t_{s+1},\dots,t_r)\eta(t_{s+1},\dots,t_r)\big)^{-1}.\end{gathered}$$

By $Q_5=Q_5(x_1,\dots,x_s)$ we denote the least common multiple of the numbers $q_1(t_{s+1},\dots,t_r)$ for $0\le t_{s+1}\le n_{s+1},\dots,0\le t_r\le n_r$, $t_{s+1}+\dots+t_r\ge 1$; $m_s=(n_1+1)\dots(n_s+1)$.

Lemma 7.15. *Suppose that Q_1 is larger than $P_1^{0.04}$. By Y we denote the number of sets $(x_1,\dots,x_s)$ satisfying the conditions*

$$\begin{gathered}Q_5\le H=P_1^a,\quad a=20(r-s)m_sn^3\rho_r,\quad \rho_r=\frac{c}{(2n)^{2r}\log n},\\ |\beta_1(t_{s+1},\dots,t_r)|\le\Delta(t_{s+1},\dots,t_r)=P_{s+1}^{-t_{s+1}}\dots P_r^{-t_r}P_1^{2n\rho_r},\\ 1\le x_1\le P_1,\dots,1\le x_s\le P_s.\end{gathered}$$

Then the variable Y satisfies the estimate $Y\ll P_1\dots P_sP_1^{-\rho}$; the constant in $\ll$ depends only on n and r.

Proof. The proof of the lemma is similar to that of Lemma 7.6. In the m_s-dimensional space, we consider the set Ω_0 of points g with coordinates

$$\begin{gathered}\{g_{t_{s+1},\dots,t_r}(x_1,\dots,x_s)\},\\ 0\le t_{s+1}\le n_{s+1},\dots,0\le t_r\le n_r,\quad t_{s+1}+\dots+t_r\ge 1,\end{gathered}$$

under the condition that $1 \le x_1 \le P_1, \dots, 1 \le x_s \le P_s$. We also assume that the coordinates of points from Ω_0 satisfy the assumptions of the lemma. We show that the set Ω_0 can intersect only one domain $\Omega_1 = \Omega_1(b, h)$ of the first class, which is determined as follows. A point α belongs to the domain Ω_1 if its coordinates $\alpha(t_1, \dots, t_r)$ can be represented as

$$\alpha(t_1, \dots, t_r) = \frac{b(t_{s+1}, \dots, t_r)}{h(t_{s+1}, \dots, t_r)} + z(t_{s+1}, \dots, t_r) = \frac{b}{h} + z,$$
$$(b, h) = 1, \quad 1 \le h \le \tau(0, \dots, 0, t_{s+1}, \dots, t_r),$$
$$|z| \le \big(h\tau(0, \dots, 0, t_{s+1}, \dots, t_r)\big)^{-1},$$
$$0 \le t_{s+1} \le n_{s+1}, \dots, 0 \le t_r \le n_r, \quad t_{s+1} + \dots + t_r \ge 1,$$

and the least common multiple of the numbers $h(t_{s+1}, \dots, t_r)$ does not exceed H. The modulus of the difference between the corresponding coordinates of the centers of the domain Ω_1 and any other domain Ω_2 of the first class is no less than H^{-2}. Hence the modulus of the difference between the corresponding coordinates of points of these domains is no less than

$$H^{-2} - 2\tau^{-1}(0, \dots, 0, t_{s+1}, \dots, t_r) \ge 0.5H^{-2},$$
$$0 \le t_{s+1} \le n_{s+1}, \dots, 0 \le t_r \le n_r, \quad t_{s+1} + \dots + t_r \ge 1.$$

Each coordinate $\{g_{t_{s+1},\dots,t_r}(x_1, \dots, x_s)\}$ of a point G from the set Ω_0 and the corresponding coordinate $\alpha(0, \dots, 0, t_{s+1}, \dots, t_r)q^{-1}(0, \dots, 0, t_{s+1}, \dots, t_r)$ of a fixed point differ by a value that does not exceed

$$\sum_{t_1=0}^{n_1} \cdots \sum_{t_s=0}^{n_s} |\beta(t_1, \dots, t_r)| P_1^{t_1} \dots P_r^{t_r} \le m_s \tau^{-1}(0, \dots, 0, t_{s+1}, \dots, t_r).$$

Hence we see that if Ω_0 intersects a domain Ω_1 of the first class, then Ω_0 intersects only one domain. Obviously, if Ω_0 and Ω_1 do not intersect, then $Y = 0$. We consider the case in which Ω_0 and Ω_1 do intersect. Then for all y satisfying the assumptions of the lemma, we have

$$\frac{a_1(t_{s+1}, \dots, t_r)}{q_1(t_{s+1}, \dots, t_r)} = \frac{b(t_{s+1}, \dots, t_r)}{h(t_{s+1}, \dots, t_r)},$$
$$\left| g_{t_{s+1},\dots,t_r}(x_1, \dots, x_s) - \frac{b(t_{s+1}, \dots, t_r)}{h(t_{s+1}, \dots, t_r)} \right| \le \Delta(t_{s+1}, \dots, t_r), \tag{7.28}$$
$$0 \le t_{s+1} \le n_{s+1}, \dots, 0 \le t_r \le n_r, \quad t_{s+1} + \dots + t_r \ge 1.$$

Since $Q_1 \ge P_1^{0.04}$, there exists a $q(0, \dots, 0, t_{s+1}, \dots, t_r)$ satisfying the inequality

$$q(0, \dots, 0, t_{s+1}, \dots, t_r) \ge P_1^u > H,$$

where $u = 0.04m_r^{-1}m_s$. Hence for the set $(t_{s+1}, \ldots, t_r)$ we have

$$\frac{a(0, \ldots, 0, t_{s+1}, \ldots, t_r)}{q(0, \ldots, 0, t_{s+1}, \ldots, t_r)} \neq \frac{b(t_{s+1}, \ldots, t_r)}{h(t_{s+1}, \ldots, t_r)}.$$

For brevity, in what follows, we denote these fractions by a/q and b/h. Moreover, we introduce the new notation

$$B(x_1, \ldots, x_s) = G - \frac{a}{q} = \sum_{t_1=0}^{n_1} \cdots \sum_{t_s=0}^{n_s} \beta(t_1, \ldots, t_s, \ldots, t_r) x_1^{t_1} \ldots x_s^{t_s}.$$

We rewrite the inequality in (7.28) corresponding to the set $(t_{s+1}, \ldots, t_r)$ as

$$\left| B(x_1, \ldots, x_s) - \frac{b}{h} + \frac{a}{q} \right| \leq \Delta = \Delta(t_{s+1}, \ldots, t_r). \tag{7.29}$$

By Y_1 we denote the number of sets $(x_1, \ldots, x_s)$ corresponding to (7.29) under the condition $1 \leq x_1 \leq P_1, \ldots, 1 \leq x_s \leq P_s$. We introduce the function $\psi(x) = \chi(x + a/q - b/h)$, where $\chi(x)$ is the function in Lemma 7.5. Then

$$Y \leq Y_1 \leq \sum_{x_1=1}^{P_1} \cdots \sum_{x_s=1}^{P_s} \psi\big(B(x_1, \ldots, x_s)\big) = Y_2.$$

Expanding the function $\psi(x)$ in the Fourier series and passing to inequalities, we obtain

$$Y_2 \ll P_1 \ldots P_s \Delta + \sum_{1 \leq \nu < M} \Delta |T(\nu)| + \sum_{M \leq \nu < M_1} \Delta^{-1} \nu^{-2} |T(\nu)| \tag{7.30}$$
$$+ P_1 \ldots P_s P_1^{-\rho_r},$$

where

$$T(\nu) = \sum_{x_1=1}^{P_1} \cdots \sum_{x_s=1}^{P_s} \exp\{2\pi i \nu B(x_1, \ldots, x_s)\}, \quad M = \Delta^{-1}, \quad M_1 = M P_1^{\rho_r}.$$

For $1 \leq \nu \leq M$, we give an estimate from above for the modulus of the partial derivatives of the polynomial $\nu B(x_1, \ldots, x_s)$. We have

$$\left| \nu \frac{\partial B(x_1, \ldots, x_s)}{\partial x_\mu} \right| \leq \nu \sum_{t_1=0}^{n_1} \ldots \sum_{t_\mu=1}^{n_\mu} \ldots \sum_{t_s=0}^{n_s} t_\mu \tau^{-1}(t_1, \ldots, t_r) P_1^{t_1} \ldots P_\mu^{t_\mu - 1} \ldots P_s^{t_s}$$
$$\leq 0.5 n_\mu m_s P_1^{1/6 - 2n\rho_r} P_\mu^{-1} \leq 0.5.$$

This implies that Lemma 5.4 (Chapter 5) can be applied to the sum $T(\nu)$. Therefore,

$$T(\nu) = \int_0^{P_1} \cdots \int_0^{P_s} \exp\{2\pi i \nu A(y_1, \ldots, y_s)\} \, dy_1 \ldots dy_s + O(P_2 \ldots P_s),$$

In the last integral we perform a change of the variables of integration of the form $y_\mu = P_\mu x_\mu$ $(\mu = 1, \dots, s)$ and pass to estimates. We obtain

$$T(\nu) \ll P_1 \dots P_s \left| \int_0^1 \cdots \int_0^1 \exp\{2\pi i \nu A(y_1, \dots, y_s)\}\, dy_1 \dots dy_s \right| + P_2 \dots P_s, \tag{7.31}$$

where

$$A(y_1, \dots, y_s) = \sum_{\substack{t_1=0 \\ t_1+\dots+t_s \ge 1}}^{n_1} \cdots \sum_{t_s=0}^{n_s} \delta(t_1, \dots, t_s) y_1^{t_1} \dots x_s^{t_s},$$

$$\delta(t_1, \dots, t_s) = \beta(t_1, \dots, t_s, \dots, t_r) P_1^{t_1} \dots P_s^{t_s}.$$

Now we give an estimate from below for $\delta = \max_{t_1,\dots,t_s} |\delta(t_1, \dots, t_s)|$. First, we have

$$\left| \frac{a}{q} - \frac{b}{h} + \sum_{t_1=0}^{n_1} \cdots \sum_{t_s=0}^{n_s} \beta(t_1, \dots, t_s, \dots, t_r) \right| \ge \left| \frac{a}{q} - \frac{b}{h} \right| - \sum_{t_1=0}^{n_1} \cdots \sum_{t_s=0}^{n_s} |\beta(t_1, \dots, t_r)|$$

$$\ge \frac{1}{qh} - \frac{1}{q\tau(0, \dots, 0, t_{s+1}, \dots, t_r)} - \sum_{\substack{t_1=0 \\ t_1+\dots+t_s \ge 1}}^{n_1} \cdots \sum_{t_s=0}^{n_s} |\beta(t_1, \dots, t_r)|$$

$$\ge 0.25\big(H\tau(0, \dots, 0, t_{s+1}, \dots, t_r)\big)^{-1}.$$

Hence, for any set $(x_1, \dots, x_s)$ satisfying inequality (7.29), we obtain

$$\left| \sum_{\substack{t_1=0 \\ t_1+\dots+t_s \ge 1}}^{n_1} \cdots \sum_{t_s=0}^{n_s} \beta(t_1, \dots, t_r)(x_1^{t_1} \dots x_s^{t_s} - 1) \right|$$

$$\ge \left| \frac{a}{q} - \frac{b}{h} + \sum_{t_1=0}^{n_1} \cdots \sum_{t_s=0}^{n_s} \beta(t_1, \dots, t_s, \dots, t_r) \right|$$

$$- \left| \frac{a}{q} - \frac{b}{h} + \sum_{t_1=0}^{n_1} \cdots \sum_{t_s=0}^{n_s} \beta(t_1, \dots, t_s, \dots, t_r) x_1^{t_1} \dots x_s^{t_s} \right|$$

$$\ge 0.25\big(H\tau(0, \dots, 0, t_{s+1}, \dots, t_r)\big)^{-1} - \Delta(t_{s+1}, \dots, t_r)$$

$$\ge \big(8H\tau(0, \dots, 0, t_{s+1}, \dots, t_r)\big)^{-1}.$$

Hence

$$\big(8H\tau(0, \dots, 0, t_{s+1}, \dots, t_r)\big)^{-1} \le \left| \sum_{\substack{t_1=0 \\ t_1+\dots+t_s \ge 1}}^{n_1} \cdots \sum_{t_s=0}^{n_s} \beta(t_1, \dots, t_r)(x_1^{t_1} \dots x_s^{t_s} - 1) \right| \le \delta m_s,$$

$$\delta \geq \left(8m_s H\tau(0,\dots,0,t_{s+1},\dots,t_r)\right)^{-1} = (8m_s H\tau)^{-1}.$$

Applying Theorem 1.6 (Chapter 1) to the integral in (7.31), we obtain the following estimate for the sum $T(\nu)$:

$$|T(\nu)| \ll P_1 \dots P_s \min\left(1, (\nu^{-1}H\tau)^{1/(2n)}\right) + P_2 \dots P_s.$$

Substituting this estimate into (7.30), we have

$$Y_2 \ll P_1 \dots P_s \Delta + \Delta \sum_{1\leq\nu<M} (\nu^{-1}H\tau)^{1/(2n)} + \Delta^{-1} \sum_{M\leq\nu<M_1} \nu^{-2}(\nu^{-1}H\tau)^{1/(2n)}$$
$$+ P_1 \dots P_s P_1^{-\rho_r} \ll P_1 \dots P_s P_1^{-\rho_r}.$$

It follows from (7.30) and (7.31) that

$$Y \leq Y_1 \leq Y_2 \ll P_1 \dots P_s P_1^{-\rho_r}.$$

The proof of the lemma is complete. □

Next, we consider the polynomial

$$\psi_1(x_1,\dots,x_s,\dots,x_r) = \sum_{\substack{t_1=0\\ t_1+\dots+t_s\geq 1}}^{n_1} \cdots \sum_{t_s=0}^{n_s} G(t_1,\dots,t_s) x_1^{t_1} \dots x_s^{t_s},$$

$$G(t_1,\dots,t_s) = \frac{a(t_1,\dots,t_s,0,\dots,0)}{q(t_1,\dots,t_s,0,\dots,0)} + \sum_{t_{s+1}=0}^{n_{s+1}} \cdots \sum_{t_r=0}^{n_r} \beta(t_1,\dots,t_r) x_{s+1}^{t_{s+1}} \dots x_r^{t_r},$$

where

$$\left(a(t_1,\dots,t_s,0,\dots,0), q(t_1,\dots,t_s,0,\dots,0)\right) = 1,$$
$$1 \leq q(t_1,\dots,t_s,0,\dots,0) \leq \tau(t_1,\dots,t_s,0,\dots,0),$$
$$|\beta(t_1,\dots,t_s,0,\dots,0)| \leq \left(q(t_1,\dots,t_s,0,\dots,0)\tau(t_1,\dots,t_s,0,\dots,0)\right)^{-1} = 1,$$
$$|\beta(t_1,\dots,t_r)| \leq \tau^{-1}(t_1,\dots,t_r),$$
$$\tau(t_1,\dots,t_r) = P_1^{t_1-1/6} P_2^{t_2} \dots P_r^{t_r},$$
$$0 \leq t_1 \leq n_1,\dots,0 \leq t_r \leq n_r, \quad t_1+\dots+t_s \geq 1.$$

By Q_2 we denote the least common multiple of $q(t_1,\dots,t_s,0,\dots,0)$ $(0 \leq t_1 \leq n_1,\dots,0 \leq t_s \leq n_s, t_1+\dots+t_s \geq 1)$. For each set $(x_{s+1},\dots,x_r)$ $(1 \leq x_{s+1} \leq P_{s+1},\dots,1 \leq x_r \leq P_r)$, we consider the D-approximations of the fractional parts of the polynomials $G = G(t_1,\dots,t_s)$ corresponding to $\tau(t_1,\dots,t_s)$

$$\tau(t_1,\dots,t_s) = P_1^{t_1-1/6} P_2^{t_2} \dots P_s^{t_s}.$$

In other words, we consider relations of the form

$$\{G\} = \frac{a_2(t_1, \dots, t_s)}{q_2(t_1, \dots, t_s)} + \beta_2(t_1, \dots, t_s),$$

where

$$\big(a_2(t_1, \dots, t_s), q_2(t_1, \dots, t_s)\big) = 1, \quad 1 \le q_2(t_1, \dots, t_s) \le \tau(t_1, \dots, t_s),$$
$$|\beta_2(t_1, \dots, t_s)| \le \big(q_2(t_1, \dots, t_s)\tau(t_1, \dots, t_s)\big)^{-1}.$$

By $Q_2(x_{s+1}, \dots, x_r)$ we denote the least common multiple of the numbers $q_2(t_1, \dots, t_s)$ $(0 \le t_1 \le n_1, \dots, 0 \le t_s \le n_s,\ t_1 + \dots + t_s \ge 1)$; $k_s = m_r m_s^{-1} = (n_{s+1} + 1) \dots (n_r + 1)$.

Lemma 7.16. *Suppose that Q_2 is larger than $P_1^{0.04}$ and Y is the number of sets $(x_{s+1}, \dots, x_r)$ satisfying the conditions*

$$Q_2(x_{s+1}, \dots, x_r) \le H = P_1^u, \quad u = 20k_s n^3 s p_r,$$
$$|\beta(t_1, \dots, t_s)| \le \Delta(t_1, \dots, t_s) = P_1^{-t_1} \dots P_s^{-t_s} P_1^{2n\rho_r},$$
$$1 \le x_{s+1} \le P_{s+1}, \dots, 1 \le x_r \le P_r.$$

Then the variable Y satisfies the estimate

$$Y \ll P_{s+1} \dots P_r P_1^{-\rho_r}, \quad \rho_r = \frac{c}{(2n)^{2r} \log n};$$

the constant in $\ll$ depends only on n and r.

Proof. The proof of this lemma is similar to that of Lemma 7.15, and we omit it here. □

Now we formulate two more lemmas.

We assume that the coefficients of the polynomial

$$f(x_1, \dots, x_k) = \sum_{t_1=0}^{n_1} \cdots \sum_{t_k=0}^{n_k} \alpha(t_1, \dots, t_k) x_1^{t_1} \dots x_k^{t_k},$$

can be represented as

$$\alpha(t_1, \dots, t_k) = \alpha = \frac{a}{q} + \beta,$$

where β is a real number and a and q are integers ($a \ge 0$, $q \ge 1$, $(a, q) = 1$). We also assume that

$$Q = \underset{t_1+\dots+t_k \ge 1}{\text{l.c.m.}} (q), \quad \delta = P_1^{t_1} \dots P_r^{t_r} \beta,$$

$$\Delta = \max_{t_1+\cdots+t_k\geq 1} |\delta|, \quad 1 \leq P_1 \leq \cdots \leq P_k.$$

We introduce a polynomial $g(x_1, \ldots, x_k) = f(x_1 + y_1, \ldots, x_k + y_k)$, where $y_1, \ldots, y_k$ are integers, $|y_s| \leq P_s$ $(s = 1, \ldots, k)$. By $\alpha_0 = \alpha_0(t_1, \ldots, t_k)$ we denote the coefficients of the polynomial $g(x_1, \ldots, x_k)$.

Lemma 7.17. *It is possible to choose integer numbers a_0, q_0 and real numbers β_0 such that for all $t_1, \ldots, t_k$, the following relations hold:*

$$\alpha_0 = \frac{a_0}{q_0} + \beta_0,$$

and moreover, $Q_0 = Q$, $\Delta \ll \Delta_0 \ll \Delta$, and the numbers Q_0 and Δ_0 are determined similarly to Q and Δ, but with α, a, q, β replaced by α_0, a_0, q_0, β_0. The constants in $\ll$ depend only on n and k.

Let us consider the polynomial

$$G(x_1, \ldots, x_k) = G = \sum_{t_1=0}^{l_1} \cdots \sum_{t_k=0}^{l_k} \alpha(t_1, \ldots, t_k) x_1^{t_1} \ldots x_k^{t_k}.$$

The set A_0 of coefficients $\alpha_0(t_1, \ldots, t_k)$ of this polynomial is a point in the m_k-dimensional Euclidean space, $m_k = (l_1 + 1) \ldots (l_k + 1)$.

Suppose that q is a natural number, $y_1, \ldots, y_k$ are nonnegative integers each of which does not exceed q. By $S_q(A_0)$ we denote the trigonometric sum

$$S_q(A_0) = \sum_{x_1\leq R_1}{}' \ldots \sum_{x_k\leq R_k}{}' \exp\{2\pi i G\}, \quad 1 \leq R_1 \leq \cdots \leq R_k,$$

where the prime on each of the summation signs means that the variables of summation $x_1, \ldots, x_k$ belong to progressions of the form

$$x_1 + y_1 \equiv 0 \pmod q, \quad \ldots, \quad x_k + y_k \equiv 0 \pmod q.$$

Suppose that Ω_1 and Ω_2 are domains of points A of the first and second classes with respect to the parameters $R_1, \ldots, R_k$. Suppose that the variables Q_0 and δ_0 are defined for a point A of the first class similarly to the variables Q and δ for a point A in Lemma 7.10, but with the parameters $P_1, \ldots, P_r$ replaced by $R_1, \ldots, R_k$.

Lemma 7.18. *Suppose that the number q satisfies the inequality*

$$q^{\mathcal{L}} \leq P_1^{0.05}, \quad \mathcal{L} = l_1 + \cdots + l_k.$$

Suppose also that, for points of the second class, the sum

$$S(A_0) = \sum_{x_1=1}^{R_1} \cdots \sum_{x_k=1}^{R_k} \exp\{2\pi i G\}$$

can be estimates as

$$|S(A_0)| \ll R_1 \dots R_k R_1^{-\rho_k},$$

where ρ_k is a positive number such that $\rho_k \le 0.02l^{-2}$, $l = \max(l_1, \dots, l_k)$. Then the sum $S_q(A_0)$ satisfies the following estimates:

(1) *If a point A_0 belongs to the second class, then*

$$|S_q(A_0)| \ll R_1 \dots R_k q^{-k} R_1^{-0.5\rho_k}.$$

(2) *If a point A_0 belongs to the first class, and moreover, $Q \le R_1^{0.07}$ and $\delta \le R_1^{0.04}$, then*

(a) $|S_q(A_0)| \ll R_1 \dots R_k q^{-k} Q_1^{-1/n+\varepsilon}$;

(b) $|S_q(A_0)| \ll R_1 \dots R_k q^{-k} (Q_1\delta)^{-1/n+\varepsilon}$ *for* $\delta \ge 1$,

where $Q_1 = Q/(Q, q^{\mathcal{L}})$ and the constants in $\ll$ depend only on k, l, and ε.

(3) *The estimate given in item* (1) *holds for the remaining points A_0 of the first class.*

In fact, the proofs of Lemmas 7.17 and 7.18 do not differ from those of Lemmas 7.8 and 7.9. Only in the proof of Lemma 7.18 we must use the result of Lemma 7.17, while in the proof of Lemma 7.9 we applied Lemma 7.8.

7.2.2 The second main lemma

The second main lemma. *Suppose that $F(x_1, \dots, x_r)$ is a polynomial with real coefficients $\alpha(t_1, \dots, t_r)$ of the form*

$$F(x_1, \dots, x_r) = \sum_{t_1=0}^{n_1} \cdots \sum_{t_r=0}^{n_r} \alpha(t_1, \dots, t_r) x_1^{t_1} \dots x_r^{t_r}, \quad \alpha(0, \dots, 0) = 0;$$

here $P_1, \dots, P_r$ are natural numbers, $P_1 \le \dots \le P_r$, $P_1 \to +\infty$.

Consider the D-approximations of the numbers $\alpha(t_1, \dots, t_r)$ corresponding to $\tau(t_1, \dots, t_r) = P_1^{t_1-1/6} P_2^{t_2} \dots P_r^{t_r}$:

$$\alpha(t_1, \dots, t_r) = \frac{a(t_1, \dots, t_r)}{q(t_1, \dots, t_r)} + \beta(t_1, \dots, t_r),$$
$$1 \le q(t_1, \dots, t_r) \le \tau(t_1, \dots, t_r), \quad \big(a(t_1, \dots, t_r), q(t_1, \dots, t_r)\big) = 1,$$
$$|\beta(t_1, \dots, t_r)| \le \big(q(t_1, \dots, t_r)\tau(t_1, \dots, t_r)\big)^{-1},$$
$$0 \le t_1 \le n_1, \dots, 0 \le t_r \le n_r.$$

Let Q be the least common multiple of the numbers $q(t_1, \dots, t_r)$. Then for $Q > P_1^{0.1}$ the trigonometric sum

$$S = S(A) = \sum_{x_1=1}^{P_1} \cdots \sum_{x_r=1}^{P_r} \exp\{2\pi i F(x_1, \dots, x_r)\}$$

satisfies the estimate

$$|S| \le cP_1 \dots P_r P_1^{-\rho},$$

where $c = c(n_1, \dots, n_r) > 0$, $\rho = \rho_r = \gamma/((2n)^{2r} \log n)$, $\gamma > 0$ *is an absolute constant.*

Prior to proving the second main lemma, we dwell upon some of its characteristic features. It should be noted that the main case of the lemma, i.e., the case in which the intervals of summation in the trigonometric sum are essentially different, can be proved by an induction approach to sums with a fewer number of variables. Recall that, in the first main lemma, we reduce estimating double trigonometric sums to estimating one-dimensional "inner" sums and to estimating the number of the fractional parts of a polynomial in a single variable contained in intervals of some special form. We shall use a similar approach to estimate sums of larger multiplicity. However, the situation is more complicated because in this case, in general, it is possible to pass to sums of lesser multiplicity in several different ways and the number of cases which we must study may depend on the number of variables r and the powers $n_1, \dots, n_r$ in the polynomial in the exponent of the r-fold sum. We overcome this difficulty by choosing a special method for passing from sums of larger multiplicity to sums of lesser multiplicity. Moreover, in fact, we establish an almost perfect correspondence between the scheme for deriving an estimate for the double sum and a similar scheme for the r-fold sum. To stress this fact, whenever possible, we consciously use the same or closely related terminology and argument.

The correspondence mentioned above can be established as follows. First, we exclude the case of sums whose intervals of summation $P_1, \dots, P_r$ do not differ very much, i.e., the case of sums that with an appropriate accuracy can be estimated by the corresponding theorems from Chapter 5. Next, starting from the assumption that the parameters $P_1, \dots, P_r$ are significantly different and using a special method, we find the index s that is less than r. Then we associate the group of variables $x_1, \dots, x_s$ with the variable x_1 in the two-fold case, and the group of variables $x_{s+1}, \dots, x_r$ with the variable x_2 in the same case. Moreover, we can write the sum S as

$$S = \sum_{x_{s+1}=1}^{P_1} \dots \sum_{x_r=1}^{P_r} \left(\sum_{x_1=1}^{P_1} \dots \sum_{x_s=1}^{P_s} \exp\left\{ 2\pi i \left(\sum_{t_1=0}^{n_1} \dots \sum_{t_s=0}^{n_s} A(t_1, \dots, t_s) x_1^{t_1} \dots x_s^{t_s} \right) \right\} \right),$$

where

$$A(t_1, \dots, t_s) = g_{t_1,\dots,t_s}(x_{s+1}, \dots, x_r).$$

Here $A(t_1, \dots, t_s)$ plays the same role as A at the beginning of the proof of the first main lemma, while the role of the inner sum over x_1 is played by the sum in parentheses in the above formula for the sum S. After this, the general scheme of reasoning in the r-fold case resembles the two-fold case very much. In particular, it is possible to establish a correspondence between the sets of indices E_0, E_1, E_2 and a sets of indices in the multiple case.

Because of this similarity, we do not further describe the scheme of the proof of the second main lemma. We only note that, in principle, it is possible to start the induction from $r = 1$ rather than from $r = 2$, as it is done here. However, this leads to significant additional technical difficulties, in particular, we must almost everywhere consider the case $r = 2$ as an exceptional case.

Proof. We assume that $r \geq 3$, since, for $r = 1$, the statement of the lemma follows from Lemma 7.1 (a) and for $r = 2$, from the first main lemma. First, we estimate the sum $S = S(A)$ under the condition that the following inequalities hold:

$$\begin{aligned} &P_2^{5/6} \leq P_1^{n_1+1}, \\ &\vdots \\ &P_{s+1}^{5/6} \leq P_1^{n_1+1} P_2^{n_2} \dots P_s^{n_s}, \\ &\vdots \\ &P_r^{5/6} \leq P_1^{n_1+1} P_2^{n_2} \dots P_{r-1}^{n_{r-1}}. \end{aligned} \tag{7.32}$$

If a point A with coordinates $\alpha(t_1, \dots, t_r)$ $(0 \leq t_1 \leq n_1, \dots, 0 \leq t_r \leq n_r, t_1 + \dots + t_r \geq 1)$ that are the coefficients of the polynomial

$$F(x_1, \dots, x_r) = \sum_{t_1=0}^{n_1} \cdots \sum_{t_s=0}^{n_s} \alpha(t_1, \dots, t_r) x_1^{t_1} \dots x_r^{t_r},$$

is a point of the second class Ω_2, then, by Lemma 7.12, the sum

$$S(A) = \sum_{x_1=1}^{P_1} \cdots \sum_{x_r=1}^{P_r} \exp\{2\pi i F(x_1, \dots, x_r)\}$$

satisfies the estimate

$$|S(A)| \ll \exp\{32\varkappa\} P_1 \dots P_r P_1^{-\rho}, \tag{7.33}$$

where

$$\rho = c(m\varkappa \log m\varkappa)^{-1}, \quad \varkappa = n_1 + \nu_2 n_2 + \dots + \nu_r n_r,$$

and the natural numbers ν_s $(s = 2, \dots, r)$ are determined by the inequalities

$$-1 < \ln P_s / \ln P_1 - \nu_s \leq 0.$$

We show that in the case under study the statement of the lemma follows from the estimate (7.33). Obviously, it suffices to prove that

$$\rho = \frac{c}{m\varkappa \log m\varkappa} \geq \frac{\gamma}{(2n)^{2r} \log n} = \rho_r.$$

By definition, the numbers ν_s $(s = 2, \dots, r)$ satisfy the inequalities $\nu_s = \ln P_s / \ln P_1 + 1$. We set

$$1 = \ln P_1 / \ln P_1 = z_1, \ \ln P_2 / \ln P_1 = z_2, \ \dots \ \ln P_r / \ln P_1 = z_r.$$

Then we have

$$\nu_s \le z_s + 1, \quad s = 2, \dots, r, \quad 1 \le z_1 \le \dots \le z_r,$$

$$\varkappa \le n_2 + \dots + n_r + n_1 z_1 + \dots + n_r z_r \le n\left(r - 1 + \sum_{s=1}^{r} z_s\right).$$

Taking the logarithm of inequality (7.32), for $s = 1, \dots, r - 1$, we obtain

$$\ln P_{s+1} \le 1.2\big((n_1 + 1) \ln P_1 + n_2 \ln P_2 + \dots + n_s \ln P_s\big).$$

This implies that the numbers $z_1, \dots, z_r$ satisfy the relations

$$
\begin{aligned}
z_1 = 1, \qquad\qquad & z_2 \le 1.2 n z_1 + 1.2,\\
& z_3 \le 1.2n(z_1 + z_2)1.2,\\
& \vdots\\
& z_r \le 1.2n(z_1 + \dots + z_{r-1}) + 1.2.
\end{aligned}
$$

To estimate $\varkappa$, we successively apply these inequalities, starting from the last, and obtain

$$
\begin{aligned}
\varkappa &\le n\left(r - 1 + \sum_{s=1}^{r} z_s\right) = \left(r - 1 + z_r + \sum_{s=1}^{r-1} z_s\right)\\
&\le n\left(r - 1 + 1.2 + (1.2n + 1) \sum_{s=1}^{r-1} z_s\right)\\
&\le n\left(r - 1 + 2 \cdot 1.2 + (1.2n + 1)^2 \sum_{s=1}^{r-2} z_s\right) \le \dots\\
&\le n\big(r - 1 + 1.2(r - 1) + (1.2n + 1)^{r-1}\big) \le 2(1.2n + 1)^r,
\end{aligned}
$$

because $2.2n(r - 1) < (1.2n + 1)^r$.

Recall that $m = (n_1 + 1) \dots (n_r + 1)$. Since $r \ge 3$ and $n \ge 2$, we obtain

$$
\begin{aligned}
m \le (n + 1)', \quad m\varkappa &\le (n + 1)^r (1.2n + 1)^r < 2(1.6n)^{2r},\\
\ln m\varkappa &< r \ln 2(n + 1)(1.2n + 1) < 5r \ln n.
\end{aligned}
$$

Hence, for $\gamma \le c/20$, we have

$$\rho = \frac{c}{m\varkappa \log m\varkappa} > \frac{c}{10r(1.6n)^{2r} \ln n} > \frac{c}{20(2n)^{2r} \ln n} \ge \frac{\gamma}{(2n)^{2r} \ln n}.$$

So we have obtained the desired estimate of the sum $S(A)$ for points of the second class under condition (7.32).

But if a point A belongs to Ω_1, then, by the definition of points of the first class, we have

$$\alpha(t_1,\dots,t_r)=\frac{a_0(t_1,\dots,t_r)}{q_0(t_1,\dots,t_r)}+\beta_0(t_1,\dots,t_r),$$
$$\big(a_0(t_1,\dots,t_r),q_0(t_1,\dots,t_r)\big)=1,$$
$$|\beta_0(t_1,\dots,t_r)|\le P_1^{-t_1+0.1}P_2^{-t_2}\dots P_r^{-t_r},$$

and the least common multiple q_0 of the numbers $q_0(t_1,\dots,t_r)$ $(0\le t_1\le n_1,\dots,0\le t_r\le n_r,\ t_1+\dots+t_r\ge 1)$ is less than $P_1^{0.1}$. By the assumptions of the lemma, Q is the least common multiple of the numbers $q(t_1,\dots,t_r)$ in the D-approximations of $\alpha(t_1,\dots,t_r)$ $(0\le t_1\le n_1,\dots,0\le t_r\le n_r,\ t_1+\dots+t_r\ge 1)$ corresponding to $\tau(t_1,\dots,t_r)=P_1^{t_1-1/6}P_2^{t_2}\dots P_r^{t_r}$, and this number Q exceed $P_1^{0.1}$. Hence there is a set $(t_1,\dots,t_r)$ such that $q(t_1,\dots,t_r)\ne q_0(t_1,\dots,t_r)$. We will show that $q_0\ge 0.5P_1^{1/15}$. Indeed, we have

$$\begin{aligned}
&\frac{1}{q(t_1,\dots,t_r)q_0(t_1,\dots,t_r)}\\
&\quad\le\left|\frac{a(t_1,\dots,t_r)}{q(t_1,\dots,t_r)}-\frac{a_0(t_1,\dots,t_r)}{q_0(t_1,\dots,t_r)}\right|\le|\beta(t_1,\dots,t_r)|+|\beta_0(t_1,\dots,t_r)|\\
&\quad\le P_1^{-t_1+0.1}P_2^{-t_2}\dots P_r^{-t_r}+P_1^{-t_1+1/6}P_2^{-t_2}\dots P_r^{-t_r}q^{-1}(t_1,\dots,t_r),\\
&\quad q_0\ge q_0(t_1,\dots,t_r)\ge 0.5P_1^{1/15}.
\end{aligned}$$

Now, to estimate the sum S, we apply Lemma 7.10. We obtain

$$|S|\ll P_1\dots P_rP_1^{-1/(30n)}\ll P_1\dots P_rP_1^{-\rho}.$$

So we have proved the statement of the lemma for the parameters $P_1,\dots,P_r$ satisfying condition (7.32). If this condition is not satisfied, then there exists an s $(1\le s\le r-1)$ for which the following inequalities hold:

$$\begin{aligned}
P_r^{5/6}&\le P_1^{n_1+1}P_2^{n_2}\dots P_{r-1}^{n_{r-1}},\\
&\ \vdots\\
P_{s+2}^{5/6}&\le P_1^{n_1+1}P_2^{n_2}\dots P_{s+1}^{n_{s+1}},\\
P_{s+1}^{5/6}&> P_1^{n_1+1}P_2^{n_2}\dots P_s^{n_s}.
\end{aligned}\tag{7.34}$$

In this case we prove the lemma by induction on the parameter r. By the induction hypothesis, the statement of the lemma holds for all r that are less than some natural number r_0. Starting from this, we prove that the statement of the lemma holds for

$r = r_0$. In what follows, for simplicity, instead of r_0, we write r. If we need to estimate the sum $S(A)$ where the number of variables is less than r, then, as if it were already proved, we use the statement of the lemma with an appropriate change of the parameter r by a smaller value.

We write the polynomial $F(x_1, \ldots, x_r)$ in the form

$$F(x_1, \ldots, x_r) = \sum_{t_1=0}^{n_1} \cdots \sum_{t_s=0}^{n_s} g_{t_1,\ldots,t_s}(x_{s+1}, \ldots, x_r) x_1^{t_1} \ldots x_s^{t_s}. \tag{7.35}$$

Recall that Q denotes the least common multiple of $q(t_1, \ldots, t_r)$ in the D-approximations of $\alpha(t_1, \ldots, t_r)$ corresponding to $\tau(t_1, \ldots, t_r)$ $(0 \le t_1 \le n_1, \ldots, 0 \le t_r \le n_r, t_1 + \cdots + t_r \ge 1)$. By Q_0 we denote the least common multiple of the numbers $q(t_1, \ldots, t_r)$ satisfying the conditions $t_1 + \cdots + t_s \ge 1$, $t_{s+1} + \cdots + t_r \ge 1$, $0 \le t_1 \le n_1, \ldots, 0 \le t_r \le n_r$, and by Q_1 we denote the least common multiple of the numbers $q(t_1, \ldots, t_r)$ satisfying the conditions $t_1 = \cdots = t_s = 0$, $t_{s+1} + \cdots + t_r \ge 1$, $0 \le t_{s+1} \le n_{s+1}, \ldots, 0 \le t_r \le n_r$. Finally, by Q_2 we denote the least common multiple of the numbers $q(t_1, \ldots, t_r)$ satisfying the conditions $t_1 + \cdots + t_s \ge 1$, $t_{s+1} = \cdots = t_r = 0$, $0 \le t_1 \le n_1, \ldots, 0 \le t_s \le n_s$.

By the assumptions of the lemma, we have $Q = [Q_0, Q_1, Q_2]$. As in the case of double sums S, we separately consider the two cases: the case of large Q_0 and the case of small Q_0.

Let $Q_0 \ge P_1^{10n^2 m_s \rho_r}$. For each set $(t_1, \ldots, t_s)$ $(0 \le t_1 \le n_1, \ldots, 0 \le t_s \le n_s, t_1 + \cdots + t_s \ge 1)$, by $Q(t_1, \ldots, t_s)$ we denote the least common multiple of the numbers $q(t_1, \ldots, t_s, \ldots, t_r)$ satisfying the conditions $0 \le t_{s+1} \le n_{s+1}, \ldots, 0 \le t_r \le n_r$, $t_{s+1} + \cdots + t_r \ge 1$. It follows from the definition of the numbers Q_0 and $Q(t_1, \ldots, t_s)$ that Q_0 is equal to the least common multiple of the numbers $Q(t_1, \ldots, t_s)$ $(0 \le t_1 \le n_1, \ldots, 0 \le t_s \le n_s, t_1 + \cdots + t_s \ge 1)$. Therefore, there exists a set $(t_1, \ldots, t_s) = N$ such that the following inequalities hold:

$$Q(t_1, \ldots, t_s) \ge Q_0^{1/m_s} \ge P_1^{10n^2 \rho_r}.$$

For this set, we assume that $Q_3 = Q(t_1, \ldots, t_3)$.

Depending on the value of Q_3, we consider the following three cases:

(a) $P_1^{10n^2 \rho_r} \le Q_3 < P_{s+1}^{0.1}$;

(b) $P_{s+1}^{0.1} \le Q_3$ and the inequalities

$$q(t_1, \ldots, t_s, \ldots, t_r) \le P_{s+1}^{t_{s+1}-1/6} P_{s+2}^{t_{s+2}} \ldots P_r^{t_r}$$

hold for $0 \le t_{s+1} \le n_{s+1}, \ldots, 0 \le t_r \le n_r$, $t_{s+1} + \cdots + t_r \ge 1$;

(c) $P_{s+1}^{0.1} \le Q_3$ and there exists a set $(t_{s+1}, \ldots, t_r)$ such that

$$q(t_1, \ldots, t_s, \ldots, t_r) \ge P_{s+1}^{t_{s+1}-1/6} P_{s+2}^{t_{s+2}} \ldots P_r^{t_r}.$$

For the set $(t_1, \dots, t_s) = N$ mentioned above, we set

$$g(x_{s+1}, \dots, x_r) = g_{t_1,\dots,t_s}(x_{s+1}, \dots, x_r)$$

in (7.35), and moreover, we have

$$\tau_1 = \tau_1(t_1, \dots, t_s) = P_1^{t_1-1/6} P_2^{t_2} \dots P_s^{t_s}.$$

We consider the D-approximations of the fractional parts of the polynomial $g(x_{s+1}, \dots, x_r)$ corresponding to τ_1, i.e., we consider the relations

$$\{g(x_{s+1}, \dots, x_r)\} = \frac{b}{l} + \frac{\theta}{l\tau_1}, \tag{7.36}$$

where $(b, l) = 1$, $1 \le l \le \tau$, and $|\theta| \le 1$.

We consider case (a). We represent the sum S in the form

$$S = S_1 + S_2 + S_3,$$

where

$$S_j = \sum \cdots \sum_{(x_{s+1},\dots,x_r) \in T_j} \sum_{x_1=1}^{P_1} \cdots \sum_{x_s=1}^{P_s} \exp\{2\pi i F(x_1, \dots, x_s, x_{s+1}, \dots, x_r)\},$$
$$j = 1, 2, 3,$$

and the domain of summation T_j over the variables $x_{s+1}, \dots, x_r$ in each of the sums S_j is its own and is determined as follows. We consider the inner sum over $x_1, \dots, x_s$:

$$S(x_{s+1}, \dots, x_r) = \sum_{x_1=1}^{P_1} \cdots \sum_{x_s=1}^{P_s} \exp\{2\pi i F(x_1, \dots, x_s, \dots, x_r)\} \tag{7.37}$$

$$= \sum_{x_1=1}^{P_1} \cdots \sum_{x_s=1}^{P_s} \exp\left\{2\pi i \left(\sum_{t_1=0}^{n_1} \cdots \sum_{t_s=0}^{n_s} G(t_1, \dots, t_s) x_1^{t_1} \dots x_s^{t_s} \right)\right\},$$

where the numbers $G(t_1, \dots, t_s)$ depend on $x_{s+1}, \dots, x_r$ and

$$G(t_1, \dots, t_s) = g_{t_1,\dots,t_s}(x_{s+1}, \dots, x_r) = g(x_{s+1}, \dots, x_r).$$

If a point G with coordinates $G(t_1, \dots, t_s)$ $(0 \le t_1 \le n_1, \dots, 0 \le t_s \le n_s,$ $t_1 + \cdots + t_s \ge 1)$ in the m_s-dimensional space is a point of the second class with respect to the parameters $P_1, \dots, P_s$, then the corresponding set $x_{s+1}, \dots, x_r$ belongs to the set T_1. If this point is a point of the first class, but its coordinate $G(t_1, \dots, t_s) = g(x_{s+1}, \dots, x_r)$ satisfies relation (7.36) with $l > H = P_1^{2n\rho_r}$, then the corresponding set $(x_{s+1}, \dots, x_r)$ belongs to the set T_2. All other sets belong to the set T_3.

For $(x_{s+1}, \dots, x_r)$ contained in the set T_1, the point G is a point of the second class and hence it belongs to the domain ω_1 or to the domain ω_2 introduced in Lemma 7.11. If the point G belongs to ω_2, then in this case the least common multiple Q^* of the denominators in the D-approximations of the numbers $g(t_1, \dots, t_s)$, corresponding to $\tau(t_1, \dots, t_s)$, is no less than $P_1^{0.1}$. Hence, by the induction assumption, we have the estimate

$$|S(x_{s+1}, \dots, x_r)| \ll P_1 \dots P_s P_1^{-\rho_s}, \quad \rho_s = \frac{\gamma}{(2n)^{2s} \log n}.$$

But if the point G belongs to ω_1, then, by the definition of the domain ω_1 and by Lemma 7.11, we obtain

$$|S(x_{s+1}, \dots, x_r)| \ll P_1 \dots P_s P_1^{-1/(20n)} \ll P_1 \dots P_s P_1^{-\rho_r}.$$

For the values contained in T_2, the least common multiple of the denominators of rational fractions in representation (7.36) is larger than H. Hence, by Lemma 7.10, the sum $S(x_{s+1}, \dots, x_r)$ satisfies the estimate

$$|S(x_{s+1}, \dots, x_r)| \ll P_1 \dots P_s H^{-1/(2n)} \ll P_1 \dots P_s P_1^{-\rho_r}.$$

Now we given an estimate from above for Y, i.e., for the number of the sets $(x_{s+1}, \dots, x_r)$ contained in T_3. In this case, the fractional parts of the polynomial $g(x_{s+1}, \dots, x_r)$ are contained at least in one of the intervals of the form $[b/l - 1/(l\tau_1), b/l + 1/(l\tau_1)]$ and $l \le H$. The number Γ of the fractional parts of the polynomial $g(x_{s+1}, \dots, x_r)$ contained in one of these intervals does not exceed

$$\Gamma_1 \ll P_{s+1} \dots P_r \big((l\tau_1)^{-1} + P_1^{-5n\rho_r}\big).$$

Indeed, the polynomial $g(x_{s+1}, \dots, x_r)$ has the form

$$g(x_{s+1}, \dots, x_r) = \sum_{t_{s+1}=0}^{n_{s+1}} \cdots \sum_{t_r=0}^{n_r} \alpha(t_1, \dots, t_s, \dots, t_r) x_{s+1}^{t_{s+1}} \dots x_r^{t_r}$$

and, by the assumption of the lemma, its coefficients satisfy the relation

$$\alpha(t_1, \dots, t_r) = \frac{a(t_1, \dots, t_r)}{q(t_1, \dots, t_r)} + \beta(t_1, \dots, t_r),$$

where

$$\begin{aligned} |\beta(t_1, \dots, t_r)| &\le \big(q(t_1, \dots, t_r)\tau(t_1, \dots, t_r)\big)^{-1} \\ &\le P_1^{-t_1+1/6} P_2^{-t_2} \dots P_r^{-t_r} \le P_{s+1}^{-t_{s+1}} \dots P_r^{-t_r}, \end{aligned}$$

and the least common multiple of the denominators $q(t_1, \dots, t_r)$, equal to Q_3, does not exceed $P_{s+1}^{0.1}$, i.e., the point with coordinates $\alpha(t_1, \dots, t_s, \dots, t_r)$ $(0 \le t_{s+1} \le$

$n_{s+1}, \dots, 0 \le t_r \le n_r, t_{s+1} + \dots + t_r \ge 1$) belongs to the first class with respect to the parameters $P_{s+1}, \dots, P_r$. Therefore, applying Lemma 7.13, item (1), (a), with Q equal to $Q_3 > P_1^{10n^3\rho_r}$ to estimate Γ_1, we obtain the estimate written above. Hence Y satisfies the estimate

$$Y \ll H^2\Gamma_1 \ll P_{s+1} \dots P_r(\tau_1^{-1} + P_1^{-5n\rho_r}) \ll P_{s+1} \dots P_r P_1^{-\rho_r}.$$

We substitute the obtained estimates for $S(x_{s+1}, \dots, x_r)$ into the sums S_1 and S_2 and estimate the sum S_3 trivially by the number of terms. We obtain

$$|S| \le |S_1| + |S_2| + |S_3| \ll P_1 \dots P_r P_1^{-\rho_r}.$$

Now we consider case (b). We represent the sum S as

$$S = S_4 + S_5 + S_6,$$

where

$$S_j = \sum \dots \sum_{(x_{s+1},\dots,x_r)\in T_j} \sum_{x_1=1}^{P_1} \dots \sum_{x_s=1}^{P_s} \exp\{2\pi i F(x_1, \dots, x_r)\}, \quad j = 4, 5, 6,$$

and the domain of summation T_j over the variables $x_{s+1}, \dots, x_r$ in each of the sums is its own and is determined as follows. We consider representation (7.37) of the inner sum over $x_1, \dots, x_s$. If a point G with coordinates $G(t_1, \dots, t_s)$ $(0 \le t_1 \le n_1, \dots, 0 \le t_s \le n_s, t_1 + \dots + t_s \ge 1)$ is a point of the second class with respect to the parameters $P_1, \dots, P_s$, then the corresponding set $(x_{s+1}, \dots, x_r)$ belongs to the set T_4. If this point is a point of the first class, but its coordinate $G(t_1, \dots, t_s) = g(x_{s+1}, \dots, x_r)$ satisfies relation (7.36) with $l \ge H_1 = P_{s+1}^{0.25\rho_{r-s}}$, then the corresponding set $(x_{s+1}, \dots, x_r)$ belongs to the set T_5. All other sets $(x_{s+1}, \dots, x_r)$ belong to the set T_6.

If the set $(x_{s+1}, \dots, x_r)$ belongs to the set T_4, the point G either belongs to the domain ω_1 (in this case we use Lemma 7.11) or to the domain ω_2 (in this case we use the induction hypothesis). We obtain

$$|S(x_{s+1}, \dots, x_r)| \ll P_1 \dots P_s P_1^{-\rho_r}.$$

Now we estimate the sum $S(x_{s+1}, \dots, x_r)$ for the sets $(x_{s+1}, \dots, x_r)$ contained in T_5. By relation (7.36), we have

$$G(t_1, \dots, t_s) = \frac{b}{l} + \frac{\theta}{l\tau_1}.$$

If $l \ge P_1^{0.1}$, then, by the induction hypothesis we have

$$|S(x_{s+1}, \dots, x_r)| \ll P_1 \dots P_s P_1^{-\rho_s}.$$

Let $l < P_1^{0.1}$. Since the point G belongs to the first class, acting similarly to the case of the sum S_5 for the double sum S, we see that the least common multiple of the denominators of the fractions determining the first class is larger than H_1. Hence we have

$$|S(x_{s+1}, \dots, x_r)| \ll P_1 \dots P_s H_1^{-1/n+\varepsilon} \ll P_1 \dots P_s P_1^{-\rho_r}.$$

Now we estimate the number of the sets $(x_{s+1}, \dots, x_r)$ contained in T_6. As in the case of the set T_3, for these sets $(x_{s+1}, \dots, x_r)$, the fractional parts of the polynomial $g(x_{s+1}, \dots, x_r)$ are contained at least in one of the intervals of the form

$$\left[\frac{b}{l} - \frac{1}{l\tau_1}, \frac{b}{l} + \frac{1}{l\tau_1}\right], \quad l \leq H_1.$$

By Lemma 7.13, item (3), the number Γ of the sets $(x_{s+1}, \dots, x_r)$ contained in one of the above intervals does not exceed

$$\Gamma_1 \ll P_{s+1} \dots P_r\bigl((l\tau_1)^{-1} + P_{s+1}^{-\rho}\bigr), \tag{7.38}$$

where the variable ρ is defined in Lemma 7.12 and is equal to

$$\rho = c(k_s\varkappa \log k_s\varkappa)^{-1}, \quad \varkappa = n_{s+1} + \nu_{s+2}n_{s+2} + \dots + \nu_r n_r,$$
$$-1 \leq \frac{\log P_t}{\log P_{s+1}} - \nu_t \leq 0, \quad t = s+2, \dots, r, \quad k_s = m_r m_s^{-1}.$$

We show that

$$\rho = \frac{c}{k_s\varkappa \log k_s\varkappa} \geq \frac{\gamma}{(2n)^{2(r-s)} \log n} = \rho_{r-s}.$$

We set

$$\frac{P_{s+1}}{\log P_{s+1}} = z_{s+1} = 1, \quad \frac{P_{s+2}}{\log P_{s+1}} = z_{s+2} = 1, \quad \dots, \quad \frac{P_r}{\log P_{s+1}} = z_r = 1.$$

Since $\nu_t \leq z_t + 1$ $(t = s+2, \dots, r)$, we have the following upper bound for $\varkappa$:

$$\varkappa \leq n\left(r - s - 1 + \sum_{t=s+1}^{r} z_t\right).$$

Relations (7.34) imply the inequalities

$$P_r^{5/6} \leq P_{s+1}^{n_{s+1}+5/6} P_{s+2}^{n_{s+2}} \dots P_{r-1}^{n_{r-1}},$$
$$\vdots$$
$$P_{s+3}^{5/6} \leq P_{s+1}^{n_{s+1}+5/6} P_{s+2}^{n_{s+2}},$$
$$P_{s+2}^{5/6} \leq P_{s+1}^{n_{s+1}+5/6}.$$

Taking logarithms of these inequalities, we obtain

$$z_{s+1} = 1, \qquad z_{s+2} \le 1.2nz_{s+1} + 1,$$
$$z_{s+3} \le 1.2n(z_{s+1} + z_{s+2}) + 1,$$
$$\vdots$$
$$z_r \le 1.2n(z_{s+1} + \cdots + z_{r-1}) + 1.$$

We use the above inequalities to estimate $\varkappa$. We obtain

$$\varkappa \le n\left(r - s - 1 + \sum_{t=s+1}^{n} z_t\right) = n\left(r - s - 1 + z_r + \sum_{t=s+1}^{r-1} z_t\right)$$
$$\le \left(r - s - 1 + 1 + (1.2n + 1)\sum_{t=s+1}^{r-1} z_t\right) \le \dots$$
$$\le \left(r - s - 1 + r - s - 1 + (1.2n + 1)^{r-s-1}\right) \le 2(1.2n + 1)^{r-s}.$$

Hence for $r \ge 3$ and $n \ge 2$, we have

$$k_s \le (n + 1)^{r-s}, \quad k_s\varkappa \le 2(n + 1)^{r-s}(1.2n + 1)^{r-s} \le 2(1.6n)^{2(r-s)}$$
$$\log k_s\varkappa \le (r - s)\log 2(n + 1)(1.2n + 1) \le 5(r - s)\log n.$$

Hence for $\gamma < c/20$, we obtain

$$\rho = \frac{c}{k_s\varkappa \log k_s\varkappa} > \frac{c}{20(2n)^{2(r-s)}\log n} \ge \rho_{r-s}.$$

Thus it follows from (7.38) that

$$\Gamma_1 \ll P_{s+1}\dots P_r\left((l\tau_1)^{-1} + P_{s+1}^{-\rho_{r-s}}\right)$$

and the number of sets Y contained in T_6 does not exceed $H_1^2\Gamma_1$,

$$Y \ll H_1^2\Gamma_1 \ll P_{s+1}\dots P_r P_{s+1}^{-0.5\rho_{r-s}} \ll P_{s+1}\dots P_r P_1^{-\rho_r}.$$

In case (b), we finally obtain

$$|S| \le |S_4| + |S_5| + |S_6| \ll P_1\dots P_r P_1^{-\rho_r}.$$

Now we study case (c). Here we have $Q_3 \ge P_{s+1}^{0.1}$, and there exists a set $(t_{s+1}, \dots, t_r)$ for which the denominator $q = q(t_1, \dots, t_s, \dots, t_r)$ of the fraction in the D-approximation of the number $\alpha(t_1, \dots, t_s, \dots, t_r)$, corresponding to $\tau = \tau(t_1, \dots, t_s, \dots, t_r)$, satisfies the condition

$$\eta = P_{s+1}^{t_{s+1}-1/6}P_{s+2}^{t_{s+2}}\dots P_r^{t_r} < q \le \tau.$$

We consider new D-approximations of the numbers $\alpha(t_1, \dots, t_s, \dots, t_r)$ for all the sets $(t_{s+1}, \dots, t_r)$ $(0 \le t_{s+1} \le n_{s+1}, \dots, 0 \le t_r \le n_r, t_{s+1} + \dots + t_r \ge 1)$ corresponding to

$$\eta(t_{s+1}, \dots, t_r) = P_{s+1}^{t_{s+1}-1/6} P_{s+2}^{t_{s+2}} \dots P_r^{t_r};$$

in other words, we consider the representations

$$\begin{gathered}\alpha(t_1, \dots, t_s, \dots, t_r) = \frac{d(t_{s+1}, \dots, t_r)}{h(t_{s+1}, \dots, t_r)} + \frac{\theta}{h(t_{s+1}, \dots, t_r)\tau(t_{s+1}, \dots, t_r)}, \\ \big(d(t_{s+1}, \dots, t_r), h(t_{s+1}, \dots, t_r)\big) = 1, \qquad (7.39) \\ 1 \le h(t_{s+1}, \dots, t_r) \le \eta(t_{s+1}, \dots, t_r), \quad |\theta| = |\theta(t_{s+1}, \dots, t_r)| \le 1.\end{gathered}$$

First, we assume that the least common multiple Q_4 of $h(t_{s+1}, \dots, t_r)$ $(0 \le t_{s+1} \le n_{s+1}, \dots, 0 \le t_r \le n_r, t_{s+1} + \dots + t_r \ge 1)$ is larger than $P_{s+1}^{0.1}$. Then, in fact, the sum S can be estimated as in case (b). We represent the sum S as

$$S = S_7 + S_8 + S_9,$$

where

$$S_j = \sum \cdots \sum_{(x_{s+1},\dots,x_r)\in T_j} \sum_{x_1=1}^{P_1} \cdots \sum_{x_s=1}^{P_s} \exp\{2\pi i F(x_1, \dots, x_r)\}, \quad j = 7, 8, 9,$$

and the domain of summation T_j over the variables $x_{s+1}, \dots, x_r$ in each of the sums is its own and is determined as follows. We consider representation (7.37) of the inner sum over $x_1, \dots, x_s$. If G is a point of the second class with respect to the parameters $P_1, \dots, P_s$, then the corresponding set $(x_{s+1}, \dots, x_r)$ belongs to T_7. If this point is a point of the first class, but its coordinate $G(t_1, \dots, t_s) = g(x_{s+1}, \dots, x_r)$ satisfies relation (7.36) with $l \ge H_1 = P_1^{0.25\rho_{r-s}}$, then the corresponding set $(x_{s+1}, \dots, x_r)$ belongs to the set T_8. All other sets $(x_{s+1}, \dots, x_r)$ belong to the set T_9.

The sums S_7 and S_8 are estimated precisely as the sums S_4 and S_5. The sum S_9 can be estimated similarly to the sum S_6; in this case the estimates are the same, but the coefficients of the polynomial $g(x_{s+1}, \dots, x_r)$ for the sum S_9 have a somewhat different representation.

Now we assume that, in the new D-approximation (7.39), the least common multiple Q_4 of the numbers $h(t_{s+1}, \dots, t_r)$ does not exceed $P_{s+1}^{0.1}$. Then the following two cases are possible:

$$P_{s+1}^{0.1} \ge Q_4 > H_2 = P_1^{10n^2\rho_r} \quad \text{and} \quad Q_4 \le H_2.$$

First, we consider the case $Q_4 > H_2$. Here the sum S is estimated similarly to the sum S in case (a) considered above. We again divide the sum S into three sums:

$$S = S_{10} + S_{11} + S_{12},$$

where

$$S_j = \sum_{(x_{s+1},\dots,x_r)\in T_j} \cdots \sum \sum_{x_1=1}^{P_1} \cdots \sum_{x_s=1}^{P_s} \exp\{2\pi i F(x_1,\dots,x_r)\}, \quad j = 10, 11, 12,$$

and the domain of summation T_j over the variables $(x_{s+1},\dots,x_r)$ in each of the sums is its own and is determined as follows. We consider representation (7.37) of the inner sum over $x_1,\dots,x_s$. If G is a point of the second class, then the corresponding set $(x_{s+1},\dots,x_r)$ belongs to the set T_{10}. If this point is a point of the first class, but its coordinate $G(t_1,\dots,t_s) = g(x_{s+1},\dots,x_r)$ satisfies relation (7.36) with $l > H = P_1^{2n\rho_r}$, then the corresponding set belongs to the set T_{11}. All other sets $(x_{s+1},\dots,x_r)$ belong to the set T_{12}.

Let us estimate the sum $S(x_{s+1},\dots,x_r)$ contained in S_{10}. In the case of points G from the set ω_1, we use Lemma 7.11, and in the case of points G from the set ω_2, we apply the induction hypothesis to the sum $S(x_{s+1},\dots,x_r)$. We obtain

$$|S(x_{s+1},\dots,x_r)| \ll P_1 \dots P_s P_1^{-\rho_s}, \qquad |S_{10}| \ll P_1 \dots P_r P_1^{-\rho_r}.$$

Now we estimate the sum S_{11}. If $l \geq P_1^{0.1}$, then, by the induction hypothesis, we obtain the estimate

$$|S(x_{s+1},\dots,x_r)| \ll P_1 \dots P_s P_1^{-\rho_s}.$$

Let $l < P_1^{0.1}$. We shall use the fact that G is a point of the first class, i.e., the relations

$$\begin{gathered} G(t_1,\dots,t_s) = \frac{a(t_1,\dots,t_s)}{q(t_1,\dots,t_s)} + \beta(t_1,\dots,t_s), \\ \big(a(t_1,\dots,t_s), q(t_1,\dots,t_s)\big) = 1, \\ |\beta(t_1,\dots,t_s)| \leq P_1^{-t_1+0.1} P_2^{-t_2} \dots P_s^{-t_s}, \\ t_1 + \dots + t_s \geq 1, \quad 0 \leq t_1 \leq n_1, \dots, 0 \leq t_s \leq n_s, \end{gathered}$$

hold and the least common multiple q of all the numbers $q(t_1,\dots,t_s)$ does not exceed $P_1^{0.1}$. As before, in estimating the double sums in the case of S_5, we show that

$$\frac{a(t_1,\dots,t_s)}{q(t_1,\dots,t_s)} = \frac{b}{l}$$

and therefore $q \geq q(t_1,\dots,t_s) = l > H$. We now apply Lemma 7.10 to the sum $S(x_{s+1},\dots,x_r)$ and obtain

$$|S(x_{s+1},\dots,x_r)| \ll P_1 \dots P_s H^{-1/n+\varepsilon} \ll P_1 \dots P_s P_1^{-\rho_r}.$$

We trivially estimate the sum S_{12} by the number of terms:

$$|S_{12}| \leq P_1 \dots P_s Y,$$

where Y is the number of sets $(x_{s+1}, \dots, x_r)$ for which the fractional parts of the polynomial $g(x_{s+1}, \dots, x_r)$ are contained at least in one of the intervals of the form

$$\left[\frac{b}{l} - \frac{1}{l\tau_1}, \frac{b}{l} + \frac{1}{l\tau_1}\right], \quad (b, l) = 1, \quad l \le H. \tag{7.40}$$

The coefficients of the polynomial $g(x_{s+1}, \dots, x_r)$ generate a point A_1 with coordinates $\alpha(t_1, \dots, t_s, t_{s+1}, \dots, t_r)$ $(0 \le t_{s+1} \le n_{s+1}, \dots, 0 \le t_r \le n_r, t_{s+1} + \dots + t_r \ge 1)$.

First, we assume that the point A_1 belongs to the second class Ω_2. The domain Ω_2 consists of domains ω_1 and ω_2. If the point A_1 belongs to the domain ω_1, then, by Lemma 7.13, item (2), the number Γ of sets $(x_{s+1}, \dots, x_r)$ contained in one of the intervals of the form (7.40) does not exceed

$$\Gamma_1 \ll P_{s+1} \dots P_r\big((l\tau_1)^{-1} + P_{s+1}^{-1/(20n)}\big).$$

If the point A_1 belongs to the domain ω_2, then, by Lemma 7.13, item (3), we have

$$\Gamma_1 \ll P_{s+1} \dots P_r\big((l\tau_1)^{-1} + \exp\{32\varkappa\}P_{s+1}^{-\rho}\big),$$

where

$$\begin{gathered}
\rho = c(k_s\varkappa \log k_s\varkappa)^{-1}, \quad k_s = m_r m_s^{-1}, \\
\varkappa = n_{s+1} + n_{s+2}\nu_{s+2} + \dots + n_r\nu_r, \\
-1 \le \log P_t / \log P_{s+1} - \nu_t \le 0, \quad t = s+2, \dots, r.
\end{gathered}$$

Repeating the argument used in estimating the sum S_6 word for word, we obtain the inequality $\rho \ge \rho_{r-s}$. Hence

$$\Gamma \ll \Gamma_1 \ll P_{s+1} \dots P_r\big((l\tau_1)^{-1} + P_{s+1}^{-\rho_{r-s}}\big).$$

Now let A_1 be a point of the first class. Then we can represent the coordinates $\alpha(t_1, \dots, t_s, t_{s+1}, \dots, t_r)$ of the point A_1 as

$$\begin{gathered}
\alpha(t_1, \dots, t_s, t_{s+1}, \dots, t_r) = \frac{a(t_{s+1}, \dots, t_r)}{q(t_{s+1}, \dots, t_r)} + \beta(t_{s+1}, \dots, t_r), \\
|\beta(t_{s+1}, \dots, t_r)| \le P_{s+1}^{-t_{s+1}+0.1} P_{s+2}^{-t_{s+2}} \dots P_r^{-t_r}, \\
0 \le t_{s+1} \le n_{s+1}, \dots, 0 \le t_r \le n_r, \quad t_{s+1} + \dots + t_r \ge 1,
\end{gathered}$$

and the least common multiple q of the numbers $q(t_{s+1}, \dots, t_r)$ $(0 \le t_{s+1} \le n_{s+1}, \dots, 0 \le t_r \le n_r, t_{s+1} + \dots + t_r \ge 1)$ does not exceed $P_{s+1}^{0.1}$. Moreover, for the coordinates $\alpha(t_1, \dots, t_r)$ of the point A_1, relations (7.39) hold for $H_2 < Q \le P_{s+1}^{0.1}$, namely,

$$\alpha(t_1, \dots, t_s, t_{s+1}, \dots, t_r) = \frac{d(t_{s+1}, \dots, t_r)}{h(t_{s+1}, \dots, t_r)} + \frac{\theta}{h(t_{s+1}, \dots, t_r)\eta(t_{s+1}, \dots, t_r)},$$

$$\eta(t_{s+1},\dots,t_r)=P_{s+1}^{t_{s+1}-1/6}P_{s+2}^{t_{s+2}}\dots P_r^{t_r},\quad |\theta|\le 1,$$
$$0\le t_{s+1}\le n_{s+1},\dots,0\le t_r\le n_r,\quad t_{s+1}+\dots+t_r\ge 1,$$
$$Q_4=\underset{t_{s+1},\dots,t_r}{\text{l.c.m.}}\big(h(t_{s+1},\dots,t_r)\big).$$

We show that the following relations hold for all the sets $(t_{s+1},\dots,t_r)$:

$$\frac{a(t_{s+1},\dots,t_r)}{q(t_{s+1},\dots,t_r)}=\frac{d(t_{s+1},\dots,t_r)}{h(t_{s+1},\dots,t_r)}.$$

Assume the contrary, i.e., assume that there is a set $(t_{s+1},\dots,t_r)$ such that

$$\frac{a(t_{s+1},\dots,t_r)}{q(t_{s+1},\dots,t_r)}\ne\frac{d(t_{s+1},\dots,t_r)}{h(t_{s+1},\dots,t_r)};$$

then, on the one hand,

$$\left|\frac{a(t_{s+1},\dots,t_r)}{q(t_{s+1},\dots,t_r)}-\frac{d(t_{s+1},\dots,t_r)}{h(t_{s+1},\dots,t_r)}\right|\ge\frac{1}{q(t_{s+1},\dots,t_r)h(t_{s+1},\dots,t_r)}\ge P_{s+1}^{-0.2},$$

since $q(t_{s+1},\dots,t_r)\le q<P_{s+1}^{0.1}$ and $h(t_{s+1},\dots,t_r)\le Q_4\le P_{s+1}^{0.1}$; on the other hand, we have

$$\begin{aligned}&\left|\frac{a(t_{s+1},\dots,t_r)}{q(t_{s+1},\dots,t_r)}-\frac{d(t_{s+1},\dots,t_r)}{h(t_{s+1},\dots,t_r)}\right|\\&\quad\le|\beta(t_1,\dots,t_r)|+\frac{1}{h(t_{s+1},\dots,t_r)\eta(t_{s+1},\dots,t_r)}\\&\quad\le P_{s+1}^{-t_{s+1}+0.1}P_{s+2}^{-t_{s+2}}\dots P_r^{-t_r}+P_{s+1}^{-t_{s+1}+1/6}P_{s+2}^{-t_{s+2}}\dots P_r^{-t_r}\le 2P_{s+1}^{-5/6},\end{aligned}$$

since

$$|\beta(t_{s+1},\dots,t_r)|\le P_{s+1}^{-t_{s+1}+0.1}P_{s+2}^{-t_{s+2}}\dots P_r^{-t_r},\quad h(t_{s+1},\dots,t_r)\ge 1,$$
$$\eta(t_{s+1},\dots,t_r)=P_{s+1}^{t_{s+1}-1/6}P_{s+2}^{t_{s+2}}\dots P_r^{t_r}.$$

The estimates obtained for

$$\left|\frac{a(t_{s+1},\dots,t_r)}{q(t_{s+1},\dots,t_r)}-\frac{d(t_{s+1},\dots,t_r)}{h(t_{s+1},\dots,t_r)}\right|$$

contradict each other. So for all the sets $(t_{s+1},\dots,t_r)$ $(0\le t_{s+1}\le n_{s+1},\dots,$ $0\le t_r\le n_r, t_{s+1}+\dots+t_r\ge 1)$ we have

$$\frac{a(t_{s+1},\dots,t_r)}{q(t_{s+1},\dots,t_r)}=\frac{d(t_{s+1},\dots,t_r)}{h(t_{s+1},\dots,t_r)}.$$

This implies $q=Q_4>H_2$.

Now we estimate Γ by Lemma 7.13, item (1), as follows:

$$\Gamma \ll P_{s+1} \dots P_r \big((l\tau_1)^{-1} + H_2^{-1/n+\varepsilon} \big).$$

So the number Γ of sets $(x_{s+1}, \dots, x_r)$ contained in one of the intervals of the form (7.40) does not exceed

$$\Gamma \ll P_{s+1} \dots P_r \big((l\tau_1)^{-1} + P_{s+1}^{-\rho_{r-s}} + P_{s+1}^{-1/(20n)} + H_2^{-1/(2n)} \big).$$

Since the number of intervals of the form (7.40) does not exceed H^2, we have

$$Y \le H^2 \Gamma \ll P_{s+1} \dots P_r H^2 H_2^{-1/(2n)} \ll P_{s+1} \dots P_r P_1^{-\rho_r}.$$

Hence

$$|S| \le |S_{10}| + |S_{11}| + |S_{12}| \ll P_1 \dots P_r P_1^{-\rho_r}.$$

Now we consider the case $Q_4 \le H_2$. We represent the sum S in the form

$$S = S_{13} + S_{14} + S_{15},$$

where

$$S_j = \sum \cdots \sum_{(x_{s+1},\dots,x_r) \in T_j} \sum_{x_1=1}^{P_1} \cdots \sum_{x_s=1}^{P_s} \exp\{2\pi i F(x_1, \dots, x_r)\}, \quad j = 13, 14, 15,$$

and the domain of summation T_j over the variables $x_{s+1}, \dots, x_r$ in each of the sums is its own and is determined as follows. We consider representation (7.37) of the inner sum over $x_1, \dots, x_s$. If G is a point of the second class, then the corresponding set $(x_{s+1}, \dots, x_r)$ belongs to the set T_{13}. If this point is a point of the first class, and moreover, in the D-representation of the fractional parts of the polynomial $g(x_{s+1}, \dots, x_r)$ of the form

$$\begin{gathered} \{g(x_{s+1}, \dots, x_r)\} = \frac{b}{l} + \beta, \quad (b, l) = 1, \quad 1 \le l \le \tau_1, \\ \tau_1 = P_1^{t_1 - 1/6} P_2^{t_2} \dots P_s^{t_s}, \quad |\beta| \le (l\tau_1)^{-1}, \quad \delta = P_1^{t_1} \dots P_s^{t_s} |\beta|, \end{gathered} \tag{7.41}$$

the variables l and δ satisfy the inequalities $l > H = P_1^{2n\rho_r}$ and $\delta > H$, then the corresponding set $(x_{s+1}, \dots, x_r)$ belongs to the set T_{14}. Finally, all the other sets $(x_{s+1}, \dots, x_r)$ belong to the set T_{15}.

We estimate each of the sums $S(x_{s+1}, \dots, x_r)$ contained in S_{13}. The second class Ω_2 consists of two sets ω_1 and ω_2. If a point G belongs to ω_1, then, by Lemma 7.11, we have

$$|S(x_{s+1}, \dots, x_r)| \ll P_1 \dots P_s P_1^{-1/(20n)}.$$

But if a point G belongs to ω_2, then, by the induction hypothesis, we obtain

$$|S(x_{s+1}, \dots, x_r)| \ll P_1 \dots P_s P_1^{-\rho_s},$$

and hence

$$|S_{13}| \ll P_1 \dots P_r P_1^{-\rho_r}.$$

Now we consider the sum S_{14}. If in representation (7.41), the variable l is larger than $P_1^{0.1}$, then, by the induction hypothesis, we obtain

$$|S(x_{s+1}, \dots, x_r)| \ll P_1 \dots P_s P_1^{-\rho_s}.$$

Now we assume that l does not exceed $P_1^{0.1}$. The point G belongs to the first class, i.e., its coordinates satisfy the relations

$$G(t_1, \dots, t_s) = \frac{a(t_1, \dots, t_s)}{q(t_1, \dots, t_s)} + \beta(t_1, \dots, t_s),$$
$$\big(a(t_1, \dots, t_s), q(t_1, \dots, t_s)\big) = 1, \quad |\beta(t_1, \dots, t_s)| \le P_1^{-t_1+0.1} P_2^{-t_2} \dots P_s^{-t_s},$$
$$0 \le t_1 \le n_1, \dots, 0 \le t_s \le n_s, \quad t_1 + \dots + t_s \ge 1,$$

and the least common multiple q of all the numbers $q(t_1, \dots, t_s)$ is less than $P_1^{0.1}$. Similarly to the case of the sum S_{12}, this implies

$$\frac{a(t_1, \dots, t_s)}{q(t_1, \dots, t_s)} = \frac{b}{l}$$

and hence $q \ge q(t_1, \dots, t_s) = l$ and $\beta(t_1, \dots, t_s) = \beta$.

By the definition of the sum S_{14}, we have either $q > H$ or $\delta > H$. Hence, by Lemma 7.10, item (b), we obtain

$$|S(x_{s+1}, \dots, x_r)| \ll P_1 \dots P_s H^{-1/n+\varepsilon} \ll P_1 \dots P_s P_1^{-\rho_r}.$$

Thus we have

$$|S_{14}| \ll P_1 \dots P_s P_1^{-\rho_r}.$$

Now we consider the sum S_{15}. We trivially estimate this sum by the number of terms as follows:

$$|S_{15}| \ll P_1 \dots P_s Y,$$

where Y is the number of sets $(x_{s+1}, \dots, x_r)$ $(1 \le x_{s+1} \le P_{s+1}, \dots, 1 \le x_r \le P_r)$ for which the fractional parts of the polynomial $g(x_{s+1}, \dots, x_r)$ are contained at least in one of the intervals $[b/l - \Delta, b/l + \Delta]$, where

$$(b, l) = 1, \quad l \le H = P_1^{2n\rho_r}, \quad \Delta = P_1^{-t_1} \dots P_s^{-t_s} P_1^{2n\rho_r}.$$

Here the variable Y is defined as in Lemma 7.14. Hence, by this lemma, we have $Y \ll P_{s+1} \dots P_r P_1^{-\rho_r}$. This implies

$$|S_{15}| \ll P_1 \dots P_s P_1^{-\rho_r},$$

and hence

$$|S| = |S_{13}| + |S_{14}| + |S_{15}| \ll P_1 \dots P_r P_1^{-\rho_r}.$$

Thus we have estimated the sum S for $Q_0 > P_1^{10m_s n^2 \rho_r}$.

Now we consider the case $Q_0 \le P_1^{10m_s n^2 \rho_r}$. Since $Q = [Q_0, Q_1, Q_2]$ and Q is larger than $P_1^{0.1}$, we have either $Q_1 \ge P_1^{0.05-5m_s n^2 \rho_r}$ or $Q_1 \le P_1^{0.05-5m_s n^2 \rho_r}$.

First, let $Q_1 \ge P_1^{0.05-5m_s n^2 \rho_r}$. We write the variables x_j $(1 \le j \le r)$ as follows:

$$x_j = Q_0 y_j + z_j, \quad 0 \le z_j \le Q_0, \quad -z_j Q_0^{-1} < y_j \le (P_j - z_j) Q_0^{-1}.$$

Recall that the D-approximations of the coefficients $\alpha(t_1, \dots, t_r)$ of the polynomial $F(x_1, \dots, x_r)$ corresponding to $\tau(t_1, \dots, t_r)$ have the form

$$\alpha(t_1, \dots, t_r) = \frac{a(t_1, \dots, t_r)}{q(t_1, \dots, t_r)} + \beta(t_1, \dots, t_r),$$
$$\big(a(t_1, \dots, t_r), q(t_1, \dots, t_r)\big) = 1, \quad 1 \le q(t_1, \dots, t_r) \le \tau(t_1, \dots, t_r),$$
$$|\beta(t_1, \dots, t_r)| \le \big(q(t_1, \dots, t_r)\tau(t_1, \dots, t_r)\big)^{-1},$$
$$0 \le t_1 \le n_1, \dots, 0 \le t_r \le n_r, \quad t_1 + \dots + t_r \ge 1.$$

It follows from these relations that the polynomial $F(x_1, \dots, x_r)$ can be written as

$$\begin{aligned} &F(Q_0 y_1 + z_1, \dots, Q_0 y_s + z_s, \dots, Q_0 y_r + z_r) \\ &\quad \equiv \Phi(z_1, \dots, z_r) + \Psi_1(Q_0 y_1 + z_1, \dots, Q_0 y_s + z_s) \\ &\qquad + \Psi(Q_0 y_1 + z_1, \dots, Q_0 y_r + z_r) \pmod 1, \end{aligned}$$

where

$$\Phi(z_1, \dots, z_s) = \sum_{\substack{t_1=0 \\ t_1+\dots+t_s \ge 1}}^{n_1} \cdots \sum_{t_s=0}^{n_s} \sum_{\substack{t_{s+1}=0 \\ t_{s+1}+\dots+t_r \ge 1}}^{n_{s+1}} \cdots \sum_{t_r=0}^{n_r} \frac{a(t_1, \dots, t_r)}{q(t_1, \dots, t_r)} z_1^{t_1} \dots z_r^{t_r},$$

$$\Psi_1(x_1, \dots, x_s) = \sum_{\substack{t_1=0 \\ t_1+\dots+t_s \ge 1}}^{n_1} \cdots \sum_{t_s=0}^{n_s} \alpha(t_1, \dots, t_s, 0, \dots, 0) x_1^{t_1} \dots x_s^{t_s},$$

$$\begin{aligned} \Psi(x_1, \dots, x_r) = \sum_{\substack{t_{s+1}=0 \\ t_{s+1}+\dots+t_r \ge 1}}^{n_{s+1}} \cdots \sum_{t_r=0}^{n_r} x_{s+1}^{t_{s+1}} \dots x_r^{t_r} \Bigg(&\frac{a(0, \dots, 0, t_{s+1}, \dots, t_r)}{q(0, \dots, 0, t_{s+1}, \dots, t_r)} \\ &+ \sum_{\substack{t_1=0 \\ t_1+\dots+t_s \ge 1}}^{n_1} \cdots \sum_{t_s=0}^{n_s} \beta(t_1, \dots, t_s, t_{s+1}, \dots, t_r) x_1^{t_1} \dots x_s^{t_s} \Bigg). \end{aligned}$$

We use the above representation of the polynomial $F(x_1, \dots, x_r)$ to transform the sum S as follows:

$$S = \sum_{x_1=1}^{P_1} \cdots \sum_{x_r=1}^{P_r} \exp\{2\pi i F(x_1, \dots, x_r)\}$$
$$= \sum_{z_1=1}^{Q_0} \cdots \sum_{z_r=1}^{Q_0} \sum_{y_1} \cdots \sum_{y_s} \exp\big\{2\pi i\big(\Phi(z_1, \dots, z_r)$$
$$+ \Psi_1(Q_0 y_1 + z_1, \dots, Q_0 y_s + z_s)\big)\big\}$$
$$\times \sum_{y_{s+1}} \cdots \sum_{y_r} \exp\big\{2\pi i \Psi(Q_0 y_1 + z_1, \dots, Q_0 y_s + z_s)\big\};$$

here the summation is taken over the variables $y_j \quad (1 \le j \le r)$ within the limits $-z_j Q_0^{-1} \le y_j \le (P_j - z_j) Q_0^{-1}$. Hence we obtain the following estimate for the sum S:

$$|S| = T_1 + \sum_{x_1=1}^{P_1} \cdots \sum_{x_s=1}^{P_s} \sum_{z_{s+1}=1}^{Q_0} \cdots \sum_{z_r=1}^{Q_0} T_2,$$

$$T_2 = \left| \sum_{y_{s+1}} \cdots \sum_{y_r} \exp\big\{2\pi i \Psi(x_1, \dots, x_s, Q_0 y_{s+1} + z_{s+1}, \dots, Q_0 y_r + z_r)\big\} \right|.$$

Therefore, to estimate the sum S, it suffices to obtain the estimate for T_1 given in the lemma. We represent the polynomial $\Psi(x_1, \dots, x_r)$ as

$$\Psi(x_1, \dots, x_r) = \sum_{\substack{t_{s+1}=0 \\ t_{s+1}+\cdots+t_r \ge 1}}^{n_{s+1}} \cdots \sum_{t_r=0}^{n_r} g_{t_{s+1},\dots,t_r}(x_1, \dots, x_s) x_{s+1}^{t_{s+1}} \dots x_r^{t_r}$$
$$= \sum_{\substack{t_{s+1}=0 \\ t_{s+1}+\cdots+t_r \ge 1}}^{n_{s+1}} \cdots \sum_{t_r=0}^{n_r} B(t_{s+1}, \dots, t_r) x_{s+1}^{t_{s+1}} \dots x_r^{t_r},$$

where

$$g_{t_{s+1},\dots,t_r}(x_1, \dots, x_s) = B(t_{s+1}, \dots, t_r) = \frac{a(0, \dots, 0, t_{s+1}, \dots, t_r)}{q(0, \dots, 0, t_{s+1}, \dots, t_r)}$$
$$+ \sum_{t_1=0}^{n_1} \cdots \sum_{t_s=0}^{n_s} \beta(t_1, \dots, t_s, \dots, t_r) x_1^{t_1} \dots x_s^{t_s}.$$

We consider the D-approximations of the fractional parts of the polynomials $g_{t_{s+1},\dots,t_r}(x_1, \dots, x_s) = B(t_{s+1}, \dots, t_r)$ corresponding to $\eta(t_{s+1}, \dots, t_r)$,

$$\eta(t_{s+1}, \dots, t_r) = P_{s+1}^{t_{s+1}-1/6} P_{s+2}^{t_{s+2}} \dots P_r^{t_r},$$

$$\{B(t_{s+1},\dots,t_s)\} = \frac{a(t_{s+1},\dots,t_r)}{q(t_{s+1},\dots,t_r)} + \beta(t_{s+1},\dots,t_r),$$
$$\big(a(t_{s+1},\dots,t_r), q(t_{s+1},\dots,t_r)\big) = 1, \quad 1 \le q(t_{s+1},\dots,t_r) \le \tau(t_{s+1},\dots,t_r),$$
$$|\beta(t_{s+1},\dots,t_r)| \le \big(q(t_{s+1},\dots,t_r)\tau(t_{s+1},\dots,t_r)\big)^{-1},$$
$$0 \le t_{s+1} \le n_{s+1},\dots,0 \le t_r \le n_r, \quad t_{s+1}+\dots+t_r \ge 1.$$

By Q_5 we denote the least common multiple of $q(t_{s+1},\dots,t_r)$, and by δ we denote the largest of the values

$$|\beta(t_{s+1},\dots,t_r)| \le P_{s+1}^{t_{s+1}}\dots P_r^{t_r},$$
$$0 \le t_{s+1} \le n_{s+1},\dots,0 \le t_r \le n_r, \quad t_{s+1}+\dots+t_r \ge 1.$$

We write the sum T_1 in the form

$$T_1 = S_{16} + S_{17} + S_{18},$$

where

$$S_j = \sum\dots\sum_{(x_{s+1},\dots,x_s)\in T_j} \sum_{z_{s+1}=1}^{Q_0}\dots\sum_{z_r=1}^{Q_0} |S(x_1,\dots,x_s,z_{s+1},\dots,z_r)|, \quad j = 16, 17, 18,$$

$$S(x_1,\dots,x_s,z_{s+1},\dots,z_r) = \sum_{y_{s+1}}\dots\sum_{y_r} \exp\big\{2\pi i\Psi(x_1,\dots,x_s,Q_0y_{s+1}+z_{s+1},\dots,Q_0y_r+z_r)\big\},$$

the summation over y_j $(s+1 \le j \le r)$ is performed within the limits $-z_jQ_0^{-1} < y_j \le (P_j - z_j)Q_0^{-1}$, and the domain of summation T_j over the variables $x_1,\dots,x_s$ in each of the sums is its own and is determined as follows. If a point B with coordinates $B(t_{s+1},\dots,t_r)$ $(0 \le t_{s+1} \le n_{s+1},\dots,0 \le t_r \le n_r, t_{s+1}+\dots+t_r \ge 1)$ is a point of the second class with respect to the parameters $P_{s+1},\dots,P_r$, then the corresponding set $(x_1,\dots,x_s)$ belongs to the sum T_{16}. If this point is a point of the first class and either $Q_5 \ge H_3 = P_1^a$ with $a = 20(r-s)n^3m_s\rho_r$ or $\delta \ge P_1^{2n\rho}$, then the corresponding set $(x_1,\dots,x_s)$ belongs to the set T_{17}. All other sets $(x_1,\dots,x_s)$ belong to the set T_{18}.

We consider the sum S_{16}. In this case, B is a point of the second class Ω_2. All points of the second class are divided into two domains ω_1 and ω_2. If the point B belongs to the domain ω_1, then the sum

$$S(B) = \sum_{x_{s+1}=1}^{P_{s+1}}\dots\sum_{x_r=1}^{P_r} \exp\{2\pi i F_1(x_{s+1},\dots,x_r)\},$$
$$F_1(x_{s+1},\dots,x_r) = \sum_{t_{s+1}=0}^{n_{s+1}}\dots\sum_{t_r=0}^{n_r} B(x_{s+1},\dots,x_r)x_{s+1}^{t_{s+1}}\dots x_r^{t_r},$$

can be estimated, by Lemma 7.11, as

$$|S(B)| \ll P_{s+1} \dots P_r P_{s+1}^{-1/(20n)}. \tag{7.42}$$

Hence, by Lemma 7.18, item (1), we have the estimate

$$|S(x_1, \dots, x_s, z_{s+1}, \dots, z_r)| \ll P_{s+1} \dots P_r P_{s+1}^{-1/(40n)} Q_0^{-r+s}.$$

If the point B belongs to the domain ω_2, then, by the induction assumption, the sum $S(B)$ satisfies the estimate

$$|S(B)| \ll P_{s+1} \dots P_r P_{s+1}^{-\rho_{r-s}}. \tag{7.43}$$

Thus, by Lemma 7.18, item (1), we have

$$|S(x_1, \dots, x_s, z_{s+1}, \dots, z_r)| \ll P_{s+1} \dots P_r P_{s+1}^{-0.5\rho_{r-s}} Q_0^{-r+s}.$$

Therefore, for the sum S_{17}, we obtain the estimate

$$|S_{16}| \ll P_1 \dots P_r P_1^{-\rho_r}.$$

Now we consider the sum S_{17}. In this case, B is a point of the first class, i.e.,

$$\begin{gathered}
B(t_{s+1}, \dots, t_r) = \frac{b(t_{s+1}, \dots, t_r)}{l(t_{s+1}, \dots, t_r)} + \beta_1(t_{s+1}, \dots, t_r), \\
\big(b(t_{s+1}, \dots, t_r), l(t_{s+1}, \dots, t_r)\big) = 1, \\
|\beta_1(t_{s+1}, \dots, t_r)| \le P_{s+1}^{-t_{s+1}+1/6} P_{s+2}^{-t_{s+2}} \dots P_r^{-t_r}, \\
0 \le t_{s+1} \le n_{s+1}, \dots, 0 \le t_r \le n_r, \quad t_{s+1} + \dots + t_r \ge 1,
\end{gathered} \tag{7.44}$$

and the least common multiple l of all the numbers $l(t_{s+1}, \dots, t_r)$ does not exceed $P_{s+1}^{0.1}$.

First, let $Q_5 \ge P_{s+1}^{0.1}$. Then we have $q(t_{s+1}, \dots, t_r) \ne l(t_{s+1}, \dots, t_r)$ for some set $(t_{s+1}, \dots, t_r)$. As before, we obtain the inequalities

$$\begin{aligned}
&\frac{1}{q(t_{s+1}, \dots, t_r)\, l(t_{s+1}, \dots, t_r)} \\
&\quad \le \left| \frac{a(t_{s+1}, \dots, t_r)}{q(t_{s+1}, \dots, t_r)} - \frac{b(t_{s+1}, \dots, t_r)}{l(t_{s+1}, \dots, t_r)} \right| \\
&\quad \le P_{s+1}^{-t_{s+1}+0.1} P_{s+2}^{-t_{s+2}} \dots P_r^{-t_r} + q^{-1}(t_{s+1}, \dots, t_r) P_{s+1}^{-t_{s+1}+1/6} P_{s+2}^{-t_{s+2}} \dots P_r^{-t_r},
\end{aligned}$$

$$\begin{aligned}
l^{-1}(t_{s+1}, \dots, t_r) &\le q(t_{s+1}, \dots, t_r) P_{s+1}^{-t_{s+1}+0.1} P_{s+2}^{-t_{s+2}} \dots P_r^{-t_r} \\
&\quad + P_{s+1}^{-t_{s+1}+1/6} P_{s+2}^{-t_{s+2}} \dots P_r^{-t_r} \le 2P_{s+1}^{-1/15}.
\end{aligned}$$

Hence $l \geq l(t_{s+1}, \dots, t_r) \geq 0.5P_{s+1}^{1/15}$. If l satisfies the inequalities $0.5P_{s+1}^{1/15} \leq l \leq P_{s+1}^{0.07}$, then, by Lemma 7.18, item (a), we have

$$|S(x_1, \dots, x_s, z_{s+1}, \dots, z_r)| \ll P_{s+1} \dots P_r L^{-1/n+\varepsilon} Q_0^{-r+s},$$
$$L = l/(l, Q_0^f), \quad f = n_{s+1} + \dots + n_r \leq (r-s)n.$$

Since

$$(l, Q_b^f) \leq P_1^b, \quad b = 10m_s n^2 f \rho_r, \quad P_1^b \leq P_1^{0.01}, \quad l \geq P_{s+1}^{0.05},$$

our sum satisfies the estimate

$$|S(x_1, \dots, x_s, z_{s+1}, \dots, z_r)| \ll P_{s+1} \dots P_r P_1^{-\rho_r} Q_0^{-r+s}.$$

Now we assume that the variable l satisfies the inequality $l > P_{s+1}^{0.07}$. Then, by Lemma 7.18, item (3), and the estimates (7.42) and (7.43), we obtain

$$|S(x_1, \dots, x_s, z_{s+1}, \dots, z_r)| \ll P_{s+1} \dots P_r P_{s+1}^{-0.5\rho_{r-s}} Q_0^{-r+s}.$$

Thus we have obtained the desired estimate of the sum S_{17} for $Q_5 \geq P_{s+1}^{0.1}$.

Now we consider the case $Q_5 < P_{s+1}^{0.1}$. The point B belongs to the first class and satisfies relations (7.44). As before, we can show that

$$\frac{a(t_{s+1}, \dots, t_r)}{q(t_{s+1}, \dots, t_r)} = \frac{b(t_{s+1}, \dots, t_r)}{l(t_{s+1}, \dots, t_r)}.$$

Hence we have $Q_5 = l$ and $\beta(t_{s+1}, \dots, t_r) = \beta_1(t_{s+1}, \dots, t_r)$.

We set

$$\delta_1(t_{s+1}, \dots, t_r) = P_{s+1}^{t_{s+1}} \dots P_r^{t_r} \beta_1(t_{s+1}, \dots, t_r).$$

If $P_{s+1}^{0.07} \geq l \geq H_3 = P_1^q$ and $q = 20(r-s)n^3 m_s \rho_r$, then, estimating the sum $S(x_1, \dots, x_s, z_{s+1}, \dots, z_r)$ by Lemma 7.18, item (2), (a), we obtain

$$|S(x_1, \dots, x_s, z_{s+1}, \dots, z_r)| \ll P_{s+1} \dots P_r L^{-1/n+\varepsilon} Q_0^{-r+s},$$
$$L = l/(l, Q_0^f), \quad f = n_{s+1} + \dots + n_r.$$

Since $(l, Q_0^f) \leq Q_0^f \leq P_1^{10m_s f n^2 \rho_r} \leq P_1^{10m_s (r-s) n^3 \rho_r} = P_1^{0.5q}$ and $l \geq P_1^{0.5q}$, we have $L \geq P_1^{0.5q}$. Hence

$$|S(x_1, \dots, x_s, z_{s+1}, \dots, z_r)| \ll P_{s+1} \dots P_r P_1^{-\rho_r} Q_0^{-r+s}.$$

If $l > P_{s+1}^{0.07}$ or $\delta > P_{s+1}^{0.04}$, then, by Lemma 7.18, item (3), and by formulas (7.42) and (7.43), we obtain

$$|S(x_1, \dots, x_s, z_{s+1}, \dots, z_r)| \ll P_{s+1} \dots P_r P_{s+1}^{-0.5\rho_{r-s}} Q_0^{-r+s}.$$

If $P_1^{2n\rho_r} \le \delta \le P_{s+1}^{0.04}$, then, by Lemma 7.18, item (2), (b), we obtain

$$|S(x_1, \dots, x_s, z_{s+1}, \dots, z_r)| \ll P_{s+1} \dots P_r \delta^{-1/n+\varepsilon} Q_0^{-r+s}.$$

So if $Q_5 \ge H_3$ or $\delta > P_1^{2n\rho_r}$, then we have the estimate

$$|S(x_1, \dots, x_s, z_{s+1}, \dots, z_r)| \ll P_{s+1} \dots P_r P_1^{-\rho_r} Q_0^{-r+s}.$$

Thus we obtain

$$|S_{17}| \ll P_1 \dots P_r P_1^{-\rho_r}.$$

Now let us estimate the sum S_{18}. We trivially estimate this sum by the number of terms as

$$|S_{18}| \ll Y P_{s+1} \dots P_r,$$

where Y is the number of sets $(x_1, \dots, x_s)$ $(1 \le x_1 \le P_1, \dots, 1 \le x_s \le P_s)$ for which we have the relations

$$|\beta(t_{s+1}, \dots, t_r)| \le \Delta(t_{s+1}, \dots, t_r) = P_{s+1}^{-t_{s+1}} \dots P_r^{-t_r} P_1^{2n\rho_r},$$
$$0 \le t_{s+1} \le n_{s+1}, \dots, 0 \le t_r \le n_r, \quad t_{s+1} + \dots + t_r \ge 1,$$
$$Q_5 \le H_3 = P_1^q, \quad q = 20(r-s)n^3 m_s \rho_r,$$

and B is a point of the first class. To estimate Y, we use Lemma 7.15 and obtain

$$Y \ll P_1 \dots P_s P_1^{-\rho_r}.$$

Hence

$$|S_{18}| \ll P_1 \dots P_r P_1^{-\rho_r},$$
$$|S| \le |S_{16}| + |S_{17}| + |S_{18}| \ll P_1 \dots P_r P_1^{-\rho_r}.$$

Thus we have estimated the sum S for $Q_1 > P_1^d$, where $d = 0.05 - 5n^3 m_s \rho_r$.

The case $Q_2 > P_1^d$ can be studied similarly to the preceding case $Q_1 > P_1^d$. The distinction is that the groups of variables $(x_1, \dots, x_s)$ and $(x_{s+1}, \dots, x_r)$ are interchanged, and hence Lemma 7.16 must be used instead of Lemma 7.15. The proof of the second main lemma is complete. □

7.2.3 Estimate for the multiple trigonometric sum

Here we state and prove a theorem about estimating a multiple trigonometric sum $S(A)$ for all points A of the unit m-dimensional cube Ω.

Theorem 7.3. *Suppose that a point A belongs to the first class Ω_1. Then the following estimate holds*:

$$|S(A)| \ll P_1 \dots P_r Q^{-1/n+\varepsilon}.$$

If, in addition, we set

$$\delta = \max_{t_1,\dots,t_r} P_1^{t_1} \dots P_r^{t_r} |\beta(t_1, \dots, t_r)|,$$

then, for $\delta > 1$, *the following estimate also holds*:

$$|S(A)| \ll P_1 \dots P_r (Q\delta)^{-1/n+\varepsilon}.$$

Suppose that the point A belongs to the second class Ω_2. *Then the following estimate holds*:

$$|S(A)| \ll P_1 \dots P_r P_1^{-\rho_r}, \quad \rho_r = \gamma (2n)^{-2r} \log^{-1} n;$$

here $\gamma > 0$ *is an absolute constant.*

The constants in $\ll$ *depend only on* r, n, *and* ε.

Proof. The proof of this theorem repeats, in fact, the proof of Theorem 7.2. The only distinction is that estimates for multiple sums are used instead of estimates for double sums and Lemmas 7.10 and 7.11 and the second main lemma are applied instead of Lemma 7.3 and the first main lemma. □

7.3 An asymptotic formula

In this section we derive an asymptotic formula for the multiple integral

$$J = J(r) = \int \cdots \int_{\Omega} |S(A)|^{2K} \, dA.$$

Its value is equal to the number of solutions of the system of Diophantine equations

$$\sum_{j=1}^{2K} (-1)^j x_{1j}^{t_1} \dots x_{rj}^{t_r} = 0, \quad 0 \le t_1 \le n_1, \dots, 0 \le t_r \le n_r,$$

where the unknowns $x_{1j}, \dots, x_{rj}$ vary within the limits

$$1 \le x_{1j} \le P_1, \dots, 1 \le x_{rj} \le P_r \quad (j = 1, 2, \dots, 2K).$$

According to the partition of the points of the cube into the classes Ω_1 and Ω_2 (see the notation in Section 7.1.1), we represent J in the form

$$J = J_1 + J_2,$$

where

$$J_1 = \int \cdots \int_{\Omega_1} |S(A)|^{2K} \, dA, \quad J_2 = \int \cdots \int_{\Omega_2} |S(A)|^{2K} \, dA.$$

We have the following assertion.

Lemma 7.19. *Suppose that $K > 2nm$. Then the variable J_1 satisfies the asymptotic formula*

$$J_1 = \theta\sigma(P_1, \dots, P_r)^{2K}(P_1^{n_1} \dots P_r^{n_r})^{-0.5m} + O\big((P_1 \dots P_r)^{2K}(P_1^{n_1} \dots P_r^{n_r})^{-0.5m} P_1^{-1/6}\big), \tag{7.45}$$

where

$$\theta = \int_{-\infty}^{+\infty} \cdots \int_{-\infty}^{+\infty} \left| \int_0^1 \cdots \int_0^1 \exp\{2\pi i F_A(x_1, \dots, x_r)\}\, dx_1 \dots dx_r \right|^{2K} dA,$$

$$F_A(x_1, \dots, x_r) = \sum_{t_1=0}^{n_1} \cdots \sum_{t_r=0}^{n_r} \alpha(t_1, \dots, t_r) x_1^{t_1} \dots x_1^{t_r}, \quad \alpha(0, \dots, 0) = 0,$$

$$\sigma = \sum_{q(0,\dots,1)=1}^{+\infty} \cdots \sum_{q(n_1,\dots,n_r)=1}^{+\infty} \sum_{\substack{a(0,\dots,1)=1 \\ (a(0,\dots,1), q(0,\dots,1))=1}}^{q(0,\dots,1)} \cdots \sum_{\substack{a(n_1,\dots,n_r)=1 \\ (a(n_1,\dots,n_r), q(n_1,\dots,n_r))=1}}^{q(n_1,\dots,n_r)} |U(a, q)|^{2K},$$

$$U(a, q) = q^{-r} \sum_{x_1=1}^{q} \cdots \sum_{x_r=1}^{q} \exp\{2\pi i \Phi(x_1, \dots, x_r)\},$$

$$\Phi(x_1, \dots, x_r) = \sum_{\substack{t_1=0 \\ t_1+\dots+t_r \geq 1}}^{n_1} \cdots \sum_{t_r=0}^{n_r} \frac{a(t_1, \dots, t_r)}{q(t_1, \dots, t_r)} x_1^{t_1} \dots x_r^{t_r},$$

$$q = q(0, \dots, 1) \dots q(n_1, \dots, n_r).$$

For the proof of this lemma, see Section 6.1 in Chapter 6.

Theorem 7.4. *Suppose that $P_1 \to +\infty$ and $P_1 \leq \dots \leq P_r$. Then there exists an absolute constant $c > 0$ such that, for $K \geq c(2n)^{2r} rnm \log n(\log P_r / \log P_1)$, the following asymptotic formula holds*:

$$J = \theta\sigma(P_1 \dots P_r)^{2K}(P_1^{n_1} \dots P_r^{n_r})^{-0.5m} + O\big((P_1 \dots P_r)^{2K}(P_1^{n_1} \dots P_r^{n_r})^{-0.5m} P_1^{-1/6}\big), \tag{7.46}$$

where θ and σ are defined in Lemma 7.19.

Proof. From Lemma 17.19, for $c \geq 1$, we have an asymptotic formula for J_1 with the desired remainder. Therefore, to prove the theorem, it suffices to estimate the integral J_2 with desired accuracy. For this, we apply Theorem 7.3 to estimate $|S(A)|$ in this integral. Since A is a point of the second class Ω_2, this theorem implies the estimate

$$|S(A)| \ll P_1 \dots P_r P_1^{-\rho_r}, \quad \rho_r = \gamma(2n)^{-2r} \log^{-1} n;$$

here $\gamma > 0$ is an absolute constant. Thus we obtain the following inequalities for J_2:

$$J_2 \ll (P_1 \dots P_r P_1^{-\rho_r})^{2K} = (P_1 \dots P_r)^{2K} P_1^{-2K\rho_r}$$
$$\ll (P_1 \dots P_r)^{2K} (P_1^{n_1} \dots P_r^{n_r})^{-0.5m} P_r^{0.5rnm} P_1^{-2K\rho_1}.$$

But, by the assumptions of the theorem, we have

$$K \geq c(2n)^{2r} rnm \log n \frac{\log P_r}{\log P_1}.$$

Thus, for $c \geq \gamma^{-1}$, we have

$$\begin{aligned} P_r^{0.5rnm} P_1^{-2K\rho_1} &= \exp\{0.5rnm \log P_r - 2K\rho_r \log P_1\} \\ &\leq \exp\{0.5rnm \log P_r - 2c(2n)^{2r} \log n \log P_r \gamma (2n)^{-2r} \log^{-1} n\} \\ &\leq P_r^{-rnm} < P_1^{-1/6}. \end{aligned}$$

Hence we can estimate the integral J_2 as follows:

$$J_2 \ll (P_1 \dots P_r)^{2K} (P_1^{n_1} \dots P_r^{n_r})^{-0.5m} P_1^{-1/6},$$

i.e., J_2 is contained in the remainder term in formula (7.46).

So we have obtained an asymptotic formula for J. The proof of the theorem is complete. □

We show that the asymptotic formula (7.46) in Theorem 7.4 cannot be obtained for K satisfying the conditions $2mn < K \leq (m/12) \cdot (\log P_r / \log P_1)$.

Suppose that $m = m_r = (n_1+1) \dots (n_r+1)$, $m_{r-1} = (n_1+1)^{-1} m_r$, and $J(r-1)$ is the number of solutions of the systems of equations

$$\sum_{j=1}^{2K} (-1)^j x_{2j}^{t_2} \dots x_{rj}^{t_r} = 0, \quad 0 \leq t_2 \leq n_2, \dots, 0 \leq t_r \leq n_r,$$

where the unknowns $x_{2j}, \dots, x_{rj}$ vary within the limits $1 \leq x_{2j} \leq P_2, \dots, 1 \leq x_{rj} \leq P_r$ $(j = 1, 2, \dots, 2K)$.

Obviously, $J(r-1) \leq J(r) = J$ for all possible values of the parameters $P_1, P_2, \dots, P_r$. Precisely as in Section 4.2.4, we obtain

$$J(r-1) \geq (2K)^{-m_{r-1}} (P_2 \dots P_r)^{2K} (P_2^{n_2} \dots P_r^{n_r})^{-0.5m_{r-1}}.$$

Suppose that the asymptotic formula (7.42) holds for some $K > 2nm$ and $P_1, \dots, P_r$ satisfying the conditions $P_1 \leq \dots \leq P_r$ as $P_1 \to +\infty$. Then we have

$$J \ll (P_1 \dots P_r)^{2K} (P_1^{n_1} \dots P_r^{n_r})^{-0.5m}.$$

In the last inequality, the constant in $\ll$ depends only on n and r, because, obviously, the singular integral θ and the singular series σ decrease with increasing parameter K.

Thus, for a sufficiently large P_1, we have

$$(2K)^{-m_{r-1}}(P_2 \dots P_r)^{2K}(P_2^{n_2} \dots P_r^{n_r})^{-0.5m_{r-1}} \\ \le J(r-1) \le J \ll (P_1 \dots P_r)^{2K}(P_2^{n_1} \dots P_r^{n_r})^{-0.5m}.$$

Taking the logarithm and passing to the limit as $P_1 \to +\infty$, we obtain

$$- m_{r-1}\log(2K) + 2K(\log P_2 + \dots + \log P_r) \\ - 0.5m_{r-1}(n_2 \log P_2 + \dots + n_r \log P_r) \le 2K(\log P_1 + \dots + \log P_r) \\ - 0.5m(n_1 \log P_1 + n_2 \log P_2 + \dots + n_r \log P_r).$$

We assume that $\log P_1 > m_{r-1}$. Then, taking into account that $m_{r-1} = m(n_1+1)^{-1} \le 0.5m$ and $\log 2K < K$, we obtain the inequalities

$$2K \log P_1 > -m_{r-1} \log 2K + 0.5(m - m_{r-1})(\log P_1 + \dots + \log P_r),$$
$$K > \frac{m}{12} \cdot \frac{\log P_r}{\log P_1},$$

as was stated above. Thus, the variable K in Theorem 7.4 has a regular order of growth with respect to the principal parameters $P_1, \dots, P_r$.

Concluding remarks on Chapter 7. The results considered in this chapter were obtained by the authors and published in [33], [34] and [35].

Chapter 8

The Hilbert–Kamke problem and its generalizations

In this chapter we present the solution of one of the classical additive problems in number theory, namely, of the Hilbert–Kamke problem, and consider some of its generalizations the most important of which is the generalization to the multidimensional case. In the Hilbert–Kamke problem it is required to represent several increasing natural numbers simultaneously as sums of the first, second, third, etc., nth powers of natural numbers. More precisely, the problem is to prove the solvability of the system of equations in the Hilbert–Kamke problem

$$x_1 + \cdots + x_k = N_1,$$
$$\vdots$$
$$x_1^n + \cdots + x_k^n = N_n$$

in natural numbers $x_1, \dots, x_k$ in the case where the number of the unknowns x is bounded and the increasing parameters $N_1, \dots, N_n$ satisfy several additional natural conditions.

It is a very interesting problem to obtain the best possible estimates for the variable $r(n)$, i.e., for the lower bound of the number k of terms for which this system is solvable. We deal with the Hilbert–Kamke problem in Sections 8.1 and 8.2, where we obtain an asymptotic approximation of the logarithm of $r(n)$.

In Section 8.3, we study the multidimensional additive problem. In this problem, instead of numbers of the form $x, x^2, \dots, x^n$ as it was in the Hilbert–Kamke problem, monomials of the form $x_1^{t_1} x_2^{t_2} \dots x_r^{t_r}$, where the exponents $t_1, \dots, t_r$ run independently through all integer values in the intervals $0 \le t_1 \le n, \dots, 0 \le t_r \le n_r$ except the values $t_1 = \cdots = t_r = 0$, are taken to be the terms in the sums.

To each set $(t_1, \dots, t_r)$, there corresponds its own equation with a right-hand side that is assumed to be a natural increasing parameter $N(t_1, \dots, t_r)$. Here we have problems of whether the number set $\{N(t_1, \dots, t_r)\}$ can be simultaneously represented as the sum of a bounded number of terms of the form mentioned above; we consider the natural conditions to which the numbers from the set $\{N(t_1, \dots, t_r)\}$ must satisfy; and we look for a function $r_0(n_1, \dots, n_r)$ similar to the variable $r(n)$ in the Hilbert–Kamke problem.

8.1 Study of the singular series in the Hilbert–Kamke problem

In Chapter 3, an asymptotic formula was obtained for the number of solutions of the system of Diophantine equations in the Hilbert–Kamke problem (Theorem 3.7). This formula is nontrivial if and only if the singular series σ and the singular integral γ in the Hilbert–Kamke problem are positive. Moreover, it turns out that for $\sigma = 0$, this system of equations does not have any solutions at all, and for $\gamma = 0$ the number of its solutions remains bounded when the numbers $N_1, \ldots, N_n$ increase. Thus the problem of the existence of $r(n)$ and of estimating $r(n)$ can be reduced to studying the singular series σ and the singular integral γ. In this section we consider the singular series σ. The next section deals with the singular integral γ.

By $W(d; k)$ we denote the number of solutions of the system of congruences

$$x_1^s + \cdots + x_k^s \equiv N_s \pmod{d}, \quad s = 1, \ldots, n, \tag{8.1}$$

for the unknowns $x_1, \ldots, x_k$. By the letter p we denote a prime number.

Lemma 8.1. *Suppose that M and m are natural numbers, $M = m!$, and $k > 0.5n(n+1) + 1$. The following relation holds*:

$$\sigma = \lim_{m \to \infty} W(M; k) M^{n-k},$$

where σ is the singular series in Theorem 3.7 *in Chapter* 3.

Proof. It follows from Theorem 3.7 (Chapter 3) that the series converges absolutely for $k > 0.5n(n+1) + 1$. Therefore, for a fixed natural number d, we have

$$\sigma = \sum\nolimits_1(d) + \sum\nolimits_2(d),$$

where

$$\sum\nolimits_1(d) = \sum_{\substack{q_1, \ldots, q_n \\ Q|d}} \sum_{a_1}^{q_1}{}' \cdots \sum_{a_n}^{q_n}{}' \left(q^{-1} \sum_{x=1}^{q} \exp\left\{ 2\pi i \left(\frac{a_1}{q_1} x + \cdots + \frac{a_n}{q_n} x^n \right) \right\} \right)^k$$
$$\times \exp\left\{ -2\pi i \left(\frac{a_1}{q_1} N_1 + \cdots + \frac{a_n}{q_n} N_n \right) \right\},$$

$$\sum\nolimits_2(d) = \sum_{\substack{1 \le q_1, \ldots, q_n < +\infty \\ Q \nmid d}} \sum_{a_1}^{q_1}{}' \cdots \sum_{a_n}^{q_n}{}' \left(q^{-1} \sum_{x=1}^{q} \exp\left\{ 2\pi i \left(\frac{a_1}{q_1} x + \cdots + \frac{a_n}{q_n} x^n \right) \right\} \right)^k$$
$$\times \exp\left\{ -2\pi i \left(\frac{a_1}{q_1} N_1 + \cdots + \frac{a_n}{q_n} N_n \right) \right\},$$

where, as above, $Q = [q_1, \dots, q_n]$ and $q = q_1, \dots, q_n$.

We set $d = M$. Since the series σ converges absolutely, the sum $\sum_2(M)$ tends to zero as $m \to \infty$. Hence we have

$$\sigma = \lim_{m\to\infty} \sum\nolimits_1(M).$$

Now we prove that $\sum_1(d) = W(d;k)d^{n-k}$. For this, we write $W(M;k)$ as a trigonometric sum. We obtain

$$W(M;k) = M^{-n} \sum_{b_1=1}^{M} \cdots \sum_{b_n=1}^{M} \left(\sum_{x=1}^{M} \exp\{2\pi i(b_1 x + \dots + b_n x^n)/M\} \right)^k \qquad (8.2)$$
$$\times \exp\{-2\pi i(b_1 N_1 + \dots + b_n N_n)/M\}.$$

We note that, for each $s = 1, \dots, n$, the sum $\sum_{b_s=1}^{M}$ can be replaced by a double sum of the form $\sum_{q_s|M} \sum_{a_s}^{\prime q_s}$. For this, we must set $q_s = (b_s, M)^{-1}M$ and $a_s = (b_s, M)^{-1}b_s$. In this notation, we also have $Q \mid M$ and, moreover,

$$\left(\sum_{x=1}^{M} \exp\{2\pi i(b_1 x + \dots + b_n x^n)/M\} \right)^k \exp\{-2\pi i(b_1 N_1 + \dots + b_n N_n)/M\}$$
$$= \left(\sum_{x=1}^{M} \exp\left\{2\pi i\left(\frac{a_1}{q_1}x + \dots + \frac{a_n}{q_n}x^n\right)\right\} \right)^k$$
$$\times \exp\left\{-2\pi i\left(\frac{a_1}{q_1}N_1 + \dots + \frac{a_n}{q_n}N_n\right)\right\},$$

$$\sum_{x=1}^{M} \exp\left\{2\pi i\left(\frac{a_1}{q_1}x + \dots + \frac{a_n}{q_n}x^n\right)\right\}$$
$$= MQ^{-1} \sum_{x=1}^{Q} \exp\left\{2\pi i\left(\frac{a_1}{q_1}x + \dots + \frac{a_n}{q_n}x^n\right)\right\}$$
$$= Mq^{-1} \sum_{x=1}^{q} \exp\left\{2\pi i\left(\frac{a_1}{q_1}x + \dots + \frac{a_n}{q_n}x^n\right)\right\}.$$

Therefore, $W(M;k) = M^{k-n} \sum_1(M)$ and hence

$$\sum\nolimits_1(M) = M^{n-k} W(M;k).$$

Passing to the limit in this relation, we obtain the statement of the lemma. The proof of the lemma is complete. □

Lemma 8.2. *The relation*

$$\sigma = \prod_p \sigma_p, \quad \sigma_p = \lim_{\alpha \to +\infty} W(p^\alpha; k) p^{\alpha(n-k)},$$

holds. The product $\prod_p \sigma_p$ *is taken over all primes* p.

Proof. First, we note that, by the well-known theorem in the elementary number theory, the number $W(d; k)$ of solutions of congruences (8.1) is multiplicative with respect to d, i.e.,

$$W(d; k) = W(d_1; k)\, W(d_2; k)$$

if only $d_1 d_2 = d$ as well as $(d_1, d_2) = 1$. Therefore, by setting

$$M = m! = \prod_{p \le m} p^{\alpha_p},$$

where α_p is the exponent of the prime p in the decomposition of the number M into prime factors, we obtain

$$W(M; k) = \prod_{p \le m} W(p^{\alpha_p}; k).$$

Hence, by Lemma 8.1, we have

$$\sigma = \lim_{m \to \infty} W(M; k) M^{n-k} = \lim_{m \to \infty} \prod_{p \le m} W(p^\alpha; k) p^{\alpha_p(n-k)}. \tag{8.3}$$

Let

$$H(d) = d^{-k} \sum_{b_1}^{d} \cdots \sum_{\substack{b_n \\ (b_1,\ldots,b_n,d)=1}}^{d} \Big(\sum_{x=1}^{d} \exp\{2\pi i (b_1 x + \cdots + b_n x^n)/d\} \Big)^k$$
$$\times \exp\{-2\pi i (b_1 N_1 + \cdots + b_n N_n)/d\}.$$

It follows from (8.2) that

$$d^n W(d; k) = \sum_{r | d} H(r) r^k;$$

hence we have (here $\mu(r)$ is the Möbius function)

$$d^k H(d) = \sum_{r | d} \mu(r) W(dr^{-1}; k)(dr^{-1})^n.$$

Since $W(d;k)$ is a multiplicative function of d, it follows from the last relation that the function $H(d)$ is also multiplicative. Therefore, expressing $W(p^\alpha;k)$ in terms of $H(d)$, we obtain

$$\prod_{p\le m} W(p^{\alpha_p};k)p^{\alpha_p(n-k)} = \prod\Big(\sum H(p^\beta)\Big) = \prod_{p\le m}\Big(\sum_{\beta=0}^{+\infty} H(p^\beta) - \sum_{\beta>\alpha_p} H(p^\beta)\Big)$$
$$= \prod_{p\le m}\sigma_p - \sum_q{}' H(q);$$

here the symbol $\sum'_q$ denotes summation over all numbers q that are not divisible by $m!$ and whose prime divisors do not exceed m. Obviously, the series $\sum'_q H(q)$ consists of a part of the terms contained in the sum $\sum_2(M)$ studied in the proof of Lemma 8.1. Since the series σ converges absolutely, the series $\sum'_q H(q)$ also converges absolutely, and moreover, as $m\to\infty$ we have

$$\sum_q{}' H(q) \to 0. \tag{8.4}$$

Thus the product $\prod_{p\le m}\sigma_p$ makes sense and from (8.3) and (8.4) we obtain

$$\sigma = \lim_{m\to\infty}\prod_{p\le m} W(p^{\alpha_p};k)p^{\alpha_p(n-k)} = \lim_{m\to\infty}\prod_{p\le m}\sigma_p + \lim_{m\to\infty}\sum_q{}' H(q) = \prod_p \sigma_p,$$

as required. □

Now we assume that the natural numbers $N_1,\dots,N_n$ $(n\ge 3)$ satisfy the condition that the system of linear equations

$$\sum_{r=1}^{n} t_r r^s = N_s, \quad s=1,\dots,n, \tag{8.5}$$

can have a solution for integer numbers $t_1,\dots,t_n$. This condition was first formulated in a somewhat different form by K. K. Mardzhanishvili (see [115], [119]). This is a necessary condition for the solvability of systems of Hilbert–Kamke equations because, for any integer x, the system

$$\sum_{r=1}^{n} t_r r^s = x^s, \quad s=1,\dots,n,$$

has an integer-valued solution of the form

$$t_r = \binom{x-1}{r-1}\binom{n-x}{n-r}, \quad r=1,\dots,n,$$

where $\binom{x}{a}$ is determined for any natural number a by the relation

$$\binom{x}{a} = \frac{x(x-1)\dots(x-a+1)}{a!}$$

and is equal to unity for $a = 0$.

Lemma 8.3. *Suppose that $p \le n$ and $k \ge n^2 2^{2n-1} = k_1$. Then the variable $W(p^k; k)$ satisfies the estimate*

$$W(p^h; k) \ge cp^{h(k-n)} > 0,$$

where $c = p^{(2\varkappa+2\delta+1)(n-k_1)}$, $\varkappa$ is the integer determined by the condition

$$p^{\varkappa} \le n < p^{\varkappa+1},$$

and δ is the exponent of p contained in the decomposition of the number $(n-1)!$ into factors.

Proof. For $s = 1, \dots, n$, we set

$$M_s = N_s - 1^s - 2^s - \dots - n^s;$$

obviously, the set of the numbers M_s satisfies condition (8.1), i.e., the system

$$\sum_{r=1}^{n} t_r r^s = M_s, \quad s = 1, \dots, n, \tag{8.6}$$

can be solved for the integers $t_1, \dots, t_n$.

Let a be a natural number, and let u_s $(s = 1, \dots, n)$ be the least nonnegative residue of the number t_s modulo p^a. Then

$$1^s + \dots + n^s + \sum_{r=1}^{n} u_r r^s \equiv N_s \pmod{p^a}.$$

We assume that $k \ge n + \sum_{r=1}^{n} u_r$. Next, we assume that

(1) $x_m = m$ if $1 \le m \le n$;

(2) $x_m = r$ if $n + u_1 + \dots + u_{r-1} < m \le n + u_1 + \dots + u_r$, $r = 1, \dots, n$;

(3) $x_m = 0$ if $u_1 + \dots + u_n + n < m \le k$.

Obviously, the number set $(x_1, \dots, x_k)$ is a solution of the system of congruences

$$\sum_{m=1}^{k} x_m^s \equiv N_s \pmod{p^a}, \quad s = 1, \dots, n;$$

by definition, the number of solutions of this system is equal to $W(p^a; k)$; hence we have

$$W(p^a; k) \geq 1.$$

Suppose that a natural number b satisfies the conditions:

(1) $2 \leq b \leq a$;

(2) $2b \geq a + 1$;

(3) $p^b > n$;

(4) $\varkappa + \delta \geq a - b$.

We show that for any integers $y_{n+1}, \dots, y_k$, there exist integers $y_1, \dots, y_n$ for which the numbers $z_1, \dots, z_k$ satisfying the conditions

$$z_m = x_m + p^b y_m, \quad m = 1, \dots, k,$$

are solutions of the system of congruences

$$\sum_{m=1}^{k} z_m^s \equiv N_s \pmod{p^{a+1}}, \quad s = 1, \dots, n. \tag{8.7}$$

Let $N_s = A_s + B_s p^a$, where A_s and B_s are nonnegative integer numbers. Since $2b \geq a + 1$, we can write system (8.7) as

$$\sum_{m=1}^{k} (x_m^s + s x_m^{s-1} y_m p^b) \equiv A_s + B_s p^a \pmod{p^{a+1}}, \quad s = 1, \dots, n.$$

Hence, for some integers D_s, we have

$$\sum_{m=1}^{n} s x_m^{s-1} y_m \equiv D_s p^{a-b} - \sum_{m=n+1}^{k} s x_m^{s-1} y_m \pmod{p^{a-b+1}}; \tag{8.8}$$

here $x_m = m$ for $1 \leq m \leq n$.

We consider the following system of linear equations for the unknowns $v_1, \dots, v_n$:

$$\sum_{m=1}^{n} s m^{s-1} v_m = D_s p^{a-b} - \sum_{m=n+1}^{k} s x_m^{s-1} y_m, \quad s = 1, \dots, n.$$

We denote the determinant of this system by α. Then

$$\alpha = \begin{vmatrix} 1 & \dots & 1 & \dots & 1 \\ 2 & \dots & 2m & \dots & 2n \\ \dots & \dots & \dots & \dots & \dots \\ n & \dots & nm^{n-1} & \dots & nn^{n-1} \end{vmatrix} = n! \begin{vmatrix} 1 & \dots & 1 \\ 1 & \dots & n^{n-1} \end{vmatrix} = n!(n-1)!\dots 2!1! \neq 0,$$

and hence this system has a unique solution of the form

$$v_r = \alpha_r/\alpha, \quad r = 1, \dots, n,$$

where α_r is the determinant obtained from the determinant α by replacing the rth column by the column composed of the numbers $F_1, \dots, F_n$, where

$$F_s = D_s p^{a-b} - \sum_{m=n+1}^{k} s x_m^{s-1} y_m, \quad s = 1, \dots, n.$$

Suppose that $\beta_r(m)$ is the determinant obtained from the determinant α_r by replacing the numbers F_s by the expressions $s x_m^{s-1} y_m$. By obvious successive transformations we reduce calculating the determinants α and $\beta_r(m)$ to calculating the Vandermonde determinants and obtain

$$\alpha^{-1}\beta_r(m) = y_m f_r(x_m),$$

where the polynomial $f_s(x)$ is given by the relation

$$f_r(x) = \frac{(n-x)\dots(r+1-x)}{(n-r)!} \cdot \frac{(x-r+1)\dots(x-1)}{(r-1)!}.$$

For an integer x, the value of $f_r(x)$ is, up to the sign, the product of binomial coefficients and hence is an integer. Hence for all m, the numbers $\alpha^{-1}\beta_r(m)$ are also integer. This implies that

$$\mu_r = \alpha_r\alpha^{-1} - \gamma_r\alpha^{-1} = \sum_{m=n+1}^{k} \alpha^{-1}\beta_r(m)$$

are also integer; here γ_s denotes the determinant obtained from the determinant α_r by replacing the numbers F_s by the numbers $D_s p^{a-b}$. Let us consider this determinant. By $\lambda(s,r)$ we denote the coefficient of x^{s-1} in the polynomial $fr(x)$. Expanding the determinant γ_r with respect to the rth column, we obtain

$$\gamma_r = \alpha p^{a-b} \sum_{s=1}^{n} s^{-1} D_s \lambda(s,r). \tag{8.9}$$

The coefficients $\lambda(s,r)$ are rational numbers, and the denominator of each of them is a divisor of the number $(n-r)!(r-1)!$ and hence of $(n-1)!$ because $(n-1)!\big((n-r)!(r-1)!\big)^{-1}$ is a binomial coefficient. This implies that the denominator Q_r of the rational number $\alpha^{-1}\gamma_r$ represented as an irreducible fraction is not divisible by p because all D_s are integers, the exponent of p contained in this representation is no less than $a - b - \varkappa - \delta$, and this number is nonnegative due to condition (4) imposed on the numbers a and b. Therefore, for the numbers v_r we have the representation

$$v_r = \alpha_r\alpha^{-1} = \mu_r + \alpha^{-1}\gamma_r = P_r Q_r^{-1},$$

where P_r is an integer and $(Q_r, p) = 1$.

We find Q'_s from the congruences

$$Q'_r Q_r \equiv 1 \pmod{p^{a-b+1}},$$

then it follows from the elementary theory of congruences that the set of numbers $y_m = Q'_m P_m$ $(m = 1, 2, \ldots, n)$ is a solution of the system of congruences (8.8). Thus, for any set of numbers $(y_{n+1}, \ldots, y_k)$, we have found the numbers $y_1, \ldots, y_n$ for which the system of congruences (8.7) is solvable. If the numbers $(y_{n+1}, \ldots, y_k)$ run independently through all the values in the complete system of residues modulo p, then the number of solutions $W(p^{a+1}; k)$ of this system satisfies the inequality

$$W(p^{a+1}; k) \geq p^{k-n}.$$

From each of the obtained solutions of system (8.7) modulo p^{a+1}, following the same scheme, we pass to solutions modulo p^{a+2}. For this, instead of the number b, we must take the number $b+1$ and verify whether the conditions imposed earlier on b with respect to the number a are satisfied for the $b+1$ with respect to the number $a+1$. Indeed, these conditions are satisfied because the inequality $2b \geq a+1$ implies $2(b+1) \geq (a+1)+1$, the inequality $b \leq a$ implies $b+1 \leq a+1$, the inequality $p^b > n$ much the more implies $p^{b+1} > n$, and moreover, $(a+1) - (b+1) = a - b \geq \varepsilon + \delta$.

The only distinction here is that, instead of $x_1 = 1, x_2 = 2, \ldots, x_n = n$, we use the numbers that are pairwise congruent to these numbers modulo p^b. The differences of these numbers are contained in the denominators of the coefficients $\lambda(s, r)$ instead of the numbers $1, \ldots, n-1$ in representation (8.9) of the determinant γ_r. But the power of the prime p, which is a divisor of these denominators, remains unchanged because the numbers s and $s + tp^b$, where t is an integer and $s \leq n-1$, are divisible by the same power of p that is less than b because $s \leq n-1 < p^b$. Because of this, the denominators Q_r of the numbers $\alpha^{-1}\alpha_r$ are not divisible by p, and hence all the other reasoning remains valid. Thus for $W(p^{a+2}; k)$ we obtain the estimate

$$W(p^{a+2}; k) \geq p^{2(k-n)}.$$

Repeating this process $h-a$ times, for $W(p^h; k)$ we obtain the estimate $W(p^h; k) \geq p^{(h-a)(k-n)}$ and hence

$$W(p^h; k) \geq p^{a(n-k)} p^{h(k-n)}. \tag{8.10}$$

Now we choose parameters a and b so that all necessary conditions be satisfied. For this, we set $a = 2\eta + 1$ and $b = \eta + 1$, where $\eta = \varepsilon + \delta$. Then we have

(1) $b = \eta + 1 \leq 2\eta + 1 = a$, i.e., $b \leq a$;

(2) $2b = 2\eta + 2 > 2\eta + 1 = a$, i.e., $2b \geq a + 1$;

(3) $p^b = pp^\eta = pp^\varkappa p^\delta \geq p^{\varkappa+1} > n$, i.e., $p^b > n$;

(4) $\varkappa + \delta = \eta = a - b$, i.e., $\varkappa + \delta \geq a - b$.

Hence, under this choice of a and b, all required conditions are satisfied, and the estimate (8.10) actually holds for

$$k \leq n + \sum_{r=1}^{n} u_r.$$

We now show that the numbers $u_1, \dots, u_n$ can be chosen so that the inequality

$$n + \sum_{r=1}^{n} u_r \leq n^2 2^{2n-1}$$

be satisfied. For this, we write system (8.6) in the equivalent form

$$\sum_{r=1}^{n} t_r r(r-1)\dots(r-s+1) = M'_s, \quad s = 1, \dots, n;$$

here M'_s are some integers and the equations in the last system are linear combinations with integer coefficients from system (8.6) and conversely. Hence the solvability conditions for both systems are equivalent. Since the quantity $r(r-1)\dots(r-s+1)$ is divisible by $s!$ without a remainder, the number M'_s is divisible by $s!$.

Suppose that the integers H_s are determined by the relations $s!H_s = M'_s$. Then the numbers $t_1, \dots, t_n$ satisfy the system of equations

$$\sum_{r=1}^{n} t_r r(r-1)\dots(r-s+1) = s!H_s,$$

and the numbers $u_1, \dots, u_n$ satisfy the system of congruences

$$\sum_{r=1}^{n} u_r r(r-1)\dots(r-s+1) \equiv s!H_s \pmod{p^a}.$$

By b_s we denote the exponent of p contained in the decomposition of $s!$ into prime divisors. Then the last system of congruences can be written as (note that $a > \delta_s$)

$$\sum_{r=1}^{n} u_r \frac{r(r-1)\dots(r-s+1)}{s!} \equiv H_s \pmod{p^{a-\delta_s}}.$$

In the determinant of this system, the units stand on the main diagonal, the zeros are below the main diagonal, and some integers are above the main diagonal. Successively solving the congruences in this system, starting from the last, we see that u_r satisfy the conditions

$$0 \leq u_r \leq p^{a-\delta_r} - 1, \quad r = n, \dots, 1.$$

Thus we have the inequality

$$n + \sum_{r=1}^{n} u_r \le \sum_{r=1}^{n} p^{a-\delta_r} \le p^a(p-1) + p^{a-1}(p-1) + \dots + (p-1) \qquad (8.11)$$
$$\le p^a(p-1)(1 + p^{-1} + p^{-2} + \dots) = p^{a+1}.$$

We write p^{a+1} as

$$p^{a+1} = p^{2\eta+2} = p^{2\varkappa+2\delta+2} = p^{2\varkappa} p^{2\delta+2}.$$

By the definition of $\varkappa$, we have $p^{2\varkappa} \le n^2$, and the definition of the number δ implies

$$p^{2\delta+2} = p^{2\left([(n-1)p^{-1}]+\dots+[(n-1)p^{-t}]\right)+2},$$

where the integer t is determined by the inequalities $p^t \le n-1 < p^{t+1}$. Hence we have

$$p^{2\delta+2} \le p^{2((n-1)/p+\dots+(n-1)/p^t)+2} \le p^{2((n-1)/(p-1))(1-p^{-t})+2}$$
$$\le p^{2((n-1)/(p-1))(1-1/(n-1))+2} \le p^{2(n-2)/(p-1)+2}.$$

Now we set $\alpha = (n-2)-(p-1) = n-p-1$. Then $n-2 = \alpha + p - 1$ and hence

$$p^{2(n-2)/(p-1)+2} = p^4 p^{2\alpha/(p-1)}.$$

If $p = 3$, then because of the inequality $n \ge 3$, we have

$$p^4 p^{2\alpha/(p-1)} = 3^n < 2^{2n-1}.$$

But if $p \ne 3$, then $p^{1/(p-1)} \le 2$ and $p^4 \le 2^{2p}$. Hence

$$p^{2\alpha/(p-1)} \le 2^{2\alpha}, \quad p^{2(n-2)/(p-1)+2} \le 2^{2p+2\alpha} = 2^{2n-2}.$$

So in both cases we have

$$p^{\alpha+1} \le 2^{2n-1}, \quad n + \sum_{r-1}^{n} u_r \le p^{a+1} \le 2^{2n-1}.$$

It follows from this and (8.11) that the estimate of this lemma holds for $k = k_1$. But if $k = k_1 + k_0$ $(k_0 > 0)$, then we can obtain the statement of the lemma by fixing the last k_0 unknowns in all possible ways. The proof of Lemma 8.3 is complete. □

Lemma 8.4. *The statement of the preceding lemma holds if we set* $k_1 = 3n^3 2^n - n$.

Proof. For $n = 3$ and $n = 4$, the statement of this lemma follows from Lemma 8.3 since in this case we have

$$n^2 2^{2n-1} < 3n^3 2^n - n;$$

hence we assume that $n \geq 5$.

For $s = 1, 2, \ldots, n$, we set

$$M_s = N_s - A \sum_{r=1}^{n} r^s,$$

where A is a natural number whose exact value will be given later. Obviously, the numbers M_s satisfy the solvability condition (8.5), and hence the system

$$\sum_{r=1}^{n} t_r r^s = M_s, \quad s = 1, \ldots, n, \tag{8.12}$$

can be solved for integers $t_1, \ldots, t_n$.

Let a be a natural number such that $a \leq 2\eta + 1$. Recall that $\eta = \varkappa + \delta$, where δ is the exponent of p contained in the decomposition of the number $(n-1)!$ into prime divisors; a natural number $\varkappa$ is determined by the condition $p^\varkappa \leq n < p^{\varkappa+1}$. It follows from (8.12) that there exist integers $u_1, \ldots, u_n$ such that

$$\sum_{r=1}^{n} u_r r^s \equiv M_s \pmod{p^a}, \quad s = 1, \ldots, n, \; 0 \leq u_r \leq p^a - 1, \; r = 1, \ldots, n.$$

Hence we have

$$A \cdot 1^s + \cdots + A \cdot n^s + \sum_{r=1}^{n} u_r r^s \equiv N \pmod{p^a}, \quad s = 1, \ldots, n,$$
$$0 \leq u_r \leq p^a - 1.$$

For $k \geq An + u_1 + \cdots + u_n$ and $r = 1, 2, \ldots, n$, we set

(1) $x_m = r$ if $A(r-1) < m \leq Ar$;

(2) $x_m = r$ if $An + u_1 + \cdots + u_{r-1} < m \leq An + u_1 + \cdots + u_r$;

(3) $x_m = 0$ if $An + u_1 + \cdots + u_n < m \leq k$.

Then the numbers $x_1, \ldots, x_k$ are solutions of the system of congruences

$$\sum_{m=1}^{k} x_m^s \equiv N_s \pmod{p^a}, \quad s = 1, \ldots, n. \tag{8.13}$$

We assume that there exists a natural number b such that

(1) $2 \leq b < a$;

$$\begin{aligned}&(2)\quad b(n_1+1)\ge a+1, \quad \text{where} \quad n_1=[0.5n]+1;\\ &(3)\quad p^b>n;\\ &(4)\quad a\ge b+2\varkappa+\delta.\end{aligned} \tag{8.14}$$

Suppose that for $m=1,\dots,k$ the integers z_m and y_m satisfy the relations

(1) $z_m=x_m+p^by_m$ if $1\le m\le nA$;

(2) $z_m=x_m$ if $nA<m\le k$.

Our nearest goal is to choose numbers $y_1,\dots,y_{nA}$ so that the following system of congruences be satisfied:

$$\sum_{m=1}^{k} z_m^s\equiv N_s \pmod{p^{a+1}}, \qquad s=1,\dots,n. \tag{8.15}$$

For $m\le nA$, $r\le n$, and $l\le A$, we set

$$z_{rl}=z_{(l-1)A+r}, \qquad x_{rl}=x_{(l-1)A+r}, \qquad y_{rl}=y_{(l-1)A+r},$$

then system (8.15) can be written as

$$\sum_{r=1}^{n}\sum_{l=1}^{A} z_{rl}^s+\sum_{m>nA}^{k} x_m^s\equiv N_s \pmod{p^{a+1}}.$$

We shall seek the numbers $y_{rt}=t_rv_l$, where t_r and v_l are some integers. Then, by the definition of z_{rl} and x_{rl}, we have

$$z_{rl}=x_{rl}+p^bt_rv_l=r+p^bt_rv_l.$$

This implies

$$z_{rl}^s=\sum_{d=0}^{s}\binom{s}{d}r^{s-d}t_r^dv_l^dp^{bd},$$

$$\sum_{r=1}^{n}\sum_{l=1}^{A} z_{rl}^s=\sum_{r=1}^{n}\sum_{l=1}^{A}\sum_{d=0}^{s}\binom{s}{d}r^{s-d}t_r^dv_l^dp^{bd}=\sum_{r=1}^{n}\sum_{d=0}^{s}\binom{s}{d}r^{s-d}t_r^d\left(\sum_{l=1}^{A}v_l^dp^{bd}\right).$$

According to the second condition imposed on the number b, we have

$$b(n_1+1)\ge a+1, \quad n_1=[0.5n]+1.$$

Therefore, after obvious transformations, system (8.15) can be written as

$$\sum_{r=1}^{n}\sum_{d=1}^{n_2}\binom{s}{d}r^{s-d}t_r^d\left(\sum_{l=1}^{A}v_l^dp^{bd}\right)\equiv\lambda_sp^a \pmod{p^{a+1}}, \quad s=1,\dots,n, \tag{8.16}$$

where $n_2 = (s, n_1)$ and λ_s are fixed integers.

Let $q = [1, \dots, n_1]$. We define the integers $M'_1, \dots, M'_{n_1}$ by the relations

$$M'_1 = q, \quad M'_2 = \cdots = M'_{n_1} = 0.$$

For these numbers, condition (8.5) is satisfied with respect to the number n_1. Indeed, this condition means that the system of equations

$$\sum_{r=1}^{n_1} t_r r^d = M'_d, \quad d = 1, \dots, n_1,$$

can be solved for the integers $t_1, \dots, t_n$. This system is equivalent to the system

$$\sum_{r=1}^{n_1} t_r f_d(r) = M''_d, \quad d = 1, \dots, n_1.$$

where

$$f_d(x) = x(x-1)\dots(x-d+1) = x^d + \alpha_{d-1}x^{d-1} + \cdots + \alpha_1 x,$$
$$M''_d = M'_d + \alpha_{d-1}M'_{d-1} + \cdots + \alpha_1 M'_1.$$

For solvability of the last system, it is necessary and sufficient that the numbers N''_d be divisible by $d!$ without a remainder. But since $M'_d = 0$ for $d \neq 1$, we have

$$M''_d = \alpha_1 M'_1 = (d-1)!q.$$

The ratio $\big((d-1)!/d!\big)q$ is an integer (because q is divisible by d by the definition) and thus condition (8.5) is satisfied for the numbers $M'_1, \dots, M'_{n_1}$.

Now we set $A = n_1^2 2^{2n_1-1}$. Then, by Lemma 8.3, there exists a number set $v_1, \dots, v_A$ for which the system of congruences

$$\sum_{l=1}^{A} v_l^d \equiv M'_d \pmod{p^{a+1}}, \quad d = 1, \dots, n_1,$$

is satisfied. Substituting this set into the congruences in system (8.16) and taking into account that $M'_2 = \cdots = M'_{n_1} = 0$, we obtain the following system (for the desired numbers $t_1, \dots, t_n$):

$$\sum_{r=1}^{n} s r^{s-1} t_r q p^b \equiv \lambda_s p^a \pmod{p^{a+1}}, \quad s = 1, \dots, n. \tag{8.17}$$

This system consists of linear congruences for the unknowns $t_1, \dots, t_n$. We show that the system has a solution. Suppose that the numbers q' and $\varkappa_1$ are determined by the conditions

$$(q', p) = 1, \quad q = q' p^{\varkappa_1},$$

then $p^{\varkappa_1} \le n_1$, and we can rewrite system (8.17) as

$$\sum_{r=1}^{n} sr^{s-1} t_r \equiv \mu_s p^{a-b-\varkappa_1} \pmod{p^{a-b-\varkappa_1+1}},$$

where μ_s are some integers ($s = 1, \dots, n$). Repeating the corresponding argument in the proof of Lemma 8.3, we first solve the system of equations

$$\sum_{r=1}^{n} sr^{s-1} w_r = \mu_r p^{a-b-\varkappa_1}, \qquad s = 1, \dots, n.$$

This system has a unique solution of the form

$$w_r = p^{a-b-\varkappa_1} \sum_{r=1}^{n} s^{-1} \mu_s \lambda(s, r),$$

where $\lambda(s, r)$ is the coefficient of x^{s-1} of the polynomial $f_r(x)$ in

$$f_r(x) = \frac{(n-x)\dots(r+1-x)}{(n-r)!} \cdot \frac{(x-1)\dots(x-r+1)}{(r-1)!}.$$

The denominator of each of the rational numbers $\lambda(s, r)$ is a divisor of the number $(n-r)!(r-1)!$ and hence of $(n-1)!$. This implies that the exponent of the prime p, which is a divisor of the denominator of the number

$$\sum_{s=1}^{n} s^{-1} \mu_s \lambda(s, r),$$

does not exceed $\varkappa + \delta$. But, since the inequality $a - b - \varkappa \ge \varkappa + \delta$ holds and, moreover, the inequality $\varkappa_1 \le \varkappa$ holds because of $n_1 \le n$, we see that, for all r, the denominator of the number w_r in its representation as an irreducible fraction is not divisible by p. Therefore, repeating the argument similarly to the corresponding argument in the proof of Lemma 8.3, we obtain the solution $t_1, \dots, t_n$ of the system of congruences (8.17). Hence the system of congruences (8.15) is solvable.

If $a + 1 < 2\eta + 1$, then we again repeat the above argument passing from the congruences modulo p^{a+1} to the congruences modulo p^{a+2} and increasing the parameter b by 1. Repeating if necessary this process several times, we can obtain the solution of the system of congruences (8.15) for $a + 1 = 2\eta + 1$ only if, instead of the numbers x_m, this system contains numbers congruent to the latter modulo p^b. But since $p^b > n$, we can further pass to large values of a similarly to the proof of Lemma 8.3. Finally, we obtain

$$W(p^h; s) \ge c_1 p^{h(k-n)}, \qquad c_1 = p^{(2\varkappa+2\delta+1)(n-k_1)}$$

if only $k \ge An + \sum_{r=1}^{n} u_r$.

To complete the proof of the lemma, it remains to choose values of the parameters a and b so that conditions (8.14) be satisfied and to estimate the variable

$$An + \sum_{r=1}^{n} u_r.$$

We set $a = 3\varkappa + \delta$ and $b = \varkappa + 1$ and verify conditions (8.14).

(1) We have $\varkappa \geq 1$, since $p \leq n, b = \varkappa + 1 \geq 2, a = 3\varkappa + \delta + 1 > \varkappa + 1 = b$, i.e., $2 \leq b < a$, and condition (8.14), (1) is satisfied.

(2) We show that $b(n_1 + 1) > a + 1$. For this, we set $t = [0.5n]$. Then for $p \neq 2$, in view of the inequality $n \geq 5$, we have $\delta \leq n - 3$, $n_1 = [0.5n] + 1 = t + 1$, $n \leq 2t + 1$, and $\delta \leq 2t - 2$. Therefore, the inequality in condition (2) follows from the inequality

$$(\varkappa + 1)(t + 2) > 3\varkappa + 2t.$$

After equivalent transformations, we thus obtain

$$\varkappa t + t + 2\varkappa + 2 > 3\varkappa + 2t, \quad (\varkappa - 1)(t - 1) + 1 > 0,$$

which indeed takes place because $\varkappa \geq 1$ and $t \geq 1$. Hence condition (2) holds for $p \neq 2$.

Now let $p = 2$. Then for $n = 5$, a straightforward verification shows that condition (2) is also satisfied. But if $n \geq 6$, then $\varkappa \geq 2, t \geq 3, \delta \leq n - 1$, and inequality (2) follows from the system of inequalities

$$(\varkappa - 1)(t - 1) > 1, \quad (\varkappa + 1)(t + 2) > 3\varkappa + 2t + 2,$$

$$b(n_1 + 1) \geq (\varkappa + 1)(t + 2) > 3\varkappa + 2t + 2 \geq 3\varkappa + n + 1 \geq 3\varkappa + \delta + 2 = a + 1.$$

Thus we have proved that condition (2) holds in all cases.

(3) The desired condition $p^b > n$ for $b = \varkappa + 1$ holds automatically according to the definition of the number $\varkappa$.

(4) If $a = 3\varkappa + \delta + 1$ and $b = \varkappa + 1$, then $b + 2\varkappa + \delta = 3\varkappa + \delta + 1$, i.e., condition (4) is also satisfied.

Now we show that the estimate

$$nA + \sum_{r=1}^{n} u_r \leq 3n^3 2^n - n$$

holds. For A, we have the relation

$$A = n_1^2 2^{2n_1 - 1}.$$

Since $n_1 = [0.5n] + 1 \ (n \geq 5)$, we have

$$A \leq 0.9n^2 2^n, \quad nA \leq 0.9n^3 2^n < n^3 2^n.$$

In the proof of Lemma 8.3, we obtained estimate (8.11) for $\sum_{r=1}^{n} u_r$ (here we also have the condition $a < \delta_s$, where δ_s is determined by the relation $p^{\delta_s} \parallel s!$). This estimate has the form

$$\sum_{r=1}^{n} u_r \le p^{a+1} - n.$$

Now we estimate

$$p^{a+1} = p^{3\varkappa+\delta+2} \le n^3 p^{\delta+2} = n^3 p^{[(n-1)/p]+\cdots+[(n-1)/p^\alpha]+2},$$

where the integer α is determined by the condition $p^\alpha \le n-1 < p^{\alpha+1}$.

For $n = p$ we have $\delta = 0$. Hence, in view of $n \ge 5$, we obtain

$$p^{a+1} = n^3 p^2 < n^3 2^n.$$

If $p < n$, then

$$\left[\frac{n-1}{p}\right] + \cdots + \left[\frac{n-1}{p^\alpha}\right] + 2 \le \frac{n-1}{p} + \cdots + \frac{n-1}{p^\alpha} + 2$$
$$= \frac{n-1}{p-1} \cdot \frac{p^\alpha - 1}{p^\alpha} + 2 \le \frac{n-2}{p-1} + 2.$$

Now for $p = 2$ we obtain $p^{a+1} \le n^3 2^{n-2+2} = n^3 2^n$. But if $2 < p < n$, then $p^{1/(p-1)} < 2$. Hence we have

$$p^{a+1} \le n^3 p^{(n-2)/(p-1)+2} \le n^3 p^3 p^{(n-p-1)/(p-1)} \le n^3 p^3 2^{-(p+2)} 2^{n+1} \le n^3 2^{n+1}.$$

From the above relations we finally obtain

$$nA + \sum_{r=1}^{n} u_r \le n^3 2^n + 2n^3 2^n - n = 3n^3 2^n - n.$$

Thus we have proved the statement of the lemma for $k_1 = 3n^3 2^n - n$, as required. □

Suppose that $n, m, k, r_1, \ldots, r_m, \lambda, N_1, \ldots, N_m$ are natural numbers ($n \ge 3$, $1 \le r_1 < \cdots < r_m = n$), p is a prime number ($p > n$), and T_λ is the number of solutions of the system of congruences

$$\left.\begin{aligned} x_1^{r_1} + \cdots + x_k^{r_1} &\equiv N_1, \\ &\vdots \\ x_1^{r_m} + \cdots + x_k^{r_m} &\equiv N_m, \end{aligned}\right\} \pmod{p^\lambda}, \quad 1 \le x_1, \ldots, x_k \le p^\lambda.$$

Further, let $k \ge 2mn \ln n$, and let $n < p \le 2n \ln n$. Then $T_1 > 1$.

Indeed, we consider the system of congruences (here g is some primitive root modulo p)

$$\left.\begin{aligned} y_1 g^{r_1} + \cdots + y_m g^{mr_1} &\equiv N_1, \\ &\vdots \\ y_1 g^{r_m} + \cdots + y_m g_k^{mr_m} &\equiv N_m, \end{aligned}\right. \quad \pmod p, \quad 1 \le y_1, \ldots, y_m \le p.$$

The determinant of this system is

$$\Delta = \begin{vmatrix} g^{r_1} & \ldots & g^{mr_1} \\ \cdots & \cdots & \cdots \\ g^{r_m} & \ldots & g^{mr_m} \end{vmatrix} \not\equiv 0 \pmod p.$$

Therefore, this system is solvable and, moreover, $1 \le y_i \le p \le 2n \ln n$ $(i = 1, \ldots, m)$. Representing each y_i as the sum of units, we obtain the following statement.

Lemma 8.5 (Yu. V. Linnik). *The number T_1 satisfies the asymptotic formula*

$$T_1 = p^{k-m} + \theta (n\sqrt{p})^k, \quad |\theta| \le 1;$$

moreover, $T_1 \ge 1$ if $p \ge 9n^2$ and $k \ge 4m \ln n$.

Proof. We have

$$T_1 = p^{-m} \sum_{a_1=0}^{p-1} \cdots \sum_{a_m=0}^{p-1} S^k(a_1, \ldots, a_m) \exp\{-2\pi i (a_1 N_1 + \cdots + a_m N_m)/p\},$$

where

$$S(a_1, \ldots, a_m) = \sum_{x=1}^{p} \exp\{2\pi i (a_1 x^{r_1} + \cdots + a_m x^{r_m})/p\},$$

Selecting the term with $a_1 = \cdots = a_m = 0$, which is equal to p^{k-m}, and applying A. Weil's estimate (Lemma A.5)

$$|S(a_1, \ldots, a_m)| \le n\sqrt{p}$$

to the other terms, we obtain the first assertion of the lemma; the second assertion follows from the first. □

Lemma 8.6. *Let $n < p \le 9n^2$, and let $k \ge [32mn \ln n]$. Then $T_1 \ge 1$.*

Proof. We have already considered the case $p \le 2n \ln n$. Therefore, we assume that $p > 2n \ln n$. We choose $Y = 2n$ and consider the following system of congruences:

$$\begin{aligned} (y_1 + z_1)x_1^{r_1} + \cdots + (y_k + z_k)x_k^{r_1} &\equiv N_1, \\ &\vdots \qquad (\text{mod } p), \\ (y_1 + z_1)x_1^{r_m} + \cdots + (y_k + z_k)x_k^{r_m} &\equiv N_m, \\ 1 \le y_1, z_1 \le Y, \quad 1 \le x_i \le p, \quad & i = 1, \ldots, k. \end{aligned}$$

We denote the number of solutions of this system by T. We have

$$T = p^{-m} \sum_{a_1=0}^{p-1} \cdots \sum_{a_m=0}^{p-1} W^k(a_1, \ldots, a_m) \exp\{-2\pi i (a_1 N_1 + \cdots + a_m N_m)/p\},$$

where

$$W(a_1, \ldots, a_m) = \sum_{y=1}^{Y} \sum_{z=1}^{Y} \sum_{x=1}^{p} \exp\{2\pi i (y+z)(a_1 x^{r_1} + \cdots + a_m x^{r_m})/p\}.$$

Let us estimate $|W(a_1, \ldots, a_m)|$ for $(a_1, \ldots, a_m) \ne (0, \ldots, 0)$. Since the congruence

$$a_1 x^{r_1} + \cdots + a_m x^{r_m} \equiv \lambda \pmod p$$

has at most n solutions for any λ, we have

$$\begin{aligned} |W(a_1, \ldots, a_m)| &\le \sum_{x=1}^{p} \left| \sum_{y=1}^{Y} \sum_{z=1}^{Y} \exp\{2\pi i (y+z)(a_1 x^{r_1} + \cdots + a_m x^{r_m})/p\} \right| \\ &\le n \sum_{\lambda=1}^{p} \left| \sum_{y=1}^{Y} \exp\{2\pi i \lambda y/p\} \right|^2 = npY. \end{aligned}$$

Hence

$$T = (Y^2 p)^k p^{-m} + \theta_1 (npY)^k = Y^{2k} p^{k-m} (1 + \theta_1 (nY^{-1})^k p^m),$$

where $|\theta_1| \le 1$. Thus, for $k \ge 8m \ln n$, we have $T \ge 1$ and, finally, for $k \ge [32mn \ln n]$, we have $T_1 \ge 1$. The proof of the lemma is complete. □

It should be noted that, for $p \ge 9n^2$, we have an asymptotic formula for T_1 and only a lower bound for T_1 for "small" p $(n < p < 9n^2)$. Moreover, from Lemma 8.5 and 8.6, we have

$$W(p; k) \ge 1$$

for $k \ge k_2 = [32n^2 \ln n] + n$, as well as

$$W(p; k) = p^{k-n} + \theta(np^{1/2})^k$$

for $p \ge 9n^2$.

Lemma 8.7. *Suppose that $p > n$, $k \geq k_2 = [32n^2 \ln n]$, and h is a natural number. Then*

$$W(p; k) \geq p^{n-k_2} p^{h(k-n)}.$$

Proof. First, we note that, without loss of generality, we can set $k = k_2$. We prove this lemma by induction on the parameter h. We consider a system of congruences of the form

$$\sum_{m=1}^{k} x_m^s \equiv N_s \pmod{p^h}, \qquad s = 1, \ldots, n, \tag{8.18}$$

$$x_m \equiv m \pmod{p}, \qquad m = 1, \ldots, n.$$

By $T(h)$ we denote the number of solutions of this system. Obviously, $W(p^h; k) \geq T(h)$. We shall prove that $T(h)$ satisfies the estimate $T(h) \geq p^{n-k_2} p^{h(k-n)}$. For $h = 1$, this statement follows from Lemma 8.5. We assume that it has already been proved for all h such that $1 \leq h \leq a$. Then we prove that this also holds for $h = a + 1$. To this end, we represent the unknowns x_m $(m = 1, \ldots, k)$ in the form $x_m = y_m + p^a z_m$.

It follows from the induction assumption that the following congruences are satisfied:

$$\sum_{m=1}^{k} y_m^s \equiv N_s \pmod{p^a}, \qquad s = 1, \ldots, n,$$

$$y_m^s \equiv m^s \pmod{p}, \qquad m = 1, \ldots, n.$$

Because of this, for $h = a+1$ and for some $\nu_1, \ldots, \nu_n$, system (8.18) is equivalent to the following linear system of congruences in the unknowns $z_1, \ldots, z_k$:

$$\sum_{m=1}^{k} y_m^{s-1} z_m \equiv \nu_s \pmod{p}, \qquad s = 1, \ldots, n.$$

If we arbitrarily choose numbers $z_{n+1}, \ldots, z_k$, then the unknowns $z_1, \ldots, z_n$ are uniquely determined, since the determinant corresponding to these unknowns is not zero because of the conditions on $y_1, \ldots, y_n$. Hence we have the relation

$$T(a + 1) = p^{k-n} T(a).$$

Thus

$$T(h) = T(1) p^{(k-n)(h-1)}, \qquad W(p^h; k) \geq p^{n-k_s} p^{(k-n)h}.$$

Lemma 8.7 is thereby proved. □

Lemma 8.8. *The following estimate holds for $p \geq 9n^2$ and $k \geq 6n^2$:*

$$W(p^h; k) \geq p^{h(k-n)}(1 - p^{-3}).$$

Proof. We divide the proof into two cases: $h = 1$ and $h > 1$. First, we consider the case $h = 1$. By Lemma 8.6, we have

$$W(p; k) \geq p^{k-n} + \theta(np^{0.5})^k = p^{k-n}(1 + \theta n^k p^{n-0.5k}), \quad |\theta| \leq 1.$$

If $p \geq n^3$, then

$$n^k p^{n-0.5k} \leq p^{n-k/6} \leq p^{n-n^2} < p^{-4},$$

but if $9n^2 < p < n^3$, then

$$n^k p^{n-0.5k} \leq p^n 3^{-k} \leq n^{3n} 3^{-6n^2} \leq p^{-4}.$$

Therefore, in both cases we have

$$W(p; k) \geq p^{k-n}(1 - p^{-4}).$$

Let now $h > 1$. By definition, $W(p^h; k)$ is the number of solutions of the system of congruences

$$\sum_{m=1}^{k} x_m^s \equiv N_m \pmod{p^h}, \quad s = 1, \ldots, n; \tag{8.19}$$

here the unknowns $x_1, \ldots, x_k$ independently run through the complete systems of residues modulo p^h.

We divide all sets of numbers $(x_1,\ldots,x_k)$ into two classes. A set $(x_1, \ldots, x_k)$ belongs to the first class if it contains at least n numbers that are pairwise noncongruent modulo p. By $W_1(h)$ we denote the number of solutions of system (8.19) that are sets of the first class. All other sets belong to the second class. By $W_2(h)$ we denote the number of solutions of system (8.19) that are sets of the second class. By this definition, we have

$$W(p^h; k) = W_1(h) + W_2(h),$$

and hence $W(p^h) \geq W(h)$. We give an estimate from below for $W_1(h)$. For $h = 1$, the variable $W_2 = W_2(h)$ satisfies the inequality

$$W_2 \leq n^k p^{n-1},$$

since in this case the number of solutions contained in the second class does not exceed the number of sets $(x_1, \ldots, x_k)$ satisfying the conditions $0 \leq x_s < p$ $(s = 1, \ldots, k)$. Moreover, for each set of numbers, there exist at most $n-1$ different numbers. Hence, for $W_1 = W_1(h)$ with $h = 1$ we have the estimate

$$W_1 \geq W(p; k) - W_2 = p^{k-n}(1 - p^{-4}) - n^k p^{n-1} \geq p^{k-n}(1 - p^{-3}).$$

But if $h > 1$, then $W_1(h)$ satisfies the relation

$$W_1(h) = p^{(k-n)(h-1)} W_1.$$

This relation can be proved by a word for word repetition of the corresponding argument in the proof of the preceding lemma. We only must replace the numbers $x_1, \dots, x_k$ in the preceding lemma by some n numbers from the set $(x_1, \dots, x_k)$ that are noncongruent modulo p. So for an arbitrary h we obtain

$$W_1(h) = p^{(k-n)(h-1)} W_1 \geq p^{h(k-n)}(1 - p^{-3}),$$

hence

$$W(p^h; k) \geq W_1(h) \geq p^{h(k-n)}(1 - p^{-3}).$$

The proof of Lemma 8.8 is complete. □

Theorem 8.1. *Let* $k \geq T = \min(n^2 2^{2n-1}, 3n^3 2^n - n)$. *Then the following inequality holds under the condition* (8.5):

$$\sigma \geq n^{-20n^4 2^n},$$

where σ *is the singular series in Lemma* 8.1.

Proof. By Lemma 8.2, we have

$$\sigma = \prod_p \sigma_p, \quad \sigma_p = \lim_{h \to \infty} W(p^h; k) p^{h(n-k)}.$$

Estimating $W(p^h; k)$ by Lemmas 8.3, 8.4, 8.7, and 8.8 for different values of p and taking into account that $k \geq T$, we obtain:

$$\begin{aligned} &(1) \quad \sigma_p \geq p^{(2\varkappa+2\delta+1)(n-T)} \quad \text{for} \quad p \leq n; \\ &(2) \quad \sigma_p \geq p^{n-k_2}, \quad k_2 = [32n^2 \ln n] + n, \quad \text{for} \quad n < p < 9n^2; \\ &(3) \quad \sigma_p \geq 1 - p^{-3} \quad \text{for} \quad p \geq 9n^2. \end{aligned} \tag{8.20}$$

We set

$$\varphi_1 = \prod_{p \leq n} \sigma_p, \quad \varphi_2 = \prod_{n<p<9n^2} \sigma_p, \quad \varphi_3 = \prod_{p \geq 9n^2} \sigma_p.$$

Then

$$\sigma = \varphi_1 \varphi_2 \varphi_3.$$

For the variable φ_3 in (8.20), we have the obvious estimate

$$\varphi_3 \geq 0.5.$$

Now we estimate φ_1 and φ_2. First, we consider φ_2. Using (8.20), we obtain

$$\varphi_2 \geq \prod_{n<p<9n^2} p^{n-k_2} > 2 \prod_{p<9n^2} p^{-32n^2 \ln n} = 2\Big(\prod_{p<9n^2} p\Big)^{-32n^2 \ln n}.$$

Next, we use the estimate $\psi(x) < x \ln 4$ (see Lemma 3.8 in Chapter 3), where $\psi(x)$ is the Chebyshev function. It follows from this estimate that

$$\prod_{p \le x} p^{\varkappa} < 4^x, \tag{8.21}$$

where $\varkappa = \varkappa(x)$ is determined by the relation $p^{\varkappa} \le x < p^{\varkappa+1}$. Hence for $x = 9n^2$, we have

$$\prod_{p<9n^2} p < \prod_{p<9n^2} p^{\varkappa} < 4^{9n^2}, \quad \varphi_2 > 2^{-2^6 3^2 n^4 \ln n + 1}.$$

Now we estimate φ_1. We have

$$\varphi_1 = \prod_{p \le n} \sigma_p \ge \prod_{p \le n} p^{(2\varkappa+2\delta+1)(n-T)} = (\varphi_4 \varphi_5)^{n-T},$$

where

$$\varphi_4 = \prod_{p \le n} p^{2\varkappa+1}, \quad \varphi_5 = \prod_{p \le n} p^{2\delta}.$$

Using (8.21), for $\varkappa = \varkappa(n)$, we obtain

$$\sigma_4 \le \prod_{p \le n} p^{3\varkappa} < 4^{3n} = 2^{6n}.$$

Next, by the definition of δ (see the assumptions of Lemma 8.3), we have

$$\varphi_5 = \prod_{p \le n} p^{2\delta} \le \left((n-1)!\right)^2$$

Obviously, $(n-1)! < 2^{-n} n^n$ and hence

$$\varphi_5 < 2^{-2n} n^{2n}, \quad \varphi_4 \varphi_5 \le 2^{4n} n^{2n}.$$

So we have $\varphi_1 > 2^{4n(n-T)} n^{2n(n-T)}$.

Using the estimates for φ_1, φ_2, and φ_3, we obtain

$$\sigma = \varphi_1 \varphi_2 \varphi_3 > 2^{4n(n-T) - 2^6 3^2 n^4 \ln n} n^{2n(n-T)} > n^{-20 n^4 2^n}.$$

The theorem is thereby proved. □

Condition (8.5) is necessary for the positiveness of the singular series σ. As Theorem 8.1 shows, this is also a sufficient condition if only the number of variables k is large; more precisely, k must be no less than T. The value of T also increases with increasing n, but it turns out that it is impossible, instead of T, to use any other variable that increases, say, slower than $T^{1-\varepsilon}$ (for any $\varepsilon > 0$). In other words, the parameter T is a variable of regular growth. The proof of this fact is the main goal in this section.

We consider a sequence of polynomials with rational coefficients

$$f_0(x) = 1, \quad f_1(x) = x, \quad \ldots, \quad f_s(x) = \frac{x(x+1)\ldots(x+s-1)}{s!}.$$

All these polynomials are integral-valued, i.e., they take integer values for integer x. For $f_0(x)$ and $f_1(x)$, this is obvious, and for all other polynomials, this follows from the property

$$f_s(x) - f_s(x-1) = f_{s-1}(x). \tag{8.22}$$

The last relation actually takes place, since

$$\begin{aligned} f_s(x) + f_s(x+1) &= \frac{x(x+1)\ldots(x+s-1)}{s!} - \frac{(x-1)x(x+1)\ldots(x+s-2)}{s!} \\ &= \frac{x(x+1)\ldots(x+s-2)}{(s-1)!} \frac{\big((x+s-1)-(x-1)\big)}{(s-1)!} \\ &= \frac{x(x+1)\ldots(x+s-2)}{(s-1)!} = f_{s-1}(x). \end{aligned}$$

Relation (8.22) also implies that if

$$Q_n(x) = a_0 f_0(x) + \cdots + a_n f_n(x),$$

then, for a natural number n, we have

$$Q_{n+1}(x) = \sum_{t=1}^{x} Q_n(t) = a_0 f_1(x) + \cdots + a_n f_{n+1}(x).$$

Further, we consider the sequence of polynomials $g_s(x)$ given by the relations

$$g_1(x) = f_1(x) = x, \qquad g_2(x) = \sum_{t=1}^{x} (2g_1(t) - 1),$$

$$\vdots$$

$$g_{s+1}(x) = \sum_{t=1}^{x} (2g_s(t) - 1).$$

It follows from the above properties of the polynomials $f_s(x)$ that

$$g_2(x) = 2f_2(x) - f_1(x), \qquad g_3(x) = 2^2 f_3(x) - 2f_2(x) - f_1(x),$$

$$\vdots$$

$$g_s(x) = 2^{s-1} f_s(x) - \sum_{r=1}^{s-1} 2^{r-1} f_r(x).$$

Lemma 8.9. *For an integer x, the polynomial $g_s(x)$ satisfies the congruence*

$$2g_s(x) \equiv 1 + (-1)^x \pmod{2^{s+1}}. \tag{8.23}$$

Proof. We prove this relation by induction on the parameter s. For $s = 1$ we have

$$2g_1(x) = 2x \equiv \begin{cases} 0 = 1 - (-1)^x & \text{if } x \text{ is even,} \\ 2 = 1 - (-1)^x & \text{if } x \text{ is odd.} \end{cases}$$

So the statement of the lemma is proved for $s = 1$. We assume that this statement holds for $s = m$, $m \geq 1$, and prove it for $s = m + 1$. By the induction hypothesis, we have

$$2g_m(x) \equiv 0 \pmod{2^{m+1}} \quad \text{if } x \text{ is even,}$$
$$2g_m(x) \equiv 2 \pmod{2^{m+1}} \quad \text{if } x \text{ is odd.}$$

For a natural number x, this implies

$$\begin{aligned} &2g_m(x) - 1 \equiv (-1)^{x+1} \pmod{2^{m+1}}, \\ &2g_{m+1}(x) = 2\sum_{t=1}^{x} \big(2g_m(t) - 1\big) \equiv 2\big(1 - 1 + \cdots + (-1)^x + (-1)^{x+1}\big) \\ &\qquad \equiv 1 - (-1)^x \pmod{2^{m+2}}. \end{aligned}$$

Thus we have proved the statement of the lemma for a natural number x. To extend the proof to all integer x, we note that, in the representation of the rational number $2^{s-1}(s!)^{-1}$ as the irreducible fraction $P_s Q_s^{-1}$, the denominator Q_s is not divisible by 2 for all natural numbers s. By this and the fact that the polynomial $g_s(x)$ is integral-valued, for all integer x, we have

$$g_s(x) \equiv G_s(x) \pmod{2^s},$$

where

$$G_s(x) = P_s Q_s' x(x+1)\ldots(x+s-1) - \sum_{r=1}^{s-1} P_r Q_r' x(x+1)\ldots(x+r-1),$$

and the integers Q_r' for $r = 1, \ldots, s$ are determined by the congruences

$$Q_r Q_r' \equiv 1 \pmod{2^s}.$$

Since the polynomial $G_s(x)$ has integer coefficients, it is periodic modulo any number, and the right-hand side of congruence (8.23) is periodic modulo any even number. Therefore, the congruences

$$2G_{m+1}(x) \equiv 2g_{m+1}(x) \equiv 1 - (-1)^x \pmod{2^{m+2}}$$

hold for all integer x, as was to be proved. □

Lemma 8.10. *Suppose that integer numbers* $a_1, \ldots, a_n$ *are coefficients of the polynomial* $G_n(x)$ *constructed in the proof of the preceding lemma. For the system of congruences*

$$\sum_{m=1}^{k} x_m^s \equiv N_s \pmod{2^n}, \quad s = 1, \ldots, n,$$

to have a solution, it is necessary that the following inequality be satisfied:

$$k \geq b_0,$$

where b_0 *is the least nonnegative residue of the number* b *modulo* 2^n,

$$b = \sum_{s=1}^{n} a_s N_s.$$

Proof. By construction, the polynomial $G_n(x)$ satisfies the congruence

$$2G_n(x) \equiv 1 - (-1)^x \pmod{2^{n+1}}.$$

This means that, for even x, the number $G_n(x)$ is congruent to zero modulo 2^n and, for odd x, with unity. Therefore, multiplying the congruence with index s by a_s and adding all congruences of the system under study, we see that the number H of odd numbers among $x_1, \ldots, x_k$ satisfies the congruence

$$H \equiv b \pmod{2^n}.$$

This readily implies the statement of the lemma. □

The solvability condition (8.5) depends only on the values of the residues of the numbers $N_1, \ldots, N_n$ modulo $n!$. Indeed, system (8.5) is equivalent to the system

$$\sum_{r=1}^{n} t_r r(r-1)\ldots(r-s+1) = M_s, \quad s = 1, \ldots, n,$$

where $(M_1, \ldots, M_n)$ is a set of integers that bijectively corresponds to the set $(N_1, \ldots, N_n)$. The solvability condition for the last system is obvious. It means that if each M_s is divisible by $s!$, then the system has an integer-valued solution. The terms $N_1, \ldots, N_n$ can be linearly and with integer coefficients expressed in terms of $M_1, \ldots, M_n$, and conversely. Therefore, in the solvability condition (8.5), it suffices, instead of the numbers N_s, to consider their residues modulo $n!$. Thus the system of equations in condition (8.5) can be replaced by a system of congruences modulo $n!$. In turn, this system is equivalent to a set of systems each of which corresponds to its own prime number p, where $p \leq n$, and the congruences are taken modulo p^δ, where δ is the exponent of the prime p contained in the decomposition of $n!$ into prime factors.

These systems are independent of one another in the sense that the unknowns in them run through their own complete systems of residues independently of one another for different primes p.

Thus the solvability condition (8.5) is equivalent to a set of independent solvability conditions for each prime p that does not exceed n. For each p, this condition written for the numbers $M_1, \dots, M_{n_1}$ means that the number M_s $(s = 1, \dots, n)$ is divisible by p^{δ_s}, where δ_s is determined by the relation $p^{\delta_s} \parallel s!$. We note that $\delta_s < n$ for all $s \leq n$. Therefore, is suffices to consider the congruences modulo p^n. All the sets $(M_1, \dots, M_n)$ and the sets $(N_1, \dots, N_n)$ satisfying the solvability condition (8.5) are now divided into classes depending on the values of the residues modulo 2^n of the numbers contained in these sets. The number of these and those sets is the same in any class. Denoting the number of classes by A, we obtain

$$A = 2^{n^2 - \delta_n - \dots - \delta_1}.$$

Further, since $\delta_1 = 0$, the residue of the number M_1 can take any value modulo 2^n independently of the other numbers M_s $(s = 2, \dots, n)$. We express the number b in Lemma 8.10 in terms of $M_1, \dots, M_n$ and obtain

$$b = P_n Q'_n M_n - \sum_{r=2}^{n-1} P_r Q'_r M_r - M_1.$$

This implies that the number of solutions of the congruence $b \equiv b_0 \pmod{2^n}$ for any fixed b_0 such that $0 \leq b_0 \leq 2^n - 1$ is equal to $2^{-n}A$. In particular, we can set $b_0 = 2^n - 1$. Then it follows from Lemma 8.10 that there exist $2^{-n}A$ sets $(N_1, \dots, N_n)$ satisfying the solvability condition and the condition that the relations

$$W(p^m; k) > 0$$

imply $k \geq 2^n - 1$. In particular, we can choose $N_1 \equiv N_2 \equiv \dots \equiv N_n \equiv 2^n - 1 \pmod{2^n}$. By setting $b_0 = 0, 1, \dots, d$ $(d \leq 2^n - 1)$, we see that there exist at least $(1 - 2^{-n}d)A$ sets $(N_1, \dots, N_n)$ satisfying the solvability condition for $p = 2$ and the condition that

$$W(2^n; k) = 0$$

if only $k < d$. Simultaneously, here we have $\sigma = 0$ because for $h \geq n$ the relation

$$W(2^n; k) = 0$$

implies that

$$W(2^h; k) = 0,$$

and hence $\sigma_2 = 0$ and $\sigma = \prod_p \sigma_p = 0$. By setting $d = 2^n - 1$, we see that the number k in Theorem 8.1 must be no less than $2^n - 1$. In other words, if T_0 is the least value of k starting from which the inequality $\sigma > 0$ holds for all $N_1, \ldots, N_n$ satisfying condition (8.1), then the following inequalities hold:

$$2^n - 1 \leq T_0 \leq T = 3n^3 2^n - n.$$

For any $\varepsilon > 0$ as $n \to \infty$, we have

$$(2^n - 1)T^{\varepsilon - 1} \to +\infty;$$

hence (here the constant in Vinogradov's symbol $\ll$ is independent of ε)

$$T^{1-\varepsilon} \ll T_0 \leq T,$$

i.e., the parameter T is a variable of regular growth. So we have proved the following theorem.

Theorem 8.2. *Let σ be the singular series in Theorem* 8.1,

$$\sigma = \prod_p \sigma_p,$$

where the numbers σ_p are defined in the assumptions of Lemma 8.2. *Then*

(1) *there are number sets $(N_1, \ldots, N_n)$ satisfying the solvability condition* (8.5) *and $\sigma = 0$ for these numbers if $k < 2^n - 1$, but $\sigma > 0$ for all sets satisfying condition* (8.5) *if*

$$k \geq T = \min(n^2 2^{2n-1}, 3n^3 2^n - n);$$

(2) *the system of equations in the solvability condition* (8.5) *can be replaced by the set of mutually independent systems of congruences; moreover, to each prime number p that does not exceed n, there corresponds its own system of congruences modulo p^n;*

(3) *among A $(A > 0)$ sets $(l_1, \ldots, l_n)$ of classes of residues modulo 2^n satisfying the solvability condition corresponding to the prime number* 2, *there exist at least $A(1 - 2^{-n}d)$ sets for which*

$$\sigma = 0 \quad \text{if } k < d < 2^n - 1, \qquad \sigma_2 > 0 \quad \text{if } k \geq T.$$

8.2 The singular integral in the Hilbert–Kamke problem

In this section we study the relation between γ, which is the value of the singular integral, and the properties of solutions of the system of equations

$$\sum_{m=1}^{k} x_m^s = \beta_s, \quad s = 1, \ldots, n, \tag{8.24}$$

where β_s are determined by the relations $\beta_s = N_s P^{-s}$, and the unknowns x_m satisfy the conditions $0 \le x_m \le 1$ for $m = 1, \dots, k$.

By Ω we denote the domain of points $(x_1, \dots, x_k)$ in the k-dimensional space for which the following inequalities hold:

$$(1)\quad 0 \le x_m \le 1, \quad m = 1, \dots, k;$$

$$(2)\quad \left|\sum_{m=1}^{k} x_m^s - \beta_s\right| \le h, \quad h > 0,\ s = 1, \dots, n.$$

We let $\mu(h)$ denote the volume of the domain Ω, i.e., we set

$$\mu(h) = \int \cdots \int_{\Omega} dx_1 \dots dx_k.$$

Lemma 8.11. *For* $k > 0.5n(n+1) + 1$, *the following relation holds*:

$$\gamma = \gamma(\beta_1, \dots, \beta_n) = \lim_{h \to 0} 2^{-n} h^{-n} \mu(h).$$

Proof. Since the integral γ converges absolutely for $k > 0.5n(n+1)+1$, this integral is a continuous function in all the variables $\beta_1, \dots, \beta_n$. We set

$$F(\beta_1, \dots, \beta_n) = \int_0^{\beta_1} \cdots \int_0^{\beta_n} \gamma(\alpha_1, \dots, \alpha_n)\, d\alpha_1 \dots d\alpha_n.$$

Then we have

$$\begin{aligned}\gamma(\beta_1, \dots, \beta_n) &= \frac{\partial F(\beta_1, \dots, \beta_n)}{\partial \beta_1, \dots, \partial \beta_n}\\ &= \lim_{h \to 0} 2^{-n} h^{-n} \int \cdots \int_{\Omega} \gamma(\alpha_1, \dots, \alpha_n)\, d\alpha_1 \dots d\alpha_n.\end{aligned}$$

Now we show that

$$F(\beta_1, \dots, \beta_n) = \int \cdots \int_{\Omega_1(\beta_1, \dots, \beta_n)} dx_1 \dots dx_k,$$

where $\Omega_1(\beta_1, \dots, \beta_n)$ denotes the domain of points $(x_1, \dots, x_k)$ determined by the conditions

$$0 < x_m < 1, \quad m = 1, \dots, k, \qquad 0 < x_1^s + \dots + x_k^s < \beta_s, \quad s = 1, \dots, n.$$

Indeed, by the definition of the functions $F(\beta_1, \dots, \beta_n)$ and $\gamma(\alpha_1, \dots, \alpha_n)$, we have

$$F(\beta_1, \dots, \beta_n) = \int_1^{\beta_1} \cdots \int_0^{\beta_n} \gamma(\alpha_1, \dots, \alpha_n)\, d\alpha_1 \dots d\alpha_n$$

$$= \int_1^{\beta_1} \cdots \int_0^{\beta_n} d\alpha_1 \ldots d\alpha_n \int_{-\infty}^{+\infty} \cdots \int_{-\infty}^{+\infty} dz_1 \ldots dz_n$$
$$\times \int_0^1 \cdots \int_0^1 \exp\left\{2\pi i \sum_{s=1}^{n} (t_s - \alpha_s) z_s\right\} dx_1 \ldots dx_k,$$

where the variables t_s are determined by the relations

$$t_s = x_1^s + \cdots + x_k^s \quad (s = 1, \ldots, n).$$

Changing the order of integration and integrating with respect to $\alpha_1, \ldots, \alpha_n$, we hence obtain

$$F(\beta_1, \ldots, \beta_n) = \int_{-\infty}^{+\infty} \cdots \int_{-\infty}^{+\infty} \left(\prod_{s=1}^{n} \frac{1 - \exp\{-2\pi i z_s \beta_s\}}{2\pi i z_s} \right) dz_1 \ldots dz_n$$
$$\times \int_0^1 \cdots \int_0^1 \exp\left\{2\pi i \sum_{s=1}^{n} t_s z_s\right\} dx_1 \ldots dx_k$$

$$= \int_0^1 \cdots \int_0^1 dx_1 \ldots dx_k$$
$$\times \int_{-\infty}^{+\infty} \cdots \int_{-\infty}^{+\infty} \prod_{s=1}^{n} \frac{\exp\{2\pi i t_s z_s\} - \exp\{2\pi i (t_s - \beta_s) z_s\}}{2\pi i z_s} dz_1 \ldots dz_n$$
$$= \pi^{-n} \int_0^1 \cdots \int_0^1 \left(\prod_{s=1}^{n} \int_0^{+\infty} \left(\frac{\sin 2\pi z_s t_s}{z_s} - \frac{\sin 2\pi z_s (t_s - \beta_s)}{z_s} \right) dz_s \right) dx_1 \ldots dx_k.$$

But since

$$\int_0^{+\infty} \frac{\sin \alpha x}{x} dx = \frac{\pi}{2} \operatorname{sgn} \alpha,$$

we have

$$F(\beta_1, \ldots, \beta_n) = \int_0^1 \cdots \int_0^1 \prod_{s=1}^{n} \left(\operatorname{sgn} t_s - \operatorname{sgn}(t_s - \beta_s) \right) dx_1 \ldots dx_k$$
$$= \underset{\substack{t_s \le \beta_s \\ 0 < x_1, \ldots, x_k < 1}}{\int \cdots \int} dx_1 \ldots dx_k = \underset{\Omega_1(\beta_1, \ldots, \beta_n)}{\int \cdots \int} dx_1 \ldots dx_k.$$

Thus we have proved the desired relation for the function $F(\beta_1, \ldots, \beta_n)$. This readily implies that the integral in the right-hand side of the above relation for the variable $\gamma(\beta_1, \ldots, \beta_n)$ is equal to $\mu(h)$. Hence

$$\gamma(\beta_1, \ldots, \beta_n) = \lim_{h \to 0} 2^{-n} h^{-n} \mu(h),$$

as required. □

We consider a system of equations for the unknowns $x_1, \dots, x_l$ of the form

$$x_1^s + \dots + x_l^s = \alpha_s, \quad s = 1, \dots, n, \ l > n. \tag{8.25}$$

We assume that the variables $x_1, \dots, x_l, \alpha_1, \dots, \alpha_n$ satisfy the conditions

$$0 \le x_m \le 1, \quad m = 1, \dots, l; \qquad 0 \le \alpha_s, \quad s = 1, \dots, n.$$

Next, we consider the set composed of l positive numbers $y_1, \dots, y_l$, where $l \ge n$. In a way, we choose n numbers with different indices among these variables. Suppose that these numbers are $z_1, \dots, z_n$. By setting $z_0 = 0$ and $z_{n+1} = 1$, we add two more numbers to this set.

The variable $\Delta(y_1, \dots, y_l)$, where

$$\Delta(y_1, \dots, y_l) = \max_{\varkappa} \min_{0 \le i < j \le n+1} |z_i - z_j|,$$

will be called the *characteristic of the set* $(y_1, \dots, y_l)$.

If the set $(y_1, \dots, y_l)$ is a solution of some system of equations, then its characteristic will be called the *characteristic of this solution of the system.*

Lemma 8.12. *Suppose that for* $l = n$ *and a positive* ε, *the characteristic* $\Delta(x_1, \dots, x_n)$ *of a fixed solution of system* (8.25) *satisfies the condition*

$$\Delta(x_1, \dots, x_n) \ge \varepsilon.$$

Suppose also that the numbers h_s $(s = 1, \dots, n)$ *satisfy the inequalities*

$$|h_s| \le H = (0.25\varepsilon)^n.$$

Then there exists a set $(y_1, \dots, y_n)$ *whose coordinates satisfy the conditions*:

(1) $\displaystyle\sum_{m=1}^{n} y_m^s = \alpha_s + h_s, \quad s = 1, \dots, n;$

(2) $\Delta(y_1, \dots, y_n) \ge 0.5\varepsilon;$

(3) $|x_m - y_m| \le H \cdot 2^{2n-2}\varepsilon^{1-n}, \quad |x_m - y_m| \le 0.25\varepsilon, \qquad m = 1, \dots, n.$

Proof. Let a be a sufficiently large natural number.

We recursively define numbers y_{mb} and z_{mb} $(m = 1, \dots, n;\ b = 0, \dots, a)$ as follows:

(1) $y_{m0} = x_m$ and $z_{m0} = 0;$

(2) $y_{m\,b+1} = y_{mb} + z_{m\,b+1};$

(3) the numbers $z_{m\,b+1}$ (for fixed numbers y_{mb}) satisfy a system of linear equations of the form

$$\sum_{m=1}^{n} z_{m\,b+1} y_{mb}^{s-1} = \frac{h_s}{as}, \quad s = 1, \dots, n. \tag{8.26}$$

We will prove that the numbers y_{mb} thus defined satisfy the inequalities

$$|y_{mb} - x_m| < 0.25\varepsilon. \tag{8.27}$$

For $b = 0$ this condition readily follows from the definition of the numbers y_{mb}. Now we assume that $d \geq 0$ and it has already been proved that this condition holds for all $b \leq d$. Then we shall prove this condition for $b = d + 1 \leq a$. Solving the linear system (8.26) for the unknowns $z_{m\,d+1}$, we obtain

$$z_{m\,d+1} = \frac{\sum_{s=1}^{n} h_s \sigma_s (as)^{-1}}{f_m(y_{md})}, \quad m = 1, \dots, n;$$

here σ_s is the coefficient of x^{s-1} in the polynomial $f_m(x)$, where

$$f_m(x) = \frac{(x - y_{1d}) \dots (x - y_{nd})}{x - y_{md}}.$$

Since $|y_{md} - x_d| < 0.25\varepsilon$ and $x_m \geq \varepsilon$, we have $y_{md} > 0$ and hence

$$\sum_{s=1}^{n} |\sigma_s| = (-1)^{n-1} f_m(-y_{md}) < 2^{n-1}.$$

Moreover, by assumption, we have $|h_s| \leq (0.25\varepsilon)^n$. Next, since $|x_m - x_r| \geq \varepsilon$, we have $|y_{md} - y_{rd}| \geq 0.5\varepsilon$. Therefore,

$$|z_{m\,d+1}| < 2^{n-1} a^{-1} h (0.5\varepsilon)^{-n+1}, \tag{8.28}$$

and hence

$$\begin{aligned} |z_{m\,d+1}| &\leq 0.25 a^{-1} \varepsilon, \\ |y_{m\,d+1} - x_m| = \left| \sum_{b=1}^{d+1} z_{mb} \right| &\leq 0.25 a^{-1} \varepsilon (d + 1) \leq 0.25\varepsilon. \end{aligned} \tag{8.29}$$

So inequality (9.27) is proved.

Now we consider the variables

$$R_s = R_s(a) = \sum_{m=1}^{n} y_{ma}^s - \sum_{m=1}^{n} x_m^s - h_s, \quad s = 1, \dots, n.$$

We show that $R_s \to 0$ as $a \to \infty$. Indeed, by the definition of y_{ma}, we have

$$R_s = \sum_{m=1}^{n}\sum_{b=0}^{a-1}(y_{m\,b+1}^s - y_{mb}^s) - h_s = \sum_{m=1}^{n}\sum_{b=0}^{a-1}\left((y_{mb} + z_{m\,b+1})^s - y_{mb}^s\right) - h_s.$$

Using the expansion of the variable $(y_{mb} + z_{m\,b+1})^s$ in the Taylor series around the point y_{mb}, we obtain

$$(y_{mb} + z_{m\,b+1})^s - y_{mb}^s = s z_{m\,b+1} y_{mb}^{s-1} + 0.5s(s-1)z_{m\,b+1}^2(y'_{mb})^{s-2},$$

where the number y'_{mb} satisfies the condition $y_{mb} \le y'_{mb} \le y_{m\,b+1}$. Hence we have $0 < y'_{mb} < 2$. Therefore, by (8.29), we obtain

$$|R_s| < 0.5s(s-1)(0.25\varepsilon)^2 a^{-1} 2^{s-2} < a^{-1}n^2 2^n.$$

We have $R_s \to 0$ as $a \to +\infty$. Hence if $(y_1, \dots, y_n)$ is the limit point for the set of points $(y_{1a}, \dots, y_{na})$, then this point satisfies condition (1) in the lemma. The second condition in the lemma is satisfied for this point because of inequality (8.28), and the third condition holds by inequality (8.29). Thus the proof of Lemma 8.12 is complete. □

We note that the system of equations in item (1) in Lemma 8.12 has only one solution that satisfies conditions (2) and (3), since the specific form of this system implies that all its solutions are permutations of a single solution and any permutation of the numbers $y_1, \dots, y_n$ does not already satisfy condition (3). Next, the numbers $y_1, \dots, y_n$ continuously depend on the variables $h_1, \dots, h_n$. To prove this fact is suffices to surround the point $(h_1, \dots, h_n)$ by a sufficiently small δ-neighborhood and to apply the already proved Lemma 8.12 to each point $(h_1+\delta_1, \dots, h_n+\delta_n)$, replacing the numbers h_s by $h_s + \delta_s$ $(s = 1, \dots, n)$.

Lemma 8.13. *Suppose that some solution* $(x_1, \dots, x_k)$ *of system* (8.24) *has a positive characteristic* ε. *Then, for a sufficiently small* h, *the volume* $\mu(h)$ *of the domain* Ω *satisfies the inequality*

$$\mu(h) \ge 2^n h^n 2^{2n(n-k)} k^{n-k} n^{-k-n} \varepsilon^{n(k-n)}.$$

Proof. Any permutation of the set $(x_1, \dots, x_k)$ is also a solution of this system. Hence we can assume that

$$x_1 \ge \varepsilon,\ x_2 - x_1 \ge \varepsilon,\ \dots\ x_n - x_{n-1} \ge \varepsilon, \quad x_n \le 1.$$

We consider the numbers $y_{n+1}, \dots, y_k$ satisfying the inequalities

$$x_m - \delta_1 \le y_m < x_m, \quad m = n+1, \dots, k,\ \delta_1 = (0.25\varepsilon)^n (kn)^{-1}.$$

For each m and for $s = 1, \dots, n$, we then have

$$|x_m^s - y_m^s| = |x_m - y_m|(x_m^{s-1} + \dots + y_m^{s-1}) \le s\delta_1 \le (0.25\varepsilon)^n k^{-1},$$

and hence, by setting

$$R_s = \sum_{m=n+1}^{k} (x_m^s - y_m^s),$$

we obtain $|R_s| \le (0.25\varepsilon)^n$.

By Lemma 8.12, there exist numbers $y_1, \dots, y_n$ satisfying the condition

$$\sum_{m=1}^{n} y_m^s = \beta_s + R_s - \sum_{m=n+1}^{k} x_m^s, \quad s = 1, \dots, n,$$

and uniquely determined by the variables R_s, i.e., by the set $(y_{n+1}, \dots, y_k)$. This implies that the point $(y_1, \dots, y_k)$ is a solution of Eq. (8.24).

Now we note that, for a sufficiently small h and for $|z_m - y_m| < hn^{-2} = \delta_2$, we have the inequality

$$|z_m^s - y_m^s| < hn^{-1},$$

and in this case the set $(z_1, \dots, z_n, y_{n+1}, \dots, y_k)$ belongs to the domain Ω. We denote the set of all such sets by Ω_1. The volume of the domain Ω_1 can be written as the multiple integral

$$\int \cdots \int_{\Omega_1} dz_1 \dots dz_n dy_{n+1} \dots du_k.$$

Since Ω_1 is contained in Ω, the volume of Ω equal to $\mu(h)$ satisfies the inequality

$$\mu(h) \ge \int \cdots \int_{\Omega_1} dz_1 \dots dz_n dy_{n+1} \dots du_k,$$

hence

$$\begin{aligned}\mu(h) &\ge \int_{x_{n+1}-\delta_1}^{x_{n+1}} \cdots \int_{x_k-\delta_1}^{x_k} dy_{n+1} \dots du_k \int_{y_1-\delta_2}^{y_1+\delta_2} \cdots \int_{y_n-\delta_2}^{y_n+\delta_2} dz_1 \dots dz_n \\ &= \delta_1^{k-n}(2\delta_2)^n \ge (0.25\varepsilon)^{n(k-n)} 2^n h^n k^{n-k} n^{-k-n}.\end{aligned}$$

This implies the statement of the lemma. □

Lemma 8.14. *Suppose that the characteristic of each solution to system* (8.24) *does not exceed* ε. *Then, for* $k \ge 2n^2$ *and a sufficiently small* h, *the volume* $\mu(h)$ *of the domain* Ω *satisfies the estimate*

$$\mu(h) \le 2^n h^n 2^{2n^2} k^{2n} n^{k-2n} \varepsilon^{k-3n-n^2}.$$

Proof. First, we prove that if the characteristic of each point $(x_1, \dots, x_l)$ in some domain ω lying in the l-dimensional unit cube ($l > n$) does not exceed a, then the volume of this domain μ_a satisfies the inequality

$$\mu_a < l^{n-1}(na)^{l-n}. \tag{8.30}$$

Without loss of generality, we can assume that all coordinates of each point from ω are distinct. We divide these points into classes as follows. To each point $\alpha = (x_1, \dots, x_l)$, we assign a set of indices $j_1, \dots, j_r$ ($r \le n-1$; $j_{m_1} \ne j_{m_2}$ for $m_1 \ne m_2$, $1 \le j_1, \dots, j_r \le k$). Suppose that the characteristic of the point α is equal to δ. Among the numbers $x_1, \dots, x_l$, there exist numbers that are larger than δ. We take the index of the smallest of them to be j_1, the index of the smallest of the numbers x_m ($m = 1, \dots, l$) satisfying the inequality

$$x_m > x_{j_1} + \delta$$

to be j_2, etc., i.e., we take the index of the smallest of the numbers x_m satisfying the inequality

$$x_m > x_{j_{r-1}} + \delta$$

to be j_r. Note that the number r for which this process stops does not exceed $n-1$. Otherwise, the characteristic of the set $(x_{j_1}, \dots, x_{j_r})$ as well as the characteristic of the point α will be larger than δ.

We assign the set of indices $j_1, \dots, j_{r-1}$ thus constructed to the point α. All the points for which the set of these indices is the same will belong to the same class. Obviously, the total number of such classes is $l + l(l-1) + \dots + l(l-1)\dots(l-r+1)$, which does not exceed l^{n-1}. These classes do not intersect. Each coordinate x_m of the point α (contained in a class) that corresponds to the indices $j_1, \dots, j_r$ lies in one of the intervals $0 \le x_m \le \delta$, $x_{j_1} < x_m \le x_{j_1} + \delta, \dots, x_{j_r} < x_m \le x_{j_r} + \delta$. Since $\delta \le \alpha$, the volume of the domain of points assigned to the same fixed class does not exceed $(na)^{l-r}$. If the coordinates with numbers $j_1, \dots, j_r$ are fixed arbitrarily, then all the remaining coordinates belong to fixed intervals whose total length does not exceed na. Multiplying this number by the number of different classes, we obtain the desired estimate for μ_a because $na < 1$ and $r < n$. Thus we have proved inequality (8.30).

Now we divide the domain Ω into two parts Ω_1 and Ω_2. The first part Ω_1 contains the points α whose characteristic $\Delta(\alpha)$ satisfies the inequality

$$(\Delta(\alpha)/8)^n \ge h. \tag{8.31}$$

The second part Ω_2 contains all the remaining points of the domain Ω. We estimate the volume $\mu(\Omega_2)$ of the domain Ω_2 using inequality (8.30). The points $\alpha \in \Omega_2$ satisfy the condition $(\Delta(\alpha)/8)^n < h$, and hence $\Delta(\alpha) < 8h^{1/n}$. By setting $a = 8h^{1/n}$, from inequality (8.30) for $l = k$, we obtain

$$\mu(\Omega_2) < k^{n-1}(8nh^{1/n})^{k-n} \le k^{n-1}(8n)^{k-n}h^{-1+k/n}.$$

Since $k \geq 2n^2$, we have $\mu(\Omega_2) < k^{n-1}(8n)^{k-n}h^{2n-1}$.

Now we estimate the volume $\mu(\Omega_1)$ of the set Ω_1. For this, we first divide the set Ω_1 into subsets $\omega_1, \ldots, \omega_T$ so that ω_t will contain the points α satisfying the condition

$$\varepsilon 2^{-t} < \Delta(\alpha) \leq \varepsilon 2^{1-t}, \quad t = 1, \ldots, T-1.$$

The last set ω_T contains the points such that

$$\varepsilon 2^{-T} < 8h^{1/n} \leq \Delta(\alpha) < \varepsilon 2^{1-T}.$$

Let us estimate the volume $\mu(\omega_t)$ of the set ω_t for $t = 1, \ldots, T$. Now we divide already the sets ω_j into classes. To each class, we assign a set of indices $j_1, \ldots, j_n$ according to the following rule. If the characteristic of a point $\alpha = (x_1, \ldots, x_k)$ is equal to the characteristic of the set of numbers $(x_{j_1}, \ldots, x_{j_n})$, then the point α is contained in the class corresponding to the indices $j_1, \ldots, j_n$. Of course, these classes can intersect. We note that the volume of the set of points for each class is the same. To verify this, it suffices to renumber the variables. The number of all classes is equal to $\binom{k}{n}$. Hence if $\mu(V)$ is the volume of the set of points of the class V corresponding to the indices $1, 2, \ldots, n$, then

$$\mu(\omega_t) \leq \binom{k}{n}\mu(V).$$

Now we shall estimate $\mu(V)$. The characteristic of the set of numbers $(x_{n+1}, \ldots, x_k)$ for each point α from V does not exceed $\varepsilon 2^{1-t}$, while the characteristic of the point α itself is no less than $\varepsilon 2^{-t}$. Therefore, the set V belongs to the set W consisting of the points $\alpha = (x_1, \ldots, x_k)$ satisfying the conditions

(1) $\alpha \in \Omega$;

(2) $\Delta(x_1, \ldots, x_n) \geq \varepsilon_t = \varepsilon 2^{-t}, \quad x_1 < \cdots < x_n$;

(3) $\Delta(x_{n+1}, \ldots, x_k) \leq 2\varepsilon_t = \varepsilon 2^{1-t}$.

We give an estimate from above for the volume $\mu(W)$ of the set W. This can be done as follows: first, we estimate the $(k-n)$-dimensional volume μ_1 of points $(x_{n+1}, \ldots, x_k)$ satisfying condition (3); then, for fixed values of the variables $x_{n+1}, \ldots, x_k$, we estimate the n-dimensional volume μ_2 of points $(x_1, \ldots, x_n)$ satisfying conditions (1) and (2). Then for $\mu(W)$ we have the estimate

$$\mu(W) \leq \mu_1\mu_2.$$

To estimate μ_2, we use inequality (8.30) with $l = k-n$ and $a = \varepsilon 2^{1-t}$. We obtain

$$\mu_2 \leq k^{n-1}(n\varepsilon 2^{1-t})^{k-2n}.$$

Now we estimate μ_1. By condition (1), each point $\alpha = (x_1, \ldots, x_k)$ of the set W satisfies the system of inequalities

$$\left|\sum_{m=1}^{n} x_m^s + \sum_{m=n+1}^{k} x_m^s - \beta_s\right| \leq h, \quad s = 1, \ldots, n.$$

Since condition (2) implies the relations

$$\big(\Delta(x_1,\dots,x_n)/8\big)^n = \big(\Delta(\alpha)/8\big)^n \geq h, \quad x_1 \leq \dots \leq x_n,$$

applying Lemma 8.12 with $h = H$ and $\varepsilon = \varepsilon_t$, we see that there exists a set of numbers $(y_1,\dots,y_n)$ such that the system of equations

$$\sum_{m=1}^{n} y_m^s + \sum_{m=n+1}^{k} x_m^s = \gamma_s, \quad s = 1,\dots,n,$$

is satisfied and the following inequalities hold:

$$|x_m - y_m| \leq h2^{2n-2}(\varepsilon 2^{1-t})^{1-n} \leq 0.25(\varepsilon 2^{-t}), \quad \Delta(y_1,\dots,y_n) \geq 0.5(\varepsilon 2^{-t}).$$

Since $x_1 \leq \dots \leq x_n$, we also have $y_1 \leq \dots \leq y_n$, which implies that the set $(y_1,\dots,y_n)$ is the same for all the numbers $x_1,\dots,x_n$ (if only the numbers $x_{n+1},\dots,x_k$ are fixed). Therefore, the entire set of points $(x_1,\dots,x_n)$ considered is contained in the n-dimensional cube centered at the point $(y_1,\dots,y_n)$; the side of this cube is equal to $2h2^{2n-2}(\varepsilon 2^{1-t})^{1-n}$. Therefore, μ_2 satisfies the estimate

$$\mu_2 \leq (2h)^n 2^{(2n-2)n}(\varepsilon 2^{1-t})^{(1-n)n}.$$

Multiplying the estimates obtained for μ_1 and μ_2, we obtain an estimate for $\mu(W)$, as well as estimates for $\mu(V)$ and $\mu(\omega_t)$. Then summing $\mu(\omega_t)$ over all $t = 1,\dots,T$ and adding the result to the estimate obtained earlier for $\mu(\Omega_2)$, after several obvious transformations, we arrive at the statement of the lemma. □

Theorem 8.3. *Denote by ε the maximal value of the characteristic of the solution $(x_1,\dots,x_n)$ of system* (8.24) *of equations. Then the following inequalities hold*:

$$2^{2n(n-k)}k^{n-k}n^{-k-n}\varepsilon^{n(k-n)} \leq \gamma \leq 2^{2n^2}k^{2n}n^{k-2n}\varepsilon^{k-3n-n^2},$$

where γ is the singular integral in the Hilbert–Kamke problem.

Proof. By Lemma 8.11, we have

$$\gamma = \lim_{h\to 0} 2^{-n}h^{-n}\mu(h).$$

Estimating $\mu(h)$ from above and below by Lemmas 8.13 and 8.14 and passing in these inequalities to the limit as $h \to 0$, we obtain the statement of the theorem. □

Remark 8.1. It follows from Theorem 8.3 that the relations $\gamma = 0$ and $\varepsilon = 0$ are equivalent. The same holds for the inequalities $\gamma > 0$ and $\varepsilon > 0$. Now we assume that the parameters $k, N_1,\dots,N_n$ in the system of Hilbert–Kamke equations take the values for which the singular series σ in the asymptotic formula for the number of solutions of this system is positive. Thus if $\varepsilon > 0$, then $\gamma > 0$ and the above asymptotic formula together with Theorem 8.3 allows us explicitly to obtain, depending on σ, k, n, and ε, the bound P_0 such that for $P \geq P_0$ the system has at least one solution. But if $\gamma = 0$, then $\varepsilon = 0$, i.e., the characteristic of any solution of Eqs. (8.24) (in real numbers) is equal to zero. This means that, among these numbers, there are at most $n - 1$ different numbers. The same can be said about the system of Hilbert–Kamke equations, since if we divide all its unknowns by P, then we obtain a solution of system (8.24). But the number of such solutions is finite for both these systems since the number of real solutions of the following system is also finite:

$$\sum_{s=1}^{r} k_m y_m^s = N_s, \quad s = 1, \dots, n,$$

where $y_1, \dots, y_r$ are unknowns, $k_1, \dots, k_r$ are fixed natural numbers such that $k_1 + \dots + k_r \leq k$, and $y_i \neq y_j$ for $i \neq j$, $1 \leq i, j \leq r$, and $r \leq n - 1$ (the number of solutions of this system does not exceed $r!$).

Thus, for $\varepsilon = \gamma = 0$, system (8.24) has only finitely many solutions. We also note that if system (8.24) is solvable and $\varepsilon > 0$, then Lemma 8.12 implies that, since the numbers $x_1, \dots, x_n$ depend on $x_{n+1}, \dots, x_k, \beta_1, \dots, \beta_n$ continuously, this system has a solution such that there are at most n distinct numbers among the numbers $x_1, \dots, x_k$. If we denote these distinct values of the numbers x_m by $y_1, \dots, y_n$, then the characteristic of the set $(y_1, \dots, y_n)$ satisfies the inequality

$$0 \leq \Delta(y_1, \dots, y_n) \leq \varepsilon.$$

In this case, for fixed natural numbers $k_1, \dots, k_n$, $k_1 + \dots + k_n \leq k$, the numbers $y_1, \dots, y_n$ satisfy the system of equations

$$\sum_{m=1}^{n} k_m y_m^s = \beta_s, \quad s = 1, \dots, n.$$

Thus we have obtained another criterion, which is numerically somewhat less precise but allows us to study the problem of whether the quantity γ is different from zero.

8.3 Multidimensional additive problem

This section is devoted to several generalizations of the Hilbert–Kamke problem. The most important of them is the multidimensional additive problem, i.e., the problem of representing a set of increasing natural numbers $N(t_1, \dots, t_r)$ simultaneously by finitely many terms of the form $x_1^{t_1}, \dots, x_r^{t_r}$. Here the exponents $t_1, \dots, t_r$ take all possible values in the intervals $[0, n_1], \dots, [0, n_r]$, respectively. Moreover, it is assumed that $t_1 + \dots + t_r \geq 1$. For $r = 1$, this problem is precisely the Hilbert–Kamke problem.

In the multidimensional case, i.e., for $r > 1$, this problem was first formulated in the monograph [29], in the section called "Problems."

The general scheme for solving the multidimensional problem is the same as in the one-dimensional case, however, new problems must be solved at each of its stages. There are three such stages. The first consists in obtaining an asymptotic formula for J, i.e., for the number of representations of the numbers $N(t_1, \dots, t_r)$. We obtained this formula in Chapter 6 (Theorem 6.2) using the results of Chapters 4 and 5. This formula can be written as

$$\begin{aligned} J &= J\big(P_1, \dots, P_r; n_1, \dots, n_r; k; N(1, 0, \dots, 0), \dots, N(n_1, \dots, n_r)\big) \\ &= \sigma\gamma P_1^{k-mn_1/2} \dots P_r^{k-mn_r/2} + O\big(P_1^{k-mn_1/2-0.1} \dots P_r^{k-mn_r/2}\big), \end{aligned} \tag{8.32}$$

where $m = (n_1 + 1) \dots (n_r + 1)$.

The quantity J is the number of solutions of the system of Diophantine equations

$$\begin{gathered} \sum_{j=1}^{k} x_{1j}^{t_1} \dots x_{rj}^{t_r} = N(t_1, \dots, t_r), \\ 0 \le t_1 \le n_1, \dots, 0 \le t_r \le n_r, \quad t_1 + \dots + t_r \ge 1, \\ 1 \le x_{1j} \le P_1, \ \dots, \ 1 \le x_{rj} \le P_r. \end{gathered} \tag{8.33}$$

In system (8.33), as in formula (8.32), the variables $P_1, \dots, P_r$ are, in fact, increasing free parameters, and their values are chosen according to the mutual orders of growth of the right-hand sides $N(t_1, \dots, t_r)$ in each of the equations in this system. The quantity σ in formula (8.32) is the value of the singular series, and the quantity γ is the value of the singular integral of the multidimensional problem.

The second stage of solving this problem is to prove that the singular series σ is positive, and the third stage is to prove that the singular integral γ is positive. It is clear that if it is proved that σ and γ are positive for some k, then formula (8.32) implies that, for increasing $P_1, \dots, P_r$, system (8.33) has solutions and there exists a simultaneous representation of $N(t_1, \dots, t_r)$ by a bounded number (by k) of terms of the required form, i.e., the complete solution of the problem under study is obtained.

Precisely as in the one-dimensional case, system (8.33) is solvable if conditions of the following two types are satisfied: arithmetic conditions related to the fact that the singular series σ is positive, and ordering conditions related to the fact that the singular integral σ is positive. Moreover, the arithmetic conditions are equivalent to the solvability conditions for the system of congruences of the form

$$\begin{gathered} \sum_{j=1}^{k} x_{1j}^{t_1} \dots x_{rj}^{t_r} = N(t_1, \dots, t_r) \pmod{q}, \\ 0 \le t_1 \le n_1, \ \dots, \ 0 \le t_r \le n_r, \end{gathered} \tag{8.34}$$

for all moduli q that do not exceed T, where $T = T(n_1, \dots, n_r)$ is a positive constant. In Theorem 8.5, we establish relations between the solvability of systems of congruences (8.34) and the solvability of some system of linear equations in integers. It turns out that the solvability of this linear system of equations in integers is precisely the required arithmetic condition.

We also note that the ordering conditions are the conditions that there exists a solution of Eqs. (8.33) in real numbers such that the Jacobi matrix corresponding to this solution has maximal rank.

Both the ordering conditions and the arithmetic conditions presented in this section are generalizations of the corresponding conditions in Sections 8.1 and 8.2 for the Hilbert–Kamke problem. It is possible to show that, for a sufficiently large number of terms, it follows from these conditions that σ and γ are positive in the asymptotic formula (8.32) for the number of solutions to Eqs. (8.33). Thus it is possible to solve the multidimensional additive problem completely. Simultaneously, we give similar conditions for some other additive problems.

We consider the system of Diophantine equations

$$\begin{aligned} \varepsilon_1 + \cdots + \varepsilon_s &= M_0, \\ \varepsilon_1 x_1 + \cdots + \varepsilon_s x_s &= M_1, \\ &\vdots \\ \varepsilon_1 x_1^n + \cdots + \varepsilon_s x_s^n &= M_n, \end{aligned} \tag{8.35}$$

where $M_0, M_1, \dots, M_n$ are fixed natural numbers; here the unknowns are the variables $x_1, \dots, x_s, \varepsilon_1, \dots, \varepsilon_s$, and $x_1, \dots, x_s$ take nonnegative integer values, while $\varepsilon_1, \dots, \varepsilon_s$ take the values ± 1.

Next, we consider the system of linear equations for the integer numbers $t_0, t_1, t_2, \dots, t_n$:

$$\begin{aligned} t_0 + t_1 + t_2 + \cdots + t_n &= M_0, \\ t_1 + 2t_2 + \cdots + nt_n &= M_1, \\ &\vdots \\ t_1 + 2^n t_2 + \cdots + n^n t_n &= M_n. \end{aligned} \tag{8.36}$$

Lemma 8.15. *The solvability of system* (8.35) *implies the solvability of system* (8.36), *and conversely, the solvability of system* (8.36) *of equations implies that there exists an s for which system* (8.35) *has a solution.*

Proof. First, for any integer x, we find a solution in the integers $T_0, T_1, \dots, T_n$ of the system of equations

$$\sum_{i=0}^{n} T_i i^s = x^s, \quad s = 1, \dots, n. \tag{8.37}$$

The variables $T_i = T_i(x)$ as functions of x are nth-degree polynomials, and moreover,

$$T_i = T_i(x) = (-1)^{n-i}\frac{x(x-1)\dots(x-i-1)(x-i+1)\dots(x-n)}{i!(n-s)!}.$$

This implies that T_i are integers. Now we assume that system (8.35) is solvable and the set $x_1, \dots, x_s, \varepsilon_1, \dots, \varepsilon_s$ is its solution. In system (8.37), we set $x = x_\nu$ $(\nu = 1, \dots, s)$. Instead of $1, x_\nu, \dots, x_\nu^n$, into (8.35), we substitute the left-hand sides of the equations in (8.37). Collecting the similar terms with $1, i, \dots, i^n$ $(i = 0, 1, \dots, n)$, we obtain the following solution of system (8.36):

$$t_i = \sum_{\nu=1}^{s} T_i(x_\nu), \quad i = 0, 1, \dots, n.$$

The first part of the lemma is proved.

Now we assume that the numbers $t_0, t_1, \dots, t_n$ give a solution of system (8.36). We set $s = |t_0| + |t_1| + \dots + |t_n|$,

$$\begin{aligned}
x_1 = \dots = x_{|t_0|} &= 0, & \varepsilon_1 = \dots = \varepsilon_{|t_0|} &= \operatorname{sgn} t_0,\\
x_{|t_0|+1} = \dots = x_{|t_0|+|t_1|} &= 1, & \varepsilon_{|t_0|+1} = \dots = \varepsilon_{|t_0|+|t_1|} &= \operatorname{sgn} t_1,\\
&\vdots & &\vdots\\
x_{|t_0|+\dots+|t_{n-1}|+1} = \dots = x_s &= n, & \varepsilon_{|t_0|+\dots+|t_{n-1}|+1} = \dots = \varepsilon_s &= \operatorname{sgn} t_n.
\end{aligned}$$

These $x_1, \dots, x_s, \varepsilon_1, \dots, \varepsilon_s$ give a solution of system (8.35) of equations. The proof of the lemma is complete. □

Suppose that $N_1, \dots, N_k$, and k are natural numbers, $f_1(x), \dots, f_k(x)$ are polynomials with integer coefficients, and n is the maximal degree of the polynomials $f_1(x), \dots, f_k(x)$.

We consider the system of equations

$$\sum_{r=1}^{s} \varepsilon_r f_\nu(x_r) = N_\nu, \quad \nu = 1, \dots, k, \tag{8.38}$$

where the unknowns are the variables $x_1, \dots, x_s, \varepsilon_1, \dots, \varepsilon_s$ and, moreover, $x_1, \dots, x_s$ take nonnegative integer values, while $\varepsilon_1, \dots, \varepsilon_s$ take the values ± 1. Next, we consider the system of linear equations for the integers $t_0, \dots, t_n$:

$$\sum_{r=0}^{n} t_r f_\nu(r) = N_\nu, \quad \nu = 1, \dots, k. \tag{8.39}$$

Theorem 8.4. *The solvability of system* (8.38) *implies the solvability of system* (8.39), *and conversely, the solvability of system* (8.39) *of equations implies that there exists an s for which system* (8.38) *has a solution.*

Proof. Obviously, the solvability of (8.39) implies that system (8.38) of equations has a solution. Now let $x_1, \dots, x_s, \varepsilon_1, \dots, \varepsilon_s$ be a solution of Eqs. (8.38), $f_\nu(x) = a_0^{(\nu)} + a_1^{(\nu)} x + \dots + a_n^{(\nu)} x^n$. Then, using Lemma 8.15, we can represent the numbers N_ν ($\nu = 1, \dots, k$) as

$$N_\nu = \sum_{r=1}^{s} \varepsilon_r f_\nu(x_r) = \sum_{r=1}^{s} \varepsilon_r \sum_{j=0}^{n} a_j^{(\nu)} x_r^j = \sum_{j=0}^{n} a_j^{(\nu)} \sum_{r=1}^{s} \varepsilon_r x_r^j$$
$$= \sum_{j=0}^{n} a_j^{(\nu)} \sum_{i=0}^{n} t_i i^j = \sum_{i=0}^{n} t_i \sum_{j=0}^{n} a_j^{(\nu)} i^j = \sum_{i=0}^{n} t_i f_\nu(i).$$

The obtained numbers $t_0, t_1, \dots, t_n$ form a solution of system (8.39).

Let $1 \le l < \dots < m < n$ be natural numbers. We consider the system of equations

$$\begin{aligned} \varepsilon_1 x_1^l + \dots + \varepsilon_s x_s^l &= N_l, \\ &\vdots \\ \varepsilon_1 x_1^m + \dots + \varepsilon_s x_s^m &= N_m, \\ \varepsilon_1 x_1^n + \dots + \varepsilon_s x_s^n &= N_n, \end{aligned} \tag{8.40}$$

where the unknowns $x_1, \dots, x_s$ take integer nonnegative values, while $\varepsilon_1, \dots, \varepsilon_s$ take the values ± 1. Further, we consider the system of linear equations for integers $t_1, \dots, t_n$:

$$\begin{aligned} t_1 + t_2 2^l + \dots + t_n n^l &= N_l, \\ &\vdots \\ t_1 + t_2 2^m + \dots + t_n n^m &= N_m, \\ t_1 + t_2 2^n + \dots + t_n n^n &= N_n. \end{aligned} \tag{8.41}$$

□

Corollary 8.1. *The solvability of system* (8.40) *implies the solvability of system* (8.41), *and conversely, the solvability of* (8.41) *implies that there exists an s for which system* (8.40) *has a solution.*

Theorem 8.1 gives this result if $f_1(x) = x^l, \dots, f_{k-1}(x) = x^m, f_k(x) = x^n$.

We consider the system of equations

$$\sum_{j=1}^{s} \varepsilon_j x_{1j}^{t_1} \dots x_{rj}^{t_r} = N(t_1, \dots, t_r), \quad 0 \le t_1 \le n_1, \dots, 0 \le t_r \le n_r, \tag{8.42}$$

where the unknowns $x_{\nu j}$ ($\nu = 1, \dots, r, j = 1, \dots, s$) take nonnegative integer values, while $\varepsilon_1, \dots, \varepsilon_s$ take the values ± 1. We consider the system of linear equations for integers $c(i_1, \dots, i_r)$ $(0 \le i_1 \le n_1, \dots, 0 \le i_r \le n_r)$:

$$\sum_{i_1=0}^{n_1} \cdots \sum_{i_r=0}^{n_r} c(i_1, \dots, i_r) i_1^{t_1} \dots i_r^{t_r} = N(t_1, \dots, t_r), \tag{8.43}$$
$$0 \le t_1 \le n_1, \dots, 0 \le t_r \le n_r.$$

Theorem 8.5. *The solvability of system* (8.43) *implies the solvability of system* (8.42), *and conversely, the solvability of system* (8.42) *implies that there exists an s for which system* (8.43) *has a solution.*

Proof. Clearly, the solvability of (8.43) implies the solvability of (8.42). Now we show that the following system of equations for the integers $c_{i_1,\dots,i_r} = c_{i_1,\dots,i_r}(x_1, \dots, x_r)$ is solvable:

$$\sum_{i_1=0}^{n_1} \cdots \sum_{i_r=0}^{n_r} c_{i_1,\dots,i_r} i_1^{t_1} \dots i_r^{t_r} = x_1^{t_1} \dots x_r^{t_r}, \tag{8.44}$$
$$0 \le t_1 \le n_1, \dots, 0 \le t_r \le n_r.$$

For a solution of the equations (8.44), we take the numbers $c_{i_1,\dots,i_r}$, where $c_{i_1,\dots,i_r} = T_{i_1}(x_1) \dots T_{i_r}(x_r)$ and the set $T_i(x)$ is a solution of system (8.37). We obtain

$$\sum_{i_1=0}^{n_1} \cdots \sum_{i_r=0}^{n_r} c_{i_1,\dots,i_r} i_1^{t_1} \dots i_r^{t_r}$$
$$= \Big(\sum_{i_1=0}^{n_1} T_{i_1}(x_1) i_1^{t_1}\Big) \dots \Big(\sum_{i_r=0}^{n_r} T_{i_r}(x_r) i_r^{t_r}\Big) = x_1^{t_1} \dots x_r^{t_r}.$$

Next, similarly to the proof of Lemma 8.15, substituting the solution of (8.44) into (8.42), we obtain a solution of system (8.43). □

Suppose that $N_1, \dots, N_k$, and k are natural numbers, $f_1(x_1, \dots, x_r), \dots, f_k(x_1, \dots, x_r)$ are polynomials with integer coefficients, and n_l is the maximal degree of the polynomials $f_1, \dots, f_k$ with respect to the variable x_l $(l = 1, \dots, r)$. We consider the system of equations

$$\sum_{j=1}^{s} \varepsilon_j f_t(x_{1j}, \dots, x_{rj}) = N_t, \quad 1 \le t \le k, \tag{8.45}$$

where the unknowns $x_{1j}, \dots, x_{rj}$ $(j = 1, \dots, s)$ take nonnegative integer values, while $\varepsilon_1, \dots, \varepsilon_s$ take the values ± 1. Further, we consider the system of linear equations for integers $c(i_1, \dots, i_r)$ $(0 \le i_1 \le n_1, \dots, 0 \le i_r \le n_r)$:

$$\sum_{i_1=0}^{n_1} \cdots \sum_{i_r=0}^{n_r} c(i_1, \dots, i_r) f_t(i_1 \dots i_r) = N_t, \quad 1 \le t \le k. \tag{8.46}$$

Theorem 8.6. *The solvability of system* (8.45) *implies the solvability of system* (8.46), *and conversely, the solvability of system* (8.46) *implies that there exists an s for which system* (8.45) *has a solution.*

Proof. The proof is similar to that of Theorem 8.4; in fact, the only distinction is that, instead of Lemma 8.15, we use Theorem 8.5. Hence we do not repeat this proof here. □

Theorem 8.7 given below is a distinctive consequence of Theorem 8.5.

Let $f(x) = f(x; a_0, \dots, a_n) = a_0 + a_1 x + \dots + a_n x^n$ be a polynomial with integer coefficients. We consider the equation in polynomials

$$g(x) = \varepsilon_1 f_1^k(x) + \dots + \varepsilon_s f_s^k(x), \tag{8.47}$$

where $f_1(x), \dots, f_s(x)$ are unknown polynomials whose degrees do not exceed n, and the unknowns $\varepsilon_1, \dots, \varepsilon_s$ take the values ± 1. We also consider the linear equation with integral-valued unknowns $t(i_0, \dots, i_n)$:

$$g(x) = \sum_{i_0=0}^{k} \cdots \sum_{i_n=0}^{k} t(i_0, \dots, i_n) f^k(x; i_0, \dots, i_n), \quad 0 \le t_0, \dots, t_n \le k. \tag{8.48}$$

Theorem 8.7. *The solvability of system* (8.47) *implies the solvability of system* (8.48), *and conversely, the solvability of system* (8.48) *implies that there exists an s for which system* (8.47) *has a solution.*

Proof. It is obvious that the solvability of Eq. (8.4) implies the solvability of Eq. (8.47). First, we show that the following equation is solvable for the integers $c(i_0, \dots, i_n)$ $(0 \le i_0, \dots, i_n \le k)$:

$$\sum_{i_0=0}^{k} \cdots \sum_{i_n=0}^{k} c(i_0, \dots, i_n) f^k(x; i_0, \dots, i_n) = f^k(x; a_0, \dots, a_n). \tag{8.49}$$

Indeed, equating the coefficients of equal powers of x, we obtain the following system of equations equivalent to the preceding equation $(0 \le s \le k)$:

$$\sum_{i_0=0}^{k} \cdots \sum_{i_n=0}^{k} c(i_0, \dots, i_n) \sum_{\substack{t_0=0 \\ t_0+\dots+t_n=k \\ t_1+\dots+t_n=s}}^{k} \cdots \sum_{t_n=0}^{k} i_0^{t_0}, \dots, i_n^{t_n} = \sum_{\substack{t_0=0 \\ t_0+\dots+t_n=k \\ t_1+\dots+t_n=s}}^{k} \cdots \sum_{t_n=0}^{k} a_0^{t_0}, \dots, a_n^{t_n}. \tag{8.50}$$

By Theorem 8.5, the following system of equations for integers $c(i_0, \dots, i_n)$ is solvable:

$$\sum_{i_0=0}^{k} \cdots \sum_{i_n=0}^{k} c(i_0, \dots, i_n)\, i_0^{t_0}, \dots, i_n^{t_n} = a_0^{t_0}, \dots, a_n^{t_n}, \quad 0 \le t_0, \dots, t_n \le k. \tag{8.51}$$

The solvability of system (8.51) implies the solvability of system (8.50), and hence the solvability of Eq. (8.49). Substituting the obtained solutions into Eq. (8.47), we obtain a solution of Eq. (8.48). The theorem is proved. □

In fact, Theorem 8.7 gives arithmetic solvability conditions for the "Waring problem in polynomials with natural (integer) coefficients." This additive problem is similar to the Waring problem. This problem studies the possibility for representing a polynomial with increasing natural (integer) coefficients as the sum of a bounded number of polynomials so that each of them be the degree (one and the same) of another polynomial again with natural (integer) coefficients.

Concluding remarks on Chapter 8. 1. The main results considered in Sections 8.1 and 8.2 were obtained by G. I. Arkhipov in [7], [8], [9].

2. Lemma 8.6 in Section 8.1 was proved by A. A. Karatsuba in [89] (see also [120], [121]).

3. G. I. Arkhipov and A. A. Karatsuba in [22] proposed a multidimensional analog of Waring's problem.

4. The results discussed in Section 8.3 were obtained by G. I. Arkhipov and V. N. Chubarikov [12].

5. Exact estimates for the number of terms in the Hilbert–Kamke problem were obtained by D. A. Mit'kin [121], [122].

Chapter 9

The *p*-adic method in three problems of number theory

In this chapter we consider the application of the p-adic method to three well-known problems in number theory. One problem concerns the algebraic number theory and is related to the problem of finding a local representation of zero by an integer-valued form in several variables. The other problem concerns the analytic number theory and is related to the problems of estimating the function $G(n)$ in the Waring problem already considered in Chapter 3. The third problem studies the behavior of fractional parts of functions increasing faster than polynomials.

9.1 The Artin problem of finding a local representation of zero by a form

The principally new results obtained in the Artin problem of finding a p-adic representation of zero by a form of an arbitrary degree continue the studies of the Hilbert–Kamke problem given in Chapter 8.

Suppose that p is a prime number, $F(x_1, \dots, x_k)$ is a form of degree n in k variables $x_1, \dots, x_k$ with integer coefficients over the field Q_p of p-adic numbers. If there exist integer p-adic numbers $x_1, \dots, x_k$ at least one of which is not zero and

$$F(x_1, \dots, x_k) = 0,$$

then it will be said that *there exists a nontrivial representation of zero by the form F in the field Q_p*.

Artin's conjecture stated that, for any $p, n \geq 1$, and $k > n^2$, any nth-degree form $F(x_1, \dots, x_k)$ can nontrivially represent zero in the field Q_p. This hypothesis was disproved in 1966 by G. Terzhanian who constructed nth-degree forms in k variables ($k \geq n^\alpha$, where $\alpha = \log_4 20$) that represent zero in Q_p only trivially. The same year, this result was sharpened by I. Brovkin who constructed forms that only trivially represent zero in Q_p with $k \geq n^{3-\varepsilon}$, where $\varepsilon > 0$ is an arbitrary, but small, fixed number.

In this section, we prove a principally stronger assertion. Theorems 9.1 and 9.2, as well as the main lemmas, which are also of interest in themselves, are stated in the language of the theory of congruences. In this statement, the assertions become more precise. Lemmas 9.6 and 9.7 are original assertions.

9.1.1 Definitions and the simplest lemmas

In what follows, we consider the nth-degree forms $F = F(x_1, \ldots, x_k)$ in k variables $x_1, \ldots, x_k$ with integer coefficients; p is a prime number.

Definition 9.1. We shall say that a form F *does not represent zero modulo* p if, for some natural number r, the congruence

$$F(x_1, \ldots, x_k) = 0 \pmod{p^r}$$

implies

$$x_1 \equiv \cdots \equiv x_k \equiv 0 \pmod{p}.$$

Lemma 9.1. *Suppose that the form $F = F(x_1, \ldots, x_k)$ is of degree n and does not represent zero modulo p. Then, for any $m \geq 1$, there exists a K such that the congruence*

$$F(x_1, \ldots, x_k) = 0 \pmod{p^K}$$

implies

$$x_1 \equiv \cdots \equiv x_k \equiv 0 \pmod{p^m}.$$

Proof. If r is the natural number in Definition 9.1, then we can take $mn + r$ to be K. □

Definition 9.2. A form F that does not represent zero modulo p is said to be *p-singular* or simply *singular*.

Definition 9.3. By $\delta_p(a)$ for $a \neq 0$ we denote the maximal degree p that divides a, i.e., $\delta_p(a) = \alpha$, where $a \equiv 0 \pmod{p^\alpha}$, but $a \neq 0 \pmod{p^{\alpha+1}}$. Moreover, we set $\delta_2(a) = \delta(a)$.

Lemma 9.2. *For any natural number n, the following relation holds*:

$$\delta(5^n - 1) = \delta(n) + 2.$$

Proof. Let $m = \delta(n)$, i.e., let

$$n = 2^m n_1, \quad (n-1, 2) = 1;$$

then

$$\begin{aligned} 5^n = (1+4)^n &= 1 + \frac{n}{1}2^2 + \frac{n(n-1)}{1 \cdot 2}2^{2\cdot 2} + \frac{n(n-1)(n-2)}{1 \cdot 2 \cdot 3}2^{2\cdot 3} + \cdots \\ &= 1 + 2^{m+2}(n_1 + 2M), \end{aligned}$$

where M is a natural number. This implies the statement of the lemma. □

Lemma 9.3. *Suppose that p is an odd prime number, $\alpha \geq 2$, g is a primitive root modulo p^α. Then the relation*

$$\delta_p(g^{n(p-1)} - 1) = \delta_p(n) + 1$$

holds for any $n \geq 1$.

Proof. The primitive root g modulo p^α $(\alpha \geq 2)$ satisfies the relation

$$g^{p-1} = 1 + pM, \quad (M, p) = 1.$$

By setting $n = p^m n_1$, $m = \delta_p(n)$, and $(n_1, p) = 1$, we hence obtain

$$g^{n(p-1)} = (1+pM)^n = 1+\frac{n}{1}pM+\frac{n(n-1)}{1 \cdot 2}p^2M^2+\cdots = 1+p^{1+m}(n_1M+pM_1),$$

i.e.,

$$\delta_p(g^{n(p-1)} - 1) = m + 1 = \delta_p(n) + 1,$$

as required. The proof of the lemma is complete. □

Lemma 9.4. *Suppose that $n = 2^h$ $(h \geq 2)$ is an integer; then for any $k < 4n$ the form $F = F(x_1, \ldots, x_k) = x_1^n + \cdots + x_k^n$ is singular modulo 2.*

Proof. We will prove that the congruence

$$x_1^n + \cdots + x_k^n \equiv 0 \pmod{2^{h+2}} \tag{9.1}$$

implies the congruence

$$x_1 \equiv \cdots \equiv x_k \equiv 0 \pmod{2}.$$

Indeed, since $h \geq 2$, we have $n = 2^h \geq h + 2$. Therefore, we can assume that in (9.1) all x_j are odd numbers, i.e., $x_j = 4m_j \pm 1$. But then

$$x_j^n = (\pm 1 + 4m_j)^n = 1 + 4nN_j \equiv 1 \pmod{2^{h+2}},$$

and the inequality

$$k \geq 2^{h+2} = 4n$$

is a necessary condition for (9.1) to be satisfied. The lemma is proved. □

Lemma 9.5. *Suppose that p is an odd prime number, $h = p^t$, and t is a natural number. Then for $k < p^{t+1}$ the form*

$$F = F(x_1, \ldots, x_k) = x_1^{h(p-1)} + \cdots + x_k^{h(p-1)}$$

is a singular form modulo p.

Proof. We prove that, for $k < p^{t+1}$, the congruence

$$x_1^{h(p-1)} + \cdots + x_k^{h(p-1)} \equiv 0 \pmod{p^{h(p-1)}} \tag{9.2}$$

implies the congruences

$$x_1 \equiv \cdots \equiv x_k \equiv 0 \pmod{p}.$$

Indeed, if the congruence (9.1) contains at least one x_j such that $(x_j, p) = 1$, then, by Lemma 9.3,

$$x_1^{h(p-1)} \equiv 0 \pmod{p^{t+1}}.$$

Since $t+1 \le h(p-1) = p^t(p-1)$, passing from (9.2) to the congruence modulo p^{t+1}, we obtain

$$k_1 \equiv 0 \pmod{p^{t+1}}, \quad k_1 \le k,$$

which is impossible. Hence all x_j $(j = 1, \dots, k)$ in (9.2) are multiples of p. The lemma is proved. □

9.1.2 Main lemmas

As we noted above, the lemmas in this subsection are original and of interest in themselves.

Lemma 9.6. *Let* $1024 \le n \le 96m$, *and let* $j_1, \dots, j_m$ *be arbitrary integers such that*

$$3n/16 < j_1 < j_2 < \cdots < j_m \le n/2.$$

We consider the system of congruences

$$\begin{aligned} x_1^{2j_1} + \cdots + x_k^{2j_1} &\equiv 0 \pmod{2^{2j_1}}, \\ &\vdots \\ x_1^{2j_m} + \cdots + x_k^{2j_m} &\equiv 0 \pmod{2^{2j_m}} \end{aligned} \tag{9.3}$$

under the condition that there are odd numbers among the unknowns in this system. Then the solvability of this system implies that

$$k \ge 2^u, \quad u = n/32.$$

Proof. Without loss of generality, we can assume that all $x_1, \dots, x_k$ in system (9.3) are odd numbers. We represent each x_j $(j = 1, \dots, k)$ as

$$x_j \equiv \pm 5^{\alpha_j} \pmod{2^n}.$$

Further, we define the polynomial $f(t)$ by the relation $f(t) = t^{\alpha_1} + \cdots + t^{\alpha_k}$. We have $f(1) = k \geq 1$. Now we prove that

$$f(1) \equiv 0 \pmod{2^u} \tag{9.4}$$

for some integer $u \geq n/32$. This will imply the statement of the lemma. First, we note that the definition and Lemma 9.3 imply the relations

$$f(5^{2j_r}) = (5^{\alpha_1})^{2j_r} + \cdots + (5^{\alpha_k})^{2j_r} \equiv x_1^{2j_r} + \cdots + x_k^{2j_r} \equiv 0 \pmod{2^{2j_r}} \tag{9.5}$$

for $r = 1, \dots, m$. We divide $f(t)$ by the product $(t - 5^{2j_1}) \dots (t - 5^{2j_m})$ with a remainder and find

$$f(t) = \varphi(t) + g(t)(t - 5^{2j_1}) \dots (t - 5^{2j_m}), \tag{9.6}$$

where $\varphi(t)$ and $g(t)$ are polynomials with integer coefficients and the degree of $\varphi(t)$ does not exceed $m - 1$. Indeed, the polynomial $\varphi(t)$ can be written as

$$\begin{aligned}\varphi(t) = a_0 + a_1(t - 5^{2j_m}) + a_2(t - 5^{2j_m})(t - 5^{2j_{m-1}}) + \cdots \\ + a_{m-1}(t - 5^{2j_m})(t - 5^{2j_{m-1}}) \dots (t - 5^{2j_2}).\end{aligned}$$

We can also take $a_0 = f(5^{2j_m})$, a_1 to be equal to the value (for $t = 5^{2j_{m-1}}$) of the quotient obtained dividing $\varphi(t) - a_0$ by $t - 5^{2j_m}$, a_2 to be equal to the value (for $t = 5^{2j_{m-2}}$) of the quotient obtained dividing $\varphi(t) - a_0 - a_1(t - 5^{2j_m})$ by $(t - 5^{2j_m})(t - 5^{2j_{m-1}})$, etc. We note that (9.5) and (9.6) imply the congruence

$$\varphi(5^{2j_r}) \equiv 0 \pmod{2^{2j_r}}, \quad r = 1, \dots, m. \tag{9.7}$$

For brevity, we denote the numbers 5^{2j_r} by the letters t_r, i.e., we assume

$$t_r = 5^{2j_r}, \quad r = 1, \dots, m.$$

Then we can represent $\varphi(t)$ as the Lagrange polynomial

$$\varphi(t) = \sum_{r=1}^{m} \varphi(t_r) \frac{(t - t_1) \dots (t - t_{r-1})(t - t_{r+1}) \dots (t - t_m)}{(t_r - t_1) \dots (t_r - t_{r-1})(t_r - t_{r+1}) \dots (t_r - t_m)}.$$

Applying Lemma 9.2, we easily obtain

$$\begin{aligned}&\delta\left(\varphi(5^{2j_r}) \frac{(1 - 5^{2j_1}) \dots (1 - 5^{2j_{r-1}})(1 - 5^{2j_{r+1}}) \dots (1 - 5^{2j_m})}{(1-5^{2(j_r - j_1)}) \dots (1-5^{2(j_r - j_{r-1})})(1-5^{2(j_{r+1} - j_r)}) \dots (1-5^{2(j_m - j_r)})}\right) \\ &= \delta\big(\varphi(5^{2j_r})\big) + 3(m-1) + \delta(j_1) + \cdots + \delta(j_{r-1}) \\ &\quad + \delta(j_{r+1}) + \cdots + \delta(j_m) - 3(m-1) - \delta(j_r - j_1) - \cdots - \delta(j_r - j_{r-1}) \\ &\quad - \delta(j_{r+1} - j_r) - \cdots - \delta(j_m - j_r) \qquad (9.8) \\ &\geq 2j_r - \delta(j_r - j_1) - \cdots - \delta(j_r - j_{r-1}) - \delta(j_{r+1} - j_r) - \cdots - \delta(j_m - j_r) \\ &\geq 2j_r - \delta\big((j_r - j_1)!\big) - \delta\big((j_m - j_r)!\big).\end{aligned}$$

But for $\delta(a!)$ $(a \geq 1)$, we have the formula

$$\delta(a!) = \left[\frac{a}{2}\right] + \left[\frac{a}{2^2}\right] + \cdots < \frac{a}{2} + \frac{a}{2^2} + \cdots = a,$$

hence the right-hand side of (9.8) is larger than

$$2j_r - (j_r - j_1) - (j_m - j_r) = 2j_r + j_1 - j_m \geq n/32.$$

This implies

$$\delta\big(\varphi(1)\big) \geq n/32. \tag{9.9}$$

Moreover, we have

$$\delta\big(g(1)(1 - 5^{2j_1}) \dots (1 - 5^{2j_m})\big) \geq 3m \geq n/32. \tag{9.10}$$

The above estimates (9.9), (9.10), and (9.6) imply (9.4). The proof of the lemma is complete. □

Lemma 9.7. *Suppose that p is an odd prime number, $n \geq 16(\rho - 1)$, $m \geq n/(16(\rho - 1))$, and $j_1, \dots, j_m$ are arbitrary integers satisfying the relations*

$$\frac{3n}{8(p-1)} < j_1 < \cdots < j_m \leq \frac{n}{p-1}.$$

We consider the system of congruences

$$\begin{aligned} x_1^{j_1(p-1)} + \cdots + x_k^{j_1(p-1)} &\equiv 0 \pmod{p^{j_1(p-1)}}, \\ &\vdots \\ x_1^{j_m(p-1)} + \cdots + x_k^{j_m(p-1)} &\equiv 0 \pmod{p^{j_m(p-1)}} \end{aligned} \tag{9.11}$$

under the condition that there are numbers indivisible by p among the unknowns in this system. Then it follows from the solvability of the system that

$$k \geq p^u, \quad u = \frac{n}{16(p-1)}.$$

Proof. We follows the argument in Lemma 9.6. Without loss of generality, we assume that all $x_1, \dots, x_k$ in system (9.11) are indivisible by p.

Let g be a primitive root modulo p^n. We represent each x_j $(j = 1, \dots, k)$ as

$$x_j \equiv g^{\alpha_j} \pmod{p^n}, \quad j = 1, \dots, k,$$

and define the polynomial $f(t) = t^{\alpha_1} + \cdots + t^{\alpha_k}$. Obviously, $f(1) = k \geq 1$; we shall prove that

$$f(1) \equiv 0 \pmod{p^u} \tag{9.12}$$

for some integer $u \geq n/(16(p-1))$. This will imply the statement of the lemma. We first note that the definition of $f(t)$ and system (9.11) imply the relations

$$\begin{aligned} f(g^{j_r(p-1)}) &= (g^{\alpha_1})^{j_r(p-1)} + \cdots + (g^{\alpha_k})^{j_r(p-1)} \\ &\equiv x_1^{j_r(p-1)} + \cdots + x_k^{j_r(p-1)} \equiv 0 \pmod{p^{j_r(p-1)}}, \\ r &= 1, \ldots, m. \end{aligned} \tag{9.13}$$

We divide $f(t)$ by the product $(t - g^{j_1(p-1)}) \ldots (t - g^{j_m(p-1)})$ with a remainder and obtain

$$f(t) = \varphi(t) + g(t)(t - g^{j_1(p-1)}) \ldots (t - g^{j_m(p-1)}), \tag{9.14}$$

where $\varphi(t)$ and $g(t)$ are polynomials with integer coefficients and the degree of $\varphi(t)$ does not exceed $m-1$.

We note that relations (9.13) and (9.14) imply the congruences

$$\varphi(g^{j_r(p-1)}) \equiv 0 \pmod{p^{j_r(p-1)}}, \quad r = 1, \ldots, m. \tag{9.15}$$

By setting

$$t_r = g^{j_r(p-1)}, \quad r = 1, \ldots, m,$$

we obtain the following Lagrangian representation for $\varphi(t)$:

$$\varphi(t) = \sum_{r=1}^{m} \varphi(t_r) \frac{(t - t_1) \ldots (t - t_{r-1})(t - t_{r+1}) \ldots (t - t_m)}{(t_r - t_1) \ldots (t_r - t_{r-1})(t_r - t_{r+1}) \ldots (t_r - t_m)}.$$

Using Lemma 9.3, we easily find

$$\begin{aligned} &\delta_p\left(\varphi(t_r) \frac{(1 - t_1) \ldots (1 - t_{r-1})(1 - t_{r+1}) \ldots (1 - t_m)}{(t_r - t_1) \ldots (t_r - t_{r-1})(t_r - t_{r+1}) \ldots (t_r - t_m)}\right) \\ &\quad = \delta_p\big(\varphi(t_r)\big) + (m-1) + \delta_p(j_1) + \cdots + \delta_p(j_{r-1}) \\ &\qquad + \delta_p(j_{r+1}) + \cdots + \delta_p(j_m) - (m-1) - \delta_p(j_r - j_1) - \cdots \\ &\qquad - \delta_p(j_r - j_{r-1}) - \delta_p(j_{r+1} - j_r) - \cdots - \delta_p(j_m - j_r) \\ &\quad \geq \delta_p(\varphi(t_r)) - \delta_p((j_r - j_1)!) - \delta_p((j_m - j_r)!). \end{aligned} \tag{9.16}$$

But for $\delta_p(a!)$, we have the formula

$$\delta_p(a!) = \left[\frac{a}{p}\right] + \left[\frac{a}{p^2}\right] + \cdots < \frac{a}{p} + \frac{a}{p^2} + \cdots = \frac{a}{p-1};$$

moreover, we obtain $\delta_p(\varphi(t_r)) \geq j_r(p-1)$ from (9.14). Therefore, the right-hand side of (9.16) is larger than

$$j_r(p-1) - \frac{j_r - j_1}{p-1} - \frac{j_m - j_r}{p-1} = j_r(p-1) - \frac{j_m - j_1}{p-1} > \frac{n}{16(p-1)}.$$

This implies

$$\delta_p(\varphi(1)) > \frac{n}{16(p-1)}, \tag{9.17}$$

and moreover,

$$\delta\big(g(1)(1-g^{j_1(p-1)})\dots(1-g^{j_m(p-1)})\big) \geq m \geq \frac{n}{16(p-1)}. \tag{9.18}$$

The estimates (9.17), (9.18), and (9.14) imply (9.12). The proof of the lemma is complete. □

9.1.3 Theorems

We state and prove theorems about forms (in a large number of variables) that do not represent zero modulo p.

Theorem 9.1. *For any natural number r, there exists an infinite sequence of natural numbers $n_1, n_2, \dots$ such that, for any $n = n_j$ $(j \geq 1)$, there is an nth-degree form $F(x_1, \dots, x_n)$ with integer coefficients that does not represent zero modulo 2; the number of its variables is k,*

$$k \geq 2^u, \quad u = \frac{n}{(\log_2 n)(\log_2\log_2 n)\dots\underbrace{(\log_2\dots\log_2 n)}_{r}\underbrace{(\log_2\dots\log_2^3 n)}_{r+1}}.$$

Proof. Suppose that h ia an arbitrary natural number, $h \geq 8$, $t = 2^h$, and $k_1 = 2^{1/16}$. We consider a form in k_1 variables of the form

$$F_1 = F_1(x_0, x_1, \dots, x_{k_1-1}) = y_0^t + \dots + y_{t-1}^t,$$

where

$$y_j = s_{2j}s_{4t-2j}, \quad j = 0, 1, \dots, t-1;$$
$$s_\nu = x_0^\nu + x_1^\nu + \dots + x_{k_1-1}^\nu, \quad \nu = 0, \dots, 4t.$$

We need to prove that F_1 is a singular form modulo 2. Indeed, the form $y_0^t + \dots + y_{t-1}^t$ in the variables $y_0, \dots, y_{t-1}$ is a singular form modulo 2 (Lemma 9.4). Hence, by Lemma 9.1, for some M, the congruence

$$y_0^t + \dots + y_{t-1}^t \equiv 0 \pmod{2^M}$$

implies the congruences

$$y_j = s_{2j}s_{4t-2j} \equiv 0 \pmod{2^{8t}}, \quad j = 0, 1, \dots, t-1.$$

The last congruences show that there are t numbers j, say, $j_1, j_2, \dots, j_t$, $0 < j_1 < j_2 < \dots < j_t \le 2t$, for which the following congruence holds:

$$s_{2j_1} \equiv \dots \equiv s_{2j_t} \equiv 0 \pmod{2^{4t}}.$$

Now let $j'_1 = j_{3t/4+1}, j'_2 = j_{3t/4+2}, \dots, j'_m = j_t$, and let $m = t/4$. Since $j_{3t/4+1} > 3t/4$, by setting $4t = n$, we obtain

$$3n/16 < j'_1 < \dots < j'_m \le n/2, \quad m = n/16, \quad n = 2^{h+2} \ge 1024,$$

and moreover,

$$s_{2j'_1} \equiv \dots \equiv s_{2j'_m} \equiv 0 \pmod{2^n}. \tag{9.19}$$

All conditions of Lemma 9.6 are satisfied, and $k_1 = 2^{t/16} = 2^{n/64} < 2^{n/32}$. Therefore, it follows from (9.19) that

$$x_0 \equiv x_1 \equiv \dots \equiv x_{k_1-1} \equiv 0 \pmod 2,$$

i.e., the form $F_1(x_0, x_1, \dots, x_{k_2-1})$ is singular modulo 2. Note that the degree of F is equal to $4t^2 = 2^{2h+2}$.

Now we consider a form $F_2 = F_2(x_0, x_1, \dots, x_{k_2-1})$ in $k_2 = 2^{k_1/16}$ variables of the form

$$F_2 = F_2(x_0, x_1, \dots, x_{k_2-1}) = F_1(y_0, y_1, \dots, y_{k_1-1}),$$

where

$$y_j = s_{2j}s_{4t-2j}, \quad j = 0, 1, \dots, t-1, \quad 4t = k_1;$$
$$s_\nu = x_0^\nu + x_1^\nu + \dots + x_{k_2-1}^\nu.$$

We prove that $F_2(x_0, x_1, \dots, x_{k_2-1})$ is a singular form modulo 2. Since $F_1(y_0, y_1, \dots, y_{k_1-1})$ is a singular form modulo 2, for some M, the congruence

$$F_1(y_0, y_1, \dots, y_{k_1-1}) \equiv 0 \pmod{2^M}$$

implies (Lemma 9.1) the relations

$$y_0 \equiv y_1 \equiv \dots \equiv y_{k_1-1} \equiv 0 \pmod{2^{8t}},$$
$$y_j = s_{2j}s_{4t-2j} = 0 \pmod{2^{8t}}, \quad j = 0, 1, \dots, t-1.$$

Thus all conditions (including the notation) for $F_1(x_0, x_1, \dots, x_{k_1-1})$ to be a singular form are satisfied. Hence, from the last congruences, repeating the above argument, we obtain

$$x_0 \equiv x_1 \equiv \dots \equiv x_{k_2-1} \equiv 0 \pmod 2,$$

namely, the form $F_2 = F_2(x_0, x_1, \dots, x_{k_2-1})$ is singular modulo 2. We note that degree of $F_2 = k_1 \cdot$ degree of $F_1 = k_1 \cdot 4t^2$. Moreover, the number k_2 of the variables in the form F_2 is equal to $2^{k_1/16}$.

If the singular form $F_s = F_s + (x_0, x_1, \ldots, x_{k_s-1})$ ($s \geq 1$) in the k_s variables $x_0, x_1, \ldots, x_{k_s-1}$ has already been constructed, then the form F_{s+1} can be constructed as follows. We take $F_{s+1} = F_{s+1}(x_0, x_1, \ldots, x_{k_{s+1}-1})$, where $k_{s+1} = 2^{k_s/16}$, of the form

$$F_{s+1} = F_{s+1}(x_0, x_1, \ldots, x_{k_{s+1}-1}) = F_s(y_0, y_1, \ldots, y_{k_s-1}),$$

where

$$y_j = s_{2j} s_{4t-2j}, \quad j = 0, 1, \ldots, t-1, \quad 4t = k_s;$$

$$s_\nu = x_0^\nu + x_1^\nu + \cdots + x_{k_{s+1}-1}^\nu.$$

The preceding argument shows that F_{s+1} is a singular form in k_{s+1} variables $x_0, x_1, \ldots, x_{k_{s+1}-1}$; we have: degree of $F_{s+1} = k_s k_{s-1} \ldots k_1 2^{2h+2}$, where $h \geq 8$ and $k_{s+1} = 2^{k_s/16}$.

We take the following numbers n_j ($j = 1, 2, \ldots$) whose existence is stated in the theorem:

$$n_j = 2^{18} k_{2r+j+3} k_{2r+j+2} \ldots k_1;$$

the singular n_jth-degree forms F_{2r+j+4} with $h = 8$ in k_{2r+j+4} variables correspond to these numbers. We set $n = n_j$, $k = k_{2r+j+4}$ and prove that the inequality

$$k \geq 2^u, \quad u = \frac{n}{(\log_2 n)(\log_2 \log_2 n) \ldots \underbrace{(\log_2 \ldots \log_2 n)}_{r} \underbrace{(\log_2 \ldots \log_2^3 n)}_{r+1}}$$

holds for $j > 0$.

First, for any $s \geq 2$, we have

$$\log_2 k_s - 5 \geq 2^{-5} k_{s-1}. \tag{9.20}$$

Indeed, $k_s = 2^{k_{s-1}/16}$ and hence

$$\log_2 k_s = \frac{k_{s-1}}{16}, \quad \frac{k_{s-1}}{16} - \frac{k_{s-1}}{32} = \frac{k_{s-1}}{32} \geq 5$$

because $k_1 = 2^{16}$.

From (9.20), for any natural numbers s and m, we obtain

$$\begin{aligned} k_{s+m} &= 2^{k_{s+m-1}/16}; \\ \log_2 k_{s+m} &= 2^{-4} k_{s+m-1}, \\ \log_2 \log_2 k_{s+m} &= \log_2 k_{s+m-1} - 4 > \log_2 k_{s+m-1} - 5 \geq 2^{-5} k_{s+m-2}, \\ \log_2 \log_2 \log_2 k_{s+m} &= \log_2 k_{s+m-2} - 5 \geq 2^{-5} k_{s+m-3}, \\ &\vdots \\ \underbrace{\log_2 \ldots \log_2}_{m} k_{s+m} &> 2^{-5} k_s \geq 2^{-5} k_1 > 5. \end{aligned}$$

It follows from the definition of k that

$$\begin{aligned} k_{2r+j+3} &= 16 \log_2 k_{2r+j+4} \leq 16 \log_2 k < 2^5 \log_2 k, \\ k_{2r+j+2} &= 16 \log_2 k_{2r+j+3} \leq 16(\log_2 \log_2 k + 5) < 2^5 \log_2 \log_2 k, \\ &\vdots \\ k_{r+j+2} &= 16 \log_2 k_{r+j+3} < \cdots < 2^5 \underbrace{\log_2 \ldots \log_2}_{r+2} k. \end{aligned} \tag{9.21}$$

Next, for any natural number $s \geq 2$, we have

$$2^{18} k_{s-1} \ldots k_1 < k_s. \tag{9.22}$$

Indeed, for $s = 2$, we have

$$k_2 = 2^{k_1/16} = 2^{2^{12}} > 2^{18+16} = 2^{34}.$$

If (9.22) takes place for $s = m \geq 2$, then

$$k_{m+1} = 2^{k_m/16} > 2^{2^{14} k_m \ldots k_1} > 2^{18} 2^{k_{m-1}/16} k_{m-1} \ldots k_1 = 2^{18} k_m k_{m-1} \ldots k_1;$$

therefore, using (9.21) for $s = 2r + j + 2$, we obtain

$$\begin{aligned} n = n_1 &= 2^{18} k_{2r+j+3} k_{2r+j+2} \ldots k_1 \\ &< 2^{5r+10} \log_2 k \cdot \log_2 \log_2 k \ldots \underbrace{\log_2 \ldots \log_2}_{r+2} k \cdot 5 \underbrace{\log_2 \ldots \log_2}_{r+2} k \\ &< 2^{5r+15} \log_2 k \cdot \log_2 \log_2 k \ldots \underbrace{\log_2 \ldots \log_2}_{r+1} k \cdot \underbrace{\log_2 \ldots \log_2^2}_{r+2} k. \end{aligned}$$

Next, from (9.22) we obtain

$$\underbrace{\log_2 \ldots \log_2}_{r+2} k = \underbrace{\log_2 \ldots \log_2}_{r+2} k_{2r+j+4} > 2^{-5} k_{r+j+2} \geq 2^{-5} k_{r+3}.$$

Moreover, for any natural number r, we have

$$2^{-5}k_{r+3} > 2^{5r+15}. \tag{9.23}$$

Indeed, for $r = 1$ we have

$$k_4 > k_2 = 2^{2^{12}} > 2^{24}.$$

If (9.23) holds for $r = m \geq 1$, then

$$2^{-5}k_{m+4} = 2^{-5}2^{k_{m+3}/16} > 2^{-5}2^{2^{5m+16}} > 2^{5m+20},$$

because $2^{5m+16} > 5m + 25$. So we have

$$\underbrace{\log_2 \ldots \log_2}_{r+2} k > 2^{5r+15},$$

and hence

$$n < \log_2 k \cdot \log_2 \log_2 k \ldots \underbrace{\log_2 \ldots \log_2}_{r+1} k \cdot \underbrace{\log_2 \ldots \log_2^3}_{r+2} k. \tag{9.24}$$

Suppose that the inequality

$$k < 2^u, \quad u = \frac{n}{\log_2 n \cdot \log_2 \log_2 n \ldots \underbrace{\log_2 \ldots \log_2}_{r} n \cdot \underbrace{\log_2 \ldots \log_2^3}_{r+1} n}$$

holds. Then we successively find

$$\begin{aligned} \log_2 k &< u, \\ \log_2 \log_2 k &< \log_2 n, \\ &\vdots \\ \underbrace{\log_2 \ldots \log_2}_{r+1} k &< \underbrace{\log_2 \ldots \log_2}_{r} n, \\ \underbrace{\log_2 \ldots \log_2}_{r+2} k &< \underbrace{\log_2 \ldots \log_2^3}_{r+1} n. \end{aligned}$$

Multiplying the left- and right-hand sides of these inequalities and using (9.24), we obtain the contradiction:

$$n < \log_2 k \cdot \log_2 \log_2 k \ldots \underbrace{\log_2 \ldots \log_2}_{r+1} k \cdot \underbrace{\log_2 \ldots \log_2^3}_{r+2} k < n.$$

Thus the assumption that $k < 2^u$ does not hold; namely, we have $k \geq 2^u$. The theorem is proved. □

Theorem 9.2. *For any natural number r, there exists an infinite sequence of natural numbers $n_1, n_2, \dots$ such that, for any $n = n_j$ $(j \geq 1)$, there is an nth-degree form $F(x_1, \dots, x_k)$ with integer coefficients that does not represent zero modulo 2; the number of variables in this form is k,*

$$k < p^u, \quad u = \frac{n}{\log_p n \cdot \log_p \log_p n \dots \underbrace{\log_p \dots \log_p}_{r} n \cdot \underbrace{\log_p \dots \log_p^3}_{r+1} n}.$$

Proof. We follow the argument in Theorem 9.1. Let $F_s = F_s(x_0, x_1, \dots, x_{k_s-1})$ be an n_sth-degree singular form in k_s variables modulo p. We shall construct F_{s+1}. By construction, we set

$$F_{s+1} = F_{s+1}(x_0, x_1, \dots, x_{k_{s+1}-1}) = F_s(y_0, y_1, \dots, y_{k_s-1}),$$

where

$$y_j = s_{j(p-1)} s_{n-j(p-1)}, \quad n = 2(p-1)k_s, \quad j = 0, 1, \dots, k_s - 1;$$
$$s_\nu = x_0^\nu + x_1^\nu + \dots + x_{k_{s+1}-1}^\nu.$$

We prove that F_{s+1} is a singular form modulo p. By Lemma 9.1, the congruence

$$F_{s+1} = F_s(y_0, y_1, \dots, y_{k_s-1}) \equiv 0 \pmod{p^M}$$

implies the congruences

$$y_j = s_{j(p-1)} s_{n-j(p-1)} \equiv 0 \pmod{p^{2n}}, \quad j = 1, \dots, k_s = t.$$

Therefore, there are numbers $1 \leq j_1 < \dots < j_t \leq 2k_s$ such that

$$s_{j_1(p-1)} \equiv \dots \equiv s_{j_t(p-1)} \equiv 0 \pmod{p^n}. \tag{9.25}$$

We choose a natural number m from the conditions $0.25t - 1 < m \leq 0.25t$. Now we assume that $n = 2(p-1)k_s = 2(p-1)t \geq 16(p-1)$ and set

$$j_t = j'_m, \; j_{t-1} = j', \; \dots, \; j_{t-m+1} = j'_1.$$

Then

$$t - m < j'_1 < \dots < j'_m \leq 2k_s = n/(p-1).$$

Moreover,

$$t - m \geq \frac{3}{4}t = \frac{3n}{8(p-1)}, \qquad m = \frac{t}{4} - 1 \geq \frac{t}{8} = \frac{n}{16(p-1)}.$$

Hence, from (8.25), we obtain the system of congruences studied in Lemma 9.7:

$$s_{j'_1(p-1)} \equiv \dots \equiv s_{j'_m(p-1)} \equiv 0 \pmod{p^n}.$$

Thus the form F_{s+1} is singular for

$$k_{s+1} < p^{n/(16(p-1))} = p^{k_s/8}.$$

We take $k_{s+1} = p^{k_1/p^2} < p^{k_s/8}$ and assume that k_s is divisible by p^2. Then it follows from the above that the form $F_{s+1} = F_{s+1}(x_0, x_1, \dots, x_{k_{s+1}-1})$ is a singular form modulo p; its degree is

$$n_{s+1} = nn_s = 2(p-1)k_s n_s$$

and the number of variables is $k_{s+1} = p^{k_s/p^2}$.

It remains to find F_1. We set

$$F_1 = F_1(x_0, x_1, \dots, x_{k_1-1}) = x_0^{h(p-1)} + x_1^{h(p-1)} + \dots + x_{k_1-1}^{h(p-1)},$$
$$h = p^6, \quad k_1 = p^6.$$

By Lemma 9.5, the form F_1 is singular. The degree of F_1 is equal to $n_1 = p^6(p-1)$. All the conditions we used above hold for all n_s and k_s $(s = 1, 2, \dots)$. Therefore, we have

$$n_{s+1} = 2^s(p-1)^s k_s k_{s-1} \dots k_1 n_1 = 2^s(p-1)^{s+1} p^6 k_s \dots k_1. \tag{9.26}$$

To obtain the statement of the theorem, we first perform simple calculations. For any $s \geq 1$, the inequality

$$\underbrace{\log_p \dots \log_p}_{s} k_{s+1} \geq 3 \tag{9.27}$$

holds. Indeed, $\log_p k_{s+1} = k_s/p^2 = p^2 \geq 3$ if $s = 1$. For $s > 1$ we have (using the fact that $\log_p k_{s-1} \geq 6$)

$$\log_p \log_p k_{s+1} = \log_p k_s - 2 \geq 0.5 \log_p k_s > k_{s-1}/p^3,$$
$$\log_p \log_p \log_p k_{s+1} > \log_p k_{s-1} - 3 \geq 0.5 \log_p k_{s-1} > k_{s-2}/p^3,$$

and so on. Finally, we obtain

$$\underbrace{\log_p \dots \log_p}_{s} k_{s+1} > k_1/p^3 = p \geq 3.$$

Further, using this inequality, for any $m \leq s-1$ we find (for brevity, we set $k_{s+1}=k$)

$$\begin{aligned} k_s &= p^2 \log_p k_{s+1} < p^3 \log_p k, \\ k_{s-1} &= p^2 \log_p k_s < p^2(\log_p \log_p k + 3) < p^3 \log_p \log_p k, \\ k_{s-2} &= p^2 \log_p k_{s-1} < p^2(\log_p \log_p \log_p k + 3) < p^3 \log_p \log_p \log_p k, \\ &\vdots \\ k_{s-m} &< p^3 \underbrace{\log_p \ldots \log_p}_{m+1} k. \end{aligned} \tag{9.28}$$

Finally, the inequality

$$p^{20s} k_1 \ldots k_{s-1} < k_s \tag{9.29}$$

holds for any $s \geq 2$. Indeed, for $s=2$ we have

$$k_2 = p^{k_1/p^2} = p^{p^4} > p^{40} k_1 = p^{46}, \quad p^4 > 46.$$

Suppose that (9.29) has already been proved for $s = m \geq 2$; we prove (9.29) for $s = m+1$. Using the assumption, we obtain $k_{m+1} = p^{k_m/p^2} > p^{p^{20m-2} k_1 \ldots k_{m-1}}$. Moreover, $p^{20(m+1)} k_1 \ldots k_m = k_1 \ldots k_{m-1} p^{k_{m-1}/p^2 + 20(m+1)}$, and thus it suffices to show that

$$p^{p^{20m-2} k_1 \ldots k_{m-1} - k_{m-1}/p^2 - 20(m+1)} > k_1 \ldots k_{m-1}.$$

But this inequality is obvious, since

$$p^{20m-2} \geq 20(m+1) + 2.$$

Hence the inequality (9.29) is always satisfied and

$$k_1 \ldots k_{s-1} < p^{-20s} k_s. \tag{9.30}$$

Now we find the sequence of numbers n'_j $(j = 1, 2, \ldots)$ whose existence is stated in the theorem. We set

$$n'_j = n_{2r+4+j}, \quad j = 1, 2, \ldots;$$

denoting $s+1 = 2r+4+j$ and $n_{s+1} = n$, we obtain from (9.26)

$$n = 2^s (p-1)^{s+1} p^6 k_s \ldots k_1. \tag{9.31}$$

Next, it follows from (9.29) that $k_1 \ldots k_{s-r-2} < p^{-20(s-r-1)} k_{s-r-1}$, hence we have

$$n < 2^s (p-1)^{s+1} p^6 p^{-20(s-r-1)} k_s k_{s-1} \ldots k_{s-r} k_{s-r-1}^2.$$

Now it follows from (9.31) and the estimates for $k_s, k_{s-1}, \dots$ that

$$n < 2^s p^{s+7-20(s-r-1)} p^{3(r+2)+3} \log_p k \cdot \log_p \log_p k \dots \underbrace{\log_p \dots \log_p}_{r+1} k \cdot \underbrace{\log_p \dots \log_p^2}_{r+2} k.$$

Since $s = 2r + 3 + j$, we readily see that

$$2^s p^{s+7-20(s-r-1)+3(r+2)+3} < p^{4r+6+2j+7+3r+6+3} p^{-20r-40-20j} < 1.$$

Thus the inequality

$$n < \log_p k \cdot \log_p \log_p k \dots \underbrace{\log_p \dots \log_p}_{r+1} k \cdot \underbrace{\log_p \dots \log_p^2}_{r+2} k \tag{9.32}$$

holds. Hence we shall prove that

$$k > p^u, \quad u = \frac{n}{\log_p n \cdot \log_p \log_p n \dots \underbrace{\log_p \dots \log_p}_{r} n \cdot \underbrace{\log_p \dots \log_p^2}_{r+1} n}.$$

Indeed, if we had $k \le p^u$, then we would obtain the inequalities

$$\begin{aligned}
\log_p k &\le u, \\
\log_p \log_p k &< \log_p n, \\
&\vdots \\
\underbrace{\log_p \dots \log_p}_{r+1} k &< \underbrace{\log_p \dots \log_p}_{r} n, \\
\underbrace{\log_p \dots \log_p^2}_{r+2} k &< \underbrace{\log_p \dots \log_p^2}_{r+1} n.
\end{aligned}$$

Multiplying the left- and right-hand sides of the inequalities and taking (9.32) into account, we obtain the contradiction: $n < n$. Thus the statement of the theorem is proved completely. □

The sequences of degrees n_j $(j = 1, 2, \dots)$ in Theorems 9.1 and 9.2 are very sparse. However, it is possible to shift these sequences in the sequence of natural numbers, which allows one to prove somewhat more general assertions. Here we state these assertions, since their proofs basically coincide with the proofs of Theorems 9.1 and 9.2 (the forms should be constructed in the reverse order, i.e., the form of the maximal degree must be constructed first; we recommend the reader to read the original papers [19], [21]).

Theorem 9.3. *For any natural number r, there exists a number $n_0 = n_0(r)$ such that, for any $n \geq n_0$, there is a form $F(x_1, \dots, x_k)$ of degree $\leq n$ with integer coefficients that does not represent zero modulo 2; the number of its variables is k,*

$$k > 2^u, \quad u = \frac{n}{\log_2 n \cdot \log_2 \log_2 n \dots \underbrace{\log_2 \dots \log_2}_{r} n \cdot \underbrace{\log_2 \dots \log_2^2}_{r+1} n}.$$

Theorem 9.4. *Let p be an odd prime number. For any natural number r, there exists a number $n_1 = n_1(r)$ such that, for any $n \geq n_1$, there is a form $F(x_1, \dots, x_k)$ of degree $\leq n$ with integer coefficients that does not represent zero modulo p; the number of its variables is k,*

$$k > p^u, \quad u = \frac{n}{\log_p n \cdot \log_p \log_p n \dots \underbrace{\log_p \dots \log_p}_{r} n \cdot \underbrace{\log_p \dots \log_p^2}_{r+1} n}.$$

9.1.4 Hypotheses

Now we state several hypotheses which, as we believe, belong to the set of problems under study (these hypotheses were stated in [19]).

Hypothesis 9.1. *If p is a prime number, $p > n$, then for any nth-degree form F in k variables, $k > n^2$, the congruence*

$$F(x_1, \dots, x_k) \equiv 0 \pmod{p^m},$$

where m is an arbitrary number, has a solution $x_1, \dots, x_k$ satisfying the condition that not all $x_1, \dots, x_k$ are multiples of p.

Hypothesis 9.2. *Let $n_1, \dots, n_s$ be natural numbers, let $p > \max(n_1, \dots, n_s)$, and let $F_1, \dots, F_s$ be forms of degrees $n_1, \dots, n_s$ in k variables. Then for any m and $k > n_1^2 + \dots + n_s^2$, the system of congruences*

$$\left.\begin{aligned} F_1(x_1, \dots, x_k) &\equiv 0, \\ &\vdots \\ F_s(x_1, \dots, x_k) &\equiv 0 \end{aligned}\right\} \pmod{p^m},$$

has the solution $x_1, \dots, x_k$ satisfying the condition that not all $x_1, \dots, x_k$ are multiples of p.

Hypothesis 9.3. *Suppose that $2 \leq p \leq n$, p is a prime number, and $\alpha_p(n)$ is the exponent of p contained in the canonical decomposition of $n!$, i.e.,*

$$\alpha_p(n) = \left[\frac{n}{p}\right] + \left[\frac{n}{p^2}\right] + \cdots$$

For any $\varepsilon > 0$, there is an $n_0 = n_0(\varepsilon)$ such that, for all $n > n_0$, for any nth-degree form F in k variables,

$$k \geq p^{(1+\varepsilon)(\alpha_p(n)+2)},$$

for any m, the congruence

$$F = F_1(x_1, \ldots, x_k) \equiv 0 \pmod{p^m}$$

has a solution $x_1, \ldots, x_k$ satisfying the condition that not all $x_1, \ldots, x_k$ are multiples of p. Moreover, for any $\varepsilon > 0$, there exists an infinite sequence of natural numbers n such that, for each of them and for any $p \leq n$, there exists an nth-degree form $F(x_1, \ldots, x_k)$ in k variables, where $k \geq p^{(1-\varepsilon)(\alpha_p(n)+2)}$, that does not represent zero modulo p.

9.2 The p-adic proof of Vinogradov's theorem on estimating $G(n)$ in the Waring problem

In what follows, we propose a distinctive application of the p-adic method in the Waring problem, namely, the proof of the well-known Vinogradov's estimate

$$G(n) \leq n(2 \ln n + 4 \ln \ln n + 2 \ln \ln \ln n + 13). \tag{9.33}$$

The estimate (9.33) has recently been the best possible for large values of n. Our proof is based on "good" estimates of trigonometric sums $A(\alpha)$ for α contained in "small" arcs,

$$S(\alpha) = \sum_x^X \exp\{2\pi i \alpha x^n\}.$$

Here x runs through X values of natural numbers of a special form, and the estimates have the form

$$S(\alpha) \ll X^{1-c/(n \log n)},$$

where $c > 0$ is an absolute constant and the constant in $\ll$ depends only on n (see Lemma 9.9 below [91]).

9.2.1 The v-numbers

In what follows, n and l are natural numbers ($n \geq 3$, $l \geq 2$) and R is an increasing parameter,

$$R \geq R_0 = R_0(n, l) = (64nl)^{64n^8 l^8 (1+1(n-1))^l};$$

for $j = 1, 2, \ldots, l$, the letters P_j, X_j, and Y_j denote the variables

$$R_j = R^{(1/n)(1-1/n)^{j-1}}, \quad X_j = 0.25 R_j, \quad Y_j = 0.6 R_j;$$

the letters p_j denote the variables running though the values of primes in their intervals $X_j < p_j \le Y_j$, and $(p_j - 1, n) = 1$.

We define the *v-numbers* corresponding to the parameters (l, R) by the relation $v = p_1 \dots p_l$. It is easy to see that the number V of such v-numbers corresponding to (l, R) satisfies the relation

$$V \asymp R^{1-(1-1/n)^l} (\ln R)^{-l},$$

while the numbers themselves vary within the limits

$$4^{-l} R^{1-(1-1/n)^l} \le v \le 2^{-l} R^{1-(1-1/n)^l}.$$

Note that the v-numbers are similar to Linnik's numbers (see Section 3.5, Chapter 3).

Lemma 9.8. *Let J be the number of solutions of the equation*

$$x_1^n + \dots + x_l^n = y_1^n + \dots + y_l^n,$$

where $x_1, \dots, x_l, y_1, \dots, y_l$ take the values of the v-numbers corresponding to the parameters (l, R). Then the following estimate holds:

$$J \le (2ln)^{20l^2 n} \prod_{s=1}^{l} (Y_s - X_s)^{2(l-s)} \prod_{j=s}^{l} (Y_j - X_j) \le (2ln)^{20l^2 n} R^{2l-n+(n-l)(1-1/n)^l}.$$

Proof. We represent J as an integral of a power of the modulus of a trigonometric sum. We have

$$J = \int_0^1 |S(\alpha)|^{2l} \, d\alpha, \tag{9.34}$$

where

$$S(\alpha) = \sum_{X_1 < p_1 \le Y_1} S(\alpha; p_1),$$

$$S(\alpha; p_1) = \sum_{X_2 < p_2 \le Y_2} \cdots \sum_{X_l < p_l \le Y_l} \exp\{2\pi i \alpha (p_1 p_2 \dots p_l)^n\}.$$

We take $H = (2l)^{8nl}$ and $H_1 = (Y_1 - X_1) H^{-1}$ and divide the interval $(X_1, Y_1]$ into H intervals of the form

$$\big(X_1 + (r-1)H_1, X_1 + rH_1\big], \quad r = 1, \dots, H.$$

According to this partition, $S(\alpha)$ is the sum of H terms:

$$S(\alpha) = \sum_{r=1}^{H} S_r(\alpha), \tag{9.35}$$

where

$$S_r(\alpha) = \sum_{X_1+(r-1)H_1<p_1\le X_1+rH_1} S(\alpha;\, p_1).$$

Raising (9.35) to the power l, we obtain

$$S^l(\alpha) = \sum_{r_1=1}^{H} \cdots \sum_{r_l=1}^{H} S_{r_1}(\alpha) \ldots S_{r_l}(\alpha).$$

We divide all sets $(r_1, \ldots, r_l)$ in the last multiple sum into two classes A and B. The class A contains the sets for which there is an r_j different from all other r_s, i.e., $r_j \ne r_s$ $(s \ne j,\ 1 \le s \le l)$. The class B contains all other sets. According to this partition, we obtain

$$S^l(\alpha) = \sum\nolimits_1 + \sum\nolimits_2, \tag{9.36}$$

where

$$\sum\nolimits_1 = \sum_{(r_1,\ldots,r_l)\in A}\cdots\sum S_{r_1}(\alpha)\ldots S_{r_l}(\alpha), \qquad \sum\nolimits_2 = \sum_{(r_1,\ldots,r_l)\in B}\cdots\sum S_{r_1}(\alpha)\ldots S_{r_l}(\alpha).$$

The terms in $\sum_1$ have the form (after $r_1, \ldots, r_l$ are reindexed)

$$S_{r_1}(\alpha) S_{r_2}^{\beta_2}(\alpha) \ldots S_{r_t}^{\beta_t}(\alpha), \quad 2 \le t \le l;$$

here $r_1 \ne r_j$ $(j = 2, \ldots, t)$ and $\beta_2 \ge 1, \ldots, \beta_t \ge 1$. The terms in $\sum_2$ have the form

$$S_{r_1}^{\beta_1}(\alpha) S_{r_2}^{\beta_2}(\alpha) \ldots S_{r_t}^{\beta_t}(\alpha);$$

here $\beta_1 \ge 2, \ldots, \beta_t \ge 2$, $r_1 < r_2 < \cdots < r_t$, and $\beta_1 + \cdots + \beta_t = l$.

The number $\mu(A)$ of terms in $\sum_1$ does not exceed H^l. We estimate the number $\mu(B)$ of terms in $\sum_2$ as

$$\mu(B) \le \sum_{t\ge 1} l^l H^t \varkappa(t), \tag{9.37}$$

where $\varkappa(t)$ is the number of solutions of the equation $\beta_1 + \cdots + \beta_t = l$ $(\beta_j \ge 2)$ or of the equation $\beta_1' + \cdots + \beta_t' = l$ $(\beta_j' \ge 0)$. Hence $t \le l/2$ and

$$\varkappa(t) = \frac{(l-t-1)!}{(l-2t)!(l-1)!} = \binom{l-t-1}{l-1} \le \binom{l-1}{t-1}. \tag{9.38}$$

From (9.37) and (9.38) we find

$$\mu(B) \le (2l)^l 2^{-2} H^{l/2}.$$

We square (9.26) and apply the Cauchy inequality. Then we obtain

$$|S(\alpha)|^{2l} \le 2\mu(A) \sum_{(r_1,\ldots,r_t)\in A}\cdots\sum |S_{r_1}(\alpha)|^2 |S_{r_2}(\alpha)|^{2\beta_2} \ldots |S_{r_t}(\alpha)|^{2\beta_t} + \tag{9.39}$$

$$+2\mu(B)\sum_{(r_1,\ldots,r_t)\in B}\cdots\sum|S_{r_1}(\alpha)|^{2\beta_1}\ldots|S_{r_t}(\alpha)|^{2\beta_t}.$$

To the terms in the first and second sums, we apply the inequality relating the geometric mean and the arithmetic mean. For the first sum we have

$$|S_{r_2}(\alpha)|^{2\beta_2}\ldots|S_{r_t}(\alpha)|^{2\beta_t}\le\frac{\beta_2|S_{r_2}(\alpha)|^{2(l-1)}+\cdots+\beta_t|S_{r_t}(\alpha)|^{2(t-1)}}{l-1};$$

for the second sum we have

$$|S_{r_1}(\alpha)|^2\ldots|S_{r_l}(\alpha)|^2\le\frac{|S_{r_1}(\alpha)|^{2l}+\cdots+|S_{r_l}(\alpha)|^{2l}}{l}.$$

Substituting these inequalities into (9.39) and then substituting (9.39) into (9.34), we obtain

$$J\le 2\mu^2(A)J_1+2\mu^2(B)J_2^{(1)},\tag{9.40}$$

where

$$J_1=\int_0^1|S_{r_1}(\alpha)|^2|S_r(\alpha)|^{2l-2}\,d\alpha,\quad J_2^{(1)}=\int_0^1|S_r(\alpha)|^{2l}\,d\alpha,$$

where $r_1\ne r$ and r_1 and r are fixed numbers ($1\le r_1, r\le H$) such that J_1 and $J_2^{(1)}$ take maximal possible values.

We note that the only difference between $J_2^{(1)}$ and J is that the interval of variation of p_1 $(X_1<p_1\le Y_1)$ in J is replaced in $J_2^{(1)}$ by the interval

$$X_1^{(1)}=X_1+(r-1)H_1<p_1\le X_1+rH_1=Y_1^{(1)},$$

whose length is $H_1=Y_1^{(1)}-X_1^{(1)}=(Y_1-X_1)H^{-1}$.

Let us estimate J_1. Applying Hölder's inequality to $|S_r(\alpha)|^{2l-2}$, we find

$$|S_r(\alpha)|^{2l-2}=\Bigg|\sum_{X_1^{(1)}<p_1\le Y_1^{(1)}}S(\alpha;p_1)\Bigg|^{2l-2}\le H_1^{2l-3}\sum_{X_1^{(1)}<p_1\le Y_1^{(1)}}|S(\alpha;p_1)|^{2l-2},$$

$$J_1=H_1^{2l-2}\int_0^1|S_{r_1}(\alpha)|^2|S(\alpha;p_1)|^{2l-2}\,d\alpha,\tag{9.41}$$

where p is a fixed prime number and $X_1+(r-1)H_1<p\le X_1+rH_1$ $(r\ne r_1)$. The last integral is equal to J_3, which is the number of solutions of the equation

$$x_1^n-y_1^n=p^n(x_2^n+\cdots+x_l^n-y_2^n-\cdots-y_l^n),\tag{9.42}$$

where x_1, y_1 take values of the form $p_1p_2\ldots p_l$, while $x_2,\ldots,y_l$ take values of the form $p_2\ldots p_l$, and

$$X_1+(r_1-1)H_1<p_1\le X_1+r_1H_1,\quad X_j<p_j\le Y_j,\quad j\ge 2.$$

Since $r \neq r_1$, we have $p \neq p_1$. It follows from (9.42) that $J_3 \leq TJ_4$, where T is the number of solutions of the congruences

$$x_1^n \equiv y_1^n \pmod{p^n}, \tag{9.43}$$

and J_4 is the number of solutions of the equation

$$x_2^n + \cdots + x_l^n = y_2^n + \cdots + y_l^n \tag{9.44}$$

in numbers $x_2, \dots, y_l$ of the form $p_2, \dots, p_l$.

By the definition of the numbers x_1 and y_1, these numbers are coprime to the modulus p^n in the congruence (9.43).

The numbers x_1 and y_1 do not exceed $2^{-l}R_1 \dots R_l = 2^{-l}R^{1-(1-1/n)^l}$, while $p > 2^{-2n}R$. Hence it follows from (9.43) that $x_1 = y_1$, and

$$T \leq n \prod_{j=1}^{l}(Y_j - X_j) = n(Y_1 - X_1)T_1. \tag{9.45}$$

Substituting the estimates (9.45) and (9.41) into (9.40), we obtain

$$\begin{aligned} J &\leq 2n\mu^2(A)H_1^{2l-2}(Y_1 - X_1)T_1J_4 + 2\mu^2(B)J_2^{(1)} \\ &\leq 2nH^{2l}H_1^{2l-2}(Y_1 - X_1)T_1J_4 + (2l)^{2l}2^{-3}H^lJ_2^{(1)}; \end{aligned} \tag{9.46}$$

here $T_1 = \prod_{j=2}^{l}(Y_j - X_j)$, the number J_4 of solutions of Eq. (9.44) in numbers $x_2, \dots, y_l$ of the form $p_2 \dots p_l$, $J_2^{(1)}$ has the same form as J, only the variables X_1, Y_1 are replaced by the variables $X_1^{(1)}, Y_1^{(1)}$.

For convenience, we rewrite inequality (9.46) as

$$J^{(0)} \leq 2nH^{2l}H_1^{2l-2}(Y_1^{(0)} - X_1^{(0)})T_1J_4 + (2l)^{2l}2^{-3}H^lJ^{(2)}, \tag{9.47}$$

where we introduce the following notation:

$$J^{(0)} = J, \quad J^{(2)} = J_2^{(1)}, \quad Y_1^{(0)} = Y_1, \quad X_1^{(0)} = X_1.$$

Now, applying (9.47) to estimate $J^{(1)}$, we obtain

$$J^{(1)} \leq 2H^{2l}H_2^{2l-2}n(Y_1^{(1)} - X_1^{(1)})T_1J_4 + (2l)^{2l}2^{-3}H^lJ^{(2)},$$

where

$$Y_1^{(1)} - X_1^{(1)} = (Y_1^{(0)} - X_1^{(0)})H^{-1}, \quad H_2 = (Y_1^{(1)} - X_1^{(1)})H^{-1};$$

here $J^{(2)}$ has the same form as J, only the variables X_1, Y_1 are replaced by the variables $X_1^{(2)}, Y_1^{(2)}$.

Let us find the number μ from the relations

$$H^{\mu+1} < Y_1 - X_1 \le H^{\mu+3}.$$

For $j = 0, 1, \ldots, \mu$, we successively find

$$J^{(j)} \le 2H^{2l} H_{j+1}^{2l-2} n(Y_1^{(j)} - X_1^{(j)}) T_1 J_4 + (2l)^{2l} 2^{-3} H^l J^{(j+1)}, \tag{9.48}$$

where

$$Y_1^{(j)} - X_1^{(j)} = (Y^{(j-1)} - X^{(j-1)})H^{-1} = (Y_1 - X_1)H^{-1},$$
$$H_{j+1} = (Y_1^{(j)} - X_1^{(j)})H^{-1} = (Y_1 - X_1)H^{-j-1};$$

the range of p_1 in $J^{(j+1)}$ has the form

$$X_1^{(j+1)} + 1 < p_1 \le Y_1^{(j+1)}.$$

Finally, we estimate $J^{\mu+1}$ as

$$J^{(\mu+1)} = \int_0^1 \Bigg| \sum_{X_1^{(\mu+1)}+1<p_1\le Y_1^{(\mu+1)}} S(\alpha; p_1)\Bigg|^{2l} d\alpha$$
$$\le (Y_1^{(\mu+1)} - X_1^{(\mu+1)})^{2l} \int_0^1 |S(\alpha; p_1)|^{2l}\, d\alpha = (Y_1 - X_1)^{2l} H^{-2l(\mu+1)} I_{l-1},$$

where I_{l-1} is the number of solutions of the equation in the lemma under the condition that $x_1, \ldots, x_l, y_1, \ldots, y_l$ take values of the form $p_2 \ldots p_l$. For I_{l-1}, we easily find the estimate

$$I_{l-1} \le (Y_2 - X_2)^2 \ldots (Y_l - X_l)^2 J_4,$$

where J_4 is the number of solutions of Eq. (9.44). We rewrite formula (9.48) as

$$J^{(j)} \le 2H^{-j(2l-1)+2}(Y_1 - X_1)^{2l-1} T_1 J_4 + aJ^{(j+1)}, \tag{9.49}$$

where $a = (2l)^{2l} 2^{-3} H^l$. Multiplying both sides of (9.49) by a^j and summing over $j = 0, 2, \ldots, \mu$, we find

$$J^{(0)} + \sum_{j=1}^{\mu} a^j J^{(j)} \le 2H^2 (Y_1 - X_1)^{2l-1} T_1 J_4 \sum_{j=0}^{\mu} a^j H^{-j(2l-1)}$$
$$+ \sum_{j=1}^{\mu} a^j J^{(j)} + a^{\mu+1} J^{(\mu+1)},$$

$$J = J^{(0)} \le 4H^2(Y_1 - X_1)^{2l-2}\prod_{j=1}^{l}(Y_j - X_j)J_4 + a^{\mu+1}\big((Y_1 - X_1)H^{-\mu-1}\big)^{2l}(Y_2 - X_2)^2 \dots (Y_l - X_l)^2 J_4.$$

Since $H^{\mu+3} \ge Y_1 - X_1$, we have

$$(Y_1 - X_1)H^{-\mu-1} \le H^2.$$

Moreover,

$$(Y_1-X_1)^{2l-2}\prod_{j=1}^{l}(Y_1-X_j) \ge a^{\mu+1}\big((Y_1-X_1)H^{-\mu-1}\big)^{2l}\prod_{j=2}^{l}(Y_1-X_j)^2. \quad (9.50)$$

Thus we obtain the main recurrence inequality for J:

$$J \le 5H^2 n(Y_1 - X_1)^{2l-2}\prod_{j=1}^{l}(Y_j - X_j)J_4^{(1)},$$

where $J_4^{(1)}$ is the number of solutions of Eq. (9.44). To $J_4^{(1)}$, we apply the same argument, and so on. After the $(s+1)$th step $(s+1 < l)$, we obtain an inequality of the form

$$J_4^{(1)} \le 5H^2 n(Y_{s+1} - X_{s+1})^{2(l-s)-2}\prod_{j=s+1}^{l}(Y_j - X_j)J_4^{(s+1)}, \quad (9.51)$$

where $J_4^{(s+1)}$ is the number of solutions of the equation

$$x_{s+2}^n + \dots + x_l^n = y_{s+2}^n + \dots + y_l^n$$

in numbers $x_{s+1}, \dots, y_l$ of the form $p_{s+2}\dots p_l$.

We note that all conditions on the parameters in deriving (9.51) are satisfied automatically; an analog of relation (9.50) looks as $(s \le l-2)$:

$$(Y_{s+1} - X_{s+1})^{2(l-s)-2}\prod_{j=s+1}^{l}(Y_j - X_j) \ge a^{\mu+1}\big((Y_{s+1} - X_{s+1})H^{-\mu-1}\big)^{2(l-s)}\prod_{j=s+2}^{l}(Y_j - X_j)^2$$

and can be verified easily. For $s = l-1$, the variable $J_4^{(s+1)}$ can be estimated trivially:

$$J_4^{(l-1)} \le Y_l - X_l.$$

Collecting all the above estimates, we obtain the statement of the lemma. □

9.2.2 Estimates of special trigonometric sums on small arcs

Throughout this subsection, we assume that $n \geq 4$ and $N \geq N_0(n)$, as well as $P = [N^{1/n}]$ and $\tau = 2nP^{n-1}$. We represent each α in the interval $-1/\tau \leq \alpha < 1 - 1/\tau$ as

$$\alpha = \frac{a}{q} + z, \quad 1 \leq q \leq \tau, \quad (a, q) = 1, \quad |z| \leq (q\tau)^{-1}. \tag{9.52}$$

The set E_1 (the set of "large arcs") contains α for which $q \leq P^{1/6}$ in representation (9.52). The set E_2 (the set of "small arcs") contains all other α in the above interval. Thus, if $\alpha \in E_2$, then $q > P^{1/6}$ in (9.52).

Next, we choose the numbers

$$X = 10^{-4}P^{1/2}, \quad Y = 10^{-2}P^{1/2}, \quad l_1 = [2n \ln n],$$

and consider the trigonometric sum

$$W(\alpha) = \sum_{X<x\leq 2X}{}' \sum_{y}{}' \exp\{2\pi i \alpha x^n y^n\},$$

where x runs through the values of the prime numbers in the above interval, while y runs through the values of the υ-numbers corresponding to the parameters (l, R), where $l = l_1$ and $R = Y$.

Lemma 9.9. *Suppose that $\alpha \in E_2$. Then $|W(\alpha)|$ satisfies the estimate*

$$|W(\alpha)| \ll P^{1-(24n \ln n)^{-1}},$$

where the constant in $\ll$ depends only on n.

Proof. We choose $\tau_1 = (4X)^n$ and represent α as

$$\alpha = \frac{a_1}{q_1} + z_1, \quad 1 \leq q_1 \leq \tau_1, \quad (a_1, q_1) = 1, \quad |z_1| \leq (q_1\tau_1)^{-1}. \tag{9.53}$$

Let us consider two cases.

1. $q_1 > X^{1/3}$. Since, in the sum $W(\alpha)$, the number of numbers x such that $(x, q_1) > 1$ does not exceed $c(\varepsilon)q_1^{\varepsilon}$, where $\varepsilon > 0$ is an arbitrarily small constant, we have

$$|W(\alpha)| \ll X^{1/2}Y + |W_1(\alpha)|,$$

where

$$|W_1(\alpha)| \leq \sum_{\substack{X<x\leq 2X \\ (x,q_1)=1}}{}' \left| \sum_{Y_1<y\leq 2^l Y_1}{}' \exp\{2\pi i \alpha x^n y^n\} \right|, \tag{9.54}$$

$$Y_1 = 4^{-l_1} Y^{1-(1-1/n)^l} < Y.$$

Let δ be an arbitrary real number satisfying the condition $|\delta| < Y^{-n}$. Then we successively obtain

$$\begin{aligned}\sum_{Y_1<y\le 2^{l_1}Y_1}{}' \exp\{2\pi i \alpha x^n y^n\} &= \sum_{Y_1<y\le 2^{l_1}Y_1}{}' \exp\{2\pi i(\alpha x^n+\delta)y^n\} \\ &+ \sum_{Y_1<y\le 2^{l_1}Y_1}{}' \exp\{2\pi i(\alpha x^n+\delta)y^n\}\big(\exp\{-2\pi i\delta y^n\}-1\big) \\ &= \mathbf{C}(2^{l_1}Y_1)\exp\{-2\pi i\delta 2^{l_1}Y_1\} - \int_{Y_1}^{2^{l_1}Y_1} \mathbf{C}(u)\, d\exp\{-2\pi i\delta u^n\},\end{aligned}$$

where

$$\mathbf{C}(u) = \sum_{Y_1<y\le u}{}' \exp\{2\pi i(\alpha x^n+\delta)y^n\},$$

$$\Big|\sum_{Y_1<y\le 2^{l_1}Y_1}{}' \exp\{2\pi i\alpha x^n y^n\}\Big| \ll |\mathbf{C}(2^{l_1}Y_1)| + |\delta| Y_1^{n-1}\int_{Y_1}^{2^{l_1}Y_1} |\mathbf{C}(u)|\, du.$$

Summing both sides of the last inequality over x and then integrating over δ within the limits $-Y^{-n} \le \delta \le Y^{-n}$, we obtain the relation

$$\begin{aligned}W &= \sum_{\substack{X<x\le 2X\\(x,q_1)=1}}{}' \Big|\sum_{Y_1<y\le 2^{l_1}Y_1}{}' \exp\{2\pi i\alpha x^n y^n\}\Big| \\ &\ll Y^n \sum_{\substack{X<x\le 2X\\(x,q_1)=1}}{}' \int_{-Y^{-n}}^{Y^{-n}} \Big|\sum_{Y_1<y\le Y_2}{}' \exp\{2\pi i(\alpha x^n+\delta)y^n\}\Big|\, d\delta,\end{aligned}$$

where Y_2 is some fixed number that does not exceed $2^{l_1}Y_1$. Further, applying Hölder's inequality, we obtain

$$\begin{aligned}W^{2l_1} &\le Y^n (X\ln^{-1} X)^{2l_1-1} \\ &\times \sum_{\substack{X<x\le 2X\\(x,q_1)=1}}{}' \int_{-Y^{-n}}^{Y^{-n}} \Big|\sum_{Y_1<y\le Y_2}{}' \exp\{2\pi i(\alpha x^n+\delta)y^n\}\Big|^{2l_1} d\delta.\end{aligned} \tag{9.55}$$

Let $x^n \not\equiv x_1^n \pmod{q_1}$. Then, using representation (9.53), we obtain

$$\|\alpha x^n - \alpha x_1^n\| = \Big\|\frac{a_1(x^n-x_1^n)}{q_1} + z_1(x^n-x_1^n)\Big\| \ge \frac{1}{q_1} - |z_1|(2X)^n > \frac{1}{2q_1} > \frac{1}{Y^n},$$

since $q_1 \le \tau = (4X)^n$.

The congruence $x^n \equiv \lambda \pmod{q_1}$ ($X < x \le 2X$, $(x, q_1) = 1$) has at most $c(\varepsilon)(X/q_1 + 1)q_1^{\varepsilon}$ solutions. Therefore, from (9.55), Lemma 9.8, and (9.54) we obtain the inequalities

$$W^{2l_1} \ll Y^n (X \ln^{-1} X)^{2l_1 - 1}\left(\frac{X}{q_1} + 1\right) q_1^{\varepsilon} \int_{-0.5}^{0.5} \Big| \sum_{Y_1 < y \le Y_2}{}' \exp\{2\pi \alpha y^n\}\Big|^{2l_1} d\alpha$$

$$\ll Y^n (X \ln^{-1} X)^{2l_1 - 1}\left(\frac{X}{q_1} + 1\right) q_1^{\varepsilon} Y^{2l_1 - n - (l_1 - n)(1 - 1/n)^{l_1}}$$

$$\ll (XY)^{2l_1}\left(\frac{1}{q_1} + \frac{1}{X}\right) q_1^{\varepsilon} Y^{-(l_1 - n)(1 - 1/n)^{l_1}} \ll P^{2l_1 - 1/6},$$

$$W \ll P^{l - 1/(12 l_1)}, \quad W(\alpha) \ll P^{l - 1/(12 l_1)} < P^{l - (24 n \ln n)^{-1}}.$$

So we have proved the statement of the lemma for $q_1 > X^{1/3}$.

2. $q_1 \le X^{1/3}$. Applying Hölder's inequality, we obtain

$$|W(\alpha)|^{2l_1} \ll X^{2l_1 - 1} \sum_{X < x \le 2X}{}' \Big| \sum_{Y_1 < y \le 2^{l_1} Y_1}{}' \exp\{2\pi i \alpha x^n y^n\}\Big|^{2l_1}$$

$$\ll X^{2l_1 - 1} \sum_{|\lambda| \ll Y_1^n} J_{l_1}(\lambda) \Big| \sum_{X < x \le 2X}{}' \exp\{2\pi i \alpha \lambda x^n\}\Big|,$$

where $J_{l_1}(\lambda)$ denotes the number of solutions of the equation

$$y_1^n + \cdots + y_{l_1}^n - y_{l_1+1}^n - \cdots - y_{2l_1}^n = \lambda$$

in the v-numbers y_j ($j = 1, \ldots, 2l_1$). We again use Hölder's inequality and the inequality $J_{l_1}(\lambda) \le J_{l_1}(0)$ and thus find

$$|W(\alpha)|^{4l_1} \ll X^{4l_1 - 2}\Big(\sum_{|\lambda| \ll Y_1^n} J_{l_1}(\lambda)\Big) \tag{9.56}$$

$$\times \Big(\sum_{|\lambda| \ll Y_1^n} J_{l_1}(\lambda) \Big| \sum_{X < x \le 2X}{}' \exp\{2\pi i \alpha \lambda x^n\}\Big|^2\Big)$$

$$\ll X^{4l_1 - 2} Y^{2l_1(1 - (1 - 1/n)^{l_1})} J_{l_1}(0) \sum_{|\lambda| \ll Y_1^n} \Big| \sum_{X < x \le 2X}{}' \exp\{2\pi i \alpha \lambda x^n\}\Big|^2$$

$$\ll X^{4l_1 - 2} Y^{4l_1 - (3l_1 - n)(1 - 1/n)^{l_1} - n} \sigma,$$

where

$$\sigma = \sum_{|\lambda| \ll Y_1^n} \Big| \sum_{X < x \le 2X}{}' \exp\{2\pi i \alpha \lambda x^n\}\Big|^2.$$

Now we square the modulus of the sum over x in σ, collect the terms with $x = x_1$, and sum the remaining terms over λ. Thus we obtain

$$\sigma \ll Y_1^n X + \sum_{x \neq x_1}{}' \min\left(Y_1^n, \frac{1}{\|\alpha(x^n - x_1^n)\|}\right) \tag{9.57}$$

$$\ll Y_1^n X + X \sum_{X < x \leq 2X} \min\left(Y_1^n, \frac{1}{\|\alpha x^n + \beta\|}\right).$$

Without loss of generality, we assume that, in representation (9.53) of the number α, the variable z_1 is nonnegative, i.e., $0 \leq z_1 \leq (q_1\tau_1)^{-1}$. In the case under study, we have $q_1 \leq X^{1/3} < P^{1/6}$; therefore, z_1 must exceed $(q_1\tau)^{-1}$, since, otherwise, α would belong to E_1, which is the set of "large arcs." So we have $(q_1\tau)^{-1} < z_1 \leq (q_1\tau_1)^{-1}$.

In the last sum, we represent the numbers x as

$$x = q_1 s + t, \quad 0 < t \leq q_1, \quad (X - t)q_1^{-1} < s \leq (2X - t)q_1^{-1}.$$

Then we have

$$\|\alpha x^n + \beta\| = \left\| \frac{a_1 t^n}{q_1} + z_1(q_1 s + t)^n + \beta \right\|.$$

If s increases by 1, then the function

$$\Phi(s) = z_1(q_1 s + t)^n + \frac{a_1 t^n}{q_1} + \beta$$

varies by a value d, where $nz_1 X^{n-1} q_1 < d < n2^n z_1 X^{n-1} q_1$.

Thus the values of $\Phi(s)$ can be represented as kd $(K < k \leq K + Xq_1^{-1} + 1)$. Since $z_1(q_1 s + t)^n < z_1(2X)^n \leq 2^{-n}$, the number kd can be integer only for a single value of k. Therefore, we have the following estimate for the sum over x:

$$\sum_{X < x \leq 2X} \min\left(Y_1^n, \frac{1}{\|\alpha x^n + \beta\|}\right) \ll q_1 \sum_{s} \min\left(Y_1^n, \frac{1}{\|\Phi(s)\|}\right) \tag{9.58}$$

$$\ll q_1\left(Y_1^n + \sum_{k \geq 1}^{s} \frac{1}{kd}\right) \ll q_1(Y_1^n + \tau X^{-n+1} \ln P).$$

From (9.58), (9.57), and (9.56) we successively find

$$\sigma \ll Y_1^n X q_1(1 + P^{n-1} Y_1^{-n} X^{-n+1}) \ll Y_1^n X^{4/3},$$

$$|W(\alpha)|^{4l_1} \ll (XY)^{4l_1} X^{-2/3} Y^{-3l_1(1-1/n)^{l_1}} \ll P^{4l_1 - 1/3},$$

$$|W(\alpha)| \ll P^{1-1/(12l_1)} \ll P^{1-(24n \ln n)^{-1}}.$$

The statement of the lemma is also proved in the second case. The proof of the lemma is complete. □

9.2.3 The u-numbers

We shall use the parameters $R,\ R_j\ (j = 1, \dots, l)$ introduced in Section 9.2.1 and introduce some new parameters; $p_1, p_2, \dots, p_l$ are arbitrary, but fixed, prime numbers, $0.5R_j < p_j \le R_j$; further, $x_1, x_2, \dots, x_l$ denote the variables running through the values of the natural numbers in the intervals $(0.25R_j)^n < x_j \le (0.5R_j)^n$ such that $(x_j, p_j) = 1\ (j = 1, \dots, l-1)$ and x_j^n takes only a single value modulo p_j.

We define the *u-numbers* corresponding to the parameters (l, R) by the relation

$$u = x_1^n + p_1^n x_2^n + (p_1 p_2)^n x_3^n + \cdots + (p_1 p_2 \dots p_{l-1})^n x_l^n.$$

The u-numbers vary within the limits $2^{-2n^2} R^n \le u \le l 2^{-n^2} R^n$. The number U of u-numbers corresponding to the parameters (l, R) satisfies the relation $U \asymp R^{n-n(1-1/n)^l}$.

If u_1, u_2 are two u-numbers,

$$u_1 = x_1^n + p_1^n x_2^n + \cdots + (p_1 \dots p_{l-1})^n x_l^n,$$
$$u_2 = y_1^n + p_1^n y_2^n + \cdots + (p_1 \dots p_{l-1})^n y_l^n,$$

then the relation $u_1 = u_2$ implies $x_1^n \equiv y_1^n \pmod{p_1^n}$; since $p_1^n > 2^{-n} R_1^n \ge x_1$ and $\ge y_1$, this implies that $x_1 = y_1$; similarly, we obtain $x_2 = y_2, \dots, x_l = y_l$. Thus for a fixed λ, the equation $u = \lambda$ in the u-numbers has at most one solution.

We note that the u-numbers constructed here are the p-adic analogs of the u-numbers constructed in Chapter 3, Sections 3.2 and 3.3.

9.2.4 The theorem

Now we state and prove an assertion that is somewhat more strict than (9.33). We assume that $n \ge 4000$.

Theorem 9.5. *The variable $G(n)$ satisfies the estimate*

$$G(n) < 2n \ln n + 2n \ln \ln n + 12n.$$

Proof. We consider the equation

$$z_1^n + \cdots + z_{4n}^n + u_1 + u_2 + (x_1 y_1)^n + \cdots + (x_k y_k)^n = N, \tag{9.59}$$

where $N \ge N_0(n)$, $z_1, \dots, z_{4n}$ are natural numbers; $P = [N^{1/n}]$, u_1, u_2 are the u-numbers corresponding to the parameters (l, R), where $l = l_2 = [n \ln n + n \ln \ln n + 2.6n]$ and $R = P$; $k = 2n$; the x_j run through the values of the prime numbers in the interval $(X, 2X)$, $X = 10^{-4} P^{1/2}$; and the y_j run through the values of the v-numbers corresponding to the parameters (l, R), $l = l_1 = [2n \ln n]$, and $R = Y = 10^{-2} P^{1/2}$.

Let J be the number of solutions of (9.59). Then we have

$$J = \int_{-1/\tau}^{1-1/\tau} S^{4n}(\alpha) T^2(\alpha) W^k(\alpha) \exp\{-2\pi i \alpha N\}\, d\alpha,$$

where

$$S(\alpha) = \sum_{z_1=1}^{P} \exp\{2\pi i \alpha z^n\}, \qquad T(\alpha) = \sum_{u} \exp\{2\pi i \alpha u\},$$
$$W(\alpha) = \sum_{x,y} \exp\{2\pi i \alpha x^n y^n\}, \qquad \tau = 2nP^{n-1}.$$

We divide the interval $-1/\tau \le \alpha < 1 - 1/\tau$ into two sets E_1 and E_2 (the sets of "large and small arcs") as in Section 9.2.2. Then we present J as the sum of two integrals: $J = J_1 + J_2$, where $J_1 = \int_{E_1}$ and $J_2 = \int_{E_2}$.

Let us estimate J_1 from below. We have

$$J_1 = \sum_{u_1,u_2} \sum_{x_1,y_1} \cdots \sum_{x_k,y_k} \int_{E_1} S^{4n}(\alpha) \exp\{-2\pi i \alpha N_1\}\, d\alpha.$$

where $N_1 = N - u_1^n - u_2^n - (x_1 y_1)^n - \cdots - (x_k y_k)^n$.

From the definition of $u_1,\ u_2,\ x_j,\ y_j$, we easily find the range of variation of N_1, namely, $0.5N < N_1 < N$. Applying Lemma 9.7 (see [162], p. 43) to estimate the last integral, we obtain

$$\int_{E_1} S^{4n}(\alpha) \exp\{-2\pi i \alpha N_1\}\, d\alpha \gg N^3,$$
$$J_1 \gg N^3 P^{2(n-n(1-1/n)^{l_2})} X^k (\ln X)^{-k} Y^{k(1\ \ (1-1/n)^{l_1})} (\ln Y)^{-kl_1}. \tag{9.60}$$

Now let us estimate J_2 from above. Applying Lemma 9.9 and using the property of the u-numbers, we find

$$J_2 \ll P^{k(1-(24n\ln n)^{-1})} N^4 P^{n-n(1-1/n)^{l_2}}. \tag{9.61}$$

It follows from (9.59) and (9.61) that $J = J_1 + J_2 > 0$, i.e., $G(n) \le 4n + 2l_2 + k < 2n \ln n + 2n \ln\ln n + 12n$. The proof of the theorem is complete. □

9.3 Fractional parts of rapidly growing functions

The old still unsolved problems on the behavior of fractional parts of the functions $(3/2)^x$, $\alpha 2^x$, $x = 1, 2, \ldots$, are well known. A distinctive record is established by Vinogradov's theorem on the uniform distribution of fractional parts of the function $\exp\{\log^c x\}$, where c is a fixed number, $1 < c < 3/2$, $x = 1, 2, \ldots$ (see [84]). Here we show that the p-adic method allows one to obtain meaningful assertions about the behavior of the fractional parts of functions that are growing even faster. We introduce a new notion, which allows us to formulate the result in a more convenient form.

Let A be a subset of the set of real numbers, and let B be an infinite subset of the set of natural numbers. We shall say that A is a regular set with respect to B if, for any $\alpha \in A$, the sequence $\{\alpha n\}$ is everywhere dense on $[0, 1)$ provided that n runs through all the values in B. For example, if A is the set of irrational numbers and B is the set of values attained by the polynomial $f(x)$ with integer positive coefficients, $x = 1, 2, \ldots$, then A is regular with respect to B. But if B is the set of numbers of the form 2^x, $x = 1, 2, \ldots$, and A is the set of irrational numbers, then A is not a regular set with respect to B. However, it is possible to assume the following: if A is the set of algebraic numbers of degree $n \geq 2$, then, in this case, A is regular with respect to B. The goal in this section is to prove the following Theorem 9.6.

Theorem 9.6. *Suppose that A is the set of real algebraic numbers of degree $n \geq 2$ and B is the set of natural numbers of the form $x^{[\log^c x]}$, $x = 1, 2, \ldots$, where c is an arbitrary fixed number in the interval $0 < c < 1$. Then A is a regular set with respect to B.*

Theorem 9.6 is an obvious consequence of the following theorem.

Theorem 9.7. *Let α be an arbitrary real algebraic number of degree $n \geq 2$, and let*

$$f(x) = \alpha \exp\{[\log^c x] \log x\},$$

where c is a constant, $0 < c < 1$, and $[\log^c x]$ is the integral part of $\log^c x$. Then there exists a number $X_1 = X_1(c) > 0$ and constants $c_1 > 0$ and $c_2 > 0$ such that, for $X \geq X_1$ and any real number ξ, the number of solutions of the inequality

$$\|\xi - f(x)\| \leq \exp\{-c_1 \log^{1-c} X\}$$

in positive integers x, $x \leq X$, is larger than or equal to

$$X \exp\{-c_2(\log^{1-c} X + \log^c X \log\log X)\}.$$

In particular,

$$\min_{x \leq X} \|\xi - f(x)\| \leq \exp\{-c_1 \log^{1-c} X\}.$$

The method used to prove Theorem 9.7 essentially repeats the method used in the preceding section. However, the estimates for the sums $S(\alpha)$ cannot be used directly in the proof of Theorem 9.7, because X is a parameter strongly increasing with n; this increase is determined by the variable R_0 introduced above in Section 9.2.1. Here by the letters $c, c_1, c_2, \ldots$ we denote absolute positive constants such that $0 < c < 1$ and $c_1, c_2, \ldots$ can be different in different formulas; X is the principal increasing parameter, $X \geq X_1(c) > 0$; $\log N \asymp \log X$; n and k are natural numbers such that $n \asymp \log^c X$ and $k \asymp \log^c X$; $\theta, \theta_1, \ldots$ are some functions satisfying the conditions $|\theta| \leq 1$, $|\theta_1| \leq 1$, etc.; and the constants in the signs O are absolute constants.

First, precisely as in the preceding section, we prove the main lemma on the upper bound for the number of solutions of the Diophantine equation

$$x_1^n + x_2^n + \cdots + x_k^n = y_1^n + y_2^n + \cdots + y_k^n. \tag{9.62}$$

The unknowns $x_1, \dots, x_k, y_1, \dots, y_k$ in Eq. (62) take values in the set of integers of a special form, which will be called *v-numbers*. To define the set V of integers corresponding to the parameters N, n, and k whose elements will be called v-numbers, we define the numbers X_j and Y_j for $j = 1, 2, \dots, k$ by the relations

$$X_j = \left(1 - \frac{1}{4k}\right) Y_j, \quad Y_j = N^{(1/n)(1-1/n)^{j-1}}.$$

For each j, $1 \le j \le k$, we make the parameter p_j range over the set of all primes in $(X_j, Y_j]$, i.e., the p_j are prime numbers such that

$$X_j < p_j \le Y_j.$$

Definition 9.4. The V-set corresponding to the parameters N, n, and k or, more briefly, the V-set, is defined to be the set of numbers of the form $v = p_1 p_2 \dots p_k$.

It is obvious that $v \ne v'$ if $p_j \ne p'_j$ for at least one j, $1 \le j \le k$. Hence the number of solutions of the equation $xy = z$, where x and y are unknown v-numbers and z is a given positive number, does not exceed 2^k. Let $\|V\|$ be the number of elements of V, i.e., the number of all v-numbers. We have

$$\|V\| = \prod_{j=1}^{k} \big(\pi(Y_j) - \pi(X_j)\big).$$

Using de la Vallée-Poussin's asymptotic formula

$$\pi(x) = \int_2^x \frac{dt}{\log t} + O\big(x \exp\big\{ -c_1 \sqrt{\log x}\,\big\}\big),$$

we obtain

$$\frac{1}{8k} \frac{Y_j}{\log Y_j} \le \pi(Y_j) - \pi(X_j) \le \frac{1}{2k} \frac{Y_j}{\log Y_j}, \qquad 4^{-k} M \le \|V\| \le M,$$

where

$$M = \left(\frac{n}{2k \log N}\right)^k \left(1 + \frac{1}{n-1}\right)^{k(k-1)/2} N^{1-(1-1/n)^k}.$$

An upper bound V_2 and a lower bound V_1 of v are given by the formula

$$V_1 = \left(1 - \frac{1}{4k}\right)^k N^{-1-(1-1/n)^k} \le v \le N^{1-(1-1/n)^k} = V_2.$$

Lemma 9.10. *Let I be the number of solutions of Eq.* (62) *under the condition that $x_1, \dots, x_k, y_1, \dots, y_k$ take values in the set of v-numbers, and let $2n + 6 < k \ll n$. Then*

$$I \le a(n; k) N^{2k-n+\omega},$$

where

$$a(n;k) = (4k)^{-(k-2n-5)(1.5k+n-41)-(2n+5)(4n+9)},$$
$$\omega = \omega(n;k) = n\left(1-\frac{1}{n}\right)^{k-2n-5} - (k+2n+4)\left(1-\frac{1}{n}\right)^{k} - 1.$$

Proof. 1. For any integer s, $0 \le s \le s_1 = k - 2(n+3)$, we denote by $I^{(s)}$ the number of solutions of the equation

$$x_{s+1}^n + \cdots + x_k^n = y_{s+1}^n + \cdots + y_k^n. \tag{9.63}$$

Here x_j and y_j, $j = s+1, \ldots, k$, take values in the set of numbers $p_{s+1} \ldots p_k$, where p_j ranges over all prime numbers in $(X_j, Y_j]$. It is obvious that $I^{(0)} = I$.

2. We claim that the following recurrent inequality holds for $I^{(s)}$:

$$I^{(s)} \le 5nH^2(Y_{s+1} - X_{s+1})^{2(k-s-1)} I^{(s+1)} \prod_{j=2}^{k-s}(Y_{s+j} - X_{s+j}), \tag{9.64}$$

where $H = (2k^5)^4$.

3. We write $I^{(s)}$ as an integral of the corresponding trigonometric sum. Let

$$S(\alpha) = \sum_{X_{s+1}<p_{s+1}\le Y_{s+1}} \cdots \sum_{X_k<p_k\le Y_k} \exp\{2\pi i\alpha(p_{s+1}\ldots p_k)^n\}.$$

Then

$$I^{(s)} = \int_0^1 |S(\alpha)|^{2(k-s)}\, d\alpha. \tag{9.65}$$

4. We transform $S(\alpha)$. Introducing the new trigonometric sums

$$S(\alpha; p_{s+1}) = \sum_{X_{s+2}<p_{s+2}\le Y_{s+2}} \cdots \sum_{X_k<p_k\le Y_k} \exp\{2\pi i\alpha(p_{s+1}\ldots p_k)^n\},$$

we obtain

$$S(\alpha) = \sum_{X_{s+1}<p_{s+1}\le Y_{s+1}} S(\alpha; p_{s+1}).$$

We put $H = (2k^5)^4$, $H_{s+1} = (Y_{s+1} - X_{s+1})H^{-1}$, and partition the interval $(X_{s+1}, Y_{s+1}]$ into H intervals j_r of the form

$$j_r : \big(X_{s+1} + (r-1)H_{s+1}, X_{s+1} + rH_{s+1}\big], \quad r = 1, 2, \ldots, H.$$

According to this partition, we represent $S(\alpha)$ as a sum of H terms,

$$S(\alpha) = \sum_{r=1}^{H} S_r(\alpha), \tag{9.66}$$

where

$$S_r(\alpha) = \sum_{p_{s+1} \in j_r} S(\alpha;\, p_{s+1}). \tag{9.67}$$

5. We transform $S^{k-s}(\alpha)$. Raising both sides of (9.66) to the power $k-s$, we obtain the equality

$$S^{k-s}(\alpha) = \sum_{r_1=1}^{H} \cdots \sum_{r_{k-s}=1}^{H} S_{r_1}(\alpha) \dots S_{r_{k-s}}(\alpha). \tag{9.68}$$

We partition all sets of numbers of the form $(r_1, \dots, r_{k-s})$, $1 \le r_1, \dots, r_{k-s} \le H$, into classes A and B as follows. Class A contains the sets $(r_1, \dots, r_{k-s})$ in which there is an r_j that is different from all other r_ν, $j \ne \nu$, i.e., $r_j \ne r_\nu$ if $\nu \ne j$. Class B contains all other sets. By $\sum_1$ we denote the sum on the right-hand side of (9.68) taken over the sets $(r_1, \dots, r_{k-s})$ of class A. The symbol $\sum_2$ will stand for the sum on the right-hand side of (9.68) taken over the sets $(r_1, \dots, r_{k-s})$ of class B. Then

$$\sum\nolimits_1 = \sum_{(r_1,\dots,r_{k-s}) \in A} \cdots \sum S_{r_1}(\alpha) \dots S_{r_{k-s}}(\alpha),$$

$$\sum\nolimits_2 = \sum_{(r_1,\dots,r_{k-s}) \in B} \cdots \sum S_{r_1}(\alpha) \dots S_{r_{k-s}}(\alpha).$$

Relation (9.68) can be written as

$$S^{k-s}(\alpha) = \sum\nolimits_1 + \sum\nolimits_2. \tag{9.69}$$

6. We transform the terms in $\sum_1$ and $\sum_2$. Changing the indexing of $(r_1, \dots, r_{k-s})$, we can write the terms of $\sum_1$ as

$$S_{r_1}(\alpha) S_{r_2}(\alpha) \dots S_{r_{k-s}}(\alpha),$$

where $r_1 \ne r_\nu$, $\nu = 2, \dots, k-s$. Changing the indexing of $(r_1, \dots, r_{k-s})$, we can write the terms of $\sum_2$ as

$$S_{r_1}^{\beta_1}(\alpha) \dots S_{r_t}^{\beta_t}(\alpha), \tag{9.70}$$

where $r_\nu \ne r_j$, $\nu \ne j$, $\beta_1 \ge 2, \dots, \beta_t \ge 2$, and

$$\beta_1 + \dots + \beta_t = k - s. \tag{9.71}$$

7. Let $\|A\|$ and $\|B\|$ be the number of sets $(r_1, \dots, r_{k-s})$ of class A and of class B, respectively. We have the following trivial upper bound for $\|A\|$: $\|A\| \le H^{k-s}$.

Let us estimate $\|B\|$. Since the numbers $\beta_1, \dots, \beta_t$ in (9.71) are larger than or equal to 2, we have $t \le 0.5(k-s)$. Let $r_1, \dots, r_t$ and $\beta_1, \dots, \beta_t$ be given. This means that in the set $(r_1, r_2, \dots, r_{k-s})$, r_1 occurs β_1 times, r_2 occurs β_2 times, ..., and r_t occurs β_t times. Therefore, in $k-s$ places, r_1 occupies β_1 places, which can be done in

$$(k-s)(k-s-1)\dots(k-s-\beta_1+1) < (k-s)^{\beta_1}$$

ways, r_2 occupies β_2 places, which can be done in at most $(k-s)^{\beta_2}$ ways, ..., and, finally, r_t occupies β_t places, which can be done in at most $(k-s)^{\beta_t}$ ways. Hence the number of terms of the form (9.70) for given $r_1, \dots, r_t$ and $\beta_1, \dots, \beta_t$ is less than or equal to

$$(k-s)!(k-s)^{\beta_1+\dots+\beta_t} < (k-s)^{2(k-s)}.$$

Further, if $r(t)$ is the number of solutions of Eq. (9.71) in $\beta_1, \dots, \beta_t$, then $r(t) \le (k-s)^{t-1}$. Hence

$$\|B\| \le \sum_{1\le t\le 0.5(k-s)} H^t r(t)(k-s)^{2(k-s)} \le 2(k-s)^{2.5(k-s)} H^{0.5(k-s)}.$$

8. We pass to inequalities in (9.69). Squaring both sides of this inequality and using the Cauchy inequality, we obtain

$$\begin{aligned} |S(\alpha)|^{2(k-s)} &\le 2\|A\| \sum_{(r_1,\dots,r_{k-s})\in A}\cdots\sum |S_{r_1}(\alpha)|^2 \dots |S_{r_{k-s}}(\alpha)|^2 \\ &\quad + 2\|B\| \sum_{(r_1,\dots,r_{k-s})\in B}\cdots\sum |S_{r_1}(\alpha)|^2 \dots |S_{r_{k-s}}(\alpha)|^2. \end{aligned} \tag{9.72}$$

We transform the terms on the right-hand side of this inequality as follows. We apply the inequality between arithmetic and geometric means to the products in the second multiple sum. Applying this inequality to the products of all factors but the first in the terms of the first multiple sum, we obtain the following inequality for the terms of the first sum:

$$\begin{aligned} &|S_{r_1}(\alpha)|^2 |S_{r_2}(\alpha)|^2 \dots |S_{r_{k-s}}(\alpha)|^2 \\ &\quad \le \frac{1}{k-s-1} |S_{r_1}(\alpha)|^2 \left(|S_{r_2}(\alpha)|^{2(k-s-1)} + \dots + |S_{r_{k-s}}(\alpha)|^{2(k-s-1)}\right). \end{aligned} \tag{9.73}$$

Here $r_1 \ne r_2$, $j = 2, \dots, k-s$.

We obtain the following inequality for the terms of the second sum:

$$\begin{aligned} &|S_{r_1}(\alpha)|^2 \dots |S_{r_{k-s}}(\alpha)|^2 \\ &\quad \le \frac{1}{k-s} \left(|S_{r_1}(\alpha)|^{2(k-s)} + \dots + |S_{r_{k-s}}(\alpha)|^{2(k-s)}\right). \end{aligned} \tag{9.74}$$

Substituting (9.73) and (9.74) into (9.72) and then substituting (9.72) into (9.65), we obtain the inequality

$$I^{(s)} \le 2\|A\|^2 I_1^{(s)} + 2\|B\|^2 I_2^{(s)}, \tag{9.75}$$

where

$$I_1^{(s)} = \int_0^1 |S_{r_1}(\alpha)|^2 |S_{r_2}(\alpha)|^{2(k-s-1)}\, d\alpha, \tag{9.76}$$

$$I_2^{(s)} = \int_0^1 |S_r(\alpha)|^{2(k-s)}\, d\alpha. \tag{9.77}$$

The r_1, r_2, and r in (9.75) are fixed positive integers such that $r_1 \neq r_2$, $1 \leq r_1, r_2, r_3 \leq H$, and the integrals $I_1^{(s)}$ and $I_2^{(s)}$ take their maximal values. Let us also note that $I_2^{(s)}$ has the same form as $I^{(s)}$, with the only difference that the range of p_{s+1} in $S(\alpha)$, which has the form

$$X_{s+1} < p_{s+1} \leq Y_{s+1},$$

is replaced in $S_r(\alpha)$ by the shorter interval

$$X_{s+1}^{(1)} = X_{s+1} + (r-1)H_{s+1} < p_{s+1} \leq X_{s+1} + rH_{s+1} = Y_{s+1}^{(1)},$$

whose length is equal to

$$H_{s+1} = Y_{s+1}^{(1)} - X_{s+1}^{(1)} = (Y_{s+1} - X_{s+1})H^{-1}.$$

9. We transform the integral $I_1^{(s)}$. Since

$$S_{r_2}(\alpha) = \sum_{X_{s+1}^{(1)} < p_{s+1} \leq Y_{s+1}^{(1)}} S(\alpha;\, p_{s+1}),$$

we can pass to inequalities. Applying Hölder's inequality, we obtain

$$\begin{aligned} |S_{r_2}(\alpha)|^{2(k-s-1)} &\leq H_{s+1}^{2(k-s-1)-1} \sum_{X_{s+1}^{(1)} < p_{s+1} \leq Y_{s+1}^{(1)}} |S(\alpha;\, p_{s+1})|^{2(k-s-1)}, \\ I_1^{(s)} &\leq H_{s+1}^{2(k-s-1)} \int_0^1 |S_{r_1}(\alpha)|^2 |S(\alpha;\, p)|^{2(k-s-1)}\, d\alpha. \end{aligned} \tag{9.78}$$

In the last integral p stands for some fixed prime number in $(X_{s+1}^{(1)}, Y_{s+1}^{(1)}]$, i.e.,

$$X_{s+1}^{(1)} = X_{s+1} + (r_2 - 1)H_{s+1} < p \leq X_{s+1} + r_2 H_{s+1} = Y_{s+1}^{(1)},$$

where $r_2 \neq r_1$ and p is such that this integral takes its maximal value.

10. Let

$$I = \int_0^1 |S_{r_1}(\alpha)|^2 |S(\alpha;\, p)|^{2(k-s-1)}\, d\alpha.$$

It is obvious that I is equal to the number of solutions of the equation

$$x_{s+1}^n - y_{s+1}^n = p^n (x_{s+2}^n + \cdots + x_k^n - y_{s+2}^n - \cdots - y_k^n).$$

In this equation the unknowns x_{s+1} and y_{s+1} take values in the set of numbers of the form $p_{s+1} \dots p_k$, and the unknowns $x_{s+2}, \dots, x_k, y_{s+2}, \dots, y_k$ take value in the set of numbers of the form $p_{s+2} \dots p_k$. Here p_{s+1} ranges over the interval

$$X_{s+1} + (r_1 - 1)H_{s+1} < p_{s+1} \le X_{s+1} + r_1 H_{s+1},$$

and $p_{s+2}, \dots, p_k$ range over the intervals

$$X_j < p_j \le Y_j, \quad j = s+2, \dots, k.$$

Since $r_2 \ne r_1$, we have $p \ne p_{s+1}$.

Let T be the number of solutions of the congruence

$$x_{s+1}^n \equiv y_{s+1}^n \pmod{p^n} \tag{9.79}$$

in x_{s+1}, y_{s+1}. Then

$$I \le T I_1, \tag{9.80}$$

where I_1 is the number of solutions of the form $p_{s+2} \dots p_k$ of the equation

$$x_{s+2}^n + \dots + x_k^n = y_{s+2}^n + \dots + y_k^n$$

in $x_j, y_j, j = s+2, \dots, k$. It is obvious that $I_1 = I^{(s+1)}$.

11. We find an upper bound for T. By definition, x_{s+1} and y_{s+1} have the form $p_{s+1} \dots p_k$, where $(p, x_{s+1}) = (p, y_{s+1}) = 1$ and $X_{s+1} < p \le Y_{s+1}$. The upper bounds of the range of $p_{s+1}, \dots, p_k$ imply the following upper bound for x_{s+1} and y_{s+1}:

$$\max(x_{s+1}, y_{s+1}) \le Y_{s+1} \dots Y_k = N^{(1-1/n)^s - (1-1/n)^k}.$$

The following lower estimate holds for p^n:

$$p^n > X_{s+1}^n = \left(1 - \frac{1}{4k}\right)^n N^{(1-1/n)^s}$$

The assumptions on N, k, n and s imply that

$$N^{(1-1/n)^s - (1-1/n)^k} < \left(1 - \frac{1}{4k}\right)^n N^{(1-1/n)^s}.$$

Hence the unknowns x_{s+1} and y_{s+1} in the congruence (9.79) take values in a subset of the reduced residue system modulo p^n. Therefore, for fixed y_{s+1}, the congruence (9.79) has at most n solutions. Hence,

$$T \le n(Y_{s+1} - X_{s+1}) \dots (Y_k - X_k). \tag{9.81}$$

12. We continue the estimates. Formulas (9.80) and (9.81) imply

$$I \le n(Y_{s+1} - X_{s+1}) \dots (Y_k - X_k) I^{(s+1)}.$$

Formula (9.78) implies

$$I_1^{(s)} \le H_{s+1}^{2(k-s-1)} I \le n H_{s+1}^{2(k-s-1)} (Y_{s+1} - X_{s+1}) \dots (Y_k - X_k) I^{(s+1)}. \tag{9.82}$$

Finally, combining the estimates for $\|A\|$ and $\|B\|$ with formulas (9.75)–(9.78) and (9.82), we obtain the first main estimate for $I^{(s)}$:

$$\begin{aligned} I^{(s)} &\le 2nH^{2(k-s)} H_{s+1}^{2(k-s-1)} (Y_{s+1} - X_{s+1}) \dots (Y_k - X_k) I^{(s+1)} \\ &\quad + 2(k-s)^{5(k-s)} H^{k-s} I_2^{(s)} \\ &= 2nH^{2(k-s)} H_{s+1}^{2(k-s-1)} (Y_{s+1} - X_{s+1}) L I^{(s+1)} \\ &\quad + 2(k-s)^{5(k-s)} H^{k-s} I_2^{(s)}. \end{aligned} \tag{9.83}$$

In this inequality H_{s+1} and L are given by the formulas

$$H_{s+1} = (Y_{s+1} - X_{s+1}) H^{-1}, \quad L = (Y_{s+2} - X_{s+2}) \dots (Y_k - X_k),$$

and $I_2^{(s)}$ has the same form as $I^{(s)}$, with the only difference that the numbers X_{s+1} and Y_{s+1} that occur in the definition of $I^{(s)}$ (the bounds of the range of the primes p_{s+1}) are replaced by new ones, namely, by $X_{s+1}^{(1)}$ and $Y_{s+1}^{(1)}$, where

$$X_{s+1}^{(1)} = X_{r+1} + (r-1) H_{s+1}, \quad Y_{s+1}^{(1)} = X_{s+1} + r X_{s+1},$$

and r is some fixed integer, $1 < r \le H$.

13. Using the notation

$$\begin{gathered} I^{(s)} = I_{2,0}^{(s)}, \quad I_2^{(s)} = I_{2,1}^{(s)}, \quad Y_{s+1} = Y_{s+1}^{(0)}, \quad X_{s+1} = X_{s+1}^{(0)}, \\ H_{s+1} = H_{s+1}^{(0)} = (Y_{s+1}^{(0)} - X_{s+1}^{(0)}) H^{-1}, \quad X_{s+1}^{(1)} = X_{s+1}^{(0)} + (r_0 - 1) H_{s+1}^{(0)}, \\ Y_{s+1}^{(1)} = X_{s+1}^{(0)} + r_0 H_{s+1}^{(0)}, \end{gathered}$$

we write inequality (9.83) as

$$\begin{aligned} I_{2,0}^{(s)} &\le 2nH^{2(k-s)} (H_{s+1}^{(0)})^{2(k-s-1)} (Y_{s+1}^{(0)} - X_{s+1}^{(0)}) L I^{(s+1)} \\ &\quad + 2(k-s)^{5(k-s)} H^{k-s} I_{2,1}^{(s)}. \end{aligned} \tag{9.84}$$

14. We now define a positive integer μ by the inequalities

$$H^{\mu+2} < Y_{s+1} - X_{s+1} \le H^{\mu+3}.$$

For $j = 0, 1, \dots, \mu$, we define the parameters $X_{s+1}^{(j+1)}$, $Y_{s+1}^{(j+1)}$, and $H_{s+1}^{(j)}$ by the relations

$$X_{s+1}^{(j+1)} = X_{s+1}^{(j)} + (r_j - 1) H_{s+1}^{(j)}, \quad Y_{s+1}^{(j+1)} = X_{s+1}^{(j)} + r_j H_{s+1}^{(j)},$$

$$H_{s+1}^{(j)} = (Y_{s+1}^{(j)} - X_{s+1}^{(j)})H^{-1}.$$

Here the r_j are integers (for example, $r_0 = r$) that occur in the corresponding iteration process, $1 \le r_j \le H$. Repeating the arguments used in items 3–12, we obtain a relation similar to (9.84):

$$I_{2,j}^{(s)} \le 2nH^{2(k-s)}(H_{s+1}^{(j)})^{2(k-s-1)}(Y_{s+1}^{(j)} - X_{s+1}^{(j)})LI^{(s+1)} + 2(k-s)^{5(k-s)}H^{k-s}I_{2,j+1}^{(s)}. \tag{9.85}$$

15. We find an upper bound for $I_{2,\mu+1}^{(s)}$. We estimate this integral in the following trivial way:

$$\begin{aligned} I_{2,\mu+1}^{(s)} &= \int_0^1 \Bigg| \sum_{X_{s+1}^{(\mu+1)} < p_{s+1} \le Y_{s+1}^{(\mu+1)}} S(\alpha; p_{s+1}) \Bigg|^{2(k-s)} d\alpha \\ &\le (Y_{s+1}^{(\mu+1)} - X_{s+1}^{(\mu+1)})^{2(k-s)} \int_0^1 |S(\alpha; p)|^{2(k-s)}\, d\alpha. \end{aligned} \tag{9.86}$$

The last integral is equal to the number of solutions of the equation

$$x_{s+1}^n + \cdots + x_k^n = y_{s+1}^n + \cdots + y_k^n,$$

where the unknowns x_j and y_j, $j = s+1, \ldots, k$, take values in the set of numbers of the form $p_{s+2} \ldots p_k$. It is obvious that

$$\int_0^1 |S(\alpha; p)|^{2(k-s)}\, d\alpha \le (Y_{s+1} - X_{s+1})^2 \ldots (Y_k - X_k)^2 I^{(s+1)}. \tag{9.87}$$

The definition of the parameters μ, $Y_{s+1}^{(j)}$, and $X_{s+1}^{(j)}$ implies that

$$Y_{s+1}^{(\mu+1)} - X_{s+1}^{(\mu+1)} = (Y_{s+1} - X_{s+1})H^{-\mu-1} \le H^2.$$

Formulas (9.86) and (9.87) imply the desired estimate for $I_{2,\mu+1}^{(s)}$:

$$I_{2,\mu+1}^{(s)} \le H^{4(k-s)}(Y_{s+1} - X_{s+1})^2 \ldots (Y_k - X_k)^2 I^{(s+1)}. \tag{9.88}$$

16. Let a be the coefficient of $I_{2,j+1}^{(s)}$ on the right-hand side of inequality (9.85):

$$a = 2(k-s)^{5(k-s)}H^{k-s}.$$

Multiplying both sides of (9.85) by a^j, we obtain

$$a^j I_{2,j}^{(s)} \le 2nH^{2(k-s)}LI^{(s+1)}a^j(H_{s+1}^{(j)})^{2(k-s-1)} + a^{j+1}I_{2,j+1}^{(s)}. \tag{9.89}$$

Summing both sides of (9.89) over $j = 0, s, \dots, \mu$, we obtain the inequality

$$I_{2,0}^{(s)} + \sum_{j=1}^{\mu} a^j I_{2,j}^{(s)} \le 2nH^{2(k-s)} L I^{s+1} \sum_{j=0}^{\mu} a^j (H_{s+1}^{(j)})^{2(k-s-1)} + \sum_{j=1}^{\mu} a^j I_{2,j}^{(s)} + a^{\mu+1} I_{2,\mu+1}^{(s)}. \tag{9.90}$$

Hence, using the formula $I_{2,0}^{(s)} = I^{(s)}$ and the estimate (9.88), we obtain

$$I^{(s)} \le (2nH^{2(k-s)} L V_1 + V_2) I^{(s+1)}, \tag{9.91}$$

where

$$V_1 = \sum_{j=0}^{\mu} a^j (H_{s+1}^{(j)})^{2(k-s-1)}, \quad V_2 = a^{\mu+2} H^{4(k-s)} (Y_{s+1} - X_{s+1})^2 \dots (Y_k - X_k)^2.$$

17. We find an upper bound for V_1. Since

$$H = (2k^5)^4, \quad a = 2(k-s)^{5(k-s)} H^{k-s} \le (2k^5)^{k-s} H^{k-s},$$
$$H_{s+1}^{(j)} = (Y_{s+1} - X_{s+1}) H^{-j-1},$$

we can estimate the summands in V_1 as follows:

$$\begin{aligned}
&a^j (H_{s+1}^{(j)})^{2(k-s-1)} \\
&\quad \le (2k^5)^{j(k-s)} H^{j(k-s)} (Y_{s+1} - X_{s+1})^{2(k-s-1)} H^{-2(k-s-1)} H^{-j(2k-2s-2)} \\
&\quad = (Y_{s+1} - X_{s+1})^{2(k-s-1)} H^{-2(k-s-1)} \big((2k^5)^{k-s} H^{-(k-s)+2}\big)^j \\
&\quad = (Y_{s+1} - X_{s+1})^{2(k-s-1)} H^{-2(k-s-1)} \big((2k^5)^{-3(k-s)+8}\big)^j.
\end{aligned}$$

Since $k - s \ge 2(n+3)$, we have

$$V_1 \le 2H^{-2(k-s-1)} (Y_{s+1} - X_{s+1})^{2(k-s-1)}.$$

We claim that

$$2nH^{2(k-s)} L \cdot 2H^{-2(k-s-1)} (Y_{s+1} - X_{s+1})^{2(k-s-1)} \le 4nH^2 (Y_{s+1} - X_{s+1})^{2(k-s-1)} \prod_{j=2}^{k-s} (Y_{s+j} - X_{s+j}), \tag{9.92}$$

$$a^{\mu+2} H^{4(k-s)} (Y_{s+1} - X_{s+1})^2 \dots (Y_k - X_k)^2 \le nH^2 (Y_{s+1} - X_{s+1})^{2(k-s-1)} \prod_{j=2}^{k-s} (Y_{s+j} - X_{s+j}). \tag{9.93}$$

Let us recall that

$$L = (Y_{s+2} - X_{s+2}) \dots (Y_k - X_k),$$
$$H^{\mu+2} < Y_{s+1} - X_{s+1} \le H^{\mu+3}, \quad a = 2(k-s)^{5(k-s)} H^{k-s}.$$

It is obvious that relation (9.92) is an equality. Dividing inequality (9.93) by the common factor, we obtain the inequality

$$a^{\mu+2} H^{4(k-s)-2} (Y_{s+2} - X_{s+2}) \dots (Y_k - X_k) \le n(Y_{s+1} - X_{s+1})^{2(k-s)-4}.$$

Replacing $a^{\mu+2}$ by a greater quantity, we claim that a stronger inequality holds:

$$\big(2(k-s)^{5(k-s)}\big)^{\mu+2} (Y_{s+1} - X_{s+1})^{k-s} H^{4(k-s)-2} (Y_{s+2} - X_{s+2}) \dots (Y_k - X_k)$$
$$\le n(Y_{s+1} - X_{s+1})^{2(k-s)-4}.$$

Since $2k^5 = H^{1/4}$, we have $(2(k-s)^5)^{k-s} \le H^{(k-s)/4}$. Besides, $H^{\mu+2} < Y_{s+1} - X_{s+1}$. We claim that an even stronger inequality holds:

$$(Y_{s+1} - X_{s+1})^{5(k-s)/4} H^{4(k-s)-2} (Y_{s+2} - X_{s+2}) \dots (Y_k - X_k)$$
$$\le n(Y_{s+2} - X_{s+2})^{2(k-s)-4}.$$

We even claim that

$$H^{4(k-s)} (Y_{s+2} - X_{s+2}) \dots (Y_k - X_k) \le (Y_{s+1} - X_{s+1})^{3(k-s)/4-4}. \tag{9.94}$$

Replacing Y_j and X_j by their values, we obtain the formulas

$$Y_j - X_j = (4k)^{-1} Y_j = (4k)^{-1} N^{(1/n)(1-1/n)^{j-1}},$$
$$(Y_{s+2} - X_{s+2}) \dots (Y_k - X_k) = \big((4k)^{-1}\big)^{k-s-1} N^{(1/n)\left((1-1/n)^{s+1} + \dots + (1-1/n)^{k-1}\right)},$$
$$(Y_{s+1} - X_{s+1})^{3(k-s)/4-4} = \big((4k)^{-1}\big)^{3(k-s)/4-4} N^{(1/n)(1-1/n)^s (3(k-s)/4-4)}.$$

Now we can rewrite formula (9.94) as

$$H^{4(k-s)} \big((4k)^{-1}\big)^{(k-s)/4+3} \le N^{\varkappa}, \tag{9.95}$$

where

$$\varkappa = \frac{1}{n}\left(1 - \frac{1}{n}\right)^s \left(\frac{3}{4}(k-s) - 4\right) - \frac{1}{n}\left(\left(1 - \frac{1}{n}\right)^{s+1} + \dots + \left(1 - \frac{1}{n}\right)^{k-1}\right)$$
$$= \left(1 - \frac{1}{n}\right)^s \left(\frac{3(k-s)}{4n} - \frac{4}{n} - 1 + \frac{1}{n} + \left(1 - \frac{1}{n}\right)^{k-s}\right).$$

We strengthen (9.95) once more:

$$H^{4(k-s)} \le N^{(1-1/n)^s (3(k-s)/(4n) - (n+3)/n)}.$$

Hence it is sufficient to prove the inequality

$$H^4 \le N^{(1-1/n)^s\,(3/(4n)-(n+3)(n(k-s)))}.$$

Taking into account that $k-s \ge 2(n+3)$, we claim that

$$H^4 \le N^{(1-1/n)^k\,1/(4n)}. \tag{9.96}$$

Recalling that $N \asymp \log X$, $k \asymp \log^c X$, $n \asymp \log^c X$, $H = (2k^5)^4$, and $0 < c < 1$, we establish that (9.96) holds with $X \ge X_1(c) > 0$.

Hence inequalities (9.92) and (9.93) hold. Combining them with formula (9.91), we obtain the second main formula

$$I^{(s)} \le 5nH^2(Y_{s+1}-X_{s+1})^{2(k-s-1)}I^{(s+1)}\prod_{j=2}^{k-s}(Y_{s+j}-X_{s+j}),$$

which coincides with inequality (9.64) in item 2.

18. Taking the product of (9.64) over $s = 0, 1, \dots, s_1$, we obtain the inequality

$$I^{(0)}\prod_{s=1}^{s_1}I^{(s)} \le I^{s_1+1}\Big(\prod_{s=1}^{s_1}I^{(s)}\Big)(5nH^2)^{s_1+1}$$
$$\times\Big(\prod_{s=0}^{s_1}(Y_{s+1}-X_{s+1})^{2(k-s-1)}\Big)\Big(\prod_{s=0}^{s_1}\prod_{j=2}^{k-s}(Y_{s+j}-X_{s+j})\Big).$$

Dividing both sides of this inequality by the corresponding product, we obtain the estimate

$$I^{(0)} \le ABCD, \tag{9.97}$$

where $A = (5nH^2)^{s_1+1}$,

$$B = \prod_{s=0}^{s_1}(Y_{s+1}-X_{s+1})^{2(k-s-1)},\quad C = \prod_{s=0}^{s_1}\prod_{j=2}^{k-s}(Y_{s+j}-X_{s+j}),$$

and $D = I^{(s_1+1)}$. Using the trivial estimate (all unknowns except one are fixed)

$$I^{(s)} \le \Big(\prod_{j=s+1}^{k}(Y_j-X_j)\Big)^{2(k-s)-1},$$

we obtain the inequality

$$D = I^{(s_1+1)} \le \Big(\prod_{j=s_1+2}^{k}(Y_j-X_j)\Big)^{2(k-s_1-1)-1}.$$

Since we have $H = (2k^5)^4$, $s_1 = k - 2(n+3) \geq 1$, and

$$Y_j - X_j = (4k)^{-1} N^{(1/n)\,(1-1/n)^{j-1}},$$

we obtain the following inequalities for A, B, C, and D:

$$A < (1280k^{41})^{k-2n-5} < 4^{6(k-2n-5)} k^{41(k-2n-5)},$$

$$B = \prod_{s=0}^{s_1} (4k)^{-2(k-s-1)} N^{2(k-s-1)/n(1-1/n)^s}$$

$$= (4k)^{-2k(s_1+1)+(s_1+1)(s_1+2)} N^{2k-2n-(2n+10)(1-1/n)^{s_1+1}},$$

$$C = \prod_{s=0}^{s_1} \prod_{j=2}^{k-s} (4k)^{-1} N^{(1/n)\,(1-1/n)^{s+j-1}}$$

$$= \prod_{s=0}^{s_1} (4k)^{-(k-s-1)} N^{(1-1/n)^{s+1}-(1-1/n)^k}$$

$$= (4k)^{-k(s_1+1)+(s_1+1)(s_1+2)/2} N^{n-1-n(1-1/n)^{s_1+2}-(s_1+1)(1-1/n)^k},$$

$$D = \Big(\prod_{j=s_1+2}^{k} (4k)^{-1} N^{(1/n)\,(1-1/n)^{j-1}} \Big)^{4n+9}$$

$$= (4k)^{-(2n+5)(4n+9)} N^{(4n+9)\left((1-1/n)^{s_1+1}-(1-1/n)^k\right)}.$$

Combining these estimates for A, B, C, and D with formula (9.97), we obtain the final inequality for $I^{(0)} = I$:

$$I = I^{(0)} < a(k;n) N^{2k-n+\omega},$$

where

$$a(k;n) = (4k)^{-(k-2n-5)(1.5k+n-41)-(2n+5)(4n+9)},$$

$$\omega = \omega(k;n) = n\left(1 - \frac{1}{n}\right)^{k-2n-5} - (k+2n+4)\left(1 - \frac{1}{n}\right)^k - 1,$$

which completes the proof of the lemma. □

Remark 9.1. Taking into account the assumptions $k > 2(n+3)$ and $n \asymp \log^c X$, we obtain a simple version of the estimate for $a(k;n)$:

$$a(k;n) < \exp\Big\{ -\frac{1}{8} k^2 \log k \Big\}.$$

Applying Lemma 9.10, we estimate the double trigonometric sum $W(\alpha)$ defined in the following lemma.

Lemma 9.11. *Let α be a real number, and let*

$$\alpha = \frac{a}{q} + \frac{\theta}{q^2},$$

where $q \geq 3$, $(a, q) = 1$, and $|\theta| \leq 1$. Consider the double trigonometric sum

$$W(\alpha) = \sum_{x\in V} \sum_{y\in V} \exp\{2\pi i\alpha m x^n y^n\},$$

where x and y range over the V-set corresponding to the parameters N, n, and k; m is a positive integer. Then

$$|W(\alpha)| \ll \|V^2\|^2 \Delta,$$

where

$$\Delta_0 = (m \log q)^{1/(4k^2)} (N^{\omega_1} q^{-1/(4k^2)} + N^{\omega_2} q^{1/(4k^2)}) \log N,$$

$$\omega_1 = -\frac{1}{2k^2} + \frac{n}{2k^2}\left(1 - \frac{1}{n}\right)^{k-2n-5} + \left(\frac{1}{2k} - \frac{3n+4}{2k^2}\right)\left(1 - \frac{1}{n}\right)^k,$$

$$\omega_2 = -\frac{n+1}{2k^2} + \frac{n}{2k^2}\left(1 - \frac{1}{n}\right)^{k-2n-5} + \left(\frac{1}{2k} - \frac{n+2}{k^2}\right)\left(1 - \frac{1}{n}\right)^k.$$

Proof. Using Hölder's inequality, we obtain

$$\begin{aligned} |W(\alpha)|^{2k} &\leq \Big(\sum_{x\in V} \Big| \sum_{y\in V} \exp\{2\pi i\alpha m x^n y^n\}\Big|\Big)^{2k} \qquad (9.98) \\ &\leq \|V\|^{2k-1} \sum_{x\in V} \Big| \sum_{y\in V} \exp\{2\pi i\alpha m x^n y^n\}\Big|^{2k} \\ &= \|V\|^{2k-1} \sum_{x\in V} \sum_{\lambda} I_k(\lambda) \exp\{2\pi i\alpha m x^n \lambda\}, \end{aligned}$$

where $I_k(\lambda)$ stands for the number of solutions of the equation

$$y_1^n + \cdots + y_k^n = y_{k+1}^n + \cdots + y_{2k}^n + \lambda \qquad (9.99)$$

in $y_1, \ldots, y_{2k} \in V$. Since $V_1 \leq y_j \leq V_2$ for $1 \leq j \leq 2k$, $I_k(\lambda)$ is equal to zero if $|\lambda| > \Lambda = k(V_2^n - V_1^n)$. For this reason, we assume that $|\lambda| \leq \Lambda$ in (9.98). Interchanging the order of summation in (9.98), we obtain the formula

$$|W(\alpha)|^{2k} \leq \|V\|^{2k-1} \sum_{|\lambda|\leq\Lambda} I_k(\lambda) \Big| \sum_{x\in V} \exp\{2\pi i\alpha m\lambda x^n\}\Big|.$$

Raising both sides of this inequality to the $2k$th power and applying Hölder's inequality, we obtain

$$|W(\alpha)|^{4k^2} \leq \|V\|^{2k(2k-1)} \Big(\sum_{|\lambda|\leq\Lambda} I_k(\lambda)\Big)^{2k-1} \times \qquad (9.100)$$

$$\times \sum_{|\lambda|\le\Lambda} I_k(\lambda)\Big|\sum_{x\in V}\exp\{2\pi i\alpha m\lambda x^n\}\Big|^{2k} = \|V\|^{4k^2-2k} W_1 W_2,$$

where

$$W_1 = \Big(\sum_{|\lambda|\le\Lambda} I_k(\lambda)\Big)^{2k-1}, \tag{9.101}$$

$$W_2 = \sum_{|\lambda|\le\Lambda} I_k(\lambda)\Big|\sum_{x\in V}\exp\{2\pi i\alpha m\lambda x^n\}\Big|^{2k}. \tag{9.102}$$

Let us estimate the sums W_1 and W_2. The sum of $I_k(\lambda)$ over all λ is equal to the number of all sets of unknowns $y_1, \dots, y_{2k}$ in Eq. (9.99), i.e.,

$$\sum_{|\lambda|\le\Lambda} I_k(\lambda) = \|V\|^{2k}, \quad W_1 = \|V\|^{4k^2-2k}.$$

Since

$$\begin{aligned} I_k(\lambda) &= \int_0^1 \Big|\sum_{x\in V}\exp\{2\pi i\alpha x^n\}\Big|^{2k}\exp\{-2\pi i\alpha\lambda\}\,d\alpha \\ &\le \int_0^1 \Big|\sum_{x\in V}\exp\{2\pi i\alpha x^n\}\Big|^{2k} d\alpha = I_k(0), \end{aligned}$$

we obtain the following chain of relations for W_2:

$$\begin{aligned} W_2 &\le I_k(0)\sum_{|\lambda|\le\Lambda}\Big|\sum_{x\in V}\exp\{2\pi i\alpha m\lambda x^n\}\Big|^{2k} \\ &= I_k(0)\sum_{|\lambda|\le\Lambda}\sum_{|\mu|\le\Lambda} I_k(\mu)\exp\{2\pi i\alpha m\lambda\mu\} \\ &\le I_k(0)\sum_{|\mu|\le\Lambda} I_k(\mu)\Big|\sum_{|\lambda|\le\Lambda}\exp\{2\pi i\alpha m\mu\lambda\}\Big| \\ &\le I_k^2(0)\sum_{|\mu|\le\Lambda}\Big|\sum_{|\lambda|\le\Lambda}\exp\{2\pi i\alpha m\mu\lambda\}\Big|. \end{aligned} \tag{9.103}$$

Since

$$\Big|\sum_{|\lambda|\le\Lambda}\exp\{2\pi i\beta\lambda\}\Big| \le \min\Big(2\Lambda+1, \frac{1}{2\|\beta\|}\Big)$$

for any real number β, relations (9.103) for W_2 imply

$$W_2 \le I_k^2(0)\sum_{|\mu|\le\Lambda}\min\Big(2\Lambda+1, \frac{1}{2\|\alpha m\mu\|}\Big) \le I_k^2(0)\sum_{|\mu|\le m\Lambda}\min\Big(2\Lambda+1, \frac{1}{2\|\alpha\mu\|}\Big).$$

Applying to this sum the well-known inequality, we obtain

$$W_2 \le 6I_k^2(0)\left(\frac{2m\Lambda+1}{q}+1\right)(2\Lambda+1+q\log q) \le 36I_k^2(0)\frac{m}{q}(\Lambda+q)^2\log q. \tag{9.104}$$

Observing that $I_k(0) = I$, where I is defined and estimated in Lemma 9.10, we deduce from (9.100)–(9.104) that

$$|W(\alpha)|^{4k^2} \le 36\|V\|^{8k^2-4k}I^2\frac{m}{q}(\Lambda+q)^2\log q, \quad |W(\alpha)| \le \|V\|^2\Delta, \tag{9.105}$$

$$\Delta = \left(\frac{36m}{q}(\Lambda+q)^2\log q\right)^{1/(4k^2)} I^{1/(2k^2)}\|V\|^{-1/k}.$$

We obtain the desired estimate for Δ by combining the upper bounds for Λ and I with the lower bound for $\|V\|$:

$$\Lambda < kV_2^n = kN^{n-n(1-1/n)^k}, \quad I < a(k;n)N^{2k-n+\omega},$$

$$\|V\| \ge \left(\frac{n}{8k\log N}\right)^k\left(1+\frac{1}{n-1}\right)^{k(k-1)/2} N^{1-(1-1/n)^k}.$$

Using Vinogradov's sign $\ll$ and taking into account Remark 9.1, following Lemma 9.10, we obtain

$$I^{1/(2k^2)}\|V\|^{-1/k} \ll \frac{k\log N}{n}N^{1/k-n/(2k^2)+\omega/(2k^2)-1/k+(1/k)(1-1/n)^k}$$

$$\ll (\log N)N^{-n/(2k^2)-1/(2k^2)+((n/(2k^2))\,(1-1/n)^{k-2n-5}+(1/(2k)-(n+2)/k^2)\,(1-1/n)^k},$$

$$\Lambda^{1/(2k)}q^{-1/(4k^2)} \ll N^{n/(2k^2)-(n/(2k^2))\,(1-1/n)^k}q^{-1/(4k^2)},$$

$$\Delta \ll \Delta_1 \ll (m\log q)^{1/(4k^2)}(N^{\omega_1}q^{-1/(4k^2)} + N^{\omega_2}q^{1/(4k^2)})\log N,$$

where

$$\begin{aligned}\omega_1 &= \frac{n}{2k^2}\left(1-\frac{1}{n}\right)^{k-2n-5} - \frac{1}{2k^2} + \frac{1}{2k}\left(1-\frac{1}{n}\right)^k \\ &\quad - \frac{n+2}{k^2}\left(1-\frac{1}{n}\right)^k - \frac{n}{2k^2}\left(1-\frac{1}{n}\right)^k \\ &= \frac{n}{2k^2}\left(1-\frac{1}{n}\right)^{k-2n-5} - \frac{1}{2k^2} + \left(\frac{1}{2k}-\frac{3n+4}{2k^2}\right)\left(1-\frac{1}{n}\right)^k,\end{aligned}$$

$$\omega_2 = -\frac{n+1}{2k^2} + \frac{n}{2k^2}\left(1-\frac{1}{n}\right)^{k-2n-5} + \left(\frac{1}{2k}-\frac{n+2}{k^2}\right)\left(1-\frac{1}{n}\right)^k.$$

Combining this with (9.105), we complete the proof of the lemma. □

Remark 9.2. The estimate in Lemma 9.11 is nontrivial only if

$$N^{4k^2\omega_1} < q < N^{-4k^2\omega_2}.$$

Proof of Theorem 9.7. 1. First we claim that for $X \geq X_1(c) > 0$ the equation $[\log^c X] = [\log^c x]$ holds either for all x in $[0.5X, X]$ or for all x in $[X, 2X]$. Indeed, let $\{\log^c X\} > 4\log^{-1+c} X$. Here the symbol $\{\cdot\}$ stands for the fractional part of a number. Then

$$\begin{aligned}(\log 0.5X)^c &= (\log^c X)\left(1 - \frac{\log 2}{\log X}\right)^c = \log^c X - c\frac{\log 2}{\log X}\log^c X + \cdots \\ &= [\log^c X] + \{\log^c X\} - \cdots > [\log^c X].\end{aligned}$$

Hence

$$[\log^c X] < (\log 0.5X)^c \leq \log^c x \leq \log^c X < [\log^c X] + 1$$

for all $0.5X \leq x \leq X$, and

$$[\log^c x] = [\log^c X]$$

for all $x \in [0.5X, X]$. If $\{\log^c X\} \leq \log^{-1+c} X$, then

$$\begin{aligned}(\log 2X)^c &= (\log^c X)\left(1 + \frac{\log 2}{\log X}\right)^c = \log^c X + c(\log 2)\log^{c-1} X + \cdots \\ &= [\log^c X] + \{\log^c X\} + c(\log 2)\log^{c-1} X + \cdots < [\log^c X] + 1.\end{aligned}$$

Therefore,

$$[\log^c X] \leq \log^c X \leq \log^c x \leq (\log 2X)^c < [\log^c X] + 1$$

if $X \leq x \leq 2X$, and

$$[\log^c x] = [\log^c X]$$

if $x \in [X, 2X]$. Hence, without loss of generality, we can assume that X satisfies the condition $[\log^c x] = [\log^c X]$ for all x in $[0.5X, X]$.

2. Setting $n = [\log^c X]$ and $k = 10n$, we define the number N by the formula

$$X = N^{2-2(1-1/n)^k}.$$

Let $V = V(N; n, k)$ be the set of v-numbers corresponding to the parameters N, n, and k. For $0 \leq a < b \leq 1$, by $K = K(X; a, b)$ we denote the number of numbers $z = xy$, where x and y range independently over the subset of V such that

$$a \leq \left\{\alpha \exp\{[\log^c z]\log z\}\right\} < b.$$

3. Suppose that $r = 2[\log X]$, $\Delta = N^{-n/(k^2)}$, the numbers a_1 and b_1 satisfy the conditions $0 \leq a_1 < b_1 \leq 1$ and $\Delta \leq b_1 - a_1 \leq 1 - \Delta$, and $\psi(x)$ is Vinogradov's "cup"

corresponding to the parameters r, Δ, a_1, and b_1 (see Lemma A.3 in the Appendix). Let us find the asymptotics of the sum

$$K_1 = K_1(X; a_1, b_1) = \sum_z \psi\left(\alpha \exp\{[\log^c z] \log z\}\right).$$

4. First, $\psi(x)$ can be expanded into the Fourier series

$$\psi(x) = b_1 - a_1 + \sum_{m \neq 0} g(m) \exp\{2\pi i m x\},$$

where

$$|g(m)| \leq \min\left(b_1 - a_1, \frac{1}{\pi |m|}, \frac{1}{\pi |m|}\left(\frac{r}{\pi |m| \Delta}\right)^r\right).$$

For $|m| > m_1 = 2r\Delta^{-1}$, we use the third estimate for $|g(m)|$ to obtain

$$\Big| \sum_{|m|>m_1} g(m) \exp\{2\pi i m x\}\Big| \leq \frac{2}{\pi}\left(\frac{r}{\pi \Delta}\right)^r \sum_{m>m_1} \frac{1}{m^{r+1}} < \frac{1}{\pi}\left(\frac{r}{\pi \Delta m_1}\right)^r < X^{-3}.$$

Therefore

$$K_1 = (b_1 - a_1)\|V\|^2 + \sum_{0<|m|\leq m_1} g(m) S(m) + \theta \|V\|^2 X^{-3}, \tag{9.106}$$

where

$$S(m) = \sum_z \exp\{2\pi i m f(z)\}, \quad f(z) = \alpha \exp\{[\log^c z] \log z\}.$$

5. Since $z = xy$, $x \in V$, and $y \in V$, we obtain the upper and lower bounds for the range of z:

$$0.5X \leq \left(1 - \frac{1}{4k}\right)^{2k} N^{2-2(1-1/n)^k} \leq z \leq N^{2-2(1-1/n)^k} = X.$$

Therefore,

$$[\log^c z] = [\log^c X] = n$$

for all values of z, whence $f(z) = \alpha z^n = \alpha x^n y^n$. Therefore, $S(m)$ is the sum defined in Lemma 9.11.

6. Since α is a real algebraic number of degree at least 2, α can be represented by an infinite continued fraction. Let P_ν and Q_ν be the numerator and denominator of the νth convergent of the continued fraction for α, $\nu = 1, 2, \ldots$. It is well known that $(P_\nu, Q_\nu) = 1$ and

$$\left|\alpha - \frac{P_\nu}{Q_\nu}\right| < \frac{1}{Q_\nu Q_{\nu+1}}. \tag{9.107}$$

By the Thue–Siegel–Roth theorem, for any $\varepsilon > 0$ there is a $c_5 = c_5(\alpha; \varepsilon) > 0$ such that

$$\left|\alpha - \frac{P_\nu}{Q_\nu}\right| \geq \frac{c_5}{Q_\nu^{2+\varepsilon}} \tag{9.108}$$

for each rational fraction P_ν/Q_ν, $(P_\nu, Q_\nu) = 1$.

Inequalities (9.107) and (9.108) imply that

$$Q_\nu < Q_{\nu+1} \leq c_6 Q_\nu^{1+\varepsilon}. \tag{9.109}$$

We take $\varepsilon = 0.01$, $c_6 = c_6(\alpha) > 0$, and the positive integer ν defined by the inequalities

$$Q_\nu < N^{n-n(1-1/n)^k} \leq Q_{\nu+1}. \tag{9.110}$$

Denoting Q_ν by q and P_ν by a, we obtain

$$\alpha = \frac{a}{q} + \frac{\theta}{q^2}, \quad (a, q) = 1, \quad |\theta| \leq 1.$$

Inequalities (9.109) and (9.110) imply the desired upper and lower estimates for q:

$$c_7 N^{(100/101)\varkappa} \leq q \leq N^{\varkappa}, \quad \varkappa = n - n\left(1 - \frac{1}{n}\right)^k. \tag{9.111}$$

7. Applying Lemma 9.11 to $S(m)$, we obtain

$$|S(m)| \ll (m \log q)^{1/(4k^2)} (N^{\omega_1} q^{-1/(4k^2)} + N^{\omega_2} q^{1/(4k^2)}) \|V\|^2 \log N.$$

Using the definition of $m_1 = 2r\Delta^{-1}$, $\Delta, \omega_1, \omega_2, k, n$, and q and the estimates (9.111), we obtain the following inequality by a simple calculation:

$$\begin{aligned}
(m \log q)^{1/(4k^2)} &\leq (m_1 n \log N)^{1/(4k^2)} \ll (\Delta^{-1} \log^3 X)^{1/(4k^2)} \\
&= (N^{n/(k^2)} \log^3 X)^{1/(4k^2)} \ll N^{n/(4k^4)}, \\
N^{\omega_1} q^{-1/(4k^2)} &\ll N^{\omega_1 - (1/(4k^2))\,(1-1/101)\,(n-n(1-1/n)^k)} \leq N^{-n/(8k^2)}, \\
N^{\omega_2} q^{1/(4k^2)} &\leq N^{\omega_2 + (1/(4k^2))\,(n-n(1-1/n)^k)} \leq N^{-n/(8k^2)}, \\
N^{n/(4k^4)} N^{-n/(8k^2)} &\leq N^{-n/(9k^2)}, \quad |S(m)| \ll N^{-n/(9k^2)} \|V\|^2 \log N.
\end{aligned} \tag{9.112}$$

8. Estimating for $0 < |m| \leq m_1$ the coefficients $g(m)$ of the Fourier series for $\psi(x)$ by $1/(\pi|m|)$, we deduce from (9.106) and (9.112) the following asymptotic formula for K_1:

$$\begin{aligned}
K_1 &= (b_1 - a_1)\|V\|^2 + \theta_1 c_8 \|V\|^2 N^{-n/(9k^2)} \log N + \theta_1 \|V\|^2 X^{-3} \\
&= (b_1 - a_1)\|V\|^2 + \theta_2 \|V\|^2 N^{-n/(10k^2)}.
\end{aligned}$$

9. Using Lemma B in [90], p. 16, with $r = 2[\log X]$, $\Delta = N^{-n/(k^2)}$, $\delta_s = \{\alpha \exp\{[\log^c z]\log z\}\}$, and $s = 1, 2, \dots, \|V\|^2$, and taking into account the asymptotic formula for $K_1 = K_1(X; a_1, b_1)$, we deduce from assertion (a) of this lemma that

$$\begin{aligned} K = K(X; a, b) &= (b-a)\|V\|^2 + O(\|V\|^2 N^{-n/(10k^2)}) + O(\|V\|^2 N^{-n/(k^2)}) \\ &= (b-a)\|V\|^2 + O(\|V\|^2 N^{-n/(10k^2)}). \end{aligned}$$

Hence $K \gg (b-a)\|V\|^2$ and for any a and b, $0 \le a < b \le 1$,

$$b - a \gg N^{-n/(10k^2)} = \exp\{-c_1 \log^{1-c} X\}.$$

As mentioned above, the number of solutions of the equation $xy = z$, $x \in V$, $y \in V$, does not exceed 2^k. Therefore, the number of positive integers $z \le X$ such that

$$a \le \{\alpha \exp\{[\log^c z]\log z\}\} < b$$

is larger than or equal to

$$\begin{aligned} c_8 2^{-k}(b-a)\|V\|^2 &\ge c_9 2^{-10n} N^{-n/(10k^2)} \left(\frac{8k\log N}{n}\right)^{-2k} \\ &\quad \times \left(1 + \frac{1}{n-1}\right)^{k(k-1)} N^{2-2(1-1/n)^k} \\ &\ge X \exp\{-c_{10}(\log^{1-c} X + \log^c X \log\log X)\}. \end{aligned}$$

In particular, this implies that for any real number ξ the number of solutions in positive integers z of the system of inequalities $z \le X$,

$$\|\xi - f(z)\| \le \exp\{-c_1 \log^{1-c} X\},$$

is larger than or equal to

$$X \exp\{-c_{10}(\log^{1-c} X + \log^c X \log\log X)\},$$

where $\min_{z \le X} \|\xi - f(z)\| \le \exp\{-c_1 \log^{1-c} X\}$, which completes the proof of the theorem. □

Remark 9.3. The assertion of the theorem remains valid if α is an irrational number with bounded partial quotients or if the partial quotients α increase but not very fast.

Concluding remarks on Chapter 9. 1. Artin's conjecture on the representation of zero by an nth degree form in k variables, $k > n^2$, in the field Q_p was proved for

$n = 2$ by Minkowski and Hasse and for $n = 3$ by V. B. Dem'yanov [60] and by D. J. Lewis [111].

2. In 1965, Yu. L. Ershov [62] and, independently, J. Ax and S. Kochen [1] proved that, for a given n, Artin's conjecture is true for all p except only finitely many of them.

3. A special notion concerning the problem of representing zero by forms over a given field was introduced: a field K has property C_a if any nth degree form in k variables with coefficients from K for $k > n^a$ can nontrivially represent zero over K. Algebraic closed fields and only these fields have property C_0. Any finite field has property C_1. The field of formal power series $F_p\{t\}$, which resembles the field Q_p very much, has property C_2.

4. A conjecture similar to Artin's conjecture but for a system of forms was rejected by G. I. Arkhipov [7], [8], [9] who proved that, in this case, k must grow exponentially, namely, like 2^n.

5. The statements presented in Section 9.1 and some of their versions were proved by G. I. Arkhipov and A. A. Karatsuba in [19], [20], [21].

6. A short presentation of a version of Theorem 9.2 is can be found in the book by Z. I. Borevich and I. R. Shafarevich [41], pp. 70–73.

7. A concise history of the Waring, Hilbert–Kamke, and Artin problems and of their generalizations is contained in [92].

8. The statements considered in Section 9.2 were proved by A. A. Karatsuba [91].

9. In [91], p. 935, it was pointed out that "...this method allows one to improve the previous results for small values of n (for each particular n, the parameters in Lemmas 1 and 2 (Lemmas 9.8 and 9.9 in this chapter) must be chosen in the optimal way)."

10. If the restriction $n \geq 4000$ under which we prove Theorem 9.5 is removed, i.e., if Theorem 9.5 is proved for $n \to +\infty$, then the number 6 in the estimate $G(n) < 2n(\ln n + \ln\ln n + 6)$ can be replaced by a smaller number.

11. The method used to prove Theorem 9.5 allowed A. A. Karatsuba [93] to prove the following assertion (see [93], p. 322).

Suppose that c_1, c_2 are absolute constants such that $0 < c_1 < c_2 < 1$; $n \geq 2$; $nc_1 \geq 1$; $P \geq P_0(n) > 0$; a real number α has the form $\alpha = a/q + \theta/q^2$, $(a, q) = 1$, $|\theta| \leq 1$, and $P^{c_1 n} \leq q \leq P^{c_2 n}$. Then, for a real number A, there exist integers w and z such that the following relation holds:

$$|\alpha z^n - w - A| \ll P^{-\rho}, \quad 0 < z < P, \quad \rho = \frac{\min(c_1,\ 1 - c_2)}{8(\ln 2/(1 - c_2))^2} \cdot \frac{1}{n},$$

and the constant in $\ll$ depends only on n.

The following example is also given in this paper (see p. 324). Let $\alpha = \sqrt{2}$, then for any $P \geq P_0(n)$ and any real number A, there exist integers w and z such that

$$|\sqrt{2}z^n - w - A| \ll P^{-c/n}, \quad 0 < z < P, \quad c = 1/(16\ln^2 4).$$

12. Theorem 9.7 was proved by A. A. Karatsuba [104].

Chapter 10

Estimates of multiple trigonometric sums with prime numbers

This chapter is concerned with estimates for multiple trigonometric sums with a general polynomial in the exponent whose variables of summation take prime values. Our results generalize Vinogradov's estimates for sums with prime numbers [159], [165] to the r-dimensional case. These results are a new application of the theory of multiple trigonometric sums that was developed in [29], [32], [33] using Karatsuba's p-adic method (see the exercises in Chapter XI in [90]). The precision of our estimates is similar to that of the analogous estimates in [25]. Here we shall make use of Vinogradov's smoothing method [11] and the results in [29], [32], [33]. The chapter is also based on the p-adic method and gives a further development of this method (see also [52]).

The chapter is organized as follows. In Section 10.1 we state some well-known lemmas, and in Section 10.2 we prove some lemmas with estimates for multiple trigonometric sums with prime numbers. In Section 10.3 we state and prove Theorem 10.1, which gives an estimate for multiple trigonometric sums with prime numbers and is the main theorem of this chapter. In Section 10.4 we prove Theorem 10.2, which concerns the uniform distribution of the fractional parts of the values of polynomials in several variables which take prime number values, and we derive an asymptotic formula for the number of simultaneous representations of a set of natural numbers by terms of the form $p_1^{t_1}\dots p_r^{t_r}$, $0 \le t_1 \le n_1, \dots, 0 \le t_r \le n_r$, where $p_1, \dots, p_r$ are prime numbers.

Notation. In what follows, $r, n_1, \dots, n_r$ and $P_1, \dots, P_r$ are natural numbers, $m = (n_1+1)\dots(n_r+1)$, $\nu \max(n_1, \dots, n_r) = 1$, $P_1 = \min(P_1, \dots, P_r)$, $p_1, \dots, p_r$ are prime numbers, Ω is the m-dimensional unit cube with coordinates $\alpha(t_1, \dots, t_r)$ satisfying the conditions

$$-\big(\tau(t_1,\dots,t_r)\big)^{-1} \le \alpha(t_1,\dots,t_r) < 1 - \big(\tau(t_1,\dots,t_r)\big)^{-1},$$
$$\tau(t_1,\dots,t_r) = P_1^{t_1}\dots P_r^{t_r} P_1^{-1/6} \quad (0 \le t_1 \le n_1, \dots, 0 \le t_r \le n_r),$$

and $F(x_1, \ldots, x_r)$ is a polynomial with real coefficients $\alpha(t_1, \ldots, t_r)$,

$$F(x_1, \ldots, x_r) = \sum_{t_1=0}^{n_1} \cdots \sum_{t_r=0}^{n_r} \alpha(t_1, \ldots, t_r) x_1^{t_1} \ldots x_r^{t_r}. \tag{10.1}$$

We let $S' = S'(A) = S'_t(A)$ denote a trigonometric sum in which the variables of summation run through the prime numbers, i.e.,

$$S' = \sum_{p_1 \le P_1} \cdots \sum_{p_r \le P_r} \exp\{2\pi i t F(p_1, \ldots, p_r)\}, \tag{10.2}$$

where the coordinates $\alpha(t_1, \ldots, t_r)$ of the point A are the coefficients of the polynomial (10.1).

$[a, b, \ldots, c]$ is the least common multiple of $a, b, \ldots, c$.

$$L_s = \log P_s, \quad s = 1, \ldots, r; \qquad L = \log P, \quad P = \max(P_1, \ldots, P_r).$$

Definition 10.1. A point A with coordinates $\alpha(t_1, \ldots, t_r), 0 \le t_1 \le n_1, \ldots, 0 \le t_r \le n_r$, is called a *point of the first class* Ω_1 if $\alpha = \alpha(t_1, \ldots, t_r)$ can be represented in the form

$$\begin{gathered} \alpha = a/q + \beta, \quad (a, q) = 1, \quad 0 \le a < q, \\ |\beta| \le m^{-1} P_1^{-t_1} \ldots P_r^{-t_r} P_1^{0.1\nu} \\ (0 \le t_1 \le n_1, \ldots, 0 \le t_r \le n_r, \ t_1 + \cdots + t_r \ge 1), \end{gathered} \tag{10.3}$$

and the least common multiple Q of the numbers $q = q(t_1, \ldots, t_r)$, $0 \le t_1 \le n_1, \ldots, 0 \le t_r \le n_r$, $t_1 + \cdots + t_r \ge 1$, does not exceed $P_1^{0.1\nu}$. The other points of the cube Ω will be called *points of the second class* Ω_2.

Definition 10.2. By a D-approximation of α corresponding to τ, $\tau \ge 1$, we mean a representation of α in the form

$$\alpha = a/q + \beta, \quad (a, q) = 1, \ q \le \tau, \ |\beta| \le (q\tau)^{-1}.$$

10.1 Some well-known lemmas

Lemma 10.1. *Let the points $A = \big(\alpha(1), \ldots, \alpha(n)\big)$ of the unit cube be divided into two classes according to Definition* 10.1. *The first class consists of points for which $Q \le P^{0.1\nu}$ and $|\beta(s)| \le \nu P^{-s+0.1\nu}$. The second class consists of all other points of the unit cube. For points of the second class we set*

$$\Delta_1 = P^{-\rho_1}, \quad \rho_1 = \gamma_1/(n^2 \ln n), \quad \mu = 1,$$

where γ_1 is a positive constant, and for points of the first class we set

$$\Delta_1 = Q^{-0.5\nu+\varepsilon}, \quad \mu = (m, Q)^{0.5\nu},$$

or alternatively,

$$\Delta_1 = Q^{-0.5\nu+\varepsilon}\delta_0^{-0.5\nu}, \quad \mu = 1,$$

where $\delta_0 = \max\big(|\beta(1)|P, \ldots, |\beta(n)|P^n\big)$. *Then for* $k \le Q\Delta_1^{-2}$

$$|S'(A)| = \Big|\sum_{p\le P} \exp\{2\pi i k F(p)\}\Big| \ll P^{1+8}\Delta_1\mu,$$

where the constant in $\ll$ *depends only on* n *and* ε.

For the proof, see [165], Chapter 7, Theorem 1.

Lemma 10.2. *Points* A *of the first class* Ω_1 *satisfy the following estimate for* $k \le Q^{2\nu}$:

$$|S(A)| = \Big|\sum_{x_1\le P_1} \cdots \sum_{x_r\le P_r} \exp\{2\pi i k F(x_1, \ldots, x_r)\}\Big|$$
$$\ll P_1 \ldots P_r Q^{-\nu+\varepsilon}\mu, \quad \mu = (k, Q)^{\nu}.$$

Further, if we set

$$\delta(t_1, \ldots, t_r) = P_1^{t_1} \ldots P_r^{t_r}\beta(t_1, \ldots, t_r), \quad \delta_0 = \max_{t_1+\cdots+t_r\ge 1} |\delta(t_1, \ldots, t_r)|,$$

then for $\delta_0 > 1$ *and* $k \le (Q\delta_0)^{2\nu}$

$$|S(A)| \ll P_1 \ldots P_r (Q\delta_0)^{-\nu+\varepsilon}.$$

The constants in $\ll$ *depend only on* $n_1, \ldots, n_r$ *and* ε.

For the proof, see [32], Lemma 15.

Lemma 10.3. *Suppose that* A *is a point of the second class* Ω_2 *and* μ_s, $s = 2, \ldots, r$, *are natural numbers satisfying the conditions*

$$-1 < \frac{\log P_s}{\log P_1} - \mu_s \le 0, \quad \varkappa = n_1 + \mu_2 n_2 + \cdots + \mu_r n_r,$$
$$\Delta_r = P_1^{-\rho_r}, \quad \rho_r^{-1} = 32m\varkappa\log(8m\varkappa).$$

Then for $k \le \Delta_r^{-2}$,

$$|S(A)| = \Big|\sum_{x_1\le P_1} \cdots \sum_{x_r\le P_r} \exp\{2\pi i k F(x_1, \ldots, x_r)\}\Big| \ll e^{32\varkappa} P_1 \ldots P_r \Delta_r.$$

The constant in $\ll$ *depends only on* $n_1, \ldots, n_r$.

For the proof, see [32], Theorem 2.

Lemma 10.4. *Suppose that $F(x_1,\ldots,x_r)$ is a polynomial with integer coefficients, $F(0,\ldots,0)=0$, and the set of coefficients is prime to q. Then*

$$|S(q,F(x_1,\ldots,x_r))| = \left|\sum_{x_1=1}^{q}\cdots\sum_{x_r=1}^{q}\exp\left\{2\pi i\frac{F(x_1,\ldots,x_r)}{q}\right\}\right| \ll q^{r-\nu+\varepsilon}.$$

The constant in $\ll$ depends only on $n_1,\ldots,n_r$, and ε.

For the proof, see [29], Chapter II, Section 1.2, Lemma 8 (a).

Lemma 10.5. *Suppose that $F(x_1,\ldots,x_r)$ is a polynomial with integer real coefficients, $F(0,\ldots,0)=0$, and α is the maximum modulus of all coefficients. Then*

$$\begin{aligned}|I_r| &= \left|\int_0^1\cdots\int_0^1 \exp\{2\pi i F(x_1,\ldots,x_n)\}\,dx_1\ldots dx_r\right|\\ &\le \min\left(1,\ 32^r\alpha^{-\nu}(\ln(\alpha+3))^{r-1}\right).\end{aligned}$$

For the proof, see [29], Chapter II, Section 1.1, Lemma 2.

Lemma 10.6. *Suppose that $F(x_1,\ldots,x_r)$ is a real differentiable function for $0\le x_j\le P_j$, $j=1,\ldots,r$, where, within the domain of the variables, the function $\partial F(x_1,\ldots,x_r)/\partial x_j$, $j=1,\ldots,r$, is piecewise monotone and of constant sign with respect to each x_s, $s=1,\ldots,r$, for any fixed values of the other variables. Suppose also that the number of intervals of monotonicity and constant sign does not exceed l and that*

$$\left|\frac{\partial F(x_1,\ldots,x_r)}{\partial x_j}\right| < \delta, \quad j=1,\ldots,r,$$

for $0<\delta<1$. Then

$$\begin{aligned}&\sum_{x_1=1}^{P_1}\cdots\sum_{x_r=1}^{P_r}\exp\{2\pi i F(x_1,\ldots,x_r)\}\\ &\quad= \int_0^{P_1}\cdots\int_0^{P_r}\exp\{2\pi i F(x_1,\ldots,x_n)\}\,dx_1\ldots dx_r\\ &\qquad+\theta l P_1\ldots P_r(P_1^{-1}+\cdots+P_r^{-1})\left(3+\frac{2\delta}{1-\delta}\right), \quad |\theta|\le 1.\end{aligned}$$

For the proof, see [29], Chapter II, Section 3.2, Lemma 16.

Lemma 10.7. *Suppose that all coefficients of the polynomial*

$$f(x_1,\ldots,x_i) = \sum_{t_1=0}^{n_1}\cdots\sum_{t_l=0}^{n_l}\alpha(t_1,\ldots,t_l)x_1^{t_1}\ldots x_l^{t_l}$$

can be represented in the form

$$\alpha = \alpha(t_1, \dots, t_l) = a/q + \beta,$$

where β is a real number, a and q are integers, $a \geq 0, q \geq 1$, and $(a, q) = 1$. Suppose also that

$$Q = \underset{t_1+\cdots+t_l \geq 1}{\text{l.c.m.}}\, q, \quad \delta = P_1^{t_1} \dots P_l^{t_l} \beta, \quad \Delta = \max_{t_1+\cdots+t_l \geq 1} |\delta|,$$

where $P_1, \dots, P_l \geq 1$. Define a polynomial $g(x_1, \dots, x_l)$ by setting

$$f(x_1 + y_1, \dots, x_l + y_l) = g(x_1, \dots, x_l),$$

where $y_1, \dots, y_l$ are integers such that $|y_s| \leq P_s, s = 1, \dots, l$. Let $\alpha_0 = \alpha_0(t_1, \dots, t_l)$ denote the coefficients of $g(x_1, \dots, x_l)$. Then one can find integers a_0 and q_0, $(a_0, q_0) = 1$, and real numbers β_0 such that for all $t_0, \dots, t_l$

$$\alpha_0 = a_0/q_0 + \beta_0,$$

where $Q_0 = Q$, $\Delta \ll \Delta_0 \ll \Delta$, and the numbers Q_0 and Δ_0 are determined in the same way as Q and Δ, except with α, a, q, and β replaced by α_0, a_0, q_0, and β_0, respectively. The constant in $\ll$ depends only on $n_1, \dots, n_l$.

For the proof, see [33], Section 6, Lemma 19.

Lemma 10.8. *Suppose that $f(x)$ is a polynomial with real coefficients, $f(x) = \alpha_0 + \alpha_1 x + \cdots + \alpha_n x^n$, $A > 0$, and $\mu = \mu(A, f)$ is the measure of all points x in the interval $[0, 1]$ for which $|f(x)| \leq A$. Then*

$$\mu \leq \min\left(1,\ 4e(A\alpha^{-1})^{1/n}\right),$$

where $\alpha = \max(|\alpha_0|, |\alpha_1|, \dots, |\alpha_n|)$.

For the proof, see [90], Chapter II, Exercise 1] (this exercise does not present a proof, but only asks for a proof).

10.2 Lemmas on estimates for multiple trigonometric sums with prime numbers

In Sections 10.2 and 10.3 we shall use the following notation.

We let E denote the set of integer r-tuples $(t_1, \dots, t_r)$, $0 \leq t_1 \leq n_1, \dots, 0 \leq t_r \leq n_r$, $t_1 + \cdots + t_r \geq 1$. We let E_0 denote the set of r-tuples in E for which $t_r \geq 1$ and $t_1 + \cdots + t_{r-1} \geq 1$; we let E_1 denote the set of r-tuples for which $t_r = 0$ and $t_1 + \cdots + t_{r-1} \geq 1$; and, finally, we let E_2 denote all other r-tuples in E, i.e., the r-tuples $(t_1, \dots, t_r)$ satisfying the conditions $t_r \geq 1$ and $t_1 = \cdots = t_{r-1} = 0$.

Next, we consider the D-approximation of the numbers $\alpha = \alpha(t_1, \dots, t_r)$, $0 \le t_1 \le n_1, \dots, 0 \le t_r \le n_r$, $t_1 + \dots + t_r \ge 1$, which correspond to

$$\tau = \tau(t_1, \dots, t_r) = P_1^{t_1 - 1/6} P_2^{t_2} \dots P_r^{t_r},$$

i.e., we consider the relations

$$\alpha = a/q + \beta, \quad 1 \le q \le \tau, \quad (a, q) = 1, \quad |\beta| \le (q\tau)^{-1}. \tag{10.4}$$

We let Q, Q_0, Q_1, and Q_2 denote the least common multiples of the numbers $q = q(t_1, \dots, t_r)$ with $(t_1, \dots, t_r)$ in the respective sets E, E_0, E_1, and E_2; we further let δ denote the maximum of the numbers $|\beta(t_1, \dots, t_r)| P_1^{t_1} \dots P_r^{t_r}$ over all r-tuples $(t_1, \dots, t_r) \in E$.

Lemma 10.9. *Let Q_0 be an integer, and let $Q_0 > P_1^{\nu/80}$. Then*

$$|S'(A)| \ll e^{8\varkappa} P_1 \dots P_r \Delta_r^{1/4},$$

where $\varkappa$ and Δ_r are as in Lemma 10.3. The constant in $\ll$ depends only on $n_1, \dots, n_r$.

Proof. We have

$$|S'(A)| \le \sum_{x_r=1}^{P_r} \Big| \sum_{p_1 \le P_1} \cdots \sum_{p_{r-1} \le P_{r-1}} \exp\{2\pi i F_1(p_1, \dots, p_{r-1}, x_r)\} \Big|,$$

where

$$F_1(p_1, \dots, p_{r-1}, x_r) = \sum_{\substack{t_1=0 \\ t_1+\dots+t_{r-1} \ge 1}}^{n_1} \cdots \sum_{t_{r-1}=0}^{n_{r-1}} \sum_{t_r=0}^{n_r} \alpha(t_1, \dots, t_r) p_1^{t_1} \dots p_{r-1}^{t_{r-1}} x_r^{t_r}.$$

We take the square of this inequality and use Cauchy's inequality. We obtain

$$\begin{aligned} |S'(A)|^2 &\le P_r \sum_{x_r=1}^{P_r} \Big| \sum_{p_1 \le P_1} \cdots \sum_{p_{r-1} \le P_{r-1}} \exp\{2\pi i F_1(p_1, \dots, p_{r-1}, x_r)\} \Big|^2 \\ &\le P_r \sum_{p_1, p_1' \le P_1} \cdots \sum_{p_{r-1}, p_{r-1}' \le P_{r-1}} \Big| \sum_{x_r=1}^{P_r} \exp\big\{2\pi i t\big(F_2(p_1, \dots, p_{r-1}, x_r) \\ &\qquad - F_2(p_1', \dots, p_{r-1}', x_r)\big)\big\} \Big| \le \end{aligned}$$

$$\le P_r \sum_{x_1,x_1'\le P_1} \cdots \sum_{x_{r-1},x_{r-1}'\le P_{r-1}} \Big|\sum_{x_r=1}^{P_r} \exp\big\{2\pi i t\big(F_2(x_1,\dots,x_r) - F_2(x_1',\dots,x_{r-1}',x_r)\big)\big\}\Big|,$$

where

$$F_2(x_1,\dots,x_r) = \sum_{\substack{t_1=0\\ t_1+\dots+t_{r-1}\ge 1}}^{n_1} \cdots \sum_{t_{r-1}=0}^{n_{r-1}} \sum_{t_r=0}^{n_r} \alpha(t_1,\dots,t_r) x_1^{t_1}\dots x_r^{t_r}.$$

We take the square of this last inequality and again use Cauchy's inequality. We have

$$|S'(A)|^4 \le P_1^2\dots P_r^2 \sum_{x_1,x_1'\le P_1} \cdots \sum_{x_{r-1},x_{r-1}'\le P_{r-1}} \sum_{x_r,x_r'\le P_r} \exp\big\{2\pi i t \times \big(F_2(x_1,\dots,x_{r-1},x_r) - F_2(x_1,\dots,x_{r-1},x_r') - F_2(x_1',\dots,x_{r-1}',x_r) + F_2(x_1',\dots,x_{r-1}',x_r')\big)\big\}.$$

From this we obtain

$$|S'(A)|^4 \le P_1^2\dots P_r^2 \sum_{x_1'\le P_1}\cdots\sum_{x_r'\le P_r} \Big|\sum_{x_1\le P_1}\cdots\sum_{x_r\le P_r} \exp\big\{2\pi i t\big(F_2(x_1,\dots,x_{r-1},x_r) - F_2(x_1,\dots,x_{r-1},x_r') - F_2(x_1',\dots,x_{r-1}',x_r)\big)\big\}\Big|.$$

Suppose that the maximum modulus of the inner sum is attained at $x_1' = a_1,\dots,x_{r-1}' = a_{r-1}$ and $x_r' = a_r$. Then

$$|S'(A)|^4 \le P_1^3\dots P_r^3|W|, \tag{10.5}$$

where

$$W = \sum_{x_1\le P_1} \cdots \sum_{x_{r-1}\le P_{r-1}} \sum_{x_r\le P_r} \exp\{2\pi i t\Phi(x_1,\dots,x_{r-1},x_r)\},$$

$$\begin{aligned}\Phi(x_1,\dots,x_{r-1},x_r) &= F_2(x_1,\dots,x_{r-1},x_r) \\ &\quad - F_2(x_1,\dots,x_{r-1},a_r) - F_2(a_1,\dots,a_{r-1},x_r) \\ &= \sum_{\substack{t_1=0\\ t_1+\dots+t_r\ge 1}}^{n_1} \cdots \sum_{t_r=0}^{n_r} \gamma(t_1,\dots,t_r) x_1^{t_1}\dots x_r^{t_r}.\end{aligned}$$

We have $\gamma(t_1,\dots,t_{r-1},t_r) = \alpha(t_1,\dots,t_{r-1},t_r)$ for $(t_1,\dots,t_{r-1},t_r) \in E_0$. Consequently, the polynomials $\Phi(x_1,\dots,x_r)$ and $F(x_1,\dots,x_r)$ have the same value Q_0, which is larger than $P_1^{\nu/80}$. We also have

$$\gamma(0,\dots,0,t_r) = -\sum_{\substack{t_1=0\\ t_1+\dots+t_r\le 1}}^{n_1} \cdots \sum_{t_{r-1}=0}^{n_{r-1}} \alpha(t_1,\dots,t_r) a_1^{t_1}\dots a_{r-1}^{t_{r-1}} \quad (1\le t_r\le n_r), \tag{10.6}$$

$$\gamma(t_1, \ldots, t_{r-1}, 0) = -\sum_{t_r=1}^{n_r} \alpha(t_1, \ldots, t_{r-1}, t_r) a_r^{t_r} \tag{10.7}$$
$$(0 \le t_1 \le n_1, \ldots, 0 \le t_{r-1} \le n_{r-1},\ t_1 + \cdots + t_{r-1} \ge 1).$$

We now estimate the sum W. We let A_0 denote the point in the m-dimensional space with coordinates

$$\alpha_0 = \alpha_0(t_1, \ldots, t_r) = \begin{cases} \alpha_0(t_1, \ldots, t_r) & \text{if } (t_1, \ldots, t_r) \in E_0, \\ 0 & \text{if } (t_1, \ldots, t_r) \notin E_0, \end{cases}$$

and we let B denote the point with coordinates $\gamma(t_1, \ldots, t_r)$, $0 \le t_1 \le n_1, \ldots,$ $0 \le t_r \le n_r, t_1 + \cdots + t_r \ge 1$. There are two possible cases:

(a) A_0 is a point of the second class;

(b) A_0 is a point of the first class.

We first consider the case (a). We show that in this case B is a point of the second class. In fact, if B were in the first class, then, by Definition 10.1, we would have $\gamma = \gamma(t_1, \ldots, t_r)$ representable in the form

$$\gamma = b/s + \xi, \qquad (b, s) = 1, \quad 0 \le b \le s, \quad |\xi| < m^{-1} P_1^{-t_1} \ldots P_r^{-t_r} P_1^{0.1\nu}$$
$$(0 \le t_1 \le n_1, \ldots, 0 \le t_r \le n_r,\ t_1 + \cdots + t_r \le 1);$$

the least common multiple Q' of the numbers $s = s(t_1, \ldots, t_r)$ $(0 \le t_1 \le n_1, \ldots,$ $0 \le t_r \le n_r, t_1 + \cdots + t_r \ge 1)$ would not exceed $P_1^{0.1\nu}$. But then the coordinates $\alpha_0 = \alpha_0(t_1, \ldots, t_r)$ of A_0 could be represented in the form

$$\begin{gathered} \alpha_0 = \gamma = b/s + \xi, \\ (b, s) = 1, \quad 0 \le b \le s, \quad |\xi| < m^{-1} P_1^{-t_1} \ldots P_r^{-t_r} P_1^{0.1\nu} \end{gathered} \tag{10.8}$$

if $(t_1, \ldots, t_r) \in E_0$ and in the form $\alpha_0 = 0/1$ if $(t_1, \ldots, t_r) \notin E_0$. Thus the least common multiple Q'' of the numbers $s = s(t_1, \ldots, t_r)$, $(t_1, \ldots, t_r) \in E_0$, does not exceed Q'. This implies that A_0 is a point of the first class, which contradicts case (a). Thus B is a point of the second class, and, by Lemma 10.3,

$$|W| \ll e^{32\varkappa} P_1 \ldots P_r \Delta_r,$$

where $\varkappa$ and Δ_r are as in Lemma 10.3.

We now consider case (b). Since A_0 is a point of the first class, we have relations (10.3). We show that $Q_0 \le P_1^{0.1\nu}$. If this is not the case, we have $Q_0 > P_1^{0.1\nu}$, and then $Q_0 \ne Q''$, since $Q'' \le P_1^{0.1\nu}$. This implies that there exists an r-tuple $(t_1, \ldots, t_r) \in E_0$ such that $s(t_1, \ldots, t_r) \ne q(t_1, \ldots, t_r)$. From this and relations (10.3) and (10.8) we find that

$$\frac{1}{sq} \le \left| \frac{a}{q} - \frac{b}{s} \right| \le |\beta| + |\xi| \le (q\tau)^{-1} + \tau^{-1} P_1^{0.1\nu - 1/6};$$

$$s^{-1} \le \tau^{-1} + q\tau^{-1}P_1^{0.1\nu-1/6}; \quad s \ge 0.5P_1^{1/6-0.1\nu}.$$

On the other hand, $s \le Q'' \le P_1^{0.1\nu}$. For $\nu^{-1} \ge 2$, these inequalities for s are contradictory; hence, $Q_0 \le P_1^{0.1\nu}$.

We represent the coefficients $\gamma(0,\dots,0,t_r)$ and $\gamma(t_1,\dots,t_{r-1},0)$ given by (10.6) and (10.7) in the form

$$\gamma(0,\dots,0,t_r) = \frac{a_1(t_r)}{q_1(t_r)} + \beta_1(t_r), \quad 1 \le t_r \le n_r;$$

$$\frac{a_1(t_r)}{q_1(t_r)} = -\sum_{\substack{t_1=0\\ t_1+\dots+t_{r-1}\ge 1}}^{n_1} \cdots \sum_{t_{r-1}=0}^{n_{r-1}} \frac{a(t_1,\dots,t_r)}{q(t_1,\dots,t_r)},$$

$$\beta_1(t_r) = -\sum_{\substack{t_1=0\\ t_1+\dots+t_{r-1}\ge 1}}^{n_1} \cdots \sum_{t_{r-1}=0}^{n_{r-1}} \beta(t_1,\dots,t_r)a_1^{t_1}\dots a_{r-1}^{t_{r-1}},$$

$$\gamma(t_1,\dots,t_{r-1},0) = \frac{a_2(t_1,\dots,t_{r-1})}{q_2(t_1,\dots,t_{r-1})} + \beta_2(t_1,\dots,t_{r-1})$$
$$(0 \le t_1 \le n_1, \dots, 0 \le t_r \le n_r, \ t_1+\dots+t_r \ge 1);$$

$$\frac{a_2(t_1,\dots,t_{r-1})}{q_2(t_1,\dots,t_{r-1})} = -\sum_{t_r=1}^{n_r} \frac{a(t_1,\dots,t_r)}{q(t_1,\dots,t_r)}a_r^{t_r},$$

$$\beta_2(t_1,\dots,t_{r-1}) = -\sum_{t_r=1}^{n_r} \beta(t_1,\dots,t_r)a_r^{t_r},.$$

From this we find that $q_1(t_r) \mid Q_0$, $q_2(t_1,\dots,t_{r-1}) \mid Q_0$, and

$$|\beta_1(t_r)| \le (n_1+1)\dots(n_{r-1}+1)P_r^{-t_r}P_1^{1/6};$$
$$|\beta_2(t_1,\dots,t_{r-1})| \le (n_r+1)P_1^{-t_1}\dots P_{r-1}^{-t_{r-1}}P_1^{1/6}$$
$$(0 \le t_1 \le n_1, \dots, 0 \le t_{r-1} \le n_{r-1}, \ 1 \le t_r \le n_r, \ t_1+\dots+t_r \ge 1).$$

We transform the sum W using the substitution $x_s = Q_0y_s + z_s$, where $1 \le z_s \le Q_0$ and $-z_sQ_0^{-1} < y_s \le (P_s - z_s)Q_0^{-1}$, $s = 1,\dots,r$. We obtain

$$W = \sum_{z_1=1}^{Q_0} \cdots \sum_{z_r=1}^{Q_0} \exp\{2\pi i t\Phi_1(z_1,\dots,z_r)\}W_1,$$
$$W_1 = \sum_{y_1} \cdots \sum_{y_r} \exp\{2\pi i t\Phi_2(Q_0y_1+z_1,\dots,Q_0y_r+z_r)\}, \tag{10.9}$$

where

$$\Phi_1(z_1,\dots,z_r)=\sum_{\substack{t_1=0\\ t_1+\dots+t_{r-1}\geq 1}}^{n_1}\cdots\sum_{t_{r-1}=0}^{n_{r-1}}\sum_{t_r=1}^{n_r}\frac{a(t_1,\dots,t_r)}{q(t_1,\dots,t_r)}z_1^{t_1},\dots,z_r^{t_r}$$
$$+\sum_{\substack{t_1=0\\ t_1+\dots+t_{r-1}\geq 1}}^{n_1}\cdots\sum_{t_{r-1}=0}^{n_{r-1}}\frac{a_2(t_1,\dots,t_{r-1})}{q_2(t_1,\dots,t_{r-1})}z_1^{t_1}\dots z_{r-1}^{t_{r-1}}+\sum_{t_r=1}^{n_r}\frac{a_1(t_r)}{q_1(t_r)}z_r^{t_r}; \tag{10.10}$$

$$\Phi_2(x_1,\dots,x_r)=\sum_{\substack{t_1=0\\ t_1+\dots+t_{r-1}\geq 1}}^{n_1}\cdots\sum_{t_{r-1}=0}^{n_{r-1}}\sum_{t_r=1}^{n_r}\beta(t_1,\dots,t_r)x_1^{t_1}\dots x_r^{t_r}$$
$$+\sum_{\substack{t_1=0\\ t_1+\dots+t_{r-1}\geq 1}}^{n_1}\cdots\sum_{t_{r-1}=0}^{n_{r-1}}\beta_2(t_1,\dots,t_{r-1})x_1^{t_1}\dots x_{r-1}^{t_{r-1}}+\sum_{t_r=1}^{n_r}\beta_1(t_r)x_r^{t_r}. \tag{10.11}$$

From the estimates for $\beta(t_1,\dots,t_r)$, $\beta_1(t_r)$, $\beta_2(t_1,\dots,t_{r-1})$, and Q_0 we obtain

$$|(\partial/\partial y_s)t\Phi_2(Q_0y_1+z_1,\dots,Q_0y_r+z_r)|\leq 0.5.$$

Consequently, Lemma 10.6 can be applied to the sum W as follows:

$$W_1=\int_{-z_1Q_0^{-1}}^{(P_1-z_1)Q_0^{-1}}\cdots\int_{-z_rQ_0^{-1}}^{(P_r-z_r)Q_0^{-1}}\exp\{2\pi it\Phi_2(Q_0y_1+z_1,\dots,Q_0y_r+z_r)\}\,dy_1\dots dy_r$$
$$+O(P_2\dots P_rQ_0^{-r+1}).$$

We make a change of the variables of integration $y_1,\dots,y_r$:

$$x_s=(Q_0y_s+z_s)P_s^{-1},\quad s=1,\dots,r.$$

We obtain

$$W_1=P_1\dots P_rQ_0^{-r}I_r+O(P_2\dots P_rQ_0^{-r+1}),$$
$$I_r=\int_0^1\cdots\int_0^1\exp\{2\pi it\Phi_3(x_1,\dots,x_r)\}\,dx_1\dots dx_r,$$
$$\Phi_3(x_1,\dots,x_r)=\Phi_2(P_1x_1,\dots,P_rx_r).$$

Thus the sum W satisfies the estimate

$$|W|\leq P_1\dots P_rQ_0^{-r}\big|S\big(Q_0,Q_0\Phi_1(z_1,\dots,z_r)\big)\big|\,|I_r|+O(P_2\dots P_rQ_0^{-r}),$$

where

$$S\big(Q_0, Q_0\Phi_1(z_1, \dots, z_r)\big) = \sum_{z_1=1}^{Q_0} \cdots \sum_{z_r=1}^{Q_0} \exp\{2\pi i t \Phi_1(z_1, \dots, z_r)\}$$

and the polynomial $\Phi_1(z_1, \dots, z_r)$ is defined by (10.10). Since $Q_0 > P_1^{\nu/80}$ by assumption, it follows from Lemma 10.4 that

$$\big|S\big(Q_0, Q_0\Phi_1(z_1, \dots, z_r)\big)\big| \ll Q_0^{r-\nu+8} \ll Q_0^r P_1^{-\nu^2/80+\nu\varepsilon/80} \ll Q_0^r \Delta_r.$$

Hence, in case (b) we have

$$|W| \ll P_1 \dots P_r \Delta_r.$$

After substituting the estimate for $|W|$ in (10.5), we find

$$|S'(A)| \ll e^{8\varkappa} P_1 \dots P_r \Delta_r^{1/4}.$$

The lemma is thereby proved. □

Lemma 10.10. *Suppose that Q_0 and Q_2 are natural numbers with $Q_0 \le P_1^{\nu/80}$ and $Q_2 > P_1^{3\nu/80}$. Then*

$$|S'(A)| \ll e^{8\varkappa} P_1 \dots P_r \Delta_r^{1/4},$$

where $\varkappa$ and Δ_r are as in Lemma 10.3. *The constant in $\ll$ depends only on $n_1, \dots, n_r$.*

Proof. We obviously have the inequality

$$\begin{aligned} |S'(A)| &\le \sum_{x_1 \le P_1} \cdots \sum_{x_{r-1} \le P_{r-1}} \Big| \sum_{p \le P_r} \exp\{2\pi i t F(x_1, \dots, x_{r-1}, p)\} \Big| \\ &= \sum_{x_1 \le P_1} \cdots \sum_{x_{r-1} \le P_{r-1}} \Big| \sum_{p \le P_r} \exp\{2\pi i t F_1(x_1, \dots, x_{r-1}, p)\} \Big| = T_1, \end{aligned}$$

where

$$\begin{aligned} F_1(x_1, \dots, x_{r-1}, p) &= \sum_{t_1=0}^{n_1} \cdots \sum_{t_{r-1}=0}^{n_{r-1}} \sum_{t_r=1}^{n_r} \alpha(t_1, \dots, t_{r-1}, t_r) x_1^{t_1} \dots x_{r-1}^{t_{r-1}} p^{t_r} \\ &= \sum_{t=1}^{n_r} f_t(x_1, \dots, x_{r-1}) p^t, \end{aligned}$$

$$f_t(x_1, \dots, x_{r-1}) = \sum_{t_1=0}^{n_1} \cdots \sum_{t_{r-1}=0}^{n_{r-1}} \alpha(t_1, \dots, t_{r-1}, t) x_1^{t_1} \dots x_{r-1}^{t_{r-1}}.$$

We transform the sum T_1 using the substitution

$$x_s = Q_0 y_s + z_s,$$
$$1 \le z_s \le Q_0, \quad -z_s Q_0^{-1} < y_s \le (P_s - z_s) Q_0^{-1}, \quad s = 1, \dots, r-1,$$

and obtain the inequality $T_1 \le T_2$, where

$$T_2 = \sum_{z_s=1}^{Q_0} \cdots \sum_{z_{r-1}=1}^{Q_0} \sum_{0 \le y_1 \le P_1 Q_0^{-1}} \cdots \sum_{0 \le y_{r-1} \le P_{r-1} Q_0^{-1}} \times \left| \sum_{p \le P_r} \exp\{2\pi i t F_1(Q_0 y_1 + z_1, \dots, Q_0 y_{r-1} + z_{r-1}, p)\} \right|. \tag{10.12}$$

We represent the polynomial $F_1(Q_0 y_1 + z_1, \dots, Q_0 y_{r-1} + z_{r-1}, p)$ in the form

$$\begin{aligned} F_1(Q_0 y_1 + z_1, \dots, Q_0 y_{r-1} + z_{r-1}, p) &= \Phi(Q_0 y_1, \dots, Q_0 y_{r-1}, p) \\ &= \sum_{t_1=0}^{n_1} \cdots \sum_{t_{r-1}=0}^{n_{r-1}} \sum_{t_r=1}^{n_r} \alpha_1(t_1, \dots, t_{r-1}, t_r)(Q_0 y_1)_1^{t_1} \dots (Q_0 y_{r-1})_{r-1}^{t_{r-1}} p^{t_r} \\ &= \sum_{t_r=1}^{n_r} g_t(Q_0 y_1, \dots, Q_0 y_{r-1}) p^t \end{aligned}$$

where

$$\alpha_1(t_1, \dots, t_{r-1}, t_r) = \sum_{s_1=t_1}^{n_1} \cdots \sum_{s_{r-1}=t_{r-1}}^{n_{r-1}} \alpha_1(t, \dots, t_{r-1}, t_r) \times \binom{s_1}{t_1} \cdots \binom{s_{r-1}}{t_{r-1}} z_1^{s_1 - t_1} \dots z_{r-1}^{s_{r-1} - t_{t-1}}.$$

By Lemma 10.6, there exist integers a_1 and q_1, $(a_1, q_1) = 1$, and real numbers β_1 such that for all $t_1, \dots, t_r$, $0 \le t_1 \le n_1, \dots, 0 \le t_{r-1} \le n_{r-1}$, $1 \le t_r \le n_r$ we have

$$\alpha_1 = \alpha_1(t_1, \dots, t_r) = a_1/q_1 + \beta_1$$

with $Q_4' = Q$ and $\delta_0 \ll \delta_0' \ll \delta_0$, where

$$Q_4' = \underset{t_r \ge 1}{\text{l.c.m.}}\, q_1, \qquad Q_4 = \underset{t_r \ge 1}{\text{l.c.m.}}\, q,$$
$$\delta_0' = \max_{t_r \ge 1} P_1^{t_1} \dots P_r^{t_r} |\beta_1|, \qquad \delta_0 = \max_{t_r \ge 1} P_1^{t_1} \dots P_r^{t_r} |\beta|.$$

We let Q_j' and $Q'(t_r)$ denote the numbers

$$Q_j' = \underset{(t_1, \dots, t_r) \in E_j}{\text{l.c.m.}}\, q_1, \quad j = 0, 1, 2,$$

$$Q'(t_r) = \underset{t_1+\cdots+t_{r-1}\geq 1}{\text{l.c.m.}} q_1(t_1, \ldots, t_r), \quad Q'(t_r) = \underset{t_1+\cdots+t_{r-1}\geq 1}{\text{l.c.m.}} q(t_1, \ldots, t_r).$$

We note that

$$Q_0 = [Q(1), \ldots, Q(n_r)], \quad Q_0' = [Q'(1), \ldots, Q'(n_r)].$$

Lemma 10.6 applied to the polynomials

$$f_t(Q_0y_1 + z_1, \ldots, Q_0y_{r-1} + z_{r-1}), \quad g_t(Q_0y_1, \ldots, Q_0y_{r-1}), \quad t = 1, \ldots, n_r,$$

implies that $Q(t_r) = Q'(t_r)$. Thus we have $Q_0 = Q_0'$. Since $Q_4 = [Q_0, Q_2'] = [Q_0, Q_2] > P_1^{3\nu/80}$, we have $Q_2' \geq Q_4Q_0^{-1} \geq P_1^{\nu/40}$. Consequently, the polynomial $\Phi(Q_0y_1 \ldots, Q_0y_{r-1}, p)$ satisfies the relation

$$\begin{aligned}\Phi(Q_0y_1, \ldots, Q_0y_{r-1}, p) &= \sum_{t_1=0}^{n_1} \cdots \sum_{t_{r-1}=0}^{n_{r-1}} \sum_{s=1}^{n_r} \left(\frac{a_1(t_1, \ldots, t_{r-1}, s)}{q_1(t_1, \ldots, t_{r-1}, s)}\right. \\ &\quad \left. + \beta(t_1, \ldots, t_{r-1}, s)\right)(Q_0y_1)^{t_1} \ldots (Q_0y_{r-1})^{t_{r-1}} p^s \\ &\equiv \Phi_1(y_1, \ldots, y_{r-1}, p) \pmod 1,\end{aligned}$$

where

$$\begin{aligned}\Phi_1(y_1, \ldots, y_{r-1}, p) &= \sum_{s=1}^{n_r} \left(\frac{a_1(0, \ldots, 0, s)}{q_1(0, \ldots, 0, s)} + \sum_{t_1=0}^{n_1} \cdots \sum_{t_{r-1}=0}^{n_{r-1}} \beta_1(t_1, \ldots, t_{r-1}, s)\right. \\ &\quad \left. \times (Q_0y_1)^{t_1} \ldots (Q_0y_{r-1})^{t_{r-1}}\right) p^s \\ &= \sum_{s=1}^{n_r} B_s p^s, \quad B_s = h_s(y_1, \ldots, y_{r-1}).\end{aligned}$$

Using this and (10.12), we obtain

$$\begin{aligned}T_2 &\leq Q_0^{r-1} \sum_{0\leq y_1\leq P_1Q_0^{-1}} \cdots \sum_{0\leq y_{r-1}\leq P_{r-1}Q_0^{-1}} \left|\sum_{p\leq P_r} \exp\{2\pi it\Phi(y, \ldots, y_{r-1}, p)\}\right| \\ &= Q_0^{r-1}T_3.\end{aligned}$$

We consider the D-approximations of the fractional parts of $h_s(y_1, \ldots, y_{r-1})$, $s = 1, \ldots, n_r$, which correspond to $\tau(s) = P_r^{s-1/6}$:

$$\begin{aligned}\{h_s(y_1, \ldots, y_{r-1})\} &= \frac{a_s(y_1, \ldots, y_{r-1})}{q_s(y_1, \ldots, y_{r-1})} + \beta(y_1, \ldots, y_{r-1}), \\ \big(a_s(y_1, \ldots, y_{r-1}), q_s(y_1, \ldots, y_{r-1})\big) &= 1, \quad 1 \leq q_s(y_1, \ldots, y_{r-1}) \leq \tau(s),\end{aligned}$$

$$|\beta_s(y_1, \ldots, y_{r-1})| \leq \big(q_s(y_1, \ldots, y_{r-1})\tau(s)\big)^{-1}.$$

The least common multiple of the numbers $q_1(y_1, \ldots, y_{r-1}) \ldots, q_{n_r}(y_1, \ldots, y_{r-1})$ we denote by $Q(\overline{y})$. The largest of the numbers $|\beta_s(y_1, \ldots, y_{r-1})|P_r^s$, $s = 1, \ldots, n_r$, we denote by $\delta(\overline{y})$. We divide the sum T_3 into three parts: $T_3 = S_1 + S_2 + S_3$, where

$$S_j = \sum_{0 \leq y_1 \leq P_1 Q_0^{-1}} \cdots \sum_{0 \leq y_{r-1} \leq P_{r-1} Q_0^{-1}} |S(y_1, \ldots, y_{r-1})|, \quad j = 1, 2, 3,$$

$$S(y_1, \ldots, y_{r-1}) = \sum_{p \leq P_r} \exp\{2\pi i t \Phi(y, \ldots, y_{r-1}, p)\},$$

and each sum S_1, S_2, S_3 has its own range of summation over $y_1, \ldots, y_{r-1}$, as follows.

If $(B_1, \ldots, B_{n_r})$ is a point of the second class with respect to the parameter P_r, then we put the corresponding $(r-1)$-tuple $(y_1, \ldots, y_{r-1})$ in S_1; if it is a point of the first class and if $Q(\overline{y}) \geq H = P_1^{3n\rho}$ or $\delta(\overline{y}) \geq H$, then we put the corresponding $(r-1)$-tuple $(y_1, \ldots, y_{r-1})$ in S_2; finally, all of the remaining $(r-1)$-tuples $(y_1, \ldots, y_{r-1})$ appear in S_3, i.e., S_3 has the $(r-1)$-tuple $(y_1, \ldots, y_{r-1})$ for which $Q(\overline{y}) < H$ and $\delta(\overline{y}) < H$.

We estimate the sum S_1. If the $(r-1)$-tuple $(y_1, \ldots, y_{r-1})$ occurs in S_1, then, by Lemma 10.1,

$$|S(y_1, \ldots, y_{r-1})| \ll P_r^{1-\rho_1+\varepsilon}, \quad \rho_1 = \gamma/(n_r^2 \log n_r),$$

where $\gamma > 0$ is a constant. Consequently,

$$S_1 \ll P_1 \ldots P_{r-1} P_r^{1-\rho_1+\varepsilon} Q_0^{-r+1} \ll P_1.$$

We proceed to estimate S_2. Since $(B_1, \ldots, B_{n_r})$ is a point of the first class, it follows from Definition 10.1 that its coordinates B_s can be represented in the form

$$B_s = b_s/l_s + \beta_s, \quad (b_s, l_s) = 1, \quad |\beta_s| \leq (n_r + 1)^{-1} P_r^{-s+0.1\nu},$$
$$s = 1, \ldots, n_r, \quad l = [l_1, \ldots, l_{n_r}] \leq P_r^{0.1\nu}.$$

We show that for a point $(B_1, \ldots, B_{n_r})$ of the first class we have $Q(\overline{y}) \leq P_r^{0.1\nu}$. In fact, otherwise, we would have $Q(\overline{y}) \neq l$. Consequently, there would exist an s, $1 \leq s \leq n$, such that $q_s(y_1, \ldots, y_{r-1}) \neq l_s$. From this we obtain

$$\frac{1}{q_s(y_1, \ldots, y_{r-1})l_s} \leq \left|\frac{a_s(y_1, \ldots, y_{r-1})}{q_s(y_1, \ldots, y_{r-1})} - \frac{b_s}{l_s}\right| \leq |\beta_s(y_1, \ldots, y_{r-1})| + |\beta_s|$$
$$\leq P_r^{-s+0.1\nu} + q_s^{-1}(y_1, \ldots, y_{r-1}) P_r^{-s+1/6},$$
$$l_s^{-1} \leq q_s(y_1, \ldots, y_{r-1}) P_r^{-s+0.1\nu} + P_r^{-s+1/6} \leq 2P_r^{0.1\nu-1/6},$$
$$l_s \leq 0.5 P_r^{1/6-0.1\nu}.$$

The last inequality contradicts the fact that $l_s \le l \le P_r^{0.1\nu}$. We thus must have $Q(\overline{y}) \le P_r^{0.1\nu}$. We show that for all s, $1 \le s \le n_r$,

$$\frac{a_s(y_1, \dots, y_{r-1})}{q_s(y_1, \dots, y_{r-1})} = \frac{b_s}{l_s}.$$

In fact, otherwise, there would exist an s, $1 \le s \le n_r$, such that

$$\frac{a_s(y_1, \dots, y_{r-1})}{q_s(y_1, \dots, y_{r-1})} \ne \frac{b_s}{l_s}.$$

Then, on the one hand, we would have

$$\left|\frac{a_s(y_1, \dots, y_{r-1})}{q_s(y_1, \dots, y_{r-1})} - \frac{b_s}{l_s}\right| \ge \frac{1}{l_s q_s(y_1, \dots, y_{r-1})} \ge P_r^{-0.2\nu},$$

and, on the other hand, we would have

$$\left|\frac{a_s(y_1, \dots, y_{r-1})}{q_s(y_1, \dots, y_{r-1})} - \frac{b_s}{l_s}\right| \le |\beta_s(y_1, \dots, y_{r-1})| + |\beta_s|$$
$$\le P_r^{-s+0.1\nu} + P_r^{-s+1/6} \le 2P_r^{-s+1/6} \le 2P_r^{-5/6}.$$

From this we find that the upper and lower bounds for the number

$$\left|\frac{a_s(y_1, \dots, y_{r-1})}{q_s(y_1, \dots, y_{r-1})} - \frac{b_s}{l_s}\right|$$

are contradictory. Thus for all $s = 1, \dots, n_r$,

$$\frac{b_s}{l_s} = \frac{a_s(y_1, \dots, y_{r-1})}{q_s(y_1, \dots, y_{r-1})}, \qquad \beta_s = \beta_s(y_1, \dots, y_{r-1}).$$

From this we obtain

$$Q(\overline{y}) = l > H, \quad \delta(\overline{y}) = \delta = \max_{1 \le s \le n_r} P_r^s |\beta_s| > H.$$

Hence, if the $(r-1)$-tuple $(y_1, \dots, y_{r-1})$ appears in S_2, it follows from Lemma 10.1 that

$$|S(y_1, \dots, y_{r-1})| \ll P_r H^{-0.5\nu+\varepsilon} \ll P_r P_1^{-\rho}, \quad |S_2| \ll P_1 \dots P_r Q_r^{-r+1} P_r^{-\rho}.$$

We estimate the sums S_3. We have $|S_3| \le Y P_r$, where Y is the number of $(r-1)$-tuples $(y_1, \dots, y_{r-1})$, $0 \le y_1 \le P_1 Q_0^{-1}, \dots, 0 \le y_{r-1} \le P_{r-1} Q_0^{-1}$, for which

$$\delta(\overline{y}) \le H = P_1^{3\rho n}, \quad Q(\overline{y}) \le H,$$

and $(B_1, \dots, B_{n_r})$ is a point of the first class. We let Ω_0 denote the set of points $(B_1, \dots, B_{n_r})$ which correspond to $(r-1)$-tuples $(y_1, \dots, y_{r-1})$ occurring in S_3.

We proceed to estimate Y. We let $\Omega_1 = \Omega_1(b_1/h_1, \dots, b_{n_r}/h_{n_r})$ denote the region in the n_r-dimensional space which is defined as follows. The point $(\alpha_1, \dots, \alpha_{r-1})$ belongs to Ω_1 if

$$\begin{gathered}\alpha_s = b_s/h_s + z_s, \quad 1 \le h_s \le \tau(s), \quad (b_s, h_s) = 1,\\ |z_s| \le (h_s\tau(s))^{-1}, \quad \tau(s) = P_r^{s-1/6}, \quad s = 1, \dots, n_r,\\ [h_1, \dots, h_{n_r}] \le H.\end{gathered} \tag{10.13}$$

Let $\Omega_1' = \Omega(b_1'/h_1', \dots, b_{n_r}'/h_{n_r}')$ be a region different from Ω_1 with the condition $[h_1', \dots, h_{n_r}'] \le H$. Then there is an index s, $1 \le s \le n_r$, such that $b_s/h_s \ne b_s'/h_s'$. Consequently,

$$|b_s/h_s - b_s'/h_s'| \ge H^{-2}.$$

Hence the distance between the sth coordinates of the points of these regions is no less than

$$H^{-2} - 2\big(\tau(s)\big)^{-1} \ge 0.5H^{-2}.$$

But the difference between $a_1(0, \dots, 0, s)/q_1(0, \dots, 0, s)$ and the sth coordinate of any point of Ω_0 does not exceed

$$\sum_{t_1=0}^{n_1} \cdots \sum_{t_{r-1}=0}^{n_{r-1}} |\beta(t_1, \dots, t_{r-1}, s)| P_1^{t_1} \dots P_{r-1}^{t_{r-1}} \ll P_r^{-s} P_1^{1/6} \ll P_1^{-5/6}.$$

Therefore, the set Ω_0 intersects with at most a single region Ω_1.

If $Y \ne 0$, then all the points of Ω_0 satisfy the relations

$$\begin{gathered}\frac{a_s(y_1, \dots, y_{r-1})}{q_s(y_1, \dots, y_{r-1})} = \frac{b_s}{l_s},\\ \left|\{h_s(y_1, \dots, y_{r-1})\} - \frac{b_s}{h_s}\right| \le P_r^{-s} H \quad (s = 1, \dots, n_r).\end{gathered} \tag{10.14}$$

Since $Q(\overline{y}) \le P_1^{3\rho n}$, $Q_2' > P_1^{\nu/40}$, and $\rho < \nu^2/120$, we have $Q_2' \ne Q(\overline{y})$, and thus for some μ we obtain

$$\frac{a_1(0, \dots, 0, \mu)}{q_1(0, \dots, 0, \mu)} \ne \frac{b_\mu}{h_\mu}, \qquad q_1(0, \dots, \mu) \ne 1.$$

From (10.14) we have $Y \le Y_1$, where Y_1 is the number of $(r-1)$-tuples for which

$$|h_\mu(y_1, \dots, y_{r-1}) - b_\mu/h_\mu| \le P_r^{-\mu} H = \Delta. \tag{10.15}$$

We set

$$B(y_1, \dots, y_{r-1}) = \sum_{t_1=0}^{n_1} \cdots \sum_{t_{r-1}=0}^{n_{r-1}} \beta_1(t_1, \dots, t_{r-1}, \mu)(Q_0 y_1)^{t_1} \dots (Q_0 y_{r-1})^{t_{r-1}},$$

$$\frac{a_1(0,\dots,0,\mu)}{q_1(0,\dots,0,\mu)} = \frac{a}{q}, \quad \frac{b_\mu}{h_\mu} = \frac{b}{h}.$$

Then (10.15) takes the form

$$|B(y_1,\dots,y_{r-1}) - b/h + a/q| \le \Delta. \tag{10.16}$$

We now define a periodic function $\chi(x)$ of period one by setting

$$\chi(x) = \begin{cases} 1 & \text{if } |x| \le D, \\ (2\Delta - |x|)\Delta^{-1} & \text{if } \Delta < |x| \le 2\Delta, \\ 0 & \text{if } 2\Delta < |x| \le 0.5, \end{cases} \tag{10.17}$$

and a function $\psi(x)$ by setting $\psi(x) = \chi(x + a/q - b/h)$. Then from (10.16) we obtain

$$Y_1 \le \sum_{0\le y_1\le P_1Q_0^{-1}} \cdots \sum_{0\le y_{r-1}\le P_{r-1}Q_0^{-1}} \psi\big(B(y_1,\dots,y_{r-1})\big) = Y_2.$$

We expand $\psi(x)$ in the Fourier series

$$\psi(x) = \Delta + \sum_{s=-\infty}^{+\infty} c(s)e^{2\pi isx},$$

$$c(0) = 0, \quad |c(s)| \le \min\left(\Delta, \frac{1}{\Delta s^2}\right), \quad |s| \ge 1.$$

Consequently,

$$Y_2 \ll P_1 \dots P_{r-1}Q_0^{-r+1}\Delta + \sum_{1\le s<M} \Delta|T(s)|$$
$$+ \sum_{M\le s\le M_1} \Delta^{-1}s^{-2}|T_s| + P_1^{1-\rho}P_2\dots P_{r-1}Q_0^{-r+1},$$

where

$$T(s) = \sum_{0\le y_1\le P_1Q_0^{-1}} \cdots \sum_{0\le y_{r-1}\le P_{r-1}Q_0^{-1}} \exp\{2\pi isB(y_1,\dots,y_{r-1})\},$$
$$M = \Delta^{-1}, \quad M_1 = MP_1^\rho.$$

We estimate the sum $T(s)$. The moduli of the first partial derivatives of the polynomial $sB(y_1,\dots,y_{r-1})$ do not exceed

$$\left|s\frac{\partial}{\partial y_j}B(y_1,\dots,y_{r-1})\right| \le |s|\sum_{t_1=0}^{n_1}\cdots\sum_{t_j=1}^{n_j}\cdots\sum_{t_{r-1}=0}^{n_{r-1}} \times t_j|\beta_1(t_1,\dots,t_j,\dots,t_{r-1},\mu)|$$

$$\begin{aligned}&\times Q_0(Q_0y_1)^{t_1}\dots Q_0(Q_0y_j)^{t_j-1}\dots Q_0(Q_0y_{r-1})^{t_{r-1}}\\&\le |s|\frac{n_jm}{n_r+1}Q_0P_1^{t_1}\dots P_j^{t_j-1}\dots P_{r-1}^{t_{r-1}}P_1^{-t_1}\dots P_{r-1}^{-t_{r-1}}P_r^{-t_r}P_1^{1/6}\\&\ll P_1^{-0.5}\le 0.5,\quad j=1,\dots,r-1.\end{aligned}$$

Consequently, by Lemma 10.6

$$\begin{aligned}T(s)=&\int_0^{P_1Q_0^{-1}}\cdots\int_0^{P_{r-1}Q_0^{-1}}\exp\{2\pi isB(y_1,\dots,y_{r-1})\}\,dy_1\dots dy_{r-1}\\&+O(P_2\dots P_{r-1}Q_0^{-r+2}).\end{aligned}$$

We transform the integral in $T(s)$ by using the substitution $z_j=P_j^{-1}Q_0y_j$, $j=1,\dots,r-1$. We obtain

$$T(s)=P_1\dots P_{r-1}Q_0^{-r+1}I_{r-1}+O(P_2\dots P_rQ_0^{-r+2}),$$

where

$$I_{r-1}=\int_0^1\cdots\int_0^1\exp\{2\pi isA(z_1,\dots,z_{r-1})\}\,dz_1\dots dz_{r-1},$$

$$A(z_1,\dots,z_{r-1})=\sum_{t_1=0}^{n_1}\cdots\sum_{t_{r-1}=0}^{n_{r-1}}\delta(t_1,\dots,t_{r-1})z_1^{t_1}\dots z_{r-1}^{t_{r-1}},$$

$$\delta(t_1,\dots,t_{r-1})=\beta_1(t_1,\dots,t_{r-1},\mu)P_1^{t_1}\dots P_{r-1}^{t_{r-1}}.$$

Let

$$\delta=\max_{t_1+\dots+t_{r-1}\ge1}|\delta(t_1,\dots,t_{r-1})|.$$

Then we have the following lower bound for δ:

$$\begin{aligned}\delta&\ge\frac{n_r+1}{m}\sum_{\substack{t_1=0\\t_1+\dots+t_{r-1}\ge1}}^{n_1}\cdots\sum_{t_{r-1}=0}^{n_{r-1}}|\beta_1(t_1,\dots,t_{r-1},\mu)|P_1^{t_1}\dots P_{r-1}^{t_{r-1}}\\&\ge\frac{n_r+1}{m}\sum_{\substack{t_1=0\\t_1+\dots+t_{r-1}\ge1}}^{n_1}\cdots\sum_{t_{r-1}=0}^{n_{r-1}}|\beta_1(t_1,\dots,t_{r-1},\mu)|(Q_0y_1)^{t_1}\dots(Q_0y_{r-1})^{t_{r-1}}\\&\ge\frac{n_r+1}{m}\Bigg(\left|\frac{a}{q}-\frac{b}{h}\right|-|\beta_1(0,\dots,0,\mu)|\\&\quad-\left|\frac{a}{q}-\frac{b}{h}+\sum_{t_{r-1}=0}^{n_{r-1}}\beta_1(t_1,\dots,t_{r-1},\mu)(Q_0y_1)^{t_1}\dots(Q_0y_{r-1})^{t_{r-1}}\right|\Bigg)\ge\end{aligned}$$

$$\geq \frac{n_r+1}{m}\left(\frac{1}{Hq} - \frac{1}{q\tau(0,\ldots,0,\mu)} - \Delta\right) \geq \frac{n_r+1}{4mH\tau(0,\ldots,0,\mu)}.$$

By Lemma 10.5, we hence obtain the following relations for $T(s)$:

$$|T(s)| \ll P_1 \ldots P_{r-1} Q_0^{-r+1} \min\left(1,\ |s|^{-1/n} H^{1/n} \big(\tau(0,\ldots,0,\mu)\big)^{1/n}\right).$$

Consequently,

$$\begin{aligned} Y_2 \ll\ & P_1 \ldots P_{r-1} Q_0^{-r+1} \Delta \\ & + P_1 \ldots P_{r-1} Q_0^{-r+1} \Delta \sum_{1 \leq s < M} m^{-1/n} H^{1/n} \big(\tau(0,\ldots,0,\mu)\big)^{1/n} \\ & + P_1 \ldots P_{r-1} Q_0^{-r+1} \Delta^{-1} \sum_{M \leq s < M_1} s^{-2-1/n} H^{1/n} \big(\tau(0,\ldots,0,\mu)\big)^{1/n} \\ & + P_1^{1-\rho} P_2 \ldots P_{r-1} Q_0^{-r+1} \ll P_1^{1-\rho} P_2 \ldots P_{r-1} Q_0^{-r+1}. \end{aligned}$$

From this we conclude that

$$\begin{aligned} S_3 &\leq Y P_r \ll P_1 \ldots P_r Q_0^{-r+1} P_1^{-\rho}, \\ T_3 = S_1 + S_2 + S_3 &\ll P_1 \ldots P_r Q_0^{-r+1} P_1^{-\rho}, \\ |S'(A)| \leq Q_0^{-r+1} T_3 &\ll P_1 \ldots P_r P_1^{-\rho}. \end{aligned}$$

The lemma is proved. □

Lemma 10.11. *Suppose that the numbers Q and δ satisfy the conditions $Q \leq P_1^{0.4\nu}$ and $\delta > m^{-1} P_1^{0.4\nu}$. Then*

$$|S'(A)| \ll P_1 \ldots P_r P_1^{-\rho},$$

where

$$\rho = \gamma/(n^2 \log n), \quad n = \max(n_1, \ldots, n_r),$$

and $\gamma > 0$ is a constant. The constant in $\ll$ depends only on $n_1, \ldots, n_r$.

Proof. Let $\delta = |\delta(t_1, \ldots, t_r)|$, $t_s \geq 1$. Then

$$|S'(A)| \leq \sum_{x_1 \leq P_1} \cdots \sum_{x_{s-1} \leq P_{s-1}} \cdots \sum_{x_r \leq P_r} \left| \sum_{p_s \leq P_s} \exp\{2\pi i t F_1(x_1, \ldots, x_{s-1}, p_s, x_{s+1}, \ldots, x_r)\} \right|, \tag{10.18}$$

where

$$F_1(x_1, \ldots, x_{s-1}, p, x_{s+1}, \ldots, x_r)$$
$$= \sum_{t_1=0}^{n_1} \cdots \sum_{t_{s-1}=0}^{n_{s-1}} \sum_{t_s=1}^{n_s} \sum_{t_{s+1}=0}^{n_{s+1}} \cdots \sum_{t_r=0}^{n_r} \alpha(t_1, \ldots, t_r) x_1^{t_1} \ldots p^{t_s} \ldots x_r^{t_r}.$$

We let q denote the least common multiple of the numbers $q(t_1, \ldots, t_r)$ with the conditions $t_1 \geq 0, \ldots, t_{s-1} \geq 0$, $t_s \geq 1$, $t_{s+1} \geq 0, \ldots, t_r \geq 0$. Then $q \leq Q$. We represent the variables x_l in the form

$$x_l = qy_l + z_l, \quad l \leq z_l \leq q,$$
$$-z_l q^{-1} < y_l \leq (P_l - z_l) q^{-1}, \quad l = 1, \ldots, s-1, s+1, \ldots, r.$$

We define the polynomial

$$\Phi_1(y_1, \ldots, y_{s-1}, p, y_{s+1}, \ldots, y_r)$$
$$= F_1(qy_1 + z_1, \ldots, qy_{s-1} + z_{s-1}, p, qy_{s+1} + z_{s+1}, \ldots, qy_r + z_r)$$
$$= \sum_{t_1=0}^{n_1} \cdots \sum_{t_{s-1}=0}^{n_{s-1}} \sum_{t_s=1}^{n_s} \sum_{t_{s+1}=0}^{n_{s+1}} \cdots \sum_{t_r=0}^{n_r} \alpha(t_1, \ldots, t_r)$$
$$\times (qy_1)^{t_1} \ldots (qy_{s-1})^{t_{s-1}} p^{t_s} (qy_{s+1})^{t_{s+1}} \ldots (qy_r)^{t_r}.$$

Then, by Lemma 10.7, there exist rational approximations to the numbers $\alpha_1 = \alpha_1(t_1, \ldots, t_r)$ such that

$$\alpha_1 = \frac{\alpha_1}{q_1} + \beta_1, \quad (a_1, q_1) = 1, \quad q' = \underset{t_s \geq 1}{\text{l.c.m.}}\, q_1, \quad \delta' = \max_{t_s \geq 1} |\beta_1| P_1^{t_1} \ldots P_r^{t_r}$$

and also $q = q'$ and $\delta \ll \delta' \ll \delta$. Consequently,

$$\Phi_1(y_1, \ldots, y_{s-1}, p, y_{s+1}, \ldots, y_r)$$
$$= \sum_{t_1=0}^{n_1} \cdots \sum_{t_{s-1}=0}^{n_{s-1}} \sum_{t_s=1}^{n_s} \sum_{t_{s+1}=0}^{n_{s+1}} \cdots \sum_{t_r=0}^{n_r} \left(\frac{a_1}{q_1} + \beta_1\right) (qy_1)^{t_r} \ldots p^{t_s} \ldots (qy_r)^{t_r}$$
$$\equiv \Phi(y_1, \ldots, y_{s-1}, p, y_{s+1}, \ldots, y_r) \pmod 1,$$

where

$$\Phi = \Phi(y_1, \ldots, y_{s-1}, p, y_{s+1}, \ldots, y_r)$$
$$= \sum_{t_1=0}^{n_1} \cdots \sum_{t_{s-1}=0}^{n_{s-1}} \sum_{t_s=1}^{n_s} \sum_{t_{s+1}=0}^{n_{s+1}} \cdots \sum_{t_r=0}^{n_r} \beta_1(t_1, \ldots, t_r)$$
$$\times (qy_1)^{t_1} \ldots (qy_{s-1})^{t_{s-1}} p^{t_s} (qy_{s+1})^{t_{s+1}} \ldots (qy_r)^{t_r} = \sum_{l=1}^{n_s} B_l p^l.$$

We let $\overline{y}$ denote the $(r-1)$-tuple $(y_1, \dots, y_{s-1}, y_{s+1}, \dots, y_r)$. We represent the coefficients B_l of the polynomial Φ in the form

$$B_l = a_l(\overline{y})/q_l(\overline{y}) + \beta_l(\overline{y}), \quad \big(a_l(\overline{y}), q_l(\overline{y})\big) = 1, \quad 1 \le q_l(\overline{y}) \le \tau(l),$$
$$|\beta_l(\overline{y})| \le \big(q_l(\overline{y}), \tau(l)\big)^{-1} = 1, \quad \tau(l) = P_s^{l-1/6}, \quad l = 1, \dots, n_s.$$

We further introduce the notation

$$Q(\overline{y}) = [q_1(\overline{y}), \dots, q_{n_s}(\overline{y})], \quad \delta(\overline{y}) = \max_{1 \le l \le n_s} |\beta_l(\overline{y})| P_s^l.$$

If in (10.18) we replace the variables of summation x_l by the variables $qy_l + z_l$, $l = 1, \dots, s-1, s+1, \dots, r$, we find that

$$|S'(A)| \le q^{r-1} \sum_{0 \le y_1 \le P_1 q^{-1}} \cdots \sum_{0 \le y_{s-1} \le P_{s-1} q^{-1}} \sum_{0 \le y_{s+1} \le P_{s+1} q^{-1}} \cdots$$
$$\cdots \sum_{0 \le y_r \le P_r q^{-1}} \Big| \sum_{p \le P_s} \exp\{2\pi i t \Phi(y_1, \dots, y_{s-1}, p, y_{s+1}, \dots, y_r)\} \Big|$$
$$= q^{r-1} T_1.$$

We divide the sum T_1 into three parts $T_1 = S_1 + S_2 + S_3$, where

$$S_j = \sum_{0 \le y_1 \le P_1 q^{-1}} \cdots \sum_{0 \le y_{s-1} \le P_{s-1} q^{-1}} \sum_{0 \le y_{s+1} \le P_{s+1} q^{-1}} \cdots \sum_{0 \le y_r \le P_r q^{-1}}^{(j)} |S(\overline{y})|,$$
$$j = 1, 2, 3,$$

$$S(\overline{y}) = \sum_{p \le P_s} \exp\{2\pi i t \Phi(y_1, \dots, y_{s-1}, p, y_{s+1}, \dots, y_r)\},$$

and each of the sums S_1, S_2, and S_3 has its own range of summation of the $(r-1)$-tuples $\overline{y} = (y_1, \dots, y_{s-1}, y_{s+1}, \dots, y_r)$, as follows. If $(B_1, \dots, B_{n_s})$ is a point of the second class with respect to the parameter P_s, then the corresponding $\overline{y}$ appears in S_1; if it is a point of the first class, and if either $Q(\overline{y}) \ge H = P_1^{3n\rho}$ or $\delta(\overline{y}) \ge H$, then the corresponding $\overline{y}$ appears in S_2; and all of the remaining $\overline{y}$ appear in S_3.

In the case when $\overline{y}$ appears in either S_1 or S_2, we use Lemma 10.1 to estimate $S(\overline{y})$. We obtain $|S(\overline{y})| \ll P_s^{1-\rho}$. Hence

$$|S_1| + |S_2| \ll P_1^{1-\rho} P_2 \dots P_r q^{-r+1}.$$

We estimate S_3. We obviously have $|S_3| \le Y P_s$, where Y is the number of $(r-1)$-tuples $\overline{y} = (y_1, \dots, y_{s-1}, y_{s+1}, \dots, y_r)$, $0 \le y_1 \le P_1 q^{-1}, \dots, 0 \le y_{s-1} \le P_{s-1} q^{-1}, 0 \le y_{s+1} \le P_{s+1} q^{-1}, \dots, 0 \le y_r \le P_r q^{-1}$, for which

$$\delta(\overline{y}) \le H = P_1^{3\rho n}, \quad Q(\overline{y}) \le H, \tag{10.19}$$

and $(B_1, \dots, B_{n_s})$ is a point of the first class. The set of such $(B_1, \dots, B_{n_s})$ in the first class and satisfying (10.19) will be denoted by Ω_0. Just as in Lemma 10.10, one proves that if $Y \neq 0$, then the entire set Ω_0 is in a single region Ω_1 defined by (10.13). Hence,

$$\frac{a_l(\overline{y})}{q_l(\overline{y})} = \frac{b_l}{h_l}, \qquad \left| B_l - \frac{b_l}{h_l} \right| \leq P_s^{-l} H, \quad 1 \leq l \leq n_s.$$

Let

$$\delta' = |\beta_1(t_1, \dots, t_{s-1}, l, t_{s+1}, \dots, t_r)| P_1^{t_1} \dots P_{s-1}^{t_{s-1}} P_s^l P_{s+1}^{t_{s+1}} \dots P_r^{t_r} \tag{10.20}$$

for some $l \geq 1$. Then Y is bounded from above by the number Y_1 of $(r-1)$-tuples $\overline{y}$ for which

$$|B_l(\overline{y}) - b_l/h_l| \leq P_s^{-1} H = \Delta, \tag{10.21}$$

where $B_l(\overline{y}) = B_l$.

We now define a periodic function $\psi_1(x)$ by setting $\psi_1(x) = \chi(x - b_l/h_l)$, where $\chi(x)$ is as in (10.17). From (10.21) and the definition of $\psi_1(x)$ we have

$$Y_1 \leq \sum_{0 \leq y_1 \leq P_1 q^{-1}} \cdots \sum_{0 \leq y_{s-1} \leq P_{s-1} q^{-1}} \sum_{0 \leq y_{s+1} \leq P_{s+1} q^{-1}} \cdots \sum_{0 \leq y_r \leq P_r q^{-1}} \psi_1(B_l(\overline{y})) = Y_2.$$

From this, if we expand $\psi_1(x)$ in the Fourier series

$$\psi_1(x) = \Delta + \sum_{t=-\infty}^{+\infty} c_1(t) e^{2\pi i t x},$$

$$c_1(0) = 0, \quad c_1(t) \leq \min(\Delta, 1/\Delta t^2) \quad \text{for } |t| \geq 1,$$

we obtain

$$\begin{aligned} Y_2 \ll P_1 \dots P_{s-1} P_{s+1} \dots P_r q^{-r+1} \Delta + \sum_{1 \leq t < M} \Delta |T(t)| \\ + \sum_{M \leq t < M_1} \Delta^{-1} t^{-2} |T(t)| + P_1 \dots P_{s-1} P_{s+1} \dots P_r q^{-r+1} P_1^{-\rho}, \end{aligned} \tag{10.22}$$

where

$$T(t) = \sum_{0 \leq y_1 \leq P_1 q^{-1}} \cdots \sum_{0 \leq y_{s-1} \leq P_{s-1} q^{-1}} \sum_{0 \leq y_{s+1} \leq P_{s+1} q^{-1}} \cdots \sum_{0 \leq y_r \leq P_r q^{-1}} \exp\{2\pi i t B_l(\overline{y})\},$$

$$M = \Delta^{-1}, \quad M_1 = M P_1^{\rho}.$$

We find an upper bound for the moduli of the first partial derivatives of the polynomial $t B_l(\overline{y})$ for $|t| \leq M_1$:

$$\left| t \frac{\partial}{\partial y_k} B_l(\overline{y}) \right| \leq |t| \sum_{t_1=0}^{n_1} \cdots \sum_{t_k=1}^{n_k} \cdots \sum_{t_r=0}^{n_r} t_k |\beta_1(t_1, \dots, t_k, \dots, t_r)| \times$$

$$\times q(qy_1)^{t_1}\dots(qy_k)^{t_k-1}\dots(qy_r)^{t_r}$$
$$\ll |t|qP_s^{-1}P_k^{-1}P_1^{1/6}\le 0.5,\quad k=1,\dots,s-1,s+1,\dots,r.$$

By Lemma 6, this implies the relations

$$|T(t)|\le\left|\int_0^{P_1q^{-1}}\dots\int_0^{P_{s-1}q^{-1}}\int_0^{P_{s+1}q^{-1}}\exp\{2\pi itB_l(\overline{y})\}\,dy_1\dots dy_{s-1}dy_{s+1}\dots dy_r\right|$$
$$+O\big(P_1\dots P_{s-1}P_{s+1}\dots P_rq^{-r+1}(P_1^{-1}+\dots+P_{s-1}^{-1}P_{s+1}^{-1}+\dots+P_r^{-1})\big)$$
$$\ll P_1\dots P_{s-1}P_{s+1}\dots P_rq^{-r+1}$$
$$\times(|I_{r-1}|+P_1^{-1}+\dots+P_{s-1}^{-1}P_{s+1}^{-1}+\dots+P_r^{-1})),$$
$$I_{r-1}=\int_0^1\dots\int_0^1\exp\{2\pi itA(\overline{y})\}\,dy_1\dots dy_{s-1}dy_{s+1}\dots dy_r,\tag{10.23}$$
$$A(\overline{y})=\sum_{\substack{t_1=0\\ t_1+\dots+t_{s-1}+t_{s+1}+\dots+t_r\ge1}}^{n_1}\dots\sum_{t_{s-1}=0}^{n_{s-1}}\sum_{t_{s+1}=0}^{n_{s+1}}\dots\sum_{t_r=0}^{n_r}\gamma(t_1,\dots,t_{s-1},t_{s+1},\dots,t_r)$$
$$\times y_1^{t_1}\dots y_{s-1}^{t_{s-1}}y_{s+1}^{t_{s+1}}\dots y_r^{t_r},$$
$$\gamma(t_1,\dots,t_{s-1},t_{s+1},\dots,t_r)=\beta_1(t_1,\dots,t_{s-1},l,t_{s+1},\dots,t_r)$$
$$\times P_1^{t_1}\dots P_{s-1}^{t_{s-1}}P_{s+1}^{t_{s+1}}\dots P_r^{t_r}.$$

There are two possible cases in relation (10.20) which defines δ':

(a) $t_1+\dots+t_{s-1}+t_{s+1}+\dots+t_r\ge1$;

(b) $t_1=\dots=t_{s-1}=t_{s+1}=\dots=t_r=0$.

Let γ be the maximum of the numbers $|\gamma(t_1,\dots,t_{s-1},t_{s+1},\dots,t_r)|$ subject to the condition that $t_1+\dots+t_{s-1}+t_{s+1}+\dots+t_r\ge1$, $0\le t_1\le n_1,\dots,0\le t_{s-1}\le n_{s-1}$, $0\le t_{s+1}\le n_{s+1},\dots,0\le t_r\le n_r$. Then in case (a) we have

$$\gamma=\delta'P_s^{-l}\gg\delta P_s^{-l}\gg P_s^{-l}P_1^{0.1\nu}.$$

We derive case (b) (see (10.21)) into two subcases: (1) $h_l>1$, (2) $h_l=1$. In subcase (1) we have

$$\gamma\ge\frac{n_s+1}{m}\sum_{\substack{t_1=0\\ t_1+\dots+t_{s-1}+t_{s+1}+\dots+t_r\ge1}}^{n_1}\dots\sum_{t_{s-1}=0}^{n_{s-1}}\sum_{t_{s+1}=0}^{n_{s+1}}\dots\sum_{t_r=0}^{n_r}|\gamma(t_1,\dots,t_{s-1},t_{s+1},\dots,t_r)|$$
$$=\frac{n_s+1}{m}\sum_{\substack{t_1=0\\ t_1+\dots+t_{s-1}+t_{s+1}+\dots+t_r\ge1}}^{n_1}\dots\sum_{t_{s-1}=0}^{n_{s-1}}\sum_{t_{s+1}=0}^{n_{s+1}}\dots\sum_{t_r=0}^{n_r}|\beta_1(t_1,\dots,t_{s-1},l,t_{s+1},\dots,t_r)|$$
$$\times P_1^{t_1}\dots P_{s-1}^{t_{s-1}}P_{s+1}^{t_{s+1}}\dots P_r^{t_r}$$

$$\geq \frac{n_s+1}{m}\sum_{\substack{t_1=0\\ t_1+\dots+t_{s-1}+t_{s+1}+\dots+t_r\geq 1}}^{n_1}\dots\sum_{t_{s-1}=0}^{n_{s-1}}\sum_{t_{s+1}=0}^{n_{s+1}}\dots\sum_{t_r=0}^{n_r}|\beta_1(t_1,\dots,t_{s-1},l,t_{s+1},\dots,t_r)|$$

$$\times (qy_1)^{t_1}\dots(qy_{s-1})^{t_{s-1}}(qy_{s+1})^{t_{s+1}}\dots(qy_r)^{t_r}$$

$$\geq \frac{n_s+1}{m}\left(\frac{b_l}{h_l}-|\beta(0,\dots,0,l,0,\dots,0|-\left|\frac{b_l}{h_l}-\sum_{t_1=0}^{n_1}\dots\sum_{t_{s-1}=0}^{n_{s-1}}\sum_{t_{s+1}=0}^{n_{s+1}}\dots\right.\right.$$

$$\left.\left.\dots\sum_{t_r=0}^{n_r}\beta_1(t_1,\dots,t_{s-1},l,t_{s+1},\dots,t_r)(qy_1)^{t_1}\dots(qy_r)^{t_r}\right|\right)$$

$$\geq \frac{n_s+1}{m}\left(\frac{1}{h_l}-P_s^{-l}P^{1/6}-\Delta\right)\gg H^{-1}\gg P_s^{-l}iP_1^{0.1\nu}$$

(all of these inequalities are written out under the assumption that the $(r-1)$-tuple $(y_1,\dots,y_{s-1},y_{s+1},\dots,y_r)$ occurs in S_2, and so (10.19) holds). We now consider subcase (2). From inequalities analogous to those in subcase (1) we obtain ($h_l=1$ and $b_l=0$)

$$\gamma\geq\frac{n_s+1}{m}\Bigg(|\beta(0,\dots,0,l,0,\dots,0)|$$

$$-\left|\sum_{t_1=0}^{n_1}\dots\sum_{t_{s-1}=0}^{n_{s-1}}\sum_{t_{s+1}=0}^{n_{s+1}}\dots\sum_{t_r=0}^{n_r}\beta_1(t_1,\dots,t_{s-1},l,t_{s+1},\dots,t_r)\right.$$

$$\left.\times(qy_1)^{t_1}\dots(qy_{s-1})^{t_{s-1}}(qy_{s+1})^{t_{s+1}}\dots(qy_r)^{t_r}\right|\Bigg)$$

$$\geq\frac{n_s+1}{m}(\delta' P_s^{-l}-\Delta)\gg\delta P_s^{-l}\gg P_s^{-l}P_1^{0.1\nu}.$$

Thus the number γ satisfies $\gamma\gg P_s^{-l}P_1^{0.1\nu}$. Consequently, by Lemma 10.5

$$|I_{r-1}|\ll|t|^{-0.5\nu}\gamma^{-0.5\nu}\ll|t|^{-0.5\nu}P_s^{-0.5\nu l}P_1^{-0.5\nu^2},$$

where $\nu\max(n_1,\dots,n_r)=1$. Substituting this bound for I_{r-1} into (10.23), and then substituting the resulting inequality for $|T(t)|$ into (10.22), we find

$$\begin{aligned}Y_2&\ll P_1\dots P_{s-1}P_{s+1}\dots P_r q^{-r+1}\Delta\\&+P_1\dots P_{s-1}P_{s+1}\dots P_r q^{-r+1}\Delta\sum_{1\leq t<M}t^{-0.5\nu}P_s^{0.5\nu l}P_1^{-0.05\nu^2}\\&+P_1\dots P_{s-1}P_{s+1}\dots P_r q^{-r+1}\Delta^{-1}\sum_{M\leq t<M_1}t^{-2-0.5\nu}P_s^{0.5\nu l}P_1^{-0.05\nu^2}\\&+P_1\dots P_{s-1}P_{s+1}\dots P_r q^{-r+1}P_1^{-\rho}.\end{aligned}$$

Since $\Delta = P_s^{-l} H = P_s^{-l} P_1^{3\rho n}$, it follows that Y_2 satisfies

$$Y_1 \ll P_1 \dots P_{s-1} P_{s+1} \dots P_r q^{-r+1} P_1^{-\rho}.$$

Consequently,

$$|S_3| \le P_s Y \le P_s Y_2 \ll P_1^{1-\rho} P_2 \dots P_r q^{-r+1}.$$

We hence obtain

$$|S'(A)| \le q^{r-1}(|S_1| + |S_2| + |S_3|) \ll P_1^{1-\rho} P_2 \dots P_r.$$

The lemma is proved. □

Suppose that A is a point of the first class. Then its coordinates $\alpha = \alpha(t_1, \dots, t_r)$ satisfy (10.3). Let E_j be the sets defined at the beginning of the section. We let Q_j denote the least common multiple of the denominators $q(t_1, \dots, t_r)$ of the rational approximations to the numbers $\alpha(t_1, \dots, t_r)$ in (10.3), over all r-tuples $(t_1, \dots, t_r)$ in the set E_j. We shall make use of this notation in Lemmas 10.12 and 10.13.

Lemma 10.12. *Suppose that A is a point of the first class, and $Q_0 > Q^{0.2}$. Then*

$$|S'(A)| \ll P_1 \dots P_r Q^{-0.05\nu+\varepsilon}(|t|, Q)^{0.25\nu}.$$

The constant in $\ll$ depends only on $n_1, \dots, n_r$ and ε.

Proof. As in the proof of Lemma 10.9, if we take the fourth power of the sum $|S'(A)|$ and apply Cauchy's inequality twice for certain fixed natural numbers $a_1, \dots, a_r$, $1 \le a_1 \le P_1, \dots, 1 \le a_r \le P_r$, we obtain

$$|S'(A)|^4 \le P_1^3 \dots P_r^3 |W|, \tag{10.24}$$

$$W = \sum_{x_1 \le P_1} \cdots \sum_{x_r \le P_r} \exp\{2\pi i t \Phi(x_1, \dots, x_{r-1}, x_r)\},$$

$$\Phi(x_1, \dots, x_{r-1}, x_r) = F_2(x_1, \dots, x_{r-1}, x_r) - F_2(x_1, \dots, x_{r-1}, a_r)$$

$$-F_2(a_1, \dots, a_{r-1}, x_r) = \sum_{\substack{t_1=0 \\ t_1+\dots+t_r \ge 1}}^{n_1} \cdots \sum_{t_r=0}^{n_r} \gamma(t_1, \dots, t_r) x_1^{t_1} \dots x_r^{t_r},$$

$$F_2(x_1, \dots, x_{r-1}, x_r) = \sum_{\substack{t_1=0 \\ t_1+\dots+t_{r-1} \ge 1}}^{n_1} \cdots \sum_{t_{r-1}=0}^{n_{r-1}} \sum_{t_r=0}^{n_r} \alpha(t_1, \dots, t_{r-1}, t_r) x_1^{t_1} \dots x_{r-1}^{t_{r-1}} x_r^{t_r}.$$

From the definition of the polynomial $\Phi(x_1, \dots, x_r)$ we find that its coefficients satisfy the following relations:

$$\gamma(t_1, \dots, t_{r-1}, t_r) = \alpha(t_1, \dots, t_{r-1}, t_r)$$

for $t_1 + \cdots + t_{r-1} \geq 1$ and $t_r \geq 1$, i.e., for an r-tuple $\bar{t} \in E_0$;

$$\gamma(0, \ldots, 0, t_r) = -\sum_{t_1=0}^{n_1} \cdots \sum_{\substack{t_{r-1}=0 \\ t_1+\cdots+t_{r-1}\geq 1}}^{n_{r-1}} \alpha(t_1, \ldots, t_{r-1}, t_r) a_1^{t_1} \ldots a_{r-1}^{t_{r-1}}$$

for $t_r \geq 1$, i.e., for $\bar{t} \in E_2$;

$$\gamma(t_1, \ldots, t_{r-1}, 0) = -\sum_{t_r=1}^{n_r} \alpha(t_1, \ldots, t_{r-1}, t_r) s_r^{t_r}$$

for $t_1 + \cdots + t_{r-1} \geq 1$, i.e., for $\bar{t} \in E_1$. Hence, if we use relations (10.3), which define the points in the first class Ω_1, we obtain

$$\gamma(\bar{t}) = a_1(\bar{t})/q_1(\bar{t}) + \beta_1(\bar{t}),$$

where

$$a_1(\bar{t})/q_1(\bar{t}) = a(\bar{t})/q(\bar{t}), \quad \beta_1(\bar{t}) = \beta(\bar{t}) \quad \text{for} \quad \bar{t} \in E_0, \tag{10.25}$$

$$\frac{a_1(\bar{t})}{q_1(\bar{t})} = -\sum_{t_r=1}^{n_r} \frac{a(t_1, \ldots, t_r)}{q(t_1, \ldots, t_r)} a_r^{t_r}, \quad \beta_1(\bar{t}) = -\sum_{t_r=1}^{n_r} \beta(\bar{t}) a_r^{t_r}, \quad \bar{t} \in E_1, \tag{10.26}$$

$$\begin{aligned} \frac{a_1(\bar{t})}{q_1(\bar{t})} &= -\sum_{t_r=1}^{n_r} \cdots \sum_{t_{r-1}=0}^{n_{r-1}} \frac{a(t_1, \ldots, t_r)}{q(t_1, \ldots, t_r)} a_1^{t_1} \ldots a_{r-1}^{t_{r-1}}, \\ \beta_1(\bar{t}) &= -\sum_{t_r=1}^{n_r} \cdots \sum_{\substack{t_{r-1}=0 \\ t_1+\cdots+t_{r-1}\geq 1}}^{n_{r-1}} \beta(t_1, \ldots, t_r) a_1^{t_1} \ldots a_{r-1}^{t_{r-1}}, \quad \bar{t} \in E_3. \end{aligned} \tag{10.27}$$

Consequently, $q_1(t_1, \ldots, t_r) \mid Q_0$ for $t_1, \ldots, t_r \geq 1$, and

$$|\beta_1(t_1, \ldots, t_{r-1}, 0)| \leq \frac{n_r + 1}{m} P_1^{-t_1} \ldots P_{r-1}^{-t_{r-1}} P_1^{0.1\nu}, \quad t_1, \ldots, t_{r-1} \geq 1,$$

$$|\beta_1(0, \ldots, 0, t_r)| \leq \frac{(n_1 + 1) \ldots (n_{r-1} + 1)}{m} P_r^{-t_r} P_1^{0.1\nu}, \quad t_r \geq 1.$$

We now transform the sum W using the substitution

$$x_s = Q_0 y_s + z_s, \quad 1 \leq z_s \leq Q_0,$$
$$-z_s Q_0^{-1} < y_s \leq (P_s - z_s) Q_0^{-1}, \quad s = 1, \ldots, r.$$

We obtain

$$\begin{aligned} W &= \sum_{z_1=1}^{Q_0} \cdots \sum_{z_r=1}^{Q_0} \exp\{2\pi i t \Phi_1(z_1, \ldots, z_r)\} W_1, \\ W_1 &= \sum_{y_1} \cdots \sum_{y_r} \exp\{2\pi i t \Phi_2(Q_0 y_1 + z_1, \ldots, Q_0 y_r + z_r)\}, \end{aligned} \tag{10.28}$$

where the summation with respect to the variables y_a is taken over all integers in the interval $-z_s Q_0^{-1} < y_s \le (P_s - z_s)Q_0^{-1}$, $s = 1, \ldots, r$,

$$\Phi_1(z_1, \ldots, z_r) = \sum_{\substack{t_1=0 \\ t_1+\cdots+t_{r-1}\ge 1}}^{n_1} \cdots \sum_{t_{r-1}=0}^{n_{r-1}} \sum_{t_r=1}^{n_r} \frac{a(t_1, \ldots, t_r)}{q(t_1, \ldots, t_r)} z_1^{t_1} \ldots z_r^{t_r}$$
$$+ \sum_{\substack{t_1=0 \\ t_1+\cdots+t_{r-1}\ge 1}}^{n_1} \cdots \sum_{t_{r-1}=0}^{n_{r-1}} \frac{a_1(t_1, \ldots, t_{r-1}, 0)}{q_1(t_1, \ldots, t_{r-1}, 0)} z_1^{t_1} \ldots z_{r-1}^{t_{r-1}} + \sum_{t_r=1}^{n_r} \frac{a_1(0, \ldots, 0, t_r)}{q_1(0, \ldots, 0, t_r)} z_r^{t_r},$$

$$\Phi_2(x_1, \ldots, x_r) = \sum_{\substack{t_1=0 \\ t_1+\cdots+t_{r-1}\ge 1}}^{n_1} \cdots \sum_{t_{r-1}=0}^{n_{r-1}} \sum_{t_r=1}^{n_r} \beta(t_1, \ldots, t_r) x_1^{t_1} \ldots x_r^{t_r} \tag{10.29}$$
$$+ \sum_{\substack{t_1=0 \\ t_1+\cdots+t_{r-1}\ge 1}}^{n_1} \cdots \sum_{t_{r-1}=0}^{n_{r-1}} \beta_1(t_1, \ldots, t_{r-1}, 0) x_1^{t_1} \ldots x_{r-1}^{t_{r-1}} + \sum_{t_r=1}^{n_r} \beta_1(0, \ldots, 0, t_r) x_r^{t_r}.$$

We estimate the first partial derivatives with respect to y_s of the polynomial $t\Phi_2(Q_0 y_1 + z_1, \ldots, Q_0 y_r + z_r)$ for $t \le Q^{0.2\nu}$;

$$\left| t\frac{\partial}{\partial y_s} \Phi_2(Q_0 y_1 + z_1, \ldots, Q_0 y_r + z_r) \right|$$
$$\le |t| \sum_{t_1=0}^{n_1} \cdots \sum_{t_s=1}^{n_s} \cdots \sum_{t_{r-1}=0}^{n_{r-1}} \sum_{t_r=1}^{n_r} t_s Q_0 |\beta(t_1, \ldots, t_r)|$$
$$\times (Q_0 y_1 + z_1)^{t_1} \ldots (Q_0 y_s + z_s)^{t_s - 1} \ldots (Q_0 y_r + z_r)^{t_r}$$
$$+ |t| \sum_{t_1=0}^{n_1} \cdots \sum_{t_s=1}^{n_s} \cdots \sum_{t_{r-1}=0}^{n_{r-1}} t_s Q_0 |\beta(t_1, \ldots, t_{r-1}, 0)|$$
$$\times (Q_0 y_1 + z_1)^{t_1} \ldots (Q_0 y_s + z_s)^{t_s - 1} \ldots (Q_0 y_r + z_r)^{t_r}$$
$$\le |t| Q_0 P_s^{-1} P_1^{0.1\nu} \ll P_1^{-0.5} \le 0.5$$

for $s = 1, \ldots, r-1$; and

$$|t(\partial/\partial y_r)\Phi_2(Q_0 y_1 + z_1, \ldots, Q_0 y_r + z_r)| \ll |t| Q_0 P_r^{-1} P_1^{0.1\nu} \le 0.5.$$

Consequently, by Lemma 10.6 we have

$$W_1 = \int_{-z_1 Q_0^{-1}}^{(P_1 - z_1)Q_0^{-1}} \dots$$
$$\dots \int_{-z_r Q_0^{-1}}^{(P_r - z_r)Q_0^{-1}} \exp\{2\pi i t \Phi_2(Q_0 y_1 + z_1, \dots, Q_0 y_r + z_r)\} \, dy_1 \dots dy_r$$
$$+ O(P_2 \dots P_r Q_0^{-r+1}).$$

We now make the change of variables of integration

$$x_s = P_s^{-1}(Q_0 y_s + z_s), \quad s = 1, \dots, r.$$

We obtain the equality

$$W_1 = P_1 \dots P_r Q_0^{-r} I_r + O(P_2 \dots P_r Q_0^{-r+1}),$$

where

$$I_r = \int_0^1 \dots \int_0^1 \exp\{2\pi i t \Phi_3(x_1, \dots, x_r)\} \, dx_1 \dots dx_r,$$
$$\Phi_3(x_1, \dots, x_r) = \Phi_2(P_1 x_1, \dots, P_r x_r).$$

Substituting W_1 in (10.28) and passing to inequalities, we find that

$$|W| \le P_1 \dots P_r Q_0^{-r} |S(Q_0, Q_0 \Phi_1(z_1, \dots, z_r))| \, |I_r| + O(P_2 \dots P_r Q_0^{-r+1}),$$

where

$$S(Q_0, Q_0 \Phi_1(z_1, \dots, z_r)) = \sum_{z_1=1}^{Q_0} \dots \sum_{z_r=1}^{Q_0} \exp\{2\pi i t \Phi_1(z_1, \dots, z_r)\}$$

and the polynomial $\Phi_1(z_1, \dots, z_r)$ is as in (10.29). The coefficients of $\Phi_1(z_1, \dots, z_r)$ are rational numbers whose denominators have least common multiple Q_0. Hence it follows from Lemma 10.4 that

$$|S(Q_0, Q_0 \Phi_1(z_1, \dots, z_r))| \ll Q_0^{r-\nu+4\varepsilon} (|t|, Q_0)^{\nu},$$

where $\varepsilon > 0$ is arbitrary small. Consequently,

$$|W| \ll P_1 \dots P_r Q_0^{-\nu+4\varepsilon} (|t|, Q_0)^{\nu}.$$

Substituting this in (10.24), we obtain

$$|S'(A)| \le P_1^{0.75} \dots P_r^{0.75} |W|^{0.25} \ll P_1 \dots P_r Q_0^{-0.25\nu+\varepsilon} (|t|, Q_0)^{0.25\nu}$$
$$\ll P_1 \dots P_r Q_0^{-0.05\nu+\varepsilon} (|t|, Q)^{0.25\nu}.$$

The lemma is proved. □

Lemma 10.13. *Suppose that A is a point of the first class Ω_1, and $Q_0 \le Q^{0.2}$ and $Q_2 > Q^{0.4}$. Then*

$$|S'(A)| \ll P_1 \dots P_r \Delta(|t|, Q)^{0.5\nu}, \quad \Delta = Q^{-0.2\nu}.$$

The constant in $\ll$ depends only on $n_1, \dots, n_r$ and ε.

Proof. We have

$$|S'(A)| \le T_1 \tag{10.30}$$

where

$$T_1 = \sum_{x_1 \le P_1} \cdots \sum_{x_{r-1} \le P_{r-1}} \Big| \sum_{p \le P_r} \exp\{2\pi i t F_1(x_1, \dots, x_{r-1}, p)\} \Big|,$$

$$F_1(x_1, \dots, x_{r-1}, x_r) = \sum_{t_1=0}^{n_1} \cdots \sum_{t_{r-1}=0}^{n_{r-1}} \sum_{t_r=1}^{n_r} \alpha(t_1, \dots, t_{r-1}, t_r) x_1^{t_1} \dots x_r^{t_r}.$$

We represent the variables x_s, $s = 1, \dots, r-1$, in the form

$$x_s = Q_0 y_s + z_s, \quad 1 \le z_s \le Q_0, \quad -z_s Q_0^{-1} < y_s \le (P_s - z_s) Q_0^{-1}. \tag{10.31}$$

We then obtain

$$F_1(Q_0 y_1 + z_1, \dots, Q_0 y_{r-1} + z_{r-1}, p) \equiv \Phi(p) + \Phi_1(z_1, \dots, z_r) \pmod 1,$$

where

$$\Phi(p) = \sum_{s=1}^{n_r} A_s p^s = \sum_{s=1}^{n_r} \Bigg(\frac{a(0, \dots, 0, s)}{q(0, \dots, 0, s)} + \sum_{t_1=0}^{n_r} \cdots \sum_{t_{r-1}=0}^{n_{r-1}} \beta(t_1, \dots, t_{r-1}, s)$$
$$\times (Q_0 y_1 + z_1)^{t_1} \dots (Q_0 y_{r-1} + z_{r-1})^{t_{r-1}} \Bigg) p^s$$
$$= \sum_{s=1}^{n_r} \left(\frac{a_s}{q_s} + B_s \right) p^s,$$

$$\Phi_1(z_1, \dots, z_r) = \sum_{t_1=0}^{n_r} \cdots \sum_{t_{r-1}=0}^{n_{r-1}} \sum_{t_r=0}^{n_r} \frac{a(t_1, \dots, t_r)}{q(t_1, \dots, t_r)} z_1^{t_1} \dots z_r^{t_r}.$$

For the number B_s, $1 \le s \le n_r$, we have the bound

$$|B_s| = \Big| \sum_{t_1=0}^{n_r} \cdots \sum_{t_{r-1}=0}^{n_{r-1}} \beta(t_1, \dots, t_{r-1}, s)(Q_0 y_1 + z_1)^{t_1} \dots (Q_0 y_{r-1} + z_{r-1})^{t_{r-1}} \Big|$$
$$\le (n_1 + 1) \dots (n_{r-1} + 1) \delta P_r^{-s} = (m/(n_r + 1)) \delta P_r^{-s}.$$

Since the point A is in the first class, we have $|\delta| \leq m^{-1} P_1^{0.1\nu}$. Consequently,

$$\delta' = |B_s| P_r^s \leq \frac{1}{n_r+1} P_1^{0.1\nu} \leq \frac{1}{n_r+1} P_r^{0.1/n_r}.$$

From this we find that $(A_1, \ldots, A_{n_r})$ is a point in the first class with respect to the parameter P_r. Thus, from Lemma 10.1 we obtain

$$T_2 = \Big| \sum_{p \leq P_r} \exp\{2\pi i t \Phi(p)\} \Big| \ll P_r \Delta_1(|t|, Q_2)^{0.5/n_r}, \quad \Delta_1 = Q_2^{-0.5/n_r+\varepsilon};$$

$$\Delta_1(|t|, Q_2)^{0.5/n_r} \ll Q_2^{-0.5\nu+\varepsilon}(|t|, Q_2)^{0.5\nu} = \Delta_1'.$$

We substitute this estimate into (10.30) and obtain

$$|S'(A)| \leq P_1 \ldots P_r T_2 \ll P_1 \ldots P_r (|t|, Q)^{0.5\nu}, \quad \Delta_r = Q^{-0.2\nu}. \ll P_1 \ldots P_r \Delta_r$$

The lemma is thereby proved. □

10.3 The main theorem

Theorem 10.1. *Let A be a point of the second class Ω_2. Then for $1 \leq t \leq P_1^{2\rho}$*

$$|S'(A)| \ll P_1 \ldots P_r \Delta, \quad \Delta = e^{8\varkappa} P_1^{-\rho_1},$$

where $\varkappa = n_1 + \mu_2 n_2 + \cdots + \mu_r n_r$, and $\mu_2, \ldots, \mu_r$ are natural numbers satisfying the conditions

$$-1 < \ln P_s / \ln P_1 - \mu_s \leq 0, \quad s = 2, \ldots, r, \quad \rho^{-1} = 128 m\varkappa \log(8m\varkappa).$$

Suppose that A is a point in the first class Ω_1. Then for $1 \leq t \leq Q^{0.2\nu}$

$$|S'(A)| \ll P_1^{1+\varepsilon} \ldots P_r \Delta(t, Q)^{0.25\nu}, \quad \Delta = Q^{-0.05\nu+\varepsilon}.$$

Finally, suppose that A is a point in the first class Ω_1 and

$$\delta = \max_{t_1+\cdots+t_r \geq 1} |\beta| P_1^{t_1} \ldots P_1^{t_r} > 1.$$

Then for $1 \leq t \leq P_1^{0.2\nu}$

$$|S'(A)| \ll P_1^{1+\varepsilon} \ldots P_r \Delta, \quad \Delta = \delta^{-\nu+\varepsilon}.$$

The constants in $\ll$ depend only on $n_1, \ldots, n_r$ and on a fixed arbitrary small number $\varepsilon > 0$.

Proof. The theorem is proved by induction on the number of variables r. The theorem is true for $r = 1$ (Lemma 10.1). By the induction assumption, the estimate in the theorem holds for $S'(A)$ in the case of $r-1$ variables and any point A. We now prove the theorem for r variables.

Suppose that A is a point in the second class Ω_2. There are four possible cases (concerning the numbers Q, Q_0, Q_1, and Q_2, see the notation at the beginning of Section 10.2):

(1) $Q_0 > P_1^{\nu/80}$;

(2) $Q_0 \le P_1^{\nu/80}$ and $Q_2 > P_1^{3\nu/80}$;

(3) $Q_0 \le P_1^{\nu/80}$, $Q_2 \le P_1^{3\nu/80}$, and $Q > P_1^{0.1\nu}$;

(4) $Q_0 > P_1^{0.1\nu}$ and $\delta \le m^{-1}P_1^{0.1\nu}$.

We obtained the estimate for $S'(A)$ in case (1) in Lemma 10.9, in case (2) in Lemma 10.10, and in case (4) in Lemma 10.11. It remains to consider case (3).

We obviously have the inequality

$$|S'(A)| \le T_1 = \sum_{x_r=1}^{P_r} \Bigg| \sum_{p_1 \le P_1} \cdots \sum_{p_{r-1} \le P_{r-1}} \exp\{2\pi i t F_1(p_1, \dots, p_{r-1}, x_r)\} \Bigg|,$$

$$F_1 = F_1(p_1, \dots, p_{r-1}, x_r) = \sum_{\substack{t_1=0 \\ t_1+\cdots+t_{r-1}\ge 1}}^{n_1} \cdots \sum_{t_{r-1}=0}^{n_{r-1}} \sum_{t_r=1}^{n_r} \alpha(t_1, \dots, t_{r-1}, t_r) p_1^{t_1} \dots p_{r-1}^{t_{r-1}} x_r^{t_r}$$

$$= \sum_{\substack{t_1=0 \\ t_1+\cdots+t_{r-1}\ge 1}}^{n_1} \cdots \sum_{t_{r-1}=0}^{n_{r-1}} f_{t_1,\dots,t_{r-1}}(x_r) p_{r-1}^{t_{r-1}}.$$

We divide the variable x_r into arithmetic progressions with difference Q_0:

$$x_r = Q_0 y_r + z_r, \quad 1 \le z_r \le Q_0, \quad -z_r Q_0^{-1} < y_r \le (P_r - z_r) Q_0^{-1}. \tag{10.32}$$

We have $T_1 \le T_2$, where

$$T_2 = \sum_{z_r=1}^{Q_0} \sum_{y_r} \Bigg| \sum_{p_1 \le P_1} \cdots \sum_{p_{r-1} \le P_{r-1}} \exp\{2\pi i t F_1(p_1, \dots, p_{r-1}, Q_0 y_r + z_r)\} \Bigg|,$$

and the variable y_r runs through the values in (10.32). We represent the polynomial F_1 in the form

$$F_1(p_1, \dots, p_{r-1}, Q_0 y_r + z_r) = \Phi(p_1, \dots, p_{r-1}, Q_0 y_r)$$

$$= \sum_{\substack{t_1=0 \\ t_1+\cdots+t_{r-1}\ge 1}}^{n_1} \cdots \sum_{t_{r-1}=0}^{n_{r-1}} \sum_{t_r=0}^{n_r} \alpha(t_1, \dots, t_{r-1}, t_r) p_1^{t_1} \dots p_{r-1}^{t_{r-1}} (Q_0 y_r)^{t_r}$$

$$= \sum_{\substack{t_1=0 \\ t_1+\cdots+t_{r-1}\geq 1}}^{n_1} \cdots \sum_{t_{r-1}=0}^{n_{r-1}} g_{t_1,\ldots,t_{r-1}}(Q_0 y_r) p_1^{t_1} \ldots p_{r-1}^{t_{r-1}}.$$

By Lemma 10.7,

$$\alpha_1 = \alpha_1(t_1, \ldots, t_{r-1}, t_r) = a_1/q_1 + \beta_1, \quad Q_4' = Q_4, \quad \delta_0 \ll \delta_0' \ll \delta_0, \quad (10.33)$$

where

$$Q_4' = \underset{t_1+\cdots+t_{r-1}\geq 1}{\text{l.c.m.}} q_1, \qquad Q_4 = \underset{t_1+\cdots+t_{r-1}\geq 1}{\text{l.c.m.}} q,$$
$$\delta_0' = \max_{t_1+\cdots+t_{r-1}\geq 1} P_1^{t_1} \ldots P_r^{t_r}|\beta_1|, \quad \delta_0 = \max_{t_1+\cdots+t_{r-1}\geq 1} P_1^{t_1} \ldots P_r^{t_r}|\beta|.$$

We let Q_j' and $Q'(t_1, \ldots, t_{r-1})$ denote the numbers

$$Q_j' = \underset{t_1+\cdots+t_r\in E_j}{\text{l.c.m.}} q_1, \quad j = 0, 1, 2,$$
$$Q'(t_1, \ldots, t_{r-1}) = \underset{t\geq 1}{\text{l.c.m.}}\, q_1(t_1, \ldots, t_{r-1}, t),$$
$$Q(t_1, \ldots, t_{r-1}) = \underset{t\geq 1}{\text{l.c.m.}}\, q(t_1, \ldots, t_{r-1}, t)$$
$$(0 \leq t_1 \leq n_1, \ldots, 0 \leq t_{r-1} \leq n_{r-1}, t_1 + \cdots + t_{r-1} \geq 1).$$

Then

$$Q_0 = \underset{t_1+\cdots+t_r\geq 1}{\text{l.c.m.}} Q(t_1, \ldots, t_{r-1}), \quad Q_0' = \underset{t_1+\cdots+t_r\geq 1}{\text{l.c.m.}} Q'(t_1, \ldots, t_{r-1}).$$

We again consider relations (10.33) for the coefficients α_1 and apply Lemma 10.6 to the polynomials $f_{t_1,\ldots,t_{r-1}}(Q_0 y_r + z_r)$ and $g_{t_1,\ldots,t_{r-1}}(Q_0 y_r)$. We obtain

$$Q(t_1, \ldots, t_{r-1}) = Q'(t_1, \ldots, t_{r-1}), \quad t_1 + \cdots + t_{r-1} \geq 1.$$

Consequently, $Q_0 = Q_0'$ and

$$Q_4 = [Q_0, Q_1] = Q_4' = [Q_0, Q_1'] \geq P_1^{3\nu/80}, \quad Q_1' \geq Q_4 Q_0^{-1} \geq P_1^{\nu/40}.$$

We transform the polynomial $\Phi(p_1, \ldots, p_{r-1}, Q_0 y_r)$ starting from relations (10.33) for $\alpha_1 = \alpha(t_1, \ldots, t_{r-1}, t_r)$:

$$\Phi(p_1, \ldots, p_{r-1}, Q_0 y_r) = \sum_{\substack{t_1=0 \\ t_1+\cdots+t_{r-1}\geq 1}}^{n_1} \cdots \sum_{t_{r-1}=0}^{n_{r-1}} \sum_{t_r=0}^{n_r} \left(\frac{a_1(t_1, \ldots, t_r)}{q_1(t_1, \ldots, t_r)} + \beta_1(t_1 + \cdots + t_r)\right)$$
$$\times\, p_1^{t_1} \ldots p_{r-1}^{t_{r-1}} (Q_0 y_r)^{t_r} \equiv$$

$$\equiv \Phi(p_1, \dots, p_{r-1}) \pmod 1,$$

$$\Phi(p_1, \dots, p_{r-1}, y) = \sum_{\substack{t_1=0\\ t_1+\dots+t_{r-1}\ge 1}}^{n_1} \cdots \sum_{t_{r-1}=0}^{n_{r-1}} \left(\frac{a_1(t_1, \dots, t_{r-1}, 0)}{q_1(t_1, \dots, t_{r-1}, 0)} + \sum_{t_r=0}^{n_r} \beta_1(t_1 + \dots + t_r)(Q_0 y)^{t_r} \right) p_1^{t_1} \dots p_{r-1}^{t_{r-1}}$$

$$= \sum_{\substack{t_1=0\\ t_1+\dots+t_{r-1}\ge 1}}^{n_1} \cdots \sum_{t_{r-1}=0}^{n_{r-1}} B(t_1 + \dots + t_{r-1}) p_1^{t_1} \dots p_{r-1}^{t_{r-1}}.$$

Consequently,

$$T_1 \le Q_0 \sum_{0 \le y_r \le P_r Q_0^{-1}} \left| \sum_{p_1 \le P_1} \cdots \sum_{p_{r-1} \le P_{r-1}} \exp\{2\pi i t F_1(p_1, \dots, p_{r-1}, y)\} \right| = Q_0 T_3.$$

The estimates for T_3 is similar to that for the corresponding sum in T_3 in Lemma 10.10. The polynomials B_s are replaced by $B(t_1, \dots, t_{r-1})$, the numbers $\tau(s) = P_r^{s-1/6}$ are replaced by $\tau(t_1, \dots, t_{r-1}) = P_1^{t_1} \dots P_{r-1}^{t_{r-1}} P_1^{-1/6}$, $Q(\overline{y}) = Q(y_1, \dots, y_{r-1})$ is replaced by $Q(y_r)$, $\delta(\overline{y})$ is replaced by $\delta(y_r)$, the $(r-1)$-tuples $\overline{y} = (y_1, \dots, t_{r-1})$ are replaced by the variable y_r, and instead of the estimate for a simple trigonometric sum (Lemma 10.1), in the corresponding places, one must apply the estimate for an $(r-1)$-dimensional sum, which holds by the induction assumption.

We have obtained the estimate for $S'(a)$ for points A in the second class Ω_2.

Now suppose that A is a point in the first class Ω_1. One has the following possible cases (the numbers Q, Q_0, Q_1, and Q_2 were defined right before Lemma 10.12):

(1) $Q_0 > Q^{0.2}$;

(2) $Q_0 \le Q^{0.2}$ and $Q_2 > Q^{0.4}$;

(3) $Q_0 \le Q^{0.2}$ and $Q_2 \le Q^{0.4}$;

(4) $Q \le P_1^{0.1\nu}$ and $1 \le \delta \le m^{-1} P_1^{0.1\nu}$.

In case (1) the estimate for $S'(A)$ was obtained in Lemma 10.12, and in case (2) it was obtained in Lemma 10.13. The derivation of the estimate for $S'(A)$ in case (3) is similar to the derivation in case (2). Namely, in the statement and proof of Lemma 10.13, one must replace Q_2 by Q_1, replace the $(r-1)$-tuple of unknowns $(x_1, \dots, x_{r-1})$ by the variable x_r, replace the variable p by the $(r-1)$-tuple $(p_1, \dots, p_{r-1})$, replace B_s by $B(t_1, \dots, t_{r-1})$, and instead of Lemma 10.1, in the corresponding places, one must use the estimate for an $(r-1)$-dimensional sum over prime numbers, which holds by the induction assumption.

It remains to consider case (4). Since A is a point in the first class, its coordinates satisfy the relations

$$\alpha = \alpha(t_1, \dots, t_r) = a/q + \beta, \quad (a, q) = 1,$$

$$\delta = \max_{t_1+\cdots+t_r\geq 1} |\beta| P_1^{t_1} \dots P_r^{t_r} \leq m^{-1} P_1^{0.1\nu}, \quad Q = \operatorname*{l.c.m.}_{t_1+\cdots+t_r\geq 1} q \leq P_1^{0.1\nu}.$$

In the case under study we have $\delta \geq 1$. We set

$$\delta = |\beta(t_1, \dots, t_r)| P_1^{t_1} \dots P_r^{t_r}, \quad t_s \geq 1. \tag{10.34}$$

Then for $|S'(A)|$ we have (for $q \neq s$)

$$|S'(A)| \leq \sum_{x_q \leq P_q} |T_1|,$$

$$T_1 = \sum_{p_1\leq P_1} \cdots \sum_{p_{q-1}\leq P_{q-1}} \sum_{p_{q+1}\leq P_{q+1}} \cdots$$
$$\cdots \sum_{p_r\leq P_r} \exp\{2\pi i t F_1(p_1, \dots, p_{q-1}, x_q, p_{q+1}, \dots, p_r)\},$$

$$F_1(x_1, \dots, x_r) = \sum_{t_q=0}^{n_q} \sum_{t_1=0}^{n_1} \cdots \sum_{t_{q-1}=0}^{n_{q-1}} \sum_{t_{q+1}=0}^{n_{q+1}} \cdots \sum_{\substack{t_r=0 \\ t_1+\cdots+t_{q-1}+t_{q+1}+\cdots+t_r\geq 1}}^{n_r} \alpha(t_1 + \cdots + t_r) x_1^{t_1} \dots x_r^{t_r}.$$

We now represent the variable x_q in the form

$$x_q = Qy + z, \quad 1 \leq z \leq Q, \quad -zQ^{-1} < y \leq (P_q - z)Q^{-1}.$$

Then

$$F_1(x_1, \dots, Qy + z, \dots, x_r) \equiv \Phi(x_1, \dots, x_{q-1}, x_{q+1}, \dots, x_r) \pmod 1,$$
$$\Phi(x_1, \dots, x_{q-1}, x_{q+1}, \dots, x_r)$$
$$= \sum_{t_1=0}^{n_1} \cdots \sum_{t_{q-1}=0}^{n_{q-1}} \sum_{t_{q+1}=0}^{n_{q+1}} \cdots \sum_{\substack{t_r=0 \\ t_1+\cdots+t_{q-1}+t_{q+1}+\cdots+t_r\geq 1}}^{n_r} A(t_1, \dots, t_{q-1}, t_{q+1}, \dots, t_r) x_1^{t_1} \dots x_r^{t_r},$$
$$A(t_1, \dots, t_{q-1}, t_{q+1}, \dots, t_r)$$
$$= \sum_{t_q=0}^{n_q} \frac{a(t_1, \dots, t_r)}{q(t_1, \dots, t_r)} z^{t_q} + \sum_{t_q=0}^{n_q} \beta(t_1, \dots, t_r)(Qy + z)^{t_q}$$

We fix z. Then we need to show that a point A_1 with coordinates

$$A(t_1, \dots, t_{q-1}, t_{q+1}, \dots, t_r)$$

belongs to the first class. In fact, the denominator of the irreducible fraction

$$\frac{a}{h} = \sum_{t_q=0}^{n_q} \frac{a(t_1, \dots, t_r)}{q(t_1, \dots, t_r)} z^{t_q}$$

divides the least common multiple Q of all the numbers $q(t_1, \dots, t_r), t_1+\dots+t_r \ge 1$. Hence the least common multiple of the numbers h does not exceed $P_1^{0.1\nu}$. The number

$$B = B(t_1, \dots, t_{q-1}, t_{q+1}, \dots, t_r) = \sum_{t_q=0}^{n_q} \beta(t_1, \dots, t_r)(Qy+z)^{t_q}$$

satisfies the inequality

$$\begin{aligned}|B| &\le (n_q+1)\delta P_1^{-t_1} \dots P_{q-1}^{-t_{q-1}} P_{q+1}^{-t_{q+1}} \dots P_r^{-t_r} \\ &\le \frac{n_q+1}{m} P_1^{-t_1} \dots P_{q-1}^{-t_{q-1}} P_{q+1}^{-t_{q+1}} \dots P_r^{-t_r} P_1^{0.1\nu}.\end{aligned}$$

We thus have

$$\delta = \max_{t_1+\dots+t_{q-1}+t_{q+1}+\dots+t_r\ge 1} |B| P_1^{t_1} \dots P_{q-1}^{t_{q-1}} P_{q+1}^{t_{q+1}} \dots P_r^{t_r} \le \frac{n_q+1}{m} P_1^{0.1\nu}.$$

The point A_1 thus belongs to the first class. Let $(t_1, \dots, t_r)$ be an r-tuple for which (10.34) holds. We let φ and g denote the polynomials

$$\varphi(Qy+z) = B(t_1, \dots, t_{q-1}, t_{q+1}, \dots, t_r), \quad g(v) = \varphi(P_q v).$$

Then the maximum modulus of the coefficients of $g(x)$ is equal to

$$\delta P_1^{-t_1} \dots P_{q-1}^{-t_{q-1}} P_{q+1}^{-t_{q+1}} \dots P_r^{-t_r}.$$

Let $K(G)$ be the number of integers y for which

$$G < |\varphi(Qy+z)| \le 2G. \tag{10.35}$$

Then, by the induction assumption,

$$\begin{aligned}|S'(A)| &\le BP_1 \dots P_{q-1}P_{q+1} \dots P_r \\ &\quad + \sum_{j=0}^{[\log P_q]+1} K(2^j D) P_1 \dots P_{q-1}P_{q+1} \dots P_r 2^{-\nu j} (\log P_r)^{r-2}, \quad (10.36)\\ D &= P_1^{-t_1} \dots P_{q-1}^{-t_{q-1}} P_{q+1}^{-t_{q+1}} \dots P_r^{-t_r},\end{aligned}$$

where B is the number of integers y that satisfy $|\varphi(Qy+z)| \le D$. By Lemma 10.8, the measure μ of the set of points x for which $|g(x)| \le 2^j D$ does not exceed

$$\ll \min\left(1,\ (2^j\delta^{-1})^{1/n_q}\right). \tag{10.37}$$

Since this set consists of at most n intervals, it follows from the definition of $g(x)$ and $\varphi(Qy+z)$ and inequality (10.37) that

$$K(2^j D) \ll P_q Q^{-1}(2^j\delta^{-1})^{1/n_q} + 1 \ll P_q Q^{-1}(2^j\delta^{-1})^{\nu} + 1 \quad \text{if} \quad 2^j\delta < 1,$$

$$K(2^j D) \ll P_q Q^{-1} + 1 \quad \text{if} \quad 2^j \delta \geq 1.$$

We substitute these estimates into (10.36). If we replace 2^j by δ in (10.36) for $2^j \delta^{-1} \geq 1$, we obtain

$$|S'(A)| \ll P_1 \dots P_r \delta^{-\nu} (\log P_r)^{r-1},$$

This completes the proof of the theorem. □

10.4 Applications

We prove Theorem 10.2, which concerns the distribution of the fractional parts of the values of a polynomial in several variables each of which runs through the sequence of prime numbers. This theorem generalizes a theorem of Vinogradov ([165], Chapter 10.8) to the *multidimensional* case.

Theorem 10.2. *Let $p_1, \dots, p_r$ be prime numbers, $F(x_1, \dots, x_r)$ be the polynomial defined in* (10.1), *and $D(\sigma)$ be the number of r-tuples $(p_1, \dots, p_r)$, $1 \leq p_1 \leq P_1, \dots, 1 \leq p_r \leq P_r$, satisfying the condition $\{F(p_1, \dots, p_r)\} < \sigma$. Suppose that $D(\sigma)$ can be represented in the form*

$$D(\sigma) = \pi(P_1) \dots \pi(P_r)\sigma + \lambda(P_1, \dots, P_r, \sigma),$$

where $\pi(x)$ is the number of primes not exceeding x. Then

$$|\lambda(P_1, \dots, P_r, \sigma)| \ll P_1^{1+\varepsilon} \dots P_r \Delta_1,$$

where Δ_1 is defined as follows: for polynomials $F(x_1, \dots, x_r)$ in the second class,

$$\Delta_1 = e^{8\varkappa} P_1^{-\rho}, \quad \rho^{-1} = 130 m\varkappa \log 8m\varkappa,$$

for polynomials $F(x_1, \dots, x_r)$ in the first class,

$$\Delta_1 = Q^{-0.05\nu+\varepsilon},$$

and, finally, if also $\delta > 1$, then

$$\Delta_1 = \delta^{-\nu+\varepsilon},$$

where $\varepsilon > 0$ is an arbitrary small constant.

Proof. Without loss of generality, we may assume that $\Delta_1 < 0.1$ and $2\Delta_1 < \sigma < 1 - \Delta_1/2$. We consider the function $\psi(x)$ in [165], Chapter II, Lemma 2, with $r = 1$, $\Delta = \Delta_1$, $\alpha = 0.5\Delta_1$, and $\beta = \sigma$. We obtain

$$D(\sigma) = N(\sigma, \Delta_1) + O(P_1 \dots P_r \Delta_1),$$

where

$$N(\sigma, \Delta_1) = \sum_{p_1 \le P_1} \cdots \sum_{p_r \le P_r} \psi\bigl(F(p_1, \ldots, p_r)\bigr),$$

We set

$$\Delta_2^{-1} = \begin{cases} P_1^{\rho} & \text{if } A \in \Omega_2, \\ Q^{0.1\nu} & \text{if } A \in \Omega_1, \\ P_1^{0.1\nu} & \text{if } A \in \Omega_1,\ \delta > 1. \end{cases}$$

Then, after expanding $\psi(x)$ in Fourier series, we have

$$N(\sigma, \Delta_1) = \pi(P_1) \ldots \pi(P_r)\sigma + H + O(P_1 \ldots P_r \Delta_1),$$

where

$$H \ll \sum_{0<t<\Delta_2^{-1}} \frac{|S_t'(A)|}{t} + \sum_{\Delta_2^{-1} \le t < \Delta_2^{-2}} \frac{|S_t'(A)|}{\Delta_1 t^2} + \sum_{t \ge \Delta_2^{-2}} \frac{|S_t'(A)|}{\Delta_1 t^2}$$
$$= \sum\nolimits_1 + \sum\nolimits_2 + \sum\nolimits_3 .$$

To estimate $\sum_1$ and $\sum_2$ we make use of the estimate for $S_t'(A)$ in Theorem 10.1, and we use the trivial estimate of $P_1 \ldots P_r$ for $\sum_3$. We obtain $H \ll P_1 \ldots P_r \Delta_1$. Consequently,

$$D(\sigma) = \pi(P_1) \ldots \pi(P_r)\sigma + O(P_1 \ldots P_r \Delta_1),$$

as was to be proved. □

There is one other important application of our estimates for multidimensional trigonometric sums with prime numbers, in the problem of simultaneously representing a set of natural numbers by terms of the form $p_1^{t_1} \ldots p_r^{t_r}$, $0 \le t_1 \le n_1, \ldots, 0 \le t_r \le n_r$. In what follows, we shall derive an asymptotic formula for the number of such representations. As in the case of the corresponding problem for one-dimensional trigonometric sums with prime numbers (see [117] and [67]), the asymptotic formula turns out to be nontrivial for large k and only if certain arithmetic conditions and order conditions are fulfilled. They are similar to the conditions in [12]; the only difference is that the variables of summation take values in a reduced system of residues modulo a certain number. The derivation of these results was stimulated by the work of Arkhipov [8] on the Hilbert–Kamke problem and the joint work by Arkhipov and Chubarikov [12] on generalizing this problem to the case of multidimensional sums in which the summation is over solid intervals of integers (not just the primes).

Let $J(M)$ denote the number of representations of the set

$$M = \bigl(M(0, \ldots, 1),\ \ldots,\ M(n_1, \ldots, n_r)\bigr)$$

in the form

$$\sum_{j=1}^{k} p_{1,j}^{t_1} \dots p_{r,j}^{t_r} = M(t_1, \dots, t_r)$$
$$(0 \le t_1 \le n_1, \dots, 0 \le t_r \le n_r,\ t_1 + \dots + t_r \ge 1),$$

where the unknowns $p_{s,j}$ are prime numbers with $2 \le p_{s,j} \le P_s$, $s = 1, \dots, r$, $j = 1, \dots, k$. The letters θ and σ denote, respectively, the singular integral and the singular series of the problem under study,

$$\theta = \int_{-\infty}^{+\infty} \cdots \int_{-\infty}^{+\infty} W^k(A) e^{-2\pi i B}\, dA,$$

$$W(A) = \int_0^1 \cdots \int_0^1 \exp\{2\pi i F_A(x_1, \dots, x_r)\}\, dx_1 \dots dx_r,$$

$$F_A(x_1, \dots, x_r) = \sum_{t_1=0}^{n_1} \cdots \sum_{t_r=0}^{n_r} \alpha(t_1, \dots, t_r) x_1^{t_1} \dots x_r^{t_r}, \quad \alpha(0, \dots, 0) = 0,$$

$$B = \sum_{t_1=0}^{n_1} \cdots \sum_{t_r=0}^{n_r} \alpha(t_1, \dots, t_r) M(t_1, \dots, t_r) P_1^{-t_1} \dots P_r^{-t_r},$$

$$\sigma = \sum_{q(0,\dots,1)=1}^{+\infty} \cdots \sum_{q(n_1,\dots,n_r)=1}^{+\infty} \cdots \sum_{\substack{a(0,\dots,1)=1 \\ (a(0,\dots,1),q(0,\dots,1))=1}}^{q(0,\dots,1)=1} \cdots \sum_{\substack{a(n_1,\dots,n_r)=1 \\ (a(n_1,\dots,n_r),q(n_1,\dots,n_r))=1}} U^k(a,q) e^{-2\pi i D},$$

$$U(a,q) = \varphi(Q)^{-r} \sum_{\substack{x_1=1 \\ (x_1,Q)=1}}^{Q} \cdots \sum_{\substack{x_r=1 \\ (x_r,Q)=1}}^{Q} \exp\{2\pi i \Phi_{a,q}(x_1, \dots, x_r)\},$$

$$\Phi_{a,q}(x_1, \dots, x_r) = \sum_{t_1=0}^{n_1} \cdots \sum_{t_r=0}^{n_r} \frac{a(t_1, \dots, t_r)}{q(t_1, \dots, t_r)} x_1^{t_1} \dots x_r^{t_t},$$

$$D = \sum_{t_1=0}^{n_1} \cdots \sum_{t_r=0}^{n_r} \frac{a(t_1, \dots, t_r)}{q(t_1, \dots, t_r)} M(t_1, \dots, t_r);$$

and Q denotes the least common multiple of the numbers $q(t_1, \dots, t_r)$, $0 \le t_1 \le n_1, \dots, 0 \le t_r \le n_1, t_1 + \dots + t_r \ge 1$. Finally, suppose that the conditions $L_s L_1^{-1} \ll 1$, $s = 2, \dots, r$, are fulfilled, and let $\varkappa$ denote the number defined in Theorem 10.1.

Theorem 10.3. *For $k > 16m\varkappa \log 16m\varkappa + 3$ one has the asymptotic formula*

$$J(M) = \sigma\theta (P_1 \dots P_r L_1^{-1} \dots L_r^{-1})^k (P_1^{n_1} \dots P_r^{n_r})^{-m/2} + O\big((P_1 \dots P_r L_1^{-1} \dots L_r^{-1})^k (P_1^{n_1} \dots P_r^{n_r})^{-m/2} L^{-1} \log L\big).$$

Proof. We represent each coordinate $\alpha = \alpha(t_1, \dots, t_r)$, $0 \le t_1 \le n_1, \dots, 0 \le t_r \le n_1$, of the m-dimensional unit cube Ω in the form

$$\alpha = a/q + \beta, \quad (a, q) = 1,\ 0 \le a < q,\ \beta = \delta P_1^{-t_1} \dots P_r^{-t_r}, \tag{10.38}$$

we let Q denote the least common multiple of the numbers q, and we let δ_0 denote the maximum of the numbers $|\delta|$ with $0 \le t_1 \le n_1, \dots, 0 \le t_r \le n_r$ and $t_1 + \dots + t_r \ge 1$. Then we let the first class Ω_1' consist of the points A for which $Q \le L^H$, $\delta_0 \le L^B$, $H = 60r\nu^{-1}$, and $B = 3r\nu^{-1}$. The remaining points A of the cube Ω are put in the second class Ω_2'. We obviously have

$$J(M) = \int \underset{\Omega}{\cdots} \int S^k(A) e^{-2\pi i (A \times M)}\, dA,$$

where

$$S(A) = \sum_{p_1 \le P_1} \cdots \sum_{p_r \le P_r} \exp\{2\pi i F_A(p_1, \dots, p_r)\},$$

$$F_A(p_1, \dots, p_r) = \sum_{t_1=0}^{n_1} \cdots \sum_{t_r=0}^{n_r} \alpha(t_1, \dots, t_r) p_1^{t_1} \dots p_r^{t_r}, \quad \alpha(0, \dots, 0) = 0,$$

$$A \times M = \sum_{t_1-0}^{n_1} \cdots \sum_{t_r=0}^{n_r} \alpha(t_1, \dots, t_r) M(t_1, \dots, t_r).$$

Corresponding to the partition of Ω into the two classes Ω_1' and Ω_2', we split the integral $J(M)$ into two parts: $J(M) = J_1 + J_2$. If A is any point in the second class Ω_2', Theorem 10.1 implies

$$|S(A)| \ll P_1 \dots P_r (L^{-0.05\nu H + \varepsilon H} + L^{-\nu B + \varepsilon B}) \ll P_1 \dots P_r L^{-2r}. \tag{10.39}$$

We set $k_0 = [8m\varkappa \log 16m\varkappa] + 1$, where the number $\varkappa$ is defined in Theorem 10.1. Then, by Theorem 3 of [32],

$$\int \underset{\Omega}{\cdots} \int |S(A)|^{k_0}\, dA \ll (P_1 \dots P_r)^{k_0} (P_1^{n_1} \dots P_r^{n_r})^{-m/2} \tag{10.40}$$

since

$$\ln P_s / \ln P_1 \ll 1, \quad s = 1, \dots, r,\ \varkappa \ll 1$$

(the constants in $\ll$ depend only on n and r). Consequently, if we use (10.39) and (10.40) and take into account that $k \ge 2k_0 + 1$, we obtain

$$J_2 \le \max_{A \in \Omega_2} |S(A)|^{k-k_0} \int \underset{\Omega}{\cdots} \int |S(A)|^{k_0}\, dA \tag{10.41}$$

$$\ll (P_1 \dots P_r L_1^{-1} \dots L_r^{-1})^k (P_1^{n_1} \dots P_r^{n_r})^{-m/2} L^{-1}.$$

We now derive an asymptotic formula for J_1. We let $\omega(a, q)$ denote the region corresponding to fixed a and q in (10.38), and we let $\delta_0 \le L^B$. If a, q and a', q' are distinct pairs, the regions $\omega(a, q)$ and $\omega(a', q')$ are disjoint. Hence

$$J_1 = \sum_{Q \le L^H} \sum_{\text{l.c.m.}(q)=Q} \cdots \sum \sum_{a \bmod q}{}' \cdots \sum{}' J_{aq,q}, \tag{10.42}$$

where $(a(t_1, \dots, t_r), q(t_1, \dots, t_r)) = 1$, $0 \le a(t_1, \dots, t_r) < q(t_1, \dots, t_r)$, $0 \le t_1 \le n_1, \dots, 0 \le t_r \le n_r$, and

$$J_{a,q} = \int \cdots \int_{\omega(a,q)} S^k(A) e^{-2\pi i (A \times M)}\, dA. \tag{10.43}$$

For the sum $S(A)$ with $A \in \omega(a, q)$ we obtain the asymptotic expression

$$\begin{aligned} S(A) &= \sum_{\sqrt{P_1} < p_1 \le P_1} \cdots \sum_{\sqrt{P_r} < p_r \le P_r} \exp\{2\pi i F_A(p_1, \dots, p_r)\} + O(P_1^{1/2} P_2 \dots P_r) \\ &= \sum_{\substack{l_1=0 \\ (l_1, Q)=1}}^{Q-1} \cdots \sum_{\substack{l_r=0 \\ (l_r, Q)=1}}^{Q-1} \exp\{2\pi i F_{a,q}(l_1, \dots, l_r)\} T(l_1, \dots, l_r) \\ &\quad + O(P_1^{1/2} P_2 \dots P_r), \end{aligned} \tag{10.44}$$

where

$$T(l_1, \dots, l_r) = \sum_{\substack{p_1 \equiv l_1 \pmod Q \\ \sqrt{P_1} < p_1 \le P_1}} \cdots \sum_{\substack{p_r \equiv l_r \pmod Q \\ \sqrt{P_r} < p_r \le P_r}} \exp\{2\pi i F_\beta(p_1, \dots, p_r)\}.$$

We rewrite $T(l_1, \dots, l_r)$ in the form

$$\begin{aligned} T(l_1, \dots, l_r) = \sum_{\sqrt{P_1} < n_1 \le P_1} \cdots \sum_{\sqrt{P_r} < n_r \le P_r} &\big(\pi(n_1, Q, l_1) - \pi(n_1 - 1, Q, l_1)\big) \dots \\ &\times \big(\pi(n_r, Q, l_r) - \pi(n_r - 1, Q, l_r)\big) \exp\{2\pi i F_\beta(n_1, \dots, n_r)\}, \end{aligned}$$

and we make an Abel transformation with respect to each variables in succession. We obtain

$$\begin{aligned} T(l_1, \dots, l_r) = \sum_{s=1}^{r} (-1)^s \sum_{1 \le j_1 < \dots < j_s \le r} &\sum_{\sqrt{P_{j_1}} < Q_{j_1} \le P_{j_1} - 1} \cdots \sum_{\sqrt{P_{j_s}} < Q_{j_s} \le P_{j_s} - 1} \\ \times \sum_{\sqrt{P_1} < n_1 \le P_1'} \cdots \sum_{\sqrt{P_r} < n_r \le P_r'} &\big(\pi(n_1, Q, l_1) - \pi(n_1 - 1, Q, l_1)\big) \dots \end{aligned}$$

$$\times \big(\pi(n_r, Q, l_r) - \pi(n_r - 1, Q, l_r)\big)\Delta_{j_1,\dots,j_s} \exp\{2\pi i F_\beta(P'_1, \dots, P'_r)\}$$
$$+ \exp\{2\pi i F_\beta(P_1, \dots, P_r)\} \sum_{\sqrt{P_1}<n_1\le P_1} \cdots \sum_{\sqrt{P_r}<n_r\le P_r} \big(\pi(n_1, Q, l_1)$$
$$- \pi(n_1 - 1, Q, l_1)\big) \dots \big(\pi(n_r, Q, l_r) - \pi(n_r - 1, Q, l_r)\big),$$
$$P'_l = \begin{cases} Q_l & \text{if } l = j_1, \dots, j_s, \\ P_l & \text{if } l \ne j_1, \dots, j_s. \end{cases}$$

Since $Q < L^H$, it follows from Siegel's theorem that $\pi(x, Q, l)$ satisfies the asymptotic formula

$$\pi(x, Q, l) = \frac{1}{\varphi(Q)} \operatorname{li} x + O(xe^{-c\sqrt{\ln x}}), \quad \operatorname{li} x = \int_2^x \frac{dt}{\ln t}.$$

We substitute this formula in the last expression for $T(l_1, \dots, l_r)$. Then

$$T(l_1, \dots, l_r) = \varphi(Q)^{-r} \sum_{\sqrt{P_1}<n_1\le P_1} \cdots \sum_{\sqrt{P_r}<n_r\le P_r} \exp\{2\pi i F_\beta(n_1, \dots, n_r)\}$$
$$\times \int_{n_1-1}^{n_1} \frac{dt_1}{\ln t_1} \cdots \int_{n_r-1}^{n_r} \frac{dt_r}{\ln t_r}$$
$$+ O(|\beta| P_1^{n_1+1} \dots P_r^{n_r+1} L_1^{-1} \dots L_r^{-1} e^{-c\sqrt{L}}) + O(P_1 \dots P_r L_1^{-1} \dots L_r^{-1} e^{-c\sqrt{L}})$$
$$= \varphi(Q)^{-r} \int_2^{P_1} \cdots \int_2^{P_r} \frac{\exp\{2\pi i F_z(t_1, \dots, t_r)\}}{\ln t_1 \dots \ln t_r} \, dt_1 \dots dt_r$$
$$+ O(P_1 \dots P_r L_1^{-1} \dots L_r^{-1} L^{-H})$$
$$= \varphi(Q)^{-r} P_1 \dots P_r L_1^{-1} \dots L_r^{-1} \int_0^1 \cdots \int_0^1 \exp\{2\pi i F_\delta(t_1, \dots, t_r)\} \, dt_1 \dots dt_r$$
$$+ O(P_1 \dots P_r L_1^{-1} \dots L_r^{-1} L^{-1} \log L).$$

From this and (10.44) we obtain

$$S(A) = U(a, q) W(\delta) + O(P_1 \dots P_r L_1^{-1} \dots L_r^{-1} L^{-1} \log L).$$

If we substitute this into (10.43), extend the integration onto the entire m-dimensional space, and then substitute the resulting expression into (10.44) and take the summation there over all $Q \ge 1$, we find that

$$J_1 = \sigma\theta(P_1 \dots P_r L_1^{-1} \dots L_r^{-1})^k (P_1^{n_1} \dots P_r^{n_r})^{-m/2}$$
$$+ O\big((P_1 \dots P_r L_1^{-1} \dots L_r^{-1})^k (P_1^{n_1} \dots P_r^{n_r})^{-m/2} L^{-1} \log L\big).$$

The last formula together with (10.41) gives the desired asymptotic formula for $J(M)$. □

10.5 On Vinogradov's problems in the theory of prime numbers

In this section we consider several number theory problems related to Vinogradov's method of trigonometric sums. By a trigonometric sum we mean a sum of the form

$$S = \sum_{x \in \Omega} e^{2\pi i f(x)},$$

where $f(x)$ is a real function of one or several variables defined on a discrete set Ω. Such sums have proved to be a very useful instrument in the study of the distribution of values of functions of several variables defined on a discrete set.

Developing the method of trigonometric sums, Vinogradov arrived at the contemporary understanding of the essence of this class of problems. His outlook is presented in [153], [158], [160], [150], [151], [154], [155], [156], [157], [165], [162] and [159]. Vinogradov has shown, in essence, how wide a class of arithmetic problems can be formulated in the language of trigonometric sums. Such a reformulation has proved to be useful, since it allows the introduction of elements of infinitesimal analysis in their solution. Thus along with arithmetic problems there have arisen problems dual to them, connected with integration of trigonometric sums. In view of this duality, Vinogradov looked at the latter problems as equally important as the initial direct problems. We note that advance in the solution of the direct problem brings something to the dual problem and conversely, but these problems are not identical.

The first problem in the theory of trigonometric sums is that of obtaining an upper bound for the modulus of the sum (see [158], Introduction, Section 1). Closely related with this bound are the problems of the distribution of values of the fractional part of a real function $F(\overline{x}) = F(x_1, \dots, x_r)$ on a discrete set Ω (see [158], Introduction, Section 2) and the distribution of values of a function $f(\overline{x}) = f(x_1, \dots, x_r)$ taking integer values on Ω (see [158], Introduction, Section 3).

It should be noted that, in general, these problems, formulated by Vinogradov, are of interest only in the case where $F(\overline{x})$, $f(\overline{x})$, and Ω have some arithmetic properties. Choosing the functions $F(\overline{x})$ and $f(\overline{x})$ and the domain Ω appropriately, we arrive at the problems of Goldbach, Warning, Goldbach–Warning, Hilbert–Kamke, the problem of estimating Weyl's sums, etc.

Problems related to multiple trigonometric sums over prime numbers belong to this class of problems. More precisely, if we take for Ω the set of points $(p_1, \dots, p_r)$ with coordinates p_s, $1 \le s \le r$, running independently over the set of consecutive primes, and for $F(x_1, \dots, x_r)$ a polynomial with arbitrary real coefficients, we arrive at the problem of obtaining an upper bound for the modulus of a multiple trigonometric sum with prime numbers, i.e., of a sum of the form

$$S = S(A) = \sum_{p_1 \le P_1} \cdots \sum_{p_r \le P_r} \exp\{2\pi i F(p_1, \dots, p_r)\},$$

where A is a point with real coordinates $\alpha(t_1, \dots, t_r)$ in the m-dimensional space, $m = (n_1 + 1)\dots(n_r + 1)$, $P_1, \dots, P_r \geq 1$,

$$F(\overline{x}) = F(x_1, \dots, x_r) = \sum_{t_1=0}^{n_1} \cdots \sum_{t_r=0}^{n_r} \alpha(t_1, \dots, t_r) x_1^{t_1} \dots x_r^{t_r}.$$

For $r = 1$, S becomes a simple trigonometric sum with prime numbers. We note that getting estimates for such sums even in the case of a linear polynomial $F(p) = \alpha p$ is fraught with great difficulties. For the first time such estimates were obtained by Vinogradov in 1937 in [154]; this allowed him to solve the Goldbach problem. The problem of finding bounds for sums with an arbitrary polynomial $F(p)$ of higher degree is substantially more complicated. Its complete solution is due to Vinogradov [165]. We outline Vinogradov's scheme [18] for estimating the sums S. By means of the sieve method he reduced it to estimating a small number of double sums W of the form

$$W = \sum_d \xi(d) \sum_m \eta(m) \exp\{2\pi i f(md)\},$$

where $\xi(d)$ and $\eta(m)$ are certain complex-valued functions.

The estimate of the sum W is obtained similarly to that found by Vinogradov in his earlier paper [153] for a sum S_1 of the form

$$S_1 = \sum_x \sum_y e^{2\pi i f(x,y)},$$

where the unknowns x and y run over certain sequences of natural numbers within the limits, respectively, from $M + 1$ to $M + X$ and from $N + 1$ to $N + Y$.

Obviously,

$$|S_l| \leq \sum_{u=M+1}^{M+X} \left| \sum_y e^{2\pi i f(u,y)} \right|,$$

where u runs through all the integers from $M + 1$ to $M + X$ (the smoothing method).

Further, applying the Cauchy inequality, we get

$$|S_l|^2 \leq X \sum_{v=N+1}^{N+Y} \sum_{v_1=N+1}^{N+Y} |T(v, v_1)|,$$

where

$$T = T(v, v_1) = \sum_{u=M+1}^{M+X} e^{2\pi i \varphi(u)}, \quad \varphi(u) = f(u, v) - f(u, v_1).$$

If for any fixed values of v, v_1 satisfactory estimates of the sums T are known, then it is possible to obtain a nontrivial estimate of the sum S_1 by such a method.

A combination of Vinogradov's method for estimating trigonometric sums with prime numbers and the theory of multiple trigonometric sums, created in [35], has made it possible to obtain estimates of multiple trigonometric sums with prime numbers [53] (see Section 10.3).

Here we find estimates of multiple trigonometric sums in which the summation is over arbitrary sequences of integers. For a nontrivial estimate of such sums it suffices that the sequences be "dense," have "small" multiplicity of repetition of their terms, and, in addition, that the corresponding simple trigonometric sums be estimated nontrivially. It should be stressed that in the derivation of estimates the key moment in the reasoning is the use of results of the theory of multiple trigonometric sums [35].

Let $t, r, n_1, \dots, n_r, P_1, \dots, P_r$ be natural numbers, $P_1 = \min(P_1, \dots, P_r)$, $n = \max(n_1, \dots, n_r)$, $\nu n = 2$, let

$$F(x_1, \dots, x_r) = \sum_{t_1=0}^{n_1} \cdots \sum_{t_r=0}^{n_r} \alpha(t_1, \dots, t_r) x_1^{t_1} \dots x_r^{t_r}$$

be a polynomial with real coefficients $\alpha(t_1, \dots, t_r)$, and let $a_1(x_1), \dots, a_r(x_r)$ be complex-valued functions of a natural argument.

Our problem is to estimate the multiple trigonometric sum

$$T = T_r(A) = \sum_{x_1 \le P_1} \cdots \sum_{x_r \le P_r} a_1(x_1), \dots, a_r(x_r) \exp\{2\pi i t F(x_1, \dots, x_r)\}$$

for any point A in the m-dimensional space, i.e., for a polynomial $F(x_1, \dots, x_r)$ of general form. We note that if $a_s(x) = 1$ for x prime, and $a_s(x) = 0$ otherwise, for $s = 1, \dots, r$, then we obtain the sum $S(A)$.

Later we shall need the following notation and definition:

Ω is an $(m-1)$-dimensional cube with coordinates $\alpha(t_1, \dots, t_r)$ satisfying

$$-\big(\tau(t_1, \dots, t_r)\big)^{-1} \le \alpha(t_1, \dots, t_r) \le 1 - \big(\tau(t_1, \dots, t_r)\big)^{-1},$$
$$\tau(t_1, \dots, t_r) = P_1^{t_1} \dots P_r^{t_r} P_1^{-1/6}$$
$$(0 \le t_1 \le n_1, \dots, 0 \le t_r \le n_r, t_1 + \cdots + t_r \ge 1), \quad \alpha(0, \dots, 0) = 0.$$

Definition 10.3. A point A with coordinates $\alpha(t_1, \dots, t_r), 0 \le t_1 \le n_1, \dots, 0 \le t_r \le n_r, t_1 + \cdots + t_r \ge 1, \alpha(0, \dots, 0)$, will be called a point of the first class Ω_1 if

$$\alpha = \alpha(t_1, \dots, t_r)$$

can be represented in the form

$$\alpha = a/q + \beta, \quad (a, q) = 1, \quad 0 \le a < q,$$
$$|\beta| \le m^{-1} P_1^{-t_1} \dots P_r^{-t_r} P_1^{0.1\nu}$$
$$(0 \le t_1 \le n_1, \dots, 0 \le t_r \le n_r, t_1 + \cdots + t_r \ge 1),$$

and the least common multiple Q of the numbers $q(t_1, \ldots, t_r), 0 \le t_1 \le n_1, \ldots, 0 \le t_r \le n_r, t_1 + \cdots + t_r \ge 1$, does not exceed $P_1^{0.1\nu}$. The remaining points of the cube Ω will be called points of the second class Ω_2.

Now we formulate a hypothesis concerning estimates of a simple trigonometric sum.

Hypothesis 10.1. *Let the point $A = (\alpha(1), \ldots, \alpha(n))$ of the unit n-dimensional cube be divided into two classes according to the above definition. Also, let*

$$R_s = \sum_{x_1 \le P_1} |a_s(x_s)|^2,$$

and let $\gamma > 0$, $c > 0$ be absolute constants. Then for $1 \le t \le \Delta_1^{-2}$ the following estimate holds:

$$|T_1(A)| = \Big| \sum_{x \le P_s} a_s(x) \exp\{2\pi i t F(x)\} \Big| \ll (R_s P_s)^{1/2} \Delta_1,$$

where

$$\Delta_1 = \begin{cases} P_s^{-\rho_1}, \ \rho_1 = \gamma (n^2 \log n)^{-1} & \text{if } A \in \Omega_2, \\ Q^{-c\nu} (tmQ)^{c\nu} & \text{if } A \in \Omega_1, \\ (Q\delta_0)^{-c\nu} & \text{if } A \in \Omega_1 \text{ and } \delta_0 \ge 1. \end{cases}$$

In the next lemma we define parameters needed to formulate the theorems in this section.

Lemma 10.14. *Let a point A belong to the second class Ω_2 and let $\mu_s, s = 2, \ldots, r$, be natural numbers satisfying the conditions*

$$-1 < \log P_s / \log P_1 - \mu_s \le 0, \quad \varkappa = n_1 + \mu_2 n_2 + \cdots + \mu_r n_r,$$
$$\Delta_r = P_1^{-\rho_r}, \quad \rho_r^{-1} = 32 m \varkappa \log(8 m \varkappa).$$

The for $1 \le t \le \Delta_r^{-2}$

$$\Big| \sum_{x_1 \le P_1} \cdots \sum_{x_r \le P_r} \exp\{2\pi i t F(x_1, \ldots, x_r)\} \Big| \ll e^{32\varkappa} P_1 \ldots P_r \Delta_r.$$

The constant in $\ll$ depends only on $n_1, \ldots, n_r$.

For the proof see [35] p. 198, Theorem 2.

Theorem 10.4. *Let a point A belong to the second class Ω_2. If Hypothesis 10.1 holds for $1 \le t \le \Delta_r^{-2}$, then*

$$|T_r(A)| \ll (R_1 \ldots R_r P_1 \ldots P_r)^{1/2} \Delta, \quad \Delta = e^{8\varkappa} \Delta_r^{1/4}.$$

The constant in $\ll$ depends only on $n_1, \ldots, n_r$.

Proof. Denote by E the set of all r-tuples of integers $(t_1, \dots, t_r)$ with the condition $0 \le t_1 \le n_1, \dots, 0 \le t_r \le n_r$, $t_1 + \dots + t_r \ge 1$. We define a subset E_0 of E by the condition $t_r \ge 1$, $t_1 + \dots + t_r \ge 1$, a subset E_1 by the condition $t_r = 0$, $t_1 + \dots + t_r \ge 1$, and finally a subset E_2 by the condition $t_r \ge 1$, $t_1 = \dots = t_r = 0$.

Now consider the numbers $\alpha = \alpha(t_1, \dots, t_r)$, $(t_1, \dots, t_r) \in E$. By the Dirichlet theorem they can be represented in the form

$$\alpha = a/q + \beta, \quad (a, q) = 1, \quad 0 \le q \le \tau, \quad |\beta| \le (q\tau)^{-1},$$
$$\tau = \tau(t_1, \dots, t_r) = P_1^{t_1} \dots P_r^{t_r} P_1^{-1/6}.$$

Further, denote by Q, Q_0, Q_1, Q_2 the least common multiples of the numbers $q = q(t_1, \dots, t_r)$ with r-tuples $(t_1, \dots, t_r)$ in E, E_0, E_1, E_2, respectively; denote by δ the maximum of the quantities

$$|\beta(t_1, \dots, t_r)| P_1^{t_1} \dots P_r^{t_r}$$

over all $(t_1, \dots, t_r) \in E$.

We partition the points A of the second class Q_2 into four subclasses Q_{21}, Q_{22}, Q_{23}, Q_{24} according to the values of the quantities Q, Q_0, Q_1, Q_2, and δ. We put in the subclass Ω_{21} those points for which $Q_0 > P_1^{\nu/80}$, in the subclass Ω_{22} the points for which $Q_0 \le P_1^{\nu/80}$ and $Q_2 > P_1^{3\nu/80}$, in the subclass Ω_{23} the points for which $Q_0 \le P_1^{\nu/80}$, $Q_2 \le P_1^{3\nu/80}$, and $Q > P_1^{0.1\nu}$, and finally, in the subclass Ω_{24} the points for which $Q > P_1^{0.1\nu}$ and $\delta \le m^{-1} P_1^{0.1\nu}$.

Let $A \in \Omega_{21}$. Then, by the Cauchy inequality, we obtain

$$\begin{aligned}
|T(A)|^2 &\le \sum_{x_r \le P_r} |a_r(x_r)|^2 \sum_{x_r \le P_r} \Bigg| \sum_{x_1 \le P_1} \cdots \sum_{x_{r-1} \le P_{r-1}} a_1(x_1) \dots a_{r-1}(x_{r-1}) \\
&\qquad \times \exp\{2\pi i t F(\overline{x})\} \Bigg|^2 \\
&\le R_r \sum_{x_1, x_1' \le P_1} |a_1(x_1)| \cdot |a_1(x_1')| \dots \sum_{x_{r-1}, x_{r-1}' \le P_{r-1}} |a_{r-1}(x_{r-1})| \cdot |a_{r-1}(x_{r-1}')| \\
&\qquad \times \Bigg| \sum_{x_r \le P_r} \exp\big\{2\pi i t \big(F_1(x_1, \dots, x_{r-1}, x_r) - F_1(x_1', \dots, x_{r-1}', x_r')\big)\big\} \Bigg|,
\end{aligned}$$

$$F_1(x_1, \dots, x_{r-1}, x_r) = \sum_{\substack{t_1=0 \\ t_1+\dots+t_{r-1} \ge 1}}^{n_1} \cdots \sum_{t_{r-1}=0}^{n_{r-1}} \sum_{t_r=0}^{n_r} \alpha(t_1, \dots, t_{r-1}, t_r) x_1^{t_1} \dots x_{r-1}^{t_{r-1}} x_r^{t_r}.$$

We use the Cauchy inequality one more time:

$$|T(A)|^4 \le R_1^2 \dots R_r^2 \sum_{x_1, x_1' \le P_1} \cdots \sum_{x_r, x_r' \le P_r} \exp\bigl\{2\pi i t\bigl(F_1(x_1, \dots, x_{r-1}, x_r) - F_1(x_1, \dots, x_{r-1}, x_r') - F_1(x_1', \dots, x_{r-1}', x_r) + F_1(x_1', \dots, x_{r-1}', x_r')\bigr)\bigr\}.$$

From this, for some fixed values $x_1' = a_1, \dots, x_r' = a_r$, we shall have

$$|T(A)|^4 \le R_1^2 \dots R_r^2 P_1 \dots P_r |W|,$$

$$W = \sum_{x_1 \le P_1} \cdots \sum_{x_r \le P_r} \exp\{2\pi i t \Phi(x_1, \dots, x_r)\},$$

$$\Phi(x_1, \dots, x_r) = F_1(x_1, \dots, x_{r-1}, x_r) - F_1(x_1, \dots, x_{r-1}, a_r) - F_1(a_1, \dots, a_{r-1}, x_r).$$

We obtained earlier in Section 10.2, Lemma 10.9, the following estimate of W:

$$|W| \ll e^{32\varkappa} P_1 \dots P_r \Delta_r,$$

where $\varkappa$ and Δ_r are defined in Lemma 10.1. Consequently,

$$|T(A)| \ll (R_1 \dots R_r P_1 \dots P_r)^{1/2} \Delta, \qquad \Delta = e^{8\varkappa} \Delta_r^{1/4}.$$

Thus the assertion of the theorem is proved for the points $A \in \Omega_{21}$.

Now let $A \in \Omega_{22}$. By the Cauchy inequality, we have

$$|T(A)|^2 \le R_1 \dots R_{r-1} \sum_{x_1 \le P_1} \cdots \sum_{x_{r-1} \le P_{r-1}} \Bigl| \sum_{x_r \le P_r} a_r(x_r) \exp\{2\pi i t F_2(x_1, \dots, x_r)\} \Bigr|^2,$$

where

$$F_2(x_1, \dots, x_{r-1}, x_r) = \sum_{t_1=0}^{n_1} \cdots \sum_{t_{r-1}=0}^{n_{r-1}} \sum_{t_r=1}^{n_r} \alpha(t_1, \dots, t_{r-1}, t_r) x_1^{t_1} \dots x_{r-1}^{t_{r-1}} x_r^{t_r} = \sum_{t=1}^{n_r} f_t(x_1, \dots, x_{r-1}) x_r^t.$$

After the change of variables $x_s = Q_0 y_s + z_s$, $1 \le z_s \le Q_0$, $-z_s Q_0^{-1} < y_s \le (P_s - z_s)Q_0^{-1}$ $(s = 1, \dots, r-1)$, we obtain

$$|T(A)|^2 \le R_1 \dots R_{r-1} \sum_{z_1=1}^{Q_0} \dots \sum_{z_{r-1}=1}^{Q_0} \sum_{0 \le y_1 \le P_1 Q_0^{-1}} \dots \sum_{0 \le y_{r-1} \le P_{r-1} Q_0^{-1}} \Big| \sum_{x_r \le P_r} a_r(x_r)$$
$$\times \exp\{2\pi i t F_1(Q_0 y_1 + z_1, \dots, Q_0 y_{r-1} + z_{r-1}, x_r)\} \Big|^2.$$

Hence for some fixed $z_1, \dots, z_{r-1}$, we shall have

$$|T(A)|^2 \le R_1 \dots R_{r-1} Q_0^{r-1} T_1,$$

where

$$T_1 = \sum_{0 \le y_1 \le P_1 Q_0^{-1}} \dots \sum_{0 \le y_{r-1} \le P_{r-1} Q_0^{-1}} \Big| \sum_{x_r \le P_r} a_r(x_r) \exp\{2\pi i t \Phi_1(y_1, \dots, y_{r-1}, x_r)\} \Big|^2,$$
$$\Phi_1(y_1, \dots, y_{r-1}, x_r) \equiv F_1(Q_0 y_1 + z_1, \dots, Q_0 y_{r-1} + z_{r-1}, x_r) \pmod 1.$$

Then, reasoning as in the proof of Lemma 10.10 in Section 10.2 and using Hypothesis 10.1 instead of Lemma 10.1 at appropriate places, we obtain

$$|T_1| \le P_1 \dots P_{r-1} Q_0^{-r+1} R_r P_r \Delta_1^2.$$

Consequently,

$$|T(A)|^2 \ll R_1 \dots R_r P_1 \dots P_r \Delta_1^2,$$

from which the estimate

$$|T(A)| \ll (R_1 \dots R_r P_1 \dots P_r)^{1/2} \ll (R_1 \dots R_r P_1 \dots P_r)^{1/2} \Delta$$

follows. For the points A from the subclass Ω_{22} the assertion of Theorem 10.4 is proved.

Let $A \in \Omega_{24}$. In this case the conditions $Q \le P_1^{0.1\nu}$, $\delta > m^{-1} P_1^{0.1\nu}$ are fulfilled. As in the proof of Lemma 10.11 in Section 10.2, $|\delta(t_1, \dots, t_r)| = \delta$, $t_s \ge 1$, and let q denote the least common multiple of the numbers $q(t_1, \dots, t_r)$ under the condition that the indices $t_1, \dots, t_r$ satisfy the inequalities

$$t_1 \ge 0, \dots, t_{s-1} \ge 0, \;\; t_s \ge 1, \;\; t_{s+1} \ge 0, \dots, t_r \ge 0.$$

Representing the variables x_l in the form

$$x_l = q y_l + z_l, \quad 1 \le z_l \le q, \quad -z_l q^{-1} < y \le (P_l - z_l) q^{-1}$$
$$(l = 1, \dots, s-1, s+1, \dots, r)$$

and applying the Cauchy inequality, we obtain

$$|T(A)|^2 \leq R_1 \dots R_{s-1} R_{s+1} \dots R_r q^{r-1} T_2,$$

$$T_2 = \sum_{0 \leq y_1 \leq P_1 q^{-1}} \cdots \sum_{0 \leq y_{s-1} \leq P_{s-1} q^{-1}} \sum_{0 \leq y_{s+1} \leq P_{s+1} q^{-1}} \cdots$$
$$\cdots \sum_{0 \leq y_r \leq P_r q^{-1}} \left| \sum_{x \leq P_s} a_s(x) \exp\{2\pi i t \Phi_2(y_1, \dots, y_{s-1}, x, y_{s+1}, \dots, y_r)\} \right|^2,$$

where

$$\Phi_2(y_1, \dots, y_{s-1}, x, y_{s+1}, \dots, y_r)$$
$$\equiv F_3(qy_1 + z_1, \dots, qy_{s-1} + z_{s-1}, x, qy_{s+1} + z_{s+1}, \dots, qy_r + z_r) \pmod 1,$$

$$F_3(x_1, \dots, x_r) = \sum_{t_1=0}^{n_1} \cdots \sum_{t_{s-1}=0}^{n_{s-1}} \sum_{t_s=1}^{n_s} \sum_{t_{s+1}=0}^{n_{s+1}} \cdots \sum_{t_r=0}^{n_r} \alpha(t_1, \dots, t_r) x_1^{t_1} \dots x_r^{t_r}.$$

The estimation of the sum T_2 is carried out similarly to that of T_1 in Lemma 10.11 in Section 10.2, with the replacement of Lemma 10.1 by Hypothesis 10.1 at appropriate places. We have

$$T_2 \ll P_1 \dots P_{s-1} P_{s+1} \dots P_r q^{-1} P_s R_s \Delta_1^2.$$

Consequently, in the case $A \in \Omega_{24}$ we obtain

$$|T(A)| \ll (R_1 \dots R_s P_1 \dots P_r)^{1/2} \Delta_1 \ll (R_1 \dots R_s P_1 \dots P_r)^{1/2} \Delta.$$

Finally, we consider the case $A \in \Omega_{23}$. In this case we use induction. Theorem 10.1 holds for $r = 1$ (Hypothesis 10.1). Assume that the hypothesis holds for a polynomial $F(x_1, \dots, x_{r-1})$ provided that the point A whose coordinates are the coefficients of the polynomial belongs to the second class. We prove the theorem for the case of r variables.

After an application of the Cauchy inequality and a change of the variables x_r of the form

$$x_r = Q_0 y_r + z_r, \quad 1 \leq z_r \leq Q_0, \quad -z_r Q_0^{-1} < y_r \leq (P_r - z_r) Q_0^{-1},$$

for some fixed z_r we obtain

$$|T(A)|^2 \ll R_r Q_0 T_3,$$

where

$$T_3 = \sum_{0 \leq y \leq P_r Q_0^{-1}} \left| \sum_{x_1 \leq P_1} \cdots \sum_{x_{r-1} \leq P_{r-1}} a_1(x_1) \dots a_{r-1}(x_{r-1}) \right.$$
$$\left. \times \exp\{2\pi i t \Phi_3(x_1, \dots, x_{r-1}, y)\} \right|^2,$$

$$\Phi_3(x_1, \ldots, x_{r-1}, y) \equiv F_4(x_1, \ldots, x_{r-1}, Q_0 y + z_r) \pmod 1,$$

$$F_4(x_1, \ldots, x_{r-1}, x_r) = \sum_{\substack{t_1=0 \\ t_1+\cdots+t_{r-1}\geq 1}}^{n_1} \cdots \sum_{t_{r-1}=0}^{n_{r-1}} \sum_{t_r=0}^{n_r} \alpha(t_1, \ldots, t_r) x_1^{t_1} \ldots x_r^{t_r}.$$

Further, acting similarly as in the case $A \in \Omega_{22}$ (see Section 10.3, proof of Theorem 10.1), we find that

$$|T(A)|^2 \ll e^{16\varkappa_{r-1}} R_1 \ldots R_r P_1 \ldots P_r \Delta_{r-1}^2,$$
$$|T(A)| \ll (R_1 \ldots R_s P_1 \ldots P_r)^{1/2} \Delta,$$

since, obviously,

$$e^{8\varkappa_{r-1}} \Delta_{r-1} \ll e^{8\varkappa} \Delta_r = \Delta.$$

Theorem 10.4 is completely proved. □

Theorem 10.5. *Let a point A belong to the first class Ω_1. For $1 \leq t \leq Q^{0.2\nu}$,*

$$|T(A)| \ll (R_1 \ldots R_r P_1 \ldots P_r)^{1/2} \Delta,$$

where

$$\Delta = \begin{cases} Q^{-c\nu}(t, Q)^{c\nu} & \text{if } A \in \Omega_1, \\ \delta^{-\nu+\varepsilon} & \text{if } A \in \Omega_1,\ \delta > 1. \end{cases}$$

The constants in $\ll$ depend only on $n_1, \ldots, n_r, \varepsilon$ and the number $\varepsilon > 0$ is arbitrarily small, but fixed.

Proof. We partition the points $A \in \Omega_1$ into three subclasses Ω_{11}, Ω_{12}, and Ω_{13}, according to the conditions:

(1) $Q_0 \geq Q^{0.2}$;

(2) $Q_0 \leq Q^{0.2}$ and $Q_2 > Q^{0.4}$;

(3) $Q_0 \leq Q^{0.2}$ and $Q_2 \leq Q^{0.4}$.

The case $A \in \Omega_{11}$ is considered as in Lemma 10.12 in Section 10.2 and the case $A \in \Omega_{12}$ as in Lemma 10.13 with the replacement, at appropriate places in the proof, of Lemma 10.1 by Hypothesis 10.1, and finally the case $A \in \Omega_{13}$ is proved by induction on the number of variables of the polynomial $F(x_1, \ldots, x_r)$, similarly to the case $A \in \Omega_{23}$. Theorem 10.5 is proved. □

Now we turn to applications of our estimates for multiple trigonometric sums $T(A)$. A similar way of applying trigonometric sums, nowadays called "Vinogradov's method of goblets or cups," has been used by him to solve problems on distribution of the fractional parts of a real function and on the number of lattice points in domains (see [158], [160], [150], [159]).

Let r sequences of natural numbers x_{ij}, $i = 1, \dots, r$, $j = 1, 2, \dots$, be given. We introduce functions $a_i(x)$ defining the multiplicity of repetition of terms of the sequence x_{ij}, i.e., $a_{ij}(x)$ is equal to the number of the x_{ij} such that $x_{ij} = x$, $j = 1, 2, \dots$. Denote by $D(\sigma)$ the number of r-tuples $(x_{1j_1}, \dots, x_{rj_r})$, $1 \le x_{1j_1} \le P_1, \dots, 1 \le x_{rj_r} \le P_r$, satisfying the condition

$$\{F(x_{1j_1}, \dots, x_{rj_r})\} < \sigma.$$

We represent $D(\sigma)$ in the form

$$D(\sigma) = \sigma \sum_{x_1 \le P_1} a_1(x_1) \cdots \sum_{x_r \le P_r} a_r(x_r) + \lambda(\overline{a}; \overline{P}; \sigma).$$

Theorem 10.6. *The following estimate holds*:

$$|\lambda(\overline{a}; \overline{P}; \sigma)| \ll (R_1 \dots R_r P_1 \dots P_r)^{1/2} \Delta \log(\Delta^{-1} + 1),$$

where Δ *is defined in Theorems* 10.4 *and* 10.5.

We omit the proof of Theorem 10.6, since it is similar to that of Theorem 10.2 in Section 10.4.

We formulate a theorem on joint distributions of the values of fractional parts of s polynomials $F_k(x_1, \dots, x_r)$, $k = 1, \dots, s$ in r variables with real coefficients,

$$F_k(x_1, \dots, x_r) = \sum_{t_1=0}^{n_1} \cdots \sum_{t_r=0}^{n_r} \alpha(t_1, \dots, t_r) x_1^{t_1} \dots x_r^{t_r},$$

where the variables $x_1, \dots, x_r$ run through the sequences $x_{1j_1}, \dots, x_{rj_r}$, respectively, considered in the preceding theorem.

Let $d_1, \dots, d_s$ be integers subject to the conditions

$$|d_k| \le \Delta^{-2}, \quad k = 1, \dots, s.$$

We define real numbers B by the equalities

$$B = B(t_1, \dots, t_r, d_1, \dots, d_s) = d_1\alpha_1(t_1, \dots, t_r) + \dots + d_s\alpha_s(t_1, \dots, t_r).$$

It follows from Dirichlet's theorem that an integer a and a natural q can be found satisfying the conditions

$$B = \frac{a}{q} + z, \quad (a, q) = 1, \quad 1 \le q \le \tau, \quad |z| \le (q\tau)^{-1},$$
$$\tau = \tau(t_1, \dots, t_r) = P_1^{t_1} \dots P_r^{t_r} P_1^{-1/6}.$$

For fixed $d_1, \dots, d_s$ we denote by $Q = Q(d_1, \dots, d_s)$ the least common multiple of the numbers $q(t_1, \dots, t_r)$ with the conditions

$$t_1 + \dots + t_r \ge 1, \quad 0 \le t_1 \le n_1, \dots, 0 \le t_r \le n_r.$$

Put

$$Q_0 = \min_{d_1,\ldots,d_s \le \Delta^{-2}} Q(d_1, \ldots, d_s), \quad \delta_0 = \min_{d_1,\ldots,d_s \le \Delta^{-2}} \delta(d_1, \ldots, d_s),$$

where

$$\delta(d_1, \ldots, d_s) = \max_{t_1,\ldots,t_r} P_1^{t_1} \ldots P_r^{t_r} |z(t_1, \ldots, t_r)|$$
$$(t_1 + \cdots + t_r \ge 1,\ 0 \le t_1 \le n_1, \ldots, 0 \le t_r \le n_r).$$

Further, we partition the collections of polynomials $(F_1, \ldots, F_s)$ into two classes E_1 and E_2. We put in the first class E_1 collections $(F_1, \ldots, F_s)$ for which

$$Q_0 \le P_1^{0.1\nu}, \quad \delta_0 \le m^{-1} P_1^{0.1\nu}.$$

The remaining collections $(F_1, \ldots, F_s)$ are put in the second class E_2.

Denote by $D(\sigma_1, \ldots, \sigma_s)$ the number of r-tuples

$$(x_{1j_1}, \ldots, x_{rj_r}), \quad 1 \le x_{1j_1} \le P_1, \ldots, 1 \le x_{rj_r} \le P_r,$$

satisfying the condition

$$\{F_1(x_{1j_1}, \ldots, x_{rj_r})\} < \sigma_1, \ \ldots, \ \{F_s(x_{1j_1}, \ldots, x_{rj_r})\} < \sigma_s.$$

We represent $D(\sigma_1, \ldots, \sigma_s)$ in the form

$$D(\sigma_1, \ldots, \sigma_s) = \sigma_1 \ldots \sigma_s \sum_{x_1 \le P_1} a_1(x_1) \cdots \sum_{x_r \le P_r} a_r(x_r) + \lambda(\overline{a}; \overline{P}; \overline{\sigma}).$$

Theorem 10.7. *The estimate*

$$|\lambda(\overline{a}; \overline{P}; \overline{\sigma})| \ll (R_1 \ldots R_r P_1 \ldots P_r)^{1/2} \Delta \log^s(\Delta^{-1} + 1)$$

holds; Δ *is defined in Theorems* 10.4 *and* 10.5.

The proof is analogous to that of Theorem 10.2 in Section 10.4, with the estimates of the trigonometric sums replaced by the estimate of the sum $T(A)$ at the appropriate places.

In Section 10.4 we obtained an asymptotic formula for the number of representations of a collection of natural numbers $N(t_1, \ldots, t_r)$ by additive terms of the form $p_1^{t_1} \ldots p_r^{t_r}$, $0 \le t_1 \le n_1, \ldots, 0 \le t_r \le n_r$, $t_1 + \cdots + t_r \ge 1$, i.e., an asymptotic formula for the number $J(\overline{N})$ of solutions of the system of equations

$$p_{11}^{t_1} \cdots p_{r1}^{t_r} + \cdots + p_{1k}^{t_1} \cdots p_{rk}^{t_r} = N(t_1, \ldots, t_r)$$
$$(0 \le t_1 \le n_1, \ldots, 0 \le t_r \le n_r,\ t_1 + \cdots + t_r \ge 1),$$

where $p_1, \dots, p_r, p_{11}, \dots, p_{rk}$ are primes. Let $N(0, \dots, 1, \dots, 0)$ with 1 at the sth place be denoted by P_s. Then for $k > 16m\varkappa \log(16m\varkappa) + 3$ we have the asymptotic formula

$$J(\overline{N}) = \sigma\theta(P_1 \dots P_r L_1^{-1} \dots L_r^{-1})^k (P_1^{n_1} \dots P_r^{n_r})^{-m/2} + O\big((P_1 \dots P_r L_1^{-1} \dots L_r^{-1})^k (P_1^{n_1} \dots P_r^{n_r})^{-m/2} L^{-1} \log L\big),$$

where $L_s = \log P_s$, $L = \max(L_1, \dots, L_r)$, $\varkappa$ is defined in Lemma 10.15 and σ and θ stand for a singular series and a singular integral.

The singular series σ was studied in [55]. Here we find a condition for the singular integral θ to be positive, depending on solvability in real numbers of the system

$$\sum_{j=1}^{k} x_{1j}^{t_1} \dots x_{rj}^{t_r} = \beta(t_1, \dots, t_r) \quad (0 \le t_1 \le n_1, \dots, 0 \le t_r \le n_r,\ t_1 + \dots + t_r \ge 1),$$

where $\beta(t_1, \dots, t_r)$ are defined by the equalities

$$\beta(t_1, \dots, t_r) = N(t_1, \dots, t_r) P_1^{-t_1} \dots P_r^{-t_r},$$

and the unknowns $x_{s,j}$ $(1 \le s \le r, 1 \le j \le k)$ satisfy the conditions $0 \le x_{s,j} \le 1$.

Denote by $\omega = \omega(h)$ the domain of the points $x_{s,j}$, $1 \le s \le r$, $1 \le j \le k$, satisfying the inequalities

(1) $0 \le x_{s,j} \le 1, s = 1, \dots, r, j = 1, \dots, k$;

(2) $|\sum_{j=1}^{k} x_{1j}^{t_1} \dots x_{rj}^{t_r} - \beta(t_1, \dots, t_r)| \le h$, $h > 0$ $(0 \le t_1 \le n_1, \dots, 0 \le t_r \le n_r,\ t_1 + \dots + t_r \ge 1)$.

We denote the volume of the domain ω by $\mu(\omega)$, i.e., we put

$$\mu(h) = \int \underset{\omega}{\cdots} \int dx_{11} \dots dx_{rk}.$$

Lemma 10.15. *For $k > nm$*

$$\theta = \theta(\overline{\beta}) = \lim_{h \to 0+} 2^{-m+1} h^{-m+1} \mu(h).$$

Proof. Since for $k > nm$ the integral converges absolutely, it is a function continuous jointly in the variables $\beta(t_1, \dots, t_r)$,

$$0 \le t_1 \le n_1, \dots, 0 \le t_r \le n_r, \quad t_1 + \dots + t_r \ge 1.$$

Put

$$F(\overline{\beta}) = \int_0^{\beta(n_1, \dots, n_r)} \dots \int_0^{\beta(0, \dots, 1)} \theta\big(\alpha(n_1, \dots, n_r), \dots, \alpha(0, \dots, 1)\big)\, d\overline{\alpha}.$$

From this we have

$$\theta(\overline{\beta}) = \frac{\partial^{m-1} F(\overline{\beta})}{\partial \overline{\beta}} = \lim_{h\to 0+} 2^{-m+1} h^{-m+1} \int \cdots \int_{\omega} \theta(\overline{\beta})\, d\overline{\alpha}. \tag{10.45}$$

Let us show that $F(\overline{\beta})$ can be represented in the form

$$F(\overline{\beta}) = \int \cdots \int_{\omega_1(\overline{\beta})} dx_{11} \dots dx_{rk},$$

where $\omega_1(\overline{\beta})$ is the domain of the points satisfying the conditions

$$\begin{gathered} 0 \le x_{s,j} \le 1, \quad 1 \le s \le r, \quad 1 \le j \le k, \\ 0 \le x_{11}^{t_1} \dots x_{r1}^{t_r} + \cdots + x_{1k}^{t_1} \dots x_{rk}^{t_r} < \beta(t_1, \dots, t_r) \\ (0 \le t_1 \le n_1, \dots, 0 \le t_r \le n_r,\ t_1 + \cdots + t_r \ge 1). \end{gathered}$$

Indeed, according to the definition of the functions $F(\overline{\beta})$ and $\theta(\overline{\alpha})$

$$\begin{aligned} F(\overline{\beta}) &= \int_0^{\beta(n_1,\dots,n_r)} \cdots \int_0^{\beta(0,\dots,1)} \theta(\overline{\alpha})\, d\overline{\alpha} \\ &= \int_0^{\beta(n_1,\dots,n_r)} \cdots \int_0^{\beta(0,\dots,1)} d\overline{\alpha} \int_{-\infty}^{+\infty} \cdots \int_{-\infty}^{+\infty} d\overline{z} \\ &\quad \times \int_0^1 \cdots \int_0^1 \exp\Bigg\{ 2\pi i \sum_{t_1=0}^{n_1} \cdots \sum_{t_r=0}^{n_r} \big(u(t_1, \dots, t_r) \\ &\quad - \alpha(t_1, \dots, t_r) \big) z(t_1, \dots, t_r) \Bigg\}\, d\overline{\alpha} \end{aligned}$$

where the $u(t_1, \dots, t_r)$ are defined by the equalities

$$\begin{gathered} u(t_1, \dots, t_r) = x_{11}^{t_1} \dots x_{r1}^{t_r} + \cdots + x_{1k}^{t_1} \dots x_{rk}^{t_r} \\ (0 \le t_1 \le n_1, \dots, 0 \le t_r \le n_r,\ t_1 + \cdots + t_r \ge 1). \end{gathered}$$

From this, changing the order of integration and integrating with respect to $\overline{\alpha}$, we obtain

$$\begin{aligned} F(\overline{\beta}) = \int_{-\infty}^{+\infty} \cdots \int_{-\infty}^{+\infty} \Bigg(\prod_{\substack{t_1=0 \\ t_1+\cdots+t_r\ge 1}}^{n_1} \cdots \prod_{t_r=0}^{n_r} \frac{1 - \exp\{-2\pi i z(t_1, \dots, t_r) \beta(t_1, \dots, t_r)\}}{2\pi i z(t_1, \dots, t_r)} \\ \times \int_0^1 \cdots \int_0^1 \exp\{2\pi i u(t_1, \dots, t_r) z(t_1, \dots, t_r)\}\, dx_{11} \dots dx_{rk} \Bigg)\, d\overline{z} \end{aligned}$$

$$= \int_0^1 \cdots \int_0^1 d\overline{x}\Bigg(\int_{-\infty}^{+\infty} \cdots \int_{-\infty}^{+\infty} \prod_{\substack{t_1=0 \\ t_1+\cdots+t_r\ge 1}}^{n_1} \cdots \prod_{t_r=0}^{n_r} \Bigg(\frac{1}{2\pi i z(t_1,\dots,t_r)}$$

$$\times \exp\{2\pi i u(t_1,\dots,t_r) z(t_1,\dots,t_r)\}$$

$$- \exp\big\{2\pi i \big(u(t_1,\dots,t_r) - \beta(t_1,\dots,t_r)\big) z(t_1,\dots,t_r)\big\}\Bigg) d\overline{z}\Bigg)$$

$$= \pi^{-m+1} \int_0^1 \cdots \int_0^1 d\overline{x} \prod_{\substack{t_1=0 \\ t_1+\cdots+t_r\ge 1}}^{n_1} \cdots \prod_{t_r=0}^{n_r} \Bigg(\int_{-\infty}^{+\infty} \Bigg(\frac{\sin 2\pi z(\overline{t}) u(\overline{t})}{z(\overline{t})}$$

$$- \frac{\sin 2\pi z(\overline{t})\big(u(\overline{t}) - \beta(\overline{t})\big)}{z(\overline{t})} \Bigg) dz(\overline{t}) \Bigg).$$

By the equality

$$\int_0^{+\infty} \frac{\sin \alpha x}{x}\, dx = \frac{\pi}{2} \operatorname{sgn} \alpha,$$

we have

$$F(\overline{\beta}) = 2^{-m+1} \int_0^1 \cdots \int_0^1 \prod_{\substack{t_1=0 \\ t_1+\cdots+t_r\ge 1}}^{n_1} \cdots \prod_{t_r=0}^{n_r} \big(\operatorname{sgn} z(\overline{t}) - \operatorname{sgn}(z(\overline{t}) - \beta(\overline{t}))\big)\, dx_{11} \dots dx_{rk}$$

$$= \int \cdots \int_{\substack{u(\overline{t}) \le \beta(\overline{t}) \\ 0 \le x_{11},\dots,x_{rk} \ge 1}} dx_{11} \dots dx_{rk} = \int \cdots \int_{\omega_1(\overline{\beta})} d\overline{x}.$$

Thus, the required equality for the function $F(\overline{\beta})$ is proved. Using the last equality and formula (10.45), we obtain

$$\theta(\overline{\beta}) = \lim_{h \to 0+} 2^{-m+1} h^{-m+1} \mu(h).$$

The lemma is proved. □

Definition 10.4. Consider the system of equations

$$F_{t_1,\dots,t_r}(\overline{x}) = \sum_{j=1}^{k} x_{1j}^{t_1} \dots x_{rj}^{t_r} = \beta(t_1,\dots,t_r) \tag{10.46}$$

$$(0 \le t_1 \le n_1, \dots, 0 \le t_r \le n_r,\ t_1 + \cdots + t_r \ge 1),$$

where $\overline{x}$ is the aggregate $(x_{11},\dots,x_{r1},\dots,x_{1k},\dots,x_{rk})$ of real numbers. The Jacobian matrix of the solution $\overline{x}$ of this system is the matrix

$$\left(\frac{\partial}{\partial x_{sj}} F_{t_1,\dots,t_r}(\overline{x}) \right),$$

the rows of which are indexed by $(t_1, \dots, t_r)$, $0 \le t_1 \le n_1, \dots, 0 \le t_r \le n_r$, $t_1 + \dots + t_r \ge 1$, ordered in some fashion, and the columns are indexed as follows: $s + r(j-1)$, $1 \le s \le r$, $1 \le j \le k$.

Theorem 10.8. *Suppose that for some solution $\overline{x}$ of system* (10.46) *for $k \ge m$ its Jacobi matrix has maximal rank, equal to $m-1$, and the modulus of the minor of order $m-1$ is equal to ε. Then*

(1) *for a sufficiently small $h > 0$, the volume $\mu(h)$ of the domain ω satisfies the estimate*

$$\mu(h) \ge c_1(\varepsilon) 2^{m-1} h^{m-1},$$

where $c_1(\varepsilon)$ is some positive constant;

(2) *for $k \ge mn$, the singular integral $theta$ satisfies the estimate*

$$\theta \ge c_1(\varepsilon) > 0.$$

Proof. Denote by $y_1, \dots, y_{m-1}$ the variables for which the determinant of the Jacobian matrix of the functions $F_{t_1,\dots,t_r}(\overline{y})$ is larger than $\varepsilon > 0$. Let $y_m, \dots, y_{kr}$ be the remainder variables. Take any real numbers $z_m, \dots, z_{kr}$ that satisfy the inequalities

$$|z_s - y_z| < \delta_1, \quad s = m, \dots, kr.$$

By the implicit function theorem there is $\delta_1 = \delta_1(\varepsilon) > 0$ such that the function

$$\begin{aligned} z_1 &= z_1(z_m, \dots, z_{kr}), \\ &\vdots \\ z_{m-1} &= z_{m-1}(z_m, \dots, z_{kr}), \end{aligned}$$

is a solution of the system of equations

$$F_{t_1,\dots,t_r}(z_1, \dots, z_{rk}) = \beta(t_1, \dots, t_r)$$
$$(0 \le t_1 \le n_1, \dots, 0 \le t_r \le n_r,\ t_1 + \dots + t_r \ge 1).$$

Take $h > 0$ sufficiently small and any $u_1, \dots, u_m$ satisfying the conditions

$$|u_s - z_s| < hr^{-1}n^{-1}m^{-1} = \delta_2, \quad s = 1, \dots, m-1.$$

Then

$$|u_{s_1}^{t_1} \dots u_{s_r}^{t_r} - y_{s_1}^{t_1} \dots y_{s_r}^{t_r}| < hm^{-1},$$

and $(u_1, \dots, u_{m-1}, z_m, \dots, z_{rk})$ belongs to the domain ω.

Consequently, we obtain

$$\mu(h) \ge \int \cdots \int_{\omega_1} du_1 \dots du_{m-1} dz_m \dots dz_{kr}$$

$$= \delta_1^{kr-m+1}(2\delta_2)^{m-1} = c_1(\varepsilon)2^{m-1}h^{m-1},$$

where

$$c_1(\varepsilon) = \delta_1^{kr-m+1}(rmn)^{-m+1} > 0.$$

By the preceding lemma, from this inequality we have

$$\theta = \theta(\overline{\beta}) \geq c_1(\varepsilon) > 0.$$

Theorem 10.8 is proved. □

We note that if for any solution of system (10.46) the rank of the Jacobian matrix is less than $m - 1$, then from Lemma 10.15 and dimensional considerations for the domain ω it follows that $\theta = 0$.

Concluding remarks on Chapter 10. The results considered in this chapter were obtained in [52], [53], [54], [55].

Chapter 11

Some applications of trigonometric sums and integrals

Let r be a natural number. Let E denote the r-dimensional space of vectors $\alpha = (\alpha_1, \dots, \alpha_r)$ with real coordinates α_j, and let $\mathcal{P}_r$ denote the class of algebraic polynomials of degree r with real coefficients,

$$\mathcal{P}_r = \Big\{P(x) : P(x) = P(\alpha, x) = \sum_{j=1}^{r} \alpha_j x^j,\ \alpha = (\alpha_1, \dots, \alpha_r) \in E^r,\ x \in E^1\Big\}.$$

For a fixed polynomial $P \in \mathcal{P}_r$, we consider the series

$$h(P) = \sum_{n \neq 0} \frac{e^{2\pi i P(n)}}{n} \tag{11.1}$$

(the sum is taken over all integers $n \neq 0$) and the symmetric partial sums

$$h_N(P) = \sum_{1 \le |n| \le N} \frac{e^{2\pi i P(n)}}{n}, \quad N = 1, 2, \dots,$$

of this series. Obviously,

$$|h_N(P)| \le \sum_{1 \le |n| \le N} \frac{1}{|n|} \sim 2 \log N \to \infty \quad (N \to \infty). \tag{11.2}$$

This is a trivial estimate of h_N. At the same time, $h_N(P)$ is, by its structure, the Hilbert transform of the sequence $\{e^{2\pi i P(n)}\}$, and the algebraically regular character of this sequence allows one to obtain a substantially better result. For example, for $r = 1$ and $P(x) = \alpha x$, $\alpha, x \in E^1$, the following relations are well known in the theory of trigonometric series (see, e.g., [170], Chapter II, Section 9):

$$h(P) = \sum_{n \neq 0} \frac{e^{2\pi i \alpha n}}{n} = 2i \sum_{n=1}^{\infty} \frac{\sin 2\pi \alpha n}{n} = 2\pi i \Big(\frac{1}{2} - \{\alpha\}\Big),$$

where $\{\alpha\}$ is the fractional part of the number $\alpha \neq 0, \pm 1, \dots$, and

$$h_N(P) = 2i \sum_{n=1}^{N} \frac{\sin 2\pi\alpha n}{n}, \qquad \sup_{P \in \mathcal{P}_1} \sup_{N=1,2,\dots} |h_N(P)| < \infty. \tag{11.3}$$

By using Vinogradov's method of trigonometric sums [165] (in a number of cases, it will be convenient for us to use the results of this method in the exposition of [8]), in this chapter we prove the following fact, which demonstrates that the uniform boundedness of the symmetric partial sums $h_N(P)$ for $N = 1, 2, \dots$ and $P \in \mathcal{P}_r$ is also true for $r \geq 2$.

Theorem 11.1. *Let $r \geq 2$. Then*

$$\sup_{N=1,2,\dots} \sup_{P \in \mathcal{P}_r} |h_N(P)| = g_r < \infty. \tag{11.4}$$

Further, for each polynomial $P \in \mathcal{P}_r$, the sequence $\{h_N(P)\}$ converges as $N \to \infty$, and so the sum of the series (11.1)*, regarded as the limit of its symmetric partial sums $h_N(P)$, is defined and bounded everywhere on $\mathcal{P}_r$.*

We treat $h_N(P)$ and $h(P)$ as functions of the coefficients $(\alpha_1, \dots, \alpha_r) = \alpha$ of the polynomial $P(\alpha, x) \in \mathcal{P}_r$ and note that Theorem 11.1 shows the properties of the special trigonometric series

$$H(\alpha) = H(\alpha_1, \dots, \alpha_r) = \sum_{n \neq 0} \frac{e^{2\pi i(\alpha_1 n + \dots + \alpha_r n^r)}}{n} = \sum_{n \neq 0} \frac{e^{2\pi i P(\alpha, n)}}{n}$$

all whose partial sums

$$H_N(\alpha) = \sum_{1 \leq |n| \leq N} \frac{e^{2\pi i P(\alpha, n)}}{n}$$

are uniformly bounded with respect to $\alpha \in E^r$, $N = 1, 2, \dots$, and converge to $H(\alpha)$ as $N \to \infty$ at each point $\alpha \in E^r$.

First we deduce some corollaries of Theorem 11.1 and then the assertion itself.

The first corollary relates to the subject of spectra of uniform convergence, i.e., a class of problems the general setting of which is due to Ul'yanov [147].

We give the corresponding definitions. Let $\mathcal{K} = \{k_n\}_0^\infty$ be a sequence of distinct integers. By $C(\mathcal{K})$ we denote the subspace of those continuous 1-periodic functions $f(\theta)$ of a single variable θ with Chebyshev norm $\|f\| = \max_\theta |f(\theta)|$ the spectrum of which is contained in $\mathcal{K}$, i.e.,

$$C(\mathcal{K}) = \Big\{ f(\theta) : f(\theta + 1) \equiv f(\theta) \in C, \ \hat{f}_k = \int_0^1 f(\theta) e^{-2\pi i k\theta} \, d\theta = 0, \ k \not\equiv \mathcal{K} \Big\}.$$

In other words, $C(\mathcal{K})$ consists of those and only those continuous 1-periodic functions whose Fourier series have the form

$$f(\theta) \sim \sum_{n=0}^{\infty} \hat{f}_{k_n} e^{2\pi i k_n \theta}. \tag{11.5}$$

For a given natural number N, by $S_N(f) = S_N(f,\theta)$ we denote the Nth partial sum of the series (11.5), and by $L_N(\mathcal{K})$ the Nth Lebesgue constant of the spectrum $\mathcal{K}$, i.e.,

$$S_N(f,\theta) = \sum_{n=0}^{N} \hat{f}_{k_n} e^{2\pi i k_n \theta}, \quad L_N(\mathcal{K}) = \sup_{f \in C(\mathcal{K}),\, f \not\equiv 0} \frac{\|S_N(f)\|}{\|f\|}.$$

A sequence $\mathcal{K}$ is called the *spectrum of uniform convergence* if for each function f from $C(\mathcal{K})$, the sequence $S_N(f,\theta)$ converges to $f(\theta)$ uniformly with respect to θ as $N \to \infty$. We denote the class of all spectra of uniform convergence by UC.

It follows from the Banach–Steinhaus theorem that a criterion for the sequence $\mathcal{K}$ to belong to the class UC is the boundedness of the corresponding Lebesgue constants:

$$L_N(\mathcal{K}) = O(1) \quad (N \to \infty). \tag{11.6}$$

However, it is impossible to consider this criterion as effective, since the principal difficulty lies precisely in obtaining estimates for $L_N(\mathcal{K})$, in explicit terms of the given sequence $\mathcal{K}$. For a long time, it remained unknown whether or not power sequences $\mathcal{K}$ (i.e., sequence of the form $\{n^2\}$, $\{n^3\}$, ...) are spectra of uniform convergence. A solution of this spectral version of Ul'yanov's problem was obtained in [129] (there one can also find a survey of results on spectra of uniform convergence). It turned out that the class UC contains not a single power sequence, and more generally, not a single polynomial sequence, i.e., a sequence of the form

$$\mathcal{K} = \mathcal{K}(P) = \{k_n : k_n = P(n),\, n = 0, 1, \ldots\}, \quad P \in \bigcup_{r \ge 1} \mathcal{P}_r.$$

An estimate from below for the Lebesgue constants of polynomial spectra is established in [129] in the form

$$L_N\big(\mathcal{K}(P)\big) \ge a_r (\log N)^{\varepsilon_r} \quad (P \in \mathcal{P}_r, \quad \varepsilon_r = 2^{-r+1}, \quad N = 1, 2, \ldots), \tag{11.7}$$

where the factor a_r is positive and depends only on the degree r of the polynomial P defining the spectrum.

In turn, this estimate was deduced from the following one:

$$|h_N(P)| \ll_r (\log N)^{1-\varepsilon_r} \quad (N = 2, 3, \ldots; \quad \varepsilon_r = 2^{-r+1}, \quad P \in \mathcal{P}_r) \tag{11.8}$$

(here and below, relations of the form $A \ll_r B$ and $A \ll_{r,\varepsilon} B$ between positive quantities A and B mean that $A \le c_r B$ and $A \le_{r,\varepsilon} B$, respectively, where the positive factors c_r and $c_{r,\varepsilon}$ depend only on the parameters indicated; if $A \leq cB$, where c is a positive absolute constant, then the notation $A \ll B$ is used).

As compared to the trivial estimate (11.2), the estimate (11.8) contains the reducing factor $(\log N)^{-\varepsilon_r}$. The method used in [129] to prove (11.8) consists in the following. Squaring $|h_N(P)|$ gives rise to the double sum

$$|h_N(P)|^2 = \sum_{1\leq|n|,|m|\leq N}\sum \frac{e^{2\pi i(P(n)-P(m))}}{nm}.$$

After the summation variable $\nu = n - m$ is introduced, and elementary estimates are performed, a relation of the form

$$|h_N(P)|^2 \ll \sum_{1\leq|\nu|\leq N} \frac{|h_N(P_\nu)|}{|\nu|} + 1 \tag{11.9}$$

is obtained, where $P_\nu(x) = P(x+\nu) - P(x)$ $(\nu = \pm1, \pm2, \dots)$. If $P \in \mathcal{P}_r$, then for each fixed ν, $P_\nu(x)$ is a polynomial of reduced degree with respect to the variable x, namely, $P_\nu \in \mathcal{P}_{r-1}$. Hence (11.8) easily follows from (11.9) by induction on r.

The idea in this method of squaring, and subsequently lowering the degree of the polynomial in the exponent, and also the character of the reducing factor, can be regarded as going back to the investigation of Gauss and Weyl (see [165], Russian pp. 6 and 8; English pp. 183–185).

The following final result strengthening (11.7) follows from Theorem 11.1.

Theorem 11.2. *Let $r = 2, 3, \dots$, and let P be a polynomial of degree r having natural numbers for its coefficients. Then the Lebesgue constants corresponding to the spectrum $\mathcal{K}(P)$ satisfy the estimate*

$$\log N \underset{r}{\ll} L_N(\mathcal{K}(P)) \quad (N = 2, 3, \dots). \tag{11.10}$$

On the other hand, since $k_n = O(n^r)$ $(n \to \infty)$, we obviously have

$$L_N(\mathcal{K}(P)) = O(\log N) \quad (N \to \infty)$$

(see, e.g., [170], Chapter II, Section 12).

Hence it follows from Theorem 11.2 that the Lebesgue constants of any polynomial spectrum have precisely logarithmic growth:

$$L_N(\mathcal{K}(P)) \asymp \log N \quad (N \to \infty), \tag{11.11}$$

the same as the Lebesgue constants of the entire trigonometric system.

Theorem 11.2 is deduced from Theorem 11.1 by using the following simple lemma, which may be used to obtain estimates from below for the Lebesgue constants of an arbitrary (nonpolynomial) spectrum.

Lemma 11.1. *Let N and M be natural numbers, $M \leq N$, and let*

$$\tau_{N,M}(\mathcal{K},\theta) = \sum_{1\leq|n|\leq M} \frac{\exp\{2\pi i k_{n+N}\,\theta\}}{n}, \quad \varkappa_{N,M}(\mathcal{K}) = \max_{\theta} |\tau_{N,M}(\mathcal{K},\theta)|.$$

Then

$$L_N(\mathcal{K}) \gg \max\left\{\frac{\log M}{\varkappa_{N,M}(\mathcal{K})} : M \in [1, N]\right\}, \tag{11.12}$$

and, in particular,

$$L_N(\mathcal{K}) \gg \frac{\log N}{\varkappa_{N,M}(\mathcal{K})}. \tag{11.13}$$

To verify this assertion, it is sufficient to note that $\tau_{N,M} \in C(\mathcal{K})$, $\|\tau_{N,M}\| = \varkappa_{N,M}(\mathcal{K})$, and that

$$\|S_N(\tau_{N,M})\| \geq |S_N(\tau_{N,M}, 0)| = \sum_{n=1}^{M} \frac{1}{n}.$$

Now if $\mathcal{K} = \mathcal{K}(P)$, where $P(x) \in \mathcal{P}_r$, then for all fixed real numbers θ and N, we have $\theta P(x+N) \in \mathcal{P}_r$, and consequently, in view of (11.4), we also have $\varkappa_{N,M} \ll_r 1$. Inequality (11.10) follows from this and (11.13).

In the case where $k_n = n$, the polynomials $\tau_{N,M}$ coincide with the so-called Fejér polynomials

$$\tau_{N,M}(\theta) = 2ie^{2\pi i N\theta} \sum_{n=1}^{M} \frac{\sin 2\pi n\theta}{n}$$

(see, e.g., [170], Chapter VIII). These polynomials have been used by many authors to construct examples of continuous functions with a "bad" sequence of Fourier sums.

Moreover, obviously (see (11.2)), for each sequence $\mathcal{K}$ we have $\varkappa_{N,M}(\mathcal{K}) = O(\log N)$. Hence it follows from Lemma 11.1 that for a sequence $\mathcal{K}$ to be a spectrum of uniform convergence, it is necessary that the quantities $\varkappa_{N,M}(\mathcal{K})$ have precisely logarithmic growth as $N \to \infty$:

$$\varkappa_{N,M}(\mathcal{K}) \approx \log N \quad (N \to \infty). \tag{11.14}$$

It would be interesting to find out how slowly the sequences $\mathcal{K}$ satisfying condition (11.4) can grow. As demonstrated in [128], the solution of this problem can turn out to be useful in connection with determining the precise order of growth of the partial sums of one-dimensional trigonometric Fourier series of class L^1 (without spectral restrictions) on a set of full measure. On the whole, here it is necessary to establish estimates for quantities of the form

$$G_{N,M} = \max_{x\in\Omega_M} \max_{y\in\Omega_M} \frac{1}{M} \sum_{n=1}^{M} \left| \sum_{m=1}^{M} \frac{\exp\{2\pi i N x_n y_m\}}{m-n+0.5} \right|,$$

where M and N are natural numbers, $M \le N$, $x = \{x_1, \dots, x_M\}$, $y = \{y_1, \dots, y_M\}$, and Ω_M is the unit cube

$$\Omega_M = \big\{x = \{x_1, \dots, x_M\},\ 0 \le x_m \le 1,\ m = 1, \dots, M\big\}$$

in E^M. Namely, it is necessary to obtain estimates as precise as possible for the lower bounds $N = N(M)$ of those numbers N for which the quantities $G_{N,M}$ have precisely logarithmic growth as $M \to \infty$:

$$N(M) = \min\{N : G_{N,M} \ge \alpha \log M\}$$

(α is a small fixed positive number).

Now we turn to other corollaries of Theorem 11.1, which apparently are themselves independent results in the theory of trigonometric series.

Corollary 11.1. *Let $P^+(x)$ and $P^-(x)$ be algebraic polynomials with real coefficients, in which*

$$P^+(-x) \equiv P^+(x), \quad P^-(x) \equiv -P^-(-x),$$

i.e., P^+ is even and P^- is odd. Then the series

$$H = \sum_{n=1}^{\infty} \frac{e^{2\pi i P^+(n)} \sin 2\pi P^-(n)}{n}$$

converges, and its partial sums

$$H_N = \sum_{n=1}^{N} \frac{e^{2\pi i P^+(n)} \sin 2\pi P^-(n)}{n}$$

are bounded above in magnitude by a quantity depending only on the degrees of the polynomials P^+ and P^-, but not on their coefficients. In particular, each series of the form

$$F_{2j+1}(\theta) = \sum_{n=1}^{\infty} \frac{\sin 2\pi n^{2j+1}\theta}{n}, \quad j = 0, 1, 2, \dots, \tag{11.15}$$

has uniformly bounded partial sums and converges for all real θ.

As already noted, for $j = 0$ (see (11.3)) the last assertion is a well-known fact in the theory of trigonometric series.

It is curious to note that the oddness of the exponent in the powers of n in (11.15) is an essential condition and no series of the form

$$F_{2j}(\theta) = \sum_{n=1}^{\infty} \frac{\sin 2\pi n^{2j}\theta}{n}, \quad j = 1, 2, \dots,$$

has the properties of boundedness that hold in the case of the series F_{2j+1}. Indeed, if $\theta = 1/3$, then, taking into account that $n^{2j} \equiv 0 \pmod 3$ for $n \equiv 0 \pmod 3$ and $n^{2j} \equiv 1 \pmod 3$ for $n \equiv \pm 1 \pmod 3$, we see that

$$S_N\left(F_{2j}, \frac{1}{3}\right) = \sin\frac{2\pi}{3} \sum_{n=3m\pm 1 \le N} \frac{1}{n} \sim 3^{-1/2} \log N \quad (N \to \infty).$$

Hardy and Littlewood [64] especially investigated the properties of the series $F_2(\theta)$, and also of the corresponding cosine series

$$\widetilde{F}_2(\theta) = \sum_{n=1}^{\infty} \frac{\cos 2\pi n^2\theta}{n},$$

and established, in particular, that both these series diverge also at certain irrational points θ. The properties of convergence and divergence of series of the more general form

$$\sum_{n=1}^{\infty} c_n \sin 2\pi n^2\theta, \quad \sum_{n=1}^{\infty} c_n \cos 2\pi n^2\theta$$

were also investigated.

Coupled with simple considerations, Theorem 11.1 allows one to go from the special trigonometric series (11.1) with coefficients $1/n$ to a wider class of trigonometric series. The corresponding assertion can be conveniently formulated in terms of multipliers of Fourier series (see [170], Chapter IV, Section 11).

Let us recall the definition. Let $\mathfrak{M}_1$ and $\mathfrak{M}_2$ be any two sets of (summable) 1-periodic functions, and let $\Lambda = \{\lambda(n)\}_{-\infty}^{+\infty}$ be a sequence of complex numbers. This sequence is said to be a *multiplier* of the class $(\mathfrak{M}_1, \mathfrak{M}_2)$ (we use the notation $\Lambda \in (\mathfrak{M}_1, \mathfrak{M}_2)$) if the Fourier series of each function of the class $\mathfrak{M}_1$ after the introduction of the multipliers $\lambda(n)$ becomes the Fourier series of some function of the class $\mathfrak{M}_2$, i.e., if

$$f(\theta) \sim \sum_n \hat{f}_n e^{2\pi i n\theta} \in \mathfrak{M}_1 \Longrightarrow f_\Lambda(\theta) \sim \sum_n \lambda(n) \hat{f}_n e^{2\pi i n\theta} \in \mathfrak{M}_2.$$

We consider the following three sets of 1-periodic (complex-valued) functions:

(1) the set V of functions $f(\theta)$ having bounded total variation in the period

$$V = \left\{f : \bigvee_0^1 f = \bigvee_0^1 \operatorname{Re} f + \bigvee_0^1 \operatorname{Im} f < \infty\right\};$$

(2) the set U^* of functions $f(\theta)$ having uniformly bounded sequence of symmetric partial sums of trigonometric Fourier series

$$\left\{f : f(\theta) \sim \sum_n \hat{f}_n e^{2\pi i n\theta}, \ \|f\|_{U^*} = \sup_{N=0,1,\dots} \sup_\theta \left| \sum_{|n|\le N} \hat{f}_n e^{2\pi i n\theta} \right| < \infty\right\};$$

(3) the set L^∞ of essentially bounded functions $f(\theta)$ with the usual norm

$$\|f\|_\infty = \operatorname*{ess\,sup}_\theta |f(\theta)|.$$

Corollary 11.2. *Let $P(x)$ be an algebraic polynomial with real coefficients. Then the sequence $e^{2\pi i P} = \{e^{2\pi i P(n)}\}_{n=-\infty}^{+\infty}$ is a multiplier of the class (V, U^*), and all the more of the class (V, L^∞):*

$$e^{2\pi i P} \in (V_1, U^*), \quad e^{2\pi i P} \in (V, U^*). \tag{11.16}$$

This follows at once from (11.4) if one notes that

$$\sum_{|n|\le N} e^{2\pi i P(n)} \hat{f}_n e^{2\pi i n\theta} = e^{2\pi i P(0)} \hat{f}_0 + \frac{1}{2\pi i}\int_0^1 h_N(P_{\theta-\varphi})\, df(\varphi),$$

where $P_\xi(x) = P(x) + \xi x$.

Proof of Theorem 11.1. As we have already mentioned above, the proof of Theorem 11.1 is based on Vinogradov's estimates [165] for the trigonometric sums $S_n(\alpha) = (1/n)\sum_{x=1}^n e^{2\pi i P(x)}$ of Weyl, where $P(x) = P(\alpha, x) \in \mathcal{P}_r$ (the factors $1/n$ are introduced for convenience of the further exposition). Here, without loss of generality, we assume that $r \ge 3$. By $I_n(\alpha)$ we denote the trigonometric integral corresponding to the polynomial $P(\alpha, x)$

$$I_n(\alpha) = \frac{1}{n}\int_0^n e^{2\pi i P(\alpha,x)}\, dx. \tag{11.17}$$

First, we sum over positive n in the sums $h_N(P) = H_N(\alpha)$. Then for a given vector $\alpha = (\alpha_1, \ldots, \alpha_r) \in E^r$, setting

$$\alpha^* = (\alpha_1^*, \ldots, \alpha_r^*), \quad \alpha_j^* = (-1)^j \alpha_j, \quad T_n(\alpha) = S_n(\alpha) - S_n(\alpha^*) \tag{11.18}$$

and taking into account that $P(\alpha, -x) \equiv P(\alpha^*, x)$, by an Abel transformation we obtain

$$H_N(\alpha) = \sum_{n=1}^{N-1} \frac{T_n(\alpha)}{n+1} + T_N(\alpha).$$

We shall prove that, for each $\alpha \in E^r$,

$$\sum_{n=1}^{\infty} \frac{|T_n(\alpha)|}{n+1} \ll_r 1, \quad T_n(\alpha) \to 0 \quad (n \to \infty). \tag{11.19}$$

Since, obviously, $|T_n(\alpha)| \le 2$, Theorem 11.1 is a consequence of (11.19).

Moreover, concerning the convergence of the series in (11.19), we prove the following strengthened assertion.

Lemma 11.2. *Let* $\varepsilon > 0$. *Then*

$$\sup_{\alpha \in E^r} \sum_{n=1}^{\infty} \frac{|T_n(\alpha)|^{\varepsilon}}{n+1} \ll_{r,\varepsilon} 1. \qquad (11.20)$$

By R^r we denote the set of rational points in E^r, i.e., the set of vectors $\beta \in E^r$ having the form

$$\beta = (a_1/q_1, \ldots, a_r/q_r),$$

where a_s and q_s are relatively prime integers, i.e., $(a_s, q_s) = 1$ and $q_s > 0$. Moreover, for $\beta \in R^r$, by $Q(\beta)$ we denote the least common multiple $[q_1, \ldots, q_r]$ of the denominators of the coordinates of this vector.

In accordance with Vinogradov's method, for a given natural number n, we split the entire space E^r into two classes, relative to the approximation of its points α by the rational points β. To the first class, which we denote by $(\mathrm{I})_n$, we allot those points $\alpha \in E^r$ admitting the representation

$$\alpha = \beta + \gamma, \quad \beta \in R^r, \qquad (11.21)$$

where the number $Q(\beta)$ and the vector of errors $\gamma = (\gamma_1, \ldots, \gamma_r)$ satisfy

$$Q(\beta) \le n^{0.3}, \quad \delta_n \le n^{0.3}, \quad \delta_n = \max_{1 \le s \le r} |\gamma_s n^s|. \qquad (11.22)$$

We allot the remaining points $\alpha \in E^r$ to the second class $(\mathrm{II})_n$, i.e. $(\mathrm{II})_n = E^r \setminus (\mathrm{I})_n$. If n is sufficiently large, say,

$$n > n_0 = 2^{10} = 1024 \qquad (11.23)$$

and $\alpha \in (\mathrm{I})_n$, then the representation of α in the form (11.21) with conditions (11.22) is unique. Indeed, assume that contrary. Then in R^r there are two vectors β^1 and β^2, $\beta^1 \ne \beta^2$, such that for the corresponding numbers $Q_1 = Q(\beta^1)$ and $Q_2 = Q(\beta^2)$ and vectors of errors $\gamma^1 = (\gamma_1^1, \ldots, \gamma_r^1)$ and $\gamma^2 = (\gamma_1^2, \ldots, \gamma_r^2)$, we have

$$\max(Q_1, Q_2) \le n^{0.3}, \quad \max(|\gamma_s^1|, |\gamma_s^2|) \le n^{0.3-s} \quad (s = 1, \ldots, r). \qquad (11.24)$$

Since $\beta^1 \ne \beta^2$, taking into account the fact that $q_s^1 \le Q_1$ and $q_s^2 \le Q_2$ $(s = 1, \ldots, r)$, we see that there exists an s, $1 \le s \le r$, such that

$$\frac{1}{Q_1 Q_2} \le \frac{1}{q_s^1 q_s^2} \le \left| \frac{a_s^1}{q_s^1} - \frac{a_s^2}{q_s^2} \right| = |\gamma_s^1 - \gamma_s^2| \le 2n^{0.3-s}. \qquad (11.25)$$

But $s \ge 1$; so it follows from (11.25) that $Q_1 Q_2 \ge 0.5 n^{0.7}$, but this, under condition (11.23), contradicts the first of conditions (11.24), which implies $Q_1 Q_2 \le n^{0.6}$.

In what follows, we assume that the natural numbers n, m, and N are larger than $n_0 = 2^{10}$; for smaller values, it is sufficient to use the trivial estimate $|T_n(\alpha)| \le 2$.

For a given vector $\alpha \in E^r$, we set

$$N_1(\alpha) = \{n : n > n_0,\ \alpha \in (\mathrm{I})_n\}, \quad N_2(\alpha) = \{n : n > n_0,\ \alpha \in (\mathrm{II})_n\}.$$

If $N_1(\alpha) \neq \emptyset$, then by $B(\alpha) = \{\beta^1, \beta^2, \dots\}$ we denote the collection of distinct positions in E^r of the rational point β in representation (11.21), (11.22) taken successively as the natural number n increases on the set $N_1(\alpha)$. We note that if $\alpha \in R^r$, then this collection is finite, and in the contrary case it is an infinite sequence.

We further set

$$Q_j = Q(\beta^j), \quad \gamma^j = \alpha - \beta^j,$$

and let $\omega_j(\alpha)$ be that segment of the series of natural numbers $n > n_0$ on which the vector $\beta \in R^r$ defined by (11.21) and (11.22) remains constant and coincides with β^j. Obviously, we have

$$\begin{gathered} N_1(\alpha) = \bigcup_j \omega_j(\alpha), \quad \omega_j(\alpha) \cap \omega_k(\alpha) = \emptyset, \\ \beta^j \neq \beta^k, \quad \gamma^j \neq \gamma^k \quad (j \neq k). \end{gathered} \tag{11.26}$$

The following lemma shows that the numbers Q_j are also distinct, and grow very fast. In addition, estimates of certain sums are given that are needed for the proof.

Lemma 11.3. *The estimates*

$$Q_{j+1} \geq 0.5 Q^{4/3} \tag{11.27}$$

hold. Further, let $\varepsilon > 0$ *and* $\varphi_n = \left(\min(\delta_n^{-1/r}, \delta_n)\right)^{\varepsilon}$ *(see* (11.22)*. Then*

$$\sup_{\alpha, j} \sum_{n \in \omega_j(\alpha)} \frac{\varphi_n}{n+1} \ll_{r,\varepsilon} 1, \tag{11.28}$$

and so

$$\sup_{\alpha} \sum_j Q_j^{-\varepsilon/r} \sum_{n \in \omega_j(\alpha)} \frac{\varphi_n}{n+1} \ll_{r,\varepsilon} 1. \tag{11.29}$$

Proof. Let n and m be natural numbers, $n \in \omega_j(\alpha)$ and $m \in \omega_{j+1}(\alpha)$. Since (see (11.26)) $\beta^j, \beta^{j+1} \in R^r$, $\beta^j \neq \beta^{j+1}$, as in (11.25), we see that there exists an s, $1 \leq s \leq r$, such that

$$\frac{1}{Q_j Q_{j+1}} \leq |\gamma_s^j - \gamma_s^{j+1}|. \tag{11.30}$$

But, in view of (11.22),

$$\begin{gathered} Q_j \leq n^{0.3}, \quad Q_{j+1} \leq n^{0.3}, \\ |\gamma_s^j| \leq n^{0.3-s} \leq n^{-0.7}, \quad |\gamma_s^{j+1}| \leq m^{0.3-s} \leq m^{-0.7}, \end{gathered}$$

so that, taking into account that $m > n$, we obtain from (11.30)

$$Q_j Q_{j+1} \geq 0.5n^{0.7} \geq 0.5Q_j^{7/3},$$

whence (11.27) follows. We now prove (11.28). We note that for a change of n in the segment $\omega_j(\alpha)$, the vector $\gamma = (\gamma_1, \dots, \gamma_r) = \gamma^j$, like the vector β, remains constant, and, as is easily seen (see (11.22)),

$$\varphi_n \leq \sum_{s=1}^{r} \varphi_{n,s}, \quad \text{where} \quad \varphi_{n,s} = \left(\min(|\gamma_s n^s|^{-1/r}, |\gamma_s n^s|)\right)^{\varepsilon}. \tag{11.31}$$

But for each $s > 0$ and a real number γ_s, we have

$$\sum_{n=1}^{\infty} \frac{\varphi_{n,s}}{n+1} \leq |\gamma_s|^{-\varepsilon/r} \sum_{n>|\gamma_s|^{-1/s}} n^{-1-s\varepsilon/r} + |\gamma_s|^{\varepsilon} \sum_{n\leq|\gamma_s|^{-1/s}} n^{-1+s\varepsilon} \underset{\varepsilon,s,r}{\ll} 1.$$

Inequality (11.28) follows from this and (11.31), while (11.29) is a consequence of (11.27) and (11.28). The lemma is proved. □

Now we present, in the form of lemmas, the estimates of trigonometric sums and integrals which are used for the proof.

Lemma 11.4 (Vinogradov [165]). *If $n \in N_2(\alpha)$, i.e., if for the n under consideration the point α belongs to the first class $(II)_n$, then*

$$|S_n(\alpha)| \underset{r}{\ll} n^{-\rho}, \quad \rho = \left(8r^2(\log r + 1.5 \log\log r + 4.2)\right)^{-1}. \tag{11.32}$$

For the proof, see [8], Lemma 7.

Lemma 11.5 (Vinogradov [165]). *Suppose that $n \in N_1(\alpha)$, i.e., for the n under consideration the point α belongs to the first class $(I)_n$. Suppose also that β and γ are defined by* (11.21) *and* (11.22). *Let $Q = Q(\beta)$. Then the estimate (asymptotic formula)*

$$S_n(\alpha) = S_n(\beta + \gamma) = S_Q(\beta) I_n(\gamma) + \Delta \tag{11.33}$$

holds, where

$$|\Delta| \leq 9rQn^{-1} \underset{r}{\ll} n^{-0.7}. \tag{11.34}$$

For the proof, see [8], Lemma 7.

Lemma 11.6 (Hua Loo-Keng). *Let $\beta \in R^r$, and let $Q = Q(\beta)$. Then the complete rational sum $S_Q(\beta)$ satisfies the estimate*

$$|S_Q(\beta)| \underset{r}{\ll} Q^{-1/r}. \tag{11.35}$$

For the proof, see [45] or [144].

Lemma 11.7 (Vinogradov [165], Chapter 2, Lemma 4). *Suppose that $\gamma = (\gamma_1, \dots, \gamma_r) \in E^r$ and*

$$\delta_n = \max_{1 \le s \le r} |\gamma_s n^s|.$$

Then the trigonometric integral $I_n(\gamma)$ satisfies

$$|I_n(\gamma)| \le \min(1,\ 32\delta_n^{-1/r}). \tag{11.36}$$

Now from Lemmas 11.4–11.7 we deduce the estimates for $T_n(\alpha)$.

Lemma 11.8. *If $n \in N_2(\alpha)$ and ρ has the same meaning as in Lemma* 11.4, *then*

$$|T_n(\alpha)| \ll_r n^{-\rho}. \tag{11.37}$$

This estimate follows at once from (11.32) if we note that (see (11.18))

$$N_1(\alpha) = N_1(\alpha^*), \quad N_2(\alpha) = N_2(\alpha^*). \tag{11.38}$$

Lemma 11.9. *Let $n \in N_1(\alpha)$, let the points β and γ be defined by* (11.21) *and* (11.22), *and let $Q = Q(\beta)$. Then*

$$|T_n(\alpha)| \ll_r Q^{-1/r} \min(\delta_n^{-1/r}, \delta_n) + n^{-0.7}. \tag{11.39}$$

Proof. If $\beta \in R^r$ and $Q = Q(\beta)$, then

$$S_Q(\beta^*) = S_Q(\beta). \tag{11.40}$$

In fact, for each $s = 1, 2, \dots$ and integer x, the congruence $(-x)^s \equiv (Q - x)^s \pmod{Q}$ is satisfied, and since $Q \equiv 0 \pmod{q_s}$, where q_s is the denominator of the sth coordinate of the vector β, we have $(-x)^s \equiv (Q - x)^s \pmod{q_s}$, $1 \le s \le r$. Hence

$$P(\beta^*, x) = P(\beta, -x) \equiv P(\beta, Q - x) \pmod{1}, \quad x = 0, \pm 1, \dots,$$

for which (11.40) follows.

Moreover, from (11.40) and (11.33) it follows that

$$S_n(\alpha^*) = S_n(\beta^* + \gamma^*) = S_Q(\beta) I_n(\gamma^*) + \Delta^*, \quad |\Delta^*| \ll_r n^{-0.7},$$

so that, in view of (11.35), we have

$$|T_n(\alpha)| = |S_n(\alpha) - S_n(\alpha^*)| \ll_r Q^{-1/r} |I_n(\gamma) - I_n(\gamma^*)| + n^{-0.7}. \tag{11.41}$$

Splitting the polynomial P into the even and odd parts

$$P^+(\alpha, x) = P\Big(\frac{\alpha+\alpha^*}{2}, x\Big), \quad P^-(\alpha, x) = P\Big(\frac{\alpha-\alpha^*}{2}, x\Big),$$

we obtain

$$I_n(\gamma) - I_n(\gamma^*) = \frac{2i}{n}\int_0^n e^{2\pi i P^+(\gamma,x)} \sin 2\pi P^-(\gamma, x)\, dx. \tag{11.42}$$

Now using (11.36) and the inequality $|\sin u| \le |u|$ ($\operatorname{Im} u = 0$), we obtain

$$|I_n(\gamma) - I_n(\gamma^*)| \ll \min\Big(\delta_n^{-1/r}, \sum_{s=1}^{r} |\gamma_s n^s|\Big) \ll_r \min(\delta_n^{-1/r}, \delta_n). \tag{11.43}$$

Inequality (11.39) follows from this and (11.41). The lemma is proved. □

In view of (11.39) and the definitions of the numbers Q_j and of the segments $\omega_j(\alpha)$ of natural numbers, we have

$$|T_n(\alpha)| \ll_r Q_j^{-1/r} \min(\delta_n^{-1/r}, \delta_n) + n^{-0.7} \quad (n \in \omega_j(\alpha)). \tag{11.44}$$

The second relation in (11.19) follows from (11.37) and (11.44) (it is necessary to observe that if $\alpha \in R^r$, then $\min(\delta_n^{-1/r}, \delta_n) = 0$ for all sufficiently large n (see also (11.27))). Moreover, from (11.37), (11.44), and (11.29) it follows that

$$\begin{aligned}
\sum_{n=1}^{\infty} \frac{|T_n(\alpha)|^\varepsilon}{n+1} &\le 2^\varepsilon \sum_{n=1}^{1024} \frac{1}{n+1} + \sum_{n\in N_1(\alpha)} \frac{|T_n(\alpha)|^\varepsilon}{n+1} + \sum_{n\in N_2(\alpha)} \frac{|T_n(\alpha)|^\varepsilon}{n+1} \\
&\ll_\varepsilon 1 + \sum_j \sum_{n\in\omega_j(\alpha)} \frac{|T_n(\alpha)|^\varepsilon}{n+1} + \sum_{n\in N_2(\alpha)} \frac{|T_n(\alpha)|^\varepsilon}{n+1} \\
&\ll_{r,\varepsilon} 1 + \sum_j Q_j^{-\varepsilon/r} \sum_{n\in\omega_j(\alpha)} \frac{\varphi_n}{n+1} + \sum_{n=1}^{\infty} n^{-1-0.7\varepsilon} + \sum_{n=1}^{\infty} n^{-1-\beta\varepsilon} \\
&\ll_{r,\varepsilon} 1,
\end{aligned}$$

which completes the proof of Lemma 11.2 and hence that of Theorem 11.1. □

Concluding remarks on Chapter 11. The results of this chapter were obtained by G. I. Arkhipov and K. I. Oskolkov in [36]. Some applications of estimates for trigonometric sums and trigonometric integrals to the Schrödinger equation are given in [131]. Some other applications can be found in [128], [129], [130].

Chapter 12

Short Kloosterman sums

In this chapter we study the distribution of reciprocal values modulo a given number. The quantities themselves take values that are relatively small (in contrast to the modulus). Our study is based on estimates for short Kloosterman sums.

Let $m > 1$, where m is an integer. For an integer x that is coprime to m, by the symbol x^* we denote a natural number that does not exceed m and satisfies the condition $x^*x \equiv 1 \pmod{m}$. Thus for a given m, the variable x^* is an integral-valued function of x, i.e., $x^* = y(x)$, $1 \leq y(x) < m$. Now suppose that x takes integer values that range from 1 to $X < m$ and are coprime to m. The problem is to study the behavior of $y = y(x)$. In particular, if a and b are fixed numbers, $1 \leq a < b < m$, then we have the question as to whether and how many times $y(x)$ hits the interval $[a, b)$. Another problem is to find how close the function approaches a given number a for $1 \leq x \leq X$. These problems are equivalent to the classical problems concerning the distribution of fractional parts of the function $y(x)/m$. The first of them is equivalent to the problem on the number of hits of the fractions $\{y(x)/m\}$ in the interval $[\alpha, \beta)$, $0 \leq \alpha < \beta < 1$. The second is equivalent to the problem of approximating a given fraction ξ, $0 \leq \xi < 1$, by $\{y(x)/m\}$.

The complexity of the study of the behavior of the function $y(x)$ significantly depends on the value of X. For example, if $X = m - 1$, then for $x = 1, 2, \ldots, X$, $(x, m) = 1$, $y(x)$ takes all the values from the reduced systems of residues modulo m. Next, if $X \geq m^{0.5+\varepsilon}$, $0 < \varepsilon < 1/2$, $m \geq m_1(\varepsilon)$, then sufficiently exact answers to the above questions about the behavior of $y(x)$ were obtained by estimating the Kloosterman sums by the Hasse–Weil method. However, for $X < m^{0.5}$, no approaches to these problems were found until 1993. The sole exception is provided by special moduli of the form $m = p^\alpha$, where p is a fixed prime number and $\alpha \to +\infty$ (or by similar moduli). Some problems for such moduli were posed and solved by A. G. Postnikov [138]. Namely, if $m = p^\alpha$, then the behavior of the fractional parts of $y(x)/m$ can be studied by Weyl's or by Vinogradov's methods, since

$$(1 + px)^* \equiv 1 - px + p^2x^2 - \cdots + (-p)^{\alpha-1}x^{\alpha-1} \pmod{m}$$

(see also [76]).

Here we consider the solution of the above problems for any moduli m under the condition that $X \geq m^\varepsilon$, where $\varepsilon > 0$ is an arbitrary fixed number and $m \geq m_1(\varepsilon)$, and even for somewhat lesser X, namely, for

$$X \geq \exp\{c \log^{2/3} m\},$$

where $c > 0$ is an absolute constant and $m > 1$.

The method used to solve these problems allowed one to proceed in solving some other problems related to the function $y(x)$ in one way or another (e.g., see [63]).

This chapter is organized as follows. In Section 12.1, we prove original theorems on the number of solutions to congruences of a special form. The results obtained are close to the final results. The numbers of solutions to these congruences themselves are the mean values of some power of the modulus of the corresponding trigonometric sums similar to short Kloosterman sums. So the theorems in this section are similar to Vinogradov's mean value theorem. In Section 12.2, we prove theorems on the estimates for trigonometric sums of a special form that are similar to incomplete (short) Kloosterman sums. In this case, we essentially use the theorems proved in Section 12.1. The fact that so short trigonometric sums of such a form admit nontrivial estimates is studied in the second original part of this chapter. In Section 12.3, we use the estimates for trigonometric sums obtained in Section 12.1 to solve problems on the distribution of fractional parts of functions of the form $(ay(x) + b)/n$. In Section 12.4, we study the mean value of the function $\alpha_k(n)$, which is essentially used in the proof of Lemma 12.1. Section 12.5 is devoted to the double Kloosterman sums with weights. Finally, in Section 12.6, we estimate a short Kloosterman sum and show how this estimate can be used.

12.1 Mean value theorems

In this section, we consider two versions of the method proposed later. The results obtained by the second version of this method are less precise, but this version is used more frequently than the first one. For natural numbers $x_1, \dots, x_k$, we denote the least common multiple of $x_1, \dots, x_k$ by the symbol $[x_1, \dots, x_k]$.

Theorem 12.1. *Suppose that $m > 1$, m is an integer, k is a natural number, and X, X_1 are real numbers such that*

$$k < X < X_1 \leq 2X, \quad k \cdot 2^{2k-1} X^{2k-1} < m.$$

Consider the congruence

$$p_1^* + \cdots + p_k^* \equiv p_{k+1}^* + \cdots + p_{2k}^* \pmod{m}, \tag{12.1}$$

where $p_1, \dots, p_{2k}$ are prime numbers that do not divide m and lie in the interval $(X, X_1]$. Then the number I of solutions to this congruence satisfies the estimate

$$I \leq k!\bigl(\pi_1(X)\bigr)^k, \tag{12.2}$$

where

$$\pi_1(X) = \pi(X_1) - \pi(X) - \sum_{p|m,\ X<p\le X_1} 1.$$

Proof. It suffices to prove that if (12.1) holds, then p_1 is equal to one of the numbers p_j, $k < j \le 2k$. Let us assume the contrary, i.e., assume that $p_1 \ne p_j$, $j = k+1, \dots, 2k$, and thus obtain a contradiction. For this, we collect together all the terms with the same p_j in the left-hand side of (12.1). Renaming the variables if necessary, we obtain the congruence

$$a_1 p_1^* + \cdots + a_t p_t^* \equiv p_{k+1}^* + \cdots + p_{2k}^* \pmod m. \tag{12.3}$$

In this congruence we have $t \ge 1$, $a_1 \ge 1, \dots, a_t \ge 1$, $a_1 + \cdots + a_t = k$,

$$X < p_j \le X_1, \quad (p_j, m) = 1, \quad j = 1, \dots, t, k+1, \dots, 2k, \tag{12.4}$$

and moreover, $p_1 \ne p_j$, $j \ge 2$.

We define the integers A and A_j by the relations

$$A = p_1 \dots p_t p_{k+1} \dots p_{2k}, \quad A_j p_j = A, \quad j = 1, \dots, t, k+1, \dots, 2k.$$

Multiplying both sides of the congruence (12.3) by A and using that

$$p_j^* p_j \equiv 1 \pmod m,$$

we obtain

$$a_1 A_1 + \cdots + a_t A_t \equiv A_{k+1} + \cdots + A_{2k} \pmod m. \tag{12.5}$$

From inequalities (12.4) for p_j, we readily find

$$0 < A_j \le (2X)^{2k-1}, \quad j = 1, \dots, t, k+1, \dots, 2k,$$

and hence we have

$$0 < a_1 A_1 + \cdots + a_t A_t \le k(2X)^{2k-1} < m,$$
$$0 < A_{k+1} + \cdots + A_{2k} \le k(2X)^{2k-1} < m.$$

Therefore, the congruence (12.5) is an equation of the form

$$a_1 A_1 + \cdots + a_t A_t = A_{k+1} + \cdots + A_{2k}. \tag{12.6}$$

It follows from the definition of the numbers A_j that, for $j \ge 2$, each A_j is divisible by p_1 and, moreover, $A_1 = p_2 \dots p_t p_{k+1} \dots p_{2k}$ is not divisible by p_1. Finally, we have the inequality $1 \le a_1 \le k < X < p_1$ for the number a_1. So from (12.6) we obtain the contradictory relation

$$a_1 A_1 \equiv 0 \pmod{p_1}.$$

The proof of the theorem is complete. □

Remark 12.1. 1. The quantity I has the obvious lower bound

$$\left(\pi_1(X)\right)^k \le I,$$

i.e., the estimate in Theorem 12.1 is sharp in order as $X \to +\infty$ for a constant k.

2. Obviously, we have

$$I = \frac{1}{m}\sum_{a=0}^{m-1}\left|\sum_{X<p\le X_1}{}' \exp\left\{2\pi i\frac{ap^*}{m}\right\}\right|^{2k},$$

where the prime on the sum over p mean the summation over prime p such that $(p,m)=1$. Thus the quantity I is the mean value of the $2k$th power of the modulus of the sum $S(a)$,

$$S(a) = \sum_{X<p\le X_1}{}' \exp\left\{2\pi i\frac{ap^*}{m}\right\}.$$

In turn, $S(a)$ is an analog of the short (incomplete) Kloosterman sum.

To prove Theorem 12.2, we need the following auxiliary lemma.

Lemma 12.1. *Let J_k be the number of solutions of the congruence*

$$\begin{gathered} x_1\dots x_k \equiv 0 \pmod{[x_1^1,\dots,x_k^2]}, \\ 0 < x_j \le X_j, \quad X_j \ge 3, \quad j=1,\dots,k. \end{gathered} \tag{12.7}$$

Then J_k satisfies the estimate

$$J_k \le (2k^8)^{k^3}\sqrt{X}(\log X)^{k^2}, \tag{12.8}$$

where $X = X_1\dots X_k$.

Proof. We introduce a new function $\alpha_k(n)$. By definition, for any natural number n, the function $\alpha_k(n)$ is equal to the number of solutions of the system

$$n = x_1\dots x_k, \quad n \equiv 0 \pmod{[x_1^2,\dots,x_k^2]}. \tag{12.9}$$

Then, obviously, the inequality

$$J_k \le \sum_{n\le X}\alpha_k(n) \tag{12.10}$$

holds. By the definition of $\alpha_k(n)$, we have

$$0 \le \alpha_k(n) \le \tau_k(n). \tag{12.11}$$

First, we prove that the function $\alpha_k(n)$ is multiplicative, i.e.,

$$\alpha_k(n, m) = \alpha_k(n)\alpha_k(m) \quad \text{for} \quad (n, m) = 1.$$

Indeed, suppose that $(n, m) = 1$ and (12.9) holds, and moreover,

$$m = y_1 \dots y_k, \quad m \equiv 0 \pmod{[y_1^2, \dots, y_k^2]}. \tag{12.12}$$

To each pair of the sets $\overline{x} = (x_1, \dots, x_k)$ and $\overline{y} = (y_1, \dots, y_k)$ that satisfy systems (12.9) and (12.12), respectively, there corresponds a set $\overline{z} = (z_1, \dots, z_k)$, $z_j = x_j t_j$, $j = 1, \dots, k$, such that

$$mn = z_1 \dots z_k, \quad mn = 0 \pmod{[z_1^2, \dots, z_k^2]}, \tag{12.13}$$

since we have $[x_1^2 y_1^2, \dots, x_k^2 y_k^2] = [x_1^2, \dots, x_k^2]\,[y_1^2, \dots, y_k^2]$ due to the fact that $(x_1 \dots x_k, y_1 \dots y_k) = 1$. Different pairs $(\overline{x}, \overline{y})$ and $(\overline{x}', \overline{y}')$ are assigned different $\overline{z}$ and $\overline{z}'$. Indeed, otherwise, we would have

$$x_j y_j = x_j' y_j', \quad j = 1, \dots, k.$$

Since $(y_j, x_j') = (x_j, y_j') = 1$, we have

$$x_j \equiv 0 \pmod{x_j'}, \quad x_j' \equiv 0 \pmod{x_j},$$

i.e., $x_j = x_j'$, $y_j = y_j'$, $j = 1, \dots, k$, and hence $(\overline{x}, \overline{y}) = (\overline{x}', \overline{y}')$. On the other hand, if $\overline{z} = (z_1, \dots, z_k)$ satisfies (12.13), then determining x_j, y_j by the relations $x_j = (n, z_j)$ and $y_j = (m, z_j)$, we obtain

$$x_j y_j = z_j.$$

Hence $\overline{z}$ is assigned the pair $\overline{x} = (x_1, \dots, x_k)$, $\overline{y} = (y_1, \dots, y_k)$ that satisfies (12.9) and (12.12), respectively. Thus we have $\alpha_k(nm) = \alpha_k(n)\alpha_k(m)$. Therefore, if $n = p_1^{\beta_1} \dots p_s^{\beta_s}$ is the canonical decomposition of n into prime factors, then

$$\alpha_k(n) = \alpha_k(p_1^{\beta_1}) \dots \alpha_k(p_s^{\beta_s}).$$

From the definition of $\alpha_k(n)$, it is easy to see that $\alpha_k(p) = 0$, where p is a prime, i.e., only the terms for which n has the form

$$n = p_1^{\beta_1} \dots p_s^{\beta_s}, \quad \beta_1 \ge 2, \dots, \beta_s \ge 2,$$

remain in the sum (12.10). Dividing the factors into two groups (with even and with odd β_j) and renaming the factors, we obtain

$$n = p_1^{\beta_1} \dots p_t^{\beta_t} p_{t+1}^{\beta_{t+1}} \dots p_s^{\beta_s},$$

where β_j are even numbers for $j \le t$ and β_j are odd numbers for $j > t$. Hence

$$n = p_1^{\beta_1} \dots p_t^{\beta_t} p_{t+1}^{\beta_{t+1}-3} \dots p_s^{\beta_s-3} p_{t+1}^3 \dots p_s^3 = m^2 r^3.$$

Thus from (12.10) and (12.11) we arrive at the inequality

$$J_k \le \sum_{m^2r^3 \le X} \alpha_k(m^2r^3) \le \sum_{m^2r^3 \le X} \tau_k(m^2r^3).$$

Since $\tau_k(ab) \le \tau_k(a)\tau_k(b)$, we have

$$J_k \le \sum_{\tau \le \sqrt[3]{X}} \tau_k^3(r) \sum_{m \le \sqrt{Xr^{-3}}} \tau_k^2(m).$$

To estimate the last double sum, we use Mardzhanishvili's inequality

$$\sum_{n=1}^{N} \tau_k^\ell(n) \le A_k^{(\ell)} N(\log N + k^\ell - 1)^{k^\ell - 1}, \quad A_k^{(\ell)} = k^\ell (k!)^{-(k^\ell-1)/(k-1)}.$$

We have

$$\begin{aligned}
\sum_{m \le \sqrt{Xr^{-3}}} \tau_k^2(m) &< k^2(k!)^{-(k+1)} \sqrt{Xr^{-3}} (\log \sqrt{Xr^{-3}} + k^2 - 1)^{k^2-1} \\
&< k^{2k^2} \sqrt{X} (\log X)^{k^2} r^{-3/2}, \\
\sum_{r \le \sqrt[3]{X}} \tau_k^3(r) r^{-3/2} &< 3k^2 + \sum_{3 < r \le \sqrt[3]{X}} \tau_k^3(r) r^{-3/2} \\
&= 3k^2 + \frac{3}{2} \int_3^{\sqrt[3]{X}} \mathbb{C}(u) u^{-5/2}\, du + \mathbb{C}(\sqrt[3]{X}) X^{-3/2},
\end{aligned}$$

where

$$\mathbb{C}(u) = \sum_{3 < r \le u} \tau_k^3(r) < u(\log u + k^3 - 1)^{k^3-1} < k^{3k^3} u (\log u)^{k^3-1},$$

$$\begin{aligned}
\int_1^{\sqrt[3]{X}} u^{-3/2} (\log u)^{k^3-1}\, du &< \int_1^\infty u^{-1/2} (\log u)^{k^3-1}\, d\log u \\
= \int_0^\infty e^{-v/2} v^{k^3-1}\, dv &< 2^{k^3} \int_0^\infty e^{-t} t^{k^3}\, dt = 2^{k^3}(k^3)!.
\end{aligned}$$

Thus for J_k we obtain

$$J_k < (2k^8)^{k^3} \sqrt{X} (\log X)^{k^2}.$$

The proof of the lemma is complete. □

Remark 12.2. In Section 12.4, by the complex integration method, for the summatory value $A_k(X)$ of the function $\alpha_k(n)$,

$$A_k(X) = \sum_{n \le X} \alpha_k(n),$$

we shall obtain an asymptotic formula similar to the formula for $T_k(X)$ in [85],

$$T_k(X) = \sum_{n \le X} \tau_k(n).$$

Theorem 12.2. *Suppose that $m > 1$, m is an integer, k is a natural number, and X, X_1 satisfy the conditions*

$$3 \le X, \quad k < X < X_1 \le 2X, \quad k 2^{2k-1} X^{2k-1} < m.$$

Consider the congruence

$$x_1^* + \cdots + x_k^* \equiv x_{k+1}^* + \cdots + x_{2k}^* \pmod{m}, \tag{12.14}$$

where $X < x_j \le X_1$, $(x_j, m) = 1$, $j = 1, \ldots, 2k$.

Then the number $I = I_k(X)$ of solutions of this congruence satisfies the estimate

$$I \le (2k)^{80k^3} X^k (\log X)^{4k^2}.$$

Proof. We consider the case $k \ge 2$, since the assertion of the theorem is trivial for $k = 1$. We assume that the set $(x_1, \ldots, x_{2k})$ satisfies (12.14). Multiplying both sides of (12.14) by the product $x_1 \ldots x_{2k}$ and using the fact that $x_j x_j^* \equiv 1 \pmod{m}$, we obtain

$$y_1 + \cdots + y_k \equiv y_{k+1} + \cdots + y_{2k} \pmod{m}, \tag{12.15}$$

where $y_j x_j = x_1 \ldots x_{2k}$, $j = 1, \ldots, 2k$. It follows from the conditions on x_j that

$$0 < y_j \le X_1^{2k-1} \le 2^{2k-1} X^{2k-1},$$

i.e., the left- and right-hand sides of Eq. (12.15) are positive and do not exceed

$$k 2^{2k-1} X^{2k-1} < m.$$

Hence the congruence (12.15) is an equality, i.e.,

$$y_1 + \cdots + y_k = y_{k+1} + \cdots + y_{2k}. \tag{12.16}$$

It follows from the definition of the variables y_j that each y_ν, $\nu \ne j$, is divisible by x_j. Therefore, according to (12.16), y_j is also divisible by x_j, i.e.,

$$y_j \equiv 0 \pmod{x_j}, \quad j = 1, \ldots, 2k. \tag{12.17}$$

Multiplying both sides of the congruence (12.17) and its modulus by x_j, we obtain

$$x_1 \dots x_{2k} \equiv 0 \pmod{x_j^2}, \quad j = 1, \dots, 2k,$$

i.e.,

$$x_1 \dots x_{2k} \equiv 0 \pmod{[x_1^2, \dots, x_{2k}^2]}. \tag{12.18}$$

Thus each set $(x_1 \dots x_{2k})$ satisfying (12.14) also satisfies (12.18), i.e.,

$$I \leq J_{2k},$$

where J_{2k} is the number of solutions of the congruence (12.17) in the lemma with obvious replacements of the corresponding parameters. Using the estimate in the lemma, we obtain the statement of the theorem:

$$I \leq \left(2(2k)^8\right)^{8k^3} 2^k X^k (\log 2X)^{4k^2} < (2k)^{80k^3} X^k (\log X)^{4k^2}.$$

□

Remark 12.3. 1. The estimate in Theorem 12.2 is close to the final estimate, since

$$I \geq \Big(X_1 - X - \sum_{(n,m)=1,\ X<n\leq X_1} 1\Big).$$

2. The quantity $I = I_k(X)$ is the mean value of the $2k$th power of the modulus of the short (incomplete) Kloosterman sum.

12.2 Analogs of incomplete Kloosterman sums and their estimates

In this section in Theorems 12.3 and 12.4, we derive and estimate trigonometric sums similar to incomplete Kloosterman sums. Namely, we consider sums S of the form

$$S = \sum_{n\leq N}{}' \exp\left\{2\pi i \frac{an^* + bn}{m}\right\},$$

where $N < m$, the prime means that the summation is performed over numbers n from some set A of natural numbers, $(n, m) = 1$, and the number of elements in A is equal to $\|A\| < \sqrt{m}$. In some cases, the estimate of $|S|$ is sufficiently sharp.

Theorem 12.3. *Suppose that $m > 1$, m is an integer, k, s are natural numbers, and real numbers X, X_1, Y, Y_1 satisfy the inequalities*

$$k < X < X_1 \leq 2X, \qquad k2^{2k-1}X^{2k-1} < m,$$
$$s < Y < Y_1 \leq 2Y, \qquad s2^{2s-1}Y^{2s-1} < m,$$

the parameter n_1 runs through N_1 values of natural numbers such that $(n_1, m) = 1$, a and b are integers such that $(a, m) = d$, p and q take values of successive prime numbers, and $(p, m) = (q, m) = 1$. Denote the set of numbers of the form $n = n_1 pq$, $X < p \le X_1$, $Y < q \le Y_1$, by the letter A and the number of elements in A by the symbol $\|A\|$. Then the trigonometric sum S,

$$S = \sum_{n \in A} \exp\left\{2\pi i \frac{an^* + bn}{m}\right\}, \tag{12.19}$$

satisfies the estimate

$$|S| \le \|A\|\Delta, \tag{12.20}$$

where

$$\Delta = (s!k!)^{1/(2sk)}(\pi_1(X))^{-1/(2s)}(\pi_1(Y))^{-1/(2k)}(sdmY)^{1/(2sk)}. \tag{12.21}$$

Proof. First, we note that

$$\|A\| = N_1\pi_1(X)\pi_1(Y).$$

Moreover, it suffices to prove the corresponding estimate for $|S_1|$,

$$S_1 = \sum_{X<p\le X_1} \sum_{Y<q\le Y_1} \exp\left\{2\pi i \frac{ap^*q^* + bpq}{m}\right\}.$$

Passing to estimates and applying Hölder's inequality, we obtain

$$\begin{aligned} |S_1|^s &\le (\pi_1(X))^{s-1} \sum_{X<p\le X_1} \left| \sum_{Y<q\le Y_1} \exp\left\{2\pi i \frac{ap^*q^* + bpq}{m}\right\}\right|^s \\ &= (\pi_1(X))^{s-1} \sum_{X<p\le X_1} \left| \sum_{\lambda=1}^{m} \sum_{sY<\mu\le sY_1} J_s(\lambda, \mu) \exp\left\{2\pi i \frac{a\lambda p^* + b\mu p}{m}\right\}\right|, \end{aligned} \tag{12.22}$$

where the symbol $J_s(\lambda, \mu)$ denotes the number of solutions of a system of congruences of the form

$$\begin{cases} q_1^* + \cdots + q_s^* \equiv \lambda, \\ q_1 + \cdots + q_s \equiv \mu, \end{cases} \pmod m,$$
$$Y < q_j \le Y_1, \quad (q_j, m) = 1, \quad j = 1, \ldots, s.$$

Let $\theta = \theta(p)$ be the argument in the sum over λ, μ in (12.22), i.e., let

$$\theta = \theta(p) = \arg \sum_{\lambda=1}^{m} \sum_{sY<\mu\le sY_1} J_s(\lambda, \mu) \exp\left\{2\pi i \frac{a\lambda p^* + b\mu p}{m}\right\}.$$

Then we have the relation

$$\left|\sum_{\lambda=1}^{m}\sum_{sY<\mu\le sY_1} J_s(\lambda,\mu)\exp\left\{2\pi i\frac{a\lambda p^*+b\mu p}{m}\right\}\right|$$
$$= e^{-i\theta(p)}\sum_{\lambda=1}^{m}\sum_{sY<\mu\le sY_1} J_s(\lambda,\mu)\exp\left\{2\pi i\frac{a\lambda p^*+b\mu p}{m}\right\}.$$

Therefore, from (12.22) we obtain the inequality

$$|S_1|^s \le (\pi_s(X))^{s-1}\sum_{\lambda=1}^{m}\sum_{sY<\mu\le sY_1} J_s(\lambda,\mu) \tag{12.23}$$
$$\times\left|\sum_{X<p\le X_1} e^{-i\theta(p)}\exp\left\{2\pi i\frac{a\lambda p^*+b\mu p}{m}\right\}\right|.$$

Raising both sides of (12.23) to the power k and again applying Hölder's inequality and then Cauchy's inequality, we successively obtain

$$|S_1|^{sk} \le (\pi_s(X))^{k(s-1)}\Big(\sum_{\lambda=1}^{m}\sum_{\mu} J_s(\lambda,\mu)\Big)^{k-1} \tag{12.24}$$
$$\times\sum_{\lambda=1}^{m}\sum_{\mu} J_s(\lambda,\mu)\left|\sum_{X<p\le X_1} e^{-i\theta(p)}\exp\left\{2\pi i\frac{a\lambda p^*+b\mu p}{m}\right\}\right|^k$$
$$\le (\pi_s(X))^{k(s-1)}\Big(\sum_{\lambda=1}^{m}\sum_{\mu} J_s(\lambda,\mu)\Big)^{k-1}\Big(\sum_{\lambda=1}^{m}\sum_{\mu} J_s^2(\lambda,\mu)\Big)^{1/2}$$
$$\times\left(\sum_{\lambda=1}^{m}\sum_{sY<\mu\le sY_1}\left|\sum_{X<p\le X_1} e^{-i\theta(p)}\exp\left\{2\pi i\frac{a\lambda p^*+b\mu p}{m}\right\}\right|^{2k}\right)^{1/2}.$$

Next, we have

$$\sum_{\lambda=1}^{m}\sum_{\mu} J_s(\lambda,\mu) = (\pi_1(Y))^s,\qquad \sum_{\lambda=1}^{m}\sum_{\mu} J_s^2(\lambda,\mu) = I(Y;m;s), \tag{12.25}$$

where the symbol $I(Y;m;s)$ denotes the number of solutions of the system of equations

$$\begin{cases} q_1^*+\cdots+q_s^* \equiv q_{s+1}^*+\cdots+q_{2s}^*, \\ q_1+\cdots+q_s \equiv q_{s+1}+\cdots+q_{2s}, \end{cases} \pmod m,$$
$$Y<q_j\le Y_1,\quad (q_j,m)=1,\quad j=1,\dots,2s.$$

Clearly, we have

$$I(Y;m;s) \le I_s(Y), \tag{12.26}$$

where $I_s(Y)$ is the number of solutions of the congruence (12.1) under the assumption that $k = s$, $X = Y$ and $X_1 = Y_1$. Finally, the sum over λ, μ under the second radical sign in (12.24), which we denote by σ, can be estimated as

$$\begin{aligned}
\sigma &= \sum_{\lambda=1}^{m} \sum_{sY<\mu\le sY_1} \left| \sum_{X<p\le X_1} e^{-i\theta(p)} \exp\left\{2\pi i \frac{a\lambda p^* + b\mu p}{m}\right\}\right|^{2k} \\
&= \sum_{sY<\mu\le sY_1} \sum_{X<p_1\le X_1} \cdots \sum_{X<p_{2k}\le X_1} \exp\{-i(\theta(p_1) - \cdots - i\theta(p_{2k}))\} \\
&\quad \times \exp\left\{2\pi i \frac{b\mu(p_1 + \cdots - p_{2k}}{m}\right\} \sum_{\lambda=1}^{m} \exp\left\{2\pi i \frac{a\lambda(p_1^* + \cdots - p_{2k}^*)}{m}\right\} \\
&\le sY \sum_{X<p_1\le X_1} \cdots \sum_{X<p_{2k}\le X_1} \sum_{\lambda=1}^{m} \exp\left\{2\pi i \frac{a\lambda(p_1^* + \cdots - p_{2k}^*)}{m}\right\} \\
&= sY \sum_{\nu=1}^{m} I(\nu) \sum_{\lambda=1}^{m} \exp\left\{2\pi i \frac{a\lambda\nu}{m}\right\},
\end{aligned}$$

where $I(\nu)$ is the number of solutions of the congruence

$$\begin{gathered}
p_1^* + \cdots - p_{2k}^* \equiv \nu \pmod m, \\
Y < p_j \le Y_1, \quad (p_j, m) = 1, \quad j = 1, \ldots, 2k.
\end{gathered}$$

Obviously, we have

$$I(\nu) \le I(0) = I = I_k(X).$$

Hence we find the final estimate for σ $((a,m) = d, a = a_1 d, m = m_1 d, (a_1, m_1) = 1)$:

$$\begin{aligned}
\sigma &\le sYI \sum_{\nu=1}^{m} \sum_{\lambda=1}^{m} \exp\left\{2\pi i \frac{a\lambda\nu}{m}\right\} = sYI \sum_{\nu=1}^{m} \sum_{\lambda=1}^{m} \exp\left\{2\pi i \frac{a_1\lambda\nu}{m_1}\right\} \qquad (12.27) \\
&= sYId \sum_{\nu=1}^{m} \sum_{\lambda_1=1}^{m_1} \exp\left\{2\pi i \frac{a_1\lambda_1\nu}{m_1}\right\} = sYIdm_1 d = sdmYI_k(X).
\end{aligned}$$

From (12.24)–(12.27) we obtain

$$|S_1|^{sk} \le (\pi_1(X))^{k(s-1)} (\pi_1(Y))^{s(k-1)} (sdmY)^{1/2} (I_s(Y) I_k(S))^{1/2}.$$

Finally, applying the estimates in Theorem 12.1 to $I_s(Y)$ and $I_k(X)$, we obtain

$$I_s(Y) \le s!(\pi_1(Y))^s, \quad I_k(X) \le k!(\pi_1(X))^k.$$

Hence

$$|S_1|^{sk} \le (s!k!)^{1/2}(\pi_1(X)\pi_1(Y))^{ks}(\pi_1(X))^{-k/2}(\pi_1(Y))^{-s/2}(sdmY)^{1/2},$$
$$|S_1| \le \pi_1(X)\pi_1(Y)\Delta,$$

where

$$\Delta = (s!k!)^{1/(2sk)}(\pi_1(X))^{-1/(2s)}(\pi_1(Y))^{-1/(2k)}(sdmY)^{1/(2sk)},$$

as required. The proof of the theorem is complete. □

Remark 12.4. In the special case of sums S, namely, in the case $b = 0$, the estimate (12.20) becomes somewhat sharper. However, this refinement is not important for the applications considered below.

Theorem 12.4. *Suppose that $m > 1$, m is an integer, k, s are natural numbers, and real numbers X, X_1, Y, Y_1 satisfy the inequalities*

$$3 \le X, \qquad k < X < X_1 \le 2X, \qquad k2^{2k-1}X^{2k-1} < m,$$
$$3 \le Y, \qquad s < Y < Y_1 \le 2Y, \qquad s2^{2s-1}Y^{2s-1} < m,$$

a and b are integers such that $(a, m) = d \ge 1$. Denote the set of natural numbers n of the form $n = xy$, where $X < x \le X_1$, $Y < y \le Y_1$, and $(xy, m) = 1$, by the letter A. Then the trigonometric sum S,

$$S = \sum_{n\in A} \exp\left\{2\pi i \frac{an^* + bn}{m}\right\},$$

satisfies the estimate

$$|S| \le XY\Delta, \tag{12.28}$$

where

$$\Delta = (2s)^{40s^2/k}(2k)^{40k^2/s}(sdmY)^{1/(2sk)}(\log Y)^{2s/k}(\log X)^{2k/s}X^{-1/(2s)}Y^{-1/(2k)}.$$

The proof of Theorem 12.4 repeats the proof of Theorem 12.3 where the estimates from Theorem 12.1 are replaced by the corresponding estimates from Theorem 12.2.

12.3 Fractional parts of functions related to reciprocal values modulo a given number

The estimates in Theorem 12.3 and 12.4 can be used successfully in different problems related to analogs of incomplete Kloosterman sums. We consider only one of them, namely, the problem on distributions of the fractional parts of functions of the form $(an^* + bn)/m$, $n \le N$.

Theorem 12.5. *Suppose that* $m \geq m_1$, $(a, m) = 1$, *b is an integer,* $1 \leq N \leq m^{4/7}$, *and* $0 \leq \alpha < \beta < 1$. *Denote the number of solutions of the system of inequalities*

$$\alpha \leq \left\{ \frac{an^* + bn}{m} \right\} < \beta, \quad n \leq N, \tag{12.29}$$

by the symbol $K = K(N; m; \alpha, \beta)$. *Then* K *has the lower bound*

$$K \geq \frac{cN}{(\log N)^{3.5}} \left((\beta - \alpha) - \exp\left\{ - \frac{\log^3 N}{320 \log^2 m} \right) \right\}, \tag{12.30}$$

where $c > 0$ *is an absolute constant.*

Proof. The estimate (12.30) is meaningful if

$$\beta - \alpha > \exp\left\{ - \frac{\log^3 N}{320 \log^2 m} \right\}$$

and, moreover,

$$N \geq \exp\{a_1 \log^{2/3} m\}.$$

Therefore, in what follows, we assume that

$$\exp\{a_1 \log^{2/3} m\} \leq N \leq m^{4/7}, \tag{12.31}$$

where $a_1 \geq 7$ and a_1 is a constant. Let us find the integer k from the inequalities

$$m^{1/(2k-1)+1/(4k-1)} \leq N < m^{1/(2k-3)+1/(4k-s)}. \tag{12.32}$$

Inequalities (12.1) imply $k \geq 2$. Moreover, it follows from (12.31) and (12.32) that

$$k \leq a_2 \log^{1/3} m, \tag{12.33}$$

and, by increasing a_1, we can obtain an arbitrarily small constant a_2. We choose

$$4X = m^{1/(2k-1)}, \qquad 4Y = m^{1/(4k-1)}, \qquad X_1 = 2X,$$
$$Y_1 = 2Y, \qquad N_1 = Nm^{-1/(2k-1)-1/(4k-1)}.$$

By the letter A we denote the set of natural numbers n of the form $n = rpq$, where r takes either the value 1 or the values of prime numbers that do not exceed N_1 and do not divide m, while p and q take the values of prime numbers that do not divide m and lie in the intervals $X < p \leq X_1$ and $Y < q \leq Y_1$. By the symbol $\|A\|$ we denote the number of numbers in the set A. Obviously, $n \in A$ satisfies the inequality $n \leq N$. It is easy to show that

$$\|A\| \asymp \pi_1(N_1)\pi_1(X)\pi_1(Y) \gg \frac{N}{(\log N)^{3.5}}.$$

We consider two integers $\alpha_1 = [\alpha m]$ and $\beta_1 = [\beta m]$. Let K_1 be equal to the number of solutions of the congruence

$$an^* + bn \equiv \ell \pmod{m} \tag{12.34}$$

under the assumption that $n \in A$ and $\alpha_1 < \ell < \beta_1$. If (12.34) is satisfied, then

$$an^* + bn = ms + \ell, \qquad \left\{\frac{\alpha n^* + bn}{m}\right\} = \frac{\ell}{m}.$$

From the inequality $\alpha_1 + 1 \le \ell \le \beta_1$ for ℓ, we find

$$\alpha m < \ell < \beta m, \quad \alpha < \ell/m < \beta.$$

This implies that $K \ge K_1$. Finally, we set $4L = \beta_1 - \alpha_1$ and consider the congruence

$$an^* + bn \equiv \alpha_1 + \ell_1 + \ell_2 + \ell_3 + \ell_4 \pmod{m}, \tag{12.35}$$

where $n \in A$ and $0 < \ell_1, \ell_2, \ell_3, \ell_4 \le L$. It follows from the conditions imposed on ℓ_j that $\ell = \alpha_1 + \ell_1 + \ell_2 + \ell_3 + \ell_4$ satisfies the inequalities

$$0 \le \alpha_1 < \ell \le \beta_1 < m.$$

Moreover, the equation

$$\ell = \alpha_1 + \ell_1 + \ell_2 + \ell_3 + \ell_4$$

in the numbers ℓ_j for a fixed ℓ has at most L^3 solutions. Therefore, if K_2 is the number of solutions of the congruence (12.35) in the numbers $n, \ell_1, \ell_2, \ell_3, \ell_4$, then

$$K_1 \ge L^{-3} K_2. \tag{12.36}$$

Using the known discontinuous factor, we write the quantity K_2 as the trigonometric sum

$$\begin{aligned} K_2 &= \frac{1}{m}\sum_{t=0}^{m-1}\sum_{n\in A}\sum_{0<\ell_1,\dots,\ell_4\le L} \exp\left\{2\pi i\frac{t(an^* + bn - \ell_1 - \cdots - \ell_4)}{m}\right\} \qquad (12.37)\\ &= \frac{1}{m}\sum_{t=0}^{m-1}\left(\sum_{n\in A}\exp\left\{2\pi i\frac{t(an^*+bn)}{m}\right\}\right)\\ &\quad\times\left(\sum_{0<\ell\le L}\exp\left\{-2\pi i\frac{t\ell}{m}\right\}\right)^4 \exp\left\{-2\pi i\frac{t\alpha_1}{m}\right\}. \end{aligned}$$

We represent the right-hand side of (12.37) as the sum of two terms the first of which is obtained for $t = 0$ and the second is the remaining part of the sum. We have

$$K_1 = \frac{1}{m}\|A\|L^4 + R,$$

$$R = \frac{1}{m}\sum_{t=1}^{m-1}\Bigl(\sum_{n\in A}\exp\Bigl\{2\pi i\frac{t(an^*+bn)}{m}\Bigr\}\Bigr) \times \Bigl(\sum_{0<\ell\le L}\exp\Bigl\{-2\pi i\frac{t\ell}{m}\Bigr\}\Bigr)^4\exp\Bigl\{-2\pi i\frac{t\alpha_1}{m}\Bigr\}. \tag{12.38}$$

We shall transform R. Let $d=(m,t)$, $1\le t<m$. Then $m=m_1d$, $t=t_1d$, $(m_1,t_1)=1$, $1\le t_1<m$. Hence for each d, $d\mid m$, $1\le d<m$, any t can be represented as $t=t_1d$, $1\le t_1<m$, $(t_1,m_1)=1$, where $m_1=md^{-1}$, and this representation is unique. Indeed, if $t_1d=t_1'd'$, where $(md^{-1},t_1)=(md'^{-1},t_1')=1$, then the relation $mt_1d=mt_1'd'$ implies

$$\frac{m}{d'}t_1=\frac{m}{d}t_1',\quad t_1\equiv 0 \pmod{t_1'},\quad t_1'\equiv 0 \pmod{t_1},$$

i.e., $t_1=t_1'$ and $d=d'$. Hence we have

$$R=\frac{1}{m}\sum_{\substack{d\mid m\\1\le d<m}}\sum_{\substack{t=1\\(t,m)=d}}^{m-1}\Bigl(\sum_{n\in A}\exp\Bigl\{2\pi i\frac{t(an^*+bn)}{m}\Bigr\}\Bigr) \times\Bigl(\sum_{0<\ell\le L}\exp\Bigl\{-2\pi i\frac{t\ell}{m}\Bigr\}\Bigr)^4\exp\Bigl\{-2\pi i\frac{t\alpha_1}{m}\Bigr\}.$$

Finally, we represent R as the sum of two terms R_1 and R_2: $R=R_1+R_2$. The term R_1 contains summands with "small" d, namely, with $1\le d\le m^{1/(32k)}$. The term R_2 contains summands with $d>m^{1/(32k)}$.

Let us estimate $|R_1|$ from above. To estimate the sum over n, we use Theorem 11.3, setting $s=2k$ in it and replacing a by at. We have

$$k2^{2k-1}X^{2k-1}=k2^{2k-1}4^{-2k+1}m<m,$$
$$2k2^{4k-1}Y^{4k-1}=2k2^{4k-1}4^{-4k+1}m<m,$$
$$k<X=\frac{1}{4}m^{1/(2k-1)},\quad k<Y=\frac{1}{4}m^{1/(4k-1)}.$$

Hence

$$|R_1|=\|A\|\Delta L^2\frac{1}{m}\sum_{t=0}^{m-1}\Bigl|\sum_{0<\ell\le L}\exp\Bigl\{2\pi i\frac{t\ell}{m}\Bigr\}\Bigr|=\|A\|\Delta L^3, \tag{12.39}$$

where

$$\Delta=\bigl((2k)!k!\bigr)^{1/(4k^2)}\bigl(\pi_1(X)\bigr)^{-1/(4k)}\bigl(\pi_1(Y)\bigr)^{-1/(2k)}\bigl(2kYm^{1+1/(32k)}\bigr)^{1/(4k^2)}.$$

Let us estimate $|R_2|$ from above. The sum over n can be trivially estimated by the number $\|A\|$. Since

$$\frac{t}{m} = \frac{t'}{m'}, \quad (t_1, m_1) = 1, \quad 1 < m_1 = \frac{m}{d} \le m^{1-1/(32k)},$$

the two sums over ℓ have the upper bound $m^{1-1/(32k)}$. We obtain

$$|R_2| \le \|A\| m^{2-1/(16k)} L. \tag{12.40}$$

Thus it follows from (12.39) and (12.40) that

$$|R| \le \|A\|(\Delta L^3 + m^{2-1/(16k)} L). \tag{12.41}$$

Therefore, from (12.36), (12.38), and (12.41), we obtain the inequality

$$K \ge L^{-3}\|A\|\Big(\frac{1}{m}L^4 - \Delta L^3 - m^{2-1/(16k)}L\Big) = \|A\|\Big(\frac{1}{m}L - \Delta - (m^{1-1/(32k)}L^{-1})^2\Big).$$

It remains to transform the last estimate to the form stated in the theorem. By assumption, we have

$$L = \frac{\beta_1 - \alpha_1}{4} \ge \frac{(\beta - \alpha)m - 1}{4}.$$

We can assume that $\beta - \alpha \ge m^{-1/(64k)}$, since, otherwise, the statement of the theorem becomes trivial. This follows from the fact that the inequality

$$m^{-1/(64k)} \le \exp\Big\{-\frac{\log^3 N}{320 \log^2 m}\Big\},$$

i.e., $k \le (5 \log^3 m)/\log^3 N$, must hold. But, it follows from (12.32) that

$$N \le m^{2/(2k-3)},$$

i.e.,

$$k - \frac{3}{2} \le \frac{\log m}{\log N}, \quad k \le \frac{3}{2} + \frac{\log m}{\log N} \le \frac{5 \log^3 m}{\log^3 N}.$$

Therefore, we have

$$L \ge \frac{1}{5}(\beta - \alpha)m \ge \frac{1}{5}m^{1-1/(64k)}, \quad m^{1-1/(32k)}L^{-1} \le 5m^{-1/(64k)},$$

$$\frac{1}{2m}L \ge \frac{1}{10}m^{1-1/(64k)} \ge (5m^{-1/(64k)})^2,$$

which, obviously, holds, since $k \le a_2 \log^{1/3} m$, $m \ge m_1$. We obtain

$$K \ge \|A\|\Big(\frac{1}{2m}L - \Delta\Big) \ge \|A\|\Big(\frac{\beta - \alpha}{10} - \Delta\Big) = \frac{1}{10}\|A\|(\beta - \alpha - 10\Delta). \tag{12.42}$$

It remains to prove the inequality

$$10((2k)!k!)^{1/(4k^2)}(\pi_1(X))^{-1/(4k)}(\pi_1(Y))^{-1/(2k)} \times (2kYm^{1+1/(32k)})^{1/(4k^2)} \le \exp\left\{-\frac{\log^3 N}{320\log^2 m}\right\}. \tag{12.43}$$

Let us estimate the left-hand side of the last inequality. First, we have

$$\log X \asymp \log Y \asymp \log N, \quad \pi_1(X) \asymp \frac{X}{\log N}, \quad \pi_1(Y) \asymp \frac{Y}{\log N},$$

i.e., the left-hand side of (12.43) is of order

$$X^{-1/(4k)}Y^{-1/(2k)}(\log N)^{1/(4k)+1/(2k)}m^{(1/(4k^2))(1+1/(32k))+1/(4k^2(4k-1))}.$$

We recall the definition of X and Y and readily see that the exponent m in the last relation is equal to δ:

$$\begin{aligned}\delta &= -\frac{1}{4k(2k-1)} - \frac{1}{2k(4k-1)} + \frac{1}{4k^2}\left(1+\frac{1}{32k}\right) + \frac{1}{4k^2(4k-1)} \\ &= -\frac{1}{4k(4k-1)(2k-1)} + \frac{1}{128k^3} \le -\frac{1}{32k^3} + \frac{1}{128k^3} \le -\frac{1}{64k^3}.\end{aligned}$$

Finally, we have

$$c_1(\log N)^{3/(4k)} \le m^{1/(128k^3)},$$

which follows from the inequalities

$$N < m, \quad k \le a_2 \log^{1/3} m.$$

Thus the left-hand side of (12.43) does not exceed

$$m^{-1/(128k^3)} \exp\left\{-\frac{\log m}{128k^3}\right\}.$$

From (12.32) we again obtain

$$N < m^{2/(2k-3)}, \quad k - \frac{3}{2} < \frac{\log m}{\log N}, \quad k \le \frac{2\log m}{\log N},$$

i.e.,

$$\exp\left\{-\frac{\log m}{128k^3}\right\} \le \exp\left\{-\frac{\log^3 N}{320\log^2 m}\right\},$$

which proves inequality (12.43).

From (12.42) and (12.43) we obtain the statement of the theorem:

$$K \ge c_1\|A\|\left(\beta - \alpha - \exp\left\{-\frac{\log^3 N}{320\log^2 m}\right)\right\}$$

$$\geq \frac{c_2 N}{(\log N)^{3.5}}\left(\beta - \alpha - \exp\left\{-\frac{\log^3 N}{320\log^2 m}\right\}\right).$$

□

Now we formulate several obvious corollaries of this theorem.

Corollary 12.1. *Suppose that $m \geq m_1$, $N < m$, $(a, m) = 1$, b is an integer, $0 \leq \alpha < \beta < 1$, and*

$$\exp\left\{-\frac{\log^3 N}{320\log^2 m}\right\} < \beta - \alpha < e^{-1}.$$

Then the interval $[\alpha, \beta]$ contains a number of the form

$$\left\{\frac{an^* + bn}{m}\right\}, \quad n \leq N.$$

Corollary 12.2. *If ξ is a real number, then for $1 < N < m$,*

$$\min_{n\leq N}\left\|\xi - \frac{an^* + bn}{m}\right\| \ll \exp\left\{-\frac{\log^3 N}{320\log^2 m}\right)\Bigg\}.$$

Prior to formulating the next theorem, we introduced some notions. Suppose that $m \geq m_1 > 1$, $(a, m) = 1$, b is an integer, $a_1 \geq 10$, and

$$\exp\{a_1 \log^{2/3} m \log^{1/4} \log m\} \leq N \leq m^{4/7}.$$

A natural number k is determined by the inequalities

$$m^{1/(2k-1)+1/(4k-1)} \leq N < m^{1/(2k-3)+1/(4k-5)}.$$

The conditions imposed on N imply that $k \geq 2$. We choose

$$4X = m^{1/(2k-1)}, \quad 4Y = m^{1/(4k-1)},$$
$$X_1 = 2X, \quad Y_1 = 2Y, \quad Z = Nm^{-1/(2k-1)-1/(4k-1)},$$

and consider the set A of natural numbers of the form $n = xyz$, where $X < x \leq X_1$, $Y < y \leq Y_1$, $z \leq Z$, and $(xyz, m) = 1$.

We denote the number of elements in the set A by the symbol $\|A\|$. Obviously, $\|A\| \leq N$ and the inequality $n \leq N$ holds for $n \in A$.

Theorem 12.6. *Let $0 \leq \alpha < \beta < 1$, and let $K = K(A; \alpha, \beta)$ be the number of solutions of the system of inequalities*

$$\alpha \leq \left\{\frac{an^* + bn}{m}\right\} < \beta, \quad n \in A.$$

Then K satisfies the asymptotic formula

$$K = (\beta - \alpha)\|A\| + O(R)$$

where

$$R = (4k)^{180k} N^{1-1/(320k^2)}.$$

Proof. We choose $r = 2[\log N]$ and assume that

$$\Delta_1 = m^{-1/(30k)} < \frac{1}{16}, \quad 2\Delta_1 \le \beta - \alpha < 1 - 2\Delta_1.$$

For given r, α, β and Δ_1, we define the Vinogradov "cup" (or "goblet") $\psi(x)$ as follows (see, e.g., Lemma A.3)

(1) $\psi(x+1) = \psi(x)$,
(2) $\psi(x) = 1$ and $\alpha + \Delta_1 \le x \le \beta - \Delta_1$,
(3) $0 < \psi(x) < 1$, $\alpha - \Delta_1 < x < \alpha + \Delta_1$, and $\beta - \Delta_1 < x < \beta + \Delta_1$,
(4) $\psi(x) = 0$ and $\beta + \Delta_1 \le x \le 1 + \alpha - \Delta_1$,
(5) $\psi(x) = \beta - \alpha + \sum_{|f|>0} g(f) e^{2\pi i f x}$,

where

$$|g(f)| \le c(f) = \min\left(\beta - \alpha, \frac{1}{|f|}, \frac{1}{|f|}\left(\frac{r}{\Delta_1 |f|}\right)^r\right).$$

By $U(\alpha, \beta)$ we denote the sum

$$U(\alpha, \beta) = \sum_{n \in A} \psi\left(\frac{an^* + bn}{m}\right).$$

Then the number K satisfies the inequalities

$$U\left(\alpha + \frac{\Delta_1}{2}, \beta - \frac{\Delta_1}{2}\right) \le K \le U\left(\alpha - \frac{\Delta_1}{2}, \beta + \frac{\Delta_1}{2}\right).$$

From the definition of $U(\alpha, \beta)$, we obtain

$$U\left(\alpha + \frac{\Delta_1}{2}, \beta - \frac{\Delta_1}{2}\right) = ((\beta - \alpha) - \Delta_1)\|A\| + O(R_1),$$
$$U\left(\alpha - \frac{\Delta_1}{2}, \beta + \frac{\Delta_1}{2}\right) = ((\beta - \alpha) + \Delta_1)\|A\| + O(R_1),$$

where

$$R_1 = \sum_{f=1}^{\infty} c(f) \left| \sum_{n \in A} \exp\left\{2\pi i \frac{f(an^* + bn)}{m}\right\}\right|.$$

We divide the sum in the last relation into two sums:

$$R_1 = R_2 + R_3,$$

where

$$R_2 = \sum_{f \le m^{1/(32k)}}, \quad R_3 = \sum_{f > m^{1/(32k)}}.$$

Since for $f \le m^{1/(32k)}$ we have

$$d = (af, m) \le f \le m^{1/(32k)},$$

we can apply the estimate in Theorem 12.4, where we set $s = 2k$, to the sum over n and thus obtain

$$\left| \sum_{n \in A} \exp\left\{2\pi i \frac{f(an^* + bn)}{m}\right\} \right| \le XYZ\Delta,$$

where

$$\Delta = (4k)^{160k}(2k)^{20k}(2kYm^{1+1/(32k)})^{1/(4k^2)}(\log Y)^4(\log X)X^{-1/(4k)}Y^{-1/(2k)}.$$

Hence we have

$$R_2 \le \sum_{0 < f \le m^{1/(32k)}} \frac{1}{f} N\Delta < (4k)^{180k} N m^{-1/(64k^3)} < (4k)^{180k} N^{1-1/(320k^2)}.$$

The sum R_3 can be estimated trivially as follows:

$$R_3 \le \sum_{f > m^{1/(32k)}} \frac{1}{f}\left(\frac{r}{\Delta_1 f}\right)^r N < N^{-1}.$$

Thus for K we obtain the asymptotic formula

$$K - (\beta - \alpha)\|A\| + O(R), \quad 2\Delta_1 \le \beta - \alpha < 1 - 2\Delta_1. \tag{12.44}$$

But if $0 < \beta - \alpha < 2\Delta_1$, then we have

$$K = K(A; \alpha, \beta) = K(A; \alpha, \alpha + 1 - 2\Delta_1) - K(A; \beta, \alpha + 1 - 2\Delta_1);$$

if $1 - 2\Delta_1 \le \beta - \alpha < 1$, then

$$K = K(A; \alpha, \beta) = K\left(A; \alpha, \alpha + \frac{1}{2}\right) + K\left(A; \alpha + \frac{1}{2}, \beta\right).$$

Therefore, the last formulas and (12.44) imply relation (12.44) already for any $0 \le \alpha < \beta < 1$. The proof of the theorem is complete. □

The following assertion can be proved on the basis of Theorem 12.6.

Theorem 12.7. *Under the assumptions of Theorem* 12.6, *the asymptotic formula*

$$\sum_{n\in A}\left\{\frac{an^*+bn}{m}\right\}=\frac{1}{2}\|A\|+O(R),$$

holds for $R=(4k)^{180k}N^{1-1/(320k^2)}$.

12.4 The function $\alpha_k(n)$ and its mean value

In the proof of Lemma 12.1, we introduced a function $\alpha_k(n)$ and proved that this function is multiplicative. In this section we study the function $\alpha_k(n)$ in more detail. In particular, we prove the mean value theorem for $\alpha_k(n)$ similarly to the mean value theorem for $\tau_k(n)$.

Lemmas 12.2 and 12.3 present assertions concerning $\alpha_k(n)$ and $A_k(n)$, which we have already proved in Lemma 12.1.

Lemma 12.2. *The function* $\alpha_k(n)$ *is multiplicative, i.e.,*

$$\alpha_k(n,m)=\alpha_k(n)\alpha_k(m)\quad \textit{for } (n,m)=1.$$

Lemma 12.3. *For* $A_k(X)$, $X\geq 3$, *and* $A_k(X)=\sum_{n\leq X}\alpha_k(n)$, *the following estimate holds*:

$$A_k(X)\leq (2k^8)^{k^3}\cdot\sqrt{X}(\log X)^{k^2}.$$

Lemma 12.4. *For a prime number p and a natural number m, the relation*

$$\alpha_k(p^m)=\frac{k(k+1)\dots(k+m-1)}{m!}\big(1-\Delta(k,m)\big)$$

holds, where

$$\Delta(k,m)=k\frac{m(m-1)\dots(m-m_1)}{(k+m-m_1-1)\dots(k+m-1)},\qquad m_1=\left[\frac{m}{2}\right].$$

Proof. By definition, $\alpha_k(p^m)$ is equal to the number of solutions of the system

$$\begin{cases} p^m=x_1\dots x_k,\\ p^m\equiv 0 \pmod{[x_1^2,\dots,x_k^2]}.\end{cases}\tag{12.45}$$

It follows from the first relation in (12.45), namely, from the equation, that each x is a nonnegative power of p, i.e., $x_j=p^{\beta_j}$, $0\leq\beta_j\leq m$, and hence

$$\beta_1+\dots+\beta_l=m.$$

The second relation in (12.45), namely, the congruence, implies that $\beta_j \le m_1$, since this congruence does not hold for $\beta_j > m_1$. But if $\beta_j \le m_1$, $j = 1, \ldots, k$, then

$$[x_1^2, \ldots, x_k^2] = [p^{2\beta_1}, \ldots, p^{2\beta_k}] = p^{\beta}, \quad \beta \le m,$$

i.e., the set $(x_1, \ldots, x_k) = (p^{\beta_1}, \ldots, p^{\beta_k})$ is a solution of (12.45). Thus we see that $\alpha_k(p^m)$ is equal to the number of solutions of the equation

$$\beta_1 + \cdots + \beta_k = m, \quad 0 \le \beta_1, \ldots, \beta_k \le m_1,$$

in nonnegative integers $\beta_1, \ldots, \beta_k$.

We define a function $f(x)$ by the relation

$$f(x) = \Big(\sum_{\beta=0}^{m_1} x^{\beta}\Big)^k = \Big(\frac{1 - x^{m_1+1}}{1 - x}\Big)^k.$$

Then we have

$$\begin{aligned}
\alpha_k(p^m) &= \frac{1}{m!}\frac{d^m}{dx^m} f(x)\Big|_{x=0} \\
&= \frac{1}{m!}\sum_{j=0}^{m} C_m^j \frac{d^j}{dx^j}(1 - x^{m_1+1})^k \frac{d^{m-j}}{dx^{m-j}}(1 - x)^{-k}\Big|_{x=0} \\
&= \frac{1}{m!}\Big(\frac{d^m}{dx^m}(1 - x)^{-k}\Big|_{x=0} \\
&\quad + C_m^{m_1+1}\frac{d^{m_1+1}}{dx^{m_1+1}}(1 - kx^{m_1+1})\frac{d^{m-m_1+1}}{dx^{m-m_1+1}}(1 - x)^{-k}\Big|_{x=0}\Big) \\
&= \frac{k(k+1)\ldots(k+m-1)}{m!} - kC_m^{m_1+1}\frac{(m_1+1)!k\ldots(k+m-m_1-2)}{m!} \\
&= \frac{k(k+1)\ldots(k+m-1)}{m!}\Big(1 - k\frac{m(m-1)\ldots(m-m_1)}{(k+m-m_1-1)\ldots(k+m-1)}\Big),
\end{aligned}$$

as required. The proof of Lemma 12.4 is complete. □

We note that $\Delta(k, m)$ satisfies the inequality

$$0 \le \Delta(k, m) \le 1.$$

Lemma 12.4 implies (1) $\alpha_k(p) = 0$; (2) $\alpha_k(p^2) = k(k-1)/2$, and (3) $\alpha_k(p^m) = \tau_k(p^m)(1 - \Delta(k, m))$.

For $\operatorname{Re} s > 1$, we define the Dirichlet series $F_k(s)$ corresponding to $\alpha_k(n)$:

$$F_k(s) = \sum_{n=1}^{\infty} \frac{\alpha_k(n)}{n^s}.$$

Since $0 \leq \alpha_k(n) \leq \tau_k(n)$, the series $F_k(s)$ converges absolutely and uniformly in the half-plane $\operatorname{Re} s \geq 1+\varepsilon$, where $\varepsilon > 0$ is arbitrary. We have the following more precise assertion about the behavior of $F_k(s)$.

Theorem 12.8. *For* $\operatorname{Re} s = \sigma > 1/3$, *the relation*

$$F_k(s) = (\zeta(2s))^{k(k-1)/2}\Phi(s)$$

holds, where $\Phi(s)$ *is a function regular in the half-plane under consideration, and* $|\Phi(s)|$ *satisfies the estimate*

$$|\Phi(s)| \leq (2k^5)^{9(2k)^5}\left(\sigma - \frac{1}{3}\right)^{-(2k)^3}.$$

Proof. Using the inequalities $0 \leq \alpha_k(n) \leq \tau_k(n)$ and the fact that $\alpha_k(n)$ is multiplicative, for $\operatorname{Re} s > 1$, we obtain

$$F_k(s) = \prod\left(1 + \frac{a_2}{p^{2s}} + \frac{a_3}{p^{3s}} + \cdots\right), \tag{12.46}$$

where

$$a_m = \alpha_k(p^m) = \frac{k(k+1)\dots(k+m-1)}{m!}(1-\Delta(k,m)).$$

Let

$$B_1(S) = \prod_{p\leq k^3}\left(1 + \frac{a_2}{p^{2s}} + \frac{a_2}{p^{3s}} + \cdots\right),$$

$$B_2(S) = \prod_{p>k^3}\left(1 + \left(\frac{a_3}{p^{3s}} + \frac{a_4}{p^{4s}} + \cdots\right)\left(1 + \frac{a_2}{p^{2s}}\right)^{-1}\right).$$

Then (12.46) implies

$$F_k(s) = B_1(s)B_2(s)\prod_{p>k^3}\left(1 + \frac{a_2}{p^{2s}}\right). \tag{12.47}$$

Further, we have

$$\frac{\zeta(2s)}{\zeta(4s)} = \prod_{p\leq k^3}\left(1 + \frac{1}{p^{2s}}\right)\prod_{p>k^3}\left(1 + \frac{1}{p^{2s}}\right), \tag{12.48}$$

$$\begin{aligned}\left(\frac{\zeta(2s)}{\zeta(4s)}\right)^{a_2} &= \prod_{p\leq k^3}\left(1 + \frac{1}{p^{2s}}\right)^{a_2}\prod_{p>k^3}\left(1 + \frac{a_2}{p^{2s}} + \binom{a_2}{2}\frac{1}{p^{4s}} + \cdots\right)\\ &= \prod_{p\leq k^3}\left(1 + \frac{1}{p^{2s}}\right)^{a_2}\prod_{p>k^3}\left(1 + \frac{a_2}{p^{2s}}\right)B_3(s),\end{aligned}$$

where

$$B_3(s) = \prod_{p>k^3} \left(1 + \frac{1}{p^{2s}}\right)^{a_2} \left(1 + \frac{a_2}{p^{2s}}\right)^{-1}. \tag{12.49}$$

From (12.47) and (12.48) we obtain

$$F_k(s) = B_1(s)B_2(s)B_3^{-1}(s)\left(\frac{\zeta(2s)}{\zeta(4s)}\right)^{a_2} \prod_{p\le k^3} \left(1 + \frac{1}{p^{2s}}\right)^{-a_2} = (\zeta(2s))^{a_2}\Phi(s),$$

where

$$\Phi(s) = B_1(s)B_2(s)B_3^{-1}(s)\zeta^{-a_2}(4s) \prod_{p\le k^3} \left(1 + \frac{1}{p^{2s}}\right)^{-a_2}.$$

For Res $> 1/3$, the function $\Phi(s)$ is regular, since each of the factors contained in the definition of $\Phi(s)$ is a regular function for Res $> 1/3$. Let $\sigma > 1/3$. We shall estimate $|\Phi(\sigma + it)|$ from above. First, we have

$$\alpha_k(p^m) = a_m \le \frac{k(k+1)\dots(k+m-1)}{m!} = \tau_k(p^m) \le m^k,$$

since $m \ge 2$ and $k \ge 2$. Hence

$$\left|1 + \frac{a_2}{p^{2s}} + \frac{a_3}{p^{3s}} + \cdots\right| \le 1 + \sum_{m=2}^{\infty} \frac{m^k}{p^{m\sigma}} \le 1 + \sum_{m=2}^{\infty} \frac{m^k}{(\sqrt[3]{p})^m} \le (6k)^{6k},$$

$$|B_1(\sigma + it)| \le \prod_{p\le k^3} (6k)^{6k} \le (6k)^{6k^4}.$$

Let us estimate $|B_2(\sigma+it)|$. Since $a_2 = k(k-1)/2 \le k^2/2$, for $\sigma > 1/3$ and $p > k^3$, we have

$$\left|1 + \frac{a_2}{p^{2s}}\right| \ge 1 - \frac{a_2}{p^{2/3}} \ge 1 - \frac{a_2}{k^2} \ge \frac{1}{2},$$

$$|B_2(\sigma + it)| \le \prod_{p>k^3} \left(1 + \frac{2a_3}{p^{3\sigma}} + \frac{2a_4}{p^{4\sigma}} + \cdots\right)$$

$$\le \prod_{p>k^3} \left(1 + \frac{\tau_{2k}(p^3)}{p^{3\sigma}} + \frac{\tau_{2k}(p^4)}{p^{4\sigma}} + \cdots\right) \le {\sum_{n=1}^{\infty}}' \frac{\tau_{2k}(n)}{n^\sigma},$$

and the prime on the last sum means the following: if $p\,|\,n$, then $p^3\,|\,n$; in other words, if $n = p_1^{\beta_1}\dots p_r^{\beta_r}$ is the canonical decomposition of n into prime factors, then $\beta_j \ge 3$, $j = 1, \dots, r$. It is easy to show that such n can be represented as

$$n = x^3y^4z^5,$$

where $x,\ y,\ z$ are natural numbers. Therefore, for $N > 3$, we have

$$\sum_{3<n\le N}{}' \frac{\tau_{2k}(n)}{n^\sigma} = \sigma \int_3^N u^{-\sigma-1}\mathbb{C}(u)\,du + N^{-\sigma}\mathbb{C}(N),$$

where

$$\mathbb{C}(u) \le \sum_{n\le u}{}' \tau_{2k}(n) \le \sum_{x^3y^4z^5\le N} \tau_{2k}^3(x)\tau_{2k}^4(y)\tau_{2k}^5(z)$$
$$= \sum_{y^4z^5\le u} \tau_{2k}^4(y)\tau_{2k}^5(z) \sum_{x\le(uy^{-4}z^{-5})^{1/3}} \tau_{2k}^3(x).$$

Using Mardzhanishvili's inequality (see Section 12.1), we obtain

$$\sum_{x\le(uy^{-4}z^{-5})^{1/3}} \tau_{2k}^3(x) < (2k^3)^{(2k)^3-1}(uy^{-4}z^{-5})^{1/3}(\log u)^{(2k)^3-1},$$

$$\mathbb{C}(u) < \sqrt[3]{u}(\log u)^{(2k)^3-1}\mathcal{D},$$

where

$$\mathcal{D} = (2k^3)^{(2k)^3-1}\Big(\sum_{y=1}^{\infty} \tau_{2k}^4(y)y^{-4/3}\Big)\Big(\sum_{z=1}^{\infty} \tau_{2k}^5(z)z^{-5/3}\Big).$$

Using the formula of partial summation and Mardzhanishvili's inequality, we can easily estimate $\mathcal{D}$ as follows:

$$\mathcal{D} \le (2k^4)^{(2k)^4}\Big(\int_1^\infty y^{-4/3}(\log y)^{(2k)^4}\,dy\Big)(2k^5)^{(2k)^5}$$
$$\times\Big(\int_1^\infty y^{-5/3}(\log z)^{(2k)^5}\,dz\Big)(2k^3)^{(2k)^3}$$
$$< (2k^4)^{2(2k)^4}(2k^5)^{2(2k)^5}(2k^3)^{(2k)^3} < (2k^5)^{5(2k)^5} = \mathcal{D}_k.$$

Hence for $1/3 < \delta \le 2$ we have

$$\sum_{3<n\le N}{}' \frac{\tau_{2k}(n)}{n^\sigma} < \sigma\,\mathcal{D}_k \int_3^N u^{-\sigma-2/3}(\log u)^{(2k)^3-1}\,du + N^{-\sigma}\mathbb{C}(N).$$

Passing to the limit as $N \to +\infty$ in the last inequality, we obtain

$$\sum_{3<n}{}' \frac{\tau_{2k}(n)}{n^\sigma} \le \sigma\,\mathcal{D}_k \int_1^\infty u^{-\sigma-2/3}(\log z)^{(2k)^3-1}\,du$$
$$= \sigma\,\mathcal{D}_k\Big(\sigma - \frac{1}{3}\Big)^{-(2k)^3} \int_0^\infty v^{(2k)^3-1}e^{-v}\,du$$

$$< \sigma \mathcal{D}_k (2k^3)^{(2k)^3} \left(\sigma - \frac{1}{3}\right)^{-(2k)^3} < (2k^5)^{6(2k)^5} \left(\sigma - \frac{1}{3}\right)^{-(2k)^3}.$$

Finally, we obtain the following estimate for $1/3 < \sigma \le 2$:

$$|B_2(\sigma + it)| < (2k^5)^{6(2k)^5} \left(\sigma - \frac{1}{3}\right)^{-(2k)^5}.$$

Let us estimate $|B_3(\sigma + it)|^{-1}$ from above. From the definition (12.49) of the function $B_3(s)$, for $\operatorname{Re} s > 1/3$, we have

$$\begin{aligned}\log B_3(s) &= \sum_{p>k^3} \left(a_2 \log\left(1 + \frac{1}{p^{2s}}\right) - \log\left(1 + \frac{a_2}{p^{2s}}\right)\right) \\ &= \sum_{p>k^3} \sum_{m=2}^{\infty} \frac{(-1)^{m-1}}{mp^{2sm}} (a_2 - a_2^m).\end{aligned}$$

The last double series converges absolutely for $\operatorname{Re} s > 1/4$. Hence for $\operatorname{Re} s > 1/3$ we have

$$|\log B_3(s)| \le \sum_{p>k^3} \sum_{m=2}^{\infty} \frac{a_2^m - a_2}{mp^{2sm}} < \sum_{p>k^3} \sum_{m=2}^{\infty} \left(\frac{a_2}{p^{2/3}}\right)^m < 2 \sum_{p>k^3} \frac{a_2}{p^{4/3}} < \frac{6}{k} a_2^2,$$

since $a_2 = k(k-1)/2 \le k^2/2$. Thus, studying the principal branch of the logarithm, we obtain

$$|\log B_3(s)| = \big|\log|B_3(s)| + i\varphi\big| \ge \big|\log|B_3(s)|\big|,$$

i.e.,

$$\big|\log|B_3(\sigma + it)|\big| < \frac{3}{2} k^3 < 2k^3,$$

$$\log|B_3(\sigma + it) > -2k^3, \quad |B_3(\sigma + it)|^{-1} < e^{2k^3}.$$

The other factors can simply be estimated for $\operatorname{Re} s \ge 1/3$ as

$$\left|\prod_{p \le k^3} \left(1 + \frac{1}{p^{2s}}\right)^{-a_2}\right| < \exp\left\{4a_2 \sum_{p \le k^3} p^{-2/3}\right\} < e^{4k^3},$$

$$|\zeta(4s)|^{-1} \le \sum_{n=1}^{\infty} n^{-4/3} < 4, \quad |\zeta(4s)|^{-a_2} < 2^{k^2}.$$

Collecting these estimates together, for $\operatorname{Re} s = \sigma > 1/3$, we obtain

$$|\Phi(s)| = |\Phi(\sigma + it)| < (2k^5)^{9(2k)^5} \left(\sigma - \frac{1}{3}\right)^{-(2k)^3}.$$

The proof of the theorem is complete. □

Theorem 12.9. *The following asymptotic formula holds*:

$$A_k(X) = \sqrt{X}\,P_m(\log X) + R(X),$$

where $P_m(u)$ *is an mth-degree polynomial in* u, $m < K = k(k-1)/2$,

$$|R(X)| < k^{c_1 k^5}(\sqrt{X})^{1-cK^{-2/3}}(\log X)^K,$$

$c > 0$ *and* $c_1 > 0$ *are absolute constants, and* $X \geq 3$.

The proof of this theorem is similar to that of the theorem in [85]. The only difference is that, whenever necessary, we must use the results of Lemma 12.3 and Theorem 12.8.

A generalization of the function $\alpha_k(n)$ is the function $\alpha_{k,m}(n)$ that, by definition, is equal to the number of solution of the system

$$\begin{cases} n = x_1 \ldots x_k, \\ n \equiv 0 \pmod{[x_1^m, \ldots, x_k^m]}. \end{cases}$$

In particular, we have $\alpha_{k,1}(n) = \tau_k(n)$ and $\alpha_{k,2}(n) = \alpha_k(n)$. Assertions similar to the corresponding assertions for $\alpha_k(n)$ also hold for the function $\alpha_{k,m}(n)$.

12.5 Double Kloosterman sums

In this section $\xi(x)$ and $\eta(y)$ are arbitrary complex-valued functions of arguments x and y; $0 < X < X_1 \leq 2X$, $0 < Y < Y_1 \leq 2Y$; the positive numbers ξ, η, ξ_0, η_0, ξ_1 are determined by the relations:

$$\xi = \max_{X<x\leq X_1} |\xi(x)|; \quad \eta = \max_{Y<y\leq Y_1} |\eta(y)|;$$
$$\xi_0 = \sum_{X<x\leq X_1} |\xi(x)|; \quad \eta_0 = \sum_{Y<y\leq Y_1} |\eta(y)|; \quad \xi_1 = \sum_{X<x\leq X_1} |\xi(x)|^2.$$

We consider the multiple trigonometric sum $W = W(a, b)$,

$$W = \sum_{X<x\leq X_1} \sum_{Y<y\leq Y_1} \xi(x)\eta(y) \exp\left\{2\pi i \frac{ax^*y^* + bxy}{m}\right\}.$$

It is natural to call the sum W a *multiple* (*double*) *Kloosterman sum with weights*. If the product $(X_1 - X)(Y_1 - Y)$ is less than m, then W is called a *short* or *incomplete sum*.

Theorem 12.10. *Suppose that* k *and* s *are positive integers, the numbers* X, X_1, Y, *and* Y_1 *satisfy the inequalities*

$$3 \leq X, \quad k < X < X_1 \leq 2X, \quad k2^{2k-1}X^{2k-1} < m,$$

$$3 \le Y, \quad s < Y < Y_1 \le 2Y, \quad s2^{2s-1}Y^{2s-1} < m,$$

and a and b are integers; moreover, $(a, m) = d \ge 1$. *Then* $|W|$ *satisfies the estimate*

$$|W| \le \xi_0 \eta_0 \Delta,$$

where

$$\Delta = (2k)^{4k^2/s}(2s)^{4s^2/k}(\xi_0^{-1}\xi\sqrt{X})^{1/s} \times (\eta_0^{-1}\eta\sqrt{Y})^{1/k}(sdmY)^{1/2ks}(\log X)^{2k/s}(\log Y)^{2s/k}.$$

Proof. We shall follow the arguments of Theorem 12.3. Passing to the inequalities, we obtain

$$|W| \le \sum_{X<x\le X_1} |\xi(x)| \left| \sum_{Y<y\le Y_1} \eta(y) \exp\left\{2\pi i \frac{ax^*y^* + bxy}{m}\right\}\right|.$$

Let us raise both parts of this inequality to the sth power and use Hölder's inequality; we obtain

$$|W|^s \le A^{s-1} \sum_{X<x\le X_1} |\xi(x)| \left| \sum_{Y<y\le Y_1} \eta(y) \exp\left\{2\pi i \frac{ax^*y^* + bxy}{m}\right\}\right|^s,$$

where

$$A = \sum_{X<x\le X_1} |\xi(x)|.$$

Using the definition of the numbers ξ_0 and ξ, we obtain the inequality

$$|W|^s \le \xi_0^{s-1}\xi W_1, \tag{12.50}$$

where

$$W_1 = \sum_{X<x\le X_1} \left| \sum_{Y<y\le Y_1} \eta(y) \exp\left\{2\pi i \frac{ax^*y^* + bxy}{m}\right\}\right|^s.$$

Let us raise the sum over y to the sth power. To do this, we define the function $J_s(\lambda, \mu)$ by the relation

$$J_s(\lambda, \mu) = \sum{}' \eta(y_1)\dots\eta(y_s),$$

where the prime on the sum indicates that summation is carried out over the sets $y_1, \dots, y_s$ satisfying the system of congruences

$$\begin{cases} y_1^* + \dots + y_s^* \equiv \lambda, \\ y_1 + \dots + y_s \equiv \mu \pmod m), \end{cases} \quad Y < y_1, \dots, y_s \le Y_1.$$

We obtain

$$\Big(\sum_{Y<y\le Y_1}\eta(y)\exp\Big\{2\pi i\frac{ax^*y^*+bxy}{m}\Big\}\Big)^s = \sum_{\lambda=1}^{m}\sum_{\mu} J_s(\lambda,\mu)\exp\{2\pi i(ax^*\lambda+bx\mu)\}.$$

Note that the parameter μ in the last relation ranges over the interval

$$sY<\mu\le sY_1<2sY\le s2^{2s-1}Y^{2s-1}<m.$$

Now suppose that $\theta(x)$ is the argument of the sum in question, i.e.,

$$\theta(x)=\arg\sum_{\lambda=1}^{m}\sum_{\mu} J_s(\lambda,\mu)\exp\Big\{2\pi i\frac{ax^*\lambda+bx\mu}{m}\Big\}.$$

Then we obtain

$$\Big|\sum_{Y<y\le Y_1}\eta(y)\exp\Big\{2\pi i\frac{ax^*y^*+bxy}{m}\Big\}\Big|^s = e^{-i\theta(x)}\sum_{\lambda=1}^{m}\sum_{\mu} J_s(\lambda,\mu)\exp\Big\{2\pi i\frac{a\lambda x^*+b\mu x}{m}\Big\};$$

$$W_1\le\sum_{\lambda=1}^{m}\sum_{\mu}|J_s(\lambda,\mu)|\Big|\sum_{X<x\le X_1} e^{-i\theta(x)}\exp\Big\{2\pi i\frac{a\lambda x^*+b\mu x}{m}\Big\}\Big|.$$

Let us raise this inequality to the kth power and again use Hölder's inequality; we obtain

$$W_1^k\le\Big(\sum_{\lambda=1}^{m}\sum_{\mu}|J_s(\lambda,\mu)|\Big)^{k-1} \times\sum_{\lambda=1}^{m}\sum_{\mu}\Big|\sum_{X<x\le X_1} e^{-i\theta(x)}\exp\Big\{2\pi i\frac{a\lambda x^*+b\mu x}{m}\Big\}\Big|^k|J_s(\lambda,\mu)|.$$

Finally, we apply Cauchy's inequality to the last sum and obtain

$$W_1^k\le B^{k-1}\sqrt{CD}, \tag{12.51}$$

where the following notation was used:

$$B=\sum_{\lambda=1}^{m}\sum_{\mu}|J_s(\lambda,\mu)|;\quad C=\sum_{\lambda=1}^{m}\sum_{\mu}|J_s(\lambda,\mu)|^2;$$

$$D = \sum_{\lambda=1}^{m} \sum_{\mu} \left| \sum_{X<x\le X_1} e^{-i\theta(x)} \exp\left\{2\pi i \frac{a\lambda x^* + b\mu x}{m}\right\}\right|^{2k}.$$

Since the parameters λ and μ in the sum B take arbitrary values, we have

$$B \le \sum_{Y<y_1,\ldots,y_s\le Y_1} |\eta(y_1)|\ldots|\eta(y_s)| = \Big(\sum_{Y<y\le Y_1} |\eta(y)|\Big)^s = \eta_0^s. \tag{12.52}$$

Similarly, for C we obtain the estimate

$$C \le \sum{}'' |\eta(y_1)|\ldots|\eta(y_{2s})|,$$

where the primes on the last sum indicate summation over the sets $y_1, \ldots, y_{2s}$ satisfying the system of congruences

$$\begin{cases} y_1^* + \cdots + y_s^* \equiv y_{s+1}^* + \cdots + y_{2s}^*, \\ y_1 + \cdots + y_s \equiv y_{s+1} + \cdots + y_{2s} \pmod m, \quad Y < y_1, \ldots, y_{2s} \le Y_1. \end{cases}$$

Obviously,

$$C \le \eta^{2s} I_s(Y), \tag{12.53}$$

where $I_s(Y)$ is the number of solutions of the congruence from Theorem 12.2 (in this theorem, we must set $k = s$ and $X = Y$). The sum D can be estimated as follows:

$$\begin{aligned} D &\le sY \sum_{\lambda=1}^{m} \left| \sum_{X<x\le X_1} \exp\left\{2\pi i \frac{a\lambda x^*}{m}\right\} \exp\{i\theta_1(x)\}\right|^{2k} \\ &= sY \sum_{x_1,\ldots,x_{2k}} \exp\big\{i\big(\theta_1(x_1) + \cdots - \theta_1(x_{2k})\big)\big\} \\ &\quad\times \sum_{\lambda=1}^{m} \exp\left\{2\pi i \frac{a\lambda(x_1^* + \cdots - x_{2k}^*)}{m}\right\} \\ &\le sY \sum_{x_1,\ldots,x_{2k}} \sum_{\lambda=1}^{m} \exp\left\{2\pi i \frac{a\lambda(x_1^* + \cdots - x_{2k}^*)}{m}\right\} \\ &= sY \sum_{\mu=1}^{m} I(\mu) \sum_{\lambda=1}^{m} \exp\left\{2\pi i \frac{a\lambda\mu}{m}\right\}, \end{aligned}$$

where $I(\mu)$ is the number of solutions of the congruence

$$x_1^* + \cdots - x_{2k}^* \equiv \mu \pmod m, \quad X < x_1, \ldots, x_{2k} \le X_1.$$

Obviously,

$$I(\mu) \le I(0) \le I_k(X).$$

Therefore, recalling that $(a, m) = d$, i.e., $a = a_1 d$, $m = m_1 d$, $(a_1, m_1) = 1$, for D we obtain the final estimate

$$D \le sYI_k(X) \sum_{\mu=1}^{m} \sum_{\lambda=1}^{m} \exp\left\{2\pi i \frac{a\lambda\mu}{m}\right\} = sdmYI_k(X). \tag{12.54}$$

From (12.50)–(12.54), we successively obtain

$$W_1^k \le \eta_0^{s(k-1)} \sqrt{\eta^{2s} I_s(Y)} \sqrt{sdmYI_k(X)}; \tag{12.55}$$

$$|W|^{ks} \le \xi_0^{ks} (\xi_0^{-1}\xi)^k \eta_0^{sk-s} \eta^s \sqrt{I_s(Y)} \sqrt{sdmY} \sqrt{I_k(X)}. \tag{12.56}$$

Finally, using the estimate for $I_k(X)$ and $I_s(Y)$ in Theorem 12.2, from (12.56) we obtain the assertion of the theorem. □

Theorem 12.11. *Suppose that the assumptions of Theorem* 12.10 *are satisfied, and moreover, $s \ge 2$. Then for $|W|$ the following estimate holds*:

$$|W| \le \xi_0 \eta_0 \Delta_1,$$

where

$$\Delta_1 = (\xi_0^{-2} \xi_1 \sqrt{X})^{1/s} (\eta_0^{-1} \eta \sqrt{Y})^{1/k} (sdmY)^{1/(2ks)} (\log X)^{2k/s} \times (\log Y)^{2s/k} (2k)^{4k^2/s} (2s)^{4s^2/k}.$$

Proof. Passing to the inequalities and using Hölder's and Cauchy's inequalities, we successively obtain

$$\begin{aligned} |W|^{s/2} &\le \left(\sum_{X<x\le X_1} |\xi(x)| \right)^{s/2-1} \\ &\quad \times \sum_{X<x\le X_1} |\xi(x)| \left| \sum_{Y<y\le Y_1} \eta(y) \exp\left\{2\pi i \frac{ax^*y^* + bxy}{m}\right\} \right|^{s/2} \\ &\le \xi_0^{s/2-1} \sqrt{\sum_{X<x\le X_1} |\xi(x)|^2} \\ &\quad \times \sqrt{\sum_{X<x\le X_1} \left| \sum_{Y<y\le Y_1} \eta(y) \exp\left\{2\pi i \frac{ax^*y^* + bxy}{m}\right\} \right|^s} \\ &= \xi_0^{s/2-1} \xi_1^{1/2} \sqrt{W_1}, \end{aligned}$$

where

$$W_1 = \sum_{X<x\le X_1} \left| \sum_{Y<y\le Y_1} \eta(y) \exp\left\{2\pi i \frac{ax^*y^* + bxy}{m}\right\} \right|^s.$$

Using estimate (12.55) and Theorem 12.11, we successively obtain

$$
\begin{aligned}
|W|^{ks} &\le \xi_0^{ks-2k}\xi_1^k W_1^k \le \xi_0^{ks-2k}\xi_1^k \eta_0^{s(k-1)}\eta^s \sqrt{I_s(Y)}\sqrt{I_k(X)}\sqrt{csdmY} \\
&\le (\xi_0\eta_0)^{ks}(\xi_0^{-2}\xi_1)^k(\eta_0^{-1}\eta)^s(2s)^{4s^3}(2k)^{4k^3}Y^{s/2}(\log Y)^{2s^2}X^{k/2}(\log X)^{2k^2}; \\
|W| &\le \xi_0\eta_0\Delta_1.
\end{aligned}
$$

The theorem is proved. □

If we set $b = 0$ in the sum $W(a,b)$, then we obtain more refined estimates. Let us only cite the statements of the corresponding results, since their proofs repeat word-for-word those of Theorems 12.10 and 12.11.

Theorem 12.12. *Assume the conditions and the notation of Theorem* 12.10. *Then the following estimate holds*:

$$|W(a,0)| \le \xi_0\eta_0\Delta(sY)^{-1/(2ks)}.$$

Theorem 12.13. *Assume the conditions and the notation of Theorem* 12.11. *Then the following estimate holds*:

$$|W(a,0)| \le \xi_0\eta_0\Delta_1(sY)^{-1/(2ks)}.$$

12.6 Short Kloosterman sums and their applications

First, we present some auxiliary lemmas, which we need to prove the main Lemma 12.A and the theorems.

Lemma 12.5. *For $x \ge 2$, the following relation holds*:

$$\sum_{p\le x}\frac{1}{p} = \log\log x + c + \frac{\theta}{\log x},$$

where $c = 0.26\ldots$ is an absolute constant and $|\theta| < 10$.

Lemma 12.6. *Suppose that $a \ge 2$, $b - a > 1$, $r \ge 0$ is an integer, and J is a set of prime numbers in the interval $(a, b]$. Suppose also that w runs through the values that are products of the form $w = p_1p_2\ldots p_r$, where $p_i \in J$, $i = 1, 2, \ldots, r$, and $p_1 < p_2 < \cdots < p_r$. Then the following inequality holds*:

$$\sum_w \frac{1}{w} \le \frac{1}{r!}\Big(\sum_{p\in J}\frac{1}{p}\Big)^r.$$

Proof. Let $q_1, \dots, q_s$ be all the prime numbers in the set J. If $s < r$, then the statement of the lemma is obvious, since in this case the sum in the left-hand side of the inequality is zero. Let $r \le s$. Each of the numbers w can be written as $w = q_1^{\alpha_1} \dots q_s^{\alpha_s}$, where α_i is equal either to 0 or to 1, $i = 1, 2, \dots, s$, and $\alpha_1 + \dots + \alpha_s = r$. Calculating the expression in the right-hand side, we obtain terms of the form

$$\frac{r!}{\alpha_1! \dots \alpha_s!}(q_1^{\alpha_1} \dots q_s^{\alpha_s}) = r!w^{-1}. \qquad \square$$

Let $x \ge y \ge 2$. By $N(x, y)$ we denote the number of numbers in the series $1, 2, \dots, [x]$ all whose prime divisors do not exceed y. We define the quantity $\Phi(x, y)$ as the number of numbers in the same series all whose prime divisors are larger than y. Lemmas 12.6 and 12.7 provide upper bounds for N and Φ, which we need in the following.

Lemma 12.7. *Suppose that $y = y(x) \le x$ and $y(x) \to +\infty$ as $x \to +\infty$. Then the following estimate*

$$N(x, y) < x \exp\left\{ -\log x \frac{\log\log\log y}{\log y} + \log\log y + O\left(\frac{\log\log y}{\log\log\log y}\right)\right\}$$

holds and all constants in the O-symbol are absolute constants.

This lemma is proved, e.g., in [139].

Lemma 12.8. *Let $x \ge y \ge y_1 > 1$. Then the following inequality holds*:

$$\Phi(x, y) < \frac{3x\log x}{(\log y)^2}.$$

Proof. We divide $\Phi(x, y)$ into two classes. The first class contains all square-free numbers. The second class contains all the other numbers. Let Φ_1 be the number of numbers in the first class, and let Φ_2 be the number of numbers in the second class.

We choose an integer $t \ge 0$. By $\Phi_1(t)$ we denote the number of n that belong to the first class and can be represented as $n = p_1 \dots p_{t+1}$, where $y < p_1 < \dots < p_{t+1} \le x$. Then

$$\Phi_1(t) = \sum_{p_1 \dots p_{t+1} \le x} 1 = \sum_{p_1 \dots p_t \le xy^{-1}} \sum_{y < p_{t+1} \le x(p_1 \dots p_t)^{-1}} 1 \le \sum_{p_1 \dots p_t \le xy^{-1}} \pi\left(\frac{x}{p_1 \dots p_t}\right).$$

For sufficiently large h, we have the inequality $\pi(h) \le 2h(\log h)^{-1}$. Using this inequality and the assertions of Lemmas 12.5 and 12.6, we obtain

$$\Phi_1(t) \le 2 \sum_{p_1 \dots p_t \le xy^{-1}} \frac{x}{p_1 \dots p_t} \pi\left(\log \frac{x}{p_1 \dots p_t}\right)^{-1}$$

$$\le 2\frac{x}{\log y}\sum_{p_1\dots p_t\le xy^{-1}}\frac{1}{p_1\dots p_t}\le 2\frac{x}{\log y}\frac{1}{t!}\Big(\sum_{y<p\le x}\frac{1}{p}\Big)^t$$
$$\le 2\frac{x}{\log y}\frac{1}{t!}\Big(\log\log x-\log\log y+\frac{20}{\log y}\Big)^t$$
$$\le 2\frac{x}{\log y}\frac{1}{t!}(\log\log x-\log\log y+0.1)^t.$$

So for Φ_1, we obtain the estimate

$$\Phi_1=\sum_{t=0}^{+\infty}\Phi_1(t)\le 2\frac{x}{\log y}\sum_{t=0}^{+\infty}\frac{1}{t!}(\log\log x-\log\log y+0.1)^t=2e^{0.1}\frac{x\log x}{(\log y)^2}.$$

It is easy to see that Φ_2 does not exceed

$$\sum_{l>y}\frac{x}{l^2}<\frac{2x}{y}.$$

Thus we have

$$\Phi(x,y)=\Phi_1+\Phi_2\le 2e^{0.1}\frac{x\log x}{(\log y)^2}+\frac{2x}{y}<\frac{3x\log x}{(\log y)^2}.$$

The proof of the lemma is complete. □

Lemma 12.9. *Suppose that $m>1$ is an integer, k and s are natural numbers, real numbers X, X_1, Y, Y_1 satisfy the inequalities*

$$k<X<X_1,\quad kX_1^{2k-1}<m,$$
$$s<Y<Y_1,\quad sY_1^{2s-1}<m,$$

and a parameter n takes N values of natural numbers such that $(n,m)=1$; a and b are integers such that $(a,m)=d$; and p and q take successive values of primes such that $(p,m)=(q,m)=1$. The set of numbers of the form npq, where $X<p\le X_1$ and $Y<q\le Y_1$ is denoted by the letter A, and the number of elements in A is denoted by the symbol $\|A\|$. Then the trigonometric sum

$$S=\sum_{n\in A}\exp\Big\{2\pi i\frac{an^*+bn}{m}\Big\}$$

satisfies the estimate

$$|S|\le\|A\|\Delta,$$

where

$$\Delta=(s!k!)^{1/(2sk)}\big(\pi_1(X)\big)^{-1/(2s)}\big(\pi_1(Y)\big)^{-1/(2k)}(smdY)^{1/(2sk)},$$
$$\pi_1(X)=\pi(X_1)-\pi(X)-\sum_{X<p\le X_1,\ p|m}1.$$

This lemma is a version of Theorem 12.3.

Lemma 12.10. *Suppose that ρ is a natural number, α and β are real numbers, $0 < \delta < 1/16$, and $2\delta < \beta - \alpha < 1 - 2\delta$. Then there exists a periodic function $\psi(x)$ with period 1 that has the following properties:*

(1) $\psi(x) = 1$ *for* $\alpha + \delta \le x \le \beta - \delta$;

(2) $0 < \psi(x) < 1$ *for* $\alpha - \delta < x < \alpha + \delta$ *and* $\beta - \delta < x < \beta + \delta$;

(3) $\psi(x) = 0$ *for* $\beta + \delta \le x \le 1 + \alpha - \delta$;

(4) $\psi(x)$ *can be expanded in the Fourier series*

$$\psi(x) = \beta - \alpha + \sum_{|r|>0} g(r) \exp\{2\pi i r x\},$$

where

$$|g(r)| \le \min\left(\beta - \alpha; \frac{1}{\pi |r|}; \frac{1}{\pi |r|}\left(\frac{\rho}{\pi |r| \delta}\right)^{\rho}\right) = c(r). \tag{12.57}$$

This lemma is a version of Lemma A.3.

Lemma 12.11. *Suppose that $\lambda_1, \lambda_2, \ldots, \lambda_Q$ are real numbers such that $0 \le \lambda_s < 1$, $s = 1, 2, \ldots, Q$. Suppose also that $\rho \ge 1$ is an integer, $0 < \delta < 1/16$, α, β, R are real numbers, $2\delta < \beta - \alpha < 1 - 2\delta$, $\psi(x)$ is the function in the preceding lemma and this functions corresponds to some given ρ, δ, α and β. Suppose that for any admissible values of α and β, the sum*

$$U(\alpha, \beta) = \sum_{s=1}^{Q} \psi(\lambda_s)$$

satisfies the relation

$$U(\alpha, \beta) = (\beta - \alpha)Q + O(R).$$

Then:

(1) *for any σ, $0 \le \sigma < 1$, the number A_σ of values of λ_s such that $0 \le \lambda_s < \sigma$ is given by the formula*

$$A_\sigma = \sigma Q + R_\sigma, \quad R_\sigma = O(R) + O(Q\delta);$$

(2) $\sum_{s=1}^{Q} \lambda_s = \frac{1}{2} Q + O(R) + O(Q\delta)$.

This lemma is a version of Lemma 3 in [162], p. 18.

Prior to stating the main Lemma 12.A, we introduce the necessary notation.

Suppose that $m \ge m_1 > 1$ is a natural number, x is a real number satisfying the inequalities

$$\exp\{(\log m)^{4/5} (\log\log m)^5\} < x \le m^{4/7}, \tag{12.58}$$

and $C > 1$ is an absolute constant whose value will be chosen later. We also set

$$U_k = \frac{1}{4}m^{1/(2k-1)}, \quad V_k = m^{1/(2k-1)-1/(4k^2-1)},$$

$$\gamma = \frac{5\log\log x - 4\log\log m - \log\log\log m - \log C}{24\log\log x}, \quad \varepsilon = \frac{\log x}{\log m},$$

$$Q_1 = \frac{1}{4}\exp\left\{\frac{1}{4}\varepsilon(\log x)^{1-10\gamma}\right\}, \qquad Q_2 = \frac{1}{4}\exp\left\{\frac{1}{4}\varepsilon(\log x)^{1-8\gamma}\right\},$$

$$Q_3 = \frac{1}{4}\exp\{(\log x)^{1-4\gamma}\}, \qquad Q_3 = \frac{1}{4}\exp\{(\log x)^{1-2\gamma}\}.$$

It is easy to see that

$$\frac{\log\log\log m}{2\log\log m} < \gamma < \frac{5}{96}, \quad \frac{(\log\log m)^5}{(\log m)^{1/5}} < \varepsilon \le \frac{4}{7}.$$

The integers k_1, k_2, s_1, and s_2 determined by the conditions

$$U_{k_1} \le Q_4 < U_{k_1-1}, \qquad U_{k_2+2} < Q_3 \le U_{k_2+1},$$
$$U_{s_1} \le Q_2 < U_{s_1-1}, \qquad U_{s_2+1} < Q_1 \le U_{s_2},$$

satisfy the inequalities

$$\begin{aligned}
\frac{1}{2\varepsilon}(\log x)^{2\gamma} + \frac{1}{2} &\le k_1 < \frac{1}{2\varepsilon}(\log x)^{2\gamma} + \frac{3}{2},\\
\frac{1}{2\varepsilon}(\log x)^{4\gamma} - \frac{3}{2} &\le k_2 < \frac{1}{2\varepsilon}(\log x)^{4\gamma} - \frac{1}{2},\\
\frac{2}{\varepsilon^2}(\log x)^{8\gamma} + \frac{1}{2} &\le s_1 < \frac{2}{\varepsilon^2}(\log x)^{8\gamma} + \frac{3}{2},\\
\frac{2}{\varepsilon^2}(\log x)^{10\gamma} - \frac{5}{2} &\le s_2 < \frac{2}{\varepsilon^2}(\log x)^{10\gamma} - \frac{3}{2}.
\end{aligned} \tag{12.59}$$

Lemma 12.A. *All natural numbers n that do not exceed x and are coprime to m, except at most*

$$\frac{60x\log\log x}{(\log x)^\gamma}$$

numbers, can be represented as

$$n = l^2 pqh, \tag{12.60}$$

where l and h are integers, p and q are prime numbers, and

(1) $\mu(pqh) \ne 0$;

(2) $1 \le l \le \log x$;

(3) *for some k and s such that $k_1 \le k \le k_2$ and $s_1 \le s \le s_2$, the following inequalities hold*:

$$V_k < p \le U_k, \quad U_{s+1} < q \le U_s.$$

Proof. First, we outline the subsequent considerations. From the series

$$1, 2, 3, \dots, n, \dots, [x], \quad (n, m) = 1, \tag{12.61}$$

we delete the numbers that cannot be represented in the form (12.60). This procedure involves several steps. At the first step (item 1°), we delete the numbers divisible by squared "large" integers from the series (12.61). At the second step (item 2°), we choose $Q = Q(x)$ in a special way and delete the numbers all whose prime divisors are larger than Q, as well as the numbers all whose prime divisors do not exceed Q. Thus in the series (12.61), all the numbers remaining after the first two steps can be represented as $n = l^2 n_1$, where n_1 is a square-free number and l is a small integer. In this case, the number n has at least one prime divisor that does not exceed Q and at least one prime divisor that is larger than Q.

The last third stage (item 3°) is the central part of the proof. Here the elimination is performed as follows. First, we delete the numbers n none of whose prime divisors belongs to the interval $(U_{s_2+1}; U_{s_1}]$. Then each of the remaining n has at least one prime divisor q that satisfies one of the inequalities $U_{s+1} < q \le U_s$, $s = s_1, s_1 + 1, \dots, s_2$. Finally, from the resulting set we delete those n none of whose prime divisors belongs to the union of the intervals $(V_k; U_k]$, $k = k_1, k_1+1, \dots, k_2$. It is easy to verify that all the numbers remaining in the series (12.61) admit the representation given in the statement of the lemma.

It should be noted that most of the numbers were deleted at the last stage. Now we prove the lemma following this plan.

1°. In the series (12.61), the number of numbers n that can be divided by a squared "large" integer $l > \log x$, i.e., that can be represented as $n = l^2 n_1, l > \log x$, does not exceed the number

$$\sum_{l > \log x} \frac{x}{l^2} < \frac{2x}{\log x}.$$

2°. We set $Q = \exp\{\log x / \log\log x\}$. According to Lemmas 12.7 and 12.8, in the series (12.61), the number of numbers n all whose prime divisors either do not exceed Q or are larger than Q is bounded above by the quantity

$$N(x, Q) + \Phi(x, Q) < \frac{4x(\log\log x)^2}{\log x}.$$

Thus throwing out

$$< \frac{2x}{\log x} + \frac{4x(\log\log)^2}{\log x} < \frac{5x(\log\log x)^2}{\log x}$$

numbers, we can restrict our consideration to the numbers $n \le x$ satisfying the conditions $n = l^2 n_1$, $1 \le l \le \log x$, $\mu(n_1) \ne 0$, and the condition that n can be divided by a prime number that does not exceed Q.

3°. For convenience of exposition, we introduce the following notation for intervals:

$$J = (1;\, U_{s_2+1}], \quad J' = (U_{s_2+1};\, U_{s_1}], \quad J'' = (U_{s_1};\, Q], \quad J_0 = (1;\, V_{k_2}],$$
$$J_i = (U_{k_2-i+1};\, V_{k_2-i}], \quad i = 1, 2, \ldots, \nu, \quad \nu = k_2 - k_1, \quad J_{\nu+1} = (U_{k_1};\, Q].$$

In what follows, i.e., in items 4° and 5°, our argument is based on the fact that each of these intervals, as well as each of the intervals $(V_k;\, U_k]$, $k = k_1, \ldots, k_2$, contains sufficiently many prime numbers that do not divide m. We show that this is the case.

Let $(u, v]$ be any of the above intervals. First, we show that

$$\frac{u}{v} \le \exp\{-(\log m)^{1/5}\}.$$

If $(u;\, v]$ is either J or J_0, then $u = 1$,

$$v \ge U_{s_2+1} = \frac{1}{4} m^{1/(2s_2+1)}, \quad \text{and} \quad \frac{u}{v} \le m^{-1/(2s_2+1)}.$$

If $(u;\, v]$ is either J'' or $J_{\nu+1}$, then $v = Q$ and $u \le U_{k_1} \le Q_4$ so that $u/v \le Q_4/Q$. Finally, it follows from the inequalities

$$U_{s_2+1} < \frac{1}{4} m^{1/(2s_1+1)} = U_{s_1} m^{-2/(4s_1^2-1)} < U_{s_1} m^{-1/(4s_1^2-1)},$$
$$U_{k+1} < \frac{1}{4} V_k m^{-1/(4k^2-1)} V_k m^{-2/(4k_2^2-1)} < V_k m^{-1/(4s_1^2-1)}, \quad k = k_1, \ldots, k_2 - 1,$$
$$V_k = 4U_k m^{-1/(4k^2-1)} \le 4U_k m^{-1/(4k_2^2-1)} < 4U_k m^{-1/(4s_1^2-1)}, \quad k = k_1, \ldots, k_2,$$

that the inequality

$$\frac{u}{v} \le 4m^{-1/(4s_1^2-1)}$$

holds for each of the remaining intervals. Thus it suffices to prove that

$$\max\left(m^{-1/(4s_2-1)};\, Q_4 Q^{-1};\, 4m^{-1/(4s_1^2-1)}\right) < \exp\{-(\log m)^{1/5}\}.$$

Using the definition of γ and ε, as well as inequalities (12.58) and (12.59), we obtain

$$2s_2 + 1 < \frac{4}{\varepsilon^2} (\log x)^{10\gamma} = 4C^{-5/12} (\log m)^{1/3} (\log x)^{1/12} (\log\log m)^{-5/12}$$
$$< 4C^{-5/12} (\log m)^{1/4} (\log\log m)^{-5/12},$$
$$m^{-1/(2s_2+1)} < \exp\left\{ -\frac{C^{5/12}}{4} (\log m)^{3/4} (\log\log m)^{5/12} \right\} < \exp\{-(\log m)^{1/5}\};$$
$$(\log x)^{2\gamma} = C^{-1/12} (\log\log m)^2 > 2\log\log x;$$
$$Q_4 Q^{-1} = \frac{1}{4} \exp\left\{ -\frac{\log x}{\log\log x} \left(1 - \frac{\log\log x}{(\log x)^{2\gamma}}\right) \right\} \le$$

$$\leq \frac{1}{4}\exp\left\{-\frac{\log x}{2\log\log x}\right\} < \exp\{-(\log m)^{1/5}\};$$

and finally, we have

$$s_1 < \frac{2}{\varepsilon^2}(\log x)^{8\gamma} + \frac{3}{2},$$
$$4s_1^2 - 1 < \frac{48}{\varepsilon^4}(\log x)^{16\gamma} = 48C^{-2/3}(\log m)^{4/5}(\log\log m)^{-4},$$
$$4m^{-1/(4s_1^2-1)} < 4\exp\left\{-\frac{C^{2/3}}{48}(\log m)^{1/5}(\log\log m)^4\right\} < \exp\{-(\log m)^{1/5}\}.$$

Next, for each of the intervals under study, we have the inequality

$$v \geq U_{s_2+1} = \frac{1}{4}m^{1/(2s_2+1)} > \frac{1}{4}\exp\left\{\frac{C^{5/12}}{4}(\log m)^{3/4}(\log\log m)^{5/12}\right\}$$
$$> \exp\{(\log m)^{3/4}(\log\log m)^{1/3}\}.$$

Since

$$\frac{1}{2}\cdot\frac{h}{\log h} \leq \pi(h) \leq 2\cdot\frac{h}{\log h}$$

for large h, for $m \geq m_1$, we have

$$\pi(v) - \pi(u) \geq \frac{v}{2\log v} - \frac{2u}{\log u} = \frac{v}{2\log v}\left(1 - 4\frac{u\log v}{v\log u}\right)$$
$$> \frac{v}{2\log v}\left(1 - 4(-(\log m)^{1/5})\log m\right) > \frac{v}{4\log v} \geq \frac{U_{s_2+1}}{4\log U_{s_2+1}}$$
$$> \frac{\exp\{(\log m)^{3/4}(\log\log m)^{1/3}\}}{4(\log m)^{3/4}(\log\log m)^{1/3}} > 2\exp\{(\log m)^{3/4}\}.$$

In the interval $(u; v]$, the number of primes that divide m does not exceed the number of all prime divisors of m, which, in turn, does not exceed $2\log m$. Therefore, for the number of primes in the interval $(u; v]$ that are coprimes to m, we have the lower bound

$$\pi(v) - \pi(u) - \sum_{u<p\leq v,\ p|m} 1 > 2\exp\{(\log m)^{3/4}\} - 2\log m > \exp\{(\log m)^{3/4}\}.$$

4°. Let N_1 be the number of numbers $n = l^2 n_1$ remaining in the series (12.61) after the first two steps that do not have prime divisors in the interval J'. Let $N_1(l)$ be the number of such numbers n corresponding to a given value of $l \leq \log x$. For each of these n, the number n_1 can be written in the form $n_1 = wv$, where all the prime divisors of w do not exceed Q and all the prime divisors of v are larger than Q. Since the canonical decomposition of n does not contain prime numbers from the interval J', we see that all the prime divisors of w can be divided into two groups. The first group contains all divisors of w that belong to the interval J. The second group contains all divisors of w that belong to the interval J''. Hence the number w can be written in the form

$$w = p_1 \dots p_{r_0} q_1 \dots q_{r_1}, \tag{12.62}$$

where $r_0 \geq 0$, $r_1 \geq 0$, $p_i \in J$, $i = 1, 2, \dots, r_0$, and $q_j \in J''$, $j = 1, 2, \dots, r_1$. Since n_1 is a square-free number, all primes $p_1, \dots, p_{r_0}, q_1, \dots, q_{r_1}$ are distinct.

By $N_1(l; r_0, r_1)$ we denote the number of numbers n that correspond to different values of r_0 and r_1. For an integer $r \geq 0$, we set

$$N_1(l; r) = \sum N_1(l; r_0, r_1),$$

where the summation is over all sets of r_0 and r_1 such that $r_0 + r_1 = 1$. We choose an arbitrary number w of the form (12.62). Recalling the definition of the quantity $\Phi(x, y)$, we see that to this w there correspond

$$\Phi\left(\frac{x}{l^2 w}, Q\right) \tag{12.63}$$

numbers v such that $n = l^2 wv \leq x$. We fix r_0 and r_1. Summing (12.63) over all w corresponding to these r_0 and r_1 and using the estimate in Lemma 12.8, we obtain

$$N_1(l; r_0, r_1) = \sum_w \Phi\left(\frac{x}{l^2 w}, Q\right) < \frac{3x \log x}{(l \cdot \log Q)^2} \sum_w \frac{1}{w}.$$

To estimate the last sum over w, we use the inequality in Lemma 12.6:

$$\sum_w \frac{1}{w} \leq \frac{1}{r_0!}\Big(\sum_{p \in J} \frac{1}{p}\Big)^{r_0} \cdot \frac{1}{r_1!}\Big(\sum_{q \in J''} \frac{1}{q}\Big)^{r_1}.$$

Now we study each of the sums separately. Applying Lemma 12.5, we obtain

$$\sum_{p \in J} \frac{1}{p} = \sum_{\substack{2 \leq p \leq U_{s_2+1} \\ (p,m)=1}} \frac{1}{p} \leq \sum_{2 \leq p \leq U_{s_2+1}} \frac{1}{p} \leq \log\log U_{s_2+1} + c + \frac{10}{\log U_{s_2+1}},$$

$$\sum_{q \in J''} \frac{1}{q} = \sum_{\substack{U_{s_1} < q \leq Q \\ (q,m)=1}} \frac{1}{q} \leq \sum_{U_{s_1} < q \leq Q} \frac{1}{q} \leq \log\log Q - \log\log U_{s_1} + \frac{20}{\log U_{s_1}}.$$

Applying inequalities (12.59), we easily obtain

$$U_{s_2+1} < \exp\left\{\frac{1}{3}\varepsilon(\log x)^{1-10\gamma}\right\}, \quad U_{s_1} < \exp\left\{\frac{1}{5}\varepsilon(\log x)^{1-8\gamma}\right\}$$

as soon as $m \geq m_1$. Therefore, we have

$$\log\log U_{s_2+1} + c + \frac{10}{\log U_{s_2+1}} \leq \log\log U_{s_2+1} + \frac{1}{2}$$

$$\leq (1-10\gamma)\log\log x + \log\varepsilon - \log 3 + \frac{1}{2} = \alpha_0,$$

$$\log\log Q - \log\log U_{s_1} + \frac{20}{\log U_{s_1}} \leq \log\log Q - \log\log U_{s_1} + \frac{1}{2}$$

$$\leq 8\gamma\log\log x - \log\log\log x + \log\varepsilon + \log 5 + \frac{1}{2} = \alpha_1.$$

Collecting together all the estimates, we successively obtain

$$\sum_w \frac{1}{w} \leq \frac{\alpha_0^{r_0}}{r_0!}\cdot\frac{\alpha_1^{r_1}}{r_1!}, \quad N_1(l;r_0,r_1) < \frac{3x\log x}{(l\cdot\log Q)^2}\cdot\frac{\alpha_0^{r_0}}{r_0!}\cdot\frac{\alpha_1^{r_1}}{r_1!},$$

$$N_1(l;r) < \frac{3x\log x}{(l\cdot\log Q)^2}\cdot\frac{1}{r!}\sum_{r_0+r_1=r}\cdot\frac{r!}{r_0!r_1!}\alpha_0^{r_0}\alpha_1^{r_1} = \frac{3x\log x}{(l\cdot\log Q)^2}\cdot\frac{(\alpha_0+\alpha_1)^r}{r!},$$

$$N_1(l) = \sum_{r=0}^{+\infty} N_1(l;r) < \frac{3x\log x}{(l\cdot\log Q)^2}\sum_{r=0}^{+\infty}\frac{(\alpha_0+\alpha_1)^r}{r!}$$

$$= \frac{3x\log x}{(l\cdot\log Q)^2}\exp\{\alpha_0+\alpha_1\}.$$

Since

$$\alpha_0+\alpha_1 = (1-2\gamma)\log\log x - \log\log x + \log(5e/3),$$

we have

$$N_1(l) < \frac{3x\log x}{(l\cdot\log Q)^2}\cdot\frac{(\log x)^{1-2\gamma}}{\log\log x}\cdot\frac{5e}{3} = \frac{5ex\log\log x}{l^2(\log x)^{2\gamma}}.$$

The relation

$$N_1 = \sum_{l\leq\log x} N_1(l)$$

implies the upper bound for N_1:

$$N_1 < 5e\frac{x\log\log x}{(\log x)^{2\gamma}}\sum_{l\leq\log x}\frac{1}{l^2} < \frac{5\pi^2 e}{6}\cdot\frac{x\log\log x}{(\log x)^{2\gamma}}.$$

We exclude these N_1 numbers from the set under study. Then for each of the remaining numbers n, it is possible to find at least one prime divisor q such that $(q,m)=1$ and $q\in J'$. Since we have the relation

$$J' = (U_{s_2+1}; U_{s_1}] = \bigcup_{s=s_1}^{s_2}(U_{s+1}; U_s],$$

for this q we can uniquely determine a number s such that

$$U_{s+1} < q \le U_s, \quad s_1 \le s \le s_2. \tag{12.64}$$

5°. Let N_2 be the number of numbers $n = l^2 n_1$ (in the set obtained in item 4°) that do not have prime divisors in the union of the intervals $(V_k;\, U_k]$, $k = k_1, k_1+1, \dots, k_2$. Let N_2 be the number of numbers n that correspond to a given value of $l \le \log x$. To estimate $N_2(l)$ and N_2, we follow the argument in item 4°.

For each of the numbers n, we write the square-free factor n_1 in the form $n_1 = wv$, where all prime divisors of w (if any) belong to the union of the intervals $J_0, J_1, \dots, L_{\nu+1}$ and all prime divisors of v are larger than Q. For each i, $i = 0, 1, \dots, \nu+1$, we define a number $r_i \ge 0$ as the number of prime divisors of w that lie in the interval J_i. By $N_2(l; r_0, r_1, \dots, r_{\nu+1})$ we denote the number of numbers n that belong to the set under study and correspond to the given values of $l, r_0, r_1, \dots, r_{\nu+1}$. Finally, for an integer $r \ge 0$, we set

$$N_2(l; r) = \sum N_2(l; r_0, r_1, \dots, r_{\nu+1}),$$

where the sum is taken over all the sets of $r_0, r_1, \dots, r_{\nu+1}$ such that $r_0 + r_1 + \dots + r_{\nu+1} = r$.

Following the argument in item 4°, we obtain the inequality

$$N_2(l; r_0, r_1, \dots, r_{\nu+1}) = \sum_w \Phi\left(\frac{x}{l^2 w}, Q\right) < \frac{3x \log x}{(l \cdot \log Q)^2} \sum_w \frac{1}{w},$$

where the sum is taken over all numbers w corresponding to $r_0, r_1, \dots, r_{\nu+1}$.

Applying Lemma 12.6, we estimate the last sum over w as follows:

$$\sum_w \frac{1}{w} \le \prod_{i=0}^{\nu+1} \frac{\sigma_i^{r_i}}{r_i!}, \quad \sigma_i = \sum_{p \in J_i} \frac{1}{p}, \quad i = 0, 1, \dots, \nu + 1.$$

Now let us consider σ_i. First, from inequalities (12.59) for k_2, we derive the upper bound for V_{k_2}:

$$V_{k_2} < \exp\{2(\log x)^{1-4\gamma}\}.$$

Applying Lemma 12.4, we obtain

$$\sigma_0 = \sum_{p \in J_0} \frac{1}{p} \le \log\log V_{k_2} + c + \frac{10}{\log V_{k_2}} < \log\log V_{k_2} + 0.3$$
$$< (1 - 4\gamma)\log\log x + \log 2 + 0.3 < (1 - 4\gamma)\log\log x + 1 = \alpha_0.$$

Let us choose an i, $1 \le i \le \nu$. We set $k = k_2 - i$. Then

$$\sigma_i = \sum_{p \in J_i} \frac{1}{p} < \log\log V_k - \log\log U_{k+1} + \frac{20}{\log U_{k+1}} = \log\left(1 + \frac{1}{2k-1}\right) + \rho(k),$$

where

$$\rho(k) = \frac{20}{\log U_{k+1}} - \log\left(1 - \frac{(2k-1)\log 4}{\log m}\right).$$

It follows from (12.59) that

$$k \le k_2 < \frac{1}{10}(\log m)^{1/3} - \frac{1}{2}.$$

Hence the absolute value of $\rho(k)$ does not exceed

$$25 \cdot \frac{(2k+1)}{\log m} < \frac{1}{4(2k-1)^2}.$$

So we obtain

$$\begin{aligned}\sigma_i &< \log\left(1 + \frac{1}{2k-1}\right) + \frac{1}{4(2k-1)^2} < \frac{1}{2k-1} - \frac{1}{4(2k-1)^2} + \frac{1}{4(2k-1)^2} \\ &= \frac{1}{2k-1} = \frac{1}{2(k_2-i)-1} = \alpha_i.\end{aligned}$$

Finally, taking into account the inequality $U_{k_1} > \exp\{(\log x)^{1-2\gamma}/2\}$, we find

$$\begin{aligned}\sigma_{\nu+1} &< \log\log Q - \log\log U_{k_1} + \frac{20}{\log U_{k_1}} < \log\log Q - \log\log U_{k_1} + 0.3 \\ &< 2\gamma\log\log x - \log\log\log x + 1 = \alpha_{\nu+1}.\end{aligned}$$

Collecting together all the estimates, we obtain the inequalities

$$\sum_w \frac{1}{w} < \prod_{i=0}^{\nu+1} \frac{\alpha_i^{r_i}}{r_i!}, \quad N_2(l; r_0, r_1, \ldots, r_{\nu+1}) < \frac{3x\log x}{(l\cdot\log Q)^2} \prod_{i=0}^{\nu+1} \frac{\alpha_i^{r_i}}{r_i!},$$

$$N_2(l;r) < \frac{3x\log x}{(l\cdot\log Q)^2} \sum_{r_0+r_1+\cdots+r_{\nu+1}=r} \frac{\alpha_0^{r_0}\alpha_1^{r_1}\ldots\alpha_{\nu+1}^{r_{\nu+1}}}{r_0!r_1!\ldots r_{\nu+1}!} = \frac{3x\log x}{(l\cdot\log Q)^2}\cdot\frac{\alpha^r}{r!},$$

where $\alpha = \alpha_0 + \alpha_1 + \cdots + \alpha_{\nu+1}$, and

$$N_1(l) = \sum_{r=0}^{+\infty} N_2(l;r) < \frac{3x\log x}{(l\cdot\log Q)^2}\exp\{\alpha\},$$

$$N_2 = \sum_{l\le\log x} N_2(l) < \frac{\pi^2}{2}\cdot\frac{x\log x}{(\log Q)^2}\exp\{\alpha\}.$$

Since

$$\alpha = (1-4\gamma)\log\log x + 1 + 2\gamma\log\log x - \log\log\log x + 1 + \sum_{k=k_1}^{k_2}\frac{1}{2k-1}$$

$$< (1-\gamma)\log\log x - \log\log\log x + 2,$$

we have

$$N_2 < \frac{(\pi e)^2 x \log x}{2(\log Q)^2} \cdot \frac{(\log x)^{1-\gamma}}{\log\log x} = \frac{(\pi e)^2 x \log\log x}{2(\log x)^\gamma}.$$

Now we also exclude these N_2 numbers. Thus we have excluded at most

$$\frac{5x(\log\log x)^2}{\log x} + \frac{5\pi^2 ex \log\log x}{6(\log x)^{2\gamma}} + \frac{(\pi e)^2 x \log\log x}{2(\log x)^\gamma} < \frac{60x \log\log x}{(\log x)^\gamma}$$

numbers from the series (12.61). For each of the remaining numbers, we can find at least one prime divisor p that belongs to the union of the intervals $(V_k, U_k]$, $k = k_1, \dots, k_2$. Together with (12,64), this means that all the remaining numbers can be represented as (12.60). The proof of Lemma 12.A is complete. □

Now we estimate the short Kloosterman sum.

Suppose that the parameters m, x, C, k_1, k_2, s_1, s_2, γ and ε satisfy conditions (12.59), a and b are integers, $(a, m) = d \geq 1$, and

$$S = \sum_{n\leq x}{}' \exp\left\{2\pi i \frac{an^* + bn}{m}\right\}.$$

Theorem 12.14. *The sum S satisfies the estimate*

$$|S| \leq x(\Delta_1 + \Delta_2),$$

where

$$\Delta_1 = d^{1/(2k_1 s_1)}(\log m)^{-5}, \quad \Delta_2 = 4(\log x)^{-5/24}(\log m)^{1/6}(\log\log m)^{25/24}.$$

Proof. We divide the sum S into two parts: $S = S_1 + S_2$. The sum S_1 contains the terms corresponding to n having the form $n = l^2 n_1$, $l \leq \log x$, $\mu(n_1) \neq 0$. The sum S_2 contains all the other terms. It is easy to see that the number of terms in S_2 does not exceed the quantity

$$\sum_{\log x < l \leq \sqrt{x}} \frac{x}{l^2} < \frac{2x}{\log x}.$$

Estimating S_2 trivially, we obtain

$$|S| \leq |S_1| + \frac{2x}{\log x}.$$

By A we denote the set of all prime numbers that do not divide m and are contained in the union of the intervals $(V_k; U_k]$, $k = k_1, \dots, k_2$. By B we denote the set of all prime numbers that do not divide m and are contained in the interval $(U_{s_2+1}; U_{s_1}]$.

We divide the sum S_1 into sums $S_{\mu,\nu}$. A sum $S_{\mu,\nu}$ contains all the terms corresponding to numbers n that can be represented as $n = l^2hu = ru$, where $\mu(hu) \neq 0$ and $l \leq \log x$. Here u is the product of μ prime factor from A and ν prime factors from B: $u = p_1 \dots p_\mu q_1 \dots q_\nu$ and r does not have prime divisors from the sets A and B.

By μ_0 and ν_0 we denote the maximal possible values of μ and ν. Since the inequalities

$$p > V_{k_2} > \exp\{(\log x)^{1-4\gamma}\}, \quad q > U_{s_2+1} > \exp\left\{\frac{1}{8}\varepsilon(\log x)^{1-10\gamma}\right\}$$

hold for any $p \in A$ and $q \in B$, we have the following estimates for μ_0 and ν_0:

$$\mu_0 \leq (\log x)^{4\gamma}, \quad \nu_0 \leq \frac{8}{\varepsilon}(\log x)^{10\gamma}. \tag{12.65}$$

We can trivially estimate the sums $S_{\mu,\nu}$ in which at least one of the indices μ, ν is nonzero. The terms in such sums correspond to different numbers $n \leq x$ for which it is impossible to find a pair of prime divisors p, q such that $p \in A$ and $q \in B$. By Lemma 12.A, the total number of such n does not exceed $x\Delta$, where $\Delta = 60 \log\log x/(\log x)^\gamma$. Hence we have

$$|S_{0,0}| + \sum_{\mu=1}^{\mu+0} |S_{\mu,0}| + \sum_{\nu=1}^{\nu_0} |S_{0,\nu}| \leq x\Delta.$$

We consider the sum

$$W = \sum_r \sum_{w \leq xr^{-1}} \exp\left\{2\pi i \frac{an^* + bn}{m}\right\}, \quad n = rw.$$

Here r runs through an increasing sequence of numbers $r = l^2h$ defined above. The quantity w takes values that are products $\mu + \nu$ of prime factors each of which, independently of the others, runs through its own increasing sequence of numbers: $w = p_1 \dots p_\mu q_1 \dots q_\nu$. Namely, $p_1 \dots p_\mu$ run through the set A independently of one another and $q_1 \dots q_\nu$ run through the set B independently of one another.

The term corresponding to the value $n = rw$, where $\mu(w) \neq 0$ (i.e., to the value w for which the numbers $p_1, \dots, p_\mu, q_1, \dots, q_\nu$ are distinct) enters the sum W exactly $\mu!\nu!$ times.

Now we consider the term corresponding to the value $n = rw$, where $\mu(w) = 0$. We assume that the set of numbers $p_1, \dots, p_\mu$ contains exactly g different numbers and, moreover, the first number occurs α_1 times, the second number occurs α_2 times, $\dots$, and the gth number occurs α_g times. We also assume that the set of numbers $q_1, \dots, q_\nu$ contains exactly t different numbers and, moreover, the first number occurs β_1 times, the second number occurs β_2 times, $\dots$, and the tth number occurs β_t times.

Then we have $\alpha_1+\alpha_2+\cdots+\alpha_g=\mu$ and $\beta_1+\beta_2+\cdots+\beta_t=\nu$. Since $\mu(w)=0$, at least one of the numbers $\alpha_1,\alpha_2,\ldots,\alpha_g,\beta_1,\beta_2,\ldots,\beta_g$ is strictly larger than 1. Hence the corresponding term occurs in the sum W

$$\frac{(\alpha_1+\cdots+\alpha_g)!(\beta_1+\cdots+\beta_t)!}{\alpha_1!\ldots\alpha_g!\beta_1!\ldots\beta_t!}<\mu!\nu!$$

times. We note that all such numbers $n=rw$ have the form $n=l^2n_1$, $\mu(n_1)\neq 0$, $l>\log x$. Therefore, the number of such $n\le x$ does not exceed the quantity

$$\sum_{\log x<l\le\sqrt{x}}\frac{x}{l^2}<\frac{2x}{\log x}.$$

So we obtain

$$(\mu!\nu!)^{-1}W=S_{\mu,\nu}+2\theta\frac{x}{\log x},\qquad |\theta|<1,\qquad |S_{\mu,\nu}|<(\mu!\nu!)^{-1}W+\frac{2x}{\log x}.$$

Now we estimate the sum W. We write $u=rp_2\ldots p_\mu q_2\ldots q_\nu$, $p=p_1$, $q=q_1$, and $n=upq$. We rewrite W as

$$W=\sum_u\sum_{\substack{p\in A\\ pq\le xu^{-1}}}\sum_{q\in B}\exp\left\{2\pi i\frac{an^*+bn}{m}\right\}.$$

We divide the sets A and B of p and q into intervals as follows. We divide each of the segments $(V_k;U_k]$, $k=k_1,\ldots,k_2$, into the intervals

$$(X;X_1]=\left(\max(V_k;2^{-(\tau+1)}U_k);2^{-\tau}U_k\right],\qquad \tau=0,1,\ldots,\tau_0.$$

If we have $X_1<2X$ for the interval corresponding to the value $\tau=\tau_0$, then we combine this interval with $(2^{-\tau_0}U_k;2^{-(\tau_0-1)}U_k]$. As a result, we obtain the interval $(X';X_1']$ for which $2X'\le X_1'\le 4X'$. Thus for the interval $(X;X_1]$ thus obtained, we have the inequalities

$$2X\le X_1\le 4X.$$

Similarly, we divide each of the intervals $(U_{s+1};U_s]$, $s=s_1,\ldots,s_2$, into the intervals $(Y;Y_1]$, where $2Y\le Y_1\le 4Y$. The number of pairs X,Y obtained by this division does not exceed

$$k_2s_2\log_2(U_{k_1})\log_2(U_{s_1})<(\log m)^3.$$

According to this, we divide the sum W into $<(\log m)^3$ sums:

$$W(X,Y)=\sum_{u\le x(XY)^{-1}}\sum_{\substack{X<p\le X_1\\ XY<pq\le Z}}{}'\sum_{Y<q\le Y_1}{}'\exp\left\{2\pi i\frac{an^*+bn}{m}\right\},$$

where $Z = [\min(X_1Y_1, xu^{-1})]$. By W_1 we denote the inner sum over p and q. We define the function $\delta(z)$ as follows:

$$\delta(z) = \begin{cases} 1, & z = pq,\ (z,m) = 1,\ X < p \le X_1,\ Y < q \le Y_1, \\ 0 & \text{otherwise.} \end{cases}$$

Then W_1 can be rewritten as

$$W_1 = \sum_{XY < z \le Z} \delta(z) \exp\left\{2\pi i \frac{a_1 z^* + b_1 z}{m}\right\},$$

where $a_1 = au^*$ and $b_1 = bu$. In this case, $(a_1, m) = (a, m) = d$. Now we transform W_1 so that the interval of variation of z be independent of u:

$$\begin{aligned} W_1 &= \sum_{XY < z \le X_1Y_1} \delta(z) \left(\frac{1}{m} \sum_{f=0}^{m-1} \sum_{y=1}^{Z} \exp\left\{2\pi i \frac{f(z-y)}{m}\right\}\right) \delta(z) \\ &\quad \times \exp\left\{2\pi i \frac{a_1 z^* + b_1 z}{m}\right\} \\ &= \frac{Z}{m} \sum_{XY < z \le X_1Y_1} \delta(z) \exp\left\{2\pi i \frac{a_1 z^* + b_1 z}{m}\right\} \\ &\quad + \frac{1}{m} \sum_{f=1}^{m-1} \frac{\exp\{-2\pi i f(Z+1)/m\} - \exp\{2\pi i f/m\}}{\exp\{-2\pi i f/m\} - 1} \\ &\quad \times \sum_{XY < z \le X_1Y_1} \delta(z) \exp\left\{2\pi i \frac{a_1 z^* + (b_1 + f) z}{m}\right\}. \end{aligned}$$

Let T be the largest possible value of the modulus of the sum

$$\sum_{XY < z \le X_1Y_1} \delta(z) \exp\left\{2\pi i \frac{a_2 z^* + b_2 z}{m}\right\} = \sum_{X < p \le X_1}{}' \sum_{Y < q \le Y_1}{}' \exp\left\{2\pi i \frac{a_2 p^* q^* + b_2 pq}{m}\right\},$$

where a_2 and b_2 are integers, $(a_2, m) = d$. Then, as is easy to see,

$$|W_1| \le \left(\frac{Z}{m} + \frac{1}{m} \sum_{f=1}^{m-1} \left(\sin \frac{\pi f}{m} \right)^{-1} \right) T \le \left(\frac{Z}{m} + \frac{1}{m} \sum_{f=1}^{m-1} \left(\frac{f}{m} \right)^{-1} \right) T < 2T \log m.$$

For given $X,\ X_1,\ Y,\ Y_1$, we determine k and s so that the following inequalities hold:

$$V_k = m^{1/(2k-1) - 1/(4k^2-1)} < X < X_1 \le U_k = \frac{1}{4} m^{1/(2k-1)}, \tag{12.66}$$

$$U_{s+1} = \frac{1}{4} m^{1/(2s+1)} < Y < Y_1 \le U_s = \frac{1}{4} m^{1/(2s-1)}, \tag{12.67}$$

$$k_1 \le k \le k_2, \quad s_1 \le s \le s_2.$$

Then we have the conditions

$$kX_1^{2k-1} < m, \quad sY_1^{2s-1} < m, \quad X > k \ge 3, \quad Y > s \ge 3.$$

This allows us to apply Lemma 12.9. So we obtain

$$T \le XY\Delta',$$

where

$$\Delta' = (k!s!)^{1/(2ks)}(\pi_1(X))^{-1/(2s)}(\pi_1(Y))^{-1/(2k)}(smdY)^{1/(2ks)},$$

and

$$\pi_1(X) = \pi(X_1) - \pi(X) - \sum_{X<p\le X_1,\ p|m} 1.$$

Using the well-known inequality

$$n! < e\sqrt{n}(n/e)^n,$$

and the fact that $s > k \ge 3$, we obtain

$$\begin{aligned}
(k!s!)^{1/(2ks)} \cdot s^{1/(2ks)} &\le \left(e^2\sqrt{ks}\left(\frac{k}{e}\right)^k\left(\frac{s}{e}\right)^s\right)^{1/(2ks)} s^{1/(2ks)} \\
&= e^{-(1/(2k)+1/(2s)-1/(ks))} \cdot (ks)^{1/(4ks)} \cdot s^{1/(2ks)} \cdot k^{1/(2s)} \cdot s^{1/(2k)} \\
&< (3\cdot 3)^{1/(4\cdot 3\cdot 3)} \cdot 3^{1/(2\cdot 3\cdot 3)} \cdot 3^{1/(2\cdot 3)} \cdot s^{1/(2k)} < 3^{1/3}s^{1/(2k)}.
\end{aligned}$$

It follows from inequalities (12.59) that

$$\begin{aligned}
s^{1/(2k)} &\le s_2^{1/(2k_1)} \le \exp\left\{\frac{\log(2\varepsilon^{-2}(\log x)^{10\gamma})}{\varepsilon^{-1}(\log x)^{2\gamma}}\right\} \\
&< \exp\left\{\frac{10\gamma\log\log x - 2\log\varepsilon + \log 2}{(\log x)^{2\gamma}}\right\} \\
&\le 2\exp\left\{\frac{\frac{25}{12}\log\log x - \frac{5}{3}\log\log m - 2\log\log x + 2\log\log m}{C^{-1/12}(\log\log m)^2}\right\} \\
&= 2\exp\left\{\frac{\frac{1}{12}\log\log x + \frac{1}{3}\log\log m}{C^{-1/12}(\log\log m)^2}\right\} \le 2\left\{\frac{1}{4}C^{1/12}(\log\log)^{-1}\right\} < 3
\end{aligned}$$

for $m \ge m_1$. Thus

$$(k!s!)^{1/(2ks)}s^{1/(2ks)} < 3^{4/3}.$$

Further, since $X_1 \ge 2X$, we have the following lower bound for $\pi_1(X)$:

$$\pi_1(X) \ge \pi(2X) - \pi(X) - 2\log m.$$

Since

$$\pi(2X) = \frac{2X}{\log X} + O\left(\frac{X}{(\log X)^2}\right), \quad \pi(X) = \frac{X}{\log X} + O\left(\frac{X}{(\log X)^2}\right),$$

for sufficiently large X (which is achieved for $m \geq m_1$), we have

$$\pi(X) \geq \frac{X}{\log X} + O\left(\frac{X}{(\log X)^2}\right) - 2\log m, \quad \pi_1(X) \geq \frac{X}{2\log X}.$$

Similarly, for $m \geq m_1$

$$\pi_1(Y) \geq \frac{Y}{2\log Y}.$$

Therefore,

$$\begin{aligned}(\pi_1(X))^{-1/(2s)}(\pi_1(Y))^{-1/(2k)} &\leq 2^{(1/(2k)+1/(2s0)}(\log X)^{1/(2s)} \\ &\quad \times (\log Y)^{1/(2k)} X^{-1/(2s)} Y^{-1/(2k)}.\end{aligned}$$

Since $k \geq k_1 \geq 3$ and $s \geq s_1 > 3$, we have

$$\begin{aligned}2^{1/(2k)+1/(2s)} &\leq 2^{1/6+1/6} = 2^{1/3}, \\ (\log X)^{1/(2s)} &\leq \exp\left\{\frac{\log\log m}{2s_1}\right\} \leq \exp\left\{\frac{\log\log m}{\varepsilon^{-2}(\log x)^{8\gamma}}\right\} \\ &\leq \exp\left\{\frac{\log\log m}{(\log x)^{2\gamma}}\right\} = \exp\{C^{1/12}(\log\log m)^{-1}\} \leq 2, \\ (\log Y)^{1/(2k)} &\leq \exp\left\{\frac{\log\log m}{2k_1}\right\} \leq \exp\left\{\frac{\log\log m}{\varepsilon^{-1}(\log x)^{2\gamma}}\right\} \\ &\leq \exp\left\{\frac{\log\log m}{(\log x)^{2\gamma}}\right\} \leq 2,\end{aligned}$$

and hence

$$(\pi_1(X))^{-1/(2s)}(\pi_1(Y))^{-1/(2k)} \leq 2^{7/3} X^{-1/(2s)} Y^{-1/(2k)}.$$

Thus we have

$$\Delta' \leq 2^{7/3} 3^{4/3} d^{1/(2ks)} X^{-1/(2s)} Y^{-1/(2k)} m^{1/(2ks)}.$$

Taking inequalities (12.66) and (12.67) into account, we have

$$\begin{aligned}&m^{1/(2ks)} X^{-1/(2s)} Y^{-1/(2k)+1/(2ks)} \\ &\quad \leq m^{1/(2ks)}\left(m^{1/(2k-1)-1/(4k^2-1)}\right)^{-1/(2s)}\left(\frac{1}{4}m^{1/(2k+1)}\right)^{-1/(2k)+1/(2ks)} \\ &\quad \leq 4^{1/(2k)-1/(2ks)} m^{-\delta},\end{aligned}$$

where

$$\delta = \frac{s - 6k^2}{2ks(2s+1)(4k^2-1)}.$$

We return to (12.59) and find the lower bound for δ:

$$\delta \geq \frac{\varepsilon^5}{16(\log x)^{24\gamma}} = \frac{C}{16}\frac{\log\log m}{\log m}.$$

Collecting together all the estimates, we obtain the upper bound for Δ':

$$\Delta' \leq 2^{7/3}3^{4/3}4^{1/6}d^{1/(2ks)}(\log m)^{-C/16} < 28d^{1/(2ks)}(\log m)^{-C/16}.$$

Finally, taking into account that $ks \geq k_1 s_1$, we obtain

$$T \leq 28XYd^{1/(2k_1s_1)}(\log m)^{-C/16}$$

so that

$$|W_1| \leq 56XYd^{1/(2k_1s_1)}(\log m)^{1-C/16}.$$

Further, it follows from the definition of u that each value of u enters the sum $W(X, Y)$ at most $(\mu - 1)!(\nu - 1)! = (\mu\nu)^{-1}\mu!\nu!$ times. Hence

$$|W(X,Y)| \leq \sum_{u \leq x(XY)^{-1}}{}' (\mu\nu)^{-1}\mu!\nu!|W_1|,$$

where the prime on the sum means that this sum does not contain repeating terms. Using the above estimate for W_1, we obtain the successive inequalities

$$\begin{aligned}
|W(X,Y)| &\leq (\mu\nu)^{-1}\mu!\nu!x(XY)^{-1}\,56XYd^{1/(2k_1s_1)}(\log m)^{1-C/16}\\
&= 56\mu!\nu!(\mu\nu)^{-1}xd^{1/(2k_1s_1)}(\log m)^{1-C/16},\\
(\mu!\nu!)^{-1}|W| &\leq (\mu!\nu!)^{-1}(\log m)^3(\mu\nu)^{-1}56\mu!\nu!xd^{1/(2k_1s_1)}(\log m)^{1-C/16}\\
&= 56(\mu\nu)^{-1}xd^{1/(2k_1s_1)}(\log m)^{4-C/16},\\
|S_{\mu,\nu}| &\leq (\mu!\nu!)^{-1}|W| + \frac{2x}{\log x}\\
&\leq 56(\mu\nu)^{-1}xd^{1/(2k_1s_1)}(\log m)^{4-C/16} + \frac{2x}{\log x},
\end{aligned}$$

$$\sum_{\mu=1}^{\mu_0}\sum_{\nu=1}^{\nu_0}|S_{\mu,\nu}| \leq 2\log\mu_0\log\nu_0 \cdot 56xd^{1/(2k_1s_1)}(\log m)^{4-C/16} + \frac{2\mu_0\nu_0 x}{\log x}.$$

Taking (12.65) into account, we obtain

$$112\log\mu_0\log\nu_0 < \log m,$$

$$\frac{2\mu_0\nu_0 x}{\log x} < \frac{16x}{(\log x)^2}\log m(\log x)^{14\gamma}$$
$$= 16C^{-7/12}x(\log m)^{-4/3}(\log x)^{11/12}(\log\log m)^{-7/12} < x(\log m)^{-1/9},$$

and finally,

$$\sum_{\mu=1}^{\mu_0}\sum_{\nu=1}^{\nu_0}|S_{\mu,\nu}| < xd^{1/(2k_1s_1)}(\log m)^{5-C/16} + x(\log m)^{-1/9}.$$

If now we choose $C = 160$, then for S_1 we obtain the estimate

$$|S_1| \le \sum_{\mu=1}^{\mu_0}\sum_{\nu=1}^{\nu_0}|S_{\mu,\nu}| < x\Delta + xd^{1/(2k_1s_1)}(\log m)^{-5} + x(\log m)^{-1/9}$$
$$\le 2x\Delta + xd^{1/(2k_1s_1)}(\log m)^{-5} = x(2\Delta + \Delta_1).$$

Estimating S, we obtain

$$|S| \le |S_1| + \frac{2x}{\log x} < x\left(2\Delta + \Delta_1 + \frac{2}{\log x}\right) < x(3\Delta + \Delta_1).$$

From the relations

$$3\Delta = 3C^{1/24}(\log x)^{-5/24}(\log m)^{1/6}(\log\log m)^{1/24}\log\log x$$
$$< 4(\log x)^{-5/24}(\log m)^{1/6}(\log\log m)^{25/24} = \Delta_2,$$

we obtain

$$|S| \le x(\Delta_1 + \Delta_2).$$

The proof of Theorem 12.14 is complete. □

Precisely as above, we assume that $m \ge m_1$ is a sufficiently large natural number, x is an arbitrary number such that

$$\exp\{(\log m)^{4/5}(\log\log m)^{73/5}\} \le x \le m^{4/7},$$

a and b are integers, and $(a, m) = 1$. By the letter N we denote the number of numbers that do not exceed x and are coprime to m.

Theorem 12.15. *Suppose that α and β are real numbers such that $0 \le \alpha < \beta < 1$ and $K(\alpha, \beta)$ is the number of solutions of the system of inequalities*

$$\alpha \le \left\{\frac{an^* + bn}{m}\right\} < \beta, \quad 1 \le n \le x, \quad (n, m = 1).$$

Then the asymptotic formula

$$K(\alpha, \beta) = (\beta - \alpha)N + O(x\Delta)$$

holds, where

$$\Delta = (\log x)^{-5/24}(\log m)^{1/6}(\log\log m)^{49/24}$$

and the constant in the O-symbol is an absolute constant.

Theorem 12.16. *The asymptotic formula*

$$\sum_{n\le x}{}' \left\{\frac{an^* + bn}{m}\right\} = \frac{1}{2}N + O(x\Delta)$$

holds, where

$$\Delta = (\log x)^{-5/24}(\log m)^{1/6}(\log\log m)^{49/24}$$

and the constant in the O-symbol is an absolute constant.

Prior to proving these assertions, we point out two facts. First, if we choose some $0 < \delta < 1/16$ and an integer $\rho \ge 1$, then it suffices to prove the statement of Theorem 12.15 in the case $2\delta < \beta - \alpha < 1 - 2\delta$, since for $0 < \beta - \alpha \le 2\delta$, we have

$$K(\alpha, \beta) = K(\alpha, \alpha + 1 - 2\delta) - K(\beta, \alpha + 1 - 2\delta),$$

and for $1 - 2\delta \le \beta - \alpha < 1$, we have

$$K(\alpha, \beta) = K(\alpha, \alpha + 1/2) - K(\alpha + 1/2, \beta).$$

Second, it follows from Lemma 12.11 that both theorems (the first in the case $2\delta < \beta - \alpha < 1 - 2\delta$) will be proved if the corresponding asymptotic formula is obtained for the sum

$$U(\alpha, \beta) = \sum_{n\le x} \psi\left(\frac{an^* + bn}{m}\right),$$

where $\psi(x)$ is Vinogradov's "cup" constructed for given δ, ρ, α, and β (see Lemma 12.10). Therefore, we combine the proofs of these two assertions.

Proof. We choose $r_0 = (\log m)^{36}$, $\delta = r_0^{-1/2}$, and $\rho = [r_0^{1/4}]$ and assume that $2\delta < \beta - \alpha < 1 - 2\delta$. Using Lemma 12.10, we construct Vinogradov's "cup" $\psi(x)$ corresponding to the chosen δ, ρ, α, and β and consider the sum $U(\alpha, \beta)$. Expanding ψ into the Fourier series, we obtain

$$U(\alpha, \beta) = (\beta - \alpha)N + R,$$

where

$$R = \sum_{r\ne 0} g(r)S_s, \quad S_r = \sum_{n\le x}{}' \exp\left\{2\pi i\frac{an^* + bn}{m}\right\}.$$

Let us estimate R. First, we note that

$$R \ll \sum_{r=1}^{+\infty} c(r)|S_r|,$$

where the quantities $c(r)$ were defined in (12.12). Next, we divide the sum over r into two sums so that

$$R \ll \Big(\sum_{r\le r_0} + \sum_{r>r_0}\Big) c(r)|S_r| = R_1 + R_2,$$

and estimate each of these sums in its own way. If $r \le r_0$, then $d = (ar, m) \le r \le r_0$. Estimating S_r by Theorem 12.14, we obtain

$$|S_r| \le x(\Delta_1 + \Delta_2)$$

and, moreover,

$$\Delta_1 = r_0^{1/(2k_1s_1)}(\log m)^{-5} \le r_0^{1/18}(\log m)^{-5} = (\log m)^{-3},$$
$$\Delta_2 = 4(\log x)^{-5/24}(\log m)^{1/6}(\log\log m)^{25/24}.$$

Since $\Delta_1 < \Delta_2$, we have $|S_r| \ll x\Delta_2$. Therefore,

$$R_1 \ll \sum_{r\le r_0} c(r)x\Delta_2 \ll x\Delta_2 \sum_{r\le r_0}\frac{1}{r} \ll x\Delta_2\log r_0 \ll x\Delta_2 \log\log m \ll x\Delta,$$

where $\Delta = (\log x)^{-5/24}(\log m)^{1/6}(\log\log m)^{49/24}$.

We estimate the sum R_2 trivially:

$$R_2 \le \sum_{r\le r_0}\frac{1}{\pi r}\left(\frac{\rho}{r\delta}\right)^{\rho} x \ll \frac{x}{r_0}\left(\frac{\rho}{r_0\delta}\right)^{\rho} \ll \frac{x}{r_0} r_0^{-\rho/4} \ll x\Delta.$$

Thus we have $R \ll x\Delta$ and

$$U(\alpha,\beta) = (\beta-\alpha)N + O(x\Delta).$$

Using Lemma 12.11 and taking into account that

$$N\delta \ll x(\log m)^{-18} \ll x\Delta,$$

we arrive at the statements of Theorem 12.15 and 12.16. □

In conclusion, we find out in which case the asymptotic formulas for $K(\alpha,\beta)$ and $S = \sum'_{n\le x}\{(an^* + bn)/m\}$ are nontrivial, i.e., the remainder is less than the leading term. To this end, it is necessary to estimate N. We have

$$N = \sum_{n\le x}{}' 1 = \sum_{n\le x}\sum_{d|(n,m)}\mu(d) = \sum_{d|m,\, d\le x}\mu(d)\left[\frac{x}{d}\right] = \sum_{d|m}\mu(d)\left[\frac{x}{d}\right]$$

$$= x\sum_{d|m}\frac{\mu(d)}{d} - \sum_{d|m}\left\{\frac{x}{d}\right\} = \frac{\varphi(m)}{m}x + \theta\cdot\tau(m),$$

where $|\theta| \le 1$. Next, it is known that

$$c_1(\log\log m)^{-1} \le \frac{\varphi(m)}{m} < 1,$$

where $c_1 > 0$ is an absolute constant. Therefore, we have

$$N \ge \frac{c_1 x}{\log\log m} - \tau(m).$$

Let m satisfy the condition that $\tau(m)$ is small, i.e., $\tau(m)$ satisfies the inequality

$$\tau(m) \le \frac{c_1 x}{2\log\log m}$$

(this holds, e.g., for $m = p$, where p is a prime number). Then

$$N \ge \frac{c_1 x}{2\log\log m}.$$

The asymptotic formula for S is nontrivial if

$$x\Delta \ll x(\log\log m)^{-1},$$

i.e., if we have the inequality

$$\log x \gg (\log m)^{4/5}(\log\log m)^{73/5}.$$

Under this condition, the asymptotic formula for $K(\alpha,\beta)$ is also nontrivial if only α and β are fixed numbers. In the general case, in view of the relation

$$\limsup_{m\to\infty}\log r(m)\frac{\log\log m}{\log m} = \log 2,$$

the asymptotic formulas for S and $K(\alpha,\beta)$ are, in general, nontrivial only if

$$x \ge \exp\left\{c_2\frac{\log m}{\log\log m}\right\},$$

where $c_2 > \log 2$.

Concluding remarks on Chapter 12. 1. In number theory, the Kloosterman sums are defined to be trigonometric sums of the form

$$S = \sum_{n=1}^{m}{}' \exp\left\{2\pi i\frac{an^* + bn}{m}\right\},$$

where m is a natural number and $a,\ b$ are integers. Here the prime on the sum means that the sum is taken over the numbers n coprime to m and the symbol n^* denotes a natural number that does not exceed m and satisfies the condition $n^*n \equiv 1 \pmod{m}$. Along with S, the sums S_1,

$$S_1 = \sum_{n\le x} \exp\left\{2\pi i \frac{an^* + bn}{m}\right\},$$

where $x < m$, are also considered; the sums S_1 are called incomplete (or short) Kloosterman sums.

2. Nontrivial estimates from above for $|S|$ and $|S_1|$ are used in a great variety of problems in number theory.

3. The first nontrivial estimates for $|S|$ and $|S_1|$ were obtained by H. Kloosterman in 1926 in the paper [106]:

$$|S| \ll m^{3/4+\varepsilon} D^{1/4}, \quad |S_1| \ll m^{7/8+\varepsilon} D^{1/4},$$

where $d = (a, m)$, $D = \max(d; (b, m))$, and the constants in $\ll$ depend on $\varepsilon > 0$.

4. In 1931, H. Salie [141] proved that

$$|S| \le 3p^{\alpha/2},$$

where $m = p^\alpha$, $p > 2$, p is a prime number, $\alpha \ge 2$, and $(a, p) = (b, p) = 1$, which implies that, in the general case, the following inequality holds:

$$|S| \le \tau(p^\alpha) p^{\alpha/2} d^{1/2}.$$

5. In 1948, A. Weil [167] proved that

$$|S| \le 2p^{1/2},$$

where $m = p$, $p > 2$, p is a prime number, and $(a, p) = (b, p) = 1$.

6. In 1957, L. Carlitz and S. Uchiyama [43] obtained the estimates

$$|S| \le \tau(m) m^{1/2} d^{1/2} \ll m^{1/2+\varepsilon} d^{1/2},$$
$$|S_1| \le \tau(m) m^{1/2} d^{1/2} \log m \ll m^{1/2+\varepsilon} d^{1/2}.$$

7. Theorems 12.1–12.9 were proved by A. A. Karatsuba in [94], [95], [96], [97]. A survey of the results is also contained in [95]. Lemma 12.1 was proved by G. I. Arkhipov (see [100]).

8. The function $\alpha_k(n)$ defined in Section 12.1 was introduced by G. I. Arkhipov. In [100], this function was called Arkhipov's function (see also [97]). A generalization of Arkhipov's function, namely, the function $\alpha_{k,m}(n)$, was introduced in [100].

9. Theorems 12.10–12.13 were proved by A. A. Karatsuba in [103].

10. Theorems 12.14–12.16 were proved by M. A. Korolev in [108].

11. In [107], M. A. Korolev proved analogs of Theorems 12.8 and 12.9 for the function $\alpha_{k,m}(n)$.

12. J. Friedlander and H. Iwaniec [63] used estimates for short Kloosterman sums in the Brun–Titchmarsh theorem.

Appendix

Here we state several assertions which we used in this book. All these assertions are called lemmas.

Lemma A.1 (Hölder's inequality). *Suppose that* $u_\nu \geq 0$, $v_\nu \geq 0$, $\alpha > 0$, $\beta > 0$, *and* $\alpha + \beta = 1$. *Then*

$$\sum_{\nu=1}^{P} u_\nu v_\nu \leq \Big(\sum_{\nu=1}^{P} u_\nu^{1/\alpha}\Big)^{\alpha} \Big(\sum_{\nu=1}^{P} v_\nu^{1/\beta}\Big)^{\beta}.$$

Corollary A.1. (1) (*Cauchy's inequality*) $\Big(\sum_{\nu=1}^{P} u_\nu v_\nu\Big)^2 \leq \Big(\sum_{\nu=1}^{P} u_\nu^2\Big)\Big(\sum_{\nu=1}^{P} v_\nu^2\Big)$.

(2) $\Big(\sum_{\nu=1}^{P} u_\nu v_\nu\Big)^k \leq \Big(\sum_{\nu=1}^{P} u_\nu\Big)^{k-1} \sum_{\nu=1}^{P} u_\nu v_\nu^k$.

(3) $\Big(\sum_{\nu=1}^{P} u_\nu\Big)^k \leq P^{k-1} \sum_{\nu=1}^{P} u_\nu^k$.

(4) (*The inequality between the geometric and arithmetic means of nonnegative numbers*)

$$k u_1 \dots u_k \leq u_1^k + \cdots + u_k^k.$$

Proof. For the proof of the lemma, see, e.g., [90], p. 85. □

Lemma A.2 (van der Korput's lemma). *Suppose that* $f(x)$ *is a real differentiable function on the interval* $a < x \leq b$ *and in the interior of this interval its derivative* $f'(x)$ *is monotone and of constant sign and satisfies the inequality* $|f'(x)| \leq \delta$ *for a constant* δ *such that* $0 < \delta < 1$. *Then we have*

$$\sum_{a<x\leq b} \exp\{2\pi i f(x)\} = \int_a^b \exp\{2\pi i f(x)\}\, dx + \theta\big(3 + 2\delta/(1-\delta)\big).$$

Proof. For the proof of the lemma, see, e.g., [165], p. 25. □

Lemma A.3 (I. M. Vinogradov's lemma on "cups"). *Suppose that* r *is a positive integer,* α *and* β *are real,* $0 < \Delta < 0.25$, *and* $\Delta \leq \beta - \alpha \leq 1 - \Delta$.

Then there exists a periodic function $\psi(x)$ *with period* 1 *satisfying the conditions*:

(1) $\psi(x) = 1$ *in the interval* $\alpha + 0.5\Delta \le x \le \beta - 0.5\Delta$;

(2) $0 < \psi(x) < 1$ *in the intervals* $\alpha - 0.5\Delta < x < \alpha + 0.5\Delta$ *and* $\beta - 0.5\Delta < x < \beta + 0.5\Delta$;

(3) $\psi(x) = 0$ *in the interval* $\beta + 0.5\Delta \le x \le 1 + \alpha - 0.5\Delta$;

(4) $\psi(x)$ *can be expanded into the Fourier series*

$$\psi(x) = \beta - \alpha + \sum_{m=1}^{\infty} \big(g_m \exp\{2\pi imx\} + h_m \exp\{-2\pi imx\}\big),$$

where

$$|g_m| \le \min\left(\frac{1}{\pi m}, \beta - \alpha, \frac{1}{\pi m}\left(\frac{r}{\pi m \Delta}\right)^r\right),$$

$$|h_m| \le \min\left(\frac{1}{\pi m}, \beta - \alpha, \frac{1}{\pi m}\left(\frac{r}{\pi m \Delta}\right)^r\right).$$

Proof. For the proof of the lemma, see, e.g., [165], p. 23. □

Lemma A.4 (I. M. Vinogradov's lemma). *Suppose that* m *is a positive integer,* λ *is real, and*

$$\Phi(y) = m\frac{ay + \lambda y}{q}, \quad (a, q) = 1, \quad q > 0;$$

moreover, y *runs through at most* Y *successive integers.*

Then for $V \ge 0$, *under the condition that*

$$\|\Phi(y)\| \le V/q,$$

the number of values of y *does not exceed*

$$\lambda Ym + m + 2V \quad \text{if} \quad Y \le q, \quad 2(\lambda Ym + m + 2V)Y/q \quad \text{if} \quad Y > q.$$

Proof. For the proof of the lemma, see, e.g., [165], p. 62. □

Lemma A.5. *Suppose that* $f(x) = a_n x^n + \cdots + a_1 x$, $a_1, \ldots, a_n$ *are integers,* $(a_, \ldots, a_n, p) = 1$, *and*

$$S = \sum_{x=1}^{p} \exp\left\{2\pi i \frac{f(x)}{p}\right\}.$$

Then the following estimate holds:

$$|S| \le n\sqrt{p}.$$

Proof. For the proof of the lemma, see, e.g., [167]. □

Bibliography

[1] J. Ax and S. Kochen, *Diophantine problems over local fields*, Amer. J. Math. 57 (1965), no. 3, 605–630.

[2] G. I. Arkhipov, *Multiple trigonometric sums*, Dokl. Akad. Nauk SSSR 219 (1974), no. 5, 1036–1037; English transl.: Soviet Math. Dokl. 15 (1974), 1702–1704.

[3] G. I. Arkhipov, *Multiple Trigonometric Sums and Applications*, Candidate Dissertation, Steklov Inst. Math. Russ. Acad. Sci., Moscow, 1975.

[4] G. I. Arkhipov, *A theorem on the mean value of the modulus of a multiple trigonometric sum*, Mat. Zametki 17 (1975), no. 1, 143–153; English transl.: Math. Notes 17 (1975), 84–90.

[5] G. I. Arkhipov, *Estimates for double trigonometric Weyl sums*, Trudy Mat. Inst. Steklov, 142 (1976), 46–66; English transl.: Proc. Steklov Inst. Math. 1979, no. 3, 47–68.

[6] G. I. Arkhipov, *On the mean value of H. Weyl sums*, Mat. Zametki 23 (1978), no. 6, 785–788; English transl.: Math. Notes 23 (1978), no. 5–6, 431–433.

[7] G. I. Arkhipov, *The values of a singular series in a Hilbert–Kamke problem*, Dokl. Akad. Nauk SSSR 259 (1981), no. 2, 265–267; English transl.: Soviet Math. Dokl. 24 (1981), no. 1, 49–51.

[8] G. I. Arkhipov, *On the Hilbert–Kamke problem*, Izv. Akad. Nauk SSSR Ser. Mat. 48 (1984), no. 1, 3–52; English transl.: Math. USSR Izv. 24 (1985), 1–47.

[9] G. I. Arkhipov, *Study of the Hilbert–Kamke Problem*, Doctorate Dissertation, Steklov Inst. Math. Russ. Acad. Sci., Moscow, 1984.

[10] G. I. Arkhipov and V. N. Chubarikov, *On multiple trigonometric sums*, Dokl. Akad. Nauk SSSR 222 (1975), no. 5, 1017–1019; English transl.: Soviet Math. Dokl. 16 (1975), 705–708.

[11] G. I. Arkhipov and V. N. Chubarikov, *Multiple trigonometric sums*, Izv. Akad. Nauk SSSR Ser. Mat. 40 (1976), 209–220; English transl.: Math. USSR Izv. 10 (1976), 200–210.

[12] G. I. Arkhipov and V. N. Chubarikov, *On arithmetic conditions for solvability of nonlinear systems of Diophantine equations*, Dokl. Akad. Nauk SSSR 284 (1985), no. 1, 16–21; English transl.: Soviet Math. Dokl. 32 (1985), 348–352.

[13] G. I. Arkhipov and V. N. Chubarikov, *On the number of summands in the Hilbert–Kamke problem in prime numbers*, Dokl. Akad. Nauk SSSR 330 (1993), no. 4, 407–408; English transl.: Soviet Math. Dokl. 47 (1993), no. 3, 485–488.

[14] G. I. Arkhipov and V. N. Chubarikov, *On the asymptotics of the number of summands in a multidimensional additive problem with prime numbers*, Dokl. Akad. Nauk SSSR 331 (1993), no. 1, 5–6; English transl.: Soviet Math. Dokl. 48 (1994), no. 1, 1–3.

[15] G. I. Arkhipov and V. N. Chubarikov, *On the number of summands in the additive Vinogradov's problem and its generalizations*, in: *Modern Problems in Number Theory and Its Applications*, Proc. IVth Intern. Conference Dedicated to the 180th Anniversary of P. L. Chebyshev and the 110th Anniversary of I. M. Vinogradov, Tula, September 10–15 2001, pp. 5–38.

[16] G. I. Arkhipov and V. N. Chubarikov, *On some applications of Vinogradov's method*, Bonner Mathematische Schriften (Proc. Session in Analytic Number Theory and Diophantine Equations, Bonn, January–June 2002), 2003, no. 360, 17pp.

[17] G. I. Arkhipov and A. A. Karatsuba, *On I. M. Vinogradov's integral*, Dokl. Akad. Nauk SSSR 239 (1978), no. 4, 764–765; English transl.: Soviet Math. Dokl. 19 (1978), 389–391.

[18] G. I. Arkhipov and A. A. Karatsuba, *A new estimate for I. M. Vinogradov's integral*, Izv. Akad. Nauk SSSR Ser. Mat. 42 (1978), no. 4, 751–762; English transl.: Math. USSR Izv. 13 (1979), 52–62.

[19] G. I. Arkhipov and A. A. Karatsuba, *On a local representation of zero by a form*, Izv. Akad. Nauk SSSR Ser. Mat. 45 (1981), no. 5, 948–961; English transl.: Math. USSR Izv. 19 (1982), 231–240.

[20] G. I. Arkhipov and A. A. Karatsuba, *On a representation of zero by a form in the field of p-adic numbers*, Dokl. Akad. Nauk SSSR 262 (1982), no. 1, 11–13; English transl.: Soviet Math. Dokl. 25 (1982), no. 1, 1–3.

[21] G. I. Arkhipov and A. A. Karatsuba, *On a problem in the theory of congruences*, Uspekhi Mat. Nauk 37 (1982), no. 5, 161–162; English transl.: Russian Math. Surveys 37 (1982), 157–158.

[22] G. I. Arkhipov and A. A. Karatsuba, *A multidimensional analogue of the Waring problem*, Dokl. Akad. Nauk SSSR, 295 (1987), no. 3, 521–523; English transl.: Soviet Math. Dokl. 36 (1988), no. 1, 75–77.

[23] G. I. Arkhipov and A. N. Zhitkov, *On the Waring problem with a noninteger exponent*, Izv. Akad. Nauk SSSR Ser. Mat. 48 (1984), no. 6, 1138–1150; English transl.: Math. USSR Izv. 25 (1985), no. 3, 443–454.

[24] G. I. Arkhipov, A. A. Karatsuba, and V. N. Chubarikov, *An upper bound of the modulus of a multiple trigonometric sum*, Trudy Steklov Mat. Inst. 143 (1977), 3–31; English transl.: Proc. Steklov Inst. Math. 143 (1980), no. 1, 1–31.

[25] G. I. Arkhipov, A. A. Karatsuba, and V. N. Chubarikov, *A sharp estimate for the number of solutions of a system of Diophantine equations*, Izv. Akad. Nauk SSSR Ser. Mat. 42 (1978), no. 6, 1187–1226; English transl.: Math. USSR Izv. 13 (1979), 461–497.

[26] G. I. Arkhipov, A. A. Karatsuba, and V. N. Chubarikov, *Distribution of fractional parts of polynomials in several variables*, Mat. Zametki 25 (1979), no. 1, 3–14; English transl.: Math. Notes 25 (1979), 3–9.

[27] G. I. Arkhipov, A. A. Karatsuba, and V. N. Chubarikov, *The convergence exponent of the singular integral in Tarry's problem*, Dokl. Akad. Nauk SSSR 248 (1979), no. 2, 268–272; English transl.: Soviet Math. Dokl. 20 (1979), no. 5, 978–981.

[28] G. I. Arkhipov, A. A. Karatsuba, and V. N. Chubarikov, *Trigonometric integrals*, Izv. Akad. Nauk SSSR Ser. Mat. 43 (1979), no. 5, 971–1003; English transl.: Math. USSR Izv. 15 (1980), 211–239.

[29] G. I. Arkhipov, A. A. Karatsuba, and V. N. Chubarikov, *Multiple Trigonometric Sums*, Trudy Mat. Inst. Steklov 151 (1980); English transl.: Proc. Steklov Inst. Math. 151 (1982).

[30] G. I. Arkhipov, A. A. Karatsuba, and V. N. Chubarikov, *On a system of Diophantine equations*, Dokl. Akad. Nauk SSSR 252 (1980), no. 2, 275–276; English transl.: Soviet Math. Dokl. 21 (1980), no. 3, 715–717.

[31] G. I. Arkhipov, A. A. Karatsuba, and V. N. Chubarikov, *Uniform estimates of multiple trigonometric sums*, Dokl. Akad. Nauk SSSR 252 (1980), no. 6, 1289–1291; English transl.: Soviet Math. Dokl. 21 (1980), no. 3, 904–906.

[32] G. I. Arkhipov, A. A. Karatsuba, and V. N. Chubarikov, *Multiple trigonometric sums and their applications*, Izv. Akad. Nauk SSSR Ser. Mat. 44 (1980), 723–781; English transl.: Math. USSR Izv. 17 (1981), 1–54.

[33] G. I. Arkhipov, A. A. Karatsuba, and V. N. Chubarikov, *Special cases in the theory of multiple trigonometric sums*, Izv. Akad. Nauk SSSR Ser. Mat. 47 (1983), no. 4, 707–784; English transl.: Math. USSR Izv. 23 (1984), 17–82.

[34] G. I. Arkhipov, A. A. Karatsuba, and V. N. Chubarikov, *New uniform estimates for multiple trigonometric sums*, Dokl. Akad. Nauk SSSR 272 (1983), no. 1, 11–12; English transl.: Soviet Math. Dokl. 28 (1983), no. 2, 301–303.

[35] G. I. Arkhipov, A. A. Karatsuba, and V. N. Chubarikov, *The Theory of Multiple Trigonometric Sums*, Nauka, Moscow, 1987 (in Russian).

[36] G. I. Arkhipov and K. I. Oskolkov, *A special trigonometric series and its applications*, Mat. Sb. 134 (176) (1987), no. 2 (10), 147–157; English transl.: Math. USSR-Sb. 62 (1989), no. 2, 45–155.

[37] M. B. Barban, Yu. V. Linnik, and N. G. Chudakov, *On prime numbers in an arithmetic progression with a prime-power difference*, Acta Arith. 9 (1964), no. 4, 375–390.

[38] L. A. Bassalygo, V. A. Zinov'ev, and S. N. Litsyn, *A lower bound for complete rational trigonometric sums in terms of multiple sums*, Dokl. Akad. Nauk SSSR 300 (1988), no. 5, 1033–1036; English transl.: Soviet Math. Dokl. 37 (1988), no. 3, 756–759.

[39] E. Bombieri, *On exponential sums in finite fields*, Amer. J. Math. 88 (1966), no. 1, 71–105.

[40] E. Bombieri and H. Davenport, *On two problems of Mordell*, Amer. J. Math. 88 (1966), no. 1, 61–70.

[41] Z. I. Borevich and I. R. Shafarevich, it Number Theory, 3rd ed., Nauka, Moscow, 1985.

[42] L. Carlitz, *A note on multiple exponential sums*, Pacific J. Math. 15 (1965), 757–765.

[43] L. Carlitz and S. Uchiyama, *Bounds for exponential sums*, Duke Math. J. 24 (1957), 37–41.

[44] K. Chandrasekharan, *Arithmetic Functions*, Springer-Verlag, Berlin, 1970.

[45] Chen Jing Run, *On Professor Hua's estimate on exponential sum*, Acta Sci. Sinica 20 (1977), no. 6, 711–719.

[46] V. N. Chubarikov, A sharp estimate of the bound for zeros of Dirichlet L-series modulo the power of a prime, Vestnik Moskov. Univ. Ser. I Mat. Mekh. (1973), no. 2, 46–52; English transl.: Moscow Univ. Math. Bull 28 (1973), No.1/2, 76–81.

[47] V. N. Chubarikov, *On a multiple trigonometric integral*, Dokl. Akad. Nauk SSSR 227 (1976), no. 6, 1308–1310; English transl.: Soviet Math. Dokl. (1976), 618–620.

[48] V. N. Chubarikov, *On multiple rational trigonometric sums and multiple integrals*, Mat. Zametki 20 (1976), no. 1, 61–68; English transl.: Math. Notes 20 (1976), 589–593.

[49] V. N. Chubarikov, *Multiple Trigonometric Sums*, Candidate Dissertation, Lomonosov Moscow State University, Moscow, 1977.

[50] V. N. Chubarikov, *An asymptotic formula for the mean value of a multiple trigonometric sum*, Mat. Zametki 23 (1978), no. 6, 799–816; English transl.: Math. Notes 23 (1978), no. 5–6, 438–448.

[51] V. N. Chubarikov, *On asymptotic formulas for I. M. Vinogradov's integral and its generalizations*, Trudy Mat. Inst. Steklov 157 (1981), 214–232; English transl.: Proc. Steklov Inst. Math. 157 (1983), no. 3, 229–248.

[52] V. N. Chubarikov, *Multiple trigonometric sums with prime numbers*, Dokl. Akad. Nauk SSSR 278 (1984), no. 2, 302–304; English transl.: Soviet Math. Dokl. 30 (1984), 391–393.

[53] V. N. Chubarikov, *Estimates for multiple trigonometric sums with primes*, Izv. Akad. Nauk SSSR Ser. Mat. 49 (1985), no. 5, 1031–1067; English transl.: Math. USSR Izv. 27 (1986), no.2, 323–357.

[54] V. N. Chubarikov, *On simultaneous representation of natural numbers by sums of powers of prime numbers*, Dokl. Akad. Nauk SSSR 286 (1986), 828–831; English transl.: Soviet Math. Dokl. 33 (1986), 211–214.

[55] V. N. Chubarikov, *A multidimensional additive problem with prime numbers*, Dokl. Akad. Nauk SSSR 290 (1986), 805–808; English transl.: Soviet Math. Dokl. 34 (1987), 365–368.

[56] V. N. Chubarikov, *Multidimensional Problems in the Theory of Prime Numbers*, Doctorate Dissertation, Lomonosov Moscow State University, Moscow, 1989.

[57] H. Davenport and D. Lewis, *Nonhomogeneous cubic equations*, J. London Math. Soc. 39 (1964), no. 4, 657–671.

[58] H. Davenport and D. Lewis, *Exponential sums in many variables*, Amer. J. Math. 84 (1962), no. 4, 649–665.

[59] P. Deligne, *Applications de la formula des traces aux sommes trigonométriques, à paraitre dans SGA*, Inst. Hautes Etud. Sci. Publ. Math. (1974), no. 43, p. 273.

[60] V. B. Dem'yanov, *Cubic forms in discrete normed fields*, Dokl. Akad. Nauk SSSR 74 (1966), no. 5, 889–891 (in Russian).

[61] Y. Eda, On the mean value theorem in an algebraic number field, Japan J. Math. 36 (1967), 5–21.

[62] Yu. L. Ershov, *On elementary theories of local fields*, Algebra i Logika 4 (1965), no. 2, 5–30.

[63] J. Friedlander and H. Iwaniec, *The Brun–Titchmarsh theorem*, Proc. Taniguchi Symp., May 1996, Kioto; *Analytic Number Theory*, Y. Motohashi, ed., Lecture Notes in Statist. 247, Cambridge University Press, Cambridge 1997.

[64] G. H. Hardy and J. E. Littlewood, *Some problems of Diophantine approximation*, II. *The trigonometric series associated with the elliptic θ-functions*, Acta Math. 37 (1914), 193–239.

[65] G. H. Hardy and J. E. Littlewood, *Some problems of Diophantine approximation*, Proc. Intern. Congr. of Math. I. Cambridge, 1912, 223–229.

[66] C. Hooley, *On exponential sums and certain of their applications*, London Math. Soc. Lecture Note Ser. 56, Cambridge University Press, Cambridge–New York 1982, 92–122.

[67] Hua Loo-Keng, *Additive theory of prime numbers*, Trudy Mat. Inst. Steklov 22 (1947); English transl. of Chinese rev. ed. Amer. Math. Soc., Providence, RI, 1965.

[68] Hua Loo-Keng, *On the number of solutions of Tarry's problem*, Acta Sci. Sinica 1 (1952), no. 1, 1–76.

[69] Hua Loo-Keng, *On the Tarry problem*, Proc. Third All-Union Math. Congr. (Moscow, 1956). Vol. IV: Survey Reports, Izdat. Akad. Nauk SSSR, Moscow, 1959, 140–143.

[70] Hua Loo-Keng, *Method of Trigonometric Sums and Its Applications in Number Theory*, Nauka, Moscow 1964 (Russian translation); *Die Abschätzung von Exponentialsummen und Ihre Anwendungen in der Zahlentheorie*, Teubner, Leipzig 1959.

[71] I. A. Ikromov, *On the convergence exponent of trigonometric integrals*, Trudy Mat. Inst. Steklov 218 (1997), 179–189.

[72] A. A. Karatsuba, *Estimates for trigonometric sums of a special form and their applications*, Dokl. Akad. Nauk SSSR 137 (1961), no. 3, 513–514 (in Russian).

[73] A. A. Karatsuba, *Waring's problem for a congruence modulo the power of a prime*, Vestnik Moskov. Univ. Ser. I Mat. Mekh. (1962), no. 4, 28–38 (in Russian).

[74] A. A. Karatsuba, *An analog of Waring's problem*, Vestnik Moskov. Univ. Ser. I Mat. Mekh. (1962), no. 1, 38–46 (in Russian).

[75] A. A. Karatsuba, *Distribution of fractional parts of polynomials of a special form*, Vestnik Moskov. Univ. Ser. I Mat. Mekh. (1962), no. 3, 34–38 (in Russian).

[76] A. A. Karatsuba, *Trigonometric sums of a special form and their applications*, Izv. Akad. Nauk SSSR Ser. Mat. 28 (1964), no. 1, 237–248 (in Russian).

[77] A. A. Karatsuba, *On systems of congruences*, Izv. Akad. Nauk SSSR Ser. Mat. 29 (1965), 935–944; English transl.: Amer. Math. Soc. Transl. (2) 82 (1969), 73–84.

[78] A. A. Karatsuba, *On estimating the number of solutions of some equations*, Dokl. Akad. Nauk SSSR 165 (1966), no. 1, 31–32 (in Russian).

[79] A. A. Karatsuba, *Systems of congruences and Waring type equations*, Dokl. Akad. Nauk SSSR 165 (1965), no. 2, 274–276 (in Russian).

[80] A. A. Karatsuba, *Theorems on mean values and complete trigonometric sums*, Izv. Akad. Nauk SSSR Ser. Mat. 30 (1966), 183–206 (in Russian).

[81] A. A. Karatsuba, *Method of Trigonometric Sums and Theorems on Mean Value*, Doctorate Dissertation, Steklov Inst. Math. Russ. Acad. Sci., Moscow, 1966.

[82] A. A. Karatsuba, *The method of trigonometric sums and mean value theorems*, Mat. Zametki 1 (1967), no. 1, 99–110; English transl.: Math. Notes 1 (1967), 64–71.

[83] A. A. Karatsuba, *On estimates of complete trigonometric sums*, Mat. Zametki 1 (1967), no. 2, 199–208; English transl.: Math. Notes 1 (1967), 133–139.

[84] A. A. Karatsuba, *Estimates of trigonometric sums by I. M. Vinogradov's method*, Trudy Mat. Inst. Steklov 112 (1971), 241–255; English transl.: Proc. Steklov Inst. Math. 112 (1973), 251–265.

[85] A. A. Karatsuba, *Uniform estimate of the remainder in the problem of Dirichlet divisors*, Izv. Akad. Nauk SSSR Ser. Mat. 36 (1972), 475–483; English transl.: Math. USSR Izv. 6 (1972), 467–475.

[86] A. A. Karatsuba, *The mean value of the modulus of a trigonometric sum*, Izv. Akad. Nauk SSSR Ser. Mat. 37 (1973), 1203–1227; English transl.: Math. USSR Izv. 7 (1973), 1199–1223.

[87] A. A. Karatsuba, *Trigonometric sums and their applications*, Proc. Intern. Congr. of Math., Vancouver 1974, 365–368.

[88] A. A. Karatsuba, *On arithmetic problems with numbers that have small prime divisors*, Acta Arith. 27 (1975), 489–492.

[89] A. A. Karatsuba, *On a system of congruences*, Mat. Zametki, 19 (1976), no. 3, 389–392; English transl.: Math. Notes 19 (1976), 237–239.

[90] A. A. Karatsuba, *Principles of Analytic Number Theory*, 2nd ed., Nauka, Moscow, 1983.

[91] A. A. Karatsuba, *On the function $G(n)$ in Waring's problem*, Izv. Akad. Nauk SSSR Ser. Mat. 49 (1985), 935–947; English transl.: Math. USSR Izv. 27 (1986), 239–249.

[92] A. A. Karatsuba, *The Hilbert-Kamke problem in analytic number theory*, Mat. Zametki 41 (1987), no. 2, 272–284; English transl.: Math. Notes 41 (1987), 155–161.

[93] A. A. Karatsuba, *On a Diophantine inequality*, Acta Arith. 53 (1989), 309–324 (in Russian).

[94] A. A. Karatsuba, *Distribution of inverse values in a residue ring modulo a given number*, Dokl. Ross. Akad. Nauk 333 (1993), no. 2, 138–139; English transl.: Russian Acad. Sci. Dokl. Math. 48 (1993), 452–454.

[95] A. A. Karatsuba, *Fractional parts of functions of a special form*, Izv. Ross. Akad. Nauk Ser. Mat. 59 (1995), no. 4, 61–80; English transl.: Russian Acad. Sci. Izv. Math. 59 (1995).

[96] A. A. Karatsuba, *Analogues of Kloosterman sums*, Izv. Ross. Akad. Nauk Ser. Mat. 59 (1995), no. 5, 93–102; English transl.: Russian Acad. Sci. Izv. Math. 59 (1995).

[97] A. A. Karatsuba, *On an arithmetic function*, Uspekhi Mat. Nauk 50 (1995), no. 5, 247–248; English transl.: Russian Math. Surveys 50 (1995), 1088–1089.

[98] A. A. Karatsuba, *Complex Analysis in Number Theory*, CRC Press, 1995.

[99] A. A. Karatsuba, *Additive congruences*, Izv. Ross. Akad. Nauk Ser. Mat. 61 (1997), no. 2, 81–94; English transl.: Russian Acad. Sci. Izv. Math. 61 (1997), no. 2, 317–329.

[100] A. A. Karatsuba, *Analogues of incomplete Kloosterman sums and their applications*, Tatra Mountains Math. Publ. 11 (1997), 89–120.

[101] A. A. Karatsuba, *On regular sets in residual classes*, Mat. Zametki 64 (1998), no. 2, 224–228; English transl.: Math. Notes 64 (1998), no. 2, 190–193.

[102] A. A. Karatsuba, *Regular sets with respect to a given modulus*, Acta et Informatica Univ. Ostraviensis 6 (1998), 129–134.

[103] A. A. Karatsuba, *Kloosterman double sums*, Mat. Zametki 66 (1999), no. 5, 682–687; English transl.: Math. Notes 66 (1999), no. 5, 565–569.

[104] A. A. Karatsuba, *On fractional parts of rapidly increasing functions*, Izv. Ross. Akad. Nauk Ser. Mat. 65 (2001), no. 4, 89–110; English transl.: Russian Acad. Sci. Izv. Math. 65 (2001), no. 4, 727–748.

[105] Yu. V. Kashirskii, *The problem of disposition of zeros of Dirichlet L-series*, Dokl. Akad. Nauk SSSR 138 (1961), no. 2, 279–282 (in Russian).

[106] H. D. Kloosterman, *On the representation of numbers in the form* $ax^2 + by^2 + cz^2 + dt^2$, Acta Math. 49 (1926), 407–464.

[107] M. A. Korolev, *On a new multiplicative function*, Uspekhi Mat. Nauk 53 (1998), no. 4, 211–212; English transl.: Russian Math. Surveys 53 (1998), no. 4, 868–869.

[108] M. A. Korolev, *Incomplete Kloosterman sums and their applications*, Izv. Ross. Akad. Nauk Ser. Mat. 64 (2000), no. 6, 41–64; English transl.: Russian Acad. Sci. Izv. Math. 64 (2000), no. 6, 1129–1152.

[109] I. M. Kozlov, *I. M. Vinogradov's mean value theorem in the ring of Gaussian numbers*, in: *Modern Problems in Number Theory and Its Applications*, Proc. IVth Intern. Conference Dedicated to the 180th Anniversary of P. L. Chebyshev and the 110th Anniversary of I. M. Vinogradov, Tula, September 10–15 2001, pp. 71–86.

[110] V. I. Levenshtein, *Bounds for the maximal capacity of a code with a linear scalar product modulus*, Dokl. Akad. Nauk SSSR 263 (1982), no. 6, 1303–1308; English transl.: Soviet Math. Dokl. 25 (1982), 526–531.

[111] D. J. Lewis, *Cubic homogeneous polynomials over p-adic number fields*, Ann. Math. 56 (1952), no. 3, 473–478.

[112] Yu. V. Linnik, *Estimates for Weyl sums by I. M. Vinogradov's method*, Izv. Akad. Nauk SSSR Ser. Mat. 6 (1942), 41–70 (in Russian).

[113] Yu. V. Linnik, *Estimates for Weyl sums*, Dokl. Akad. Nauk SSSR, 34 (1942), no. 7, 201–203 (In Russian).

[114] Yu. V. Linnik, *On Weyl sums*, Mat. Sb. 12 (1942), 28–39. (In Russian).

[115] K. K. Mardzhanishvili, *On the simultaneous representation of numbers by sums of complete first, second, ..., and nth powers*, Izv. Akad. Nauk SSSR Ser. Mat. 1 (1937), 609–631 (in Russian).

[116] K. K. Mardzhanishvili, *Estimate of an arithmetic sum*, Dokl. Akad. Nauk SSSR 22 (1939), 391–393 (in Russian).

[117] K. K. Mardzhanishvili, *On an additive problem of number theory*, Izv. Akad. Nauk SSSR Ser. Mat. 4 (1940), 193–214 (in Russian).

[118] K. K. Mardzhanishvili, *On a system of equations in prime numbers*, Dokl. Akad. Nauk SSSR 70 (1950), 381–383 (in Russian).

[119] K. K. Mardzhanishvili, *On some nonlinear systems of equations in integers*, Mat. Sb. 33 (1953), 630–675 (in Russian).

[120] K. K. Mardzhanishvili, *On a singular series*, Trudy Mat. Inst. Steklov 142 (1976), 174–181; English transl.: Proc. Steklov Inst. Math. 142 (1979), 185–194.

[121] D. A. Mit'kin, *Estimate for the number of summands in the Hilbert–Kamke problem*, Mat. Sb. 129 (171) (1986), no. 4, 549–577; English transl.: Math. USSR-Sb. 57 (1987), no. 2, 561–590.

[122] D. A. Mit'kin, *Estimate for the number of summands in the Hilbert–Kamke problem*, II, Mat. Sb. 132 (174) (1987), no. 3, 345–351; English transl.: Math. USSR-Sb. 60 (1988), no. 2, 339–346.

[123] L. J. Mordell, *On a cubic exponential sum in two variables*, J. London Math. Soc. 38 (1963), 356–358.

[124] L. J. Mordell, *Some exponential sums*, Trudy Mat. Inst. Steklov 132 (1973), 30–34; English transl.: Proc. Steklov Inst. Math. 132 (1973), 29–34.

[125] M. B. Nathanson, *On a problem of Rohrbach for finite groups*, J. Number Theory 41 (1992), 69–76.

[126] V. I. Nechaev, *On the representation of natural numbers by a sum of terms of the form* $x(x+1)\dots(x+n-1)/n!$, Izv. Akad. Nauk SSSR Ser. Mat. 17 (1953), 485–498 (in Russian).

[127] V. I. Nechaev, *An estimate for a complete rational trigonometric sum*, Mat. Zametki 17 (1975), no. 6, 839–849; English transl.: Math. Notes 17 (1975), 504–511.

[128] K. I. Oskolkov, *Subsequences of Fourier sums of integrable functions*, Trudy Mat. Inst. Steklov 167 (1985), 239–260; English transl.: Proc. Steklov Inst. Math. 167 (1986), no. 2, 267–290.

[129] K. I. Oskolkov, *On spectra of uniform convergence*, Dokl. Akad. Nauk SSSR 288 (1986), no. 1, 54–58; English transl.: Soviet Math. Dokl. 33 (1986), no. 3, 616–620.

[130] K. I. Oskolkov, *Properties of a class of I. M. Vinogradov series*, Dokl. Akad. Nauk SSSR 300 (1988), no. 4, 803–807; English transl.: Soviet Math. Dokl. 37 (1988), no. 3, 737–741.

[131] K. I. Oskolkov, *I. M. Vinogradov series and integrals and their applications*, Trudy Mat. Inst. Steklov 190 (1989), 186–221; English transl.: Proc. Steklov Inst. Math. 190 (1992), 193–229.

[132] M. M. Petechuk, *The sum of values of a function of divisors in arithmetic progressions with a difference equal to a power of an odd prime number*, Izv. Akad. Nauk SSSR Ser. Mat. 43 (1979), no. 4, 892–908; English transl.: Math. USSR Izv. 15 (1980), no. 1, 145–160.

[133] G. Polya and G. Szego, *Problems and Theorems in Analysis*, vol. II, Springer-Verlag, Berlin 1972.

[134] D. A. Popov, *Spherical convergence of a series and of the Fourier integral of the indicator of a two-dimensional domain*, Trudy Mat. Inst. Steklov 218 (1997), 354–373; English transl.: Proc. Steklov Inst. Math. 218 (1997), 352–371.

[135] D. A. Popov, *Reconstruction of characteristic functions in two-dimensional Radon tomography*, Uspekhi Mat. Nauk 53 (1998), no. 1, 115–198; English transl.: Russian Math. Surveys 53 (1998), 109–194.

[136] D. A. Popov, *Uniform Estimates of Oscillating Integrals and Their Applications*, Doctorate Dissertation, Math. Inst. Russ. Acad. Sci., Moscow, 1998.

[137] D. A. Popov, *On the number of lattice points in three-dimensional solids of revolution*, Izv. Ross. Akad. Nauk Ser. Mat. 64 (2000), no. 2, 121–140; English transl.: Russian Acad. Sci. Izv. Math. 64 (2000), no. 2, 343–362.

[138] A. G. Postnikov, *On the sum of characters modulo the power of a prime*, Izv. Akad. Nauk SSSR Ser. Mat. 19 (1955), no. 1, 11–16 (in Russian).

[139] K. Prachar, *Primzahlverteilung*, Springer-Verlag, Berlin 1957.

[140] S. M. Rozin, *On the zeros of Dirichlet L-series*, Izv. Akad. Nauk SSSR Ser. Mat. 23 (1959), no. 2, 503–508 (in Russian).

[141] H. Salie, *Über die Kloostermanschen Summen $S(u, v, q)$*, Math. Z. 34 (1931), 91–109.

[142] B. M. Sidel'nikov, *On the mutual correlation of sequences*, Problemy Kibernetiki, 1974, no. 24, 15–42 (in Russian).

[143] S. B. Stechkin, *On mean values of the modulus of a trigonometric sum*, Trudy Mat. Inst. Steklov 134 (1975), 283–309; English transl.: Proc. Steklov Inst. Math. 134 (1975), 321–350.

[144] S. B. Stechkin, *Estimate of a complete rational trigonometric sum*, Trudy Mat. Inst. Steklov 143 (1977), 188–207; English transl.: Proc. Steklov Inst. Math. 143 (1980), 201–220.

[145] E. Titchmarsh, *The Theory of the Riemann Zeta-Function*, 2nd ed., The Clarendon Press, Oxford University Press, New York 1986.

[146] O. V. Tyrina, *A new estimate for I. M. Vinogradov's trigonometric integral*, Izv. Akad. Nauk SSSR Ser. Mat. 51 (1987), no. 2, 363–378; English transl.: Math. USSR Izv. 51 (1987), 363–378.

[147] P. L. Ul'yanov, *Some questions in the theory of orthogonal and biorthogonal series*, Izv. Akad. Nauk Azerbaidzhan Ser. Fiz.-Techn. i Mat. Nauk (1965), no. 6, 11–13 (in Russian).

[148] I. M. Vinogradov, *On the distribution of fractional parts of functions in two variables*, Izv. Leningrad. Politekhn. Inst. 30 (1927), 31–52 (in Russian).

[149] I. M. Vinogradov, *Representation of a number by an integer polynomial in several variables*, Izv. Akad. Nauk SSSR Ser. Phys. Mat. (1928), no. 4/5, 401–414 (in Russian).

[150] I. M. Vinogradov, *On problems of analytic number theory*, Proc. November Jubilee Session Acad. Sci. USSR (Leningrad, 1932), Izdat. Akad Nauk SSSR, Leningrad, 1933, pp. 1–11 (in Russian).

[151] I. M. Vinogradov, *Some theorems of analytic number theory*, C. R. (Dokl.) Akad. Sci URSS 4(5) (1934), 185–187 (in Russian).

[152] I. M. Vinogradov, *A new estimate for $G(n)$ in the Waring problem*, Dokl. Akad. Nauk SSSR 5 (1934), 249–253 (in Russian).

[153] I. Vinogradov, *On some new problems of the theory of numbers*, C. R. (Dokl.) Akad. Sci URSS 3 (4) (1934), 1–3; English transl.: ibid., 4–6.

[154] I. M. Vinogradov, *Representation of an odd number as the sum of three primes*, C. R. (Dokl.) Akad. Sci URSS, 15 (1937), 291–294 (also in *Selected Works*, Springer-Verlag, Berlin 1985).

[155] I. M. Vinogradov, *Some new problems of the theory of primes*, C. R. (Dokl.) Akad. Sci URSS 16 (1937), 131–132.

[156] I. M. Vinogradov, *Analytical theory of numbers*, Izv. Akad. Nauk SSSR Ser. Mat. 9 (1945), 159–164; English transl.: ibid., 165–168.

[157] I. M. Vinogradov, *Additive problems of the theory of prime numbers*, Jubilee Collection on the Thirtieth Anniversary of the Great October Socialist Revolution, Part I, Izdat. Akad. Nauk SSSR, Moscow, 1947, pp. 65–79 (in Russian).

[158] I. M. Vinogradov, *The method of trigonometric sums in the theory of numbers*, Trudy Mat. Inst. Steklov, 23 (1947); English transl.: Interscience, New York 1954.

[159] I. M. Vinogradov, *Selected Works*, Izdat. Akad. Nauk SSSR, Moscow, 1952 (in Russian).

[160] I. M. Vinogradov, *Certain problems in analytic number theory*, Proc. Third All-Union Math. Congrc. (Moscow, 1956). Vol. III: Survey Reports, Izdat. Akad. Nauk SSSR, Moscow 1958, 3–13; English transl.: *Selected Works*, Springer-Verlag, Berlin 1985.

[161] I. M. Vinogradov, *A new estimate for* $\zeta(1 + it)$, Izv. Akad. Nauk SSSR Ser. Mat. 22 (1958), no. 62, 161–164 (in Russian).

[162] I. M. Vinogradov, *Special Versions of the Method of Trigonometric Sums*, Nauka, Moscow 1976.

[163] I. M. Vinogradov, *Fundamentals of Number Theory*, Nauka, Moscow 1981.

[164] I. M. Vinogradov and A. A. Karatsuba, *The method of trigonometric sums in number theory*, Trudy Mat. Inst. Steklov 168 (1984), 4–30; English transl.: Proc. Steklov Inst. Math. 168 (1986), no. 3, 3–30.

[165] I. M. Vinogradov, *The Method of Trigonometric Sums in Number Theory*, 2nd ed., Nauka, Moscow 1980; English transl.: *Selected works*, Springer-Verlag, Berlin 1985.

[166] S. M. Voronin and A. A. Karatsuba, The Riemann Zeta Function, Fizmatlit, Moscow, 1994; English transl.: de Gruyter Exp. Math. 5, Berlin–New York 1992.

[167] A. Weil, *On some exponential sums*, Proc. Nat. Acad. Sci. USA 34 (1948), 204–207.

[168] H. Weyl, *Über die Gleichverteilung der Zahlen mod Eins*, Math. Ann. 77 (1916), 313–352.

[169] V. A. Zinov'ev and S. N. Litsyn, *Lower bounds for complete rational trigonometric sums*, Uspekhi Mat. Nauk 43 (1988), no. 1, 199–200; English transl.: Russian Math. Surveys 43 (1988), 259–260.

[170] A. Zygmund, *Trigonometric Series*, 2nd rev. ed., Vol. 1, Cambridge University Press, Cambridge 1959.

Index